Joe   551   mon 3-5

Final Dec 22nd 12-3 PM

www.duxbury.com

www.duxbury.com is the World Wide Web site for Duxbury and is your direct source to dozens of online resources.

At *www.duxbury.com* you can find out about supplements, demonstration software, and student resources. You can also send e-mail to many of our authors and preview new publications and exciting new technologies.

www.duxbury.com
Changing the way the world learns®

DUXBURY

Probability and Statistics for Engineers and Scientists

THIRD EDITION

Anthony J. Hayter

Georgia Institute of Technology

THOMSON

BROOKS/COLE ™

Australia • Brazil • Canada • Mexico • Singapore • Spain • United Kingdom • United States

THOMSON

BROOKS/COLE

Probability and Statistics for Engineers and Scientists, Third Edition

Anthony J. Hayter

Acquisitions Editor: *Carolyn Crockett*
Technology Project Manager: *Fiona Chong*
Assistant Editor: *Ann Day*
Executive Marketing Manager: *Joe Rogove*
Marketing Assistant: *Brian Smith*
Marketing Communications Manager: *Darlene Amidon-Brent*
Manager, Editorial Production: *Kelsey McGee*
Creative Director: *Rob Hugel*
Art Director: *Lee Friedman*
Print Buyer: *Judy Inouye*
Permissions Editor: *Bob Kauser*

Production Service and Composition: *Interactive Composition Corporation*
Text Designer: *Andrew Ogus*
Art Editor: *Lisa Torri*
Copy Editor: *Martha Williams*
Cover Designer: *Laurie Anderson*
Cover Image: *Louis Psihoyos/Getty Images, Science Faction Collection*
Cover Printing, Printing & Binding: *Transcontinental Printing/Louiseville*

Thomson Higher Education
10 Davis Drive
Belmont, CA 94002-3098
USA

Library of Congress Control Number: 2005930896
ISBN 0-495-10757-3

Cover Image: *Black Pearls in an Oyster Shell, Tahiti. A rare collection of perhaps some of the best Tahitian black pearls in the world, hi-graded from a 25-year harvest of 80% of the pearls that come out of Tahiti.*

ABOUT THE AUTHOR

Dr. Anthony Hayter obtained a triple first-class degree in mathematics from Cambridge University in England and a Ph.D. in statistics from Cornell University. His doctoral thesis included the proof of a famous mathematical conjecture that had remained unsolved for thirty years, and he is the author of numerous research publications in the fields of experimental design and applied data analysis.

Dr. Hayter is passionate about empowering students and researchers with the skills and knowledge that they need to perform effective and accurate data analysis. With his experience as a teacher, researcher, and consultant in a wide variety of settings, Dr. Hayter knows how essential these skills are in today's workplace. Dr. Hayter's work has shown him the substantial advantages that accrue from a clear understanding of the concepts of probability and statistics, together with the pitfalls that can arise from their misuse.

Dr. Hayter collaborates on many research projects, including work to improve wheelchair designs and to provide better assistive technologies to disabled people. In addition, he has worked on projects to improve the safety of bridges, to monitor air pollution levels, and to promote equitable taxation rates, along with various other projects that are used as examples and data sets in this textbook.

In his spare time, Dr. Hayter likes to read the detective stories of Timothy Hemion. In fact, Dr. Hayter tells his students that conducting a good data analysis is like being part of a detective story. A well-designed experiment provides pertinent evidence, and the statistician's job is to know how to extract the relevant clues from the data set. These clues can then be used to piece together a picture of the true state of affairs, and they may be used to disprove or substantiate the theories and hypotheses that have been put forward. If you don't have the proper skills at your disposal, then you're not going to be able to solve the crime!

CONTENTS

PREFACE

Unlikely events happen all the time to somebody somewhere—it's just
that it would be strange if they happened to you or me.

—*Inspector Morimoto and the Sushi Chef* by Timothy Hemion

The primary guideline governing the development of the third edition of this textbook has been
to extend the strengths of the previous editions that have resulted in its adoption worldwide for
teaching probability and statistics at both undergraduate and graduate levels. The cornerstone
of the success of this textbook has been that it is full of real examples, which I believe is the best
way to teach and capture the interests of students. It is clearly important to include examples
that are relevant to engineering. However, it is just as important to incorporate examples that
are interesting to both the instructor and the students.

This textbook has been built around three main pedagogical tenets: (1) talk to students with
a language and vocabulary that students find familiar from their other science and engineering
courses, (2) provide clear explanations and expositions of the statistical concepts that students
need for their work and research, and (3) provide a firm reinforcement of the theoretical
concepts in interesting examples to which the students can relate. Moreover, the foundation
underlying all my teaching activities is my conviction that teaching is a valuable and noble
enterprise and that it is worthwhile to devote time and energy toward doing it as well as
possible. The education of the following generations is an important way for us to pay back
something in response to the many advantages and opportunities that we have been afforded
in our own lives.

This book has been adopted for undergraduate sequences providing an introduction to
data analysis from probability theory through basic statistical techniques and leading to more
advanced statistical inference methods. The book has also been used for graduate-level service
courses, and it provides a useful handbook for researchers in engineering and the sciences. It
is intended for students with reasonable quantitative abilities, although it is designed mainly
to provide an applied rather than a theoretical exposition.

Highlights of the Book

- The book has been developed from extensive teaching experience with undergraduate
  and graduate engineering students.

- Real examples from engineering sciences are developed throughout the book.

- The applied presentation stresses the comprehension of the underlying concepts and
  the application of statistical methodologies.

- A large number of interesting data sets from a wide range of fields are included.

- A motivating case study is included at the beginning of the book and is continued at
  the ends of the first twelve chapters.

- A large number and variety of exercise problems of various levels of difficulty are included.

- The book provides a handbook of statistical methodologies for undergraduate and graduate engineering students.

- Computer notes offer help and tips for data analysis with statistical software packages.

- The composition of the book allows flexibility in the order in which the material is taught.

Motivation and Goals

The primary goal of this book is to provide a means of leading the reader through the important issues of data collection, data presentation, and data analysis. These are tremendously important topics for students and researchers today, and the book is designed to show the reader how to think about these issues properly and accurately. A key issue in this goal is illustrating to the reader why statistical analysis methods are relevant and useful for engineering and the sciences.

The topics are presented in the context of a wide range of engineering examples that provide a motivation for the development of the material. The examples show how the techniques can be used to gain an understanding of the data set under consideration. The examples are intended to be readily understood by the reader and to be interesting and thought provoking. This book can also be used as a handbook of statistical techniques and probability distributions for all scientists, engineers, and anybody involved in data analysis. Many students have commented that it is a valuable resource that they have used long after finishing their statistics course.

The book concentrates on allowing the reader to obtain an understanding of the concepts behind the methodologies presented, rather than providing an unnecessary amount of theory. The reader is encouraged to look at a formula and to understand what it is doing and how it works. The reader is then able to use statistical software packages properly and knows which analysis techniques to employ and how to use and interpret the results.

Presentation of Topics

Each of the topics presented in this book is introduced with reference to several examples from different engineering and scientific areas and with reference to some data sets. After the technical development of the topic has been described, the important points are summarized in a highlighted box.

The examples are then used to show the proper application of the new methodology. These examples are built on and developed throughout the chapters as increasingly sophisticated methodologies are considered. This presentation provides ties and connections between the different chapters, and it also shows how each of the individual topics fits into the wider range of statistical methodologies that can be applied to any particular problem. Moreover, the relevance and importance of the statistical analysis to these problems are demonstrated. A list of the examples and the sections where they appear in the text is provided on the inside of the front cover.

Continuing Case Study

A new addition to the third edition is the motivating case study at the beginning of the book that shows how the subjects of probability and statistics can be applied to the issues that arise in computer chip construction. The analysis is decomposed into several core constituents that are tied to the different chapters of the book. This case study provides students with an immediate explanation for why it is necessary and important to study probability and statistics. Instructors may choose to use the case study at the beginning of the course as motivating material and as a road map to show students where they will be going. Instructors can also refer back to the case study before each new chapter or topic is started in order to show how the different topics all fit with each other.

Composition of the Book

The figure below illustrates the composition of this book. The chapters are arranged from general probability theory and probability distributions to descriptive statistics, basic statistical inference techniques, and more advanced statistical inference techniques. The book also includes chapters on nonparametric methods, quality control methods, and reliability analysis and life testing. Instructors can be very flexible in the selection and order in which the chapters are actually taught in a course.

Composition of the Book

CASE STUDY

Probability

CHAPTER 1:   Probability Theory

CHAPTER 2:   Random Variables

CHAPTER 3:   Discrete Probability Distributions

CHAPTER 4:   Continuous Probability Distributions

CHAPTER 5:   The Normal Distribution

Basic Statistics

CHAPTER 6:   Descriptive Statistics

CHAPTER 7:   Statistical Estimation and Sampling Distributions

CHAPTER 8:   Inferences on a Population Mean

CHAPTER 9:   Comparing Two Population Means

CHAPTER 10:   Discrete Data Analysis

Advanced Statistics

CHAPTER 11:   The Analysis of Variance

CHAPTER 12:   Simple Linear Regression and Correlation

CHAPTER 13:   Multiple Linear Regression and Nonlinear Regression

CHAPTER 14:   Multifactor Experimental Design and Analysis

CHAPTER 15:   Nonparametric Statistical Analysis

CHAPTER 16:   Quality Control Methods

CHAPTER 17:   Reliability Analysis and Life Testing

Data Sets

The Data Sets CD accompanying this textbook contains all the data sets used in the book. The data sets of the worked examples are included so that readers can replicate the results with their own statistical software package. Many of the exercise problems also involve data sets included on the CD that readers can analyze with their own statistical software packages.

A list of the data sets and the sections where they are used is provided on the inside of the back cover.

Exercise Problems

This book contains a large number of exercise problems of varying difficulty levels. The problems are presented at the end of every section within the chapters, and in addition a set of supplementary problems is provided at the end of each chapter. The initial problems take the reader through the steps of the new material that has been presented and allow the reader to practice the material. The subsequent problems become more difficult and more open-ended. Most of the problems are presented in the context of engineering problems and data sets. Answers to all the odd-numbered problems at the ends of the chapter sections are given at the back of this book, and worked solutions can be found in the *Student Solution Manual* for these odd-numbered problems and in the *Instructor Solution Manual* for all of the problems.

Accompanying Materials

- All the data sets in this book are available in the Data library at www.duxbury.com.

- Worked solutions and answers to all the problems are presented in the *Instructor Solution Manual*.

- Worked solutions and answers to all the odd-numbered problems at the ends of the chapter sections are presented in the *Student Solution Manual*.

- A separate, stand-alone Instructor's Suite CD contains data sets, the complete solutions manual, and a multimedia manager of all the art figures in the text.

Acknowledgments

I would like to express my heartfelt thanks to my editor, Carolyn Crockett, and to all members of the team that have contributed to this third edition with their wonderful talents, wisdom, experience, and energy. This especially includes Fiona Chong, Jennifer Crotteau, Ann Day, Anne Draus, Daniel Geller, Christina Ho, and Kelsey McGee.

Finally, I would also like to thank various reviewers who have helped with the development of this book, especially Jackie Miller of the Ohio State University for her work on the third edition. The reviewers of the first edition include Mary R. Anderson, Arizona State University; Charles E. Antle, The Pennsylvania State University, University Park; Sant Ram Arora, University of Minnesota—Twin Cities Campus; William R. Astle, Colorado School of Mines; Lee J. Bain, University of Missouri—Rolla; Douglas M. Bates, University of Wisconsin—Madison; Rajan Batta, The State University of New York at Buffalo; Alan C. Bovik, University of Texas at Austin; Don B. Campbell, Western Illinois University; M. Jeya Chandra, Pennsylvania State University—University Park; Yueh-Jane Chang, Idaho State University; Chung-Lung Chen, Mississippi State University; Inchan Choi, Wichita State University; John R. Cook, North Dakota State University; Rianto A. Djojosugito, South Dakota School of Mines and Technology; Lucien Duckstein, University of Arizona; Earnest W. Fant, University of Arkansas;

Richard F. Feldman, Texas A & M University; Sam Gutmann, Northeastern University; Carol O'Connor Holloman, University of Louisville; Chi-Ming Ip, University of Miami; Rasul A. Khan, Cleveland State University; Stojan Kotefski, New Jersey Institute of Technology; Walter S. Kuklinski, University of Massachusetts—Lowell; S. Kumar, Rochester Institute of Technology; Gang Li, University of North Carolina at Charlotte; Jiye-Chyi Lu, North Carolina State University; Ditlev Monrad, University of Illinois at Urbana—Champaign; John Morgan, California Polytechnic State University—Pomona; Paul J. Nahin, University of New Hampshire; Larry Ringer, Texas A & M University; Paul L. Schillings, Montana State University; Ioannis Stavrakakis, The University of Vermont; and James J. Swain, University of Alabama—Huntsville.

The reviewers of the second edition were Alexander Dukhovny, San Francisco State University; Marc Genton, Massachusetts Institute of Technology; Diwakar Gupta, University of Minnesota; Joseph J. Harrington, Harvard University; and Jim Rowland, University of Kansas.

Survey respondents for the third edition include Mostafa S. Aminzadeh, Towson University; Barb Barnet, University of Wisconsin—Platteville; Ronald D. Bennett, Bethel College; Shannon Brewer, Northeast State Community College; Frank C. Castronova, Lawrence Technological University; Mike Doviak, Old Dominion University; Natarajan Gautam, Penn State University; Peggy Hart, Doane College; Wei-Min Huang, Lehigh University; Xiaoming Huo, Georgia Tech; Bruce N. Janson, University of Colorado at Denver; Scott Jilek, University of St. Thomas; Michael Kostreva, Clemson University; Paul Kvam, Georgia Tech; David W. Matolak, Ohio University; Gary C. McDonald, Oakland University; Megan Meece, University of Florida; Luke Miller, University of San Diego; Steve Patch, University of North Carolina at Asheville; Robi Polikar, Rowan University; Andrew M. Ross, Lehigh University; Manuel D. Rossetti, University of Arkansas; Robb Sinn, North Georgia College and State University; Bradley Thiessen, St. Ambrose University; Dolores Tichenor, Tri-State University; Lewis VanBrackle, Kennesaw State University; Jerry Weyand, Cleary University; Ed Wheeler, University of Tennessee at Martin; Elaine Zanutto, The Wharton School, University of Pennsylvania; and Kathy Zhong, University of Detroit Mercy.

Anthony J. Hayter

This is a running case study that continues through the first twelve chapters of the book. It demonstrates how probability and statistical inference can be applied to this important engineering problem.

CONTINUING CASE STUDY: MICROELECTRONIC SOLDER JOINTS

Solder joints are an important component of microelectronic assemblies. Figure CS.1 shows a cross-section of a typical assembly known as a flip chip in which as series of conductive bump-shaped solder joints are used to attach a silicon chip to a printed circuit board, which is known as the substrate. These solder joints provide the conductive path from the silicon ship to the substrate, and fatigue in the solder joints is responsible for almost all of the mechanical and electrical failures of the assembly.

The area surrounding the solder joints between the silicon chip and the substrate is filled with a substance known as the underfill, which is non-conductive epoxy that helps to protect the solder joints from moisture as well as adding strength to the assembly. The underfill also helps to minimize the stress in the solder joints that arises from the different thermal expansions of the silicon chip and the substrate. This helps ensure that the connections are not damaged or broken.

It is important to investigate the reasons behind the development of cracks in the solder joints which can affect the operation of the assembly. The development of these cracks can be related to the shapes of the solder joints, which are illustrated in Figure CS.2. While most of the joints turn out to be barrel shaped, some may have cylinder shapes or hourglass shapes. In addition to cracks in the solder joints, failures can also be caused by solder extrusions that connect two adjacent solder joints as shown in Figure CS.3, or which leave only a very small gap between two adjacent solder joints.

In addition, a critical component of the assembly is the bonding between the solder joint and the substrate which is achieved through suitable metallization of the substrate. As Figure CS.4 shows, a bond pad is created in the substrate made of copper, which is coated with thin layers of nickel and gold. The thickness of the gold layer has an important effect on the reliability of the electrical connection between the solder joint and the substrate.

Reliability assessment of these microelectronic assemblies is often performed with accelerated life tests. Since it is known that temperature changes cause stress in the solder joints that can ultimately lead to failure, the accelerated life tests often consist of subjecting the assembly

FIGURE CS.1

Cross section of a typical flip chip microelectronic assembly

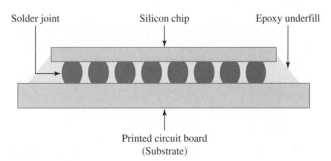

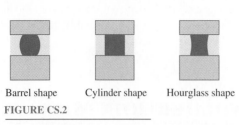

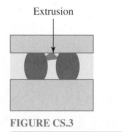

Extrusion

Barrel shape   Cylinder shape   Hourglass shape

FIGURE CS.2

Shapes of solder joints in a microelectronic assembly

FIGURE CS.3

An extrusion between solder joints in a microelectronic assembly

FIGURE CS.4

Diagram of a substrate bond pad in a microelectronic assembly

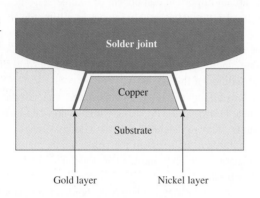

Solder joint

Copper

Substrate

Gold layer   Nickel layer

to alternate periods at low and high temperatures. For example, the assembly may be alternately immersed in liquid at −55 degrees Centigrade for five minutes, and then switched to a liquid at 125 degrees Centigrade for five minutes. This cycle is repeated many times until the assembly eventually fails. These accelerated life tests are designed to mimic (at an accelerated speed) the conditions to which the assembly will be subjected in its everyday operation, and it is anticipated that designs which survive the most cycles of the accelerated life test will have the best reliability in real world applications.

A considerable amount of research is conducted on the fatigue life of the solder joints together with their optimal production method, and the areas of probability and statistics play major roles in this research. This case study will be continued through the first twelve chapters of the book to show how probability theory and statistical inference can be applied to this important engineering problem.

Chapter 1 Probability Theory

For a given production method, it is important to know the probabilities that the solder joints will be formed according to each of the three shape profiles. Additionally, the probability that cracking occurs in each of the different solder joint shapes is investigated, and the probability of finding a given number of cracked solder joints within a random sample of solder joints is calculated.

Chapter 2 Random Variables

The number of extrusions in an assembly with a large number of solder joints is critical to the overall lifetime of the assembly. In this chapter the average number of extrusions in an assembly is investigated, and an assessment is made of the amount of variability in the number of extrusions. In addition, the total number of extrusions in a batch of 250 assemblies is considered.

Chapter 3 Discrete Probability Distributions

An analysis is conducted of the number of hourglass shaped solder joints on an assembly which contains 64 solder joints. A calculation is also performed concerning the amount of work that is necessary if a researcher intends to examine solder joints one at a time until two cracked hourglass shaped solder joints are discovered. Finally, the distribution of the number of solder joints of each shape on an assembly consisting of 16 solder joints is discussed.

Chapter 4 Continuous Probability Distributions

Accelerated life tests are considered and attention is directed to the question of how many temperature cycles an assembly can survive before it fails. A probability distribution is used to model this failure time. An investigation is also conducted into how different types of epoxies employed in the underfill can affect the reliability of the assembly.

Chapter 5 The Normal Distribution

The thickness of the gold layer at the top of the bond pad has an important effect on the reliability of the bond between the solder joint and the substrate. In this chapter these thicknesses are modelled with a normal probability distribution, and it is shown how to calculate the probability that the thickness of the gold layer lies within an optimal range. The average thickness of the gold layers on an assembly consisting of 16 bond pads is also considered. Finally, calculations are made concerning the number of hourglass shaped solder joints that will be produced on an assembly comprised of 512 solder joints.

Chapter 6 Descriptive Statistics

A data set is obtained by measuring the thicknesses of the nickel layers deposited by a new method on each of the bond pads of an assembly. Summary statistics are calculated for the data set. Also, a categorical data set is obtained from the frequencies of the solder joint shapes in a large assembly produced by a particular methodology.

Chapter 7 Statistical Estimation and Sampling Distributions

The data sets given in Chapter 6 are used to obtain estimates of the quantities of interests. The average amount of nickel deposited by the new method is estimated, and an assessment of the estimate's accuracy is made. Also, for the methodology under consideration an estimate is obtained for the probability that a solder joint will have a barrel shape, and the precision of this estimate is also considered.

Chapter 8 Inferences on a Population Mean

The data set of nickel layer thicknesses is used to examine whether the new method is applying nickel according to the desired target value. The information pertinent to this question is extracted from the data set and is summarized in various ways. Finally, calculations are performed to estimate how many additional nickel layer thickness measurements are needed in order to increase the sensitivity of the statistical inferences to a specified level.

Chapter 9 Comparing Two Population Means

A comparison is made between the original method of applying the nickel layer to the substrate bond pads and the new method. A data set of nickel layer thicknesses achieved from the original method is compared with the data set of nickel layer thicknesses achieved with the new method in order to test whether there is any evidence of a difference between the two methods. Analysis methods are introduced which allow the researcher to quantify this difference.

Chapter 10 Discrete Data Analysis

A test is performed to determine whether the frequencies of the solder joint shapes observed in the analysis of a large assembly are consistent with some theoretical probability values. The data set is also used to calculate a range for one of these probability values. In addition, an experiment is conducted to investigate which of two epoxy formulations for the underfill results in the smallest failure probability after 2000 temperature cycles of the accelerated life test, and the analysis of the resulting data set is performed.

Chapter 11 The Analysis of Variance

In this chapter it is shown how a comparison can be made between four different companies in terms of the thicknesses of the gold layers on the substrate bond pads. A random selection is made of bond pads on assemblies produced by each of the four companies, and a data set is formed by measuring the gold layer thicknesses. An analysis of the data set is then performed which allows the determination of which company has the thinnest gold layers and which company has the thickest gold layers.

Chapter 12 Simple Linear Regression and Correlation

The researcher is interested in whether the heights of the solder joints have any influence on the reliability of the microelectronic assembly. An experiment is conducted whereby assemblies with different solder joint heights are subjected to an accelerated life test until they fail. The number of temperature cycles that the assemblies can withstand before failing is measured, and an analysis is performed to investigate whether there is any evidence that the solder joint heights have any effect on the reliability of the assembly.

Probability and Statistics for Engineers and Scientists

CHAPTER ONE

Probability Theory

1.1 Probabilities

1.1.1 Introduction

Jointly with statistics, probability theory is a branch of mathematics that has been developed to deal with **uncertainty**. Classical mathematical theory had been successful in describing the world as a series of fixed and real observable events, yet before the seventeenth century it was largely inadequate in coping with processes or experiments that involved uncertain or random outcomes. Spurred initially by the mathematician's desire to analyze gambling games and later by the scientific analysis of mortality tables within the medical profession, the theory of probability has been developed as a scientific tool dealing with **chance**.

Today, probability theory is recognized as one of the most interesting and also one of the most useful areas of mathematics. It provides the basis for the science of statistical inference through experimentation and data analysis—an area of crucial importance in an increasingly quantitative world. Through its applications to problems such as the assessment of system reliability, the interpretation of measurement accuracy, and the maintenance of suitable quality controls, probability theory is particularly relevant to the engineering sciences today.

1.1.2 Sample Spaces

An **experiment** can in general be thought of as any process or procedure for which more than one **outcome** is possible. The goal of probability theory is to provide a mathematical structure for understanding or explaining the chances or likelihoods of the various outcomes actually occurring. A first step in the development of this theory is the construction of a list of the possible experimental outcomes. This collection of outcomes is called the **sample space** or **state space** and is denoted by S.

Sample Space

The **sample space** S of an experiment is a set consisting of all of the possible experimental outcomes.

The following examples help illustrate the concept of a sample space.

Example 1
Machine Breakdowns

An engineer in charge of the maintenance of a particular machine notices that its breakdowns can be characterized as due to an electrical failure within the machine, a mechanical failure of some component of the machine, or operator misuse. When the machine is running, the engineer is uncertain what will be the cause of the next breakdown. The problem can be thought of as an experiment with the sample space

$$S = \{\text{electrical, mechanical, misuse}\}$$

| Example 2
| Defective Computer
| Chips

A company sells computer chips in boxes of 500, and each chip can be classified as either satisfactory or defective. The number of defective chips in a particular box is uncertain, and the sample space is

$$S = \{0 \text{ defectives, } 1 \text{ defective, } 2 \text{ defectives, } 3 \text{ defectives, } 4 \text{ defectives, } \dots,$$
$$499 \text{ defectives, } 500 \text{ defectives}\}$$

| Example 3
| Software Errors

The control of errors in computer software products is obviously of great importance. The number of separate errors in a particular piece of software can be viewed as having a sample space

$$S = \{0 \text{ errors, } 1 \text{ error, } 2 \text{ errors, } 3 \text{ errors, } 4 \text{ errors, } 5 \text{ errors, } \dots\}$$

In practice there will be an upper bound on the possible number of errors in the software, although conceptually it is all right to allow the sample space to consist of all of the positive integers.

| Example 4
| Power Plant Operation

A manager supervises the operation of three power plants, plant X, plant Y, and plant Z. At any given time, each of the three plants can be classified as either generating electricity (1) or being idle (0). With the notation (0, 1, 0) used to represent the situation where plant Y is generating electricity but plants X and Z are both idle, the sample space for the status of the three plants at a particular point in time is

$$S = \{(0, 0, 0) \ (0, 0, 1) \ (0, 1, 0) \ (0, 1, 1) \ (1, 0, 0) \ (1, 0, 1) \ (1, 1, 0) \ (1, 1, 1)\}$$

| S
| (0, 0, 0)   (1, 0, 0)
| (0, 0, 1)   (1, 0, 1)
| (0, 1, 0)   (1, 1, 0)
| (0, 1, 1)   (1, 1, 1)

FIGURE 1.1

Sample space for power plant example

It is often helpful to portray a sample space as a diagram. Figure 1.1 shows a diagram of the sample space for this example, where the sample space is represented by a box containing the eight individual outcomes. Diagrams of this kind are known as **Venn diagrams**.

GAMES OF CHANCE

Games of chance commonly involve the toss of a coin, the roll of a die, or the use of a pack of cards. The toss of a single coin has a sample space

$$S = \{\text{head, tail}\}$$

and the toss of two coins (or one coin twice) has a sample space

$$S = \{(\text{head, head}) \ (\text{head, tail}) \ (\text{tail, head}) \ (\text{tail, tail})\}$$

where (head, tail), say, represents the event that the first coin resulted in a head and the second coin resulted in a tail. Notice that (head, tail) and (tail, head) are two distinct outcomes since observing a head on the first coin and a tail on the second coin is different from observing a tail on the first coin and a head on the second coin.

A usual six-sided die has a sample space

$$S = \{1, 2, 3, 4, 5, 6\}$$

FIGURE 1.2

Sample space for rolling two dice

					S
(1, 1)	(1, 2)	(1, 3)	(1, 4)	(1, 5)	(1, 6)
(2, 1)	(2, 2)	(2, 3)	(2, 4)	(2, 5)	(2, 6)
(3, 1)	(3, 2)	(3, 3)	(3, 4)	(3, 5)	(3, 6)
(4, 1)	(4, 2)	(4, 3)	(4, 4)	(4, 5)	(4, 6)
(5, 1)	(5, 2)	(5, 3)	(5, 4)	(5, 5)	(5, 6)
(6, 1)	(6, 2)	(6, 3)	(6, 4)	(6, 5)	(6, 6)

FIGURE 1.3

Sample space for choosing one card

												S
A♡	2♡	3♡	4♡	5♡	6♡	7♡	8♡	9♡	10♡	J♡	Q♡	K♡
A♣	2♣	3♣	4♣	5♣	6♣	7♣	8♣	9♣	10♣	J♣	Q♣	K♣
A♢	2♢	3♢	4♢	5♢	6♢	7♢	8♢	9♢	10♢	J♢	Q♢	K♢
A♠	2♠	3♠	4♠	5♠	6♠	7♠	8♠	9♠	10♠	J♠	Q♠	K♠

FIGURE 1.4

Sample space for choosing two cards with replacement

					S
(A♡, A♡)	(A♡, 2♡)	(A♡, 3♡)	⋯	(A♡, Q♠)	(A♡, K♠)
(2♡, A♡)	(2♡, 2♡)	(2♡, 3♡)	⋯	(2♡, Q♠)	(2♡, K♠)
(3♡, A♡)	(3♡, 2♡)	(3♡, 3♡)	⋯	(3♡, Q♠)	(3♡, K♠)
⋮	⋮	⋮		⋮	⋮
(Q♠, A♡)	(Q♠, 2♡)	(Q♠, 3♡)	⋯	(Q♠, Q♠)	(Q♠, K♠)
(K♠, A♡)	(K♠, 2♡)	(K♠, 3♡)	⋯	(K♠, Q♠)	(K♠, K♠)

If two dice are rolled (or, equivalently, if one die is rolled twice), then the sample space is shown in Figure 1.2, where (1, 2) represents the event that the first die recorded a 1 and the second die recorded a 2. Again, notice that the events (1, 2) and (2, 1) are both included in the sample space because they represent two distinct events. This can be seen by considering one die to be red and the other die to be blue, and by distinguishing between obtaining a 1 on the red die and a 2 on the blue die and obtaining a 2 on the red die and a 1 on the blue die.

If a card is chosen from an ordinary pack of 52 playing cards, the sample space consists of the 52 individual cards as shown in Figure 1.3. If two cards are drawn, then it is necessary to consider whether they are drawn with or without **replacement**. If the drawing is performed *with replacement*, so that the initial card drawn is returned to the pack and the second drawing is from a full pack of 52 cards, then the sample space consists of events such as (6♡, 8♣), where the first card drawn is 6♡ and the second card drawn is 8♣. Altogether there will be $52 \times 52 = 2704$ elements of the sample space, including events such as (A♡, A♡), where the A♡ is drawn twice. This sample space is shown in Figure 1.4.

FIGURE 1.5

Sample space for choosing two cards without replacement

		$(A\heartsuit, 2\heartsuit)$	$(A\heartsuit, 3\heartsuit)$	$\cdots$	$(A\heartsuit, Q\spadesuit)$	$(A\heartsuit, K\spadesuit)$
$(2\heartsuit, A\heartsuit)$			$(2\heartsuit, 3\heartsuit)$	$\cdots$	$(2\heartsuit, Q\spadesuit)$	$(2\heartsuit, K\spadesuit)$
$(3\heartsuit, A\heartsuit)$	$(3\heartsuit, 2\heartsuit)$			$\cdots$	$(3\heartsuit, Q\spadesuit)$	$(3\heartsuit, K\spadesuit)$
$\vdots$	$\vdots$	$\vdots$		$\vdots$	$\vdots$	$\vdots$
$(Q\spadesuit, A\heartsuit)$	$(Q\spadesuit, 2\heartsuit)$	$(Q\spadesuit, 3\heartsuit)$	$\cdots$			$(Q\spadesuit, K\spadesuit)$
$(K\spadesuit, A\heartsuit)$	$(K\spadesuit, 2\heartsuit)$	$(K\spadesuit, 3\heartsuit)$	$\cdots$	$(K\spadesuit, Q\spadesuit)$		

If two cards are drawn *without replacement*, so that the second card is drawn from a reduced pack of 51 cards, then the sample space will be a subset of that above, as shown in Figure 1.5. Specifically, events such as $(A\heartsuit, A\heartsuit)$, where a particular card is drawn twice, will not be in the sample space. The total number of elements in this new sample space will therefore be $2704 - 52 = 2652$.

1.1.3 Probability Values

The likelihoods of particular experimental outcomes actually occurring are found by assigning a set of **probability values** to each of the elements of the sample space. Specifically, each outcome in the sample space is assigned a probability value that is a number between zero and one. The probabilities are chosen so that the sum of the probability values over all of the elements in the sample space is one.

Probabilities

A set of **probability values** for an experiment with a sample space $S = \{O_1, O_2, \ldots, O_n\}$ consists of some probabilities

$$p_1, p_2, \ldots, p_n$$

that satisfy

$$0 \le p_1 \le 1, 0 \le p_2 \le 1, \ldots, 0 \le p_n \le 1$$

and

$$p_1 + p_2 + \cdots + p_n = 1$$

The probability of outcome O_i occurring is said to be p_i, and this is written $P(O_i) = p_i$.

An intuitive interpretation of a set of probability values is that the *larger* the probability value of a particular outcome, the *more likely* it is to happen. If two outcomes have identical probability values assigned to them, then they can be thought of as being equally likely to occur. On the other hand, if one outcome has a larger probability value assigned to it than another outcome, then the first outcome can be thought of as being more likely to occur.

FIGURE 1.6

Probability values for machine
breakdown example

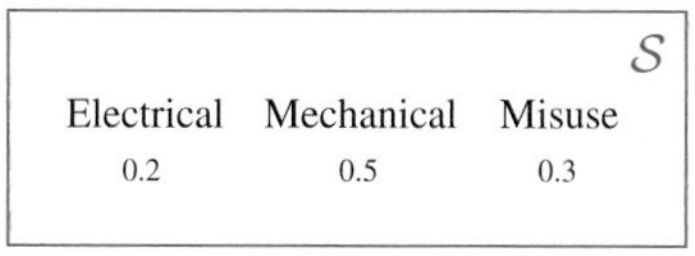

		$\mathcal{S}$
Electrical	Mechanical	Misuse
0.2	0.5	0.3

If a particular outcome has a probability value of one, then the interpretation is that it is certain to occur, so that there is actually no uncertainty in the experiment. In this case all of the other outcomes must necessarily have probability values of zero.

The following examples illustrate the assignment of probability values.

Example 1

Machine Breakdowns

Suppose that the machine breakdowns occur with probability values of $P(\text{electrical}) = 0.2$, $P(\text{mechanical}) = 0.5$, and $P(\text{misuse}) = 0.3$. This is a valid probability assignment since the three probability values 0.2, 0.5, and 0.3 are all between zero and one and they sum to one. Figure 1.6 shows a diagram of these probabilities by recording the respective probability value with each of the outcomes. These probability values indicate that mechanical failures are most likely, with misuse failures being more likely than electrical failures.

In addition, $P(\text{mechanical}) = 0.5$ indicates that about half of the failures will be attributable to mechanical causes. This does not mean that of the next two machine breakdowns, exactly one will be for mechanical reasons, or that in the next ten machine breakdowns, exactly five will be for mechanical reasons. However, it means that in the *long run*, the manager can reasonably expect that roughly half of the breakdowns will be for mechanical reasons. Similarly, in the long run, the manager will expect that about 20% of the breakdowns will be for electrical reasons, and that about 30% of the breakdowns will be attributable to operator misuse.

Example 3

Software Errors

Suppose that the number of errors in a software product has probabilities

$$P(0 \text{ errors}) = 0.05, \quad P(1 \text{ error}) = 0.08, \quad P(2 \text{ errors}) = 0.35,$$
$$P(3 \text{ errors}) = 0.20, \quad P(4 \text{ errors}) = 0.20, \quad P(5 \text{ errors}) = 0.12,$$
$$P(i \text{ errors}) = 0, \quad \text{for } i \geq 6$$

These probabilities show that there are at most five errors since the probability values are zero for six or more errors. In addition, it can be seen that the most likely number of errors is two and that three and four errors are equally likely.

It is reasonable to ask how anybody would ever know the probability assignments in the above two examples. In other words, how would the engineer know that there is a probability of 0.2 that a breakdown will be due to an electrical fault, or how would a computer programmer know that the probability of an error-free product is 0.05? In practice these probabilities would have to be estimated from a collection of data and prior experiences. Later in this book, in Chapters 7 and 10, it will be shown how statistical analysis techniques can be employed to help the engineer and programmer conduct studies to **estimate** probabilities of these kinds.

In some situations, notably games of chance, the experiments are conducted in such a way that all of the possible outcomes can be considered to be equally likely, so that they must be assigned identical probability values. If there are n outcomes in the sample space that are equally likely, then the condition that the probabilities sum to one requires that each probability value be $1/n$.

For a coin toss, the probabilities will in general be given by

$$P(\text{head}) = p, \qquad P(\text{tail}) = 1 - p$$

for some value of p with $0 \le p \le 1$. A fair coin will have $p = 0.5$ so that

$$P(\text{head}) = P(\text{tail}) = 0.5$$

with the two outcomes being equally likely. A biased coin will have $p \neq 0.5$. For example, if $p = 0.6$, then

$$P(\text{head}) = 0.6, \qquad P(\text{tail}) = 0.4$$

as shown in Figure 1.7, and the coin toss is more likely to record a head.

A fair die will have each of the six outcomes equally likely, with each being assigned the same probability. Since the six probabilities must sum to one, this implies that each of the six outcomes must have a probability of $1/6$, so that

$$P(1) = P(2) = P(3) = P(4) = P(5) = P(6) = \frac{1}{6}$$

This case is shown in Figure 1.8. An example of a biased die would be one for which

$$P(1) = 0.10, \quad P(2) = 0.15, \quad P(3) = 0.15,$$
$$P(4) = 0.15, \quad P(5) = 0.15, \quad P(6) = 0.30$$

as in Figure 1.9. In this case the die is most likely to score a 6, which will happen roughly three times out of ten as a long-run average. Scores of 2, 3, 4, and 5 are equally likely, and a score of 1 is the least likely event, happening only one time in ten on average.

If two dice are thrown and each of the 36 outcomes are equally likely (as will be the case with two fair dice that are shaken properly), the probability value of each outcome will necessarily be $1/36$. This is shown in Figure 1.10.

If a card is drawn at random from a pack of cards, then there are 52 possible outcomes in the sample space, and each one is equally likely so that each would be assigned a probability value of $1/52$. Thus, for example, $P(A\heartsuit) = 1/52$, as shown in Figure 1.11. If two cards are drawn with replacement, and if both the cards can be assumed to be chosen at random through suitable shuffling of the pack before and between the drawings, then each of the $52 \times 52 = 2704$ elements of the sample space will be equally likely and hence should each be assigned a probability value of $1/2704$. In this case $P(A\heartsuit, 2\clubsuit) = 1/2704$, for example, as shown in Figure 1.12. If the drawing is performed without replacement but again at random, then the sample space has only 2652 elements and each would have a probability of $1/2652$, as shown in Figure 1.13.

					$\mathcal{S}$
Head			Tail		
0.6			0.4		

FIGURE 1.7

Probability values for a biased coin

					$\mathcal{S}$
1	2	3	4	5	6
1/6	1/6	1/6	1/6	1/6	1/6

FIGURE 1.8

Probability values for a fair die

					$\mathcal{S}$
1	2	3	4	5	6
0.10	0.15	0.15	0.15	0.15	0.30

FIGURE 1.9

Probability values for a biased die

FIGURE 1.10

Probability values for rolling two dice

$\mathcal{S}$

(1, 1) 1/36	(1, 2) 1/36	(1, 3) 1/36	(1, 4) 1/36	(1, 5) 1/36	(1, 6) 1/36
(2, 1) 1/36	(2, 2) 1/36	(2, 3) 1/36	(2, 4) 1/36	(2, 5) 1/36	(2, 6) 1/36
(3, 1) 1/36	(3, 2) 1/36	(3, 3) 1/36	(3, 4) 1/36	(3, 5) 1/36	(3, 6) 1/36
(4, 1) 1/36	(4, 2) 1/36	(4, 3) 1/36	(4, 4) 1/36	(4, 5) 1/36	(4, 6) 1/36
(5, 1) 1/36	(5, 2) 1/36	(5, 3) 1/36	(5, 4) 1/36	(5, 5) 1/36	(5, 6) 1/36
(6, 1) 1/36	(6, 2) 1/36	(6, 3) 1/36	(6, 4) 1/36	(6, 5) 1/36	(6, 6) 1/36

FIGURE 1.11

Probability values for choosing one card

$\mathcal{S}$

A♡	2♡	3♡	4♡	5♡	6♡	7♡	8♡	9♡	10♡	J♡	Q♡	K♡
1/52	1/52	1/52	1/52	1/52	1/52	1/52	1/52	1/52	1/52	1/52	1/52	1/52
A♣	2♣	3♣	4♣	5♣	6♣	7♣	8♣	9♣	10♣	J♣	Q♣	K♣
1/52	1/52	1/52	1/52	1/52	1/52	1/52	1/52	1/52	1/52	1/52	1/52	1/52
A♢	2♢	3♢	4♢	5♢	6♢	7♢	8♢	9♢	10♢	J♢	Q♢	K♢
1/52	1/52	1/52	1/52	1/52	1/52	1/52	1/52	1/52	1/52	1/52	1/52	1/52
A♠	2♠	3♠	4♠	5♠	6♠	7♠	8♠	9♠	10♠	J♠	Q♠	K♠
1/52	1/52	1/52	1/52	1/52	1/52	1/52	1/52	1/52	1/52	1/52	1/52	1/52

FIGURE 1.12

Probability values for choosing two cards with replacement

$\mathcal{S}$

(A♡, A♡) 1/2704	(A♡, 2♡) 1/2704	(A♡, 3♡) 1/2704	$\cdots$	(A♡, Q♠) 1/2704	(A♡, K♠) 1/2704
(2♡, A♡) 1/2704	(2♡, 2♡) 1/2704	(2♡, 3♡) 1/2704	$\cdots$	(2♡, Q♠) 1/2704	(2♡, K♠) 1/2704
(3♡, A♡) 1/2704	(3♡, 2♡) 1/2704	(3♡, 3♡) 1/2704	$\cdots$	(3♡, Q♠) 1/2704	(3♡, K♠) 1/2704
$\vdots$	$\vdots$	$\vdots$		$\vdots$	$\vdots$
(Q♠, A♡) 1/2704	(Q♠, 2♡) 1/2704	(Q♠, 3♡) 1/2704	$\cdots$	(Q♠, Q♠) 1/2704	(Q♠, K♠) 1/2704
(K♠, A♡) 1/2704	(K♠, 2♡) 1/2704	(K♠, 3♡) 1/2704	$\cdots$	(K♠, Q♠) 1/2704	(K♠, K♠) 1/2704

FIGURE 1.13

Probability values for choosing two cards without replacement

$\mathcal{S}$

	(A♡, 2♡) 1/2652	(A♡, 3♡) 1/2652	... ...	(A♡, Q♠) 1/2652	(A♡, K♠) 1/2652
(2♡, A♡) 1/2652		(2♡, 3♡) 1/2652	... ...	(2♡, Q♠) 1/2652	(2♡, K♠) 1/2652
(3♡, A♡) 1/2652	(3♡, 2♡) 1/2652		... ...	(3♡, Q♠) 1/2652	(3♡, K♠) 1/2652
⋮	⋮	⋮		⋮	⋮
(Q♠, A♡) 1/2652	(Q♠, 2♡) 1/2652	(Q♠, 3♡) 1/2652	... ...		(Q♠, K♠) 1/2652
(K♠, A♡) 1/2652	(K♠, 2♡) 1/2652	(K♠, 3♡) 1/2652	... ...	(K♠, Q♠) 1/2652	

1.1.4 Problems

1.1.1 What is the sample space when a coin is tossed three times?

1.1.2 What is the sample space for counting the number of females in a group of n people?

1.1.3 What is the sample space for the number of Aces in a hand of 13 playing cards?

1.1.4 What is the sample space for a person's birthday?

1.1.5 A car repair is performed either on time or late and either satisfactorily or unsatisfactorily. What is the sample space for a car repair?

1.1.6 A bag contains balls that are either red or blue and either dull or shiny. What is the sample space when a ball is chosen from the bag?

1.1.7 A probability value p is often reported as an *odds ratio*, which is $p/(1-p)$. This is the ratio of the probability that the event happens to the probability that the event does not happen.
(a) If the odds ratio is 1, what is p?
(b) If the odds ratio is 2, what is p?
(c) If $p = 0.25$, what is the odds ratio?

1.1.8 An experiment has five outcomes, I, II, III, IV, and V. If $P(\text{I}) = 0.13$, $P(\text{II}) = 0.24$, $P(\text{III}) = 0.07$, and $P(\text{IV}) = 0.38$, what is $P(\text{V})$?

1.1.9 An experiment has five outcomes, I, II, III, IV, and V. If $P(\text{I}) = 0.08$, $P(\text{II}) = 0.20$, and $P(\text{III}) = 0.33$, what are the possible values for the probability of outcome V? If outcomes IV and V are equally likely, what are their probability values?

1.1.10 An experiment has three outcomes, I, II, and III. If outcome I is twice as likely as outcome II, and outcome II is three times as likely as outcome III, what are the probability values of the three outcomes?

1.2 Events

1.2.1 Events and Complements

Interest is often centered not so much on the individual elements of a sample space, but rather on collections of individual outcomes. These collections of outcomes are called **events**.

FIGURE 1.14

$P(A) = 0.10 + 0.15 + 0.30 = 0.55$

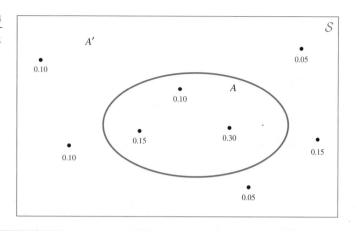

An **event** A is a subset of the sample space S. It collects outcomes of particular interest. The probability of an event A, $P(A)$, is obtained by summing the probabilities of the outcomes contained within the event A.

An event is said to occur if one of the outcomes contained within the event occurs.

Figure 1.14 shows a sample space S consisting of eight outcomes, each of which is labeled with a probability value. Three of the outcomes are contained within the event A. The probability of the event A is calculated as the sum of the probabilities of these three events, so that

$$P(A) = 0.10 + 0.15 + 0.30 = 0.55$$

The **complement** of an event A is taken to mean the event consisting of everything in the sample space S that is not contained within the event A. The notation A' is used for the complement of A. In this example, the probability of the complement of A is obtained by summing the probabilities of the five outcomes not contained within A, so that

$$P(A') = 0.10 + 0.05 + 0.05 + 0.15 + 0.10 = 0.45$$

Notice that $P(A) + P(A') = 1$, which is a general rule.

Complements of Events

The event A', the **complement** of an event A, is the event consisting of everything in the sample space S that is not contained within the event A. In all cases

$$P(A) + P(A') = 1$$

It is useful to consider both individual outcomes and the whole sample space as also being events. Events that consist of an individual outcome are sometimes referred to as **elementary events** or **simple events**. If an event is defined to be a particular single outcome, then its probability is just the probability of that outcome. If an event is defined to be the whole sample space, then obviously its probability is one.

1.2.2   Examples of Events

Example 2
Defective Computer Chips

Consider the following probability values for the number of defective chips in a box of 500 chips:

$$P(0 \text{ defectives}) = 0.02, \quad P(1 \text{ defective}) = 0.11,$$
$$P(2 \text{ defectives}) = 0.16, \quad P(3 \text{ defectives}) = 0.21,$$
$$P(4 \text{ defectives}) = 0.13, \quad P(5 \text{ defectives}) = 0.08$$

and suppose that the probabilities of the additional elements of the sample space (6 defectives, 7 defectives, ..., 500 defectives) are unknown. The company is thinking of claiming that each box has no more than 5 defective chips, and it wishes to calculate the probability that the claim is correct.

The event *correct* consists of the six outcomes listed above, so that

$$\text{correct} = \{0 \text{ defectives}, \ 1 \text{ defective}, \ 2 \text{ defectives}, \ 3 \text{ defectives},$$
$$4 \text{ defectives}, \ 5 \text{ defectives}\} \subset \mathcal{S}$$

The probability of the claim being correct is then

$$P(\text{correct}) = P(0 \text{ defectives}) + \cdots + P(5 \text{ defectives})$$
$$= 0.02 + 0.11 + 0.16 + 0.21 + 0.13 + 0.08 = 0.71$$

Consequently, on average, only about 71% of the boxes will meet the company's claim that there are no more than 5 defective chips. The complement of the event *correct* is that there will be at least 6 defective chips so that the company's claim will be incorrect. This has a probability of $1 - 0.71 = 0.29$.

Example 3
Software Errors

Consider the event A that there are no more than two errors in a software product. This event is given by

$$A = \{0 \text{ errors}, 1 \text{ error}, 2 \text{ errors}\} \subset \mathcal{S}$$

and its probability is

$$P(A) = P(0 \text{ errors}) + P(1 \text{ error}) + P(2 \text{ errors})$$
$$= 0.05 + 0.08 + 0.35 = 0.48$$

The probability of the complement of the event A is

$$P(A') = 1 - P(A) = 1 - 0.48 = 0.52$$

which is the probability that a software product has three or more errors.

Example 4
Power Plant Operation

Consider the probability values given in Figure 1.15, where, for instance, the probability that all three plants are idle is $P((0, 0, 0)) = 0.07$, and the probability that only plant X is idle is $P((0, 1, 1)) = 0.18$. The event that plant X is idle is given by

$$A = \{(0, 0, 0), (0, 0, 1), (0, 1, 0), (0, 1, 1)\}$$

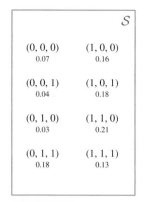

FIGURE 1.15

Probability values for power plant example

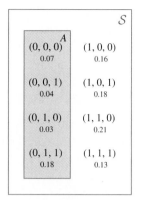

FIGURE 1.16

Event A: plant X idle

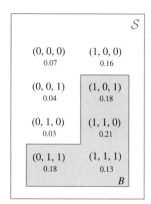

FIGURE 1.17

Event B: at least two plants generating electricity

as illustrated in Figure 1.16, and it has a probability of

$$P(A) = P((0, 0, 0)) + P((0, 0, 1)) + P((0, 1, 0)) + P((0, 1, 1))$$
$$= 0.07 + 0.04 + 0.03 + 0.18 = 0.32$$

The complement of this event is

$$A' = \{(1, 0, 0), (1, 0, 1), (1, 1, 0), (1, 1, 1)\}$$

which corresponds to plant X generating electricity, and it has a probability of

$$P(A') = 1 - P(A) = 1 - 0.32 = 0.68$$

Suppose that the manager is interested in the proportion of the time that at least two out of the three plants are generating electricity. This event is given by

$$B = \{(0, 1, 1), (1, 0, 1), (1, 1, 0), (1, 1, 1)\}$$

as illustrated in Figure 1.17, with a probability of

$$P(B) = P((0, 1, 1)) + P((1, 0, 1)) + P((1, 1, 0)) + P((1, 1, 1))$$
$$= 0.18 + 0.18 + 0.21 + 0.13 = 0.70$$

This result indicates that, on average, at least two of the plants will be generating electricity about 70% of the time. The complement of this event is

$$B' = \{(0, 0, 0), (0, 0, 1), (0, 1, 0), (1, 0, 0)\}$$

which corresponds to the situation in which at least two of the plants are idle. The probability of this is

$$P(B') = 1 - P(B) = 1 - 0.70 = 0.30$$

FIGURE 1.18

Event A: sum equal to 6

(1, 1) 1/36	(1, 2) 1/36	(1, 3) 1/36	(1, 4) 1/36	(1, 5) 1/36	(1, 6) 1/36
(2, 1) 1/36	(2, 2) 1/36	(2, 3) 1/36	(2, 4) 1/36	(2, 5) 1/36	(2, 6) 1/36
(3, 1) 1/36	(3, 2) 1/36	(3, 3) 1/36	(3, 4) 1/36	(3, 5) 1/36	(3, 6) 1/36
(4, 1) 1/36	(4, 2) 1/36	(4, 3) 1/36	(4, 4) 1/36	(4, 5) 1/36	(4, 6) 1/36
(5, 1) 1/36	(5, 2) 1/36	(5, 3) 1/36	(5, 4) 1/36	(5, 5) 1/36	(5, 6) 1/36
(6, 1) 1/36	(6, 2) 1/36	(6, 3) 1/36	(6, 4) 1/36	(6, 5) 1/36	(6, 6) 1/36

GAMES OF CHANCE

The event that an *even* score is recorded on the roll of a die is given by

$$\text{even} = \{2, 4, 6\}$$

For a fair die this event would have a probability of

$$P(\text{even}) = P(2) + P(4) + P(6) = \frac{1}{6} + \frac{1}{6} + \frac{1}{6} = \frac{1}{2}$$

Figure 1.18 shows the event that the sum of the scores of two dice is equal to 6. This event is given by

$$A = \{(1, 5), (2, 4), (3, 3), (4, 2), (5, 1)\}$$

If each outcome is equally likely with a probability of $1/36$, then this event clearly has a probability of $5/36$. A sum of 6 will be obtained with two fair dice roughly 5 times out of 36 on average, that is, on about 14% of the throws. The probabilities of obtaining other sums can be obtained in a similar manner, and it is seen that 7 is the most likely score, with a probability of $6/36 = 1/6$. The least likely scores are 2 and 12, each with a probability of $1/36$.

Figure 1.19 shows the event that at least one of the two dice records a 6, which is seen to have a probability of $11/36$. The complement of this event is the event that neither die records a 6, with a probability of $1 - 11/36 = 25/36$.

Figure 1.20 illustrates the event that a card drawn from a pack of cards belongs to the heart suit. This event consists of the 13 outcomes corresponding to the 13 cards in the heart suit. If the drawing is done at random, with each of the 52 possible outcomes being equally likely with a probability of $1/52$, then the probability of drawing a heart is clearly $13/52 = 1/4$. This result makes sense since there are four suits that are equally likely. Figure 1.21 illustrates the event that a picture card (Jack, Queen, or King) is drawn, with a probability of $12/52 = 3/13$.

FIGURE 1.19

Event *B*: at least one 6 recorded

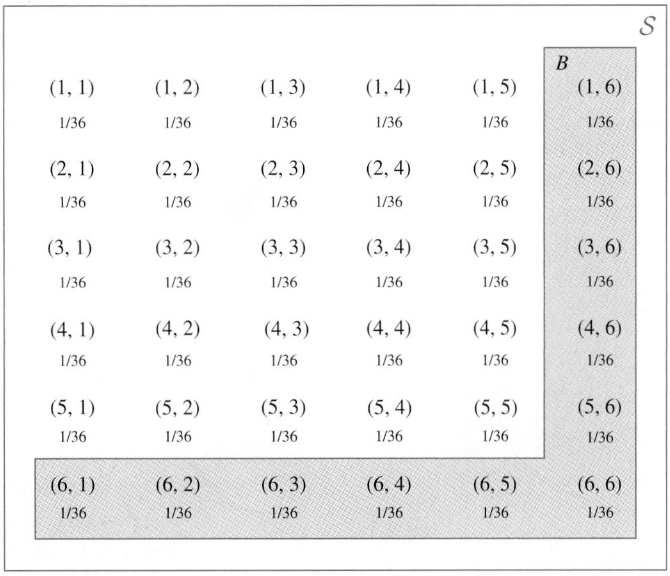

S

					B
(1, 1)	(1, 2)	(1, 3)	(1, 4)	(1, 5)	(1, 6)
1/36	1/36	1/36	1/36	1/36	1/36
(2, 1)	(2, 2)	(2, 3)	(2, 4)	(2, 5)	(2, 6)
1/36	1/36	1/36	1/36	1/36	1/36
(3, 1)	(3, 2)	(3, 3)	(3, 4)	(3, 5)	(3, 6)
1/36	1/36	1/36	1/36	1/36	1/36
(4, 1)	(4, 2)	(4, 3)	(4, 4)	(4, 5)	(4, 6)
1/36	1/36	1/36	1/36	1/36	1/36
(5, 1)	(5, 2)	(5, 3)	(5, 4)	(5, 5)	(5, 6)
1/36	1/36	1/36	1/36	1/36	1/36
(6, 1)	(6, 2)	(6, 3)	(6, 4)	(6, 5)	(6, 6)
1/36	1/36	1/36	1/36	1/36	1/36

FIGURE 1.20

Event *A*: card belongs to heart suit

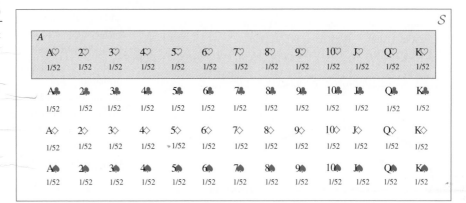

FIGURE 1.21

Event *B*: picture card is chosen

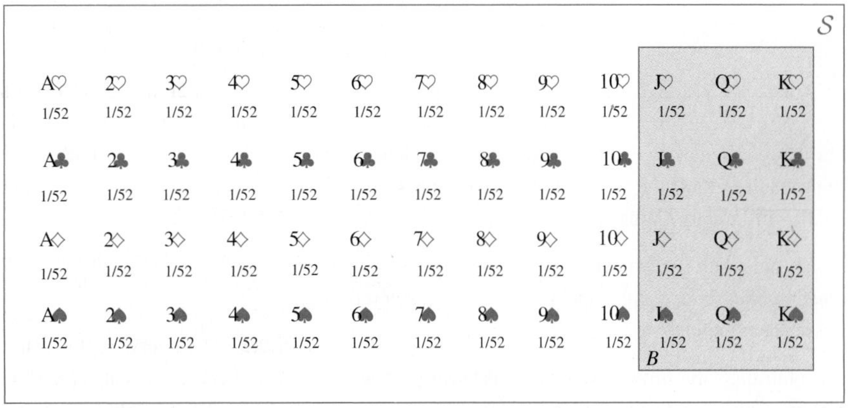

■ 1.2.3   Problems

1.2.1 Consider the sample space in Figure 1.22 with outcomes a, b, c, d, and e. Calculate:
(a) $P(b)$ (b) $P(A)$ (c) $P(A')$

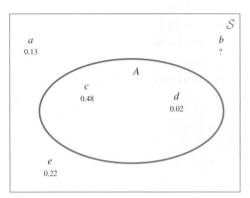

FIGURE 1.22

1.2.2 Consider the sample space in Figure 1.23 with outcomes a, b, c, d, e, and f. If $P(A) = 0.27$, calculate:
(a) $P(b)$ (b) $P(A')$ (c) $P(d)$

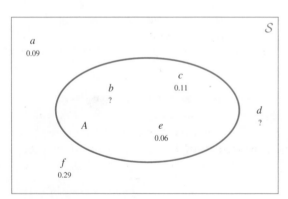

FIGURE 1.23

1.2.3 If birthdays are equally likely to fall on any day, what is the probability that a person chosen at random has a birthday in January? What about February?

1.2.4 If a fair die is thrown, what is the probability of scoring a *prime* number (suppose that the number 1 is considered to be a prime number)?

1.2.5 If two fair dice are thrown, what is the probability that at least one score is a prime number? What is the complement of this event? What is its probability?

1.2.6 Two fair dice are thrown, one red and one blue. What is the probability that the red die has a score that is *strictly greater* than the score of the blue die? Why is this probability less than 0.5? What is the complement of this event?

1.2.7 If a card is chosen at random from a pack of cards, what is the probability that the card is from one of the two black suits?

1.2.8 If a card is chosen at random from a pack of cards, what is the probability that it is an Ace?

1.2.9 A *winner* and a *runner-up* are decided in a tournament of four players, one of whom is Terica. If all the outcomes are equally likely, what is the probability that
(a) Terica is the winner?
(b) Terica is either the winner or the runner-up?

1.2.10 Three types of batteries are being tested, type I, type II, and type III. The outcome (I, II, III) denotes that the battery of type I fails first, the battery of type II next, and the battery of type III lasts the longest. The probabilities of the six outcomes are given in Figure 1.24. What is the probability that
(a) the type I battery lasts longest?
(b) the type I battery lasts shortest?
(c) the type I battery does not last longest?
(d) the type I battery lasts longer than the type II battery?
(This problem is continued in Problem 1.4.9.)

	$\mathcal{S}$
(I, II, III)	(I, III, II)
0.11	0.07
(II, I, III)	(II, III, I)
0.24	0.39
(III, I, II)	(III, II, I)
0.16	0.03

FIGURE 1.24

Probability values for battery lifetimes

1.2.11 A factory has two assembly lines, each of which is *shut down* (S), at *partial capacity* (P), or at *full capacity* (F). The sample space is given in Figure 1.25, where, for example, (S, P) denotes that the first assembly line is

		$\mathcal{S}$
(S, S)	(S, P)	(S, F)
0.02	0.06	0.05
(P, S)	(P, P)	(P, F)
0.07	0.14	0.20
(F, S)	(F, P)	(F, F)
0.06	0.21	0.19

FIGURE 1.25

Probability values for assembly line operations

shut down and the second one is operating at partial capacity. What is the probability that

(a) both assembly lines are shut down?
(b) neither assembly line is shut down?
(c) at least one assembly line is at full capacity?
(d) exactly one assembly line is at full capacity?

What is the complement of the event in part (b)? What is the complement of the event in part (c)?

(This problem is continued in Problem 1.4.10.)

1.2.12 A fair coin is tossed three times. What is the probability that two heads will be obtained *in succession*?

1.3 Combinations of Events

In general, more than one event will be of interest for a particular experiment and sample space. For two events A and B, in addition to the consideration of the probability of event A occurring and the probability of event B occurring, it is often important to consider other probabilities such as the probability of **both** events occurring simultaneously. Other quantities of interest may be the probability that **neither** event A nor event B occurs, the probability that **at least one** of the two events occurs, or the probability that event A occurs, but event B does not.

1.3.1 Intersections of Events

Consider first the calculation of the probability that both events occur simultaneously. This can be done by defining a new event to consist of the outcomes that are in both event A and event B.

> **Intersections of Events**
>
> The event $A \cap B$ is the **intersection** of the events A and B and consists of the outcomes that are contained within both events A and B. The probability of this event, $P(A \cap B)$, is the probability that both events A and B occur simultaneously.

Figure 1.26 shows a sample space $\mathcal{S}$ that consists of nine outcomes. Event A consists of three outcomes, and its probability is given by

$$P(A) = 0.01 + 0.07 + 0.19 = 0.27$$

Event B consists of five outcomes, and its probability is given by

$$P(B) = 0.07 + 0.19 + 0.04 + 0.14 + 0.12 = 0.56$$

The intersection of these two events, shown in Figure 1.27, consists of the two outcomes that are contained within both events A and B. It has a probability of

$$P(A \cap B) = 0.07 + 0.19 = 0.26$$

which is the probability that both events A and B occur simultaneously.

FIGURE 1.26

Events A and B

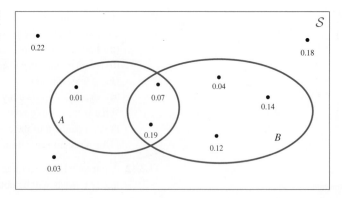

FIGURE 1.27

The event $A \cap B$

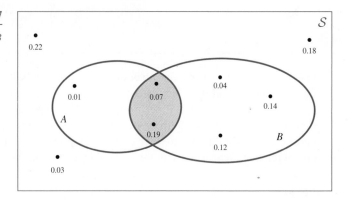

Event A', the complement of the event A, is the event consisting of the six outcomes that are not in event A. Notice that there are obviously no outcomes in $A \cap A'$, and this is written

$$A \cap A' = \emptyset$$

where $\emptyset$ is referred to as the "empty set," a set that does not contain anything. Consequently,

$$P(A \cap A') = P(\emptyset) = 0$$

and it is impossible for the event A to occur at the same time as its complement.

A more interesting event is the event $A' \cap B$ illustrated in Figure 1.28. This event consists of the three outcomes that are contained within event B but that are not contained within event A. It has a probability of

$$P(A' \cap B) = 0.04 + 0.14 + 0.12 = 0.30$$

which is the probability that event B occurs but event A does not occur. Similarly, Figure 1.29 shows the event $A \cap B'$, which has a probability of

$$P(A \cap B') = 0.01$$

This is the probability that event A occurs but event B does not.

FIGURE 1.28

The event $A' \cap B$

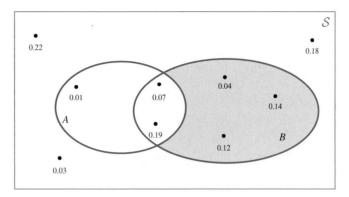

FIGURE 1.29

The event $A \cap B'$

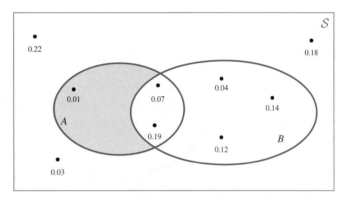

Notice that

$$P(A \cap B) + P(A \cap B') = 0.26 + 0.01 = 0.27 = P(A)$$

and similarly that

$$P(A \cap B) + P(A' \cap B) = 0.26 + 0.30 = 0.56 = P(B)$$

The following two equalities hold in general for all events A and B:

$$P(A \cap B) + P(A \cap B') = P(A) \qquad P(A \cap B) + P(A' \cap B) = P(B)$$

Two events A and B that have *no* outcomes in common are said to be **mutually exclusive** events. In this case $A \cap B = \emptyset$ and $P(A \cap B) = 0$.

Mutually Exclusive Events

Two events A and B are said to be **mutually exclusive** if $A \cap B = \emptyset$ so that they have no outcomes in common.

Figure 1.30 illustrates a sample space $\mathcal{S}$ that consists of seven outcomes, three of which are contained within event A and two of which are contained within event B. Since no outcomes are contained within both events A and B, the two events are mutually exclusive.

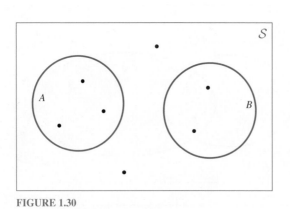

FIGURE 1.30

A and *B* are mutually exclusive events

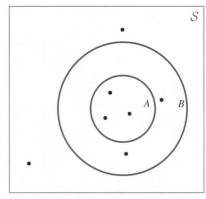

FIGURE 1.31

$A \subset B$

Finally, Figure 1.31 illustrates a situation where an event A is contained within an event B, that is, $A \subset B$. Each outcome in event A is also contained in event B. It is clear that in this case $A \cap B = A$.

Some other simple results concerning the intersections of events are as follows:

$$A \cap B = B \cap A \qquad A \cap A = A$$
$$A \cap S = A \qquad A \cap \emptyset = \emptyset$$
$$A \cap A' = \emptyset \qquad A \cap (B \cap C) = (A \cap B) \cap C$$

1.3.2   Unions of Events

The event that *at least one* out of two events A and B occurs, shown in Figure 1.32, is denoted by $A \cup B$ and is referred to as the **union** of events A and B. The probability of this event, $P(A \cup B)$, is the sum of the probability values of the outcomes that are in either of events A or B (including those events that are in both events A and B).

> **Unions of Events**
>
> The event $A \cup B$ is the **union** of events A and B and consists of the outcomes that are contained within at least one of the events A and B. The probability of this event, $P(A \cup B)$, is the probability that at least one of the events A and B occurs.

Notice that the outcomes in the event $A \cup B$ can be classified into three kinds. They are

1. in event A, but not in event B

2. in event B, but not in event A

3. in both events A and B

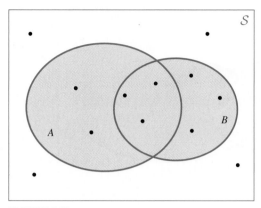

FIGURE 1.32

The event $A \cup B$

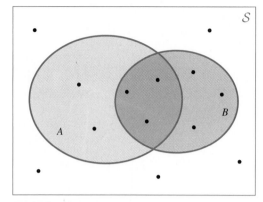

FIGURE 1.33

Decomposition of the event $A \cup B$

The outcomes of type 1 form the event $A \cap B'$, the outcomes of type 2 form the event $A' \cap B$, and the outcomes of type 3 form the event $A \cap B$, as shown in Figure 1.33. Since the probability of $A \cup B$ is obtained as the sum of the probability values of the outcomes within these three (mutually exclusive) events, the following result is obtained:

$$P(A \cup B) = P(A \cap B') + P(A' \cap B) + P(A \cap B)$$

This equality can be presented in another form using the relationships

$$P(A \cap B') = P(A) - P(A \cap B)$$

and

$$P(A' \cap B) = P(B) - P(A \cap B)$$

Substituting in these expressions for $P(A \cap B')$ and $P(A' \cap B)$ gives the following result:

$$P(A \cup B) = P(A) + P(B) - P(A \cap B)$$

This equality has the intuitive interpretation that the probability of at least one of the events A and B occurring can be obtained by adding the probabilities of the two events A and B and then subtracting the probability that both the events occur simultaneously. The probability that both events occur, $P(A \cap B)$, needs to be subtracted since the probability values of the outcomes in the intersection $A \cap B$ have been counted *twice*, once in $P(A)$ and once in $P(B)$.

Notice that if events A and B are mutually exclusive, so that no outcomes are in $A \cap B$ and $P(A \cap B) = 0$ as in Figure 1.30, then $P(A \cup B)$ can just be obtained as the sum of the probabilities of events A and B.

If the events A and B are mutually exclusive so that $P(A \cap B) = 0$, then

$$P(A \cup B) = P(A) + P(B)$$

FIGURE 1.34

The event $A \cup B$

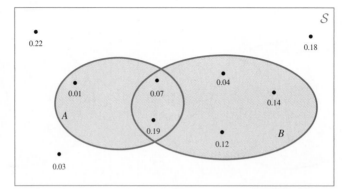

FIGURE 1.35

The event $A' \cup B'$

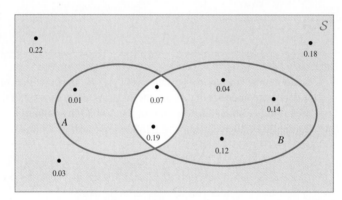

The sample space of nine outcomes illustrated in Figure 1.26 can be used to demonstrate some general relationships between unions and intersections of events. For this example, the event $A \cup B$ consists of the six outcomes illustrated in Figure 1.34, and it has a probability of

$$P(A \cup B) = 0.01 + 0.07 + 0.19 + 0.04 + 0.14 + 0.12 = 0.57$$

The event $(A \cup B)'$, which is the complement of the union of the events A and B, consists of the three outcomes that are neither in event A nor in event B. It has a probability of

$$P((A \cup B)') = 0.03 + 0.22 + 0.18 = 0.43 = 1 - P(A \cup B)$$

Notice that the event $(A \cup B)'$ can also be written as $A' \cap B'$ since it consists of those outcomes that are simultaneously neither in event A nor in event B. This is a general result:

$$(A \cup B)' = A' \cap B'$$

Furthermore, the event $A' \cup B'$ consists of the seven outcomes illustrated in Figure 1.35, and it has a probability of

$$P(A' \cup B') = 0.01 + 0.03 + 0.22 + 0.18 + 0.12 + 0.14 + 0.04 = 0.74$$

However, this event can also be written as $(A \cap B)'$ since it consists of the outcomes that are in the complement of the intersection of sets A and B. Hence, its probability could have been calculated by

$$P(A' \cup B') = P((A \cap B)') = 1 - P(A \cap B) = 1 - 0.26 = 0.74$$

Again, this is a general result:

$$(A \cap B)' = A' \cup B'$$

Finally, if event A is contained within event B, $A \subset B$, as shown in Figure 1.31, then clearly $A \cup B = B$.

Some other simple results concerning the unions of events are as follows:

$$A \cup B = B \cup A \qquad A \cup A = A \qquad A \cup \mathcal{S} = \mathcal{S}$$
$$A \cup \emptyset = A \qquad A \cup A' = \mathcal{S} \qquad A \cup (B \cup C) = (A \cup B) \cup C$$

1.3.3 Examples of Intersections and Unions

Example 4
Power Plant Operation

Consider again Figures 1.15, 1.16, and 1.17, and recall that event A, the event that plant X is idle, has a probability of 0.32, and that event B, the event that at least two out of the three plants are generating electricity, has a probability of 0.70.

The event $A \cap B$ consists of the outcomes for which plant X is idle *and* at least two out of the three plants are generating electricity. Clearly, the only outcome of this kind is the one where plant X is idle and both plants Y and Z are generating electricity, so that

$$A \cap B = \{(0, 1, 1)\}$$

as illustrated in Figure 1.36. Consequently,

$$P(A \cap B) = P((0, 1, 1)) = 0.18$$

The event $A \cup B$ consists of outcomes where *either* plant X is idle *or* at least two plants are generating electricity (or both). Seven out of the eight outcomes satisfy this condition, so that

$$A \cup B = \{(0, 0, 0), (0, 0, 1), (0, 1, 0), (0, 1, 1), (1, 0, 1), (1, 1, 0), (1, 1, 1)\}$$

as illustrated in Figure 1.37. The probability of the event $A \cup B$ is thus

$$P(A \cup B) = P((0, 0, 0)) + P((0, 0, 1)) + P((0, 1, 0)) + P((0, 1, 1))$$
$$+ P((1, 0, 1)) + P((1, 1, 0)) + P((1, 1, 1))$$
$$= 0.07 + 0.04 + 0.03 + 0.18 + 0.18 + 0.21 + 0.13 = 0.84$$

Another way of calculating this probability is

$$P(A \cup B) = P(A) + P(B) - P(A \cap B)$$
$$= 0.32 + 0.70 - 0.18 = 0.84$$

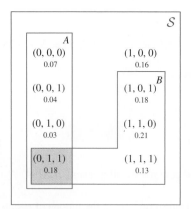

FIGURE 1.36

The event $A \cap B$

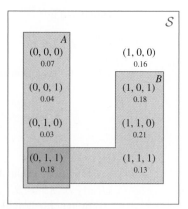

FIGURE 1.37

The event $A \cup B$

Still another way is to notice that the complement of the event $A \cup B$ consists of the single outcome $(1, 0, 0)$, which has a probability value of 0.16, so that

$$P(A \cup B) = 1 - P((A \cup B)') = 1 - P((1, 0, 0)) = 1 - 0.16 = 0.84$$

Example 5

Television Set Quality

A company that manufactures television sets performs a final quality check on each appliance before packing and shipping it. The quality check has two components, the first being an evaluation of the quality of the *picture* obtained on the television set, and the second being an evaluation of the *appearance* of the television set, which looks for scratches or other visible deformities on the appliance. Each of the two evaluations is graded as *Perfect*, *Good*, *Satisfactory*, or *Fail*. The 16 outcomes are illustrated in Figure 1.38 together with a set of probability values, where the notation (P, G), for example, means that an appliance has a *Perfect* picture and a *Good* appearance.

The company has decided that an appliance that fails on either of the two evaluations will not be shipped. Furthermore, as an additional conservative measure to safeguard its reputation, it has decided that appliances that score an evaluation of *Satisfactory* on *both* accounts will also not be shipped.

An initial question of interest concerns the probability that an appliance cannot be shipped. This event A, say, consists of the outcomes

$$A = \{(F, P), (F, G), (F, S), (F, F), (P, F), (G, F), (S, F), (S, S)\}$$

as illustrated in Figure 1.39. The probability that an appliance cannot be shipped is then

$$\begin{aligned}
P(A) &= P((F, P)) + P((F, G)) + P((F, S)) + P((F, F)) + P((P, F)) \\
&\quad + P((G, F)) + P((S, F)) + P((S, S)) \\
&= 0.004 + 0.011 + 0.009 + 0.008 + 0.007 + 0.012 + 0.010 + 0.013 \\
&= 0.074
\end{aligned}$$

In the long run about 7.4% of the television sets will fail the quality check.

From a technical point of view, the company is also interested in the probability that an appliance has a picture that is graded as either *Satisfactory* or *Fail*. This event B, say, is

(P, P)	(P, G)	(P, S)	(P, F)	$\mathcal{S}$
0.140	0.102	0.157	0.007	
(G, P)	(G, G)	(G, S)	(G, F)	
0.124	0.141	0.139	0.012	
(S, P)	(S, G)	(S, S)	(S, F)	
0.067	0.056	0.013	0.010	
(F, P)	(F, G)	(F, S)	(F, F)	
0.004	0.011	0.009	0.008	

FIGURE 1.38

Probability values for television set example

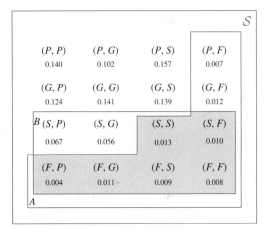

FIGURE 1.39

Event A: appliance not shipped

(P, P)	(P, G)	(P, S)	(P, F)	$\mathcal{S}$
0.140	0.102	0.157	0.007	
(G, P)	(G, G)	(G, S)	(G, F)	
0.124	0.141	0.139	0.012	
B (S, P)	(S, G)	(S, S)	(S, F)	
0.067	0.056	0.013	0.010	
(F, P)	(F, G)	(F, S)	(F, F)	
0.004	0.011	0.009	0.008	

FIGURE 1.40

Event B: picture *Satisfactory* or *Fail*

(P, P)	(P, G)	(P, S)	(P, F)	$\mathcal{S}$
0.140	0.102	0.157	0.007	
(G, P)	(G, G)	(G, S)	(G, F)	
0.124	0.141	0.139	0.012	
B (S, P)	(S, G)	(S, S)	(S, F)	
0.067	0.056	0.013	0.010	
(F, P)	(F, G)	(F, S)	(F, F)	
0.004	0.011	0.009	0.008	

FIGURE 1.41

Event $A \cap B$

illustrated in Figure 1.40, and it has a probability of

$$P(B) = P((F, P)) + P((F, G)) + P((F, S)) + P((F, F)) + P((S, P))$$
$$+ P((S, G)) + P((S, S)) + P((S, F))$$
$$= 0.004 + 0.011 + 0.009 + 0.008 + 0.067 + 0.056 + 0.013 + 0.010$$
$$= 0.178$$

The event $A \cap B$ consists of outcomes where the appliance is not shipped *and* the picture is evaluated as being either *Satisfactory* or *Fail*. It contains the six outcomes illustrated in Figure 1.41, and it has a probability of

$$P(A \cap B) = P((F, P)) + P((F, G)) + P((F, S)) + P((F, F))$$
$$+ P((S, S)) + P((S, F))$$
$$= 0.004 + 0.011 + 0.009 + 0.008 + 0.013 + 0.010 = 0.055$$

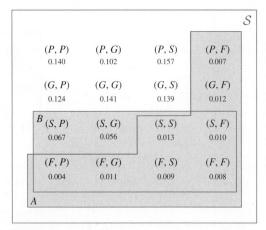

FIGURE 1.42

Event $A \cup B$

FIGURE 1.43

Event $A \cap B'$

The event $A \cup B$ consists of outcomes where the appliance was *either* not shipped *or* the picture was evaluated as being either *Satisfactory* or *Fail*. It contains the 10 outcomes illustrated in Figure 1.42, and its probability can be obtained either by summing the individual probability values of these ten outcomes or more simply as

$$P(A \cup B) = P(A) + P(B) - P(A \cap B)$$
$$= 0.074 + 0.178 - 0.055 = 0.197$$

Television sets that have a picture evaluation of either *Perfect* or *Good* but that cannot be shipped constitute the event $A \cap B'$. This event is illustrated in Figure 1.43 and consists of the outcomes

$$A \cap B' = \{(P, F), (G, F)\}$$

It has a probability of

$$P(A \cap B') = P((P, F)) + P((G, F)) = 0.007 + 0.012 = 0.019$$

Notice that

$$P(A \cap B) + P(A \cap B') = 0.055 + 0.019 = 0.074 = P(A)$$

as expected.

GAMES OF CHANCE

The event A that an even score is obtained from a roll of a die is

$$A = \{2, 4, 6\}$$

If the event B, a high score, is defined to be

$$B = \{4, 5, 6\}$$

then

$$A \cap B = \{4, 6\} \quad \text{and} \quad A \cup B = \{2, 4, 5, 6\}$$

If a fair die is used, then $P(A \cap B) = 2/6 = 1/3$, and $P(A \cup B) = 4/6 = 2/3$.

If two dice are thrown, recall that Figure 1.18 illustrates the event A, that the sum of the scores is equal to 6, and Figure 1.19 illustrates the event B, that at least one of the two dice records a 6. If all the outcomes are equally likely with a probability of $1/36$, then $P(A) = 5/36$ and $P(B) = 11/36$. Since there are no outcomes in both events A and B,

$$A \cap B = \emptyset$$

and $P(A \cap B) = 0$. Consequently, the events A and B are mutually exclusive.

The event $A \cup B$ consists of the five outcomes in event A together with the 11 outcomes in event B, and its probability is

$$P(A \cup B) = \frac{16}{36} = \frac{4}{9} = P(A) + P(B)$$

If one die is red and the other is blue, then Figure 1.44 illustrates the event C, say, that an even score is obtained on the red die, and Figure 1.45 illustrates the event D, say, that an even score is obtained on the blue die. Figure 1.46 then illustrates the event $C \cap D$, which is the event that both dice have even scores. If all outcomes are equally likely, then this event has a probability of $9/36 = 1/4$. Figure 1.47 illustrates the event $C \cup D$, the event that at least one die has an even score. This event has a probability of $27/36 = 3/4$. Notice that $(C \cup D)'$, the complement of the event $C \cup D$, is just the event that both dice have odd scores.

Recall that Figure 1.20 illustrates the event A, that a card drawn from a pack of cards belongs to the heart suit, and Figure 1.21 illustrates the event B, that a picture card is drawn.

FIGURE 1.44

Event C: even score on red die

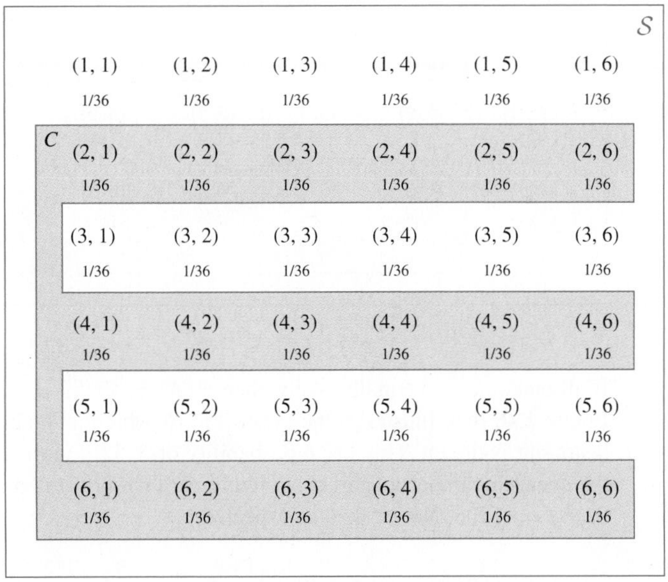

FIGURE 1.45

Event D: even score on blue die

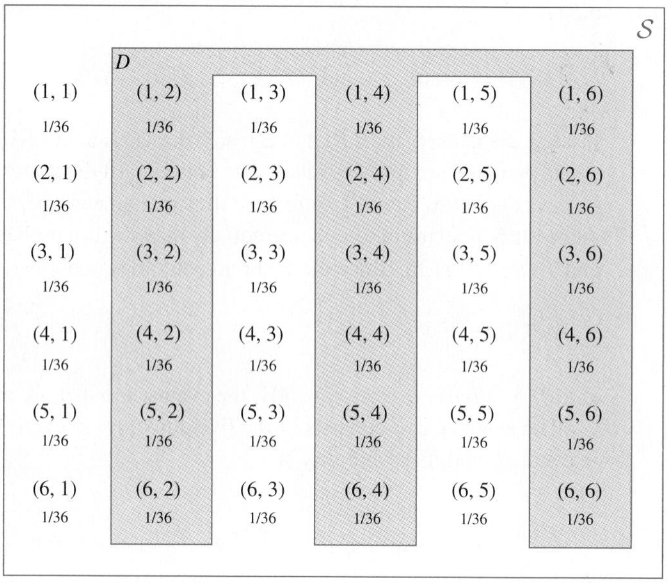

FIGURE 1.46

Event $C \cap D$

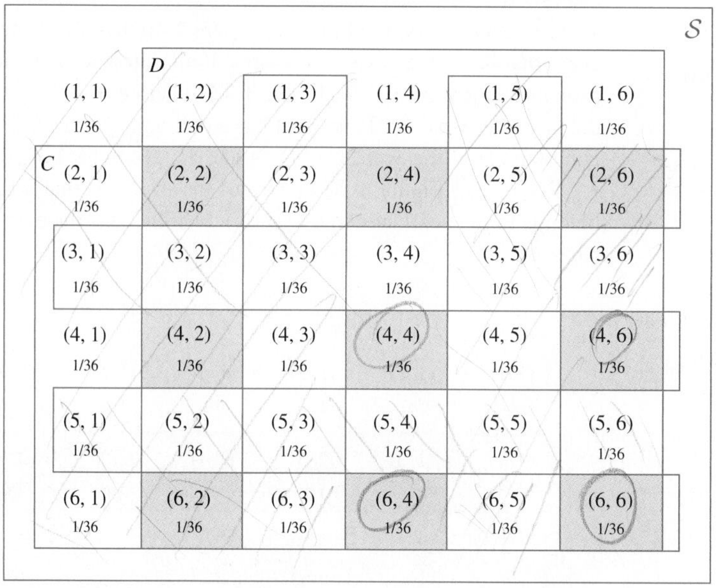

If all outcomes are equally likely, then $P(A) = 13/52 = 1/4$, and $P(B) = 12/52 = 3/13$. Figure 1.48 then illustrates the event $A \cap B$, which is the event that a picture card from the heart suit is drawn. This has a probability of $3/52$. Figure 1.49 illustrates the event $A \cup B$, the event that either a heart or a picture card (or both) is drawn, which has a probability of $22/52 = 11/26$. Notice that, as expected,

$$P(A) + P(B) - P(A \cap B) = \frac{13}{52} + \frac{12}{52} - \frac{3}{52} = \frac{22}{52} = P(A \cup B)$$

FIGURE 1.47

Event $C \cup D$

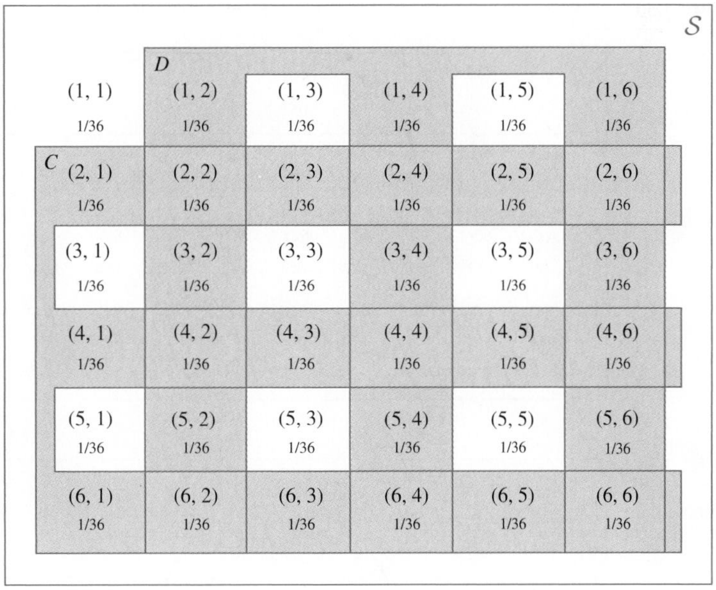

FIGURE 1.48

Event $A \cap B$

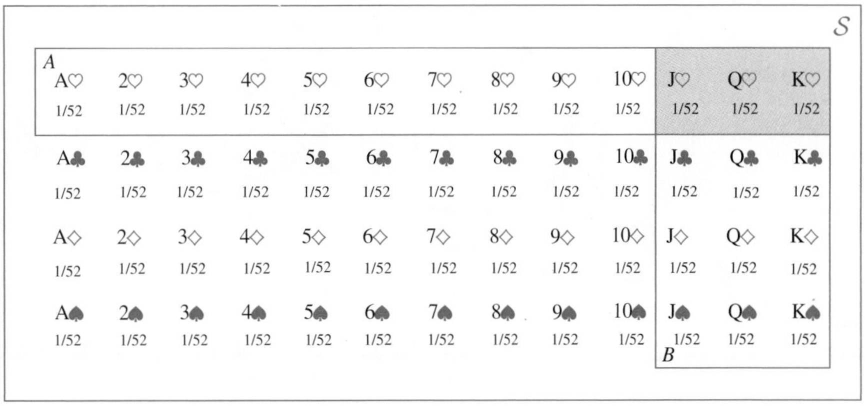

FIGURE 1.49

Event $A \cup B$

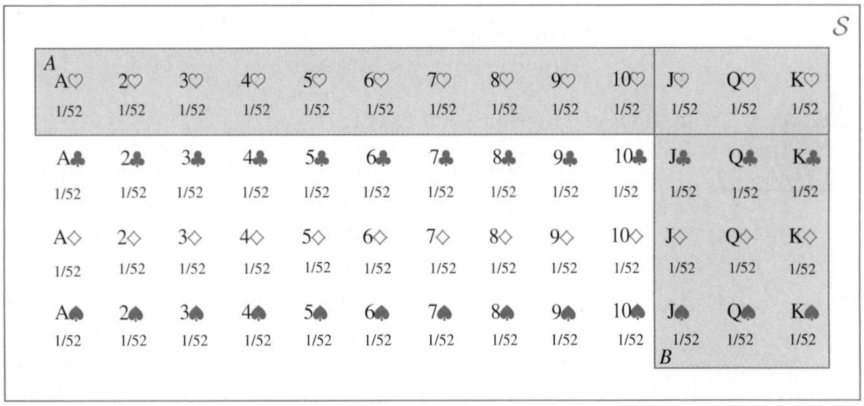

FIGURE 1.50

Event $A' \cap B$

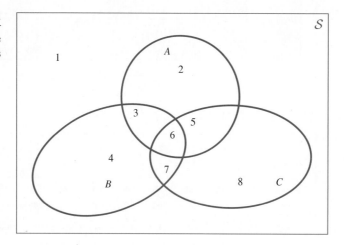

													$\mathcal{S}$
A A♡	2♡	3♡	4♡	5♡	6♡	7♡	8♡	9♡	10♡	J♡	Q♡	K♡	
1/52	1/52	1/52	1/52	1/52	1/52	1/52	1/52	1/52	1/52	1/52	1/52	1/52	
A♣	2♣	3♣	4♣	5♣	6♣	7♣	8♣	9♣	10♣	J♣	Q♣	K♣	
1/52	1/52	1/52	1/52	1/52	1/52	1/52	1/52	1/52	1/52	1/52	1/52	1/52	
A♢	2♢	3♢	4♢	5♢	6♢	7♢	8♢	9♢	10♢	J♢	Q♢	K♢	
1/52	1/52	1/52	1/52	1/52	1/52	1/52	1/52	1/52	1/52	1/52	1/52	1/52	
A♠	2♠	3♠	4♠	5♠	6♠	7♠	8♠	9♠	10♠	J♠	Q♠	K♠	
1/52	1/52	1/52	1/52	1/52	1/52	1/52	1/52	1/52	1/52	1/52 B	1/52	1/52	

FIGURE 1.51

Three events decompose the sample space into eight regions

Finally, Figure 1.50 illustrates the event $A' \cap B$, which is the event that a picture card from a suit other than the heart suit is drawn. It has a probability of 9/52. Again, notice that

$$P(A \cap B) + P(A' \cap B) = \frac{3}{52} + \frac{9}{52} = \frac{12}{52} = P(B)$$

as expected.

1.3.4   Combinations of Three or More Events

Intersections and unions can be extended in an obvious manner to three or more events. Figure 1.51 illustrates how three events A, B, and C can divide a sample space into eight distinct and separate regions. The event A, for example, is composed of the regions 2, 3, 5, and 6, and the event $A \cap B$ is composed of the regions 3 and 6.

The event $A \cap B \cap C$, the intersection of the events A, B, and C, consists of the outcomes that are simultaneously contained within all three events A, B, and C. In Figure 1.51 it corresponds to region 6. The event $A \cup B \cup C$, the union of the events A, B, and C, consists of the outcomes that are in at least one of the three events A, B, and C. In Figure 1.51 it corresponds to all of the regions except for region 1. Hence region 1 can be referred to as $(A \cup B \cup C)'$ since it is the complement of the event $A \cup B \cup C$.

In general, care must be taken to avoid ambiguities when specifying combinations of three or more events. For example, the expression

$$A \cup B \cap C$$

is ambiguous since the two events

$$A \cup (B \cap C) \quad \text{and} \quad (A \cup B) \cap C$$

are different. In Figure 1.51 the event $B \cap C$ is composed of regions 6 and 7, so $A \cup (B \cap C)$ is composed of regions 2, 3, 5, 6, and 7. In contrast, the event $A \cup B$ is composed of regions 2, 3, 4, 5, 6, and 7, so $(A \cup B) \cap C$ is composed of just regions 5, 6, and 7.

Figure 1.51 can also be used to justify the following general expression for the probability of the union of three events:

Union of Three Events

The probability of the **union of three events** A, B, and C is the sum of the probability values of the simple outcomes that are contained within at least one of the three events. It can also be calculated from the expression

$$P(A \cup B \cup C) = [P(A) + P(B) + P(C)]$$
$$- [P(A \cap B) + P(A \cap C) + P(B \cap C)] + P(A \cap B \cap C)$$

The expression for $P(A \cup B \cup C)$ can be checked by matching up the regions in Figure 1.51 with the various terms in the expression. The required probability, $P(A \cup B \cup C)$, is the sum of the probability values of the outcomes in regions 2, 3, 4, 5, 6, 7, and 8. However, the sum of the probabilities $P(A)$, $P(B)$, and $P(C)$ counts regions 3, 5, and 7 *twice*, and region 6 *three* times. Subtracting the probabilities $P(A \cap B)$, $P(A \cap C)$, and $P(B \cap C)$ removes the double counting of regions 3, 5, and 7 but also subtracts the probability of region 6 *three* times. The expression is then completed by adding back on $P(A \cap B \cap C)$, the probability of region 6.

Figure 1.52 illustrates three events A, B, and C that are mutually exclusive because no two events have any outcomes in common. In this case,

$$P(A \cup B \cup C) = P(A) + P(B) + P(C)$$

because the event intersections all have probabilities of zero. More generally, for a sequence $A_1, A_2, \ldots, A_n$ of mutually exclusive events where no two of the events have any outcomes

---

FIGURE 1.52

Three mutually exclusive events

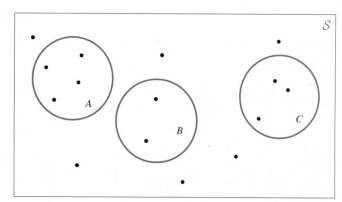

in common, the probability of the union of the events can be obtained by summing the probabilities of the individual events.

Union of Mutually Exclusive Events

For a sequence $A_1, A_2, \ldots, A_n$ of **mutually exclusive events**, the probability of the **union** of the events is given by

$$P(A_1 \cup \cdots \cup A_n) = P(A_1) + \cdots + P(A_n)$$

If a sequence $A_1, A_2, \ldots, A_n$ of mutually exclusive events has the additional property that their union consists of the whole sample space $\mathcal{S}$, then they are said to be an **exhaustive** sequence. They are also said to provide a **partition** of the sample space.

Sample Space Partitions

A **partition** of a sample space is a sequence $A_1, A_2, \ldots, A_n$ of *mutually exclusive* events for which

$$A_1 \cup \cdots \cup A_n = \mathcal{S}$$

Each outcome in the sample space is then contained within one and only one of the events A_i.

Figure 1.53 illustrates a partition of a sample space $\mathcal{S}$ into eight mutually exclusive events.

Example 5
Television Set Quality

In addition to the events A and B discussed before, consider also the event C that an appliance is of "mediocre quality." The event is defined to be appliances that score either *Satisfactory* or *Good* on each of the two evaluations, so that

$$C = \{(S, S), (S, G), (G, S), (G, G)\}$$

The three events A, B, and C are illustrated in Figure 1.54.
    Notice that

$$A \cap C = \{(S, S)\} \qquad \text{and} \qquad B \cap C = \{(S, S), (S, G)\}$$

FIGURE 1.53

A partition of the sample space

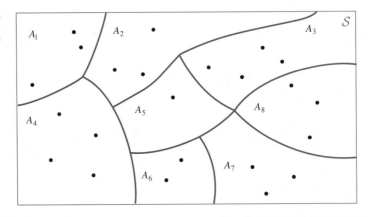

FIGURE 1.54

Events A, B, and C

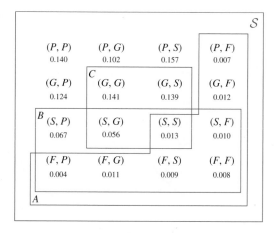

Also, the intersection of the three events is

$$A \cap B \cap C = \{(S, S)\}$$

The fact that the events $A \cap C$ and $A \cap B \cap C$ are identical, both consisting of the single outcome (S, S), is a consequence of the fact that $A \cap B' \cap C = \emptyset$. There are no outcomes that are not shipped (in event A), have a picture rating of *Good* or *Perfect* (in event B'), and are of mediocre quality (in event C).

The company may be particularly interested in the event D, that an appliance is of "high quality," defined to be the complement of the union of the events A, B, and C:

$$D = (A \cup B \cup C)'$$

Notice that this event can also be written as

$$D = A' \cap B' \cap C'$$

since it consists of the outcomes that are shipped (in event A'), have a picture rating of *Good* or *Perfect* (in event B'), and are not of mediocre quality (in event C'). Specifically, the event D consists of the outcomes

$$D = \{(G, P), (P, P), (P, G), (P, S)\}$$

and it has a probability of

$$\begin{aligned} P(D) &= P((G, P)) + P((P, P)) + P((P, G)) + P((P, S)) \\ &= 0.124 + 0.140 + 0.102 + 0.157 = 0.523 \end{aligned}$$

■ 1.3.5 Problems

1.3.1 Consider the sample space $S = \{0, 1, 2\}$ and the event $A = \{0\}$. Explain why $A \neq \emptyset$.

1.3.2 Consider the sample space and events in Figure 1.55. Calculate the probabilities of the events:
- **(a)** B
- **(b)** $B \cap C$
- **(c)** $A \cup C$
- **(d)** $A \cap B \cap C$
- **(e)** $A \cup B \cup C$
- **(f)** $A' \cap B$
- **(g)** $B' \cup C$
- **(h)** $A \cup (B \cap C)$
- **(i)** $(A \cup B) \cap C$
- **(j)** $(A' \cup C)'$

(This problem is continued in Problem 1.4.1.)

1.3.3 Use Venn diagrams to illustrate the equations:
- **(a)** $A \cup (B \cap C) = (A \cup B) \cap (A \cup C)$

FIGURE 1.55

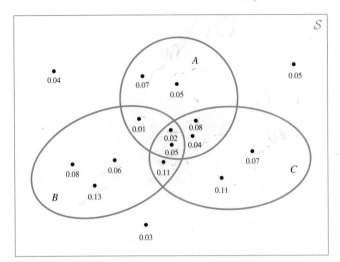

(b) $A \cap (B \cup C) = (A \cap B) \cup (A \cap C)$

(c) $(A \cap B \cap C)' = A' \cup B' \cup C'$

1.3.4 Let A be the event that a person is *female*, let B be the event that a person has *black hair*, and let C be the event that a person has *brown eyes*. Describe the kinds of people in the following events:

(a) $A \cap B$ **(b)** $A \cup C'$

(c) $A' \cap B \cap C$ **(d)** $A \cap (B \cup C)$

1.3.5 A card is chosen from a pack of cards. Are the events that a card from one of the two red suits is chosen and that a card from one of the two black suits is chosen mutually exclusive? What about the events that an Ace is chosen and that a heart is chosen?

1.3.6 If $P(A) = 0.4$ and $P(A \cap B) = 0.3$, what are the possible values for $P(B)$?

1.3.7 If $P(A) = 0.5$, $P(A \cap B) = 0.1$, and $P(A \cup B) = 0.8$, what is $P(B)$?

1.3.8 A fair die is thrown. A is the event that an *even* score is obtained, and B is the event that a *prime* score is obtained. Give the probabilities:

(a) $A \cap B$ **(b)** $A \cup B$ **(c)** $A \cap B'$

1.3.9 A card is drawn at random from a pack of cards. A is the event that a heart is obtained, B is the event that a club is obtained, and C is the event that a diamond is obtained. Are these three events mutually exclusive? What is $P(A \cup B \cup C)$? Explain why $B \subset A'$.

1.3.10 A card is drawn from a pack of cards. A is the event that an Ace is obtained, B is the event that a card from one of the two red suits is obtained, and C is the event that a picture card is obtained. What cards do the following events consist of?

(a) $A \cap B$ **(b)** $A \cup C$

(c) $B \cap C'$ **(d)** $A \cup (B' \cap C)$

1.3.11 A car repair can be performed either on time or late and either satisfactorily or unsatisfactorily. The probability of a repair being on time and satisfactory is 0.26. The probability of a repair being on time is 0.74. The probability of a repair being satisfactory is 0.41. What is the probability of a repair being late and unsatisfactory?

1.3.12 A bag contains 200 balls that are either red or blue and either dull or shiny. There are 55 shiny red balls, 91 shiny balls, and 79 red balls. If a ball is chosen at random, what is the probability that it is either a shiny ball or a red ball? What is the probability that it is a dull blue ball?

1.3.13 In a study of patients arriving at a hospital emergency room, the gender of the patients is considered, together with whether the patients are younger or older than 30 years of age, and whether or not the patients are admitted to the hospital. It is found that 45% of the patients are male, 30% of the patients are younger than 30 years of age, 15% of the patients are females older than 30 years of age who are admitted to the hospital, and 21% of the patients are females younger than 30 years of age. What proportion of the patients are females older than 30 years of age who are not admitted to the hospital?

1.4 Conditional Probability

1.4.1 Definition of Conditional Probability

For experiments with two or more events of interest, attention is often directed not only at the probabilities of individual events, but also at the probability of an event occurring **conditional** on the knowledge that another event has occurred. Probabilities such as these are important and very useful since they provide appropriate **revisions** of a set of probabilities once a particular event is known to have occurred.

The probability that event A occurs conditional on event B having occurred is written

$$P(A|B)$$

Its interpretation is that if the outcome occurring is known to be contained within the event B, then this **conditional probability** measures the probability that the outcome is also contained within the event A. Conditional probabilities can easily be obtained using the following formula:

Conditional Probability

The **conditional probability** of event A conditional on event B is

$$P(A|B) = \frac{P(A \cap B)}{P(B)}$$

for $P(B) > 0$. It measures the probability that event A occurs when it is known that event B occurs.

One simple example of conditional probability concerns the situation in which two events A and B are mutually exclusive. Since in this case events A and B have no outcomes in common, it is clear that the occurrence of event B precludes the possibility of event A occurring, so that intuitively, the probability of event A conditional on event B must be zero. Since $A \cap B = \emptyset$ for mutually exclusive events, this intuitive reasoning is in agreement with the formula

$$P(A|B) = \frac{P(A \cap B)}{P(B)} = \frac{0}{P(B)} = 0$$

Another simple example of conditional probability concerns the situation in which an event B is contained within an event A, that is $B \subset A$. Then if event B occurs, it is clear that event A must also occur, so that intuitively, the probability of event A conditional on event B must be one. Again, since $A \cap B = B$ here, this intuitive reasoning is in agreement with the formula

$$P(A|B) = \frac{P(A \cap B)}{P(B)} = \frac{P(B)}{P(B)} = 1$$

For a less obvious example of conditional probability, consider again Figure 1.26 and the events A and B shown there. Suppose that event B is known to occur. In other words, suppose that it is known that the outcome occurring is one of the five outcomes contained within the event B. What then is the conditional probability of event A occurring?

Since two of the five outcomes in event B are also in event A (that is, there are two outcomes in $A \cap B$), the conditional probability is the probability that one of these two outcomes occurs

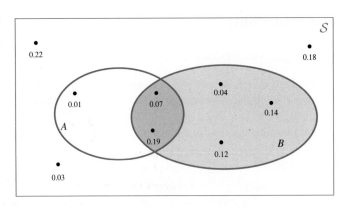

FIGURE 1.56

$P(A|B) = P(A \cap B)/P(B)$

rather than one of the other three outcomes (which are in $A' \cap B$). As Figure 1.56 shows, the conditional probability is calculated to be

$$P(A|B) = \frac{P(A \cap B)}{P(B)} = \frac{0.26}{0.56} = 0.464$$

Notice that this conditional probability is different from $P(A) = 0.27$. If the event B is known *not* to occur, then the conditional probability of event A is

$$P(A|B') = \frac{P(A \cap B')}{P(B')} = \frac{P(A) - P(A \cap B)}{1 - P(B)} = \frac{0.27 - 0.26}{1 - 0.56} = 0.023$$

In the same way that $P(A) + P(A') = 1$, it is also true that

$$P(A|B) + P(A'|B) = 1$$

This is reasonable because if event B occurs, it is still the case that either event A occurs or it does not, and so the two conditional probabilities should sum to one. Formally, this result can be shown by noting that

$$P(A|B) + P(A'|B) = \frac{P(A \cap B)}{P(B)} + \frac{P(A' \cap B)}{P(B)}$$

$$= \frac{1}{P(B)}(P(A \cap B) + P(A' \cap B)) = \frac{1}{P(B)}P(B) = 1$$

However, there is no general relationship between $P(A|B)$ and $P(A|B')$.

Finally, the event conditioned on can be represented as a combination of events. For example,

$$P(A|B \cup C)$$

represents the probability of event A conditional on the event $B \cup C$, that is conditional on either event B or C occurring. It can be calculated from the formula

$$P(A|B \cup C) = \frac{P(A \cap (B \cup C))}{P(B \cup C)}$$

FIGURE 1.57

$\mathcal{S}$

| 0 defectives |
| 0.02 |
| 1 defective |
| 0.11 |
| 2 defectives |
| 0.16 |
| 3 defectives |
| 0.21 |
| 4 defectives |
| 0.13 |
| 5 defectives |
| 0.08 |
| **correct** |
| 6 defectives |
| ? |
| ⋮ |
| 500 defectives |
| ? |

Sample space for computer chips example

1.4.2   Examples of Conditional Probabilities

Example 2

Defective Computer Chips

Consider Figure 1.57 that illustrates the sample space for the number of defective chips in a box of 500 chips, and recall that the event *correct*, with a probability of $P(\text{correct}) = 0.71$, consists of the six outcomes corresponding to no more than five defectives.

The probability that a box has no defective chips is

$$P(0 \text{ defectives}) = 0.02$$

so that if a box is chosen at random, it has a probability of only 0.02 of containing no defective chips. If the company guarantees that a box has no more than five defective chips, then customers can be classified as either satisfied or unsatisfied depending on whether the guarantee is met. Clearly, an unsatisfied customer did not purchase a box containing no defective chips. However, it is interesting to calculate the probability that a satisfied customer purchased a box that contained no defective chips. Intuitively, this conditional probability should be larger than the unconditional probability 0.02.

The required probability is the probability of no defectives conditional on there being no more than five defectives, which is calculated to be

$$P(0 \text{ defectives}|\text{correct}) = \frac{P(0 \text{ defectives} \cap \text{correct})}{P(\text{correct})} = \frac{P(0 \text{ defectives})}{P(\text{correct})}$$

$$= \frac{0.02}{0.71} = 0.028$$

This conditional probability indicates that whereas 2% of all the boxes contain no defectives, 2.8% of the satisfied customers purchased boxes that contained no defectives.

Example 4	The probability that plant X is idle is $P(A) = 0.32$. However, suppose it is known that at
Power Plant Operation	least two out of the three plants are generating electricity (event B). How does this change the

The probability that plant X is idle is $P(A) = 0.32$. However, suppose it is known that at least two out of the three plants are generating electricity (event B). How does this change the probability of plant X being idle?

The probability that plant X is idle (event A) conditional on at least two out of the three plants generating electricity (event B) is

$$P(A|B) = \frac{P(A \cap B)}{P(B)} = \frac{0.18}{0.70} = 0.257$$

as shown in Figure 1.58. Therefore, whereas plant X is idle about 32% of the time, it is idle only about 25.7% of the time when at least two of the plants are generating electricity.

Example 5
Television Set Quality

Recall that the probability that an appliance has a picture graded as either *Satisfactory* or *Fail* is $P(B) = 0.178$. However, suppose that a technician takes a television set from a pile of sets that could not be shipped. What is the probability that the appliance taken by the technician has a picture graded as either *Satisfactory* or *Fail*?

The required probability is the probability that an appliance has a picture graded as either *Satisfactory* or *Fail* (event B) conditional on the appliance not being shipped (event A). As Figure 1.59 shows, this can be calculated to be

$$P(B|A) = \frac{P(A \cap B)}{P(A)} = \frac{0.055}{0.074} = 0.743$$

so that whereas only about 17.8% of all the appliances manufactured have a picture graded as either *Satisfactory* or *Fail*, 74.3% of the appliances that cannot be shipped have a picture graded as either *Satisfactory* or *Fail*.

GAMES OF CHANCE

If a fair die is rolled the probability of scoring a 6 is $P(6) = 1/6$. If somebody rolls a die without showing you but announces that the result is *even*, then intuitively the chance that a 6 has been obtained is 1/3 since there are three equally likely even scores, one of which is a 6.

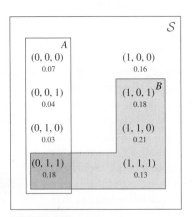

FIGURE 1.58

$P(A|B) = P(A \cap B)/P(B)$

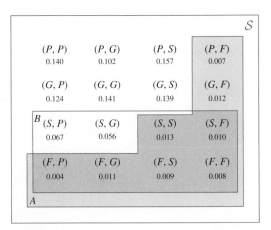

FIGURE 1.59

$P(B|A) = P(A \cap B)/P(A)$

Mathematically, this conditional probability is calculated to be

$$P(6|\text{even}) = \frac{P(6 \cap \text{even})}{P(\text{even})} = \frac{P(6)}{P(\text{even})}$$

$$= \frac{P(6)}{P(2) + P(4) + P(6)} = \frac{1/6}{1/6 + 1/6 + 1/6} = \frac{1}{3}$$

as expected.

If a red die and a blue die are thrown, with each of the 36 outcomes being equally likely, let A be the event that the red die scores a 6, so that

$$P(A) = \frac{6}{36} = \frac{1}{6}$$

Also, let B be the event that at least one 6 is obtained on the two dice (see Figure 1.19) with a probability of

$$P(B) = \frac{11}{36}$$

Suppose that somebody rolls the two dice without showing you but announces that at least one 6 has been scored. What then is the probability that the red die scored a 6? As Figure 1.60 shows, this conditional probability is calculated to be

$$P(A|B) = \frac{P(A \cap B)}{P(B)} = \frac{P(A)}{P(B)} = \frac{1/6}{11/36} = \frac{6}{11}$$

As expected, this conditional probability is larger than $P(A) = 1/6$. It is also slightly larger than 0.5, which is accounted for by the outcome $(6, 6)$ where both dice score a 6.

Contrast this problem with the situation where the announcement is that *exactly* one 6 has been scored, event C, say. In this case, it is intuitively clear that the 6 obtained is equally likely to have been scored on the red die or the blue die, so that the conditional probability $P(A|C)$

FIGURE 1.60

$P(A|B) = P(A \cap B)/P(B)$

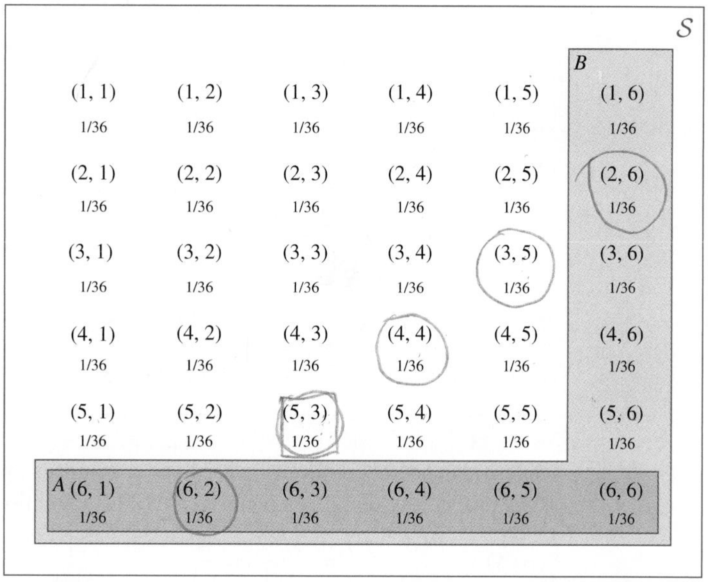

FIGURE 1.61

$P(A|C) = P(A \cap C)/P(C)$

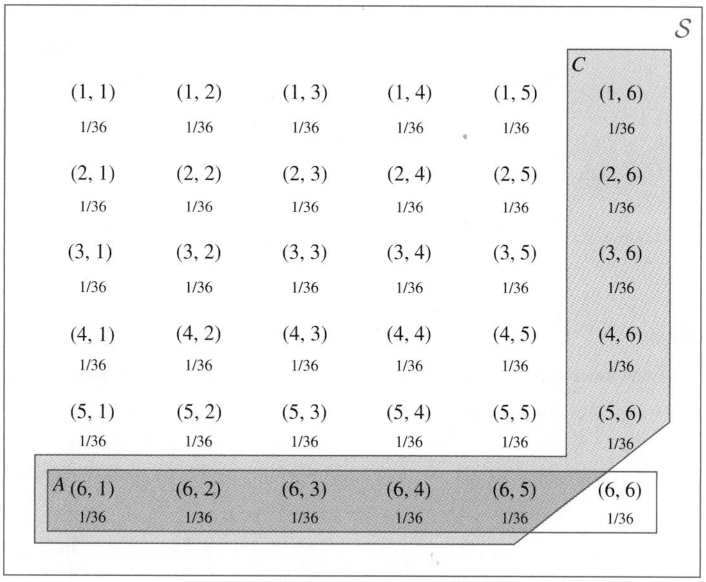

should be equal to 1/2. As Figure 1.61 shows, this is correct since

$$P(A|C) = \frac{P(A \cap C)}{P(C)} = \frac{5/36}{10/36} = \frac{1}{2}$$

If a card is drawn from a pack of cards, let A be the event that a card from the heart suit is obtained, and let B be the event that a picture card is drawn. Recall that $P(A) = 13/52 = 1/4$ and $P(B) = 12/52 = 3/13$. Also, the event $A \cap B$, the event that a picture card from the heart suit is drawn, has a probability of $P(A \cap B) = 3/52$.

FIGURE 1.62

$P(C|A) = P(C \cap A)/P(A)$

A♡ C	2♡	3♡	4♡	5♡	6♡	7♡	8♡	9♡	10♡	J♡	Q♡	K♡ A
1/52	1/52	1/52	1/52	1/52	1/52	1/52	1/52	1/52	1/52	1/52	1/52	1/52
A♣	2♣	3♣	4♣	5♣	6♣	7♣	8♣	9♣	10♣	J♣	Q♣	K♣
1/52	1/52	1/52	1/52	1/52	1/52	1/52	1/52	1/52	1/52	1/52	1/52	1/52
A♢	2♢	3♢	4♢	5♢	6♢	7♢	8♢	9♢	10♢	J♢	Q♢	K♢
1/52	1/52	1/52	1/52	1/52	1/52	1/52	1/52	1/52	1/52	1/52	1/52	1/52
A♠	2♠	3♠	4♠	5♠	6♠	7♠	8♠	9♠	10♠	J♠	Q♠	K♠
1/52	1/52	1/52	1/52	1/52	1/52	1/52	1/52	1/52	1/52	1/52	1/52	1/52

Suppose that somebody draws a card and announces that it is from the heart suit. What then is the probability that it is a picture card? This conditional probability is

$$P(B|A) = \frac{P(A \cap B)}{P(A)} = \frac{3/52}{1/4} = \frac{3}{13}$$

Notice that in this case, $P(B|A) = P(B)$ because the proportion of picture cards in the heart suit is identical to the proportion of picture cards in the whole pack. The events A and B are then said to be **independent** events, which are discussed more fully in Section 1.5.

Finally, let C be the event that the $A♡$ is chosen, with $P(C) = 1/52$. If it is known that a card from the heart suit is obtained, then intuitively the conditional probability of the card being $A♡$ is $1/13$ since there are 13 equally likely cards in the heart suit. As Figure 1.62 shows, this is correct because

$$P(C|A) = \frac{P(C \cap A)}{P(A)} = \frac{P(C)}{P(A)} = \frac{1/52}{1/4} = \frac{1}{13}$$

■ 1.4.3 Problems

1.4.1 Consider again Figure 1.55 and calculate the probabilities:
(a) $P(A|B)$
(b) $P(C|A)$
(c) $P(B|A \cap B)$
(d) $P(B|A \cup B)$
(e) $P(A|A \cup B \cup C)$
(f) $P(A \cap B|A \cup B)$

1.4.2 Let A be the event that a *prime* number is obtained from the roll of a fair die. Calculate $P(5|A)$, $P(6|A)$, and $P(A|5)$.

1.4.3 A card is drawn at random from a pack of cards. Calculate:
(a) $P(A♡|\text{card from red suit})$
(b) $P(\text{heart}|\text{card from red suit})$
(c) $P(\text{card from red suit}|\text{heart})$

(d) $P(\text{heart}|\text{card from black suit})$
(e) $P(\text{King}|\text{card from red suit})$
(f) $P(\text{King}|\text{red picture card})$

1.4.4 If $A \subset B$ and $B' \neq \emptyset$, is $P(A)$ larger or smaller than $P(A|B)$? Provide some intuitive reasoning for your answer.

1.4.5 A ball is chosen at random from a bag containing 150 balls that are either red or blue and either dull or shiny. There are 36 red shiny balls and 54 blue balls. What is the probability of the chosen ball being shiny conditional on it being red? What is the probability of the chosen ball being dull conditional on it being red?

1.4.6 A car repair is either on time or late and either satisfactory or unsatisfactory. If a repair is made on time, then there is a probability of 0.85 that it is satisfactory.

There is a probability of 0.77 that a repair will be made on time. What is the probability that a repair is made on time and is satisfactory?

1.4.7 Assess whether the probabilities of the events (i) increase, decrease, or remain unchanged when they are conditioned on the events (ii).

(**a**) (i) It rains tomorrow, (ii) it is raining today.

(**b**) (i) A lottery winner has black hair, (ii) the lottery winner has brown eyes.

(**c**) (i) A lottery winner has black hair, (ii) the lottery winner owns a red car.

(**d**) (i) A lottery winner is more than 50 years old, (ii) the lottery winner is more than 30 years old.

1.4.8 Suppose that births are equally likely to be on any day. What is the probability that somebody chosen at random has a birthday on the *first day* of a month? How does this probability change conditional on the knowledge that the person's birthday is in March? In February?

1.4.9 Consider again Figure 1.24 and the battery lifetimes. Calculate the probabilities:

(**a**) A type I battery lasts longest conditional on it not failing first

(**b**) A type I battery lasts longest conditional on a type II battery failing first

(**c**) A type I battery lasts longest conditional on a type II battery lasting the longest

(**d**) A type I battery lasts longest conditional on a type II battery not failing first

1.4.10 Consider again Figure 1.25 and the two assembly lines. Calculate the probabilities:

(**a**) Both lines are at full capacity conditional on neither line being shut down

(**b**) At least one line is at full capacity conditional on neither line being shut down

(**c**) One line is at full capacity conditional on exactly one line being shut down

(**d**) Neither line is at full capacity conditional on at least one line operating at partial capacity

1.4.11 The length, width, and height of a manufactured part are classified as being either within or outside specified tolerance limits. In a quality inspection 86% of the parts are found to be within the specified tolerance limits for width, but only 80% of the parts are within the specified tolerance limits for all three dimensions. However, 2% of the parts are within the specified tolerance limits for width and length but not for height, and 3% of the parts

are within the specified tolerance limits for width and height but not for length. Moreover, 92% of the parts are within the specified tolerance limits for either width or height, or both of these dimensions.

(**a**) If a part is within the specified tolerance limits for height, what is the probability that it will also be within the specified tolerance limits for width?

(**b**) If a part is within the specified tolerance limits for length and width, what is the probability that it will be within the specified tolerance limits for all three dimensions?

1.4.12 A gene can be either type A or type B, and it can be either dominant or recessive. If the gene is type B, then there is a probability of 0.31 that it is dominant. There is also a probability of 0.22 that a gene is type B and it is dominant. What is the probability that a gene is of type A?

1.4.13 A manufactured component has its quality graded on its performance, appearance, and cost. Each of these three characteristics is graded as either pass or fail. There is a probability of 0.40 that a component passes on both appearance and cost. There is a probability of 0.31 that a component passes on all three characteristics. There is a probability of 0.64 that a component passes on performance. There is a probability of 0.19 that a component fails on all three characteristics. There is a probability of 0.06 that a component passes on appearance but fails on both performance and cost.

(**a**) What is the probability that a component passes on cost but fails on both performance and appearance?

(**b**) If a component passes on both appearance and cost, what is the probability that it passes on all three characteristics?

1.4.14 An agricultural research establishment grows vegetables and grades each one as either good or bad for its taste, good or bad for its size, and good or bad for its appearance. Overall 78% of the vegetables have a good taste. However, only 69% of the vegetables have both a good taste and a good size. Also, 5% of the vegetables have both a good taste and a good appearance, but a bad size. Finally, 84% of the vegetables have either a good size or a good appearance.

(**a**) If a vegetable has a good taste, what is the probability that it also has a good size?

(**b**) If a vegetable has a bad size and a bad appearance, what is the probability that it has a good taste?

1.4.15 There is a 4% probability that the plane used for a commercial flight has technical problems, and this causes a delay in the flight. If there are no technical problems with the plane, then there is still a 33% probability that the flight is delayed due to all other reasons. What is the probability that the flight is delayed?

1.4.16 In a reliability test there is a 42% probability that a computer chip survives more than 500 temperature cycles. If a computer chip does not survive more than 500 temperature cycles, then there is a 73% probability that it was manufactured by company A. What is the probability that a computer chip is not manufactured by company A and does not survive more than 500 temperature cycles?

1.5   Probabilities of Event Intersections

1.5.1   General Multiplication Law

It follows from the definition of the conditional probability $P(A|B)$ that the probability of the intersection of two events $A \cap B$ can be calculated as

$$P(A \cap B) = P(B)\,P(A|B)$$

That is, the probability of events A and B both occurring can be obtained by multiplying the probability of event B by the probability of event A conditional on event B. It also follows from the definition of the conditional probability $P(B|A)$ that

$$P(A \cap B) = P(A)\,P(B|A)$$

so that the probability of events A and B both occurring can also be obtained by multiplying the probability of event A by the probability of event B conditional on event A. Therefore, it does not matter which of the two events A or B is conditioned upon.

More generally, since

$$P(C|A \cap B) = \frac{P(A \cap B \cap C)}{P(A \cap B)}$$

the probability of the intersection of three events can be calculated as

$$P(A \cap B \cap C) = P(A \cap B)\,P(C|A \cap B) = P(A)\,P(B|A)\,P(C|A \cap B)$$

Thus, the probability of all three events occurring can be obtained by multiplying together the probability of one event, the probability of a second event conditioned on the first event, and the probability of the third event conditioned on the intersection of the first and second events. This formula can be extended in an obvious way to the following **multiplication law** for the intersection of a series of events.

> **Probabilities of Event Intersections**
>
> The probability of the **intersection of a series of events** $A_1, \ldots, A_n$ can be calculated from the expression
>
> $$P(A_1 \cap \cdots \cap A_n) = P(A_1) \times P(A_2|A_1) \times P(A_3|A_1 \cap A_2) \times \cdots \times P(A_n|A_1 \cap \cdots \cap A_{n-1})$$

This expression for the probability of event intersections is particularly useful when the conditional probabilities $P(A_i|A_1 \cap \cdots \cap A_{i-1})$ are easily obtainable, as illustrated in the following example.

Suppose that two cards are drawn at random *without replacement* from a pack of cards. Let A be the event that the *first* card drawn is from the heart suit, and let B be the event that the *second* card drawn is from the heart suit. What is $P(A \cap B)$, the probability that both cards are from the heart suit?

Figure 1.13 shows the sample space for this problem, which consists of 2652 equally likely outcomes, each with a probability of $1/2652$. One way to calculate $P(A \cap B)$ is to count the number of outcomes in the sample space that are contained within the event $A \cap B$, that is, for which both cards are in the heart suit. In fact,

$$A \cap B = \{(A\heartsuit, 2\heartsuit), (A\heartsuit, 3\heartsuit), \ldots, (A\heartsuit, K\heartsuit), (2\heartsuit, A\heartsuit), (2\heartsuit, 3\heartsuit), \ldots,$$
$$(2\heartsuit, K\heartsuit), \ldots (K\heartsuit, A\heartsuit), (K\heartsuit, 2\heartsuit), \ldots, (K\heartsuit, Q\heartsuit)\}$$

which consists of $13 \times 12 = 156$ outcomes. Consequently, the required probability is

$$P(A \cap B) = \frac{156}{2652} = \frac{3}{51}$$

However, a more convenient way of calculating this probability is to note that it is the product of $P(A)$ and $P(B|A)$. When the first card is drawn, there are 13 heart cards out of a total of 52 cards, so

$$P(A) = \frac{13}{52} = \frac{1}{4}$$

Conditional on the first card being a heart (event A), when the second card is drawn there will be 12 heart cards remaining in the reduced pack of 51 cards, so that

$$P(B|A) = \frac{12}{51}$$

The required probability is then

$$P(A \cap B) = P(A)\, P(B|A) = \frac{1}{4} \times \frac{12}{51} = \frac{3}{51}$$

as before.

1.5.2 Independent Events

Two events A and B are said to be **independent** events if

$$P(B|A) = P(B)$$

so that the probability of event B remains the same whether or not the event A is conditioned upon. In other words, knowledge of the occurrence (or lack of occurrence) of event A does not affect the probability of event B. In this case

$$P(A \cap B) = P(A)\, P(B|A) = P(A)\, P(B)$$

and

$$P(A|B) = \frac{P(A \cap B)}{P(B)} = \frac{P(A)P(B)}{P(B)} = P(A)$$

Thus, in a similar way, the probability of event A remains the same whether or not event B is conditioned upon, and the probability of both events occurring, $P(A \cap B)$, is obtained simply by multiplying together the individual probabilities of the two events $P(A)$ and $P(B)$.

Independent Events

Two events A and B are said to be **independent** events if

$$P(A|B) = P(A), \quad P(B|A) = P(B), \quad \text{and} \quad P(A \cap B) = P(A)\,P(B)$$

Any one of these three conditions implies the other two. The interpretation of two events being independent is that knowledge about one event does not affect the probability of the other event.

The concept of independence is most easily understood from a practical standpoint, with two events being independent if they are "unrelated" to each other. For example, suppose that a person is chosen at random from a large group of people, as in a lottery, for instance. Let A be the event that the person is over 6 feet tall, and let B be the event that the person weighs more than 200 pounds. Intuitively, these two events are *not* independent because knowledge of one event influences our perception of the likelihood of the other event. For example, if the lottery winner is known to be over 6 feet tall, then this fact increases the likelihood that the person weighs more than 200 pounds.

On the other hand, if event C is that the person owns a red car, then intuitively the events A and C are independent, as are the events B and C. The knowledge that the lottery winner is over 6 feet tall does not change our perception of the probability that the person owns a red car. Conversely, the knowledge that the lottery winner owns a red car does not change our perception of the probability that the person is over 6 feet tall.

From a mathematical standpoint, two events A and B can be proven to be independent by establishing one of the conditions

$$P(A|B) = P(A), \quad P(B|A) = P(B), \quad \text{or} \quad P(A \cap B) = P(A)\,P(B)$$

In practice, however, an assessment of whether two events are independent or not is usually made by the practical consideration of whether the two events are "unrelated."

Events $A_1, A_2, \ldots, A_n$ are said to be independent if conditioning on combinations of some of the events does not affect the probabilities of the other events. In this case, the expression given earlier for the probability of the intersection of the events simplifies to the product of the probabilities of the individual events.

Intersections of Independent Events

The probability of the intersection of a series of **independent events** $A_1, \ldots, A_n$ is simply given by

$$P(A_1 \cap \cdots \cap A_n) = P(A_1)\,P(A_2) \cdots P(A_n)$$

Consider again the problem discussed above where two cards are drawn from a pack of cards, and where A is the event that the *first* card drawn is from the heart suit and B is the event that the *second* card drawn is from the heart suit. Suppose that now the drawings are made *with replacement*. What is $P(A \cap B)$ in this case?

Figure 1.12 shows the sample space for this problem, which consists of 2704 equally likely outcomes, each with a probability of $1/2704$. As before, one way to calculate $P(A \cap B)$ is to count the number of outcomes in the sample space that are contained within the event

$A \cap B$. This event is now

$$A \cap B = \{(A\heartsuit, A\heartsuit), (A\heartsuit, 2\heartsuit), \ldots, (A\heartsuit, K\heartsuit), (2\heartsuit, A\heartsuit), (2\heartsuit, 2\heartsuit), \ldots,$$
$$(2\heartsuit, K\heartsuit), \ldots (K\heartsuit, A\heartsuit), (K\heartsuit, 2\heartsuit), \ldots, (K\heartsuit, K\heartsuit)\}$$

It consists of $13 \times 13 = 169$ outcomes, so that the required probability is

$$P(A \cap B) = \frac{169}{2704} = \frac{1}{16}$$

However, it is easier to notice that events A and B are independent with $P(A) = P(B) = 1/4$, so that

$$P(A \cap B) = P(A)\, P(B) = \frac{1}{4} \times \frac{1}{4} = \frac{1}{16}$$

The independence follows from the fact that with the replacement of the first card and with appropriate shuffling of the pack to ensure randomness, the outcome of the second drawing is not related to the outcome of the first drawing.

If the drawings are performed without replacement, then clearly events A and B are not independent. This can be confirmed mathematically by noting that

$$P(B|A) = \frac{12}{51} \quad \text{and} \quad P(B|A') = \frac{13}{51}$$

which are different from $P(B) = 1/4$.

1.5.3   Examples and Probability Trees

Example 2

Defective Computer Chips

Suppose that 9 out of the 500 chips in a particular box are defective, and suppose that 3 chips are *sampled* at random from the box without replacement. If each of the 3 chips sampled is tested to determine whether it is defective (1) or satisfactory (0), the sample space has eight outcomes. For example, the outcome $(0, 1, 0)$ corresponds to the first and third chips being satisfactory and the second chip being defective.

The probability values of the eight outcomes can be calculated using a **probability tree** as illustrated in Figure 1.63. The events A, B, and C are, respectively, the events that the first, second, and third chips sampled are defective. These events are not independent since the sampling is conducted without replacement. The probability tree starts at the left with two branches corresponding to the events A and A'. The probabilities of these events

$$P(A) = \frac{9}{500} \quad \text{and} \quad P(A') = \frac{491}{500}$$

are recorded at the ends of the branches.

Each of these two branches then splits into two more branches corresponding to the events B and B', and the *conditional* probabilities of these events are recorded. These conditional probabilities are

$$P(B|A) = \frac{8}{499}, \qquad P(B'|A) = \frac{491}{499}, \qquad P(B|A') = \frac{9}{499}, \qquad P(B'|A') = \frac{490}{499}$$

which are constructed by considering how many of the 499 chips left in the box are defective when the second chip is chosen. For example, $P(B|A) = 8/499$ since if the first chip chosen is defective (event A), then 8 out of 499 chips in the box are defective when the second chip is chosen.

The probability tree is completed by adding additional branches for the events C and C', and by recording the probabilities of these events *conditional* on the outcomes of the first two

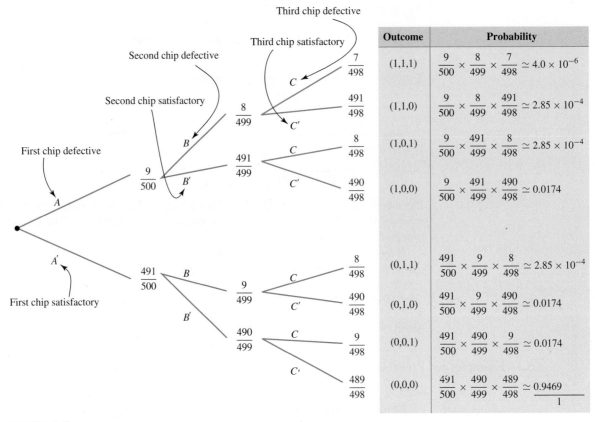

FIGURE 1.63

Probability tree for computer chip sampling

choices. For example,

$$P(C|A \cap B') = \frac{8}{498}$$

because conditional on the event $A \cap B'$ (the first choice is defective and the second is satisfactory), 8 out of the 498 chips in the box are defective when the third choice is made.

The probability values of the eight outcomes are found by *multiplying* the probabilities along the branches. Thus, the probability of choosing 3 defective chips is

$$P((1, 1, 1)) = P(A \cap B \cap C) = P(A)P(B|A)P(C|A \cap B)$$
$$= \frac{9}{500} \times \frac{8}{499} \times \frac{7}{498} = \frac{21}{5,177,125} \simeq 4.0 \times 10^{-6}$$

The probability of choosing 2 satisfactory chips followed by a defective chip is

$$P((0, 0, 1)) = P(A' \cap B' \cap C) = P(A')P(B'|A')P(C|A' \cap B')$$
$$= \frac{491}{500} \times \frac{490}{499} \times \frac{9}{498} = \frac{72,177}{4,141,700} \simeq 0.0174$$

Notice that the probabilities of the outcomes $(1, 0, 0)$, $(0, 1, 0)$, and $(0, 0, 1)$ are identical, although they are calculated in different ways. Similarly, the probabilities of the outcomes $(1, 1, 0)$, $(0, 1, 1)$, and $(1, 0, 1)$ are identical. The probability of exactly 1 defective chip being found is

$$P(1 \text{ defective}) = P((1, 0, 0)) + P((0, 1, 0)) + P((0, 0, 1))$$
$$\simeq 3 \times 0.0174 = 0.0522$$

In fact, if attention is focused solely on the number of defective chips in the sample, then the required probabilities can be found from the hypergeometric distribution which is discussed in Section 3.3. Finally, it is interesting to note that, in practice, the number of defective chips in a box will not usually be known, but probabilities of these kinds are useful in *estimating* the number of defective chips in the box. In later chapters of this book, statistical techniques will be employed to use the information provided by a random sample (here the number of defective chips found in the sample) to make inferences about the population that is sampled (here the box of chips).

Example 6	A satellite launch system is controlled by a computer (computer 1) that has two identical
Satellite Launching	backup computers (computers 2 and 3). Normally, computer 1 controls the system, but if it

has a malfunction then computer 2 automatically takes over. If computer 2 malfunctions then computer 3 automatically takes over, and if computer 3 malfunctions there is a general system shutdown.

The state space for this problem consists of the four situations

$$\mathcal{S} = \{\text{computer 1 in use, computer 2 in use, computer 3 in use, system failure}\}$$

Suppose that a computer malfunctions with a probability of 0.01 and that malfunctions of the three computers are *independent* of each other. Also, let the events A, B, and C be, respectively, the events that computers 1, 2, and 3 malfunction.

Figure 1.64 shows the probability tree for this problem, which starts at the left with two branches corresponding to the events A and A' with probabilities $P(A) = 0.01$ and

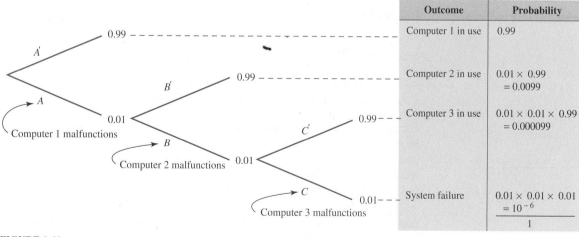

FIGURE 1.64

Probability tree for computer backup system

$P(A') = 0.99$. The top branch (event A') corresponds to computer 1 being in use, and there is no need to extend it further. However, the bottom branch (event A) extends into two further branches for the events B and B'.

Since events A and B are independent, the probabilities of these second-stage branches (events B and B') do not need to be conditioned on the first-stage branch (event A), and so their probabilities are just $P(B) = 0.01$ and $P(B') = 0.99$. The probability tree is completed by adding branches for the events C and C' following on from the events A and B.

The probability values of the four situations are obtained by multiplying the probabilities along the branches, so that

$P(\text{computer 1 in use}) = 0.99$
$P(\text{computer 2 in use}) = 0.01 \times 0.99 = 0.0099$
$P(\text{computer 3 in use}) = 0.01 \times 0.01 \times 0.99 = 0.000099$
$P(\text{system failure}) = 0.01 \times 0.01 \times 0.01 = 10^{-6}$

The design of the system backup capabilities is obviously conducted with the aim of minimizing the probability of a system failure. Notice that a key issue in the determination of this probability is the assumption that the malfunctions of the three computers are *independent* events. In other words, a malfunction in computer 1 should not affect the probabilities of the other two computers malfunctioning. An essential part of such a backup system is ensuring that these events are as independent as it is possible to make them.

In particular, it is sensible to have three teams of programmers working independently to supply software to the three computers. If only one piece of software is written and then copied onto the three machines, then the computer malfunctions will not be independent since a malfunction due to a software error in computer 1 will be repeated in the other two computers.

Finally, it is worth noting that this system can be thought of as consisting of three computers connected in parallel, as discussed in Section 17.1.2, where system reliability is considered in more detail.

Example 7

Car Warranties

A company sells a certain type of car that it assembles in one of four possible locations. Plant I supplies 20% of the cars; plant II, 24%; plant III, 25%; and plant IV, 31%. A customer buying a car does not know where the car has been assembled, and so the probabilities of a purchased car being from each of the four plants can be thought of as being 0.20, 0.24, 0.25, and 0.31.

Each new car sold carries a one-year bumper-to-bumper warranty. The company has collected data that show

$P(\text{claim}|\text{plant I}) = 0.05 \quad P(\text{claim}|\text{plant II}) = 0.11$
$P(\text{claim}|\text{plant III}) = 0.03 \quad P(\text{claim}|\text{plant IV}) = 0.08$

For example, a car assembled in plant I has a probability of 0.05 of receiving a claim on its warranty. This information, which is a closely guarded company secret, indicates which assembly plants do the best job. Plant III is seen to have the best record, and plant II the worst record. Notice that claims are clearly not independent of assembly location because these four conditional probabilities are unequal.

Figure 1.65 shows a probability tree for this problem. It is easily constructed because the probabilities of the second-stage branches are simply obtained from the conditional probabilities above. The probability that a customer purchases a car that was assembled in plant I and that does not require a claim on its warranty is seen to be

$P(\text{plant I, no claim}) = 0.20 \times 0.95 = 0.19$

FIGURE 1.65

Probability tree for car warranties example

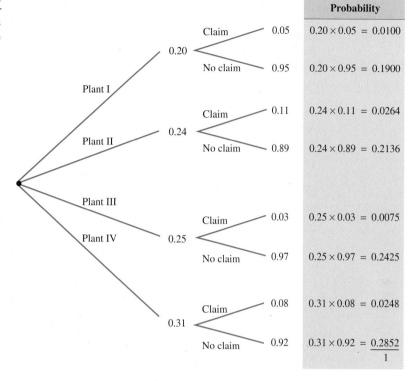

	Probability
Claim — 0.05	$0.20 \times 0.05 = 0.0100$
No claim — 0.95	$0.20 \times 0.95 = 0.1900$
Claim — 0.11	$0.24 \times 0.11 = 0.0264$
No claim — 0.89	$0.24 \times 0.89 = 0.2136$
Claim — 0.03	$0.25 \times 0.03 = 0.0075$
No claim — 0.97	$0.25 \times 0.97 = 0.2425$
Claim — 0.08	$0.31 \times 0.08 = 0.0248$
No claim — 0.92	$0.31 \times 0.92 = \underline{0.2852}$
	1

From a customer's point of view, the probability of interest is the probability that a claim on the warranty of the car will be required. This can be calculated as

$$P(\text{claim}) = P(\text{plant I, claim}) + P(\text{plant II, claim}) + P(\text{plant III, claim})$$
$$+ P(\text{plant IV, claim})$$
$$= (0.20 \times 0.05) + (0.24 \times 0.11) + (0.25 \times 0.03) + (0.31 \times 0.08)$$
$$= 0.0687$$

In other words, about 6.87% of the cars purchased will have a claim on their warranty. Notice that this overall claim rate is a *weighted* average of the four individual plant claim rates, with weights corresponding to the supply proportions of the four plants.

GAMES OF CHANCE

In the roll of a fair die, consider the events

$$\text{even} = \{2, 4, 6\} \quad \text{and} \quad \text{high score} = \{4, 5, 6\}$$

Intuitively, these two events are not independent since the knowledge that a high score is obtained increases the chances of the score being even, and vice versa, the knowledge that the score is even increases the chances of the score being high. Mathematically, this may be confirmed by noting that the probabilities

$$P(\text{even}) = \frac{1}{2} \quad \text{and} \quad P(\text{even}|\text{high score}) = \frac{2}{3}$$

are different.

If a red die and a blue die are rolled, consider the probability that both dice record even scores. In this case the scores on the two dice will be independent of each other since the score on one die does not affect the score that is obtained on the other die. If A is the event that the red die has an even score, and B is the event that the blue die has an even score, the required probability is

$$P(A \cap B) = P(A)\,P(B) = \frac{1}{2} \times \frac{1}{2} = \frac{1}{4}$$

A more tedious way of calculating this probability is to note that 9 out of the 36 outcomes in the sample space (see Figure 1.46) have both scores even, so that the required probability is $9/36 = 1/4$.

Suppose that two cards are drawn from a pack of cards *without replacement*. What is the probability that exactly one card from the heart suit is obtained? A very tedious way to solve this problem is to count the number of outcomes in the sample space (see Figure 1.13) that satisfy this condition. A better way is

$$P(\text{exactly one heart}) = P(\text{first card heart, second card not heart})$$
$$+ P(\text{first card not heart, second card heart})$$
$$= \left(\frac{13}{52} \times \frac{39}{51}\right) + \left(\frac{39}{52} \times \frac{13}{51}\right) = \frac{13}{34} = 0.382$$

Since the second drawing is made without replacement, the events "first card heart" and "second card heart" are not independent. However, notice that if the second card is drawn *with replacement*, then the two events are independent, and the required probability is

$$P(\text{exactly one heart}) = P(\text{first card heart, second card not heart})$$
$$+ P(\text{first card not heart, second card heart})$$
$$= \left(\frac{1}{4} \times \frac{3}{4}\right) + \left(\frac{3}{4} \times \frac{1}{4}\right) = \frac{3}{8} = 0.375$$

It is interesting that the probability is slightly higher when the second drawing is made without replacement.

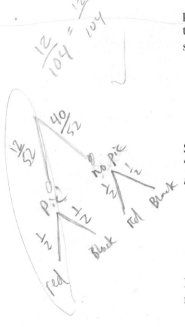

■ 1.5.4   Problems

1.5.1 Two cards are chosen from a pack of cards *without replacement*. Calculate the probabilities:
 (a) Both are picture cards.
 (b) Both are from red suits.
 (c) One card is from a red suit and one card is from a black suit.

1.5.2 Repeat Problem 1.5.1, except that the second drawing is made *with replacement*. Compare your answers with those from Problem 1.5.1.

1.5.3 Two cards are chosen from a pack of cards *without replacement*. Are the following events independent?
 (a) (i) The first card is a picture card, (ii) the second card is a picture card.

(b) (i) The first card is a heart, (ii) the second card is a picture card.
(c) (i) The first card is from a red suit, (ii) the second card is from a red suit.
(d) (i) The first card is a picture card, (ii) the second card is from a red suit.
(e) (i) The first card is a red picture card, (ii) the second card is a heart.

1.5.4 Four cards are chosen from a pack of cards *without replacement*. What is the probability that all four cards are hearts? What is the probability that all four cards are from red suits? What is the probability that all four cards are from different suits?

FIGURE 1.66

Switch diagram

1.5.5 Repeat Problem 1.5.4, except that the drawings are made *with replacement*. Compare your answers with those from Problem 1.5.4.

1.5.6 Show that if the events A and B are independent events, then so are the events
(a) A and B' (b) A' and B (c) A' and B'

1.5.7 Consider the network given in Figure 1.66 with three switches. Suppose that the switches operate independently of each other and that switch 1 allows a message through with probability 0.88, switch 2 allows a message through with probability 0.92, and switch 3 allows a message through with probability 0.90. What is the probability that a message will find a route through the network?

1.5.8 Suppose that birthdays are equally likely to be on any day of the year (ignore February 29 as a possibility). Show that the probability that two people chosen at random have different birthdays is $364/365$. Show that the probability that three people chosen at random all have different birthdays is

$$\frac{364}{365} \times \frac{363}{365}$$

and extend this pattern to show that the probability that n people chosen at random all have different birthdays is

$$\frac{364}{365} \times \cdots \times \frac{366 - n}{365}$$

What then is the probability that in a group of n people, at least two people will share the same birthday? Evaluate this probability for $n = 10$, $n = 15$, $n = 20$, $n = 25$, $n = 30$, and $n = 35$. What is the smallest value of n for which the probability is larger than a half? Do you think that birthdays are equally likely to be on any day of the year?

1.5.9 Suppose that 17 lightbulbs in a box of 100 lightbulbs are broken and that 3 are selected at random without replacement. Construct a probability tree for this problem. What is the probability that there will be no broken lightbulbs in the sample? What is the probability that there will be no more than 1 broken lightbulb in the sample? (This problem is continued in Problem 1.7.8.)

1.5.10 Repeat Problem 1.5.9, except that the drawings are made *with replacement*. Compare your answers with those from Problem 1.5.9.

1.5.11 Suppose that a bag contains 43 red balls, 54 blue balls, and 72 green balls, and that 2 balls are chosen at random without replacement. Construct a probability tree for this problem. What is the probability that 2 green balls will be chosen? What is the probability that the 2 balls chosen will have different colors?

1.5.12 Repeat Problem 1.5.11, except that the drawings are made *with replacement*. Compare your answers with those from Problem 1.5.11.

1.5.13 A biased coin has a probability p of resulting in a head. If the coin is tossed twice, what value of p minimizes the probability that the same result is obtained on both throws?

1.5.14 If a fair die is rolled six times, what is the probability that each score is obtained exactly once? If a fair die is rolled seven times, what is the probability that a 6 is not obtained at all?

1.5.15 (a) If a fair die is rolled five times, what is the probability that the numbers obtained are all even numbers?
(b) If a fair die is rolled three times, what is the probability that the three numbers obtained are all different?
(c) If three cards are taken at random from a pack of cards with replacement, what is the probability that there are two black cards and one red card?

(d) If three cards are taken at random from a pack of cards without replacement, what is the probability that there are two black cards and one red card?

1.5.16 Suppose that n components are available, and that each component has a probability of 0.90 of operating correctly, independent of the other components. What value of n is needed so that there is a probability of at least 0.995 that at least one component operates correctly?

1.5.17 Suppose that an insurance company insures its clients for flood damage to property. Can the company reasonably expect that the claims from its clients will be independent of each other?

1.5.18 A system has four computers. Computer 1 works with a probability of 0.88; computer 2 works with a probability of 0.78; computer 3 works with a probability of 0.92; computer 4 works with a probability of 0.85. Suppose that the operations of the computers are independent of each other.

(a) Suppose that the system works only when all four computers are working. What is the probability that the system works?

(b) Suppose that the system works only if at least one computer is working. What is the probability that the system works?

(c) Suppose that the system works only if at least three computers are working. What is the probability that the system works?

1.6   Posterior Probabilities

1.6.1   Law of Total Probability

Let $A_1, \ldots, A_n$ be a partition of a sample space $\mathcal{S}$ so that the events A_i are mutually exclusive with

$$\mathcal{S} = A_1 \cup \cdots \cup A_n$$

Suppose that the probabilities of these n events, $P(A_1), \ldots, P(A_n)$, are known. In addition, consider an event B as shown in Figure 1.67, and suppose that the conditional probabilities $P(B|A_1), \ldots, P(B|A_n)$ are also known.

An initial question of interest is how to use the probabilities $P(A_i)$ and $P(B|A_i)$ to calculate $P(B)$, the probability of the event B. In fact, this is easily achieved by noting that

$$B = (A_1 \cap B) \cup \cdots \cup (A_n \cap B)$$

FIGURE 1.67

A partition $A_1, \ldots, A_n$ and an event B

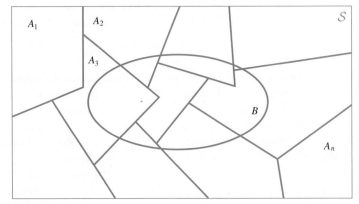

where the events $A_i \cap B$ are mutually exclusive, so that

$$P(B) = P(A_1 \cap B) + \cdots + P(A_n \cap B)$$
$$= P(A_1) P(B|A_1) + \cdots + P(A_n) P(B|A_n)$$

This result, known as the **law of total probability**, has the interpretation that if it is known that one and only one of a series of events A_i can occur, then the probability of another event B can be obtained as the weighted average of the conditional probabilities $P(B|A_i)$, with weights equal to the probabilities $P(A_i)$.

Law of Total Probability

If $A_1, \ldots, A_n$ is a partition of a sample space, then the probability of an event B can be obtained from the probabilities $P(A_i)$ and $P(B|A_i)$ using the formula

$$P(B) = P(A_1) P(B|A_1) + \cdots + P(A_n) P(B|A_n)$$

Example 7

Car Warranties

The law of total probability was tacitly used in the previous section when the probability of a claim being made on a car warranty was calculated to be 0.0687. If A_1, A_2, A_3, and A_4 are, respectively, the events that a car is assembled in plants I, II, III, and IV, then they provide a partition of the sample space, and the probabilities $P(A_i)$ are the supply proportions of the four plants.

If B is the event that a claim is made, then the conditional probabilities $P(B|A_i)$ are the claim rates for the four individual plants, so that

$$P(B) = P(A_1) P(B|A_1) + P(A_2) P(B|A_2) + P(A_3) P(B|A_3) + P(A_4) P(B|A_4)$$
$$= (0.20 \times 0.05) + (0.24 \times 0.11) + (0.25 \times 0.03) + (0.31 \times 0.08)$$
$$= 0.0687$$

as obtained before.

1.6.2 Calculation of Posterior Probabilities

An additional question of interest is how to use the probabilities $P(A_i)$ and $P(B|A_i)$ to calculate the probabilities $P(A_i|B)$, the revised probabilities of the events A_i conditional on the event B. The probabilities

$$P(A_1), \ldots, P(A_n)$$

can be thought of as being the *prior* probabilities of the events $A_1, \ldots, A_n$. However, the observation of the event B provides some additional information that allows the revision of these prior probabilities into a set of *posterior* probabilities

$$P(A_1|B), \ldots, P(A_n|B)$$

which are the probabilities of the events $A_1, \ldots, A_n$ conditional on the event B.

From the law of total probability, these posterior probabilities are calculated to be

$$P(A_i|B) = \frac{P(A_i \cap B)}{P(B)} = \frac{P(A_i) P(B|A_i)}{P(B)} = \frac{P(A_i) P(B|A_i)}{\sum_{j=1}^{n} P(A_j) P(B|A_j)}$$

which is known as **Bayes' theorem**.

HISTORICAL NOTE

Thomas Bayes was born in London, England, in 1702. He was ordained and ministered at a Presbyterian church in Tunbridge Wells, about 35 miles outside London. He was elected a Fellow of the Royal Society in 1742 and died on April 17, 1761. His work on posterior probabilities was discovered in his papers after his death.

Bayes' Theorem

If $A_1, \ldots, A_n$ is a partition of a sample space, then the **posterior probabilities** of the events A_i conditional on an event B can be obtained from the probabilities $P(A_i)$ and $P(B|A_i)$ using the formula

$$P(A_i|B) = \frac{P(A_i)\,P(B|A_i)}{\sum_{j=1}^{n} P(A_j)\,P(B|A_j)}$$

Bayes' theorem is an important result in probability theory because it shows how *new information* can properly be used to update or revise an existing set of probabilities. In some cases the prior probabilities $P(A_i)$ may have to be estimated based on very little information or on subjective feelings. It is then important to be able to improve these probabilities as more information becomes available, and Bayes' theorem provides the means to do this.

1.6.3 Examples of Posterior Probabilities

Example 7
Car Warranties

When a customer buys a car, the (prior) probabilities of it having been assembled in a particular plant are

$$P(\text{plant I}) = 0.20 \quad P(\text{plant II}) = 0.24$$
$$P(\text{plant III}) = 0.25 \quad P(\text{plant IV}) = 0.31$$

If a claim is made on the warranty of the car, how does this change these probabilities? From Bayes' theorem, the posterior probabilities are calculated to be

$$P(\text{plant I}|\text{claim}) = \frac{P(\text{plant I})\,P(\text{claim}|\text{plant I})}{P(\text{claim})} = \frac{0.20 \times 0.05}{0.0687} = 0.146$$

$$P(\text{plant II}|\text{claim}) = \frac{P(\text{plant II})\,P(\text{claim}|\text{plant II})}{P(\text{claim})} = \frac{0.24 \times 0.11}{0.0687} = 0.384$$

$$P(\text{plant III}|\text{claim}) = \frac{P(\text{plant III})\,P(\text{claim}|\text{plant III})}{P(\text{claim})} = \frac{0.25 \times 0.03}{0.0687} = 0.109$$

$$P(\text{plant IV}|\text{claim}) = \frac{P(\text{plant IV})\,P(\text{claim}|\text{plant IV})}{P(\text{claim})} = \frac{0.31 \times 0.08}{0.0687} = 0.361$$

which are tabulated in Figure 1.68. Notice that plant II has the largest claim rate (0.11), and its posterior probability 0.384 is much larger than its prior probability of 0.24. This is expected since the fact that a claim is made increases the likelihood that the car has been assembled in a plant that has a high claim rate. Similarly, plant III has the smallest claim rate (0.03), and its posterior probability 0.109 is much smaller than its prior probability of 0.25, as expected.

FIGURE 1.68

Prior and posterior probabilities for car warranties example

	Prior Probabilities	Posterior Probabilities Claim	No claim
Plant I	0.200	0.146	0.204
Plant II	0.240	0.384	0.229
Plant III	0.250	0.109	0.261
Plant IV	0.310	0.361	0.306
	1.000	1.000	1.000

On the other hand, if *no* claim is made on the warranty, the posterior probabilities are calculated to be

$$P(\text{plant I}|\text{no claim}) = \frac{P(\text{plant I})P(\text{no claim}|\text{plant I})}{P(\text{no claim})}$$

$$= \frac{0.20 \times 0.95}{0.9313} = 0.204$$

$$P(\text{plant II}|\text{no claim}) = \frac{P(\text{plant II})P(\text{no claim}|\text{plant II})}{P(\text{no claim})}$$

$$= \frac{0.24 \times 0.89}{0.9313} = 0.229$$

$$P(\text{plant III}|\text{no claim}) = \frac{P(\text{plant III})P(\text{no claim}|\text{plant III})}{P(\text{no claim})}$$

$$= \frac{0.25 \times 0.97}{0.9313} = 0.261$$

$$P(\text{plant IV}|\text{no claim}) = \frac{P(\text{plant IV})P(\text{no claim}|\text{plant IV})}{P(\text{no claim})}$$

$$= \frac{0.31 \times 0.92}{0.9313} = 0.306$$

as tabulated in Figure 1.68. In this case when no claim is made, the probabilities decrease slightly for plant II and increase slightly for plant III.

Finally, it is interesting to notice that when a claim is made the probabilities are revised quite substantially, but when no claim is made the posterior probabilities are almost the same as the prior probabilities. Intuitively, this is because the claim rates are all rather small, and so a claim is an "unusual" occurrence, which requires a more radical revision of the probabilities.

Example 8

Chemical Impurity Levels

A chemical company has to pay particular attention to the impurity levels of the chemicals that it produces. Previous experience leads the company to estimate that about *one in a hundred* of its chemical batches has an impurity level that is too high.

To ensure better quality for its products, the company has invested in a new laser-based technology for measuring impurity levels. However, this technology is not foolproof, and its manufacturers warn that it will falsely give a reading of a high impurity level for about 5% of batches that actually have satisfactory impurity levels (these are "false-positive" results). On the other hand, it will falsely indicate a satisfactory impurity level for about 2% of batches that have high impurity levels (these are "false-negative" results). With this in mind, the chemical company is interested in questions such as these:

■ If a high impurity reading is obtained, what is the probability that the impurity level really is high?

■ If a satisfactory impurity reading is obtained, what is the probability that the impurity level really is satisfactory?

To answer these questions, let A be the event that the impurity level is too high. Event A and its complement A' form a partition of the sample space, and they have prior probability values of

$$P(A) = 0.01 \quad \text{and} \quad P(A') = 0.99$$

Let B be the event that a high impurity reading is obtained. The false-negative rate then indicates that

$$P(B|A) = 0.98 \quad \text{and} \quad P(B'|A) = 0.02$$

and the false-positive rate indicates that

$$P(B|A') = 0.05 \quad \text{and} \quad P(B'|A') = 0.95$$

If a high impurity reading is obtained, Bayes' theorem gives

$$P(A|B) = \frac{P(A)\,P(B|A)}{P(A)\,P(B|A) + P(A')\,P(B|A')}$$
$$= \frac{0.01 \times 0.98}{(0.01 \times 0.98) + (0.99 \times 0.05)} = 0.165$$

and

$$P(A'|B) = \frac{P(A')\,P(B|A')}{P(A)\,P(B|A) + P(A')\,P(B|A')}$$
$$= \frac{0.99 \times 0.05}{(0.01 \times 0.98) + (0.99 \times 0.05)} = 0.835$$

If a satisfactory impurity reading is obtained, Bayes' theorem gives

$$P(A|B') = \frac{P(A)\,P(B'|A)}{P(A)\,P(B'|A) + P(A')\,P(B'|A')}$$
$$= \frac{0.01 \times 0.02}{(0.01 \times 0.02) + (0.99 \times 0.95)} = 0.0002$$

and

$$P(A'|B') = \frac{P(A')\,P(B'|A')}{P(A)\,P(B'|A) + P(A')\,P(B'|A')}$$
$$= \frac{0.99 \times 0.95}{(0.01 \times 0.02) + (0.99 \times 0.95)} = 0.9998$$

These posterior probabilities are tabulated in Figure 1.69.

FIGURE 1.69

Prior and posterior probabilities for the chemical impurities example

	Prior Probabilities	Posterior Probabilities B: high reading	Posterior Probabilities B': satisfactory reading
A: impurity level too high	0.0100	0.1650	0.0002
A': impurity level satisfactory	0.9900	0.8350	0.9998
	1.0000	1.0000	1.0000

We can see that if a satisfactory impurity reading is obtained, then the probability of the impurity level actually being too high is only 0.0002, so that on average, only 1 in 5000 batches testing satisfactory is really not satisfactory. However, if a high impurity reading is obtained, there is a probability of only 0.165 that the impurity level really is high, and the probability is 0.835 that the batch is really satisfactory. In other words, only about 1 in 6 of the batches testing high actually has a high impurity level.

At first this may appear counterintuitive. Since the false-positive and false-negative error rates are so low, why is it that most of the batches testing high are really satisfactory? The answer lies in the fact that about 99% of the batches have satisfactory impurity levels, so that 99% of the time there is an "opportunity" for a false-positive result, and only about 1% of the time is there an "opportunity" for a genuine positive result.

In conclusion, the chemical company should realize that it is wasteful to disregard off-hand batches that are indicated to have high impurity levels. Further investigation of these batches should be undertaken to identify the large proportion of them that are in fact satisfactory products.

■ 1.6.4 Problems

1.6.1 Suppose it is known that 1% of the population suffers from a particular disease. A blood test has a 97% chance of identifying the disease for diseased individuals, but also has a 6% chance of falsely indicating that a healthy person has the disease.

(a) What is the probability that a person will have a positive blood test?

(b) If your blood test is positive, what is the chance that you have the disease?

(c) If your blood test is negative, what is the chance that you do not have the disease?

1.6.2 Bag A contains 3 red balls and 7 blue balls. Bag B contains 8 red balls and 4 blue balls. Bag C contains 5 red balls and 11 blue balls. A bag is chosen at random, with each bag being equally likely to be chosen, and then a ball is chosen at random from that bag. Calculate the probabilities:

(a) A red ball is chosen.

(b) A blue ball is chosen.

(c) A red ball from bag B is chosen.

If it is known that a red ball is chosen, what is the probability that it comes from bag A? If it is known that a blue ball is chosen, what is the probability that it comes from bag B?

1.6.3 A class had two sections. Section I had 55 students of whom 10 received A grades. Section II had 45 students of whom 11 received A grades. Now 1 of the 100 students is chosen at random, with each being equally likely to be chosen.

(a) What is the probability that the student was in section I?

(b) What is the probability that the student received an A grade?

(c) What is the probability that the student received an A grade if the student is known to have been in section I?

(d) What is the probability that the student was in section I if the student is known to have received an A grade?

1.6.4 An island has three species of bird. Species 1 accounts for 45% of the birds, of which 10% have been tagged. Species 2 accounts for 38% of the birds, of which 15% have been tagged. Species 3 accounts for 17% of the birds, of which 50% have been tagged. If a tagged bird is observed, what are the probabilities that it is of species 1, of species 2, and of species 3?

1.6.5 After production, an electrical circuit is given a quality score of A, B, C, or D. Over a certain period of time, 77% of the circuits were given a quality score A, 11% were given a quality score B, 7% were given a quality score C, and 5% were given a quality score D. Furthermore, it was found that 2% of the circuits given a quality score A eventually failed, and the failure rate was 10% for circuits given a quality score B, 14% for circuits given a quality score C, and 25% for circuits given a quality score D.

(a) If a circuit failed, what is the probability that it had received a quality score either C or D?

(b) If a circuit did not fail, what is the probability that it had received a quality score A?

1.6.6 The weather on a particular day is classified as either cold, warm, or hot. There is a probability of 0.15 that it is cold and a probability of 0.25 that it is warm. In addition, on each day it may either rain or not rain. On cold days there is a probability of 0.30 that it will rain, on warm days there is a probability of 0.40 that it will rain, and on hot days there is a probability of 0.50 that it will rain. If it is not raining on a particular day, what is the probability that it is cold?

1.6.7 A valve can be used at four temperature levels. If the valve is used at a cold temperature, then there is a probability of 0.003 that it will leak. If the valve is used at a medium temperature, then there is a probability of 0.009 that it will leak. If the valve is used at a warm temperature, then there is a probability of 0.014 that it will leak. If the valve is used at a hot temperature, then there is a probability of 0.018 that it will leak. Under standard operating conditions, the valve is used at a cold temperature 12% of the time, at a medium

temperature 55% of the time, at a warm temperature 20% of the time, and at a hot temperature 13% of the time.

(a) If the valve leaks, what is the probability that it is being used at the hot temperature?

(b) If the valve does not leak, what is the probability that it is being used at the medium temperature?

1.6.8 A company sells five types of wheelchairs, with type A being 12% of the sales, type B being 34% of the sales, type C being 7% of the sales, type D being 25% of the sales, and type E being 22% of the sales. In addition, 19% of the type A wheelchair sales are motorized, 50% of the type B wheelchair sales are motorized, 4% of the type C wheelchair sales are motorized, 32% of the type D wheelchair sales are motorized, and 76% of the type E wheelchair sales are motorized.

(a) If a motorized wheelchair is sold, what is the probability that it is of type C?

(b) If a nonmotorized wheelchair is sold, what is the probability that it is of type D?

1.7   Counting Techniques

In many situations the sample space S consists of a very large number of outcomes that the experimenter will not want to list in their entirety. However, if the outcomes are equally likely, then it suffices to know the *number* of outcomes in the sample space and the *number* of outcomes contained within an event of interest. In this section, various **counting techniques** are discussed that can be used to facilitate such computations. Remember that if a sample space S consists of N equally likely outcomes, of which n are contained within the event A, then the probability of the event A is

$$P(A) = \frac{n}{N}$$

1.7.1   Multiplication Rule

Suppose that an experiment consists of k "components" and that the ith component has n_i possible outcomes. The total number of experimental outcomes will then be equal to the product

$$n_1 \times n_2 \times \cdots \times n_k$$

This is known as the **multiplication rule** and can easily be seen by referring to the tree diagram in Figure 1.70. The n_1 outcomes of the first component are represented by the n_1 branches at the beginning of the tree, each of which splits into n_2 branches corresponding to the outcomes of the second component, and so on. The total number of experimental outcomes (the size of the sample space) is equal to the number of branch ends at the end of the tree, which is equal to the product of the n_i.

FIGURE 1.70

Probability tree illustrating the
multiplication rule

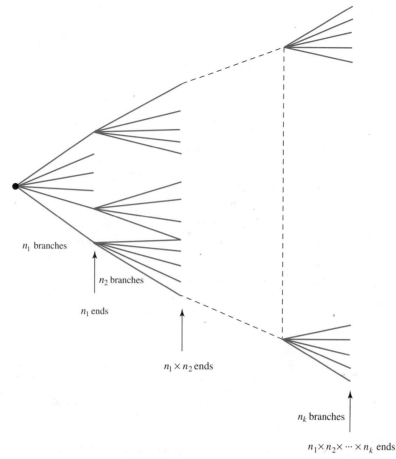

FIGURE 1.70

Probability tree illustrating the
multiplication rule

n_1 branches

n_2 branches

n_1 ends

$n_1 \times n_2$ ends

n_k branches

$n_1 \times n_2 \times \cdots \times n_k$ ends

Multiplication Rule

If an experiment consists of k components for which the number of possible outcomes
are $n_1, \ldots, n_k$, then the total number of experimental outcomes (the size of the sample
space) is equal to

$$n_1 \times n_2 \times \cdots \times n_k$$

Example 9

**Car Body Assembly
Line**

A side panel for a car is made from a sheet of metal in the following way. The metal sheet is
first sent to a cleaning machine, then to a pressing machine, and then to a cutting machine.
The process is completed by a painting machine followed by a polishing machine. Each of the
five tasks can be performed on one of several machines whose number and location within
the factory are determined by the management so as to streamline the whole manufacturing
process.

In particular, suppose that there are six cleaning machines, three pressing machines,
eight cutting machines, five painting machines, and eight polishing machines, as illustrated in
Figure 1.71. As a quality control procedure, the company attaches a bar code to each panel that

FIGURE 1.71

Manufacturing process for car side
panels

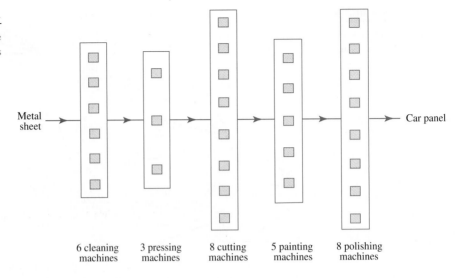

6 cleaning machines   3 pressing machines   8 cutting machines   5 painting machines   8 polishing machines

Total number of pathways is $6 \times 3 \times 8 \times 5 \times 8 = 5760$.

identifies which of the machines have been used in its construction. The number of possible "pathways" through the manufacturing process is

$$6 \times 3 \times 8 \times 5 \times 8 = 5760$$

The number of pathways that include a particular pressing machine are

$$6 \times 8 \times 5 \times 8 = 1920$$

If the 5760 pathways can be considered to be equally likely, then a panel chosen at random will have a probability of 1/5760 of having each of the pathways. However, notice that the pathways will probably not be equally likely, since, for example, the factory layout could cause a panel coming out of one pressing machine to be more likely to be passed on to a particular cutting machine than panels from another pressing machine.

Example 10
Fiber Coatings

Thin fibers are often coated by passing them through a cloud chamber containing the coating material. The fiber and the coating material are provided with opposite electrical charges to provide a means of attraction. Among other things, the quality of the coating will depend on the sizes of the *electrical charges* employed, the *density* of the coating material in the cloud chamber, the *temperature* of the cloud chamber, and the *speed* at which the fiber is passed through the chamber.

A chemical engineer wishes to conduct an experiment to determine how these four factors affect the quality of the coating. The engineer is interested in comparing three charge levels, five density levels, four temperature levels, and three speed levels, as illustrated in Figure 1.72. The total number of possible experimental conditions is then

$$3 \times 5 \times 4 \times 3 = 180$$

In other words, there are 180 different combinations of the four factors that can be investigated.

However, the cost of running 180 experiments is likely to be prohibitive, and the engineer may have a budget sufficient to investigate, say, only 30 experimental conditions. Nevertheless,

FIGURE 1.72

Experimental configurations for fiber coatings

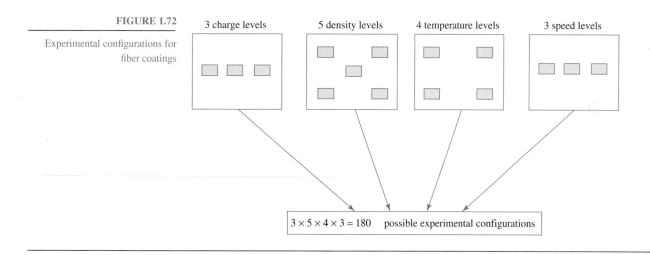

with an appropriate **experimental design** and statistical analysis, the engineer can carefully choose which experimental conditions to investigate in order to provide a maximum amount of information about the four factors and how they influence the quality of the coating. The analysis of experiments of this kind is discussed in Chapter 14.

Often the k components of an experiment are identical because they are replications of the same process. In such cases $n_1 = \cdots = n_k = n$, say, and the total number of experimental outcomes will be n^k. For example, if a die is rolled twice, there are $6 \times 6 = 36$ possible outcomes. If a die is rolled k times, there are 6^k possible outcomes.

Computer codes consist of a series of binary digits 0 and 1. The number of different strings consisting of k digits is then 2^k. For example, a string of 20 digits can have

$$2^{20} = 1,048,576$$

possible values. Calculations such as these indicate how much "information" can be carried by the strings.

Computer passwords typically consist of a string of eight characters, say, which are either 1 of the 26 letters or a numerical digit. The possible number of choices for a password is then

$$36^8 \simeq 2.82 \times 10^{12}$$

If a password is chosen at random, the chance of somebody "guessing" it is thus negligibly small. Nevertheless, a feeling of security could be an illusion for at least two reasons. First, if a computer can be programmed to search repeatedly through possible passwords quickly in an organized manner, it may not take it long to hit on the correct one. Second, few people choose passwords at random, since they themselves have to remember them.

1.7.2 Permutations and Combinations

Often it is important to be able to calculate how many ways a series of distinguishable k objects can be drawn from a pool of n objects. If the drawings are performed with replacement, then the k drawings are identical events, each with n possible outcomes, and the multiplication rule shows that there are n^k possible ways to draw the k objects.

If the drawings are made *without replacement*, then the outcome is said to be a **permutation** of k objects from the original n objects. If only one object is chosen, then clearly there are only n possible outcomes. If two objects are chosen, then there will be

$$n(n-1)$$

possible outcomes, since there are n possibilities for the first choice and then only $n-1$ possibilities for the second choice. More generally, if k objects are chosen, there will be

$$n(n-1)(n-2)\cdots(n-k+1)$$

possible outcomes, which is obtained by multiplying together the number of choices at each drawing.

For dealing with expressions such as these, it is convenient to use the following notation:

Factorials

If n is a positive integer, the quantity $n!$ called "n factorial" is defined to be

$$n! = n(n-1)(n-2)\cdots(1)$$

Also, the quantity $0!$ is taken to be equal to 1.

The number of permutations of k objects from n objects is given the notation P_k^n.

Permutations

A **permutation** of k objects from n objects ($n \geq k$) is an **ordered** sequence of k objects selected without replacement from the group of n objects. The number of possible permutations of k objects from n objects is

$$P_k^n = n(n-1)(n-2)\cdots(n-k+1) = \frac{n!}{(n-k)!}$$

Notice that if $k = n$, the number of permutations is

$$P_n^n = n(n-1)(n-2)\cdots 1 = n!$$

which is just the number of ways of ordering n objects.

Example 11

Taste Tests

A food company has four different recipes for a potential new product and wishes to compare them through consumer taste tests. In these tests, a participant is given the four types of food to taste in a random order and is asked to rank various aspects of their taste. This ranking procedure simply provides an ordering of the four products, and the number of possible ways in which it can be performed is

$$P_4^4 = 4! = 4 \times 3 \times 2 \times 1 = 24$$

In a different taste test, each participant samples eight products and is asked to pick the best, the second best, and the third best. The number of possible answers to the test is then

$$P_3^8 = 8 \times 7 \times 6 = 336$$

$$P_3^8 = \frac{8!}{(8-3)!} = \frac{8!}{5!} = \frac{8 \cdot 7 \cdot 6 \cdot 5 \cdot 4 \cdot 3 \cdot 2 \cdot 1}{5 \cdot 4 \cdot 3 \cdot 2 \cdot 1} = 8 \cdot 7 \cdot 6$$

Notice that with permutations, the *order* of the sequence is important. For example, if the eight products in the taste test are labeled *A–H*, then the permutation *ABC* (in which product *A* is judged to be best, product *B* second best, and product *C* third best) is considered to be different from the permutation *ACB*, say. That is, each of the six orderings of the products *A*, *B*, and *C* is considered to be a different permutation.

Sometimes when *k* objects are chosen from a group of *n* objects, the ordering of the drawing of the *k* objects is not of importance. In other words, interest is focused on which *k* objects are chosen, but not on the order in which they are chosen. Such a collection of objects is called a *combination* of *k* objects from *n* objects. The notation C_k^n is used for the total possible number of such combinations, and it is calculated using the formula

$$C_k^n = \frac{n!}{(n-k)! \, k!}$$

This formula for the number of combinations follows from the fact that each combination of *k* objects can be associated with the *k*! permutations that consist of those objects. Consequently,

$$P_k^n = k! \times C_k^n$$

so that

$$C_k^n = \frac{P_k^n}{k!} = \frac{n!}{(n-k)! \, k!}$$

A common alternative notation for the number of combinations is

$$C_k^n = \binom{n}{k}$$

Combinations

A **combination** of *k* objects from *n* objects ($n \geq k$) is an **unordered** collection of *k* objects selected without replacement from the group of *n* objects. The number of possible combinations of *k* objects from *n* objects is

$$C_k^n = \binom{n}{k} = \frac{n!}{(n-k)! \, k!}$$

Notice that

$$C_1^n = \binom{n}{1} = \frac{n!}{(n-1)! \, 1!} = n$$

and

$$C_2^n = \binom{n}{2} = \frac{n!}{(n-2)! \, 2!} = \frac{n(n-1)}{2}$$

so that there are *n* ways to choose one object from *n* objects, and $n(n-1)/2$ ways to choose two objects from *n* objects (without attention to order). Also,

$$C_{n-1}^n = \binom{n}{n-1} = \frac{n!}{1! \, (n-1)!} = n$$

and

$$C_n^n = \binom{n}{n} = \frac{n!}{0!\,n!} = 1$$

This last equation just indicates that there is only one way to choose all n objects. It is also useful to note that $C_k^n = C_{n-k}^n$.

Example 11

Taste Tests

Suppose that in the taste test, each participant samples eight products and is asked to select the three best products, but not in any particular order. The number of possible answers to the test is then

$$\binom{8}{3} = \frac{8!}{5!\,3!} = \frac{8 \times 7 \times 6}{3 \times 2 \times 1} = 56$$

Example 2

Defective Computer Chips

Suppose again that 9 out of 500 chips in a particular box are defective, and that 3 chips are sampled at random from the box without replacement. The total number of possible samples is

$$\binom{500}{3} = \frac{500!}{497!\,3!} = \frac{500 \times 499 \times 498}{3 \times 2 \times 1} = 20,708,500$$

which are all equally likely.

The probability of choosing 3 defective chips can be calculated by dividing the number of samples that contain 3 defective chips by the total number of samples. Since there are 9 defective chips, the number of samples that contain 3 defective chips is

$$\binom{9}{3} = \frac{9!}{6!\,3!} = \frac{9 \times 8 \times 7}{3 \times 2 \times 1} = 84$$

so that the probability of choosing 3 defective chips is

$$\frac{\binom{9}{3}}{\binom{500}{3}} = \frac{\left(\frac{9!}{6!\,3!}\right)}{\left(\frac{500!}{497!\,3!}\right)} = \frac{9 \times 8 \times 7}{500 \times 499 \times 498} \simeq 4.0 \times 10^{-6}$$

as obtained before.

Also, the number of samples that contains exactly 1 defective chip is

$$9 \times \binom{491}{2}$$

since there are 9 ways to choose the defective chip and C_2^{491} ways to choose the 2 satisfactory chips. Consequently, the probability of obtaining exactly 1 defective chip is

$$\frac{9 \times \binom{491}{2}}{\binom{500}{3}} = \frac{\left(9 \times \frac{491!}{489!\,2!}\right)}{\left(\frac{500!}{497!\,3!}\right)} = \frac{9 \times 491 \times 490 \times 3}{500 \times 499 \times 498} = 0.0522$$

as obtained before. These calculations are examples of the hypergeometric distribution that is discussed in Section 3.3.

GAMES OF CHANCE

Suppose that four cards are taken at random without replacement from a pack of cards. What is the probability that two Kings and two Queens are chosen?

The number of ways to choose four cards is

$$\binom{52}{4} = \frac{52!}{48!\,4!} = \frac{52 \times 51 \times 50 \times 49}{4 \times 3 \times 2 \times 1} = 270{,}725$$

The number of ways of choosing two Kings from the four Kings in the pack as well as the number of ways of choosing two Queens from the four Queens in the pack is

$$\binom{4}{2} = \frac{4!}{2!\,2!} = \frac{4 \times 3}{2 \times 1} = 6$$

so that the number of hands consisting of two Kings and two Queens is

$$\binom{4}{2} \times \binom{4}{2} = 36$$

The required probability is thus

$$\frac{36}{270{,}725} \simeq 1.33 \times 10^{-4}$$

which is a chance of about 13 out of 100,000.

■ 1.7.3 Problems

1.7.1 Evaluate:
(a) 7! (b) 8! (c) 4! (d) 13!

1.7.2 Evaluate:
(a) P_2^7 (b) P_5^9 (c) P_2^5 (d) P_4^{17}

1.7.3 Evaluate:
(a) C_2^6 (b) C_4^8 (c) C_2^5 (d) C_6^{14}

1.7.4 A menu has five appetizers, three soups, seven main courses, six salad dressings, and eight desserts. In how many ways can a full meal be chosen? In how many ways can a meal be chosen if either an appetizer or a soup is ordered, but not both?

1.7.5 In an experiment to test iron strengths, three different ores, four different furnace temperatures, and two different cooling methods are to be considered. Altogether, how many experimental configurations are possible?

1.7.6 Four players compete in a tournament and are ranked from 1 to 4. They then compete in another tournament and are again ranked from 1 to 4. Suppose that their performances in the second tournament are unrelated to their performances in the first tournament, so that the two sets of rankings are independent.
(a) What is the probability that each competitor receives an identical ranking in the two tournaments?

(b) What is the probability that nobody receives the same ranking twice?

1.7.7 Twenty players compete in a tournament. In how many ways can rankings be assigned to the top five competitors? In how many ways can the best five competitors be chosen (without being in any order)?

1.7.8 There are 17 broken lightbulbs in a box of 100 lightbulbs. A random sample of 3 lightbulbs is chosen without replacement.
(a) How many ways are there to choose the sample?
(b) How many samples contain no broken lightbulbs?
(c) What is the probability that the sample contains no broken lightbulbs?
(d) How many samples contain exactly 1 broken lightbulb?
(e) What is the probability that the sample contains no more than 1 broken lightbulb?

1.7.9 Show that $C_k^n = C_k^{n-1} + C_{k-1}^{n-1}$. Can you provide an interpretation of this equality?

1.7.10 A poker hand consists of five cards chosen at random from a pack of cards.
(a) How many different hands are there?
(b) How many hands consist of all hearts?

(c) How many hands consist of cards all from the same suit (a "flush")?

(d) What is the probability of being dealt a flush?

(e) How many hands contain all four Aces?

(f) How many hands contain four cards of the same number or picture?

(g) What is the probability of being dealt a hand containing four cards of the same number or picture?

1.7.11 In an arrangement of n objects in a circle, an object's neighbors are important, but an object's place in the circle is not important. Thus, rotations of a given arrangement are considered to be the same arrangement. Explain why the number of different arrangements is $(n-1)!$

1.7.12 In how many ways can six people sit in six seats in a line at a cinema? In how many ways can the six people sit around a dinner table eating pizza after the movie?

1.7.13 Repeat Problem 1.7.12 with the condition that one of the six people, Andrea, must sit next to Scott. In how many ways can the seating arrangements be made if Andrea refuses to sit next to Scott?

1.7.14 A total of n balls are to be put into k boxes with the conditions that there will be n_1 balls in box 1, n_2 balls in box 2, and so on, with n_k balls being placed in box k $(n_1 + \cdots + n_k = n)$. Explain why the number of ways of doing this is

$$\frac{n!}{n_1! \times \cdots \times n_k!}$$

Explain why this is just $C_{n_1}^n = C_{n_2}^n$ when $k = 2$.

1.7.15 Explain why the following two problems are identical and solve them.

(a) In how many ways can 12 balls be placed in 3 boxes, when the first box can hold 3 balls, the second box can hold 4 balls, and the third box can hold 5 balls.

(b) In how many ways can 3 red balls, 4 blue balls, and 5 green balls be placed in a straight line?

(See Problem 1.7.14.)

1.7.16 A garage employs 14 mechanics, of whom 3 are needed on one job and, at the same time, 4 are needed on another

job. The remaining 7 are to be kept in reserve. In how many ways can the job assignments be made? (See Problem 1.7.14.)

1.7.17 A company has 15 applicants to interview, and 3 are to be invited on each day of the working week. In how many ways can the applicants be scheduled? (See Problem 1.7.14.)

1.7.18 A quality inspector selects a sample of 12 items at random from a collection of 60 items, of which 18 have excellent quality, 25 have good quality, 12 have poor quality, and 5 are defective.

(a) What is the probability that the sample only contains items that have either excellent or good quality?

(b) What is the probability that the sample contains three items of excellent quality, three items of good quality, three items of poor quality, and three defective items?

1.7.19 A salesman has to visit 10 different cities. In how many different ways can the ordering of the visits be made? If he decides that 5 of the visits will be made one week, and the other 5 visits will be made the following week, in how many different ways can the 10 cities be split into two groups of 5 cities?

1.7.20 Suppose that 5 cards are taken without replacement from a deck of 52 cards. How many ways are there to do this so that there are 2 red cards and 3 black cards?

1.7.21 A hand of 8 cards is chosen at random from an ordinary deck of 52 playing cards without replacement.

(a) What is the probability that the hand does not have any hearts?

(b) What is the probability that the hand consists of two hearts, two diamonds, two clubs, and two spades?

1.7.22 A box contains 40 batteries, 5 of which have low lifetimes, 30 of which have average lifetimes, and 5 of which have high lifetimes. A consumer requires 8 batteries to run an appliance and randomly selects them all from the box. What is the probability that among the 8 batteries fitted into the consumer's appliance, there are exactly 2 low, 4 average and 2 high lifetimes batteries?

1.8   Case Study: Microelectronic Solder Joints

Suppose that using a particular production method there is a probability of 0.85 that a solder joint has a barrel shape, there is a probability of 0.03 that a solder joint has a cylinder shape, and there is a probability of 0.12 that a solder joint has an hourglass shape. If it is known that a particular solder joint does not have a barrel shape, what is the probability that it has a cylinder shape? This is a conditional probability that can be calculated as

$$P(\text{cylinder}|\text{not barrel}) = \frac{P(\text{cylinder and not barrel})}{P(\text{not barrel})} = \frac{P(\text{cylinder})}{P(\text{cylinder}) + P(\text{hourglass})}$$

$$= \frac{0.03}{0.03 + 0.12} = 0.2$$

Furthermore, suppose that after a certain number of temperature cycles in an accelerated life test there is a probability of 0.002 that a solder joint is cracked if it has a barrel shape, there is a probability of 0.004 that a solder joint is cracked if it has a cylinder shape, and there is a probability of 0.005 that a solder joint is cracked if it has an hourglass shape. This information can be represented by the conditional probabilities shown in Figure 1.73.

If a solder joint is known to be cracked, Bayes' theorem can be used to calculate the probabilities of it having each of the three shapes. For example, the probability that it has a barrel shape is

$P(\text{barrel}|\text{cracked})$

$$= \frac{P(\text{barrel})\,P(\text{cracked}|\text{barrel})}{\left(\begin{array}{c} P(\text{barrel})\,P(\text{cracked}|\text{barrel}) + P(\text{cylinder})\,P(\text{cracked}|\text{cylinder}) \\ + P(\text{hourglass})\,P(\text{cracked}|\text{hourglass}) \end{array}\right)}$$

$$= \frac{0.85 \times 0.002}{(0.85 \times 0.002) + (0.03 \times 0.004) + (0.12 \times 0.005)} = 0.70248$$

Similarly, if a solder joint is known not to be cracked, then Bayes' theorem can be used to calculate the probability that it has a cylinder shape, for example, as

$P(\text{cylinder}|\text{not cracked})$

$$= \frac{P(\text{cylinder})\,P(\text{not cracked}|\text{cylinder})}{\left(\begin{array}{c} P(\text{barrel})\,P(\text{not cracked}|\text{barrel}) + P(\text{cylinder})\,P(\text{not cracked}|\text{cylinder}) \\ + P(\text{hourglass})\,P(\text{not cracked}|\text{hourglass}) \end{array}\right)}$$

$$= \frac{0.03 \times 0.996}{(0.85 \times 0.998) + (0.03 \times 0.996) + (0.12 \times 0.995)} = 0.02995$$

Figure 1.74 shows all of the shape probabilities conditional on whether the solder joint is known to be cracked or not cracked. Notice that the probabilities in each column sum to one, and that whereas the knowledge that the solder joint is not cracked has little effect on the shape probabilities, the knowledge that the solder joint is cracked (which is a considerably rarer event) has much more effect on the shape probabilities.

FIGURE 1.73

Conditional probabilities of cracking for solder joints

$P(\text{cracked}|\text{barrel}) = 0.002$	$P(\text{not cracked}|\text{barrel}) = 0.998$
$P(\text{cracked}|\text{cylinder}) = 0.004$	$P(\text{not cracked}|\text{cylinder}) = 0.996$
$P(\text{cracked}|\text{hourglass}) = 0.005$	$P(\text{not cracked}|\text{hourglass}) = 0.995$

FIGURE 1.74

Shape probabilities conditional on whether the solder joint is cracked or not

No information on whether the solder joint is cracked or not	Solder joint is known to be cracked	Solder joint is known not to be cracked
$P(\text{barrel}) = 0.85$	$P(\text{barrel}|\text{cracked}) = 0.70248$	$P(\text{barrel}|\text{not cracked}) = 0.85036$
$P(\text{cylinder}) = 0.03$	$P(\text{cylinder}|\text{cracked}) = 0.04959$	$P(\text{cylinder}|\text{not cracked}) = 0.02995$
$P(\text{hourglass}) = 0.12$	$P(\text{hourglass}|\text{cracked}) = 0.24793$	$P(\text{hourglass}|\text{not cracked}) = 0.11969$

Finally, suppose that an assembly consists of 16 solder joints and that unknown to the researcher 5 of these solder joints are cracked. If the researcher randomly chooses a sample of 4 of the solder joints for inspection, then the state space of the number of cracked joints in the sample is {0, 1, 2, 3, 4}. The total number of different samples that can be chosen is

$$\binom{16}{4} = \frac{16!}{12!4!} = 1820$$

and the probability that there will be exactly two cracked solder joints in the researcher's sample is

$$\frac{\binom{5}{2} \times \binom{11}{2}}{\binom{16}{4}} = \frac{10 \times 55}{1820} = 0.302$$

This hypergeometric distribution is discussed more comprehensively in Section 3.3.

1.9   Supplementary Problems

1.9.1 What is the sample space for the average score of two dice?

1.9.2 What is the sample space when a *winner* and a *runner-up* are chosen in a tournament with four contestants.

1.9.3 A *biased* coin is known to have a greater probability of recording a head than a tail. How can it be used to determine *fairly* which team in a football game has the choice of kick-off?

1.9.4 If two fair dice are thrown, what is the probability that their two scores differ by no more than one?

1.9.5 If a card is chosen at random from a pack of cards, what is the probability of choosing a diamond picture card?

1.9.6 Two cards are drawn from a pack of cards. Is it more likely that two hearts will be drawn when the drawing is with replacement or without replacement?

1.9.7 Two fair dice are thrown. *A* is the event that the sum of the scores is no larger than 4, and *B* is the event that the two scores are identical. Calculate the probabilities:
(a) $A \cap B$ **(b)** $A \cup B$ **(c)** $A' \cup B$

1.9.8 Two fair dice are thrown, one red and one blue. Calculate:
(a) $P(\text{red die is 5} | \text{sum of scores is 8})$
(b) $P(\text{either die is 5} | \text{sum of scores is 8})$
(c) $P(\text{sum of scores is 8} | \text{either die is 5})$

1.9.9 Consider the network shown in Figure 1.75 with five switches. Suppose that the switches operate independently and that each switch allows a message through with a probability of 0.85. What is the probability that a message will find a route through the network?

1.9.10 Which is more likely: obtaining at least one head in two tosses of a fair coin, or at least two heads in four tosses of a fair coin?

FIGURE 1.75

Switch diagram

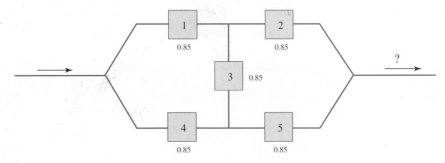

1.9.11 Bag 1 contains 6 red balls, 7 blue balls, and 3 green balls. Bag 2 contains 8 red balls, 8 blue balls, and 2 green balls. Bag 3 contains 2 red balls, 9 blue balls, and 8 green balls. Bag 4 contains 4 red balls, 7 blue balls, and no green balls. Bag 1 is chosen with a probability of 0.15, bag 2 with a probability of 0.20, bag 3 with a probability of 0.35, and bag 4 with a probability of 0.30, and then a ball is chosen at random from the bag. Calculate the probabilities:

   (a) A blue ball is chosen
   (b) Bag 4 was chosen if the ball is green
   (c) Bag 1 was chosen if the ball is blue

1.9.12 A fair die is rolled. If an even number is obtained, then that is the recorded score. However, if an odd number is obtained, then a fair coin is tossed. If a head is obtained, then the recorded score is the number on the die, but if a tail is obtained, then the recorded score is *twice* the number on the die.

   (a) Give the possible values of the recorded score.
   (b) What is the probability that a score of 10 is recorded?
   (c) What is the probability that a score of 3 is recorded?
   (d) What is the probability that a score of 6 is recorded?
   (e) What is the probability that a score of 4 is recorded if it is known that the coin is tossed?
   (f) If a score of 6 is recorded, what is the probability that an odd number was obtained on the die?

1.9.13 How many sequences of length 4 can be made when each component of the sequence can take 5 different values? How many sequences of length 5 can be made when each component of the sequence can take 4 different values? In general, if $3 \leq n_1 < n_2$, are there more sequences of length n_1 with n_2 possible values for each component, or more sequences of length n_2 with n_1 possible values for each component?

1.9.14 Twenty copying jobs need to be done. If there are four copy machines, in how many ways can five jobs be assigned to each of the four machines? If an additional copier is used, in how many ways can four jobs be assigned to each of the five machines?

1.9.15 A bag contains two counters with each independently equally likely to be either black or white. What is the distribution of X, the number of white counters in the bag? Suppose that a white counter is added to the bag and then one of the three counters is selected at random and taken out of the bag. What is the distribution of X conditional on the counter taken out being white? What if the counter taken out of the bag is black?

1.9.16 It is found that 28% of orders received by a company are from first-time customers, with the other 72% coming from repeat customers. In addition, 75% of the orders from first-time customers are dispatched within one day, and overall 30% of the company's orders are from repeat customers whose orders are not dispatched within one day. If an order is dispatched within one day, what is the probability that it was for a first-time customer?

1.9.17 When asked to select their favorite opera work, 26% of the respondents selected a piece by Puccini, and 22% of the respondents selected a piece by Verdi. Moreover, 59% of the respondents who selected a piece by Puccini were female, and 45% of the respondents who selected a piece by Verdi were female. Altogether, 62% of the respondents were female.

   (a) If a respondent selected a piece that is by neither Puccini nor Verdi, what is the probability that the respondent is female?
   (b) What proportion of males selected a piece by Puccini?

1.9.18 A random sample of 10 fibers is taken from a collection of 92 fibers that consists of 43 fibers of polymer A, 17 fibers of polymer B, and 32 fibers of polymer C.

   (a) What is the probability that the sample does not contain any fibers of polymer B?
   (b) What is the probability that the sample contains exactly one fiber of polymer B?
   (c) What is the probability that the sample contains three fibers of polymer A, three fibers of polymer B, and four fibers of polymer C?

1.9.19 A fair coin is tossed five times. What is the probability that there is not a sequence of three outcomes of the same kind?

1.9.20 Consider telephone calls made to a company's complaint line. Let A be the event that the call is answered within 10 seconds. Let B be the event that the call is answered by one of the company's experienced telephone operators. Let C be the event that the call lasts less than 5 minutes. Let D be the event that the complaint is handled successfully by the telephone operator. Describe the following events.

   (a) $B \cap C'$ **(b)** $(A \cup B') \cap D$
   (c) $A' \cap C' \cap D'$ **(d)** $(A \cap C) \cup (B \cap D)$

1.9.21 A manager has 20 different job orders, of which 7 must be assigned to production line I, 7 must be assigned to production line II, and 6 must be assigned to production line III.

(a) In how many ways can the assignments be made?

(b) If the first job and the second job must be assigned to the same production line, in how many ways can the assignments be made?

(c) If the first job and the second job cannot be assigned to the same production line, in how many ways can the assignments be made?

1.9.22 A hand of 3 cards (without replacement) is chosen at random from an ordinary deck of 52 playing cards.

(a) What is the probability that the hand contains only diamonds?

(b) What is the probability that the hand contains one Ace, one King, and one Queen?

1.9.23 A hand of 4 cards (without replacement) is chosen at random from an ordinary deck of 52 playing cards.

(a) What is the probability that the hand does not have any Aces?

(b) What is the probability that the hand has exactly one Ace?

Suppose now that the 4 cards are taken with replacement.

(c) What is the probability that the same card is obtained four times?

1.9.24 Are the following statements true or false?

(a) If a fair coin is tossed three times, the probability of obtaining two heads and one tail is the same as the probability of obtaining one head and two tails.

(b) If a card is drawn at random from a deck of cards, the probability that it is a heart increases if it is conditioned on the knowledge that it is an Ace.

(c) The number of ways of choosing five different letters from the alphabet is more than the number of seconds in a year.

(d) If two events are independent, then the probability that they both occur can be calculated by multiplying their individual probabilities.

(e) It is always true that $P(A|B) + P(A'|B) = 1$.

(f) It is always true that $P(A|B) + P(A|B') = 1$.

(g) It is always true that $P(A|B) \leq P(A)$.

1.9.25 There is a probability of 0.55 that a soccer team will win a game. There is also a probability of 0.85 that the soccer team will not have a player sent off in the game. However, if the soccer team does not have a player sent off, then there is a probability of 0.60 that the team will win the game. What is the probability that the team has a player sent off but still wins the game?

1.9.26 A warehouse contains 500 machines. Each machine is either new or used, and each machine has either good quality or bad quality. There are 120 new machines that have bad quality. There are 230 used machines. Suppose that a machine is chosen at random, with each machine being equally likely to be chosen.

(a) What is the probability that the chosen machine is a new machine with good quality?

(b) If the chosen machine is new, what is the probability that it has good quality?

1.9.27 A class has 250 students, 113 of whom are male, and 167 of whom are mechanical engineers. There are 52 female students who are not mechanical engineers. There are 19 female mechanical engineers who are seniors.

(a) If a randomly chosen student is not a mechanical engineer, what is the probability that the student is a male?

(b) If a randomly chosen student is a female mechanical engineer, what is the probability that the student is a senior?

1.9.28 A business tax form is either filed on time or late, is either from a small or a large business, and is either accurate or inaccurate. There is an 11% probability that a form is from a small business and is accurate and on time. There is a 13% probability that a form is from a small business and is accurate but is late. There is a 15% probability that a form is from a small business and is on time. There is a 21% probability that a form is from a small business and is inaccurate and is late.

(a) If a form is from a small business and is accurate, what is the probability that it was filed on time?

(b) What is the probability that a form is from a large business?

1.9.29 (a) If four cards are taken at random from a pack of cards without replacement, what is the probability of having exactly two hearts?

(b) If four cards are taken at random from a pack of cards without replacement, what is the probability of having exactly two hearts and exactly two clubs?

(c) If four cards are taken at random from a pack of cards without replacement and it is known that there are no clubs, what is the probability that there are exactly three hearts?

1.9.30 An applicant has a 0.26 probability of passing a test when they take it for the first time, and if they pass it they can move on to the next stage. However, if they fail the test the first time, they must take the test a second time, and when an applicant takes the test for the second time there is a 0.43 chance that they will pass and be allowed to move on to the next stage. The applicant is rejected if the test is failed on the second attempt.

(a) What is the probability that an applicant moves on to the next stage but needs two attempts at the test?

(b) What is the probability that an applicant moves on to the next stage?

(c) If an applicant moves on to the next stage, what is the probability that they passed the test on the first attempt?

1.9.31 A fair die is rolled five times. What is the probability that the first score is strictly larger than the second score which is strictly larger than the third score which is strictly larger than the fourth score which is strictly larger than the fifth score (i.e., the five scores are strictly decreasing).

1.9.32 A software engineer makes two backup copies of his file, one on a CD and another on a diskette. Suppose that there is a probability of 0.05% that the file is corrupted when it is backed-up onto the CD, and a probability of 0.1% that the file is corrupted when it is backed-up onto the diskette, and that these events are independent of each other. What is the probability that the engineer will have at least one uncorrupted copy of the file?

1.9.33 A warning light in the cockpit of a plane is supposed to indicate when a hydraulic pump is inoperative. If the pump is inoperative, then there is a probability of 0.992 that the warning light will come on. However, there is a probability of 0.003 that the warning light will come on even when the pump is operating correctly. Furthermore, there is a probability of 0.996 that the pump is operating correctly. If the warning light comes on, what is the probability that the pump really is inoperative?

1.9.34 A hand of 10 cards is chosen at random without replacement from a deck of 52 cards. What is the probability that the hand contains exactly 2 Aces, 2 Kings, 3 Queens, and 3 Jacks?

1.9.35 There are 11 items of a product on a shelf in a retail outlet, and unknown to the customers, 4 of the items are overage. Suppose that a customer takes 3 items at random.

(a) What is the probability that none of the overage products are selected by the customer?

(b) What is the probability that exactly 2 of the items taken by the customer are overage?

1.9.36 Among those people who are infected with a certain virus, 32% have strain A, 59% have strain B, and the remaining 9% have strain C. Furthermore, 21% of people infected with strain A of the virus exhibit symptoms, 16% of people infected with strain B of the virus exhibit symptoms, and 63% of people infected with strain C of the virus exhibit symptoms.

(a) If a person has the virus and exhibits symptoms of it, what is the probability that they have strain C?

(b) If a person has the virus but doesn't exhibit any symptoms of it, what is the probability that they have strain A?

(c) What is the probability that a person who has the virus does not exhibit any symptoms of it?

"When solving mysteries like this one, it's always a question of prior probabilities and posterior probabilities." (From *Inspector Morimoto and the Two Umbrellas*, by Timothy Hemion)

CHAPTER TWO

Random Variables

After the general discussion of probability theory presented in Chapter 1, **random variables** are introduced in this chapter. They are one of the fundamental building blocks of probability theory and statistical inference. The basic probability theory developed in Chapter 1 is extended for outcomes of a numerical nature, and distinctions are drawn between **discrete** random variables and **continuous** random variables.

Useful properties of random variables are discussed, and summary measures such as the **mean** and **variance** are considered, together with combinations of several random variables. Specific examples of common families of random variables are given in Chapters 3, 4, and 5.

2.1 Discrete Random Variables

2.1.1 Definition of a Random Variable

A random variable is formed by assigning a **numerical value** to each outcome in the sample space of a particular experiment. The state space of the random variable consists of these numerical values. Technically, a random variable can be thought of as being generated from a function that maps each outcome in a particular sample space onto the real number line $\mathbb{R}$, as illustrated in Figure 2.1.

Random Variables

A **random variable** is obtained by assigning a numerical value to each outcome of a particular experiment.

A random variable is therefore a special kind of experiment in which the outcomes are numerical values, either positive or negative or possibly zero. Sometimes the experimental outcomes are already numbers, and then these may just be used to define the random variable. In other cases the experimental outcomes are not numerical, and then a random variable may be defined by assigning "scores" or "costs" to the outcomes.

Example 1

Machine Breakdowns

The sample space for the machine breakdown problem is

$$\mathcal{S} = \{\text{electrical, mechanical, misuse}\}$$

and each of these failures may be associated with a **repair cost**. For example, suppose that electrical failures generally cost an average of \$200 to repair, mechanical failures have an average repair cost of \$350, and operator misuse failures have an average repair cost of only \$50. These repair costs generate a random variable *cost*, as illustrated in Figure 2.2, which has a state space of {50, 200, 350}.

FIGURE 2.1

A random variable is formed by assigning a numerical value to each outcome in a sample space

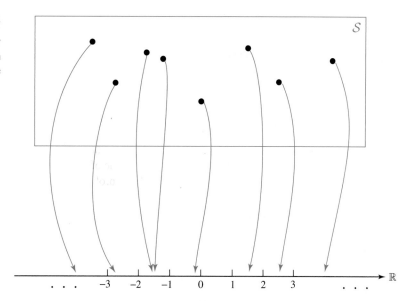

FIGURE 2.2

The random variable "cost" for machine breakdowns

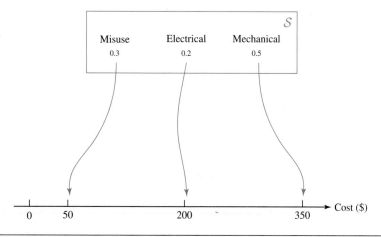

Notice that *cost* is a random variable because its values 50, 200, and 350 are numbers. The breakdown *cause*, defined to be electrical, mechanical, or operator misuse, is not considered to be a random variable because its values are not numerical.

Example 4

Power Plant Operation

Figure 1.15 illustrates the sample space for the power plant example, where the outcomes designate which of the three power plants are generating electricity (1) and which are idle (0). Suppose that interest is directed only at the number of plants that are generating electricity. This creates a random variable

$X =$ number of power plants generating electricity

which can take the values 0, 1, 2, and 3, as shown in Figure 2.3.

Example 12

Personnel Recruitment

A company has one position available for which eight applicants have made the short list. The company's strategy is to interview the applicants sequentially and to make an offer immediately

FIGURE 2.3

X = number of power plants generating electricity

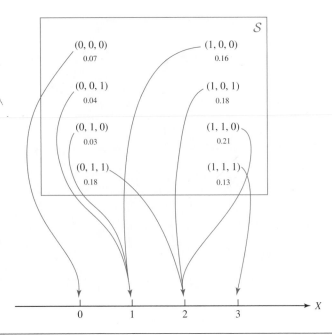

to anyone they feel is outstanding (without interviewing the additional applicants). If none of the first seven applicants interviewed is judged to be outstanding, the eighth applicant is interviewed and then the best of the eight applicants is offered the job.

The company is interested in how many applicants will need to be interviewed under this strategy. A random variable

X = number of applicants interviewed

can be defined taking the values 1, 2, 3, 4, 5, 6, 7, and 8.

Example 13
Factory Floor
Accidents

For safety and insurance purposes, a factory manager is interested in how many factory floor accidents occur in a given year. A random variable

X = number of accidents

can be defined, which can hypothetically take the value 0 or any positive integer.

GAMES OF CHANCE

The score obtained from the roll of a die can be thought of as a random variable taking the values 1 to 6. If two dice are rolled, a random variable can be defined to be the sum of the scores, taking the values 2 to 12. Figure 2.4 illustrates a random variable defined to be the positive difference between the scores obtained from two dice, taking the values 0 to 5.

It is usual to refer to random variables generically with *uppercase* letters such as X, Y, or Z. The values taken by the random variables are then labeled with *lowercase* letters. For example, it may be stated that a random variable X takes the values $x = -0.6$, $x = 2.3$, and $x = 4.0$. This custom helps clarify whether the random variable is being referred to, or a particular value taken by the random variable, and it is helpful in the subsequent mathematical discussions.

FIGURE 2.4

X = positive difference between the scores of two dice

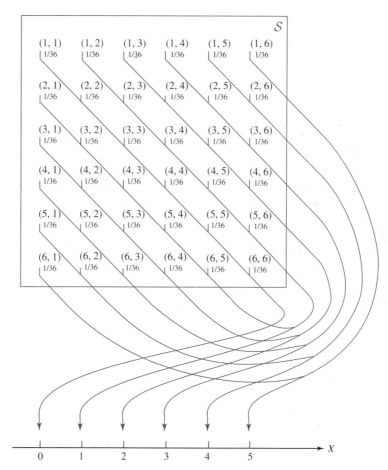

The examples given above all concern *discrete* random variables as opposed to *continuous* random variables, which are discussed in the next section. A continuous random variable is one that may take any value within a continuous interval. For example, a random variable that can take any value between 0 and 1 is a continuous random variable. Mathematically speaking, continuous random variables can take *uncountably* many values.

In contrast, discrete random variables can take only certain discrete values, as their name suggests. There may be only a finite number of values, as in Examples 1, 4, and 12, or infinitely many values, as in Example 13. The distinction between discrete and continuous random variables can most easily be understood by comparing the examples above with the examples of continuous random variables in the next section. The distinction is important since the **probability** properties of discrete and continuous random variables need to be handled in two different ways.

2.1.2   Probability Mass Function

The probability properties of a *discrete* random variable are based upon the assignment of a probability value p_i to each of the values x_i taken by the random variable. These probability values, which are known as the **probability mass function** of the random variable, must each be between 0 and 1 and must sum to 1.

Probability Mass Function

The **probability mass function (p.m.f.)** of a random variable X is a set of probability values p_i assigned to each of the values x_i taken by the *discrete* random variable. These probability values must satisfy $0 \le p_i \le 1$ and $\sum_i p_i = 1$. The probability that the random variable takes the value x_i is said to be p_i, and this is written $P(X = x_i) = p_i$.

The probabilistic properties of a discrete random variable are defined by specifying its probability mass function, that is, by specifying what values the random variable can take and the probability values of its taking each of those values. The probability mass function is also referred to as the **distribution** of the random variable, and the abbreviation p.m.f. is often used.

The probability mass function may typically be given in either tabular or graphical form as illustrated in the following examples. In addition, it will be seen in Chapter 3 that some common and useful discrete random variables have their probability mass functions specified by succinct formulas.

Example 1

Machine Breakdowns

It follows from Figure 2.2 that $P(\text{cost} = 50) = 0.3$, $P(\text{cost} = 200) = 0.2$, and $P(\text{cost} = 350) = 0.5$. This probability mass function is given in tabular form in Figure 2.5 and as a line graph in Figure 2.6.

Example 4

Power Plant Operation

The probability mass function for the number of plants generating electricity can be inferred from Figure 2.3 and is given in Figures 2.7 and 2.8. For example, the probability

x_i	50	200	350
p_i	0.3	0.2	0.5

FIGURE 2.5

Tabular presentation of the probability mass function for machine breakdown costs

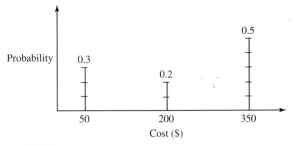

FIGURE 2.6

Line graph of the probability mass function for machine breakdown costs

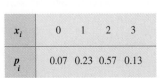

x_i	0	1	2	3
p_i	0.07	0.23	0.57	0.13

FIGURE 2.7

Tabular presentation of the probability mass function for power plant example

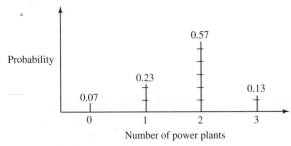

FIGURE 2.8

Line graph of the probability mass function for power plant example

that no plants are generating electricity ($X = 0$) is simply the probability of the outcome $(0, 0, 0)$, namely 0.07. The probability that exactly one plant is generating electricity ($X = 1$) is the sum of the probabilities of the outcomes $(1, 0, 0)$, $(0, 1, 0)$, and $(0, 0, 1)$, which is $0.04 + 0.03 + 0.16 = 0.23$.

Example 13
Factory Floor Accidents

Suppose that the probability of having x accidents is

$$P(X = x) = \frac{1}{2^{x+1}}$$

This is a valid probability mass function since

$$\sum_{x=0}^{\infty} P(X = x) = \sum_{x=0}^{\infty} \frac{1}{2^{x+1}} = \frac{1}{2} + \frac{1}{4} + \frac{1}{8} + \frac{1}{16} + \cdots = 1$$

so that the probability values sum to 1 (recall that in general for $0 < p < 1$, $p + p^2 + p^3 + \cdots = p/(1 - p)$). A line graph of this probability mass function is given in Figure 2.9. It is a special case of the *geometric distribution* discussed in further detail in Chapter 3.

GAMES OF CHANCE

The probability mass function for the positive difference between the scores obtained from two dice can be inferred from Figure 2.4 and is given in Figures 2.10 and 2.11. For example, since there are four outcomes, $(1, 5)$, $(5, 1)$, $(2, 6)$, and $(6, 2)$, for which the positive difference

FIGURE 2.9

Line graph of the probability mass function for factory floor accidents example

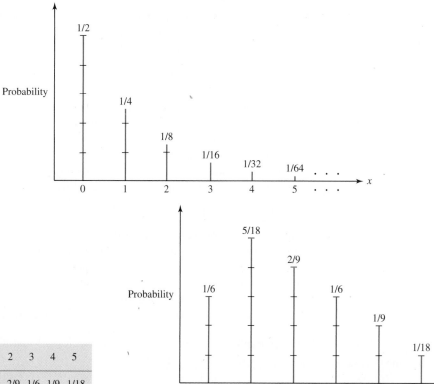

x_i	0	1	2	3	4	5
p_i	1/6	5/18	2/9	1/6	1/9	1/18

FIGURE 2.10

Tabular presentation of the probability mass function for dice example

FIGURE 2.11

Line graph of the probability mass function for dice example

is equal to 4, and each is equally likely with a probability of 1/36, the probability of having a positive difference equal to 4 is $P(X = 4) = 4/36 = 1/9$.

2.1.3   Cumulative Distribution Function

An alternative way of specifying the probabilistic properties of a random variable X is through the function

$$F(x) = P(X \leq x)$$

which is known as the **cumulative distribution function**, for which the abbreviation c.d.f. is often used. One advantage of this function is that it can be used for both discrete and continuous random variables.

Cumulative Distribution Function

The **cumulative distribution function (c.d.f.)** of a random variable X is the function

$$F(x) = P(X \leq x)$$

Like the probability mass function, the cumulative distribution function summarizes the probabilistic properties of a random variable. Knowledge of either the probability mass function or the cumulative distribution function allows the other function to be calculated.

For example, suppose that the probability mass function is known. The cumulative distribution function can then be calculated from the expression

$$F(x) = \sum_{y:y \leq x} P(X = y)$$

In other words, the value of $F(x)$ is constructed by simply adding together the probabilities $P(X = y)$ for values y that are no larger than x.

The cumulative distribution function $F(x)$ is an increasing step function with steps at the values taken by the random variable. The heights of the steps are the probabilities of taking these values. Mathematically, the probability mass function can be obtained from the cumulative distribution function through the relationship

$$P(X = x) = F(x) - F(x^-)$$

where $F(x^-)$ is the limiting value from below of the cumulative distribution function. If there is no step in the cumulative distribution function at a point x, then $F(x) = F(x^-)$ and $P(X = x) = 0$. If there is a step at a point x, then $F(x)$ is the value of the cumulative distribution function at the *top* of the step, and $F(x^-)$ is the value of the cumulative distribution function at the *bottom* of the step, so that $P(X = x)$ is the height of the step.

These relationships are illustrated in the following examples.

Example 1

Machine Breakdowns

The probability mass function given in Figures 2.5 and 2.6 can be used to construct the cumulative distribution function as follows:

$$-\infty < x < 50 \Rightarrow F(x) = P(\text{cost} \leq x) = 0$$
$$50 \leq x < 200 \Rightarrow F(x) = P(\text{cost} \leq x) = 0.3$$
$$200 \leq x < 350 \Rightarrow F(x) = P(\text{cost} \leq x) = 0.3 + 0.2 = 0.5$$
$$350 \leq x < \infty \Rightarrow F(x) = P(\text{cost} \leq x) = 0.3 + 0.2 + 0.5 = 1.0$$

This cumulative distribution function is illustrated in Figure 2.12.

FIGURE 2.12

Cumulative distribution function for machine breakdown costs

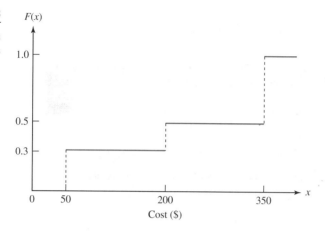

FIGURE 2.13

Cumulative distribution function for power plant example

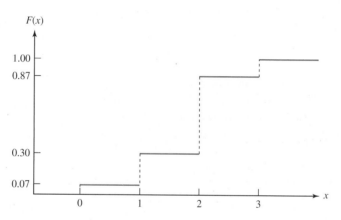

Notice that the cumulative distribution function is a step function that starts at a value of 0 for small values of x and increases to a value of 1 for large values of x. The steps occur at the points $x = 50$, $x = 200$, and $x = 350$, which are the possible values of the cost, and the sizes of the steps at these points 0.3, 0.2, and 0.5 are just the values of the probability mass function.

Example 4
Power Plant Operation

The cumulative distribution function for the number of plants generating electricity can be inferred from the probability mass function given in Figure 2.7 and is given in Figure 2.13. For example, the probability that no more than one plant is generating electricity is simply

$$F(1) = P(X \leq 1) = P(X = 0) + P(X = 1) = 0.07 + 0.23 = 0.30$$

Example 12
Personnel Recruitment

Suppose that Figure 2.14 provides the cumulative distribution function for the random variable X, the number of applicants interviewed, which is graphed in Figure 2.15. The probability mass function of X can be obtained by measuring the heights of the steps of the cumulative distribution function. For example,

$$P(X = 1) = F(1) - F(1^-) = 0.18 - 0.00 = 0.18$$
$$P(X = 2) = F(2) - F(2^-) = 0.28 - 0.18 = 0.10$$

A line graph of the probability mass function is presented in Figure 2.16.

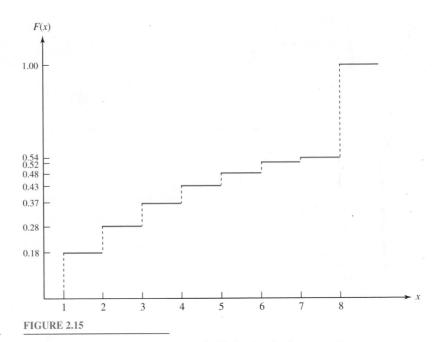

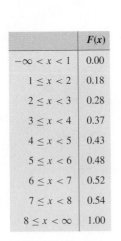

	F(x)
$-\infty < x < 1$	0.00
$1 \leq x < 2$	0.18
$2 \leq x < 3$	0.28
$3 \leq x < 4$	0.37
$4 \leq x < 5$	0.43
$5 \leq x < 6$	0.48
$6 \leq x < 7$	0.52
$7 \leq x < 8$	0.54
$8 \leq x < \infty$	1.00

FIGURE 2.14

Tabular representation of the cumulative distribution function for personnel recruitment example

FIGURE 2.15

Graphical representation of the cumulative distribution function for personnel recruitment example

FIGURE 2.16

Line graph of the probability mass function for personnel recruitment example

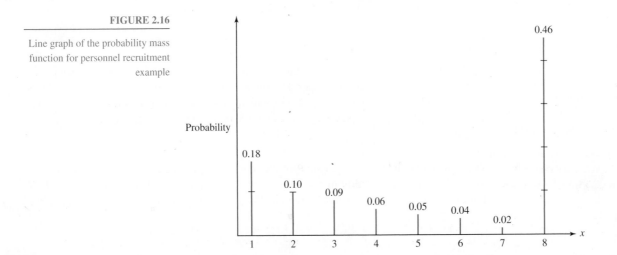

Notice that the probability values $P(X = 1)$ to $P(X = 7)$ are decreasing, but the most likely outcome is that all eight applicants need to be interviewed. This could happen if the company interviews the applicants in order of decreasing merit based upon prior information. The chance of an applicant being judged as good enough to be offered the job immediately then decreases the further down the list that he or she is.

FIGURE 2.17

Cumulative distribution function for dice example

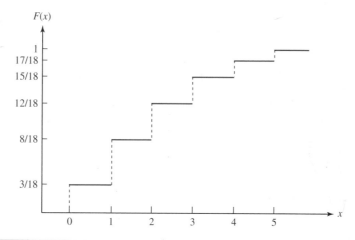

The cumulative distribution function for the positive difference between the scores obtained from two dice can be inferred from the probability mass function given in Figure 2.10 and is graphed in Figure 2.17. For example, the probability that the difference is no larger than 2 is

$$F(2) = P(X \leq 2) = P(X = 0) + P(X = 1) + P(X = 2) = \frac{1}{6} + \frac{5}{18} + \frac{2}{9} = \frac{2}{3}$$

■ 2.1.4 Problems

2.1.1 An office has four copying machines, and the random variable X measures how many of them are in use at a particular moment in time. Suppose that $P(X = 0) = 0.08$, $P(X = 1) = 0.11$, $P(X = 2) = 0.27$, and $P(X = 3) = 0.33$.
(a) What is $P(X = 4)$?
(b) Draw a line graph of the probability mass function.
(c) Construct and plot the cumulative distribution function.
(This problem is continued in Problems 2.3.1 and 2.4.2.)

2.1.2 Figure 2.18 presents the cumulative distribution function of a random variable. Make a table and line graph of its probability mass function.

2.1.3 Suppose that two fair dice are rolled and that the two numbers recorded are *multiplied* to obtain a final score. Construct and plot the probability mass function and the cumulative distribution function of the final score. (This problem is continued in Problem 2.3.2.)

2.1.4 Two cards are drawn at random from a pack of cards *with replacement*. Let the random variable X be the number of cards drawn from the heart suit.
(a) Construct the probability mass function.
(b) Construct the cumulative distribution function.

	$F(x)$
$-\infty \leq x < -4$	0.00
$-4 \leq x < -1$	0.21
$-1 \leq x < 0$	0.32
$0 \leq x < 2$	0.39
$2 \leq x < 3$	0.68
$3 \leq x < 7$	0.81
$7 \leq x < \infty$	1.00

FIGURE 2.18

Cumulative distribution function of a random variable

(c) What is the most likely value of the random variable X?
Repeat parts (a)–(c) when the second drawing is made *without replacement*. (This problem is continued in Problem 2.3.3.)

2.1.5 Two fair dice, one red and one blue, are rolled. A score is calculated to be twice the value of the blue die if the red die has an even value, and to be the value of the red die minus the value of the blue die if the red die has an odd value. Construct and plot the probability mass function and the cumulative distribution function of the score.

2.1.6 A fair coin is tossed three times. A player wins $1 if the first toss is a head, but loses $1 if the first toss is a tail. Similarly, the player wins $2 if the second toss is a head, but loses $2 if the second toss is a tail, and wins or loses $3 according to the result of the third toss. Let the random variable X be the total winnings after the three tosses (possibly a negative value if losses are incurred).
(a) Construct the probability mass function.
(b) Construct the cumulative distribution function.
(c) What is the most likely value of the random variable X?

2.1.7 Consider Example 5 and the probability values given in Figure 1.38. The company has decided that each television set should be given a *quality score* calculated in the following manner. A perfect picture scores 4, a good picture scores 2, a satisfactory picture scores 1, and a failed picture scores 0. Also, a perfect appearance scores 3, a good appearance scores 2, a satisfactory appearance scores 1, and a failed appearance scores 0. The final quality score is obtained by *multiplying* the picture score and the appearance score. For example, the outcome (P, G), where the picture is perfect and the appearance is good, would be assigned a quality score of $4 \times 2 = 8$.
(a) Construct the probability mass function of the quality score.
(b) Construct the cumulative distribution function of the quality score.
(c) What is the most likely value of the quality score? Recall that an appliance is not shipped if it fails on either the picture or the appearance evaluation, or if it is evaluated as only satisfactory in both cases. How can the probability that an appliance is not shipped be obtained from the cumulative distribution function?

2.1.8 Four cards are labeled $1, $2, $3, and $6. A player pays $4, selects two cards at random, and then receives the sum of the winnings indicated on the two cards. Calculate the probability mass function and the cumulative distribution function of the *net* winnings (that is, winnings minus the $4 payment).

2.1.9 A company has five warehouses, only two of which have a particular product in stock. A salesperson calls the five warehouses in a random order until a warehouse with the product is reached. Let the random variable X be the number of calls made by the salesperson, and calculate its probability mass function and cumulative distribution function. (This problem is continued in Problems 2.3.4 and 2.4.3.)

2.1.10 Suppose that a random variable X can take the value 1, 2, or any other positive integer. Is it possible that

$$P(X = i) = \frac{c}{i^2}$$

for some value of the constant c? Is it possible that

$$P(X = i) = \frac{c}{i}$$

for some value of c?

2.1.11 A consultant has six appointment times that are open, three on Monday and three on Tuesday. Suppose that when making an appointment a client randomly chooses one of the remaining open times, with each of those open times equally likely to be chosen. Let the random variable X be the total number of appointments that have already been made over both days at the moment when Monday's schedule has just been completely filled.
(a) What is the state space of the random variable X?
(b) Calculate the probability mass function and the cumulative distribution function of X.
(This problem is continued in Problems 2.3.16 and 2.4.12.)

2.2   Continuous Random Variables

2.2.1   Examples of Continuous Random Variables

Random variables are classified as either discrete or continuous according to the set of values that they can take. Continuous random variables can take any value within a continuous region, as illustrated in the following examples.

Example 14
Metal Cylinder Production

A company manufactures metal cylinders that are used in the construction of a particular type of engine. These cylinders, which must slide freely within an outer casing, are designed to have a diameter of 50 mm. The company discovers, however, that the cylinders it manufactures can have a diameter anywhere between 49.5 and 50.5 mm.

Suppose that the random variable X is the diameter of a randomly chosen cylinder manufactured by the company. Since this random variable can take any value between 49.5 and 50.5, it is a continuous random variable.

Example 15
Battery Failure Times

Suppose that a random variable X is the time to failure of a newly charged battery. Failure can be defined to be the moment at which the battery can no longer supply enough energy to operate a certain appliance. This random variable is continuous since it can hypothetically take any positive value. Its state space can be thought of as the interval $[0, \infty)$.

Example 16
Concrete Slab Breaking Strengths

The strength of a concrete slab is measured by a machine that applies increasing amounts of pressure to the center of the slab. The pressure at which the slab first cracks is then said to be the breaking strength of the concrete. If X is a random variable corresponding to the breaking strength of a randomly chosen concrete slab, then it is a continuous random variable taking any value between certain practical limits.

Example 17
Milk Container Contents

A machine-filled milk container is labeled as containing 2 liters. However, the actual amount of milk deposited into the container by the filling machine varies between 1.95 and 2.20 liters. If the random variable X measures the amount of milk in a randomly chosen container, it is a continuous random variable taking any value in the interval $[1.95, 2.20]$.

GAMES OF CHANCE

Suppose that a dial is spun and the angle θ that it makes with a fixed mark is measured once it has come to a halt, as illustrated in Figure 2.19. Let the angle θ be measured so that it lies between $0°$ and $180°$. The value of θ obtained from a spin is then a continuous random variable taking any value within the interval $[0, 180]$.

Also, suppose that a player spins the dial and then wins an amount corresponding to

$$\$1000 \times \frac{\theta}{180}$$

Then it is convenient to consider the amount won as a continuous random variable taking values within the interval $[0, 1000]$.

An astute reader may have noticed that the distinction between discrete and continuous random variables is sometimes not all that clear. For instance, in Example 15 above, if the failure time is recorded to the nearest minute or nearest hour, then it may be thought of as a discrete random variable taking a positive integer value. Similarly, in the spinning game above, if the angle θ is measured to the nearest degree, then it can be thought of as being a discrete random variable taking one of the values $0, 1, 2, \ldots, 180$. In addition, the monetary amount won in the game would actually be a discrete number of cents.

FIGURE 2.19

Dial-spinning game

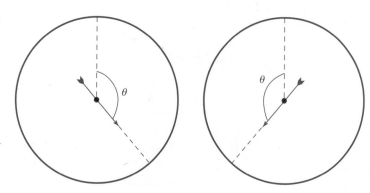

In practice, the real distinction between discrete and continuous random variables lies in how their probabilistic properties are defined. Whereas the probabilistic properties of discrete random variables are defined through a probability mass function, as described in the previous section, the probabilistic properties of a continuous random variable are defined through a **probability density function**, as described below.

Finally, it is useful to note that a discrete random variable can be obtained from a continuous random variable by grouping the elements of the state space in a certain manner. For example, the milk container contents in Example 17 may be defined to be *inadequate* if the contents are less than 2.0 liters, *adequate* if the contents are between 2.0 and 2.1 liters, and *excessive* if the contents are greater than 2.1 liters. If some scores are assigned to these three categories, then a discrete random variable has been generated that can take three values.

2.2.2   Probability Density Function

The probabilistic properties of a continuous random variable are defined through a function $f(x)$ known as the probability density function, for which the abbreviation p.d.f. is often used. The probability that the random variable lies between two values a and b is obtained by *integrating* the probability density function between these two values, so that

$$P(a \leq X \leq b) = \int_a^b f(x)\,dx$$

This probability is the *area* under the probability density function between the points a and b as illustrated in Figure 2.20. A valid probability density function $f(x)$ cannot take negative values and must integrate to one over the whole sample space, so that the total probability is equal to 1.

Probability Density Function

A **probability density function** $f(x)$ defines the probabilistic properties of a *continuous* random variable. It must satisfy $f(x) \geq 0$ and

$$\int_{\text{state space}} f(x)\,dx = 1$$

The probability that the random variable lies between two values is obtained by integrating the probability density function between the two values.

FIGURE 2.20

$P(a \leq x \leq b)$ is the area under the probability density function $f(x)$ between the points a and b

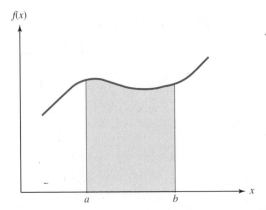

FIGURE 2.21

Probability density function for metal cylinder diameters

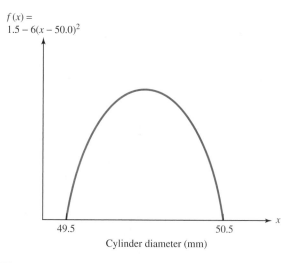

$f(x) =$
$1.5 - 6(x - 50.0)^2$

Cylinder diameter (mm)

It is useful to notice that the probability that a continuous random variable X takes any specific value a is always 0! Technically, this can be seen by noting that

$$P(X = a) = \int_a^a f(x)\,dx = 0$$

Of course, this is in contrast to discrete random variables, which can have nonzero probabilities of taking specific values. Actually, this may be the best way of distinguishing between discrete and continuous random variables. Even though continuous random variables have zero probabilities of taking specific values, they can have nonzero probabilities of falling within certain continuous regions.

Example 14

Metal Cylinder Production

Suppose that the diameter of a metal cylinder has a probability density function

$$f(x) = 1.5 - 6(x - 50.0)^2$$

for $49.5 \le x \le 50.5$ and $f(x) = 0$ elsewhere, as shown in Figure 2.21. This is a valid probability density function because it is positive within the state space $[49.5, 50.5]$ and because

$$\int_{49.5}^{50.5} (1.5 - 6(x - 50.0)^2)\,dx = [1.5x - 2(x - 50.0)^3]_{49.5}^{50.5}$$
$$= [1.5 \times 50.5 - 2(50.5 - 50.0)^3]$$
$$\quad - [1.5 \times 49.5 - 2(49.5 - 50.0)^3]$$
$$= 75.5 - 74.5 = 1.0$$

The probability that a metal cylinder has a diameter between 49.8 and 50.1 mm can be calculated to be

$$\int_{49.8}^{50.1} (1.5 - 6(x - 50.0)^2)\,dx = [1.5x - 2(x - 50.0)^3]_{49.8}^{50.1}$$
$$= [1.5 \times 50.1 - 2(50.1 - 50.0)^3]$$
$$\quad - [1.5 \times 49.8 - 2(49.8 - 50.0)^3]$$
$$= 75.148 - 74.716 = 0.432$$

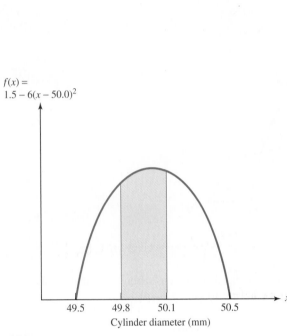

$$f(x) = \frac{2}{(x+1)^3}$$

$$f(x) = 1.5 - 6(x - 50.0)^2$$

49.5 49.8 50.1 50.5

Cylinder diameter (mm)

FIGURE 2.22

Probability that a metal cylinder diameter is between 49.8 and 50.1 mm

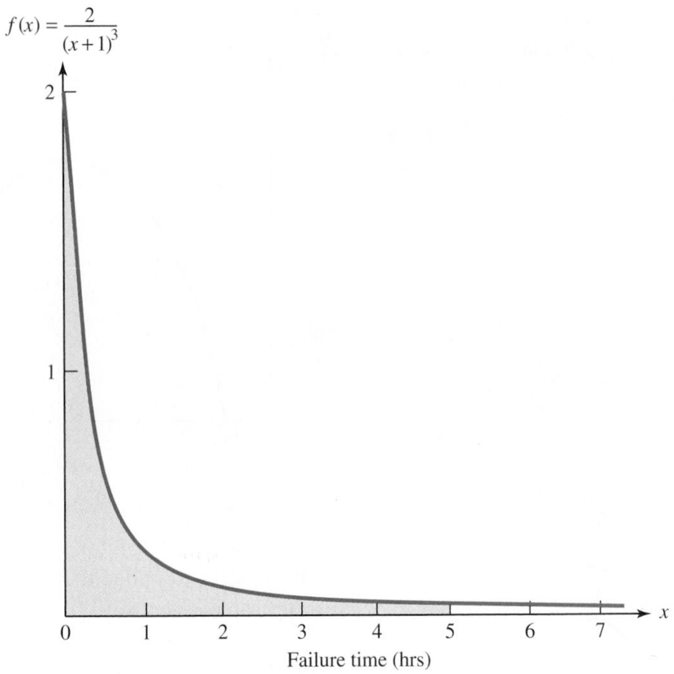

0 1 2 3 4 5 6 7

Failure time (hrs)

FIGURE 2.23

Probability density function for battery failure times and the area corresponding to the probability of failing within five hours

as illustrated in Figure 2.22. Consequently, about 43% of the cylinders will have diameters within these limits.

Example 15
Battery Failure Times

Suppose that the battery failure time, measured in hours, has a probability density function given by

$$f(x) = \frac{2}{(x+1)^3}$$

for $x \geq 0$ and $f(x) = 0$ for $x < 0$. This is a valid probability density function because it is positive and because

$$\int_0^\infty \frac{2}{(x+1)^3}\, dx = \left[\frac{-1}{(x+1)^2} \right]_0^\infty = [0] - [-1] = 1$$

The probability density function is illustrated in Figure 2.23 along with the area representing the probability that the battery fails within the first five hours. This probability can be calculated to be

$$P(0 \leq X \leq 5) = \int_0^5 \frac{2}{(x+1)^3}\, dx = \left[\frac{-1}{(x+1)^2} \right]_0^5 = \left[\frac{-1}{36} \right] - \left[\frac{-1}{1} \right] = \frac{35}{36}$$

The probability that a battery lasts *longer* than five hours is consequently 1/36.

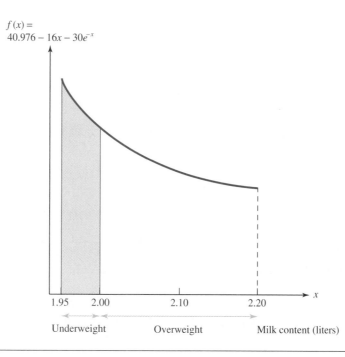

FIGURE 2.24

Probability density function for milk container contents and the area corresponding to the probability of underweight containers

$f(x) = 40.976 - 16x - 30e^{-x}$

1.95 2.00 2.10 2.20

Underweight   Overweight   Milk content (liters)

Example 17
Milk Container Contents

Suppose that the probability density function of the amount of milk deposited in a milk container is

$$f(x) = 40.976 - 16x - 30e^{-x}$$

for $1.95 \leq x \leq 2.20$ and $f(x) = 0$ elsewhere. This is a valid probability density function since it is positive within the state space $[1.95, 2.20]$ and

$$\int_{1.95}^{2.20} (40.976 - 16x - 30e^{-x})\, dx = [40.976x - 8x^2 + 30e^{-x}]_{1.95}^{2.20}$$
$$= 54.751 - 53.751 = 1$$

Figure 2.24 illustrates the area that corresponds to the probability that the actual amount of milk is less than the advertised 2.00 liters. The area can be calculated to be

$$\int_{1.95}^{2.00} (40.976 - 16x - 30e^{-x})\, dx = [40.976x - 8x^2 + 30e^{-x}]_{1.95}^{2.00}$$
$$= 54.012 - 53.751 = 0.261$$

Consequently, about 26% of the milk containers are underweight.

GAMES OF CHANCE

Consider again the game that involves the spinning of a dial. If the dial is fair, so that after spinning it is no more likely to point in any particular direction than another, then conceptually all possible values of θ within the state space $[0, 180]$ should be "equally likely." This can be achieved with a "flat" probability density function as illustrated in Figure 2.25. Since the total area under the probability density function must be equal to 1, its height must be $1/180$, so that

$$f(\theta) = \frac{1}{180}$$

FIGURE 2.25

Probability density function for
dial angle θ and area corresponding
to $P(a \leq \theta \leq b)$

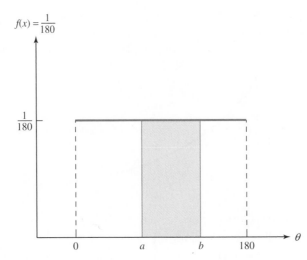

FIGURE 2.25

Probability density function for
dial angle θ and area corresponding
to $P(a \leq \theta \leq b)$

for $0 \leq \theta \leq 180$ and $f(\theta) = 0$ elsewhere. This is an example of the *uniform* probability density function, which is discussed in greater detail in Chapter 4.

Notice that the probability that the angle θ lies between two values a and b is

$$\int_a^b \frac{1}{180} \, d\theta = \frac{b-a}{180}$$

In other words, the probability that θ lies within the interval $[a, b]$ is proportional to the length of the interval $b - a$. Thus, for example, $P(0 \leq \theta \leq 10) = 10/180 = 1/18$, which is twice as large as $P(10 \leq \theta \leq 15) = 5/180 = 1/36$, which is expected for a fair dial.

The winnings

$$X = \$1000 \times \frac{\theta}{180}$$

are intuitively "equally likely" to be anywhere between \$0 and \$1000, and so the probability density function should be flat between these two limits. Since the total area must be 1, the probability density function must be

$$f(x) = \frac{1}{1000}$$

for $0 \leq x \leq 1000$ and $f(x) = 0$ elsewhere.

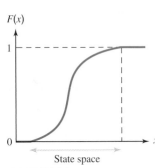

$F(x)$

State space

FIGURE 2.26

The cumulative distribution function
increases from 0 at the beginning of
the state space to 1 at the end of the
state space

2.2.3 Cumulative Distribution Function

The **cumulative distribution function** of a continuous random variable X is defined in exactly the same way as for a discrete random variable, namely

$$F(x) = P(X \leq x)$$

For a continuous random variable, the cumulative distribution function $F(x)$ is a continuous nondecreasing function that takes the value 0 prior to and at the beginning of the state space and increases to a value of 1 at the end of and after the state space, as illustrated in Figure 2.26, although the shape of the cumulative distribution function will vary.

Like the probability density function, the cumulative distribution function summarizes the probabilistic properties of a continuous random variable, and knowledge of either function allows the other function to be constructed. For example, if the probability density function $f(x)$ is known, then the cumulative distribution function can be calculated from the expression

$$F(x) = P(X \le x) = \int_{-\infty}^{x} f(y)\,dy$$

In practice, the lower integration limit $-\infty$ can be replaced by the lower endpoint of the state space since the probability density function is 0 outside the state space. The probability density function can be obtained by differentiating the cumulative distribution function

$$f(x) = \frac{dF(x)}{dx}$$

In addition, notice that the cumulative distribution function provides a convenient way of obtaining the probability that a random variable lies within a certain region, since

$$P(a \le X \le b) = P(X \le b) - P(X \le a) = F(b) - F(a)$$

Example 14
Metal Cylinder Production

The cumulative distribution function of the metal cylinder diameters can be constructed from the probability density function as

$$\begin{aligned} F(x) = P(X \le x) &= \int_{49.5}^{x} (1.5 - 6(y - 50.0)^2)\,dy \\ &= [1.5y - 2(y - 50.0)^3]_{49.5}^{x} \\ &= [1.5x - 2(x - 50.0)^3] - [1.5 \times 49.5 - 2(49.5 - 50.0)^3] \\ &= 1.5x - 2(x - 50.0)^3 - 74.5 \end{aligned}$$

for $49.5 \le x \le 50.5$. As expected, the cumulative distribution function is an increasing function between the limits $x = 49.5$ and $x = 50.5$, as illustrated in Figure 2.27, with $F(49.5) = 0$ and $F(50.5) = 1$. Technically, in addition $F(x) = 0$ for $x < 49.5$ and $F(x) = 1$ for $x > 50.5$.

FIGURE 2.27

Cumulative distribution function for metal cylinder diameters illustrating $P(49.7 \le X \le 50.0)$

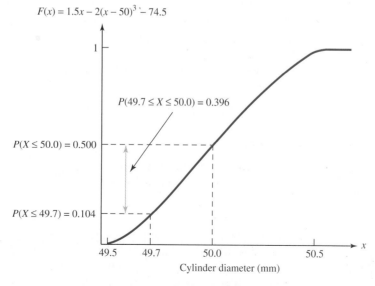

$F(x) = 1.5x - 2(x - 50)^3 - 74.5$

$P(49.7 \le X \le 50.0) = 0.396$

$P(X \le 50.0) = 0.500$

$P(X \le 49.7) = 0.104$

Cylinder diameter (mm)

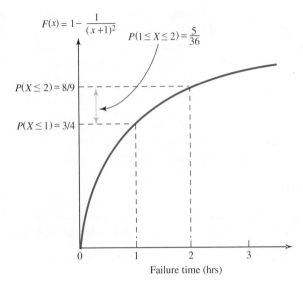

Figure 2.27 also illustrates the probability that a metal cylinder has a diameter between
49.7 and 50.0 mm, which is

$$P(49.7 \leq X \leq 50.0) = F(50.0) - F(49.7)$$
$$= (1.5 \times 50.0 - 2(50.0 - 50.0)^3 - 74.5)$$
$$- (1.5 \times 49.7 - 2(49.7 - 50.0)^3 - 74.5)$$
$$= 0.5 - 0.104 = 0.396$$

Consequently, about 40% of the cylinders have diameters between these two limits.

Example 15
Battery Failure Times

The cumulative distribution function of the battery failure times is

$$F(x) = P(X \leq x) = \int_0^x \frac{2}{(y+1)^3} \, dy$$

$$= \left[\frac{-1}{(y+1)^2} \right]_0^x = \left[\frac{-1}{(x+1)^2} \right] - [-1] = 1 - \frac{1}{(x+1)^2}$$

for $x \geq 0$. This is illustrated in Figure 2.28, together with the probability that a battery lasts
between one and two hours, which is

$$P(1 \leq X \leq 2) = F(2) - F(1)$$
$$= \left(1 - \frac{1}{(2+1)^2} \right) - \left(1 - \frac{1}{(1+1)^2} \right) = \frac{8}{9} - \frac{3}{4} = \frac{5}{36}$$

Example 16
Concrete Slab Breaking Strengths

Suppose that the concrete slab breaking strengths measured in certain units are between 120
and 150, with a cumulative distribution function of

$$F(x) = 3.856 - 12.8e^{-x/100}$$

for $120 \leq x \leq 150$. This is a valid cumulative distribution function since

$$F(120) = 3.856 - 12.8e^{-120/100} = 0$$
$$F(150) = 3.856 - 12.8e^{-150/100} = 1$$

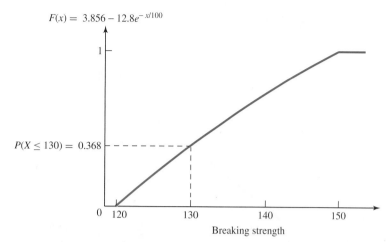

FIGURE 2.29

Cumulative distribution function for concrete breaking strengths illustrating $P(X \leq 130)$

and the function is increasing between these two values. Figure 2.29 illustrates the cumulative distribution function together with the probability that a concrete slab has a strength less than 130, which is

$$F(130) = 3.856 - 12.8e^{-130/100} = 0.368$$

The probability density function of the breaking strengths can be calculated from the cumulative distribution function to be

$$f(x) = \frac{dF(x)}{dx} = \frac{d}{dx}(3.856 - 12.8e^{-x/100}) = 0.128e^{-x/100}$$

for $120 \leq x \leq 150$ and $f(x) = 0$ elsewhere.

GAMES OF CHANCE Consider again the game that involves the spinning of a dial. The cumulative distribution function of the angle θ is

$$F(\theta) = \int_0^\theta f(y)\, dy = \int_0^\theta \frac{1}{180}\, dy = \frac{\theta}{180}$$

for $0 \leq \theta \leq 180$. Similarly, the cumulative distribution function of the winnings X is

$$F(x) = \int_0^x f(y)\, dy = \int_0^x \frac{1}{1000}\, dy = \frac{x}{1000}$$

for $0 \leq x \leq 1000$.

Figure 2.30 illustrates the cumulative distribution function of the winnings together with the probability that the winnings are between \$250 and \$750, which is

$$P(250 \leq X \leq 750) = F(750) - F(250) = \frac{750}{1000} - \frac{250}{1000} = \frac{1}{2}$$

This result is expected because these winnings correspond to values of θ between 45° and 135°, which make up half of the dial, as illustrated in Figure 2.31.

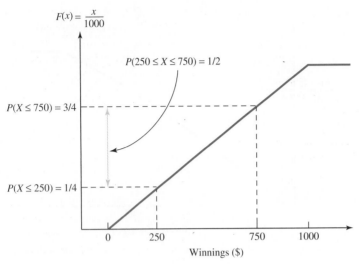

FIGURE 2.30

Cumulative distribution function for dial-spinning game illustrating
$P(250 \leq X \leq 750)$

FIGURE 2.31

Dial region corresponding to winnings
between \$250 and \$750

■ 2.2.4   Problems

2.2.1 Consider a random variable measuring the following
quantities. In each case state with reasons whether you
think it more appropriate to define the random variable as
discrete or as continuous.
(a) A person's height
(b) A student's course grade
(c) The thickness of a metal plate
(d) The purity of a chemical solution
(e) The type of personal computer a person owns
(f) A person's age

2.2.2 A random variable X takes values between 4 and 6 with a
probability density function

$$f(x) = \frac{1}{x \ln(1.5)}$$

for $4 \leq x \leq 6$ and $f(x) = 0$ elsewhere.
(a) Make a sketch of the probability density function.
(b) Check that the total area under the probability density
function is equal to 1.
(c) What is $P(4.5 \leq X \leq 5.5)$?
(d) Construct and sketch the cumulative distribution
function.
(This problem is continued in Problems 2.3.10 and 2.4.5.)

2.2.3 A random variable X takes values between -2 and 3 with
a probability density function

$$f(x) = \frac{15}{64} + \frac{x}{64}, \qquad -2 \leq x \leq 0$$
$$f(x) = \frac{3}{8} + cx, \qquad 0 \leq x \leq 3$$

and $f(x) = 0$ elsewhere.
(a) Find the value of c and sketch the probability density
function.
(b) What is $P(-1 \leq X \leq 1)$?
(c) Construct and sketch the cumulative distribution
function.

2.2.4 A random variable X takes values between 0 and 4 with a
cumulative distribution function

$$F(x) = \frac{x^2}{16}$$

for $0 \leq x \leq 4$.
(a) Sketch the cumulative distribution function.
(b) What is $P(X \leq 2)$?
(c) What is $P(1 \leq X \leq 3)$?
(d) Construct and sketch the probability density function.
(This problem is continued in Problems 2.3.11 and 2.4.6.)

2.2.5 A random variable X takes values between 0 and ∞ with
a cumulative distribution function

$$F(x) = A + Be^{-x}$$

for $0 \leq x \leq \infty$.

(a) Find the values of A and B and sketch the cumulative distribution function.

(b) What is $P(2 \leq X \leq 3)$?

(c) Construct and sketch the probability density function.

2.2.6 A car panel is spray-painted by a machine, and the technicians are particularly interested in the thickness of the resulting paint layer. Suppose that the random variable X measures the thickness of the paint in millimeters at a randomly chosen point on a randomly chosen car panel, and that X takes values between 0.125 and 0.5 mm with a probability density function of

$$f(x) = A(0.5 - (x - 0.25)^2)$$

for $0.125 \leq x \leq 0.5$ and $f(x) = 0$ elsewhere.

(a) Find the value of A and sketch the probability density function.

(b) Construct and sketch the cumulative distribution function.

(c) What is the probability that the paint thickness at a particular point is less than 0.2 mm?

(This problem is continued in Problems 2.3.12 and 2.4.7.)

2.2.7 Suppose that the random variable X is the time taken by a garage to service a car. These times are distributed between 0 and 10 hours with a cumulative distribution function

$$F(x) = A + B \ln(3x + 2)$$

for $0 \leq x \leq 10$.

(a) Find the values of A and B and sketch the cumulative distribution function.

(b) What is the probability that a repair job takes longer than two hours?

(c) Construct and sketch the probability density function.

2.2.8 The bending capabilities of plastic sheets are investigated by bending sheets at increasingly large angles until a deformity appears in the sheet. The angle θ at which the deformity first appears is then recorded. Suppose that this angle takes values between $0°$ and $10°$ with a probability density function

$$f(\theta) = A(e^{10-\theta} - 1)$$

for $0 \leq \theta \leq 10$ and $f(\theta) = 0$ elsewhere.

(a) Find the value of A and sketch the probability density function.

(b) Construct and sketch the cumulative distribution function.

(c) What is the probability that a plastic sheet can be bent up to an angle of $8°$ without deforming?

(This problem is continued in Problems 2.3.13 and 2.4.8.)

2.2.9 An archer shoots an arrow at a circular target with a radius of 50 cm. If the arrow hits the target, the distance r between the point of impact and the center of the target is measured. Suppose that this distance has a cumulative distribution function

$$F(r) = A + \frac{B}{(r + 5)^3}$$

for $0 \leq r \leq 50$.

(a) Find the values of A and B and sketch the cumulative distribution function.

(b) What is the probability that the arrow hits within 10 cm of the center of the target?

(c) What is the probability that the arrow hits more than 30 cm away from the center of the target?

(d) Construct and sketch the probability density function. (This problem is continued in Problems 2.3.14 and 2.4.9.)

2.2.10 Sometimes a random variable is a mix of discrete and continuous components. For example, suppose that the dial-spinning game is modified in the following way. First a fair coin is tossed and if a head is obtained, the player wins $500 and the dial is not spun. However, if a tail is obtained, the player spins the dial and receives winnings of

$$\$1000 \times \frac{\theta}{180}$$

as before. In this game there is a probability of 0.5 of winning $500, with all the other possible winnings between $0 and $1000 being equally likely. The coin toss provides a discrete element to the winnings, and the dial spin provides a continuous element. The best way to describe the probabilistic properties of *mixed* random variables such as this is through a cumulative distribution function. The cumulative distribution function of the winnings from this game is given in Figure 2.32.

(a) What is the probability of winning less than $200?

(b) What is the probability of winning between $400 and $700?

Interpret your answers.

2.2.11 The resistance X of an electrical component has a probability density function

$$f(x) = Ax(130 - x^2)$$

for resistance values in the range $10 \leq x \leq 11$.

(a) Calculate the value of the constant A.

(b) Calculate the cumulative distribution function.

(c) What is the probability that the electrical component has a resistance between 10.25 and 10.5?

(This problem is continued in Problems 2.3.17, 2.4.13, and 2.6.12.)

FIGURE 2.32

The cumulative distribution function of the winnings of the game in Problem 2.2.10.

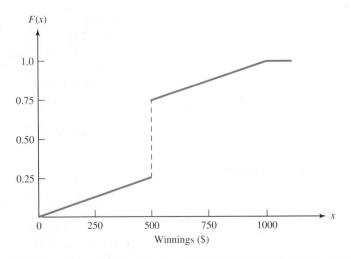

2.3 The Expectation of a Random Variable

Whereas the probability mass function or the probability density function provides complete information about the probabilistic properties of a random variable, it is often useful to employ some summary measures of these properties. One of the most basic summary measures is the **expectation** or **mean** of a random variable, which is denoted by $E(X)$ and represents an "average" value of the random variable. Two random variables with the same expected value can be thought of as having the same average value, although their probability mass functions or probability density functions may be quite different.

2.3.1 Expectations of Discrete Random Variables

A discrete random variable X taking the values x_i with probability values p_i has an expected value of

$$E(X) = \sum_i p_i x_i$$

This value can be interpreted as a *weighted* average of the values in the state space x_i, where the weights are the probability values p_i.

Expected Value of a Discrete Random Variable

The **expected value** or **expectation** of a discrete random variable with a probability mass function $P(X = x_i) = p_i$ is

$$E(X) = \sum_i p_i x_i$$

$E(X)$ provides a summary measure of the average value taken by the random variable and is also known as the **mean** of the random variable.

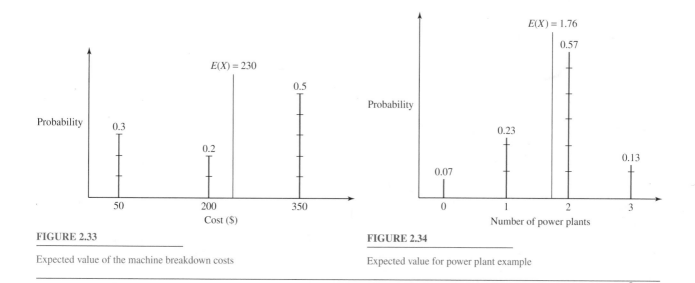

FIGURE 2.33

Expected value of the machine breakdown costs

FIGURE 2.34

Expected value for power plant example

The calculation and interpretation of the expected value are illustrated in the following examples.

Example 1
Machine Breakdowns

The expected repair cost is

$$E(\text{cost}) = (\$50 \times 0.3) + (\$200 \times 0.2) + (\$350 \times 0.5) = \$230$$

which is illustrated with the probability mass function in Figure 2.33. Over a long period of time, the repairs will cost an average of about $230 each. Notice that this expected value is not equal to one of the values $50, $200, and $350 in the state space, but it is the average of these values weighted by their probabilities.

Example 4
Power Plant Operation

The expected number of power plants generating electricity is

$$E(X) = (0 \times 0.07) + (1 \times 0.23) + (2 \times 0.57) + (3 \times 0.13) = 1.76$$

which is illustrated in Figure 2.34. This expected value, 1.76, provides a summary measure of the average number of power plants generating electricity at particular points in time.

Example 12
Personnel Recruitment

The expected number of applicants interviewed is

$$E(X) = (1 \times 0.18) + (2 \times 0.10) + (3 \times 0.09) + (4 \times 0.06) + (5 \times 0.05)$$
$$+ (6 \times 0.04) + (7 \times 0.02) + (8 \times 0.46)$$
$$= 5.20$$

which is illustrated in Figure 2.35. This expected value provides some indication of how many of the eight applicants will actually have to be interviewed under the company's interviewing strategy.

If another strategy is employed, whereby the weakest two candidates on the short list are dropped so that a maximum of only six applicants are interviewed, then the probability that $X = 6$ increases to $0.04 + 0.02 + 0.46 = 0.52$, and the expected number of applicants

FIGURE 2.35

Expected value for personnel
recruitment example

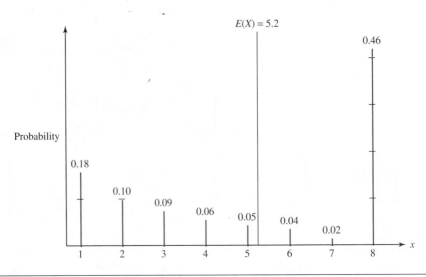

interviewed falls to

$$E(X) = (1 \times 0.18) + (2 \times 0.10) + (3 \times 0.09) + (4 \times 0.06) + (5 \times 0.05) + (6 \times 0.52)$$
$$= 4.26$$

The drop in the expected value from 5.20 to 4.26 illustrates the potential savings under the new strategy, which is at the expense of not having the chance to hire one of the two applicants dropped from the short list.

Finally, notice that if the strategy is to interview all eight applicants and then to hire the best one, then obviously $P(X = 8) = 1$, and the expected number of applicants interviewed is $E(X) = 8$.

GAMES OF CHANCE If a fair die is rolled, the expected value of the outcome is

$$E(X) = \left(1 \times \frac{1}{6}\right) + \left(2 \times \frac{1}{6}\right) + \left(3 \times \frac{1}{6}\right) + \left(4 \times \frac{1}{6}\right) + \left(5 \times \frac{1}{6}\right) + \left(6 \times \frac{1}{6}\right)$$
$$= 3.5$$

Since each outcome is equally likely, this is just the normal arithmetic average of the six outcomes.

The probability mass function of the positive difference between the scores obtained from two dice is given in Figure 2.36 together with the expected difference, which is calculated to be

$$E(X) = \left(0 \times \frac{1}{6}\right) + \left(1 \times \frac{5}{18}\right) + \left(2 \times \frac{2}{9}\right) + \left(3 \times \frac{1}{6}\right) + \left(4 \times \frac{1}{9}\right) + \left(5 \times \frac{1}{18}\right)$$
$$= \frac{35}{18} = 1.94$$

Suppose that you organize a game whereby a player rolls two dice and you pay the player the dollar amount of the difference in the scores. How much should you charge a person to play? Since your expected or long-run average payment is $1.94 per game, a "fair" charge would be $1.94. However, if you want to make a profit, you must charge more than this. If you

FIGURE 2.36

Expected value for dice example

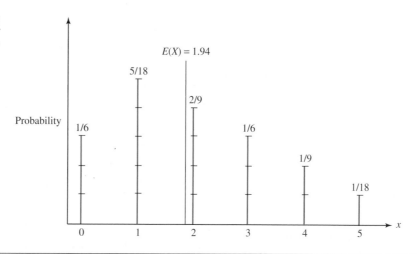

charge $2.00 per game, you will lose $3, $2, or $1, break even, or win $1 or $2 on each game, but your expected long-run profit will be $2.00 − $1.94 per game, that is, about 6 cents per game. The more people who play the better, because your total expected profits will increase and the chance of an unlucky net loss will diminish. Of course, if the dice are not fair, things may not be as they appear to be!

2.3.2 Expectations of Continuous Random Variables

The expectation of a continuous random variable X with a probability density function $f(x)$ is given by

$$E(X) = \int_{\text{state space}} x f(x)\, dx$$

Again, $E(X)$ can be interpreted as a *weighted* average of the values within the state space, with weights corresponding to the probability density function $f(x)$.

Expected Value of a Continuous Random Variable

The **expected value** or **expectation** of a continuous random variable with a probability density function $f(x)$ is

$$E(X) = \int_{\text{state space}} x f(x)\, dx$$

The expected value provides a summary measure of the average value taken by the random variable, and it is also known as the **mean** of the random variable.

The calculation and interpretation of the expected value of continuous random variables are illustrated in the following examples.

Example 14

Metal Cylinder Production

The expected diameter of a metal cylinder is

$$E(X) = \int_{49.5}^{50.5} x(1.5 - 6(x - 50.0)^2)\, dx$$

$f(x) = 1.5 - 6(x - 50.0)^2$

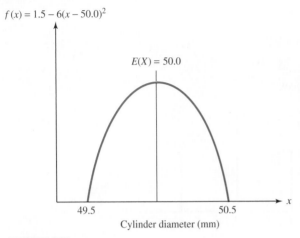

FIGURE 2.37

Expected value for metal cylinder diameters

FIGURE 2.38

Expected value for a symmetric probability density function

The evaluation of this integral can be simplified using the transformation $y = x - 50.0$, so that

$$E(X) = \int_{-0.5}^{0.5} (y + 50.0)(1.5 - 6y^2)\, dy$$

$$= \int_{-0.5}^{0.5} (-6y^3 - 300y^2 + 1.5y + 75)\, dy$$

$$= [-3y^4/2 - 100y^3 + 0.75y^2 + 75y]_{-0.5}^{0.5}$$

$$= [25.09375] - [-24.90625] = 50.0$$

Consequently, the metal cylinders have an average diameter of 50.0 mm.

The probability density function and the expected value are illustrated in Figure 2.37. Notice that the probability density function is **symmetric** about the value 50.0 mm, which is the expected value. It is a general result that a random variable with a symmetric probability density function has an expectation equal to the point of symmetry.

Symmetric Random Variables

If a continuous random variable X has a probability density function $f(x)$ that is symmetric about a point μ, so that $f(\mu + x) = f(\mu - x)$ for all $x \in \mathbb{R}$, then $E(X) = \mu$, so that the expectation of the random variable is equal to the point of symmetry.

This general result is illustrated in Figure 2.38.

Example 15
Battery Failure Times

The expected battery failure time is

$$E(X) = \int_0^\infty x \frac{2}{(x+1)^3}\, dx = \int_0^\infty \left(\frac{2}{(x+1)^2} - \frac{2}{(x+1)^3} \right) dx$$

$$= \left[\frac{-2}{(x+1)} + \frac{1}{(x+1)^2} \right]_0^\infty = [0] - [-1] = 1$$

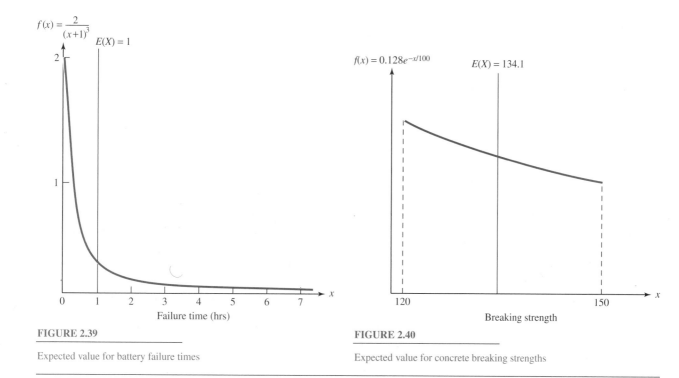

FIGURE 2.39

Expected value for battery failure times

FIGURE 2.40

Expected value for concrete breaking strengths

which is illustrated in Figure 2.39. This expected value indicates that the batteries fail on average after one hour of operation.

Example 16
Concrete Slab Breaking Strengths

Figure 2.40 illustrates the probability density function for the concrete breaking strengths together with the expected breaking strength, which is

$$E(X) = \int_{120}^{150} x 0.128 e^{-x/100} \, dx$$

$$= [-12.8(100 + x)e^{-x/100}]_{120}^{150} = [-714.02] - [-848.16] = 134.1$$

The relative advantage of another type for concrete might be judged by determining whether it has an expected breaking strength smaller or larger than this value of 134.1.

GAMES OF CHANCE

The probability density function for the winnings in the dial-spinning game is illustrated in Figure 2.41. Since it is symmetric about the middle value of $500, the expected winnings are immediately known to be $500. Formally, this can be checked by noting that

$$E(X) = \int_0^{1000} x \frac{1}{1000} \, dx = \left[\frac{x^2}{2000} \right]_0^{1000} = [500] - [0] = 500$$

A fair price for playing the game is thus $500.

Finally, the concept of expectation can be extended in a natural way from the expectation $E(X)$ of a random variable X to the expectation $E(g(X))$ of a function $g(X)$ of the random variable. An example of this, which is useful in the next section concerning the variances of random variables, is when $g(X) = X^2$. The expectation of the square of the random variable X,

FIGURE 2.41

Expected winnings for
dial-spinning game

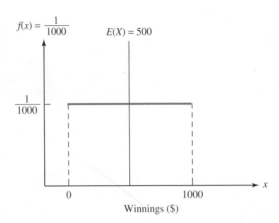

$f(x) = \dfrac{1}{1000}$ $E(X) = 500$

$\dfrac{1}{1000}$

0

1000

Winnings (\$)

x

$E(X^2)$, can be interpreted as an average value of the squares of the values taken by the random variable.

Expectations of functions of random variables can be calculated from the formulas

$$E(g(X)) = \sum_i p_i g(x_i)$$

for discrete random variables and

$$E(g(X)) = \int_{\text{state space}} g(x) f(x)\, dx$$

for continuous random variables. Notice that in general $E(g(X)) \neq g(E(X))$, and in particular $E(X^2)$ is in general not equal to $E(X)^2$. Functions and combinations of random variables are discussed in more detail in Section 2.6.

2.3.3   Medians of Random Variables

The **median** is another summary measure of the distribution of a random variable that provides information about the "middle" value of the random variable. It is defined to have the property that the random variable is equally likely to be either smaller or larger than the median. The median is most often used with continuous random variables and is the value of x for which $F(x) = 0.5$.

Median

The **median** of a continuous random variable X with a cumulative distribution function $F(x)$ is the value x in the state space for which

$$F(x) = 0.5$$

The random variable is then equally likely to fall above or below the median value.

Example 14

Metal Cylinder Production

The median value of the metal cylinder diameters is the solution to

$$F(x) = 1.5x - 2(x - 50.0)^3 - 74.5 = 0.5$$

which is $x = 50.0$. This is not really a surprise because the probability density function is symmetric about the point $x = 50.0$, so that a diameter is equally likely to be smaller than 50 mm and larger than 50 mm.

In general, it is clear that a random variable with a symmetric probability density function has a *median* as well as an *expectation* equal to the point of symmetry. The following statement generalizes the previous result on the expectation of symmetric random variables.

Symmetric Random Variables

If a continuous random variable X has a probability density function $f(x)$ that is symmetric about a point μ, then both the *median* and the *expectation* of the random variable are equal to μ.

The following examples illustrate that in general the expectation and the median of a random variable can take different values.

Example 15

Battery Failure Times

The median of the battery failure times is the solution to

$$F(x) = 1 - \frac{1}{(x+1)^2} = 0.5$$

which is $x = \sqrt{2} - 1 = 0.414$. Figure 2.42 illustrates this median value relative to the expected failure time of 1 hour. Whereas the batteries will operate for an average of 1 hour, half of them will fail before 0.414 hour, or about 25 minutes. The reason the expected lifetime is so much longer than the median lifetime is that those batteries that last longer than 25 minutes have a chance of lasting a considerable length of time.

FIGURE 2.42

Comparison of median and mean battery failure times

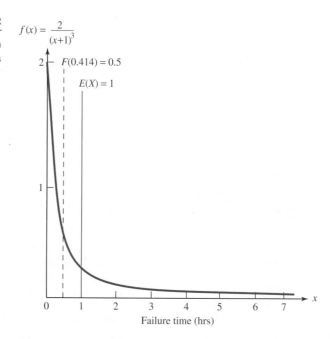

FIGURE 2.43

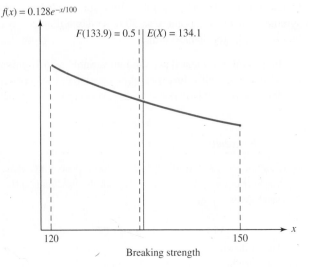

Comparison of median and mean concrete breaking strengths

This example illustrates that the median and the mean can both provide useful information about a random variable. If an engineer has one appliance and wants to know how many batteries will be required to keep it running over a long period of time, then the mean value of one hour provides the most useful information. To keep the appliance running for one day, the engineer can plan on needing about 24 batteries on average.

On the other hand, if the engineer has many appliances and wishes to know how many of them will last more than 30 minutes on newly charged batteries, then a median value of about 25 minutes indicates that the engineer can expect fewer than half of the appliances to last more than 30 minutes.

Example 16

Concrete Slab Breaking Strengths

The median concrete breaking strength is the solution to

$$F(x) = 3.856 - 12.8e^{-x/100} = 0.5$$

which is $x = 133.9$. Figure 2.43 illustrates this together with the expected breaking strength of 134.1. The median and the mean strengths are about the same here since the probability density function is almost flat.

■ 2.3.4   Problems

2.3.1 Consider again the four copying machines discussed in Problem 2.1.1. What is the expected number of copying machines in use at a particular moment in time?

2.3.2 Consider again Problem 2.1.3 where the numbers obtained on two fair dice are multiplied to obtain a final score. What is the expected value of this score?

2.3.3 Consider again Problem 2.1.4 where two cards are drawn from a pack of cards. Is the expected number of hearts drawn larger when the second drawing is made with or without replacement? Does this answer surprise you?

2.3.4 Consider again the salesperson discussed in Problem 2.1.9 who is trying to locate a particular product. What is the expected number of warehouses called by the salesperson?

2.3.5 Suppose that a player draws a card at random from a pack of cards, and wins $15 if an Ace, King, Queen, or Jack is obtained, and otherwise wins the face value of the card in dollars. What is the expected amount won by the player? Would you pay $9 to play this game?

2.3.6 Two fair dice, one red and one blue, are rolled, and a fair coin is tossed. If a head is obtained on the coin toss, then a player wins the *sum* of the scores on the two dice. If a tail is obtained on the coin toss, then the player wins the score on the red die. What are the expected winnings?

2.3.7 A player pays $1 to play a game where three fair dice are rolled. If three 6s are obtained the player wins $500, and otherwise the player wins nothing. What are the expected net winnings of this game? Would you want to play this game? Does your answer depend upon how many times you can play the game?

2.3.8 A state lottery generally consists of many tickets being sold at prices of about $1, each with a chance to win a large jackpot which is often over $1 million. Would you expect the expected net winnings on each ticket to be positive or negative? Why do people play lotteries?

2.3.9 Suppose that you are organizing the game described at the end of Section 2.3.1, where you charge players $2 to roll two dice, and then you pay them the difference in the scores. If you fix the dice so that each die has a probability of 0.2 of scoring a 3 and equal probabilities of 0.16 of scoring the other five numbers, do your expected winnings increase beyond 6 cents per game? Is this a surprise?

2.3.10 Consider again the random variable described in Problem 2.2.2 with a probability density function of

$$f(x) = \frac{1}{x \ln(1.5)}$$

for $4 \le x \le 6$ and $f(x) = 0$ elsewhere.
(a) What is the expected value of this random variable?
(b) What is the median of this random variable?

2.3.11 Consider again the random variable described in Problem 2.2.4 with a cumulative distribution function of

$$F(x) = \frac{x^2}{16}$$

for $0 \le x \le 4$.
(a) What is the expected value of this random variable?
(b) What is the median of this random variable?

2.3.12 Consider again the car panel painting machine discussed in Problem 2.2.6. What is the expected paint thickness? What is the median paint thickness?

2.3.13 Consider again the plastic bending capabilities discussed in Problem 2.2.8. What is the expected deformity angle? What is the median deformity angle?

2.3.14 Consider again the archery problem discussed in Problem 2.2.9. What is the expected deviation from the center of the target? What is the median deviation?

2.3.15 Prove that a continuous random variable with a probability density function that is symmetric about a point μ has an expected value equal to the point of symmetry μ.

2.3.16 Recall Problem 2.1.11 concerning the scheduling of appointments with a consultant. What is the expected value of the total number of appointments that have already been made over both days at the moment when Monday's schedule has just been completely filled?

2.3.17 Recall Problem 2.2.11 concerning the resistance of an electrical component.
(a) What is the expected value of the resistance?
(b) What is the median value of the resistance?

2.3.18 A random variable has a probability density function $f(x) = A(x - 1.5)$ over the state space $2 \le x \le 3$.
(a) What is the value of A?
(b) What is the median of the random variable?

2.4 The Variance of a Random Variable

2.4.1 Definition and Interpretation of Variance

Another important summary measure of the distribution of a random variable is the **variance**, which measures the *spread* or *variability* in the values taken by the random variable. Whereas the mean or expectation measures the central or average value of the random variable, the variance measures the spread or deviation of the random variable about its mean value.

Specifically, the variance of a random variable is defined as

$$\mathrm{Var}(X) = E\big((X - E(X))^2\big)$$

Thus, it is defined to be the expected value of the squares of the deviations of the random variable values about the expected value $E(X)$. Necessarily, the variance is always *positive*, and larger values of the variance indicate a greater spread in the distribution of the random variable about the mean value. An alternative and often simpler expression for calculating the variance is

$$
\begin{aligned}
\mathrm{Var}(X) &= E((X - E(X))^2) \\
&= E(X^2 - 2X E(X) + (E(X))^2) \\
&= E(X^2) - 2E(X)E(X) + (E(X))^2 \\
&= E(X^2) - (E(X))^2
\end{aligned}
$$

Variance

The **variance** of a random variable X is defined to be

$$
\mathrm{Var}(X) = E((X - E(X))^2)
$$

or equivalently

$$
\mathrm{Var}(X) = E(X^2) - (E(X))^2
$$

The variance is a positive quantity that measures the spread of the distribution of the random variable about its mean value. Larger values of the variance indicate that the distribution is more spread out.

The concept of variance can be illustrated graphically. Figure 2.44 shows two probability density functions that have different mean values, but identical variances. The variances are the same because the shape or spread of the density functions about their mean values is the same. On the other hand, Figure 2.45 shows two probability density functions that have the

FIGURE 2.44

Two distributions with different mean values but identical variances

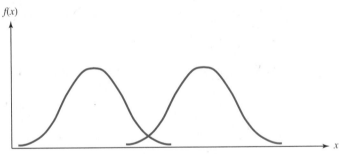

FIGURE 2.45

Two distributions with identical mean values but different variances

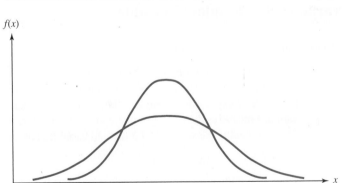

same mean values, but different variances. The density function that is flatter and more spread out has the larger variance.

It is common to use the symbol μ to denote the mean or expectation of a random variable and the symbol σ^2 to denote the variance. The square root of the variance, σ, is known as the **standard deviation** of the distribution of the random variable and is often used in place of the variance to describe the spread of the distribution.

Standard Deviation

The **standard deviation** of a random variable X is defined to be the positive square root of the variance. The symbol σ^2 is often used to denote the variance of a random variable, so that σ represents the standard deviation.

Notice that the standard deviation has the same units as the random variable X, but the variance has the square of these units. For example, if the random variable X is measured in seconds, then the standard deviation will also be measured in seconds but the variance will be measured in seconds^2.

2.4.2 Examples of Variance Calculations

Example 1
Machine Breakdowns

Recall that the repair costs are $50, $200, and $350 with respective probability values of 0.3, 0.2, and 0.5, and that the expected repair cost is $E(X) = \$230$. The variance of the repair cost can be calculated from the formula

$$\text{Var}(X) = E((X - E(X))^2) = \sum_i p_i(x_i - E(X))^2$$

and the calculations are shown in Figure 2.46. The variance is 17,100 so that the standard deviation is $\sqrt{17,100} = \$130.77$.

Alternatively, since

$$E(X^2) = \sum_i p_i x_i^2 = (0.3 \times 50^2) + (0.2 \times 200^2) + (0.5 \times 350^2) = 70,000$$

FIGURE 2.46

Mean and standard deviation of machine breakdown costs

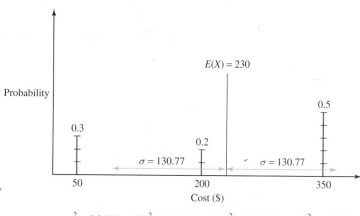

$$\sigma^2 = 0.3\,(50 - 230)^2 + 0.2\,(200 - 230)^2 + 0.5\,(350 - 230)^2 = 17100$$

the variance can be calculated from the simpler formula to be

$$\mathrm{Var}(X) = E(X^2) - (E(X))^2 = 70{,}000 - 230^2 = 17{,}100$$

as before.

Example 12
Personnel Recruitment

The expected value of the square of the number of applicants interviewed is

$$E(X^2) = \sum_i p_i x_i^2$$

$$= (0.18 \times 1^2) + (0.10 \times 2^2) + (0.09 \times 3^2) + (0.06 \times 4^2) + (0.05 \times 5^2)$$
$$+ (0.04 \times 6^2) + (0.02 \times 7^2) + (0.46 \times 8^2) = 34.96$$

Since the expected value of the number of applicants interviewed is $E(X) = 5.20$, the variance can be calculated to be

$$\mathrm{Var}(X) = E(X^2) - (E(X))^2 = 34.96 - 5.20^2 = 7.92$$

The standard deviation is therefore $\sigma = \sqrt{7.92} \doteq 2.81$ people, which is illustrated in Figure 2.47.

Example 14
Metal Cylinder
Production

Recall that the mean cylinder diameter is $E(X) = 50.0$ mm, so that the variance is

$$\mathrm{Var}(X) = E((X - E(X))^2) = \int (x - 50.0)^2 f(x)\, dx$$

$$= \int_{49.5}^{50.5} (x - 50.0)^2 (1.5 - 6(x - 50.0)^2)\, dx$$

$$= [0.5(x - 50.0)^3 - 1.2(x - 50.0)^5]_{49.5}^{50.5} = [0.025] - [-0.025] = 0.05$$

FIGURE 2.47

Mean and standard deviation for
personnel recruitment example

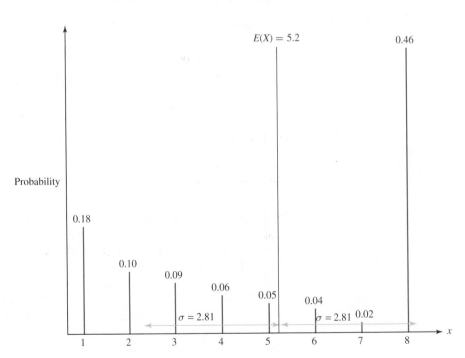

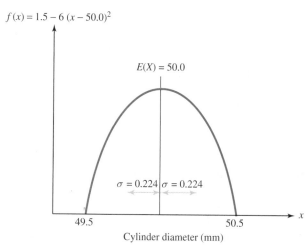

FIGURE 2.48

Mean and standard deviation of metal cylinder diameters

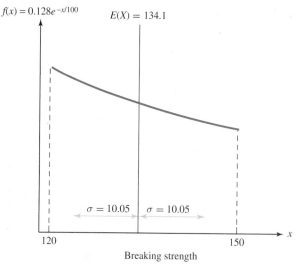

FIGURE 2.49

Mean and standard deviation of concrete breaking strengths

The standard deviation of the metal cylinder diameters is therefore $\sigma = \sqrt{0.05} = 0.224$ mm, which is illustrated in Figure 2.48.

Example 16
Concrete Slab Breaking
Strengths

The expected squared breaking strength is

$$E(X^2) = \int x^2 f(x)dx = \int_{120}^{150} x^2 0.128 e^{-x/100}dx$$
$$= [-12.8(20,000 + 200x + x^2)e^{-x/100}]_{120}^{150}$$
$$= [-207,064.79] - [-225,148.70] = 18,083.91$$

Since the mean breaking strength is $E(X) = 134.1$, the variance of the breaking strengths is

$$\text{Var}(X) = E(X^2) - (E(X))^2 = 18,083.91 - 134.1^2 = 101.10$$

The standard deviation of the breaking strengths is thus $\sigma = \sqrt{101.10} = 10.05$, which is illustrated in Figure 2.49.

GAMES OF CHANCE

Consider two games. In game I a fair die is rolled and a player wins the dollar amount of the score obtained. The probability mass function of the winnings is given in Figure 2.50, and the expected winnings are $3.50. In game II the same die is rolled, but the player wins $3 if a score of 1, 2, or 3 is obtained, and wins $4 if a score of 4, 5, or 6 is obtained. The probability mass function of the winnings from this game is given in Figure 2.51, and again the expected winnings are $3.50.

In both games the expected winnings are the same, and so both games produce the same average revenue. However, the variability in the winnings of game I is clearly larger than the variability in the winnings of game II, and consequently the winnings should have a larger variance in game I than in game II.

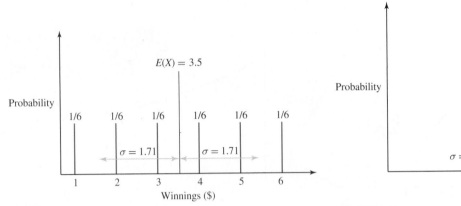

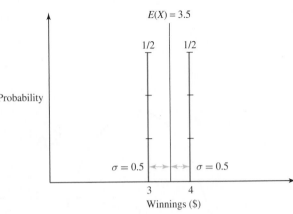

FIGURE 2.50

Winnings from game I

FIGURE 2.51

Winnings from game II

For Game I,

$$\text{Var}(X) = E\left((X - E(X))^2\right) = \sum_i p_i (x_i - 3.5)^2$$

$$= \frac{1}{6}(1 - 3.5)^2 + \frac{1}{6}(2 - 3.5)^2 + \frac{1}{6}(3 - 3.5)^2 + \frac{1}{6}(4 - 3.5)^2$$

$$+ \frac{1}{6}(5 - 3.5)^2 + \frac{1}{6}(6 - 3.5)^2 = \frac{35}{12}$$

whereas for Game II,

$$\text{Var}(X) = E\left((X - E(X))^2\right) = \sum_i p_i (x_i - 3.5)^2$$

$$= \frac{1}{2}(3 - 3.5)^2 + \frac{1}{2}(4 - 3.5)^2 = \frac{1}{4}$$

Thus, the standard deviation of the winnings for game I is $\sigma = \sqrt{35/12} = 1.71$, which, as expected, is larger than the standard deviation of the winnings for game II, which is $\sigma = \sqrt{1/4} = 0.50$.

2.4.3   Chebyshev's Inequality

Chebyshev's inequality is a general result that underlines the importance of the variance and standard deviation of a distribution. It provides general bounds on the probability that a random variable can take values greater than so many standard deviations away from its expected value.

Chebyshev's Inequality

If a random variable has a mean μ and a variance σ^2, then

$$P(\mu - c\sigma \leq X \leq \mu + c\sigma) \geq 1 - \frac{1}{c^2}$$

for $c \geq 1$.

FIGURE 2.52

Illustration of Chebyshev's inequality

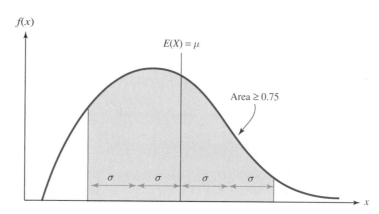

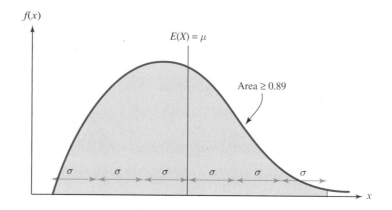

HISTORICAL NOTE

Pafnutii Lvovich Chebyshev (1821–1894) was the leader of a mathematical school in Russia known as the St. Petersburg School, which was of paramount importance in the development of mathematics in Russia. In particular this school provided a solid mathematical background for probability theory. The inequality presented here appears in Chebyshev's master's dissertation entitled "An essay on elementary analysis of probability theory" presented at Moscow University in 1846.

This result indicates that regardless of the actual distribution of a random variable, there are very large probabilities that it will take a value within a few standard deviations of its mean value. For example, taking $c = 2$ gives

$$P(\mu - 2\sigma \leq X \leq \mu + 2\sigma) \geq 1 - \frac{1}{2^2} = 0.75$$

and taking $c = 3$ gives

$$P(\mu - 3\sigma \leq X \leq \mu + 3\sigma) \geq 1 - \frac{1}{3^2} = 0.89$$

Thus any random variable has a probability of at least 75% of taking a value within two standard deviations of its mean and has a probability of at least 89% of taking a value within three standard deviations of its mean, as illustrated in Figure 2.52.

The importance of Chebyshev's inequality is that these results are true regardless of the exact distribution of a random variable, and they require knowledge of only the mean and the standard deviation of the distribution. However, if the exact distribution is known or can reasonably be approximated, then the probabilities can be calculated more exactly and may be much larger than the lower bounds provided by Chebyshev's inequality. For example, the exact values for a normal distribution are shown in Figure 5.16.

Example 18

Tomato Plant Heights

A researcher is interested in how tomato plants are affected by different growing conditions. It is found that under particular growing conditions, three weeks after planting, the heights of the plants have a mean of 29.4 cm and a standard deviation of 2.1 cm. Two standard deviations either side of the mean is

$$[29.4 - (2 \times 2.1), 29.4 + (2 \times 2.1)] = [25.2, 33.6]$$

and three standard deviations either side of the mean is

$$[29.4 - (3 \times 2.1), 29.4 + (3 \times 2.1)] = [23.1, 35.7]$$

Consequently, the researcher can infer that tomato plants grown under these conditions have a probability of at least 75% of having a height between 25.2 cm and 33.6 cm after three weeks, and have a probability of at least 89% of having a height between 23.1 cm and 35.7 cm after three weeks.

Notice that these conclusions can be drawn without knowing the actual distribution of the plant heights, since only the mean and standard deviation are required. However, an important question is: How does the researcher estimate these values? The discussion on statistical *estimation* in Chapter 7 will indicate how the researcher can estimate the mean value to be 29.4 cm and the standard deviation to be 2.1 cm.

Example 14

Metal Cylinder Production

Recall that the metal cylinder diameters have a cumulative distribution function of

$$F(x) = 1.5x - 2(x - 50.0)^3 - 74.5$$

for $49.5 \leq x \leq 50.5$, and that the diameters have a mean of 50.0 mm and a standard deviation of 0.224 mm. Two standard deviations either side of the mean is therefore

$$[50.0 - (2 \times 0.224), 50.0 + (2 \times 0.224)] = [49.552, 50.448]$$

so that Chebyshev's inequality indicates that there is at least a 75% probability that a cylinder will have a diameter between 49.552 and 50.448 mm.

However, this probability can be calculated exactly to be

$$F(50.448) - F(49.552) = 0.992 - 0.008 = 0.984$$

While this result confirms that Chebyshev's inequality is correct in this case, it also illustrates that it provides a very poor lower bound on the true probability. The estimation of the actual probability distribution can allow substantially more accurate inferences to be made. Notice also that for this example there is in fact a probability of 1 that a cylinder diameter falls within three standard deviations of the mean.

2.4.4   Quantiles of Random Variables

Quantiles of random variables are additional summary measures that can provide information about the spread or variability of the distribution of the random variable. The *p*th **quantile** of a random variable X with a cumulative distribution function $F(x)$ is defined to be the value x for which

$$F(x) = p$$

so that there is a probability of p that the random variable takes a value smaller than the *p*th quantile. The probability p is often written as a percentage, and the resulting quantiles are then called **percentiles**, so that, for instance, the 70th percentile of a distribution is the value x for which $F(x) = 0.70$, as illustrated in Figure 2.53. Notice that the 50th percentile of a distribution is the *median* value.

FIGURE 2.53

Illustration of 70th percentile

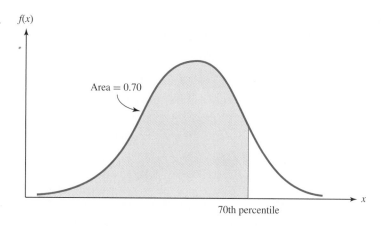

FIGURE 2.54

Illustration of quartiles and median

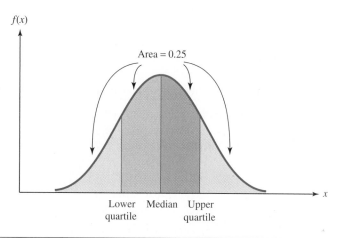

Quantiles

The pth **quantile** of a random variable X with a cumulative distribution function $F(x)$ is defined to be the value x for which

$$F(x) = p$$

This is also referred to as the **$p \times 100$th percentile** of the random variable. There is a probability of p that the random variable takes a value less than the pth quantile.

An idea of the spread of a distribution can be obtained by calculating its **quartiles**. The **upper quartile** of a distribution is defined to be the 75th percentile of the distribution, and the **lower quartile** of a distribution is defined to be the 25th percentile. Notice that the two quartiles, together with the median, partition the state space of a random variable into four "quarters," each of which has a probability of 0.25, as illustrated in Figure 2.54.

The **interquartile range**, defined to be the distance between the two quartiles as shown in Figure 2.55, can be used similar to the variance to provide an indication of the spread of a distribution. Larger values of the interquartile range obviously indicate that a random variable has a distribution that is more spread out.

FIGURE 2.55

Illustration of the interquartile range

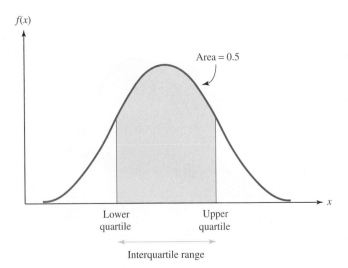

Quartiles and Interquartile Range

The **upper quartile** of a distribution is defined to be the 75th percentile of the distribution, and the **lower quartile** of a distribution is defined to be the 25th percentile. The **interquartile range** is the distance between the two quartiles and like the variance provides an indication of the spread of the distribution.

Example 14
Metal Cylinder Production

The cumulative distribution function of the metal cylinder diameters is

$$F(x) = 1.5x - 2(x - 50.0)^3 - 74.5$$

for $49.5 \leq x \leq 50.5$. The *upper quartile* of the distribution is the value of x for which

$$F(x) = 0.75$$

which is 50.17 mm. The *lower quartile* satisfies

$$F(x) = 0.25$$

and is 49.83 mm. The interquartile range is therefore $50.17 - 49.83 = 0.34$ mm, and half of the cylinders will have diameters between 49.83 mm and 50.17 mm, as illustrated in Figure 2.56.

Example 15
Battery Failure Times

The cumulative distribution function of the battery failure times is

$$F(x) = 1 - \frac{1}{(x + 1)^2}$$

The *upper quartile* of the distribution is thus seen to be 1 hour, and the *lower quartile* is seen to be $(2/\sqrt{3}) - 1 = 0.155$ hour, which is about 9.3 minutes. The interquartile range is thus $60 - 9.3 = 50.7$ minutes, and half of the batteries will fail between 9.3 minutes and 1 hour, as illustrated in Figure 2.57.

Recall that the expected failure time is also 1 hour. However, even though the average failure time is 1 hour, on average three out of four batteries will fail before 1 hour.

FIGURE 2.56

Interquartile range for metal cylinder diameters

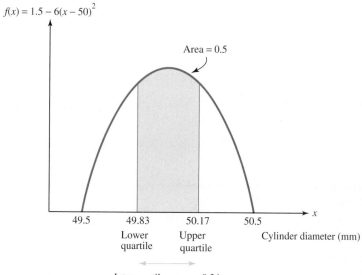

$f(x) = 1.5 - 6(x - 50)^2$

Area = 0.5

49.5   49.83   50.17   50.5 x

Lower quartile   Upper quartile   Cylinder diameter (mm)

Interquartile range = 0.34

FIGURE 2.57

Interquartile range for battery failure times

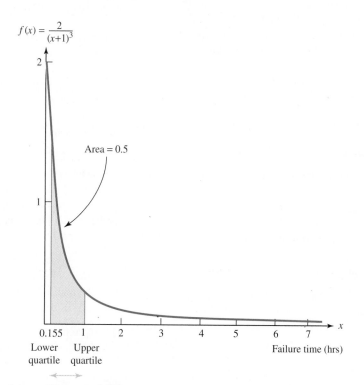

$f(x) = \dfrac{2}{(x+1)^3}$

Area = 0.5

0.155   1   2   3   4   5   6   7 x

Lower quartile   Upper quartile   Failure time (hrs)

Interquartile range = 0.845

■ 2.4.5 Problems

2.4.1 Suppose that the random variable X takes the values -2, 1, 4, and 6 with probability values 1/3, 1/6, 1/3, and 1/6, respectively.
(a) Find the expectation of X.
(b) Find the variance of X using the formula
$$\text{Var}(X) = E((X - E(X))^2)$$
(c) Find the variance of X using the formula
$$\text{Var}(X) = E(X^2) - (E(X))^2$$

2.4.2 Consider again the four copying machines discussed in Problems 2.1.1 and 2.3.1. Calculate the variance and standard deviation of the number of copying machines in use at a particular moment.

2.4.3 Consider again the salesperson discussed in Problems 2.1.9 and 2.3.4 who is trying to locate a particular product. Calculate the variance and standard deviation of the number of warehouses called by the salesperson.

2.4.4 Suppose that you are organizing the game described at the end of Section 2.3.1, where you charge players $2 to roll two dice and then you pay them the difference in the scores. What is the variance in your profit from each game? If you are playing a game in which you have positive expected winnings, would you prefer a small or a large variance in the winnings?

2.4.5 Consider again the random variable described in Problems 2.2.2 and 2.3.10 with a probability density function of
$$f(x) = \frac{1}{x \ln(1.5)}$$
for $4 \leq x \leq 6$ and $f(x) = 0$ elsewhere.
(a) What is the variance of this random variable?
(b) What is the standard deviation of this random variable?
(c) Find the upper and lower quartiles of this random variable.
(d) What is the interquartile range?

2.4.6 Consider again the random variable described in Problems 2.2.4 and 2.3.11 with a cumulative distribution function of
$$F(x) = \frac{x^2}{16}$$
for $0 \leq x \leq 4$.
(a) What is the variance of this random variable?
(b) What is the standard deviation of this random variable?

(c) Find the upper and lower quartiles of this random variable.
(d) What is the interquartile range?

2.4.7 Consider again the car panel painting machine discussed in Problems 2.2.6 and 2.3.12.
(a) What is the variance of the paint thickness?
(b) What is the standard deviation of the paint thickness?
(c) Find the upper and lower quartiles of the paint thickness.
(d) What is the interquartile range?

2.4.8 Consider again the plastic bending capabilities discussed in Problems 2.2.8 and 2.3.13.
(a) What is the variance of the deformity angle?
(b) What is the standard deviation of the deformity angle?
(c) Find the upper and lower quartiles of the deformity angle.
(d) What is the interquartile range?

2.4.9 Consider again the archery problem discussed in Problems 2.2.9 and 2.3.14.
(a) What is the variance of the deviation from the center of the target?
(b) What is the standard deviation of the deviation from the center of the target?
(c) Find the upper and lower quartiles of the deviation from the center of the target.
(d) What is the interquartile range?

2.4.10 The time taken to serve a customer at a fast-food restaurant has a mean of 75.0 seconds and a standard deviation of 7.3 seconds. Use Chebyshev's inequality to calculate time intervals that have 75% and 89% probabilities of containing a particular service time.

2.4.11 A machine produces iron bars whose lengths have a mean of 110.8 cm and a standard deviation of 0.5 cm. Use Chebyshev's inequality to obtain a lower bound on the probability that an iron bar chosen at random has a length between 109.55 cm and 112.05 cm.

2.4.12 Recall Problems 2.1.11 and 2.3.16 concerning the scheduling of appointments with a consultant. What is the standard deviation of the total number of appointments that have already been made over both days at the moment when Monday's schedule has just been completely filled?

2.4.13 Recall Problems 2.2.11 and 2.3.17 concerning the resistance of an electrical component.
  (a) What is the standard deviation of the resistance?
  (b) What is the 80th percentile of the resistance? What is the 10th percentile of the resistance?

2.4.14 A continuous random variable has a probability density function $f(x) = Ax^{2.5}$ for $2 \leq x \leq 3$.
  (a) What is the value of A?
  (b) What is the expectation of the random variable?
  (c) What is the standard deviation of the random variable?
  (d) What is the median of the random variable?

2.4.15 In a game a player either loses $1 with a probability 0.25, wins $1 with a probability 0.4, or wins $4 with a probability 0.35. What are the expectation and the standard deviation of the winnings?

2.4.16 A random variable X has a probability density function $f(x) = A/\sqrt{x}$ for $3 \leq x \leq 4$.
  (a) What is the value of A?
  (b) What is the cumulative distribution function of X?
  (c) What is the expected value of X?
  (d) What is the standard deviation of X?
  (e) What is the median of X?
  (f) What is the upper quartile of X?

2.4.17 When a construction project is opened for bidding, two proposals are received with probability 0.11, three proposals are received with probability 0.19, four proposals are received with probability 0.55, and five proposals are received with probability 0.15.
  (a) What is the expectation of the number of proposals received?
  (b) What is the standard deviation of the number of proposals received?

2.4.18 A random variable X has a distribution given by the probability density function $f(x) = (1 - x)/2$ with a state space $-1 \leq x \leq 1$.
  (a) What is the expected value of X?
  (b) What is the standard deviation of X?
  (c) What is the upper quartile of X?

2.5 Jointly Distributed Random Variables

2.5.1 Joint Probability Distributions

Instead of considering one random variable X and its probability distribution, it is often appropriate to consider two random variables X and Y and their **joint probability distribution**. If the random variables are discrete, then the **joint probability mass function** consists of probability values $P(X = x_i, Y = y_j) = p_{ij} \geq 0$ satisfying

$$\sum_i \sum_j p_{ij} = 1$$

If the random variables are continuous, then the **joint probability density function** is a function $f(x, y) \geq 0$ satisfying

$$\int \int_{\text{state space}} f(x, y) \, dx \, dy = 1$$

The probability that $a \leq X \leq b$ and $c \leq Y \leq d$ is obtained from the joint probability density function as

$$\int_{x=a}^{b} \int_{y=c}^{d} f(x, y) \, dy \, dx$$

The **joint cumulative distribution function** is defined to be

$$F(x, y) = P(X \leq x, Y \leq y),$$

which is

$$F(x, y) = \sum_{i: x_i \leq x} \sum_{j: y_j \leq y} p_{ij}$$

for discrete random variables and

$$F(x, y) = \int_{w=-\infty}^{x} \int_{z=-\infty}^{y} f(w, z) \, dz \, dw$$

for continuous random variables.

Joint Probability Distributions

The **joint probability distribution** of two random variables X and Y is specified by a set of probability values $P(X = x_i, Y = y_j) = p_{ij}$ for discrete random variables, or a joint probability density function $f(x, y)$ for continuous random variables. In either case, the **joint cumulative distribution function** is defined to be

$$F(x, y) = P(X \le x, Y \le y)$$

The following two examples illustrate jointly distributed random variables.

Example 19
Air Conditioner Maintenance

A company that services air conditioner units in residences and office blocks is interested in how to schedule its technicians in the most efficient manner. Specifically, the company is interested in how long a technician takes on a visit to a particular location, and the company recognizes that this mainly depends on the number of air conditioner units at the location that need to be serviced.

If the random variable X, taking the values 1, 2, 3, and 4, is the *service time* in hours taken at a particular location, and the random variable Y, taking the values 1, 2, and 3, is the *number of air conditioner units* at the location, then these two random variables can be thought of as jointly distributed.

Suppose that their joint probability mass function p_{ij} is given in Figure 2.58. The figure indicates, for example, that there is a probability of 0.12 that $X = 1$ and $Y = 1$, so that there is a probability of 0.12 that a particular location chosen at random has one air conditioner unit that takes a technician one hour to service. Similarly, there is a probability of 0.07 that a location has three air conditioner units that take four hours to service. Notice that this is a valid probability mass function since

$$\sum_i \sum_j p_{ij} = 0.12 + 0.08 + \cdots + 0.07 = 1.00$$

FIGURE 2.58

Joint probability mass function for air conditioner maintenance example

		$X =$ service time (hrs)			
		1	2	3	4
	1	0.12	0.08	0.07	0.05
$Y =$ number of air conditioner units	2	0.08	0.15	0.21	0.13
	3	0.01	0.01	0.02	0.07

FIGURE 2.59			X = service time (hrs)		
		1	2	3	4
Joint cumulative distribution function for air conditioner maintenance example	1	0.12	0.20	0.27	0.32
Y = number of air conditioner units	2	0.20	0.43	0.71	0.89
	3	0.21	0.45	0.75	1.00

The joint cumulative distribution function

$$F(x, y) = P(X \leq x, Y \leq y) = \sum_{i=1}^{x} \sum_{j=1}^{y} p_{ij}$$

is given in Figure 2.59. For example, the probability that a location has no more than two air conditioner units that take no more than two hours to service is

$$F(2, 2) = p_{11} + p_{12} + p_{21} + p_{22} = 0.12 + 0.08 + 0.08 + 0.15 = 0.43$$

Example 20
Mineral Deposits

In order to determine the economic viability of mining in a certain area, a mining company obtains samples of ore from the location and measures their *zinc* content and their *iron* content. Suppose that the random variable X is the zinc content of the ore, taking values between 0.5 and 1.5, and that the random variable Y is the iron content of the ore, taking values between 20.0 and 35.0. Furthermore, suppose that their joint probability density function is

$$f(x, y) = \frac{39}{400} - \frac{17(x-1)^2}{50} - \frac{(y-25)^2}{10,000}$$

for $0.5 \leq x \leq 1.5$ and $20.0 \leq y \leq 35.0$.

The validity of this joint probability density function can be checked by ascertaining that $f(x, y) \geq 0$ within the state space $0.5 \leq x \leq 1.5$ and $20.0 \leq y \leq 35.0$, and that

$$\int_{x=0.5}^{1.5} \int_{y=20.0}^{35.0} f(x, y) \, dx \, dy = 1$$

The joint probability density function provides complete information about the joint probabilistic properties of the random variables X and Y. For example, the probability that a randomly chosen sample of ore has a zinc content between 0.8 and 1.0 and an iron content between 25 and 30 is

$$\int_{x=0.8}^{1.0} \int_{y=25.0}^{30.0} f(x, y) \, dx \, dy$$

which can be calculated to be 0.092. Consequently only about 9% of the ore at the location has mineral levels within these limits.

2.5.2   Marginal Probability Distributions

Even though two random variables X and Y may be jointly distributed, if interest is focused on only one of the random variables, then it is appropriate to consider the probability distribution of that random variable alone. This is known as the **marginal distribution** of the random variable and can be obtained quite simply by summing or integrating the joint probability distribution over the values of the other random variable.

For example, for two discrete random variables X and Y, the probability values of the marginal distribution of X are

$$P(X = x_i) = p_{i+} = \sum_j p_{ij}$$

and for two continuous random variables, the probability density function of the marginal distribution of X is

$$f_X(x) = \int_{-\infty}^{\infty} f(x, y)\, dy$$

where in practice the summation and integration limits can be curtailed at the appropriate boundaries of the state space. Note that the marginal distribution of a random variable X and the marginal distribution of a random variable Y do not uniquely determine their joint distribution.

Marginal Probability Distributions

The **marginal distribution** of a random variable X is obtained from the joint probability distribution of two random variables X and Y by summing or integrating over the values of the random variable Y. The marginal distribution is the individual probability distribution of the random variable X considered alone.

The expectations and variances of the random variables X and Y can be obtained from their marginal distributions in the usual manner, as illustrated in the following examples.

Example 19
Air Conditioner
Maintenance

The marginal probability mass function of X, the time taken to service the air conditioner units at a particular location, is given in Figure 2.60 and is obtained by summing the appropriate values of the joint probability mass function. For example,

$$P(X = 1) = \sum_{j=1}^{3} p_{1j} = 0.12 + 0.08 + 0.01 = 0.21$$

The expected service time is

$$E(X) = \sum_{i=1}^{4} i\, P(X = i)$$
$$= (1 \times 0.21) + (2 \times 0.24) + (3 \times 0.30) + (4 \times 0.25) = 2.59$$

Since

$$E(X^2) = \sum_{i=1}^{4} i^2 P(X = i)$$
$$= (1 \times 0.21) + (4 \times 0.24) + (9 \times 0.30) + (16 \times 0.25) = 7.87$$

FIGURE 2.60

Marginal probability mass
functions for air conditioner
maintenance example

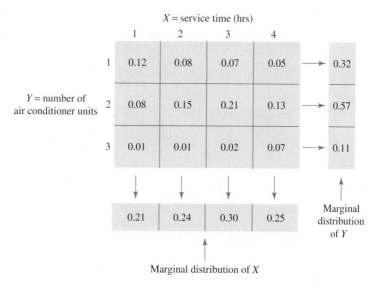

FIGURE 2.60

Marginal probability mass functions for air conditioner maintenance example

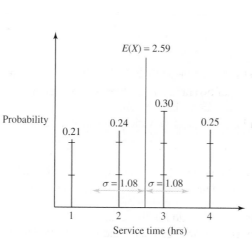

FIGURE 2.61

Marginal probability mass function of the service time

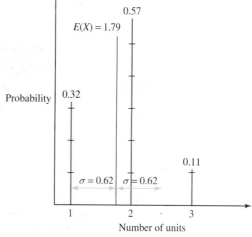

FIGURE 2.62

Marginal probability mass function of the number of air conditioner units

the variance in the service times is

$$\text{Var}(X) = E(X^2) - (E(X))^2 = 7.87 - 2.59^2 = 1.162$$

The standard deviation is therefore $\sigma = \sqrt{1.162} = 1.08$ hours, or about 65 minutes, as indicated in Figure 2.61.

The marginal probability mass function of Y, the number of air conditioner units at a particular location, is given in Figure 2.62. Here,

$$P(Y = 1) = \sum_{i=1}^{4} p_{i1} = 0.12 + 0.08 + 0.07 + 0.05 = 0.32$$

FIGURE 2.63

Marginal probability density
function of zinc content

$$f_X(x) = \frac{57}{40} - \frac{51(x-1)^2}{10}$$

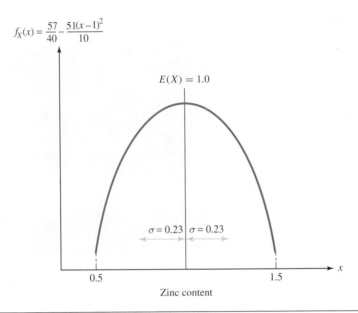

$E(X) = 1.0$

$\sigma = 0.23 \mid \sigma = 0.23$

0.5   1.5 x

Zinc content

for example. The expected number of air conditioner units can be calculated to be $E(Y) = 1.79$, and the standard deviation is $\sigma = 0.62$.

Example 20
Mineral Deposits

The marginal probability density function of X, the zinc content of the ore, is

$$f_X(x) = \int_{y=20.0}^{y=35.0} f(x, y)\, dy$$

$$= \int_{y=20.0}^{y=35.0} \left(\frac{39}{400} - \frac{17(x-1)^2}{50} - \frac{(y-25)^2}{10{,}000} \right) dy$$

$$= \left[\frac{39y}{400} - \frac{17y(x-1)^2}{50} - \frac{(y-25)^3}{30{,}000} \right]_{y=20.0}^{y=35.0} = \frac{57}{40} - \frac{51(x-1)^2}{10}$$

for $0.5 \leq x \leq 1.5$. This is shown in Figure 2.63, and since it is symmetric about the point $x = 1$, the expected zinc content is $E(X) = 1$. The variance of the zinc content is

$$\text{Var}(X) = E((X - E(X))^2)$$

$$= \int_{0.5}^{1.5} (x-1)^2 f_X(x)\,dx = \int_{0.5}^{1.5} (x-1)^2 \left(\frac{57}{40} - \frac{51(x-1)^2}{10} \right) dx$$

$$= \left[\frac{19}{40}(x-1)^3 - \frac{51}{50}(x-1)^5 \right]_{0.5}^{1.5} = [0.0275] - [-0.0275] = 0.055$$

and the standard deviation is therefore $\sigma = \sqrt{0.055} = 0.23$.

The probability that a sample of ore has a zinc content between 0.8 and 1.0 can be calculated from the marginal probability density function to be

$$P(0.8 \leq X \leq 1.0) = \int_{0.8}^{1.0} f_X(x)\,dx = \int_{0.8}^{1.0} \left(\frac{57}{40} - \frac{51(x-1)^2}{10} \right) dx$$

$$= \left[\frac{57x}{40} - \frac{17(x-1)^3}{10} \right]_{0.8}^{1.0} = [1.425] - [1.1536] = 0.2714$$

Consequently about 27% of the ore has a zinc content within these limits.

FIGURE 2.64

Marginal probability density
function of iron content

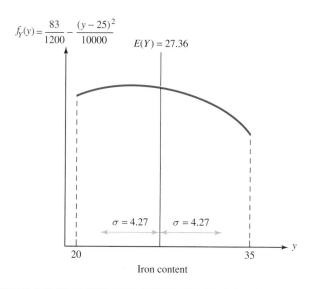

The marginal probability density function of Y, the iron content of the ore, is

$$
f_Y(y) = \int_{x=0.5}^{x=1.5} f(x, y) dx = \int_{x=0.5}^{x=1.5} \left(\frac{39}{400} - \frac{17(x-1)^2}{50} - \frac{(y-25)^2}{10,000} \right) dx
$$

$$
= \left[\frac{39x}{400} - \frac{17(x-1)^3}{150} - \frac{x(y-25)^2}{10,000} \right]_{x=0.5}^{x=1.5} = \frac{83}{1200} - \frac{(y-25)^2}{10,000}
$$

for $20.0 \le y \le 35.0$. This is shown in Figure 2.64 together with the expected iron content
and the standard deviation of the iron content, which can be calculated to be $E(Y) = 27.36$
and $\sigma = 4.27$.

2.5.3 Conditional Probability Distributions

If two random variables X and Y are jointly distributed, then it is sometimes useful to consider
the distribution of one random variable *conditional* on the other random variable having taken
a particular value. Conditional probabilities were discussed in Section 1.4, and they allow
probabilities, or more generally random variable distributions, to be revised following the
observation of a certain event.

If two discrete random variables X and Y are jointly distributed, then the **conditional
distribution** of random variable X conditional on the event $Y = y_j$ consists of the probability
values

$$
p_{i|Y=y_j} = P(X = x_i | Y = y_j) = \frac{P(X = x_i, Y = y_j)}{P(Y = y_j)} = \frac{p_{ij}}{p_{+j}}
$$

where $p_{+j} = P(Y = y_j) = \sum_i p_{ij}$. If two continuous random variables X and Y are jointly
distributed, then the conditional distribution of random variable X conditional on the event
$Y = y$ has a probability density function

$$
f_{X|Y=y}(x) = \frac{f(x, y)}{f_Y(y)}
$$

where the denominator $f_Y(y)$ is the **marginal distribution** of the random variable Y. Conditional expectations and variances can be calculated in the usual manner from these conditional
distributions.

Conditional Probability Distributions

The **conditional distribution** of a random variable X conditional on a random variable Y taking a particular value summarizes the probabilistic properties of the random variable X under the knowledge provided by the value of Y. It consists of the probability values

$$p_{i|Y=y_j} = P(X = x_i | Y = y_j) = \frac{P(X = x_i, Y = y_j)}{P(Y = y_j)} = \frac{p_{ij}}{p_{+j}}$$

for discrete random variables or the probability density function

$$f_{X|Y=y}(x) = \frac{f(x, y)}{f_Y(y)}$$

for continuous random variables, where $f_Y(y)$ is the **marginal distribution** of the random variable Y.

It is important to recognize the difference between a *marginal distribution* and a *conditional distribution*. The marginal distribution for X is the appropriate distribution for the random variable X when nothing is known about the random variable Y. In contrast, the conditional distribution for X conditional on a particular value y of Y is the appropriate distribution for the random variable X when the random variable Y is known to take the value y. This difference is illustrated in the following examples.

Example 19
Air Conditioner
Maintenance

Suppose that a technician is visiting a location that is known to have three air conditioner units, an event that has a probability of

$$P(Y = 3) = p_{+3} = 0.01 + 0.01 + 0.02 + 0.07 = 0.11$$

The conditional distribution of the service time X consists of the probability values

$$p_{1|Y=3} = P(X = 1 | Y = 3) = \frac{p_{13}}{p_{+3}} = \frac{0.01}{0.11} = 0.091$$

$$p_{2|Y=3} = P(X = 2 | Y = 3) = \frac{p_{23}}{p_{+3}} = \frac{0.01}{0.11} = 0.091$$

$$p_{3|Y=3} = P(X = 3 | Y = 3) = \frac{p_{33}}{p_{+3}} = \frac{0.02}{0.11} = 0.182$$

$$p_{4|Y=3} = P(X = 4 | Y = 3) = \frac{p_{43}}{p_{+3}} = \frac{0.07}{0.11} = 0.636$$

These values are shown in Figure 2.65, and they are clearly different from the marginal distribution of the service time given in Figure 2.61. Conditioning on a location having three air conditioner units increases the chances of a large service time being required.

The conditional expectation of the service time is

$$E(X|Y = 3) = \sum_{i=1}^{4} i p_{i|Y=3}$$
$$= (1 \times 0.091) + (2 \times 0.091) + (3 \times 0.182) + (4 \times 0.636) = 3.36$$

which, as expected, is considerably larger than the "overall" expected service time of 2.59 hours. The difference between these expected values can be interpreted in the following way.

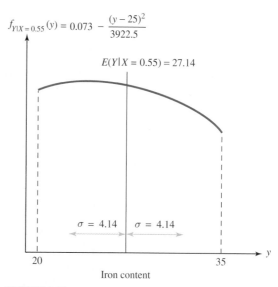

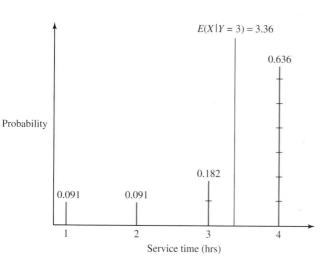

FIGURE 2.65

Conditional probability mass function of service time when $Y = 3$

FIGURE 2.66

Conditional probability density function of iron content when $X = 0.55$

If a technician sets off for a location for which the number of air conditioner units is not known, then the expected service time at the location is 2.59 hours. However, if the technician knows that there are three air conditioner units at the location that need servicing, then the expected service time is 3.36 hours.

Example 20
Mineral Deposits

Suppose that a sample of ore has a zinc content of $X = 0.55$. What is known about its iron content? The information about the iron content Y is summarized in the conditional probability density function for the iron content, which is

$$f_{Y|X=0.55}(y) = \frac{f(0.55, y)}{f_X(0.55)}$$

where the denominator is the marginal distribution of the zinc content X evaluated at 0.55. Since

$$f_X(0.55) = \frac{57}{40} - \frac{51(0.55 - 1.00)^2}{10} = 0.39225$$

the conditional probability density function is

$$f_{Y|X=0.55}(y) = \frac{f(0.55, y)}{0.39225} = \frac{39}{400 \times 0.39225} - \frac{17(0.55 - 1.00)^2}{50 \times 0.39225} - \frac{(y - 25)^2}{10,000 \times 0.39225}$$

$$= 0.073 - \frac{(y - 25)^2}{3922.5}$$

for $20.0 \le y \le 35.0$. This is shown in Figure 2.66 with the conditional expectation of the iron content, which can be calculated to be 27.14, and with the conditional standard deviation, which is 4.14.

2.5.4   Independence and Covariance

In the same way that two events A and B were said to be independent in Chapter 1 if they are "unrelated" to each other, two random variables X and Y are said to be **independent** if the value taken by one random variable is "unrelated" to the value taken by the other random variable. More specifically, the random variables are independent if the distribution of one of the random variables does not depend upon the value taken by the other random variable.

Independent Random Variables

Two random variables X and Y are defined to be **independent** if their joint probability mass function or joint probability density function is the *product* of their two marginal distributions. If the random variables are discrete, then they are independent if

$$p_{ij} = p_{i+} p_{+j}$$

for all values of x_i and y_j. If the random variables are continuous, then they are independent if

$$f(x, y) = f_X(x) f_Y(y)$$

for all values of x and y. If two random variables are independent, then the probability distribution of one of the random variables does not depend upon the value taken by the other random variable.

Notice that if the random variables X and Y are independent, then their *conditional* distributions are identical to their *marginal* distributions. If the random variables are discrete, this is because

$$p_{i|Y=y_j} = \frac{p_{ij}}{p_{+j}} = \frac{p_{i+} p_{+j}}{p_{+j}} = p_{i+}$$

and if the random variables are continuous, this is because

$$f_{X|Y=y}(x) = \frac{f(x, y)}{f_Y(y)} = \frac{f_X(x) f_Y(y)}{f_Y(y)} = f_X(x)$$

In either case, the conditional distributions do not depend upon the value conditioned upon, and they are equal to the marginal distributions. This result has the interpretation that knowledge of the value taken by the random variable Y does not influence the distribution of the random variable X, and vice versa.

As a simple example of two independent random variables, suppose that X and Y have a joint probability density function of

$$f(x, y) = 6xy^2$$

for $0 \le x \le 1$ and $0 \le y \le 1$ and $f(x, y) = 0$ elsewhere. The fact that this joint density function is a function of x *multiplied* by a function of y (and that the state spaces of the random variables [0, 1] do not depend upon each other) immediately indicates that the two random variables are independent. Specifically, the marginal distribution of X is

$$f_X(x) = \int_{y=0}^{1} 6xy^2 \, dy = 2x$$

for $0 \le x \le 1$, and the marginal distribution of Y is

$$f_Y(y) = \int_{x=0}^{1} 6xy^2 \, dx = 3y^2$$

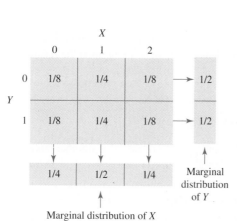

FIGURE 2.67

Joint probability mass function and marginal
probability mass functions for X and Y in
coin-tossing game

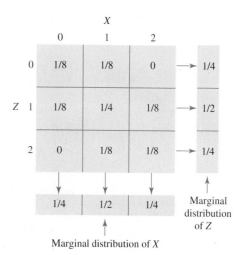

FIGURE 2.68

Joint probability mass function and marginal
probability mass functions for X and Z in
coin-tossing game

for $0 \leq y \leq 1$. The fact that $f(x, y) = f_X(x) f_Y(y)$ confirms that the random variables X and Y are independent.

GAMES OF CHANCE   Suppose that a fair coin is tossed three times so that there are eight equally likely outcomes, and that the random variable X is the number of heads obtained in the *first and second* tosses, the random variable Y is the number of heads in the *third* toss, and the random variable Z is the number of heads obtained in the *second and third* tosses.

The joint probability mass function of X and Y is given in Figure 2.67 together with the marginal distributions of X and Y. For example, $P(X = 0, Y = 0) = P(TTT) = 1/8$ and $P(X = 0) = P(TTT) + P(TTH) = 1/4$. It is easy to check that

$$P(X = i, Y = j) = P(X = i) P(Y = j)$$

for all values of $i = 0, 1, 2$ and $j = 0, 1$, so that the joint probability mass function is equal to the product of the two marginal probability mass functions. Consequently, the random variables X and Y are independent, which is not surprising since the outcome of the third coin toss should be unrelated to the outcomes of the first two coin tosses.

Figure 2.68 shows the joint probability mass function of X and Z together with the marginal distributions of X and Z. For example, $P(X = 1, Z = 1) = P(HTH) + P(THT) = 1/4$. Notice, however, that

$$P(X = 0, Z = 0) = P(TTT) = \frac{1}{8}$$
$$P(X = 0) = P(TTH) + P(TTT) = \frac{1}{4}$$

and

$$P(Z = 0) = P(HTT) + P(TTT) = \frac{1}{4}$$

so that

$$P(X = 0, Z = 0) \neq P(X = 0)P(Z = 0)$$

This result indicates that the random variables X and Z are *not* independent. In fact, their dependence is a result of their both depending upon the result of the second coin toss.

The strength of the dependence of two random variables on each other is indicated by their **covariance**, which is defined to be

$$Cov(X, Y) = E((X - E(X))(Y - E(Y)))$$

The covariance can be any positive or negative number, and independent random variables have a covariance of zero. It is often convenient to calculate the covariance from an alternative expression

$$
\begin{aligned}
Cov(X, Y) &= E((X - E(X))(Y - E(Y))) \\
&= E(XY - XE(Y) - E(X)Y + E(X)E(Y)) \\
&= E(XY) - E(X)E(Y) - E(X)E(Y) + E(X)E(Y) \\
&= E(XY) - E(X)E(Y)
\end{aligned}
$$

Covariance

The **covariance** of two random variables X and Y is defined to be

$$Cov(X, Y) = E((X - E(X))(Y - E(Y))) = E(XY) - E(X)E(Y)$$

The covariance can be any positive or negative number, and independent random variables have a covariance of 0.

In practice, the most convenient way to assess the strength of the dependence between two random variables is through their **correlation**.

Correlation

The **correlation** between two random variables X and Y is defined to be

$$Corr(X, Y) = \frac{Cov(X, Y)}{\sqrt{Var(X)Var(Y)}}$$

The correlation takes values between -1 and 1, and independent random variables have a correlation of 0.

Random variables with a positive correlation are said to be positively correlated, and in such cases there is a tendency for *high* values of one random variable to be associated with *high* values of the other random variable. Random variables with a negative correlation are said to be negatively correlated, and in such cases there is a tendency for *high* values of one random variable to be associated with *low* values of the other random variable. The strength of these tendencies increases as the correlation moves further away from 0 to 1 or to -1.

As an illustration of the calculation of a covariance, consider again the simple example where

$$f(x, y) = 6xy^2$$

for $0 \le x \le 1$ and $0 \le y \le 1$. The expectation of the random variable X can be calculated from its marginal distribution to be

$$E(X) = \int_{x=0}^{1} x f_X(x) \, dx = \int_{x=0}^{1} 2x^2 \, dx = \frac{2}{3}$$

and similarly

$$E(Y) = \int_{y=0}^{1} y f_Y(y) \, dy = \int_{y=0}^{1} 3y^3 \, dy = \frac{3}{4}$$

Also,

$$E(XY) = \int_{x=0}^{1} \int_{y=0}^{1} xy f(x, y) \, dy \, dx = \int_{x=0}^{1} \int_{y=0}^{1} 6x^2 y^3 \, dy \, dx = \frac{1}{2}$$

so that

$$\text{Cov}(X, Y) = E(XY) - E(X)E(Y) = \frac{1}{2} - \left(\frac{2}{3} \times \frac{3}{4} \right) = 0$$

This result is expected since the random variables X and Y were shown to be independent.

Example 19
Air Conditioner
Maintenance

The expected service time is $E(X) = 2.59$ hours, and the expected number of units serviced is $E(Y) = 1.79$. In addition,

$$E(XY) = \sum_{i=1}^{4} \sum_{j=1}^{3} ij \, p_{ij}$$
$$= (1 \times 1 \times 0.12) + (1 \times 2 \times 0.08) + \cdots + (4 \times 3 \times 0.07) = 4.86$$

so that the covariance is

$$\text{Cov}(X, Y) = E(XY) - E(X)E(Y) = 4.86 - (2.59 \times 1.79) = 0.224$$

Since $\text{Var}(X) = 1.162$ and $\text{Var}(Y) = 0.384$, the correlation between the service time and the number of units serviced is

$$\text{Corr}(X, Y) = \frac{\text{Cov}(X, Y)}{\sqrt{\text{Var}(X)\text{Var}(Y)}} = \frac{0.224}{\sqrt{1.162 \times 0.384}} = 0.34$$

As expected, the service time and the number of units serviced are not independent but are positively correlated. This makes sense because there is a tendency for locations with a large number of air conditioner units to require relatively long service times.

GAMES OF CHANCE

Consider again the tossing of three coins and the random variables X, Y, and Z. It is easy to check that $E(X) = 1$ and $E(Y) = 1/2$, and that

$$E(XY) = \sum_{i=0}^{2} \sum_{j=0}^{1} ij \, p_{ij} = \left(1 \times 1 \times \frac{1}{4} \right) + \left(2 \times 1 \times \frac{1}{8} \right) = \frac{1}{2}$$

Consequently,

$$\text{Cov}(X, Y) = E(XY) - E(X)E(Y) = \frac{1}{2} - \left(1 \times \frac{1}{2} \right) = 0$$

which is expected because the random variables X and Y are independent.

However, $E(Z) = 1$ and

$$E(XZ) = \sum_{i=0}^{2}\sum_{j=0}^{2} ij\, p_{ij}$$

$$= \left(1 \times 1 \times \frac{1}{4}\right) + \left(2 \times 1 \times \frac{1}{8}\right) + \left(1 \times 2 \times \frac{1}{8}\right) + \left(2 \times 2 \times \frac{1}{8}\right)$$

$$= \frac{5}{4}$$

so that

$$\text{Cov}(X, Z) = E(XZ) - E(X)E(Z) = \frac{5}{4} - (1 \times 1) = \frac{1}{4}$$

Also, since $\text{Var}(X) = \text{Var}(Z) = 1/2$,

$$\text{Corr}(X, Z) = \frac{\text{Cov}(X, Z)}{\sqrt{\text{Var}(X)\text{Var}(Z)}} = \frac{\frac{1}{4}}{\sqrt{\frac{1}{2} \times \frac{1}{2}}} = \frac{1}{2}$$

so that the random variables X and Z are positively correlated.

■ 2.5.5 Problems

2.5.1 Consider Example 20 on mineral deposits.
  (a) Show that $P(0.8 \leq X \leq 1, 25 \leq Y \leq 30) = 0.092$.
  (b) Show that the iron content has an expected value of 27.36 and a standard deviation of 4.27.
  (c) Show that conditional on $X = 0.55$, the iron content has an expected value of 27.14 and a standard deviation of 4.14.

2.5.2 Consider Example 19 on air conditioner maintenance.
  (a) Suppose that a location has only one air conditioner that needs servicing. What is the conditional probability mass function of the service time required, and the conditional expectation and standard deviation of the service time?
  (b) Suppose that a location requires a service time of two hours. What is the conditional probability mass function of the number of air conditioner units serviced, and the conditional expectation and standard deviation of the number of air conditioner units serviced?

2.5.3 Suppose that two continuous random variables X and Y have a joint probability density function

$$f(x, y) = A(x - 3)y$$

for $-2 \leq x \leq 3$ and $4 \leq y \leq 6$, and $f(x, y) = 0$ elsewhere.
  (a) What is the value of A?
  (b) What is $P(0 \leq X \leq 1, 4 \leq Y \leq 5)$?

  (c) Construct the marginal probability density functions $f_X(x)$ and $f_Y(y)$.
  (d) Are the random variables X and Y independent?
  (e) If $Y = 5$, what is the conditional probability density function of X?

2.5.4 A fair coin is tossed four times, and the random variable X is the number of heads in the *first three* tosses and the random variable Y is the number of heads in the *last three* tosses.
  (a) What is the joint probability mass function of X and Y?
  (b) What are the marginal probability mass functions of X and Y?
  (c) Are the random variables X and Y independent?
  (d) What are the expectations and variances of the random variables X and Y?
  (e) What is the covariance of X and Y?
  (f) If there is one head in the last three tosses, what is the conditional probability mass function of X? What are the conditional expectation and variance of X?

2.5.5 Suppose that two continuous random variables X and Y have a joint probability density function

$$f(x, y) = A\left(e^{x+y} + e^{2x-y}\right)$$

for $1 \leq x \leq 2$ and $0 \leq y \leq 3$, and $f(x, y) = 0$ elsewhere.

(a) What is the value of A?

(b) What is $P(1.5 \le X \le 2, 1 \le Y \le 2)$?

(c) Construct the marginal probability density functions $f_X(x)$ and $f_Y(y)$.

(d) Are the random variables X and Y independent?

(e) If $Y = 0$, what is the conditional probability density function of X?

2.5.6 Two cards are drawn *without replacement* from a pack of cards, and the random variable X measures the number of hearts drawn and the random variable Y measures the number of clubs drawn.

(a) What is the joint probability mass function of X and Y?

(b) What are the marginal probability mass functions of X and Y?

(c) Are the random variables X and Y independent?

(d) What are the expectations and variances of the random variables X and Y?

(e) What is the covariance of X and Y?

(f) What is the correlation between X and Y?

(g) If no hearts are drawn, what is the conditional probability mass function of Y? If one heart is drawn, what is the conditional probability mass function of Y?

2.5.7 Repeat Problem 2.5.6 when the second card is drawn *with replacement*.

2.5.8 The random variable X measures the concentration of ethanol in a chemical solution, and the random variable Y measures the acidity of the solution. They have a joint probability density function

$$f(x, y) = A(20 - x - 2y)$$

for $0 \le x \le 5$ and $0 \le y \le 5$, and $f(x, y) = 0$ elsewhere.

(a) What is the value of A?

(b) What is $P(1 \le X \le 2, 2 \le Y \le 3)$?

(c) Construct the marginal probability density functions $f_X(x)$ and $f_Y(y)$.

(d) Are the ethanol concentration and the acidity independent?

(e) What are the expectation and the variance of the ethanol concentration?

(f) What are the expectation and the variance of the acidity?

(g) If the ethanol concentration is 3, what is the conditional probability density function of the acidity?

(h) What is the covariance between the ethanol concentration and the acidity?

(i) What is the correlation between the ethanol concentration and the acidity?

2.5.9 Two safety inspectors inspect a new building and assign it a "safety score" of 1, 2, 3, or 4. Suppose that the random variable X is the score assigned by the first inspector and the random variable Y is the score assigned by the second inspector, and that they have a joint probability mass function given in Figure 2.69.

			X		
		1	2	3	4
	1	0.09	0.03	0.01	0.01
Y	2	0.02	0.15	0.03	0.01
	3	0.01	0.01	0.24	0.04
	4	0.00	0.01	0.02	0.32

FIGURE 2.69

Joint probability mass function for safety scores

(a) What is the probability that both inspectors assign the same safety score?

(b) What is the probability that the second inspector assigns a higher safety score than the first inspector?

(c) What are the marginal probability mass function, expectation, and variance of the score assigned by the first inspector?

(d) What are the marginal probability mass function, expectation, and variance of the score assigned by the second inspector?

(e) Are the scores assigned by the two inspectors independent of each other? Would you expect them to be independent? How would you interpret the situation if they were independent?

(f) If the first inspector assigns a score of 3, what is the marginal probability mass function of the score assigned by the second inspector?

(g) What is the covariance of the scores assigned by the two inspectors?

(h) What is the correlation between the scores assigned by the two inspectors? If you are responsible for training the safety inspectors to perform proper safety evaluations of buildings, what correlation value would you like there to be between the scores of two safety inspectors?

2.5.10 Joint probability distributions of three or more random variables can be interpreted by extending the ideas in this section. For example, suppose that three continuous random variables X, Y, and Z have a joint probability density function

$$f(x, y, z) = \frac{3xyz^2}{32}$$

for $0 \leq x \leq 2$, $0 \leq y \leq 2$, and $0 \leq z \leq 2$, and $f(x, y, z) = 0$ elsewhere.

(a) Establish that this is a valid joint probability density function by showing that it is always positive within the state space $0 \leq x \leq 2$, $0 \leq y \leq 2$, and $0 \leq z \leq 2$, and that the total probability is equal to 1.

(b) In general,

$$P(a \leq X \leq b, c \leq Y \leq d, e \leq Z \leq f)$$

$$= \int_{x=a}^{b} \int_{y=c}^{d} \int_{z=e}^{f} f(x, y, z)\, dx\, dy\, dz$$

What is $P(0 \leq X \leq 1, 0.5 \leq Y \leq 1.5, 1 \leq Z \leq 2)$?

(c) The marginal probability density function of a particular random variable can be calculated from the joint probability density function by integrating over all values of the other random variables. What is the marginal probability density function of the random variable X?

2.6   Combinations and Functions of Random Variables

2.6.1   Linear Functions of a Random Variable

A **linear function** of a random variable X is another random variable

$$Y = aX + b$$

for some numbers $a, b \in \mathbb{R}$. It is a general result that

$$E(Y) = aE(X) + b$$

so that the expectation of the random variable Y is just the same linear function of the expectation of the random variable X. However,

$$\mathrm{Var}(Y) = a^2 \mathrm{Var}(X)$$

and the standard deviation of Y, σ_Y, is given by

$$\sigma_Y = |a|\sigma_X$$

where σ_X is the standard deviation of X. Notice that the "shift" amount b does not influence the variance of Y and that the "scale" amount a is *squared* in the relationship between the two variances.

Linear Functions of a Random Variable

If X is a random variable and $Y = aX + b$ for some numbers $a, b \in \mathbb{R}$, then

$$E(Y) = aE(X) + b$$

and

$$\mathrm{Var}(Y) = a^2 \mathrm{Var}(X)$$

An important application of this result concerns the "standardization" of a random variable to have a zero mean and a unit variance. If a random variable X has an expectation of μ and a variance of σ^2, notice that the result above implies that the random variable

$$Y = \frac{X - \mu}{\sigma} = \frac{1}{\sigma}X + \left(-\frac{\mu}{\sigma}\right)$$

has an expectation of 0 and a variance of 1. This standardization of a random variable is often useful, and notice that it is performed by first subtracting the mean from the random variable and then dividing by the standard deviation of the random variable.

The actual cumulative distribution function $F_Y(y)$ of the random variable $Y = aX + b$ can be obtained from the cumulative distribution function $F_X(x)$ of the random variable X by noting that

$$F_Y(y) = P(Y \le y) = P(aX + b \le y) = P(aX \le y - b)$$

If $a > 0$, this gives

$$F_Y(y) = P\left(X \le \frac{y - b}{a}\right) = F_X\left(\frac{y - b}{a}\right)$$

and if $a < 0$, this gives

$$F_Y(y) = P\left(X \ge \frac{y - b}{a}\right) = 1 - F_X\left(\frac{y - b}{a}\right)$$

Example 21 **Test Score** **Standardization**	Suppose that the raw scores X from a particular testing procedure are distributed between -5 and 20 with an expected value of 10 and a variance of 7. In order to *standardize* the scores so that they lie between 0 and 100, the linear transformation

$$Y = 4X + 20$$

is applied to the scores. This means, for example, that a raw score of $x = 12$ corresponds to a standardized score of $y = (4 \times 12) + 20 = 68$.

The expected value of the standardized scores is then known to be

$$E(Y) = 4E(X) + 20 = (4 \times 10) + 20 = 60$$

with a variance of

$$\mathrm{Var}(Y) = 4^2 \mathrm{Var}(X) = 4^2 \times 7 = 112$$

The standard deviation of the standardized scores is $\sigma_Y = \sqrt{112} = 10.58$, which is $4 \times \sigma_X = 4 \times \sqrt{7}$.

Example 22 **Chemical Reaction** **Temperatures**	The temperature X in degrees Fahrenheit of a particular chemical reaction is known to be distributed between $220°$ and $280°$ with a probability density function of

$$f_X(x) = \frac{x - 190}{3600}$$

This is shown in Figure 2.70 together with the expectation and standard deviation of the chemical reaction temperature, which are easily calculated to be $E(X) = 255°$ and $\sigma_X = 16.58°$. The variance of the chemical reaction temperature is $\mathrm{Var}(X) = 275$. In addition, the cumulative distribution function of the chemical reaction temperature is

$$F_X(x) = \int_{z=220}^{x} f_X(z)\,dz = \frac{(x - 190)^2}{7200} - \frac{1}{8}$$

Suppose that a chemist wishes to convert the temperatures to degrees Centigrade. If the random variable Y measures the reaction temperature in degrees Centigrade, then it is obtained

$f(x) = \dfrac{x - 190}{3600}$

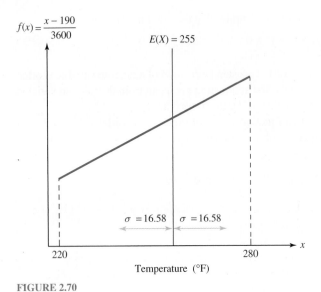

FIGURE 2.70

Probability density function for chemical reaction temperature in degrees Fahrenheit

$f(y) = \dfrac{9y}{10000} - \dfrac{79}{1000}$

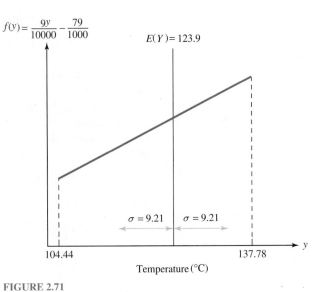

FIGURE 2.71

Probability density function for chemical reaction temperature in degrees Centigrade

as the following linear function of the random variable X

$$Y = \frac{5}{9}X - \frac{160}{9}$$

Notice that $x = 220°F$ corresponds to

$$y = \frac{5}{9} \times 220 - \frac{160}{9} = 104.44°C$$

and that $x = 280°F$ corresponds to

$$y = \frac{5}{9} \times 280 - \frac{160}{9} = 137.78°C$$

Since $a = 5/9$ is positive, the cumulative distribution function of Y is

$$F_Y(y) = F_X\left(\frac{y - b}{a}\right) = F_X\left(\frac{y + \frac{160}{9}}{\frac{5}{9}}\right) = \frac{(9y - 790)^2}{180{,}000} - \frac{1}{8}$$

The probability density function of Y is obtained by differentiation to be

$$f_Y(y) = \frac{d}{dy}F_Y(y) = \frac{9y}{10{,}000} - \frac{79}{1000}$$

for $104.44 \le y \le 137.78$, as shown in Figure 2.71. The expectation and variance of Y can be obtained directly from its probability density function or from the relationships

$$E(Y) = \frac{5}{9}E(X) - \frac{160}{9} = \frac{5}{9} \times 255 - \frac{160}{9} = 123.9$$

and

$$\mathrm{Var}(Y) = \left(\frac{5}{9}\right)^2 \mathrm{Var}(X) = \left(\frac{5}{9}\right)^2 \times 275 = 84.88$$

2.6.2   Linear Combinations of Random Variables

Given two random variables X_1 and X_2, it is often useful to consider the random variable obtained as their sum. It is a general result that

$$E(X_1 + X_2) = E(X_1) + E(X_2)$$

The expectation of the sum of two random variables is equal to the sum of the expectations of the two random variables.

Also, in general,

$$\text{Var}(X_1 + X_2) = \text{Var}(X_1) + \text{Var}(X_2) + 2\text{Cov}(X_1, X_2)$$

Notice that if the two random variables are independent so that their covariance is 0, then the variance of their sum is equal to the sum of their two variances.

The variance of the sum of two independent random variables is equal to the sum of the variances of the two random variables.

These are nice simple results, but it is important to remember that whereas the expectation of the sum of two random variables is *always* equal to the sum of the expectations of the two random variables, the variance of the sum of two random variables is equal to the sum of their variances only when they are *independent* random variables.

Sums of Random Variables

If X_1 and X_2 are two random variables, then

$$E(X_1 + X_2) = E(X_1) + E(X_2)$$

and

$$\text{Var}(X_1 + X_2) = \text{Var}(X_1) + \text{Var}(X_2) + 2\text{Cov}(X_1, X_2)$$

If X_1 and X_2 are independent random variables so that $\text{Cov}(X_1, X_2) = 0$, then

$$\text{Var}(X_1 + X_2) = \text{Var}(X_1) + \text{Var}(X_2).$$

Consider now a sequence of random variables $X_1, \ldots, X_n$ together with some constants $a_1, \ldots, a_n$ and b, and define a new random variable Y to be the *linear* combination

$$Y = a_1 X_1 + \cdots + a_n X_n + b$$

Linear combinations of random variables like this are important in many situations, and it is useful to obtain some general results for them.

The expectation of the linear combination is

$$E(Y) = a_1 E(X_1) + \cdots + a_n E(X_n) + b$$

which is just the linear combination of the expectations of the random variables X_i. Furthermore, if the random variables $X_1, \ldots, X_n$ are *independent* of one another, then

$$\text{Var}(Y) = a_1^2 \, \text{Var}(X_1) + \cdots + a_n^2 \, \text{Var}(X_n)$$

Again, notice that the "shift" amount b does not influence the variance of Y and that the coefficients a_i are *squared* in this expression.

Linear Combinations of Random Variables

If $X_1, \ldots, X_n$ is a sequence of random variables and $a_1, \ldots, a_n$ and b are constants, then

$$E(a_1 X_1 + \cdots + a_n X_n + b) = a_1 E(X_1) + \cdots + a_n E(X_n) + b$$

If, in addition, the random variables are independent, then

$$\text{Var}(a_1 X_1 + \cdots + a_n X_n + b) = a_1^2 \text{Var}(X_1) + \cdots + a_n^2 \text{Var}(X_n)$$

As an illustration of these results, suppose that two independent random variables X_1 and X_2 both have an expectation of μ and a variance of σ^2. Suppose that

$$Y_1 = X_1 + X_2$$

is the sum of the two random variables and that

$$Y_2 = X_1 - X_2$$

is their difference. Then

$$E(Y_1) = E(X_1) + E(X_2) = 2\mu \qquad \text{and} \qquad E(Y_2) = E(X_1) - E(X_2) = 0$$

so that the sum has an expectation of 2μ and the difference has an expectation of 0.

Since X_1 and X_2 are independent random variables,

$$\text{Var}(Y_1) = \text{Var}(X_1) + \text{Var}(X_2) = 2\sigma^2$$

In addition,

$$\text{Var}(Y_2) = \text{Var}(X_1 + (-1)X_2) = \text{Var}(X_1) + (-1)^2 \text{Var}(X_2) = 2\sigma^2$$

so that the variances of the sum and the difference of X_1 and X_2 are *both* equal to $2\sigma^2$. It is also interesting to note that there is "more" variability in the random variables Y_1 and Y_2 than in the random variables X_1 and X_2 since their variances are twice as large. This illustrates the general rule:

Adding or subtracting independent random variables increases variability.

As another illustration, suppose that $X_1, \ldots, X_n$ is a sequence of independent random variables each with an expectation μ and a variance σ^2, and consider the **average**

$$\bar{X} = \frac{X_1 + \cdots + X_n}{n}$$

In particular, the random variables may represent the observations in a sample of size n from some population of interest. Then

$$E(\bar{X}) = E\left(\frac{1}{n}X_1 + \cdots + \frac{1}{n}X_n\right) = \frac{1}{n}E(X_1) + \cdots + \frac{1}{n}E(X_n)$$

$$= \frac{1}{n}\mu + \cdots + \frac{1}{n}\mu = \mu$$

and

$$\text{Var}(\bar{X}) = \text{Var}\left(\frac{1}{n}X_1 + \cdots + \frac{1}{n}X_n\right) = \left(\frac{1}{n}\right)^2 \text{Var}(X_1) + \cdots + \left(\frac{1}{n}\right)^2 \text{Var}(X_n)$$

$$= \left(\frac{1}{n}\right)^2 \sigma^2 + \cdots + \left(\frac{1}{n}\right)^2 \sigma^2 = \frac{\sigma^2}{n}$$

Notice that whereas the expectation of the average is equal to the expectations of the individual random variables μ, the variance of the average is *reduced* to σ^2/n. This illustrates the general rule:

Averaging independent random variables reduces variability.

Averaging Independent Random Variables

Suppose that $X_1, \ldots, X_n$ is a sequence of independent random variables each with an expectation μ and a variance σ^2, and with an **average**

$$\bar{X} = \frac{X_1 + \cdots + X_n}{n}$$

Then

$$E(\bar{X}) = \mu$$

and

$$\text{Var}(\bar{X}) = \frac{\sigma^2}{n}$$

Figure 2.72 summarizes how the mean and the variance change for simple combinations of random variables. Notice that (1) doubling a random variable has more variability than adding two independent realizations of the random variable; (2) the sum of two independent realizations of a random variable has the same variability as the difference between the two random variables, and these are larger than the variability of the individual random variable; (3) averaging two realizations of a random variable has smaller variability than the individual random variable.

In summary, while the expectation of a combination of random variables is intuitively obvious and is easy to calculate, be careful when calculating the variance because it can be less intuitive.

Example 23

Piston Head Construction

A circular piston head is designed to slide smoothly within a cylinder. However, there is some variability in the piston head and cylinder sizes about their specified dimensions, and the company that manufactures them is interested in how well the piston heads actually fit within the cylinders.

Suppose that the random variable X_1 measures the radius of a piston head, and that it has an expected value of 30.00 mm and a standard deviation of 0.05 mm. Also, suppose that the random variable X_2 measures the inside radius of a cylinder, with an expected value of 30.25 mm and a standard deviation of 0.06 mm. The "gap" between the piston head and the cylinder is $Y = X_2 - X_1$, as illustrated in Figure 2.73.

FIGURE 2.72

Expectations and variances of combinations of a random variable

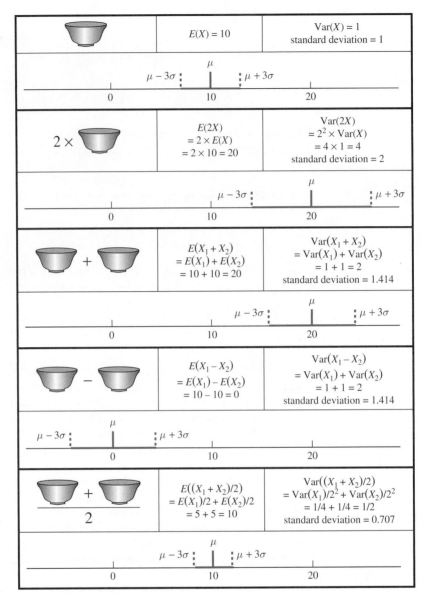

The expected value of the gap is

$$E(Y) = E(X_2) - E(X_1) = 30.25 - 30.00 = 0.25 \text{ mm}$$

Also, since the sizes of a piston head and a cylinder can reasonably be expected to be independent, the variance of the gap is

$$Var(Y) = Var(X_2 + (-1)X_1) = Var(X_2) + (-1)^2 Var(X_1)$$
$$= 0.06^2 + 0.05^2 = 0.0061$$

The standard deviation of the gap is therefore $\sigma_Y = \sqrt{0.0061} = 0.078$ mm, which is larger than the standard deviations of both the piston heads and the cylinders.

FIGURE 2.73

Cross-sectional view of piston head within cylinder

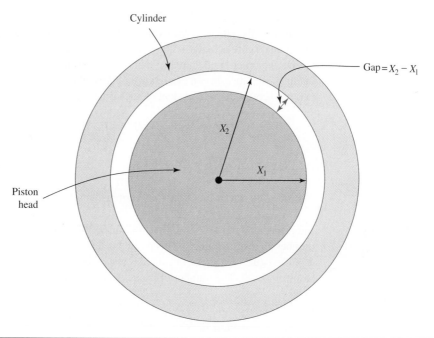

Cylinder

Gap $= X_2 - X_1$

X_2

X_1

Piston
head

Example 21

**Test Score
Standardization**

Suppose that in a certain examination procedure candidates must take two tests, and that X_1 measures the score of a candidate on test 1 and X_2 measures the score of a candidate on test 2. Furthermore, suppose that the scores on test 1 are distributed between 0 and 30 with an expectation of 18 and a variance of 24, and that the scores on test 2 are distributed between -10 and 50 with an expectation of 30 and a variance of 60.

The examining board wishes to standardize each test score to lie between 0 and 100, and to then calculate a final score out of 100 that is weighted 2/3 from test 1 and 1/3 from test 2. The standardized scores of the two tests are

$$Y_1 = \frac{10}{3} X_1 \qquad \text{and} \qquad Y_2 = \frac{5}{3} X_2 + \frac{50}{3}$$

so that the final score is

$$Z = \frac{2}{3} Y_1 + \frac{1}{3} Y_2 = \frac{20}{9} X_1 + \frac{5}{9} X_2 + \frac{50}{9}$$

Thus a candidate receiving a score of $x_1 = 11$ on test 1 and a score of $x_2 = 2$ on test 2 receives a final score of

$$z = \left(\frac{20}{9} \times 11 \right) + \left(\frac{5}{9} \times 2 \right) + \frac{50}{9} = 31.11$$

The expected value of the final score is

$$E(Z) = \frac{20}{9} E(X_1) + \frac{5}{9} E(X_2) + \frac{50}{9}$$

$$= \left(\frac{20}{9} \times 18 \right) + \left(\frac{5}{9} \times 30 \right) + \frac{50}{9} = 62.22$$

What is the variance of the final score? At this point it is important to remember that the scores a candidate receives on the two tests may not be independent. In fact, they will be independent only if they measure two *unrelated* qualities of the candidate. If, for example, test 1 measures a candidate's proficiency at *probability* and test 2 measures a candidate's proficiency at *statistics*, then the test scores should not be independent and the calculation of the variance of the final score actually requires knowledge of the covariance between the two test scores.

On the other hand, suppose that test 1 measures a candidate's proficiency at *probability* but that test 2 measures a candidate's *athletic* abilities, then it is reasonable to assume that the two test scores are independent of each other. In this case, the variance of the final score is

$$\text{Var}(Z) = \text{Var}\left(\frac{20}{9}X_1 + \frac{5}{9}X_2 + \frac{50}{9}\right) = \left(\frac{20}{9}\right)^2 \times \text{Var}(X_1) + \left(\frac{5}{9}\right)^2 \times \text{Var}(X_2)$$

$$= \left(\frac{20}{9}\right)^2 \times 24 + \left(\frac{5}{9}\right)^2 \times 60 = 137.04$$

The standard deviation of the final score is therefore $\sigma_Z = \sqrt{137.04} = 11.71$.

GAMES OF CHANCE

Suppose that a fair coin is tossed 100 times. What are the expectation and the variance of the number of heads obtained? This problem can easily be solved by defining the random variable X_i to be equal to 1 if the ith coin toss results in a head, and to be equal to 0 if the ith coin toss results in a tail. The number of heads obtained is therefore

$$Y = X_1 + \cdots + X_{100}$$

The expectation of X_i is clearly $1/2$, and the variance can easily be shown to be $1/4$. Therefore,

$$E(Y) = E(X_1) + \cdots + E(X_{100}) = \frac{1}{2} + \cdots + \frac{1}{2} = \frac{100}{2} = 50$$

and since the coin tosses are independent,

$$\text{Var}(Y) = \text{Var}(X_1) + \cdots + \text{Var}(X_{100}) = \frac{1}{4} + \cdots + \frac{1}{4} = \frac{100}{4} = 25$$

Hence, the expected number of heads obtained is 50 with a standard deviation of 5. This is an example of the binomial distribution that is discussed in Section 3.1.

Suppose now that a die is rolled 10 times and the *average* of the 10 scores is recorded. The probability mass function of the average score is tedious to compute, but again, the expectation and variance of the average score are easily calculated.

Since the score from the roll of a single die has an expected value of $\mu = 3.5$ and a variance of $\sigma^2 = 35/12$ (see Section 2.4.2), the average score also has an expected value of 3.5 but has a variance of $\sigma^2/10 = 35/120 = 7/24$. The standard deviation of the average score is $\sqrt{7/24} = 0.54$.

2.6.3   Nonlinear Functions of a Random Variable

A nonlinear function of a random variable X is another random variable $Y = g(X)$ for some nonlinear function g. Some useful examples might be $Y = X^2$, $Y = \sqrt{X}$, $Y = e^X$, and $Y = 1/(1 + X)$. There are no general results that relate the expectation and variance of the random variable Y to the expectation and variance of the random variable X. Usually, the easiest way to construct the probability distribution of the random variable Y is to construct its cumulative distribution function from the cumulative distribution function of the random variable X.

For example, suppose that the random variable X is distributed between 0 and 1 with a probability density function of

$$f_X(x) = 1$$

for $0 \leq x \leq 1$ and $f(x) = 0$ elsewhere. Since the probability density function is symmetric about $x = 0.5$, the expectation of X is 0.5. Also, the cumulative distribution function of X is

$$F_X(x) = x$$

for $0 \leq x \leq 1$.

Consider the random variable Y defined to be

$$Y = e^X$$

Clearly, Y takes values between 1 and $e = 2.718$, and its cumulative distribution function is

$$F_Y(y) = P(Y \leq y) = P(e^X \leq y) = P(X \leq \ln(y)) = F_X(\ln(y)) = \ln(y)$$

The probability density function of Y is obtained by differentiation to be

$$f_Y(y) = \frac{dF_Y(y)}{dy} = \frac{1}{y}$$

for $1 \leq y \leq 2.718$, which is shown together with the probability density function of X in Figure 2.74. Notice that

$$E(Y) = \int_{z=1}^{2.718} z \, f_Y(z) \, dz = \int_{z=1}^{2.718} 1 \, dz = 2.718 - 1 = 1.718$$

FIGURE 2.74

Comparison of the probability density functions of X and $Y = e^x$

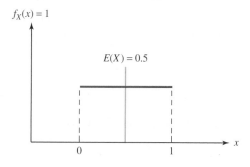

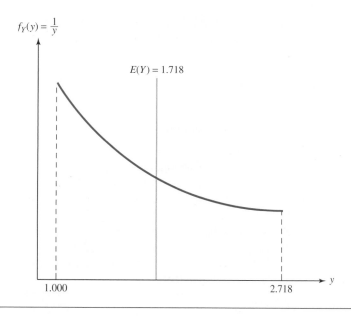

and that, even though $Y = e^X$,

$$E(Y) \neq e^{E(X)} = e^{0.5} = 1.649$$

Example 23
Piston Head
Construction

Suppose that X_1, the radius of the piston heads, actually takes values between 29.9 mm and 30.1 mm with a probability density function of

$$f_{X_1}(x) = 5$$

for $29.9 \leq x \leq 30.1$ and $f(x) = 0$ elsewhere. The cumulative distribution function of X_1 is therefore

$$F_{X_1}(x) = \int_{z=29.9}^{x} f_{X_1}(z)\, dz = 5x - 149.5$$

for $29.9 \leq x \leq 30.1$.

The piston manufacturers are particularly interested in the *area* of the piston head since this directly affects the performance of the piston. The area of the piston head is

$$Y = \pi X_1^2$$

which takes values between $\pi 29.9^2 = 2808.6$ mm^2 and $\pi 30.1^2 = 2846.3$ mm^2. The cumulative distribution function of the piston head area is

$$F_Y(y) = P(Y \leq y) = P\left(\pi X_1^2 \leq y\right)$$
$$= P\left(X_1 \leq \sqrt{y/\pi}\right) = F_{X_1}\left(\sqrt{y/\pi}\right) = 2.821\sqrt{y} - 149.5$$

so that the probability density function is

$$f_Y(y) = \frac{dF_Y(y)}{dy} = \frac{1.410}{\sqrt{y}}$$

for $2808.6 \leq y \leq 2846.3$, as shown in Figure 2.75.

FIGURE 2.75

Probability density function of piston head area

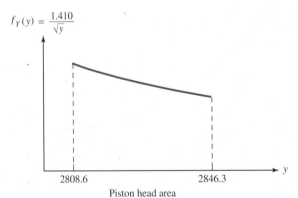

$f_Y(y) = \dfrac{1.410}{\sqrt{y}}$

2808.6   2846.3 y

Piston head area

■ 2.6.4 Problems

2.6.1 Suppose that the random variables X, Y, and Z are independent with $E(X) = 2$, $\text{Var}(X) = 4$, $E(Y) = -3$, $\text{Var}(Y) = 2$, $E(Z) = 8$, and $\text{Var}(Z) = 7$. Calculate the expectation and variance of the following random variables.
(a) $3X + 7$ (b) $5X - 9$
(c) $2X + 6Y$ (d) $4X - 3Y$
(e) $5X - 9Z + 8$ (f) $-3Y - Z - 5$
(g) $X + 2Y + 3Z$ (h) $6X + 2Y - Z + 16$

2.6.2 Recall that for any function $g(X)$ of a random variable X,

$$E(g(X)) = \int g(x) f(x) \, dx$$

where $f(x)$ is the probability density function of X. Use this result to show that

$$E(aX + b) = aE(X) + b$$

and

$$\text{Var}(aX + b) = a^2 \text{Var}(X)$$

2.6.3 Suppose that X_1, X_2, and X_3 are independent random variables each with a mean of μ and a variance of σ^2. Compare the means and variances of

$$Y = 3X_1 \quad \text{and} \quad Z = X_1 + X_2 + X_3$$

2.6.4 A machine part is assembled by fastening *two* components of type A and *one* component of type B end to end. Suppose that the lengths of components of type A have an expectation of 37.0 mm and a standard deviation of 0.7 mm, whereas the lengths of components of type B have an expectation of 24.0 mm and a standard deviation of 0.3 mm. What are the expectation and variance of the length of the machine part?

2.6.5 In a game a player either wins $10 with a probability of $1/8$ or loses $1 with a probability of $7/8$. What are the expectation and variance of the total net winnings of a player after 50 turns at the game?

2.6.6 The weight of a certain type of brick has an expectation of 1.12 kg with a standard deviation of 0.03 kg.
(a) What are the expectation and variance of the average weight of 25 bricks randomly selected?
(b) How many bricks need to be selected so that their average weight has a standard deviation of no more than 0.005 kg?

2.6.7 Ten cards are drawn *with replacement* from a pack of cards. What are the expectation and variance of the

number of Aces drawn? If the drawings are made *without replacement*, what is the expected number of Aces drawn?

2.6.8 Suppose that the random variable X has a probability density function

$$f(x) = 2x$$

for $0 \le x \le 1$. Find the probability density function and the expectation of the random variable Y in the following cases.
(a) $Y = X^3$ (b) $Y = \sqrt{X}$
(c) $Y = 1/(1 + X)$ (d) $Y = 2^X$

2.6.9 The radius of a soap bubble has a probability density function

$$f(r) = A(1 - (r - 1)^2)$$

for $0 \le r \le 2$.
(a) What is the value of A?
(b) What is the probability density function of the *volume* of the soap bubble?
(c) What is the expected value of the volume of the soap bubble?

2.6.10 A rod of length L is bent until it snaps in two. The point of breakage X, as measured from one end of the rod, has a probability density function

$$f(x) = Ax(L - x)$$

for $0 \le x \le L$.
(a) What is the value of A?
(b) What is the probability density function of the *difference* in the lengths of the two pieces of the rod?
(c) What is the expected difference in the lengths of the two pieces of the rod?

• **2.6.11** If $x is invested in mutual fund A, the annual return has an expectation of $0.1x and a standard deviation of $0.02x. If $x is invested in mutual fund B, the annual return has an expectation of $0.1x and a standard deviation of $0.03x. Suppose that the returns on the two funds are independent of each other and that I have $1000 to invest.
(a) What are the expectation and variance of my annual return if I invest all my money in fund A?
(b) What are the expectation and variance of my annual return if I invest all my money in fund B?
(c) What are the expectation and variance of my total annual return if I invest half of my money in fund A and half in fund B?

(d) Suppose I invest $$x$ in fund A and the rest of my money in fund B. What value of x *minimizes* the variance of my total annual return?

Explain why your answers illustrate the importance of *diversity* in an investment strategy.

2.6.12 Recall Problems 2.2.11, 2.3.17, and 2.4.13 concerning the resistance of an electrical component. Suppose that five of the components are used in an electrical circuit so that the total resistance is the sum of the five individual resistances. Furthermore, suppose that it is reasonable to assume that the resistances of the five components are independent of each other. What will be the expected value and the standard deviation of the total resistance?

2.6.13 Suppose that items from a manufacturing process are subject to three separate evaluations, and that the results of the first evaluation X_1 have a mean value of 59 with a standard deviation of 10, the results of the second evaluation X_2 have a mean value of 67 with a standard deviation of 13, and the results of the third evaluation X_3 have a mean value of 72 with a standard deviation of 4. In addition, suppose that the results of the three evaluations can be taken to be independent of each other.
   (a) If a final evaluation score is obtained as the average of the three evaluations $X = (X_1 + X_2 + X_3)/3$, what are the mean and the standard deviation of the final evaluation score?
   (b) If a final evaluation score is obtained as the weighted average of the three evaluations $X = 0.4X_1 + 0.4X_2 + 0.2X_3$, what are the mean and the standard deviation of the final evaluation score?

2.6.14 The random variable X has an expectation of 77 and a standard deviation of 9. Find the values of a and b such that the random variable $Y = a + bX$ has an expectation of 1000 and a standard deviation of 10.

2.6.15 Suppose that components are manufactured such that their heights are independent of each other with $\mu = 65.90$ and $\sigma = 0.32$.
   (a) What are the mean and the standard deviation of the average height of five components?
   (b) If eight components are stacked on top of each other, what are the mean and the standard deviation of the total height?

2.6.16 An object has a weight W. When it is weighed with machine 1, a value X_1 is obtained. When it is weighed with machine 2, a value X_2 is obtained. The value X_1 has a mean W and a standard deviation 3. The value X_2 has a

mean W and a standard deviation 4. The values X_1 and X_2 are independent.
   (a) Suppose that an engineer uses the value $A = (X_1 + X_2)/2$ to estimate the weight. What are the expectation and the standard deviation of A?
   (b) Suppose that an engineer uses the value $B = \delta X_1 + (1 - \delta)X_2$ to estimate the weight. What value of δ gives an estimate B with the smallest standard deviation? What is this smallest standard deviation?

2.6.17 A fair six-sided die is rolled 80 times and the sum of the 80 scores is calculated. What are the expectation and the standard deviation of this total score?

2.6.18 A relay race is run between team A and team B. Each team has four runners, who successively run around a track and pass a baton to the next runner in their team when they have finished. The team finishes when the fourth runner has completed the run. The runners on team A take times with an expectation of 33.2 seconds and a standard deviation of 1.4 seconds, while the runners on team B take times with an expectation of 33.0 seconds and a standard deviation of 1.3 seconds. The times of all eight runners are independent of each other.
   (a) What are the expectation and the standard deviation of the total time taken by team A?
   (b) An official measures the total time taken by team A and subtracts from it the total time taken by team B? What are the expectation and the standard deviation of this difference?
   (c) The manager of team A measures the time taken by the first runner on team A and subtracts from it the average of the times of the other three runners on team A. What are the expectation and the standard deviation of this difference?

2.6.19 Suppose that a temperature has a mean of 110°F and a standard deviation of 2.2°F. What are the mean and the standard deviation in degrees Centigrade? ($F = 9C/5 + 32$)

2.6.20 A person's cholesterol level C can be measured by three different tests. Test-α returns a value X_α with a mean C and a standard deviation 1.2, test-β returns a value X_β with a mean C and a standard deviation 2.4, and test-γ returns a value X_γ with a mean C and a standard deviation 3.1. Suppose that the three test results are independent. If a doctor decides to use the weighted average $0.5X_\alpha + 0.3X_\beta + 0.2X_\gamma$ what is the standard deviation of the cholesterol level obtained by the doctor?

2.6.21 Suppose that the thicknesses of electrical components have a standard deviation of 56 microns. How many components need to be taken so that the standard deviation of their average thickness is no larger than 10 microns?

2.6.22 Suppose that the impurity levels of water samples taken from a particular source are independent with a mean value of 3.87 and a standard deviation of 0.18.

(a) What are the mean and the standard deviation of the sum of the impurity levels from two water samples?

(b) What are the mean and the standard deviation of the sum of the impurity levels from three water samples?

(c) What are the mean and the standard deviation of the average of the impurity levels from four water samples?

(d) If the impurity levels of two water samples are averaged, and the result is subtracted from the impurity level of a third sample, what are the mean and the standard deviation of the resulting value?

2.7 Case Study: Microelectronic Solder Joints

Figure 2.76 shows the probability mass function of the number of extrusions on an assembly with a large number of solder joints. These extrusions are defined to be cases where the gap between the solder of two adjacent solder joints is less than a specified value, and their presence is likely to reduce the lifetime of the assembly. The expected number of extrusions is

$$E(X) = (0 \times 0.721) + (1 \times 0.133) + (2 \times 0.083) + (3 \times 0.055)$$
$$+ (4 \times 0.007) + (5 \times 0.001) = 0.497$$

and since

$$E(X^2) = (0^2 \times 0.721) + (1^2 \times 0.133) + (2^2 \times 0.083)$$
$$+ (3^2 \times 0.055) + (4^2 \times 0.007) + (5^2 \times 0.001) = 1.097$$

the variance is

$$\text{Var}(X) = 1.097 - 0.497^2 = 0.850$$

which gives a standard deviation of $\sqrt{0.850} = 0.922$. Consequently, the average number of extrusions per assembly is about 0.5, with a standard deviation of just less than 1. Furthermore, it is clear from the probability mass function that there is a probability of more than 99% that an assembly will have no more than 3 extrusions.

If a batch of these products consist of 250 assemblies, then the total number of extrusions in the complete batch has an expected value

$$E(X) = 250 \times 0.497 = 124.25$$

and a variance

$$\text{Var}(X) = 250 \times 0.850 = 212.50$$

so that the standard deviation is $\sqrt{212.50} = 14.58$.

FIGURE 2.76

The probability mass function of the number of solder joint extrusions in an assembly

x_i	0	1	2	3	4	5
p_i	0.721	0.133	0.083	0.055	0.007	0.001

2.8 Supplementary Problems

2.8.1 A box contains four red balls and two blue balls. Balls are drawn at random without replacement, and the random variable X measures the total number of balls selected up to the point when both blue balls have been selected.
(a) What is the probability mass function of X?
(b) What is the expected value of X?

2.8.2 The probability mass function of the number of calls taken by a switchboard within 1 minute is given in Figure 2.77.

x_i	0	1	2	3	4	5	6
p_i	0.21	0.39	0.18	0.16	0.03	0.02	0.01

FIGURE 2.77

Probability mass function of the number of calls taken by a switchboard

(a) Compute and sketch the cumulative distribution function of the number of calls taken within a minute.
(b) What is the expected number of calls taken within a minute?
(c) What is the variance of the number of calls taken within a minute?
(d) If the numbers of calls taken in two different minutes are independent, what are the expectation and variance of the number of calls taken within one hour?

2.8.3 A box initially contains two red balls and two blue balls. At each turn, one of the balls within the box is selected at random and then replaced together with another ball of the *opposite* color.
(a) After three turns, what is the distribution of the number of red balls in the box?
(b) After three turns, what are the expectation and variance of the number of red balls in the box?
(c) Repeat parts (a) and (b) if at each turn the selected ball is replaced with a ball of the *same* color.

2.8.4 Suppose that an Ace, a King, a Queen, and a Jack are all worth 15 points, and that other cards are all worth their face value. If a hand of 13 cards is dealt, what is the expected value of the total score of the hand?

2.8.5 The acidity level X of a soil sample has a probability density function
$$f(x) = A\left(\frac{3}{2}\right)^x$$
for $1 \le x \le 11$.

(a) What is the value of A?
(b) Compute and sketch the cumulative distribution function of X.
(c) What is the median soil acidity level?
(d) What is the interquartile range of the soil acidity levels?

2.8.6 Suppose that the random variables X and Y have a joint probability density function
$$f(x, y) = 4x(2 - y)$$
for $0 \le x \le 1$ and $1 \le y \le 2$.
(a) What is the marginal probability density function of X?
(b) Are the random variables X and Y independent?
(c) What is $\mathrm{Cov}(X, Y)$?
(d) What is the probability density function of X conditional on $Y = 1.5$?

2.8.7 The density X of a chemical solution is
$$f(x) = A\left(x + \frac{2}{x}\right)$$
for $5 \le x \le 10$.
(a) What is the value of A?
(b) Compute and sketch the cumulative distribution function of the density.
(c) What is the mean density level?
(d) What is the variance of the density level?
(e) What is the median density level?
(f) What is the interquartile range of the density level?
(g) Suppose that 10 chemical solutions are made independently of each other. What are the mean and variance of the average density of the ten solutions?

2.8.8 Recall that
$$\mathrm{Var}(aX + b) = a^2 \mathrm{Var}(X)$$
and
$$\mathrm{Var}(X_1 + X_2) = \mathrm{Var}(X_1) + \mathrm{Var}(X_1)$$
if the random variables X_1 and X_2 are independent. Use these results to show that
$$\mathrm{Var}(a_1 X_1 + \cdots + a_n X_n + b)$$
$$= a_1^2 \mathrm{Var}(X_1) + \cdots + a_n^2 \mathrm{Var}(X_n)$$
if the random variables X_i are independent.

2.8.9 Suppose that the scores from a test have a mean of 75 and a standard deviation of 12. How should the scores be standardized so that they have a mean of 100 and a standard deviation of 20?

2.8.10 If the random variables X and Y are related through the expression

$$Y = aX + b$$

show that they have a correlation of 1 if $a > 0$ and of -1 if $a < 0$.

2.8.11 An insurance company charges a customer an annual premium of \$100, and there is a probability of 0.9 that the customer will not need to make a claim. If the customer does make a claim, the amount of the claim \$$X$ has a probability density function

$$f(x) = \frac{x(1800 - x)}{972,000,000}$$

for $0 \leq x \leq 1800$. Each customer also incurs administrative costs to the insurance company of \$5. If the insurance company has 10,000 customers, what is its expected annual profit? Would you expect the customers' claims to be independent of each other?

2.8.12 Suppose that telephone calls on a particular line have an expected length of 320 seconds with a standard deviation of 63 seconds, and suppose that the call lengths are independent of each other.
(a) What are the expectation and the standard deviation of the total length of five calls?
(b) What are the expectation and the standard deviation of the average length of ten calls?

2.8.13 In an evaluation procedure, n items are ranked in order of effectiveness. No ties are allowed, so that the ranks are the positive integers from 1 to n. Suppose that all of the possible rankings are equally likely.
(a) What is the probability mass function of the rank of a particular item?
(b) What is the expectation of the rank of a particular item?
(c) What is the variance of the rank of a particular item?

2.8.14 Nancy and Tom have to take a bus ride. The time taken by a bus for the specified journey has an expectation of 87 minutes with a standard deviation of 3 minutes, and the times taken by different buses are independent of each other. Consider the random variable X, which is defined to be the *sum* of the time that Nancy spends on a bus and the time that Tom spends on a bus.

(a) What are the expectation and the standard deviation of X if Nancy and Tom ride on different buses?
(b) What are the expectation and the standard deviation of X if Nancy and Tom ride together on the same bus?

2.8.15 When a fair coin is tossed, 10 points are scored if a head is obtained and 20 points are scored if a tail is obtained. Suppose that the coin is tossed two times, and let the random variable X be the total score obtained.
(a) What is the state space of X?
(b) What is the probability mass function of X?
(c) What is the cumulative distribution function of X?
(d) What is the expectation of X?
(e) What is the standard deviation of X?

2.8.16 A continuous random variable has a probability density function

$$f(x) = Ax$$

for $5 \leq x \leq 6$.
(a) What is the value of A?
(b) What is the cumulative distribution function of the random variable?
(c) What is the expectation of the random variable?
(d) What is the standard deviation of the random variable?

2.8.17 Components have a weight with an expectation of 438 and a standard deviation of 4.
(a) What are the expectation and the standard deviation of the sum of the weights of three randomly selected components?
(b) What are the expectation and the standard deviation of the average of the weights of eight randomly selected components?

2.8.18 In a game a player rolls a fair six-sided die. If the score is even, the player receives an amount of dollars equal to the score. If the score is odd, the player receives an amount of dollars equal to the score multiplied by three. It costs the player \$5 to play the game.
(a) What is the probability mass function of the player's net winnings?
(b) What are the expectation and the standard deviation of the net winnings?
(c) Suppose that a player plays the game 10 times. What are the expectation and the standard deviation of the total net winnings from all 10 games?

2.8.19 Are the following statements true or false?
(a) The variance of a random variable is measured in the same units as the random variable.

(b) In a diving competition, the scores awarded by judges for a particular type of dive have an expected value of 78 with a standard deviation of 5. If the scores are doubled so that they can be compared with scores from an easier type of dive, the new scores will have an expected value of 156 and a standard deviation of 10.

(c) The variance of the difference between two independent random variables cannot be smaller than the larger of their two variances.

(d) If a continuous random variable has a symmetric probability density function, then the mean and the median are identical.

(e) If X is a continuous random variable, then $P(X \geq x) = P(X > x)$ for any value of x.

(f) If X is a discrete random variable, then $P(X \geq x) = P(X > x)$ for any value of x.

2.8.20 Suppose that the time taken to download a file of a certain kind onto a computer has an expected value of 22 minutes and a standard deviation of 1.8 minutes. If a technician needs to download five of these files onto a computer one after the other, and the downloading times are independent of each other, what are the expected value and the standard deviation of the total time taken to download all five files? What are the expected value and the standard deviation of the average downloading time of the five files?

2.8.21 When a computer chip is examined to discover how many of the solder joints have become cracked, there is a probability of 0.12 that none of the joints are cracked, a probability of 0.43 that one of the joints is cracked, a probability of 0.28 that two of the joints are cracked, and a probability of 0.17 that three of the joints are cracked.

(a) What is the mean number of cracked joints in a computer chip?

(b) What is the standard deviation of the number of cracked joints in a computer chip?

(c) Suppose that two different chips are randomly selected. What are the mean and the standard deviation of the total number of cracked joints in the two chips combined?

2.8.22 A discrete random variable takes the values -22, 3, 19, and 23 with probabilities 0.3, 0.2, 0.1, and 0.4 respectively.

(a) What is the expectation of the random variable?

(b) What is the standard deviation of the random variable?

2.8.23 A random variable has a probability density function $f(x) = A/x^2$ for $2 \leq x \leq 4$.

(a) What is the value of A?

(b) What is the lower quartile of the distribution?

2.8.24 Bricks have weights that have a mean 250 and a standard deviation 4.

(a) Suppose X is the weight of a randomly chosen brick. Let $Y = c + dX$. What are the values of c and d such that Y has a mean 100 and a standard deviation 1.

(b) Suppose 10 bricks are chosen at random. What are the mean and the standard deviation of the total weight of these 10 bricks?

2.8.25 An evaluation score X_1 of a candidate using method 1 has a mean of 100 and a standard deviation of 12, while an evaluation score X_2 of a candidate using method 2 has a mean of 100 and a standard deviation of 13. If the two evaluation scores are independent, what values of c_1 and c_2 can be chosen so that the combined score $Y = c_1 X_1 + c_2 X_2$ has a mean of 100 and a standard deviation of 10?

2.8.26 Wafers of type A have thicknesses with a mean of 134.9 and a standard deviation of 0.7, while wafers of type B have thicknesses with a mean of 138.2 and a standard deviation of 1.1. The thicknesses of the wafers are all independent of each other.

(a) Three wafers of type A are stacked on top of each other. What are the mean and the standard deviation of the total thickness?

(b) Two wafers of type A and two wafers of type B are stacked on top of each other. What are the mean and the standard deviation of the total thickness?

(c) Four wafers of type A and three wafers of type B are taken. What are the mean and the standard deviation of the average thickness of these seven wafers?

CHAPTER THREE

Discrete Probability Distributions

Several important discrete probability distributions are discussed in this chapter, and they allow the concise description of many commonly occurring phenomena. The probability mass functions of these distributions are described by formulas that depend upon some **parameter** values, and their expectations and variances are functions of these parameters. Familiarity with these distributions leads to an understanding of many probabilistic problems and to the development of statistical inference methods discussed in later chapters.

The binomial, geometric, negative binomial, hypergeometric, Poisson, and multinomial distributions are discussed in this chapter. With the exception of the Poisson distribution, each of these distributions is generated by the specific problem under consideration. In other words, the nature of the problem determines which probability distribution is appropriate for its analysis, and it is important to be able to recognize which probability distribution is appropriate for a particular problem. On the other hand, the Poisson distribution is chosen by the experimenter as being likely to provide a useful analysis.

3.1 The Binomial Distribution

3.1.1 Bernoulli Random Variables

The most simple discrete random variable is one that can take just two values. Such a random variable can be used to model the outcome of a coin toss, whether a valve is open or shut, whether an item is defective or not, or any other process that has only two possible outcomes. Typically, the outcomes are labeled 0 and 1, and the random variable is defined by the parameter p, $0 \leq p \leq 1$, which is the probability that the outcome is 1.

These simple random variables are known as **Bernoulli** random variables, and their expectation is

$$E(X) = (0 \times P(X = 0)) + (1 \times P(X = 1)) = (0 \times (1 - p)) + (1 \times p) = p$$

Also, since

$$E(X^2) = (0^2 \times P(X = 0)) + (1^2 \times P(X = 1)) = p$$

their variance is

$$\mathrm{Var}(X) = E(X^2) - (E(X))^2 = p - p^2 = p(1 - p)$$

Bernoulli Random Variables

A **Bernoulli** random variable with parameter p, $0 \le p \le 1$, takes the values 0 and 1 with $P(X = 0) = 1 - p$ and $P(X = 1) = p$. The expectation and variance of the random variable are

$$E(X) = p \quad \text{and} \quad \text{Var}(X) = p(1 - p)$$

An experiment that has only two outcomes is often referred to as a *Bernoulli trial*.

3.1.2   Definition of the Binomial Distribution

Many processes can be thought of as consisting of a *sequence* of Bernoulli trials, such as, for example, the repeated tossing of a coin or the repeated examination of objects to determine whether or not they are defective. In such cases, a random variable of interest is the *number of successes* obtained within a fixed number of trials n, where a "success" is defined in an appropriate manner. Such a random variable is called a **binomial** random variable and it is probably the most important of all discrete probability distributions.

Specifically, if n *independent* Bernoulli trials $X_1, \ldots, X_n$ are performed, each with a probability p of recording a 1, then the random variable

$$X = X_1 + \cdots + X_n$$

is said to have a binomial distribution with parameters n and p, which is written

$$X \sim B(n, p)$$

Notice that the random variable X takes the values $0, 1, \ldots, n$ and counts the number of the Bernoulli random variables that take the value 1 or, in other words, counts the number of "successes" in the n trials.

The probability mass function of a $B(n, p)$ random variable is given by the formula

$$P(X = x) = \binom{n}{x} p^x (1 - p)^{n-x}$$

for $x = 0, 1, \ldots, n$. This formula is illustrated by the case $n = 4$, say, where the $2^4 = 16$ possible outcomes of the four Bernoulli trials are listed in Figure 3.1. The probability that the binomial random variable takes the value 3 is the probability that exactly three of the Bernoulli random variables take the value 1, which is

$$P(X = 3) = P(0111) + P(1011) + P(1101) + P(1110)$$
$$= (1 - p)ppp + p(1 - p)pp + pp(1 - p)p + ppp(1 - p)$$
$$= 4p^3(1 - p)$$

Also, the probability that the binomial random variable takes the value 2 is the probability that exactly two of the Bernoulli random variables take the value 1, which is

$$P(X = 2) = P(0011) + P(0101) + P(0110) + P(1001) + P(1010) + P(1100)$$
$$= (1 - p)(1 - p)pp + (1 - p)p(1 - p)p + (1 - p)pp(1 - p)$$
$$\quad + p(1 - p)(1 - p)p + p(1 - p)p(1 - p) + pp(1 - p)(1 - p)$$
$$= 6p^2(1 - p)^2$$

HISTORICAL NOTE

James (Jakob, Jacques) Bernoulli (1654–1705) was one of several famous Swiss mathematicians from the same family. He obtained a degree in theology from the University of Basel in 1676 and held a chair in mathematics at the same university from 1687. One of his early interests was how to teach mathematics to the blind. He is reported to have been an excellent teacher, and students from all over Europe came to study under him. Much of his work on probability was contained in his manuscript *Ars Conjectandi* (The Art of Conjecturing), which was published (in Latin) after his death in 1713.

FIGURE 3.1

Illustration of the random variable
$X \sim B(4, p)$

Outcome of four Bernoulli trials	Probability	X = number of successes
0 0 0 0	$(1 - p)(1 - p)(1 - p)(1 - p)$	0
0 0 0 1	$(1 - p)(1 - p)(1 - p)p$	1
0 0 1 0	$(1 - p)(1 - p)p(1 - p)$	1
0 0 1 1	$(1 - p)(1 - p)pp$	2
0 1 0 0	$(1 - p)p(1 - p)(1 - p)$	1
0 1 0 1	$(1 - p)p(1 - p)p$	2
0 1 1 0	$(1 - p)pp(1 - p)$	2
0 1 1 1	$(1 - p)ppp$	3
1 0 0 0	$p(1 - p)(1 - p)(1 - p)$	1
1 0 0 1	$p(1 - p)(1 - p)p$	2
1 0 1 0	$p(1 - p)p(1 - p)$	2
1 0 1 1	$p(1 - p)pp$	3
1 1 0 0	$pp(1 - p)(1 - p)$	2
1 1 0 1	$pp(1 - p)p$	3
1 1 1 0	$ppp(1 - p)$	3
1 1 1 1	$pppp$	4

These calculations illustrate the general rule that if $X \sim B(n, p)$, then $P(X = x)$ is obtained as the sum of a number of probabilities that are each equal to $p^x(1 - p)^{n-x}$ since they correspond to exactly x of the Bernoulli random variables taking the value 1 and $(n - x)$ of the Bernoulli random variables taking the value 0. Furthermore, the number of these separate terms is simply the number of ways in which x 1s and $(n - x)$ 0s can be arranged, which is

$$\binom{n}{x}$$

so that the probability mass function is as indicated.

Notice that in particular

$$P(X = 0) = \binom{n}{0} p^0(1 - p)^{n-0} = \frac{n!}{n! \, 0!}(1 - p)^n = (1 - p)^n$$

which is the probability that each Bernoulli trial results in a 0, and

$$P(X = n) = \binom{n}{n} p^n(1 - p)^{n-n} = \frac{n!}{0! \, n!}p^n = p^n$$

which is the probability that each Bernoulli trial results in a 1.

Since the Bernoulli random variables each have an expectation of p, the expectation of a $B(n, p)$ random variable is easily calculated to be

$$E(X) = E(X_1) + \cdots + E(X_n) = p + \cdots + p = np$$

Also, because the Bernoulli random variables each have a variance of $p(1 - p)$ and are independent, the variance of a $B(n, p)$ random variable is

$$Var(X) = Var(X_1) + \cdots + Var(X_n)$$
$$= p(1 - p) + \cdots + p(1 - p) = np(1 - p)$$

The Binomial Distribution

Consider an experiment consisting of

- n Bernoulli trials

- that are independent and

- that each have a constant probability p of success.

Then the total number of successes X is a random variable that has a **binomial** distribution with parameters n and p, which is written

$$X \sim B(n, p)$$

The probability mass function of a $B(n, p)$ random variable is

$$P(X = x) = \binom{n}{x} p^x (1 - p)^{n-x}$$

for $x = 0, 1, \ldots, n$, with

$$E(X) = np \quad \text{and} \quad \text{Var}(X) = np(1 - p)$$

Figure 3.2 illustrates the probability mass function and cumulative distribution function of a $B(8, 0.5)$ random variable. In this case,

$$P(X = 3) = \binom{8}{3} \times 0.5^3 \times (1 - 0.5)^5 = \frac{8!}{3!\, 5!} \times 0.5^8 = 0.219$$

FIGURE 3.2

Probability mass function and cumulative distribution function of a $B(8, 0.5)$ random variable

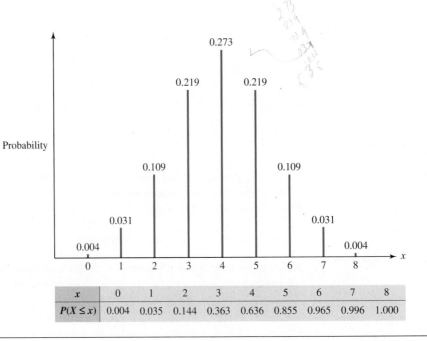

x	0	1	2	3	4	5	6	7	8
$P(X \le x)$	0.004	0.035	0.144	0.363	0.636	0.855	0.965	0.996	1.000

and

$$P(X \le 1) = P(X = 0) + P(X = 1)$$

$$= \binom{8}{0} \times 0.5^0 \times (1 - 0.5)^8 + \binom{8}{1} \times 0.5^1 \times (1 - 0.5)^7$$

$$= \frac{8!}{0! \, 8!} \times 0.5^8 + \frac{8!}{1! \, 7!} \times 0.5^8 = 0.004 + 0.031 = 0.035$$

Notice that this is a **symmetric distribution**, and in fact it is straightforward to establish that a $B(n, 0.5)$ distribution is symmetric for any value of the parameter n.

Symmetric Binomial Distributions

A $B(n, 0.5)$ distribution is a **symmetric probability distribution** for any value of the parameter n. The distribution is symmetric about the expected value $n/2$.

The $B(8, 0.5)$ distribution is, for example, the distribution of the number of heads obtained in eight tosses of a fair coin. The probability of obtaining exactly *four* heads in eight tosses is

$$P(X = 4) = 0.273$$

and the probability of obtaining at least *six* heads in eight tosses is

$$P(X \ge 6) = 1 - P(X \le 5) = 1 - 0.855 = 0.145$$

The expected number of heads obtained in eight tosses is

$$E(X) = np = 8 \times 0.5 = 4$$

and the variance is

$$\mathrm{Var}(X) = np(1 - p) = 8 \times 0.5 \times 0.5 = 2$$

Figure 3.3 illustrates the probability mass function and cumulative distribution function of a $B(8, 1/3)$ random variable. In this case,

$$P(X = 3) = \binom{8}{3} \times \left(\frac{1}{3}\right)^3 \times \left(1 - \frac{1}{3}\right)^5 = \frac{8!}{3! \, 5!} \times \left(\frac{1}{3}\right)^3 \times \left(\frac{2}{3}\right)^5 = 0.273$$

and

$$P(X \le 1) = P(X = 0) + P(X = 1)$$

$$= \binom{8}{0} \times \left(\frac{1}{3}\right)^0 \times \left(1 - \frac{1}{3}\right)^8 + \binom{8}{1} \times \left(\frac{1}{3}\right)^1 \times \left(1 - \frac{1}{3}\right)^7$$

$$= \frac{8!}{0! \, 8!} \times \left(\frac{2}{3}\right)^8 + \frac{8!}{1! \, 7!} \times \left(\frac{1}{3}\right) \times \left(\frac{2}{3}\right)^7 = 0.039 + 0.156 = 0.195$$

Suppose that a fair die is rolled eight times and that $1 is won each time a 5 or a 6 is scored. The distribution of the winnings is then given by the $B(8, 1/3)$ distribution, since there is a probability of $p = 1/3$ of winning $1 on each of the eight rolls of the die. The probability

FIGURE 3.3

Probability mass function and cumulative distribution function of a $B(8, 1/3)$ random variable

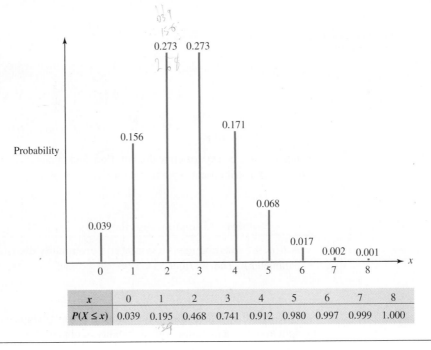

x	0	1	2	3	4	5	6	7	8
$P(X \le x)$	0.039	0.195	0.468	0.741	0.912	0.980	0.997	0.999	1.000

that exactly \$4 is won is therefore

$$P(X = 4) = 0.171$$

and the probability that no more than \$2 is won is

$$P(X \le 2) = 0.468$$

The expected winnings are

$$E(X) = np = 8 \times \frac{1}{3} = \$2.67$$

and the variance is

$$\mathrm{Var}(X) = np(1 - p) = 8 \times \frac{1}{3} \times \frac{2}{3} = 1.78$$

Figure 3.4 illustrates the probability mass function and cumulative distribution function of a $B(8, 2/3)$ random variable, which is seen to be a "reflection" of the $B(8, 1/3)$ distribution. In general, if

$$X \sim B(n, p)$$

and if the random variable Y is defined by $Y = n - X$, then

$$Y \sim B(n, 1 - p)$$

This makes sense since if X counts the number of "successes" in n Bernoulli trials where the success probability is p, then $Y = n - X$ counts the number of "failures" in n Bernoulli trials, where the failure probability is $1 - p$.

FIGURE 3.4

Probability mass function and cumulative distribution function of a $B(8, 2/3)$ random variable

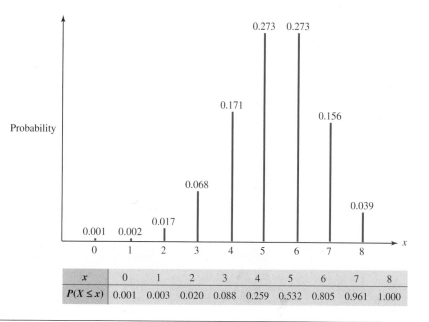

x	0	1	2	3	4	5	6	7	8
$P(X \leq x)$	0.001	0.003	0.020	0.088	0.259	0.532	0.805	0.961	1.000

COMPUTER NOTE Statistical software packages provide a convenient way of finding binomial probabilities without having to compute them by hand or to look them up in tables. You will need to indicate which parameter values n and p you are interested in, and you are usually given the option of being told the probability mass function or the cumulative distribution function. The statistical software package should also provide information on the other distributions in this chapter.

3.1.3 Examples of the Binomial Distribution

Example 17

Milk Container Contents

Recall that there is a probability of 0.261 that a milk container is underweight. Suppose that the milk containers are shipped to retail outlets in boxes of 20 containers. What is the distribution of the number of underweight containers in a box?

If the milk contents of two different milk containers are independent of each other, the number of underweight containers in a box has a binomial distribution with parameters $n = 20$ and $p = 0.261$. This is because each individual milk container represents a Bernoulli trial with a constant probability $p = 0.261$ of being underweight, as illustrated in Figure 3.5. The expected number of underweight cartons in a box is

$$E(X) = np = 20 \times 0.261 = 5.22$$

and the variance is

$$\text{Var}(X) = np(1 - p) = 20 \times 0.261 \times 0.739 = 3.86$$

so that the standard deviation is $\sigma = \sqrt{3.86} = 1.96$.

The probability mass function of the number of underweight containers is shown in Figure 3.6. The probability that a box contains exactly *seven* underweight containers, for example, is

$$P(X = 7) = \binom{20}{7} \times 0.261^7 \times 0.739^{13} = \frac{20!}{7! \, 13!} \times 0.261^7 \times 0.739^{13} = 0.125$$

FIGURE 3.5

The total number of underweight milk containers in the box has a binomial distribution with parameters $n = 20$ and $p = 0.261$

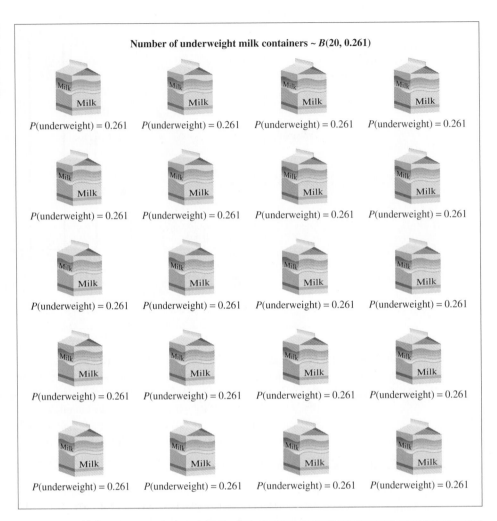

Number of underweight milk containers ~ $B(20, 0.261)$

$P(\text{underweight}) = 0.261$ $P(\text{underweight}) = 0.261$ $P(\text{underweight}) = 0.261$ $P(\text{underweight}) = 0.261$

$P(\text{underweight}) = 0.261$ $P(\text{underweight}) = 0.261$ $P(\text{underweight}) = 0.261$ $P(\text{underweight}) = 0.261$

$P(\text{underweight}) = 0.261$ $P(\text{underweight}) = 0.261$ $P(\text{underweight}) = 0.261$ $P(\text{underweight}) = 0.261$

$P(\text{underweight}) = 0.261$ $P(\text{underweight}) = 0.261$ $P(\text{underweight}) = 0.261$ $P(\text{underweight}) = 0.261$

$P(\text{underweight}) = 0.261$ $P(\text{underweight}) = 0.261$ $P(\text{underweight}) = 0.261$ $P(\text{underweight}) = 0.261$

The probability that a box contains *no more than three* underweight containers is

$$P(X \leq 3) = P(X = 0) + P(X = 1) + P(X = 2) + P(X = 3)$$

$$= \binom{20}{0} \times 0.261^0 \times 0.739^{20} + \binom{20}{1} \times 0.261^1 \times 0.739^{19}$$

$$+ \binom{20}{2} \times 0.261^2 \times 0.739^{18} + \binom{20}{3} \times 0.261^3 \times 0.739^{17}$$

$$= 0.0024 + 0.0167 + 0.0559 + 0.1185 = 0.1935$$

Example 24
Air Force Scrambles

An Air Force intercept squadron consists of 16 planes that should always be ready for immediate launch. However, a plane's engines are troublesome, and there is a probability of 0.25 that the engines of a particular plane will not start at a given attempt. If this happens, then the mechanics must wait five minutes before trying to start the engines again.

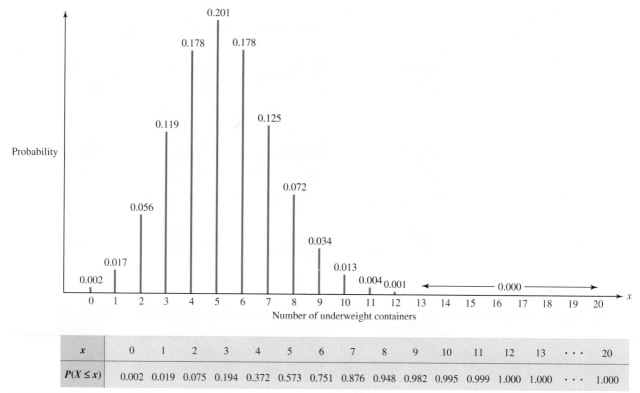

x	0	1	2	3	4	5	6	7	8	9	10	11	12	13	$\cdots$	20
$P(X \leq x)$	0.002	0.019	0.075	0.194	0.372	0.573	0.751	0.876	0.948	0.982	0.995	0.999	1.000	1.000	$\cdots$	1.000

FIGURE 3.6

Probability mass function and cumulative distribution function of the number of underweight milk containers

The squadron commander is interested in how many planes will immediately become airborne if the squadron is scrambled. As Figure 3.7 illustrates, the number of planes successfully launched has a binomial distribution with parameters $n = 16$ and $p = 0.75$. Each plane represents a Bernoulli trial with a constant probability $p = 0.75$ of launching on time.

The expected number of planes launched is

$$E(X) = np = 16 \times 0.75 = 12$$

and the variance is

$$\text{Var}(X) = np(1 - p) = 16 \times 0.75 \times 0.25 = 3$$

so that the standard deviation is $\sigma = \sqrt{3} = 1.73$.

The probability mass function of the number of planes that scramble successfully is shown in Figure 3.8. The probability that exactly 12 planes scramble successfully is

$$P(X = 12) = \binom{16}{12} \times 0.75^{12} \times 0.25^4 = \frac{16!}{12!\,4!} \times 0.75^{12} \times 0.25^4 = 0.225$$

FIGURE 3.7

The number of planes launched on
time has a binomial distribution
with parameters $n = 16$ and
$p = 0.75$

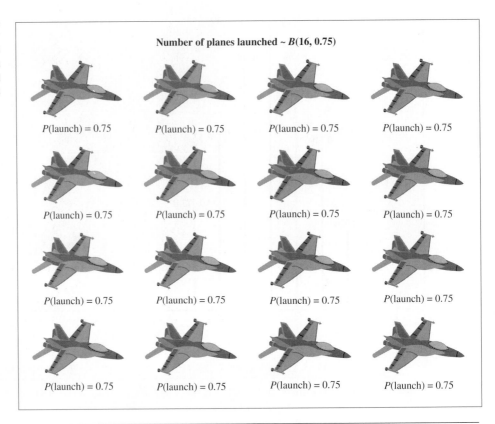

Number of planes launched ~ $B(16, 0.75)$

The probability that at least 14 planes scramble successfully is

$$P(X \geq 14) = P(X = 14) + P(X = 15) + P(X = 16)$$

$$= \binom{16}{14} \times 0.75^{14} \times 0.25^2 + \binom{16}{15} \times 0.75^{15} \times 0.25^1$$

$$+ \binom{16}{16} \times 0.75^{16} \times 0.25^0$$

$$= 0.134 + 0.054 + 0.010 = 0.198$$

GAMES OF CHANCE   Suppose that a fair coin is tossed n times. The distribution of the number of heads obtained, X, is binomial with parameters n and $p = 0.5$. The expected number of heads obtained is therefore

$$E(X) = np = \frac{n}{2}$$

and the variance is

$$\text{Var}(X) = np(1 - p) = \frac{n}{4}$$

The *proportion* of heads obtained is $Y = X/n$, which has an expectation and variance of

$$E(Y) = \frac{E(X)}{n} = \frac{1}{2} \quad \text{and} \quad \text{Var}(Y) = \frac{\text{Var}(X)}{n^2} = \frac{1}{4n}$$

Notice that the proportion of heads has an expected value equal to the probability of obtaining a head on one coin toss, which is not surprising, and that the variance of the proportion of

FIGURE 3.8

Probability mass function and
cumulative distribution function of
the number of planes launched on
time

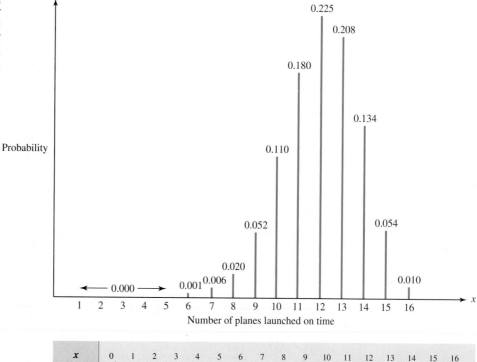

x	0	1	2	3	4	5	6	7	8	9	10	11	12	13	14	15	16
$P(X \leq x)$	0.000	0.000	0.000	0.000	0.000	0.000	0.001	0.007	0.027	0.079	0.189	0.369	0.594	0.802	0.936	0.990	1.000

heads *decreases* as the number of tosses n *increases*. In practice, as the coin is tossed more and more times, the *number* of heads may vary widely about the expected value of $n/2$, but there will be a tendency for the *proportion* of heads to become closer and closer to the value $1/2$.

Figure 3.9 shows the results of 80 coin tosses, and the proportion of heads obtained is graphed against the number of tosses in Figure 3.10. As expected, the fluctuation in the proportion of heads decreases as the number of tosses increases, and there is a tendency toward the expected value of $p = 1/2$. The calculation presented at the end of Section 5.3.1 provides further information on this point.

In general, the following result holds for the proportion of successes in a series of independent Bernoulli trials.

Proportion of Successes in Bernoulli Trials

If the random variable X counts the number of successes in n independent Bernoulli trials with a constant success probability p, so that $X \sim B(n, p)$, then the *proportion* of successes $Y = X/n$ has an expected value and variance of

$$E(Y) = p \quad \text{and} \quad \text{Var}(Y) = \frac{p(1-p)}{n}$$

The variance of the proportion Y decreases as the number of trials n increases, so that there is a tendency for Y to become closer and closer to the success probability p as the number of trials n increases.

FIGURE 3.9

Results of 80 coin tosses

n	Result of nth coin toss	Number of heads in first n coin tosses	Proportion of heads in first n coin tosses
1	tail	0	0.000
2	head	1	0.500
3	head	2	0.667
4	head	3	0.750
5	tail	3	0.600
6	head	4	0.667
7	head	5	0.714
8	head	6	0.750
9	head	7	0.778
10	tail	7	0.700
11	head	8	0.727
12	tail	8	0.667
13	tail	8	0.615
14	tail	8	0.571
15	head	9	0.600
16	head	10	0.625
17	tail	10	0.588
18	tail	10	0.556
19	tail	10	0.526
20	head	11	0.550
21	head	12	0.571
22	head	13	0.591
23	head	14	0.609
24	tail	14	0.583
25	tail	14	0.560
26	tail	14	0.538
27	tail	14	0.519
28	tail	14	0.500
29	tail	14	0.483
30	tail	14	0.467
31	tail	14	0.452
32	tail	14	0.438
33	tail	14	0.424
34	tail	14	0.412
35	head	15	0.429
36	tail	15	0.417
37	head	16	0.432
38	tail	16	0.421
39	head	17	0.436
40	tail	17	0.425
41	head	18	0.439
42	tail	18	0.429
43	tail	18	0.419
44	head	19	0.432
45	head	20	0.444
46	tail	20	0.435
47	tail	20	0.426
48	tail	20	0.417
49	tail	20	0.408
50	tail	20	0.400
51	head	21	0.412
52	tail	21	0.404
53	tail	21	0.396
54	tail	21	0.389
55	head	22	0.400
56	tail	22	0.393
57	head	23	0.404
58	tail	23	0.397
59	head	24	0.407
60	head	25	0.417
61	head	26	0.426
62	tail	26	0.419
63	head	27	0.429
64	head	28	0.438
65	head	29	0.446
66	tail	29	0.439
67	tail	29	0.433
68	head	30	0.441
69	head	31	0.449
70	head	32	0.457
71	tail	32	0.451
72	tail	32	0.444
73	head	33	0.452
74	tail	33	0.446
75	head	34	0.453
76	head	35	0.461
77	tail	35	0.455
78	head	36	0.462
79	tail	36	0.456
80	head	37	0.463

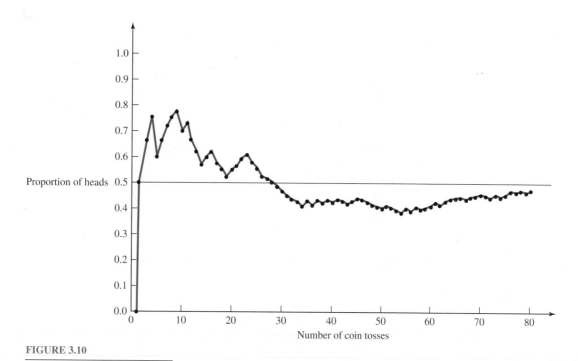

FIGURE 3.10

Graph of proportion of heads against number of coin tosses for 80 coin tosses

This result will be useful in Chapter 7 when the problem of *estimating* an unknown success probability p is addressed. Also, it should be noted that for large values of n, the probability mass function of a $B(n, p)$ random variable becomes tedious to calculate. A simple approximate method of calculating binomial probabilities for large values of n based on the *normal distribution* is discussed in Section 5.3.1.

■ 3.1.4 Problems

3.1.1 Suppose that $X \sim B(10, 0.12)$. Calculate:
   (a) $P(X = 3)$ **(b)** $P(X = 6)$
   (c) $P(X \leq 2)$ **(d)** $P(X \geq 7)$
   (e) $E(X)$ **(f)** $\mathrm{Var}(X)$

3.1.2 Suppose that $X \sim B(7, 0.8)$. Calculate:
   (a) $P(X = 4)$ **(b)** $P(X \neq 2)$
   (c) $P(X \leq 3)$ **(d)** $P(X \geq 6)$
   (e) $E(X)$ **(f)** $\mathrm{Var}(X)$

3.1.3 Draw line graphs of the probability mass functions of a $B(6, 0.5)$ distribution and a $B(6, 0.7)$ distribution. Mark the expected values of the distributions on the line graphs and calculate the standard deviations of the two distributions.

3.1.4 An archer hits a bull's-eye with a probability of 0.09, and the results of different attempts can be taken to be independent of each other. If the archer shoots *nine* arrows, calculate the probability that:
(a) Exactly two arrows score bull's-eyes.
(b) At least two arrows score bull's-eyes.
What is the expected number of bull's-eyes scored?
(This problem is continued in Problems 3.2.5 and 3.5.3.)

3.1.5 A fair die is rolled *eight* times. Calculate the probability that there are:
(a) Exactly five even numbers
(b) Exactly one 6 $n=1 \quad p=\frac{1}{6}$ $n=8$
(c) No 4s $p=\frac{1}{2}$
(d) At least six prime numbers

3.1.6 A multiple-choice quiz consists of ten questions, each with five possible answers of which only one is correct. A student passes the quiz if seven or more correct answers are obtained. What is the probability that a student who guesses blindly at all of the questions will pass the quiz? What is the probability of passing the quiz if, on each question, a student can eliminate three incorrect answers and then guesses between the remaining two?

3.1.7 A flu virus hits a company employing 180 people. Independent of the other employees, there is a probability of $p = 0.35$ that each person needs to take sick leave. What are the expectation and variance of the *proportion* of the workforce who need to take sick leave? In general, what value of the sick rate p produces the *largest* variance for this proportion?

3.1.8 Consider the two independent binomial random variables $X_1 \sim B(n_1, p)$ and $X_2 \sim B(n_2, p)$. If $Y = X_1 + X_2$, explain why $Y \sim B(n_1 + n_2, p)$.

3.1.9 A company receives 60% of its orders over the Internet. Within a collection of 18 independently placed orders, what is the probability that
  (a) between 8 and 10 of the orders are received over the Internet?
  (b) no more than 4 of the orders are received over the Internet?
(This problem is continued in Problems 3.2.9, 3.5.5, and 5.3.10.)

3.2   The Geometric and Negative Binomial Distributions

3.2.1   Definition of the Geometric Distribution

Consider a sequence of independent Bernoulli trials with a constant success probability p. Whereas the binomial distribution is the distribution of the number of successes occurring in a fixed number of trials n, it is sometimes of interest to count instead the number of trials performed until the *first* success occurs. Such a random variable is said to have a **geometric** distribution with parameter p, and it has a probability mass function given by

$$P(X = x) = (1 - p)^{x-1} \, p$$

for $x = 1, 2, 3, 4 \ldots$. It is easy to see that the probability mass function is of this form, because if the first success occurs on the xth trial, then the first $x - 1$ trials must all be failures. The probability of these $x - 1$ failures is

$$(1 - p)^{x-1}$$

which is then multiplied by p, the probability that the xth trial is a success.

It can be shown that a geometric distribution with parameter p has an expected value and a variance of

$$E(X) = \frac{1}{p} \quad \text{and} \quad \text{Var}(X) = \frac{1 - p}{p^2}$$

Also, the cumulative distribution function can be calculated to be

$$P(X \leq x) = \sum_{i=1}^{x} P(X = i) = \sum_{i=1}^{x} (1 - p)^{i-1} \, p$$
$$= p \left(1 + (1 - p) + (1 - p)^2 + \cdots + (1 - p)^{x-1} \right)$$
$$= p \times \frac{1 - (1 - p)^x}{1 - (1 - p)} = 1 - (1 - p)^x$$

Figure 3.11 illustrates a geometric distribution with parameter $p = 1/2$, and Figure 3.12 illustrates a geometric distribution with parameter $p = 1/5$. The former has an expected value of $E(X) = 2$, while the latter has an expected value of $E(X) = 5$. Notice that the probability values form a decreasing geometric series, as the name of the distribution suggests.

The Geometric Distribution

The number of trials up to and including the *first* success in a sequence of independent Bernoulli trials with a constant success probability p has a **geometric** distribution with parameter p. The probability mass function is

$$P(X = x) = (1 - p)^{x-1} \, p$$

for $x = 1, 2, 3, 4 \ldots$, and the cumulative distribution function is

$$P(X \leq x) = 1 - (1 - p)^x$$

The geometric distribution with parameter p has an expected value and a variance of

$$E(X) = \frac{1}{p} \quad \text{and} \quad \text{Var}(X) = \frac{1-p}{p^2}$$

The distribution with $p = 1/2$ is appropriate for modeling the number of tosses of a fair coin made until a head is obtained for the first time, since in this case the "success" probability (the probability of obtaining a head) is $p = 1/2$. The probability that a head is obtained for the first time on the *fourth* coin toss is therefore

$$P(X = 4) = (1 - p)^{4-1} \, p = \left(\frac{1}{2}\right)^3 \times \frac{1}{2} = \frac{1}{16}$$

which is simply the probability of obtaining three tails followed by a head.

FIGURE 3.11

Probability mass function and cumulative distribution function of a geometric distribution with parameter $p = 1/2$

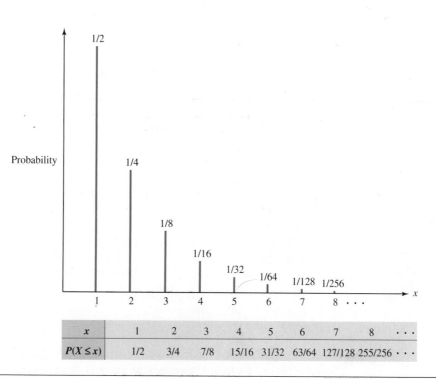

x	1	2	3	4	5	6	7	8	$\cdots$
$P(X \leq x)$	1/2	3/4	7/8	15/16	31/32	63/64	127/128	255/256	$\cdots$

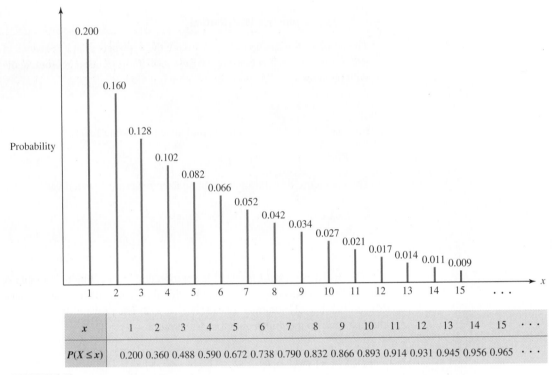

FIGURE 3.12

Probability mass function and cumulative distribution function of a geometric distribution with parameter $p = 1/5$

3.2.2   Definition of the Negative Binomial Distribution

The geometric distribution can be generalized to situations in which the quantity of interest is the number of trials up to and including the rth success. In this case, the appropriate distribution is the **negative binomial** distribution with parameters p and r, which has a probability mass function given by

$$P(X = x) = \binom{x-1}{r-1} (1-p)^{x-r} p^r$$

for $x = r, r+1, r+2, r+3, \ldots$.

This probability mass function can be understood by noting that if the rth success occurs on the xth trial, then the first $x - 1$ trials consist of $r - 1$ successes and $x - r$ failures. The probability of this event is obtained from the binomial distribution as

$$\binom{x-1}{r-1} (1-p)^{x-r} p^{r-1}$$

The required probability value of the negative binomial distribution is obtained by multiplying this probability by p, the probability that the xth trial results in a success.

The Negative Binomial Distribution

The number of trials up to and including the rth success in a sequence of independent Bernoulli trials with a constant success probability p has a **negative binomial** distribution with parameters p and r. The probability mass function is

$$P(X = x) = \binom{x-1}{r-1} (1-p)^{x-r} p^r$$

for $x = r, r+1, r+2, r+3, \ldots$, with an expected value and a variance of

$$E(X) = \frac{r}{p} \quad \text{and} \quad \text{Var}(X) = \frac{r(1-p)}{p^2}$$

Figure 3.13 illustrates a negative binomial distribution with parameter values $p = 1/2$ and $r = 2$. This distribution is appropriate for modeling the number of tosses of a fair coin made until a head is obtained for the second time. For example, the probability that only *two*

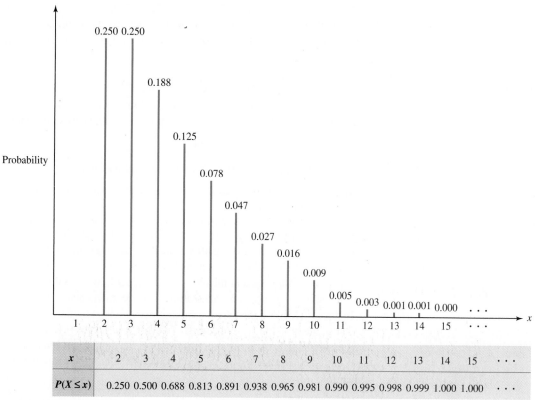

x	2	3	4	5	6	7	8	9	10	11	12	13	14	15	$\cdots$
$P(X \le x)$	0.250	0.500	0.688	0.813	0.891	0.938	0.965	0.981	0.990	0.995	0.998	0.999	1.000	1.000	$\cdots$

FIGURE 3.13

Probability mass function and cumulative distribution function of a negative binomial distribution with parameters $p = 1/2$ and $r = 2$

coin tosses are required is

$$P(X = 2) = \binom{1}{1} \times 0.5^0 \times 0.5^2 = \frac{1}{4}$$

which is simply the probability that the first two coin tosses result in heads. The probability that *five* coin tosses are required is

$$P(X = 5) = \binom{4}{1} \times 0.5^3 \times 0.5^2 = \frac{1}{8}$$

which is the sum of the probabilities of the four outcomes $TTTHH, TTHTH, THTTH,$ and $HTTTH$. The expected number of coin tosses required to obtain two heads is

$$E(X) = \frac{r}{p} = \frac{2}{0.5} = 4$$

3.2.3   Examples of the Geometric and Negative Binomial Distributions

Example 24

Air Force Scrambles

Recall that a plane's engines start successfully at a given attempt with a probability of 0.75. Any time that the mechanics are unsuccessful in starting the engines, they must wait five minutes before trying again. What is the distribution of the number of attempts needed to start a plane's engines?

A "success" here is the event that the plane's engines start, so that the success probability is $p = 0.75$. Furthermore, the geometric distribution is appropriate since attention is directed at the number of trials until the *first* success. The probability that the engines start on the third attempt is therefore

$$P(X = 3) = 0.25^2 \times 0.75 = 0.047$$

The probability that the plane is launched within 10 minutes of the first attempt to start the engines is the probability that no more than three attempts are required, which is

$$P(X \leq 3) = 1 - 0.25^3 = 0.984$$

The expected number of attempts required to start the engines is

$$E(X) = \frac{1}{p} = \frac{1}{0.75} = 1.33$$

Example 25

Telephone Ticket Sales

Telephone ticket sales for a popular event are handled by a bank of telephone salespersons who start accepting calls at a specified time. In order to get through to an operator, a caller has to be lucky enough to place a call at just the time when a salesperson has become free from a previous client. Suppose that the chance of this is 0.1. What is the distribution of the number of calls that a person needs to make until a salesperson is reached?

In this problem, the placing of a call represents a Bernoulli trial with a "success" probability, that is, the probability of reaching a salesperson, of $p = 0.1$, as illustrated in Figure 3.14. The geometric distribution is appropriate since the quantity of interest is the number of calls made until the *first* success. The probability that a caller gets through on the fifth attempt, say, is therefore

$$P(X = 5) = 0.9^4 \times 0.1 = 0.066$$

The expected number of calls needed to get through to a salesperson is

$$E(X) = \frac{1}{p} = \frac{1}{0.1} = 10$$

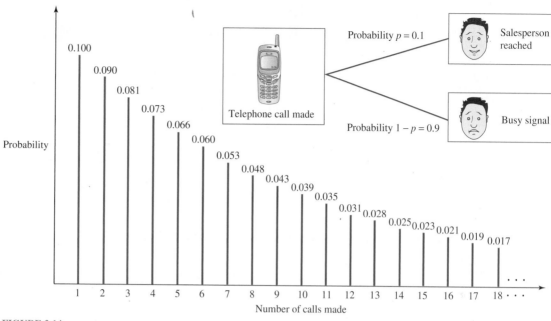

FIGURE 3.14

Probability mass function of a geometric distribution with parameter $p = 0.1$, the distribution of the number of calls made until a ticket salesperson is reached

and the probability that 15 or more calls are needed is

$$P(X \geq 15) = 1 - P(X \leq 14) = 1 - \left(1 - 0.9^{14}\right) = 0.9^{14} = 0.229$$

which is simply the probability that the first 14 calls are unsuccessful.

Example 12
Personnel Recruitment

Suppose that a company wishes to hire *three* new workers and that each applicant interviewed has a probability of 0.6 of being found acceptable, as illustrated in Figure 3.15. What is the distribution of the total number of applicants that the company needs to interview?

In this problem, an interview represents a Bernoulli trial with a "success" being that the applicant is found to be acceptable for the position. The success probability is therefore $p = 0.6$. Furthermore, the negative binomial distribution with $r = 3$ is appropriate since the quantity of interest is the number of interviews that must be undertaken until three suitable applicants have been found.

The negative binomial distribution with parameters $p = 0.6$ and $r = 3$ is illustrated in Figure 3.16. For example, the probability that exactly six applicants need to be interviewed is

$$P(X = 6) = \binom{5}{2} \times 0.4^3 \times 0.6^3 = 0.138$$

If the company has a budget that allows up to six applicants to be interviewed, then the probability that the budget is sufficient is

$$P(X \leq 6) = P(X = 3) + P(X = 4) + P(X = 5) + P(X = 6)$$
$$= 0.216 + 0.259 + 0.207 + 0.138 = 0.820$$

FIGURE 3.15

Applicants

The number of applicants who
need to be interviewed to fill three
positions has a negative binomial
distribution with parameters
$p = 0.6$ and $r = 3$

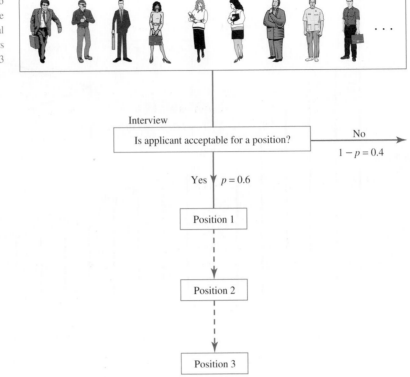

Interview

Is applicant acceptable for a position?

No
$1 - p = 0.4$

Yes $p = 0.6$

Position 1

Position 2

Position 3

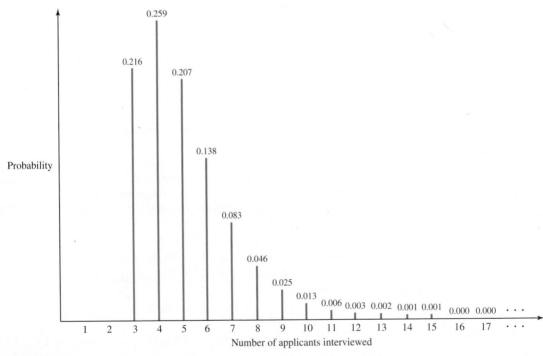

FIGURE 3.16

Probability mass function of a negative binomial distribution with parameters $p = 0.6$ and $r = 3$, the distribution of the number of applicants interviewed

The expected number of interviews required is

$$E(X) = \frac{r}{p} = \frac{3}{0.6} = 5$$

GAMES OF CHANCE If a fair die is repeatedly thrown, the number of throws made until a 6 is obtained has a geometric distribution with parameter $p = 1/6$. The expected number of throws made until a 6 is obtained is therefore

$$E(X) = \frac{1}{p} = 6$$

The probability that a 6 is *not* obtained in the first six throws is

$$P(X \geq 7) = 1 - P(X \leq 6) = 1 - \left(1 - \left(\frac{5}{6}\right)^6\right) = \left(\frac{5}{6}\right)^6 = 0.335$$

The distribution of the number of throws made until a 6 is obtained for the *second* time has a negative binomial distribution with parameters $p = 1/6$ and $r = 2$. The expected number of throws required is

$$E(X) = \frac{r}{p} = 12$$

The probability that two 6s are obtained in the first four throws is

$$P(X \leq 4) = P(X = 2) + P(X = 3) + P(X = 4)$$
$$= \binom{1}{1} \times \left(\frac{5}{6}\right)^0 \times \left(\frac{1}{6}\right)^2 + \binom{2}{1} \times \left(\frac{5}{6}\right)^1 \times \left(\frac{1}{6}\right)^2$$
$$+ \binom{3}{1} \times \left(\frac{5}{6}\right)^2 \times \left(\frac{1}{6}\right)^2$$
$$= 0.028 + 0.046 + 0.058 = 0.132$$

■ **3.2.4 Problems**

3.2.1 If X has a geometric distribution with parameter $p = 0.7$, calculate:
(a) $P(X = 4)$ (b) $P(X = 1)$
(c) $P(X \leq 5)$ (d) $P(X \geq 8)$

3.2.2 If X has a negative binomial distribution with parameters $p = 0.6$ and $r = 3$, calculate:
(a) $P(X = 5)$ (b) $P(X = 8)$
(c) $P(X \leq 7)$ (d) $P(X \geq 7)$

3.2.3 If X has a geometric distribution with parameter $p = 0.7$, show that

$$E(X) = \sum_{x=1}^{\infty} x P(X = x) = \frac{1}{p}$$

Also, show that

$$E(X^2) = \sum_{x=1}^{\infty} x^2 P(X = x) = \frac{2}{p^2} - \frac{1}{p}$$

and deduce that

$$\text{Var}(X) = \frac{1 - p}{p^2}$$

3.2.4 Suppose that $X_1, \ldots, X_r$ are independent random variables, each with a geometric distribution with parameter p. Explain why

$$Y = X_1 + \cdots + X_r$$

has a negative binomial distribution with parameters p and r. Use this relationship to establish the mean and variance of a negative binomial distribution.

3.2.5 Recall Problem 3.1.4 where an archer hits a bull's-eye with a probability of 0.09, and the results of different attempts can be taken to be independent of each other.
(a) If the archer shoots a series of arrows, what is the probability that the *first* bull's-eye is scored with the fourth arrow?

bi Geo $(1-P)^{4-1} (P)$

(b) What is the probability that the *third* bull's-eye is scored with the tenth arrow?

(c) What is the expected number of arrows shot before the *first* bull's-eye is scored?

(d) What is the expected number of arrows shot before the *third* bull's-eye is scored?

3.2.6 A supply container dropped from an aircraft by parachute hits a target with a probability of 0.37.

(a) What is the expected number of container drops needed to hit a target?

(b) If hits from three containers are required to provide sufficient supplies, what is the expected number of containers dropped before sufficient supplies have been provided?

(c) What is the probability that sufficient supplies are provided by ten container drops?

(d) What is the probability that it is the tenth container dropped that completes the provision of supplies?

3.2.7 Cards are chosen randomly from a pack of cards with replacement. Calculate the probability that:

(a) The first heart is obtained on the third drawing.

(b) The fourth heart is obtained on the tenth drawing. What is the expected number of cards drawn before the fourth heart is obtained? If the first two cards drawn are spades, what is the probability that the first heart is obtained on the fifth drawing?

3.2.8 When a fisherman catches a fish, it is a young one with a probability of 0.23 and it is returned to the water. On the other hand, an adult fish is kept to be eaten later.

(a) What is the expected number of fish caught by the fisherman before an adult fish is caught?

(b) What is the probability that the fifth fish caught is the first young fish?

Suppose that the fisherman wants three fish to eat for lunch.

(c) What is the probability that the first time the fisherman can stop for lunch is immediately after the sixth fish has been caught?

(d) If the fisherman catches eight fish, what is the probability that there are sufficient fish for lunch?

3.2.9 Recall Problem 3.1.9, in which a company receives 60% of its orders over the Internet. Within a certain period of time:

(a) What is the probability that the fifth order received is the first Internet order?

(b) What is the probability that the eighth order received is the fourth Internet order?

3.2.10 Consider a fair six-sided die. The die is rolled until a 6 is obtained for the third time. What is the expectation of the number of die rolls needed?

3.2.11 A fair coin is tossed until the fifth head is obtained. What is the probability that the coin is tossed exactly 10 times?

3.3 The Hypergeometric Distribution

3.3.1 Definition of the Hypergeometric Distribution

Consider a collection of N items of which r are of a certain kind. For example, it may be useful to think of r of the N items as being "defective." Alternatively, it may be useful to envision a box containing N balls of which exactly r are red. If one of the items is chosen at random, the probability that it is of the special kind is clearly

$$p = \frac{r}{N}$$

Consequently, if n items are chosen at random *with replacement*, then clearly the distribution of X, the number of defective items chosen, is

$$X \sim B(n, r/N)$$

However, if n items are chosen at random *without replacement*, then the binomial distribution cannot be applied because the success probability, that is, the probability of selecting an item of the special kind, is no longer constant. Instead, the appropriate distribution for the number of defective items chosen is the **hypergeometric distribution**. The probability mass function of the hypergeometric distribution is

$$P(X = x) = \frac{\binom{r}{x} \times \binom{N-r}{n-x}}{\binom{N}{n}}$$

for $\max\{0, n + r - N\} \le x \le \min\{n, r\}$. Notice that the number of defective items x must take a value at least as large as $n + r - N$ if this is positive. This is because when $n + r - N$ is positive, the sample size n is larger than the number of nondefective items $N - r$, so that the sample must contain at least $n - (N - r)$ defective items. Also, the number of defective items x cannot be larger than either the sample size n or the total number of defective items r, whichever is smaller.

The Hypergeometric Distribution

The **hypergeometric distribution** has a probability mass function given by

$$P(X = x) = \frac{\binom{r}{x} \times \binom{N-r}{n-x}}{\binom{N}{n}}$$

for $\max\{0, n + r - N\} \le x \le \min\{n, r\}$, with an expected value of

$$E(X) = \frac{nr}{N}$$

and a variance of

$$\text{Var}(X) = \left(\frac{N-n}{N-1}\right) \times n \times \frac{r}{N} \times \left(1 - \frac{r}{N}\right)$$

It represents the distribution of the number of items of a certain kind in a random sample of size n drawn *without replacement* from a population of size N that contains r items of this kind.

The hypergeometric distribution was encountered in Section 1.7 during the discussion of Example 2 on defective computer chips. In this example, a box of $N = 500$ computer chips contains $r = 9$ defective chips, and $n = 3$ chips are selected at random without replacement. The total number of possible samples that can be taken is

$$\binom{500}{3}$$

which in general is

$$\binom{N}{n}$$

The number of samples containing exactly one defective chip is

$$\binom{9}{1} \times \binom{491}{2}$$

where the first term represents the number of ways of choosing one defective chip and the second term represents the number of ways of choosing two satisfactory chips. In general, the number of samples containing exactly x defective items can similarly be shown to be

$$\binom{r}{x} \times \binom{N-r}{n-x}$$

The probability mass function of the hypergeometric distribution is then derived by noting that since all possible samples are equally likely to be chosen, the probability of choosing x defective items is obtained by dividing the number of samples containing x defective items by the total possible number of samples.

Finally, it is useful to note that if the total number of items N is *much larger* than the sample size n, then sampling without replacement is very similar to sampling with replacement. This is because the reduction of the total population by the items selected does not appreciably change the selection probabilities of additional items. In such a case, while the hypergeometric distribution is technically correct, it is usually convenient to use the binomial distribution

$$X \sim B(n, r/N)$$

to approximate the distribution of the number of defective items chosen.

3.3.2 Examples of the Hypergeometric Distribution

Example 17
Milk Container Contents

Suppose that milk is shipped to retail outlets in boxes that hold 16 milk containers. One particular box, which happens to contain 6 underweight containers, is opened for inspection, and 5 containers are chosen at random. What is the distribution of the number of underweight milk containers in the sample chosen by the inspector?

The required distribution is the hypergeometric distribution with $N = 16$, $r = 6$, and $n = 5$, which is illustrated in Figure 3.17. For example, the probability that the inspector

FIGURE 3.17

Probability mass function of the number of underweight milk containers in the inspector's sample

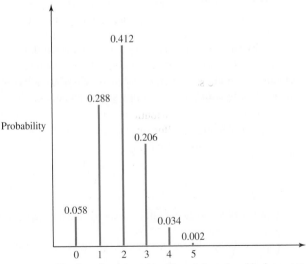

Number of underweight milk containers found by inspector

FIGURE 3.18

Probability mass function of the number of tagged fish that are recaptured

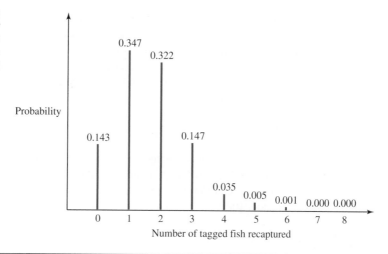

chooses exactly two underweight containers is

$$P(X = 2) = \frac{\binom{6}{2} \times \binom{10}{3}}{\binom{16}{5}} = \frac{\left(\frac{6!}{2!\,4!}\right) \times \left(\frac{10!}{3!\,7!}\right)}{\left(\frac{16!}{5!\,11!}\right)} = 0.412$$

The expected number of underweight containers chosen by the inspector is

$$E(X) = \frac{nr}{N} = \frac{5 \times 6}{16} = 1.875$$

Example 26

Fish Tagging and Recapture

A small lake contains 50 fish. One day a fisherman catches 10 of these fish and tags them so that they can be recognized if they are caught again. The tagged fish are released back into the lake. The next day the fisherman goes out and catches 8 fish, which are kept in the fishing boat until they are all released at the end of the day. What is the distribution of the number of tagged fish caught by the fisherman on the second day?

The second day's fishing can be thought of as a sample of size eight taken without replacement from the fish stock. The sample is taken without replacement since the fish that are caught are kept in the fishing boat until all 8 fish have been caught (thereby eliminating the possibility of the same fish being caught twice on the second day). Consequently, given that all 50 fish are equally likely to be caught, the number of tagged fish caught on the second day has a hypergeometric distribution with $N = 50$, $r = 10$, and $n = 8$.

This distribution is illustrated in Figure 3.18. For example, the probability that 3 tagged fish are caught on the second day is

$$P(X = 3) = \frac{\binom{10}{3} \times \binom{40}{5}}{\binom{50}{8}} = \frac{\left(\frac{10!}{3!\,7!}\right) \times \left(\frac{40!}{5!\,35!}\right)}{\left(\frac{50!}{8!\,42!}\right)} = 0.147$$

The expected number of tagged fish recaptured on the second day is

$$E(X) = \frac{nr}{N} = \frac{8 \times 10}{50} = 1.6$$

Suppose that there is also a much larger lake containing 5000 fish of which 80 have been tagged over a period of time. If a fisherman goes out one day and catches 8 fish (without replacement), the distribution of the number of tagged fish caught will be a hypergeometric distribution with $N = 5000$, $r = 80$, and $n = 8$. However, because of the large number of fish in the lake, it is appropriate to approximate the distribution as a $B(8, p)$ distribution, where the success probability is

$$p = \frac{r}{N} = \frac{80}{5000} = 0.016$$

In practice, tagging and recapture experiments of this kind can be used to *estimate* the total size of the fish stock. For example, if 100 fish have been tagged and the fishermen find that tagged fish tend to represent about 10% of their catches, this suggests that the total fish stock is about 1000 fish.

GAMES OF CHANCE

If six cards are randomly drawn without replacement from a pack of cards, the number of cards chosen from the heart suit has a hypergeometric distribution with $N = 52$, $r = 13$, and $n = 6$. The expected number of hearts chosen is

$$E(X) = \frac{nr}{N} = \frac{6 \times 13}{52} = 1.5$$

The number of Aces chosen has a hypergeometric distribution with $N = 52$, $r = 4$, and $n = 6$, with an expected value of

$$E(X) = \frac{nr}{N} = \frac{6 \times 4}{52} = 0.462$$

■ 3.3.3 Problems

3.3.1 Let X have a hypergeometric distribution with $N = 11$, $r = 6$, and $n = 7$. Calculate:
  (a) $P(X = 4)$ **(b)** $P(X = 5)$ **(c)** $P(X \le 3)$

3.3.2 A committee consists of eight right-wing members and seven left-wing members. A subcommittee is formed by randomly choosing five of the committee members. Draw a line graph of the probability mass function of the number of right-wing members serving on the subcommittee.

3.3.3 A box contains 17 balls of which 10 are red and 7 are blue. A sample of 5 balls is chosen at random and placed in a jar. Calculate the probability that:
  (a) The jar contains exactly 3 red balls.
  (b) The jar contains exactly 1 red ball.
  (c) The jar contains more blue balls than red balls.

3.3.4 A jury of 12 people is selected at random from a group of 16 men and 18 women. What is the probability that the jury contains exactly 7 women? Suppose that the jury is selected at random from a group of 1600 men and

1800 women. Use the binomial approximation to the hypergeometric distribution to calculate the probability that in this case the jury contains exactly 7 women.

3.3.5 Five cards are selected at random from a pack of cards *without replacement*. What is the probability that exactly three of them are picture cards (Kings, Queens, or Jacks)? If a hand of 13 cards is dealt from the pack, what are the expectation and variance of the number of picture cards that it contains?

3.3.6 Consider a collection of N items of which r_i are of type i, for $1 \le i \le k$, where

$$r_1 + \cdots + r_k = N$$

Suppose that a sample of size n is chosen at random from the N items without replacement. If X_i is the number of items of type i in the sample, $1 \le i \le k$, with

$$X_1 + \cdots + X_k = n$$

then they are said to have a **multivariate hypergeometric** distribution. Explain why their probability mass function is

$$P(X_1 = x_1, \ldots, X_k = x_k) = \frac{\binom{r_1}{x_1} \times \cdots \times \binom{r_k}{x_k}}{\binom{N}{n}}$$

for suitable values of $x_1, \ldots, x_k$.

A box contains four red balls, five blue balls, and six green balls. Five balls are chosen at random and placed in a jar. What is the probability that the jar contains one red ball, two blue balls, and two green balls?

3.3.7 There are 11 items of a product on a shelf in a retail outlet, and unknown to the customers, 4 of the items are outdated. Suppose that a customer takes 3 items at random.

(a) What is the probability that none of the outdated products are selected by the customer?

(b) What is the probability that exactly 2 of the items taken by the customer are outdated?

3.3.8 A plate has 15 cakes on it, of which 9 are chocolate and 6 are strawberry. A child randomly selects 5 of the cakes and eats them. What is the probability that the number of chocolate cakes remaining on the plate is between 5 and 7 inclusive?

3.3.9 (a) A box contains 8 red balls and 8 blue balls, and 4 balls are taken at random without replacement. What is the probability that 2 red balls and 2 blue balls are taken?

(b) A box contains 50,000 red balls and 50,000 blue balls, and 4 balls are taken at random without replacement. Estimate the probability that 2 red balls and 2 blue balls are taken.

3.3.10 In a ground water contamination study, the researchers identify 25 possible sites for drilling and sample collection. Unknown to the researchers, 19 of these sites have ground water with a high contamination, while the other 6 sites have ground water with a low contamination. The researchers have a budget that only allows them to drill at 5 sites, so they randomly choose these 5 sites from their list of 25 sites. What is the probability that at least 4 out of the 5 sites that the researchers examine have ground water with a high contamination?

3.4 The Poisson Distribution

3.4.1 Definition of the Poisson Distribution

It is often useful to define a random variable that *counts* the number of "events" that occur within certain specified boundaries. For example, an experimenter may be interested in the number of defects in an item, the number of radioactive particles emitted by a substance, or the number of telephone calls received by an operator within a certain time limit. The **Poisson distribution** is often appropriate to model such situations.

A random variable X with a Poisson distribution takes the values $x = 0, 1, 2, 3, \ldots$ with a probability mass function

$$P(X = x) = \frac{e^{-\lambda} \lambda^x}{x!}$$

where λ is the parameter of the distribution. This can be written

$$X \sim P(\lambda)$$

which should be read "X is distributed as a Poisson random variable with parameter λ."

Notice that the series expansion of e^λ ensures that these probability values sum to 1 since

$$\sum_{x=0}^{\infty} P(X = x) = \sum_{x=0}^{\infty} \frac{e^{-\lambda} \lambda^x}{x!} = e^{-\lambda} \left(\frac{1}{1} + \frac{\lambda}{1} + \frac{\lambda^2}{2!} + \frac{\lambda^3}{3!} + \frac{\lambda^4}{4!} + \cdots \right)$$

$$= e^{-\lambda} \times e^{\lambda} = 1$$

It can also be shown that if $X \sim P(\lambda)$, then

$$E(X) = \text{Var}(X) = \lambda$$

The Poisson Distribution

A random variable X distributed as a **Poisson** random variable with parameter λ, which is written

$$X \sim P(\lambda)$$

has a probability mass function

$$P(X = x) = \frac{e^{-\lambda} \lambda^x}{x!}$$

for $x = 0, 1, 2, 3, \ldots$. The Poisson distribution is often useful to model the number of times that a certain event occurs per unit of time, distance, or volume, and it has a mean and variance both equal to the parameter value λ.

Figures 3.19 and 3.20 illustrate and contrast the probability mass functions and cumulative distribution functions of Poisson distributions with parameters $\lambda = 2$ and $\lambda = 5$. It can be seen that since the mean and variance of the Poisson distribution are both equal to the parameter value, the distribution with the larger parameter value has a larger expected value and is more

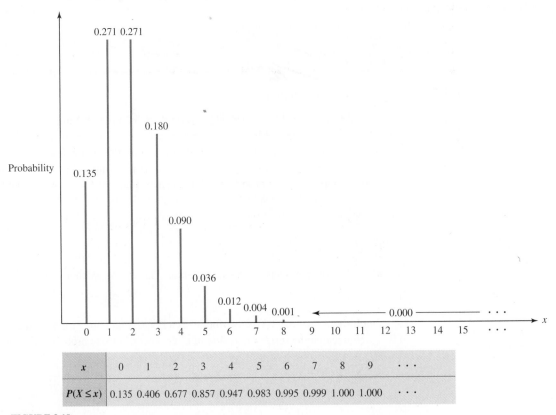

x	0	1	2	3	4	5	6	7	8	9	$\cdots$
$P(X \leq x)$	0.135	0.406	0.677	0.857	0.947	0.983	0.995	0.999	1.000	1.000	$\cdots$

FIGURE 3.19

Probability mass function and cumulative distribution function of a Poisson random variable with mean $\lambda = 2.0$

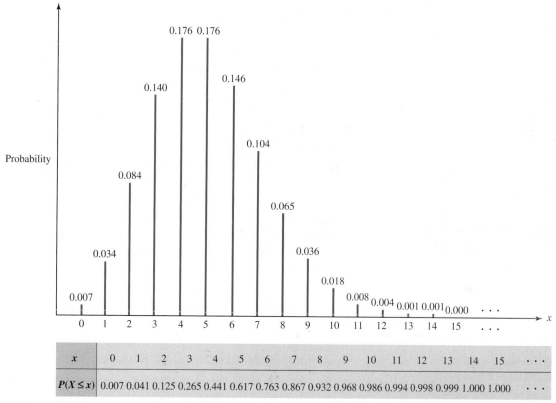

x	0	1	2	3	4	5	6	7	8	9	10	11	12	13	14	15	···
$P(X \le x)$	0.007	0.041	0.125	0.265	0.441	0.617	0.763	0.867	0.932	0.968	0.986	0.994	0.998	0.999	1.000	1.000	···

FIGURE 3.20

Probability mass function and cumulative distribution function of a Poisson random variable with mean $\lambda = 5.0$

spread out. As a check on some of the probability values given, notice that for $\lambda = 2$,

$$P(X = 3) = \frac{e^{-2} \times 2^3}{3!} = \frac{0.135 \times 8}{6} = 0.180$$

and for $\lambda = 5$,

$$\begin{aligned} P(X \le 2) &= P(X = 0) + P(X = 1) + P(X = 2) \\ &= \frac{e^{-5} \times 5^0}{0!} + \frac{e^{-5} \times 5^1}{1!} + \frac{e^{-5} \times 5^2}{2!} \\ &= e^{-5}\left(\frac{1}{1} + \frac{5}{1} + \frac{25}{2}\right) = 0.125 \end{aligned}$$

As a final point, it may be useful to note that the Poisson distribution can be used to approximate the $B(n, p)$ distribution when n is very large (larger than 150, say) and the success probability p is very small (smaller than 0.01, say). A parameter value of $\lambda = np$ should be used for the Poisson distribution, so that it has the same expected value as the binomial distribution.

3.4.2 Examples of the Poisson Distribution

<u>Example 3</u>

<u>Software Errors</u>

Suppose that the number of errors in a piece of software has a Poisson distribution with parameter $\lambda = 3$. This parameter immediately implies that the expected number of errors is three and that the variance in the number of errors is also equal to three.

The distribution of the number of errors is illustrated in Figure 3.21, and the probability that a piece of software has no errors is

$$P(X = 0) = \frac{e^{-3} \times 3^0}{0!} = e^{-3} = 0.050$$

The probability that there are three or more errors in a piece of software is

$$P(X \geq 3) = 1 - P(X = 0) - P(X = 1) - P(X = 2)$$
$$= 1 - \frac{e^{-3} \times 3^0}{0!} - \frac{e^{-3} \times 3^1}{1!} - \frac{e^{-3} \times 3^2}{2!}$$
$$= 1 - e^{-3}\left(\frac{1}{1} + \frac{3}{1} + \frac{9}{2}\right) = 1 - 0.423 = 0.577$$

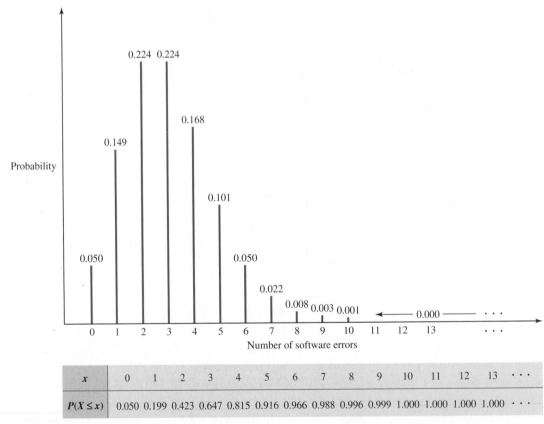

x	0	1	2	3	4	5	6	7	8	9	10	11	12	13	$\cdots$
$P(X \leq x)$	0.050	0.199	0.423	0.647	0.815	0.916	0.966	0.988	0.996	0.999	1.000	1.000	1.000	1.000	$\cdots$

FIGURE 3.21

Probability mass function and cumulative distribution function of a Poisson distribution with parameter $\lambda = 3$, the distribution of the number of software errors

Example 27
Glass Sheet Flaws

A quality inspector at a glass manufacturing company inspects sheets of glass to check for any slight imperfections. Suppose that the number of these flaws in a glass sheet has a Poisson distribution with parameter $\lambda = 0.5$. This implies that the expected number of flaws per sheet is only 0.5.

The distribution of the number of flaws per sheet is shown in Figure 3.22. The probability that there are no flaws in a sheet is

$$P(X = 0) = \frac{e^{-0.5} \times 0.5^0}{0!} = e^{-0.5} = 0.607$$

so that about 61% of the glass sheets are in "perfect" condition. Sheets with two or more flaws are scrapped by the company, and this happens with a probability of

$$P(X \geq 2) = 1 - P(X = 0) - P(X = 1) = 1 - \frac{e^{-0.5} \times 0.5^0}{0!} - \frac{e^{-0.5} \times 0.5^1}{1!}$$
$$= 1 - e^{-0.5}\left(\frac{1}{1} + \frac{0.5}{1}\right) = 1 - 0.910 = 0.090$$

Consequently, about 9% of the glass sheets need to be scrapped and recycled through the company's manufacturing process.

FIGURE 3.22

Probability mass function and cumulative distribution function of a Poisson distribution with parameter $\lambda = 0.5$, the distribution of the number of flaws in a glass sheet

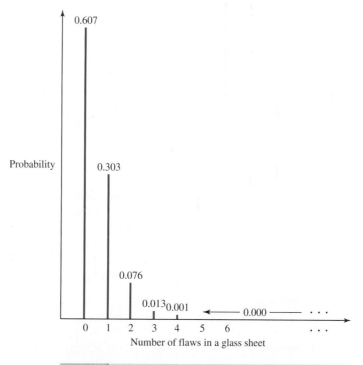

x	0	1	2	3	4	5	6	$\cdots$
$P(X \leq x)$	0.607	0.910	0.986	0.999	1.000	1.000	1.000	$\cdots$

Example 28

Hospital Emergency Room Arrivals

A hospital emergency room accepts an average of about 47 bone fracture patients per *week*. How might the number of bone fracture patients arriving in a certain *day* be modeled? Assuming that bone fracture accidents are equally likely to occur on any day of the week, it is reasonable to expect an average of $47/7 = 6.71$ patients per day. If the Poisson distribution is chosen to model the number of bone fracture patients, then it is appropriate to choose a parameter value of $\lambda = 6.71$ because this is the expected value of the distribution.

The hospital manager has decided to allocate emergency room resources that are sufficient to comfortably cope with up to 10 bone fracture patients per day. The Poisson distribution can be used to predict the probability that on any given day these resources will be inadequate. The probability that no more than 10 bone fracture patients arrive at the emergency room is predicted to be

$$P(X \le 10) = \sum_{x=0}^{10} \frac{e^{-6.71} \times 6.71^x}{x!}$$

$$= e^{-6.71} \left(\frac{1}{1} + \frac{6.71}{1} + \frac{6.71^2}{2!} + \frac{6.71^3}{3!} + \cdots + \frac{6.71^{10}}{10!} \right) = 0.921$$

Consequently the manager should expect the emergency room to require additional assistance on about 8% of days, which is an average of about 29 days per year.

■ 3.4.3   Problems

3.4.1 If $X \sim P(3.2)$, calculate:
(a) $P(X = 1)$ (b) $P(X \le 3)$
(c) $P(X \ge 6)$ (d) $P(X = 0 | X \le 3)$

3.4.2 If $X \sim P(2.1)$, calculate:
(a) $P(X = 0)$ (b) $P(X \le 2)$
(c) $P(X \ge 5)$ (d) $P(X = 1 | X \le 2)$

3.4.3 If $X \sim P(\lambda)$, show that

$$E(X) = \sum_{x=0}^{\infty} x P(X = x) = \lambda$$

Also show that

$$E(X^2) - E(X) = E(X(X - 1))$$

$$= \sum_{x=0}^{\infty} x(x-1) P(X = x) = \lambda^2$$

and use this result to show that $\mathrm{Var}(X) = \lambda$.

3.4.4 The number of cracks in a ceramic tile has a Poisson distribution with a mean of $\lambda = 2.4$. What is the probability that a tile has no cracks? What is the probability that a tile has four or more cracks?

3.4.5 On average there are about 25 imperfections in 100 meters of optical cable. Use the Poisson distribution to estimate the probability that there are no imperfections in 1 meter of

cable. What is the probability that there is no more than one imperfection in 1 meter of cable?

3.4.6 On average there are *four* traffic accidents in a city during one hour of rush-hour traffic. Use the Poisson distribution to calculate the probability that in one such hour there are
(a) no accidents
(b) at least six accidents

3.4.7 Recall that the Poisson distribution with a parameter value of $\lambda = np$ can be used to approximate the $B(n, p)$ distribution when n is very large and the success probability p is very small.

A box contains 500 electrical switches, each one of which has a probability of 0.005 of being defective. Use the Poisson distribution to make an approximate calculation of the probability that the box contains no more than 3 defective switches.

3.4.8 In a scanning process, the number of misrecorded pieces of information has a Poisson distribution with parameter $\lambda = 9.2$.
(a) What is the probability that there are between 6 and 10 misrecorded pieces of information?
(b) What is the probability that there are no more than 4 misrecorded pieces of information?
(This problem is continued in Problem 5.6.10.)

3.5 The Multinomial Distribution

3.5.1 Definition of the Multinomial Distribution

Whereas the binomial distribution is generated from a sequence of Bernoulli trials, each with only two possible outcomes, the **multinomial distribution** is a more general distribution that arises when each of the trials can have three or more possible outcomes.

The Multinomial Distribution

Consider a sequence of n independent trials where each individual trial can have k outcomes that occur with constant probability values $p_1, \ldots, p_k$ with $p_1 + \cdots + p_k = 1$. The random variables $X_1, \ldots, X_k$ that count the number of occurrences of each outcome are said to have a **multinomial distribution,** and their joint probability mass function is

$$P(X_1 = x_1, \ldots, X_k = x_k) = \frac{n!}{x_1! \cdots x_k!} \times p_1^{x_1} \times \cdots \times p_k^{x_k}$$

for nonnegative integer values of the x_i satisfying $x_1 + \cdots + x_k = n$.

The random variables $X_1, \ldots, X_k$ have expectations and variances given by

$$E(X_i) = np_i \qquad \text{and} \qquad \text{Var}(X_i) = np_i(1 - p_i)$$

but they are *not* independent.

The justification of this probability mass function is similar to that of the binomial distribution. If $X_1 = x_1, \ldots, X_k = x_k$, then this means that x_i of the trials take outcome i with probability p_i, so that a particular realization of the k trials has a probability of

$$p_1^{x_1} \times \cdots \times p_k^{x_k}$$

Since the total number of possible rearrangements of the n trial outcomes satisfying $X_1 = x_1, \ldots, X_k = x_k$ is

$$\frac{n!}{x_1! \cdots x_k!}$$

the probability mass function is as indicated.

Since the random variable X_i counts the number of trials that take outcome i, which has a constant probability of p_i, the *marginal* distribution of X_i is

$$X_i \sim B(n, p_i)$$

Consequently, $E(X_i) = np_i$ and $\text{Var}(X_i) = np_i(1 - p_i)$. However, the random variables $X_1, \ldots, X_k$ are *not* independent.

3.5.2 Examples of the Multinomial Distribution

Example 1
Machine Breakdowns

Recall that the machine breakdowns are attributable to electrical faults, mechanical faults, and operator misuse, and that these causes occur with probabilities of 0.2, 0.5, and 0.3, respectively. The engineer in charge of the maintenance of the machine is interested in predicting the causes of the next 10 breakdowns. If X_1 is the number of breakdowns due to electrical reasons, X_2 is the number of breakdowns due to mechanical reasons, and X_3 is the number of breakdowns due to operator misuse, then

$$X_1 + X_2 + X_3 = 10$$

and if the breakdown causes can be assumed to be independent of one another, the random variables X_1, X_2, and X_3 have a multinomial distribution with a probability mass function

$$P(X_1 = x_1, X_2 = x_2, X_3 = x_3) = \frac{10!}{x_1!\, x_2!\, x_3!} \times 0.2^{x_1} \times 0.5^{x_2} \times 0.3^{x_3}$$

The probability that there will be *three* electrical breakdowns, *five* mechanical breakdowns, and *two* misuse breakdowns is therefore

$$P(X_1 = 3, X_2 = 5, X_3 = 2) = \frac{10!}{3!\, 5!\, 2!} \times 0.2^3 \times 0.5^5 \times 0.3^2 = 0.057$$

The expected number of electrical breakdowns is

$$E(X_1) = np_1 = 10 \times 0.2 = 2$$

the expected number of mechanical breakdowns is

$$E(X_2) = np_2 = 10 \times 0.5 = 5$$

and the expected number of misuse breakdowns is

$$E(X_3) = np_3 = 10 \times 0.3 = 3$$

If the engineer is interested in the probability of there being no more than *two* electrical breakdowns, then this can be calculated by noting that $X_1 \sim B(10, 0.2)$, so that

$$P(X_1 \leq 2) = P(X_1 = 0) + P(X_1 = 1) + P(X_1 = 2)$$
$$= \binom{10}{0} \times 0.2^0 \times 0.8^{10} + \binom{10}{1} \times 0.2^1 \times 0.8^9$$
$$+ \binom{10}{2} \times 0.2^2 \times 0.8^8$$
$$= 0.107 + 0.268 + 0.302 = 0.677$$

Example 29
Drug Allergies

Patients being treated with a particular drug run the risk of being allergic to the drug. A patient's reaction to the drug is characterized by doctors as being *hyperallergic*, *allergic*, *mildly allergic,* or *not allergic*, and these have probability values of 0.12, 0.28, 0.33, and 0.27, respectively.

Suppose that *nine* patients are administered the drug, as illustrated in Figure 3.23. How can the doctors predict the types of allergy that will be encountered? If X_1, X_2, X_3, and X_4 are the numbers of patients exhibiting the four types of reaction, then assuming that one patient's reaction can reasonably be assumed to be independent of another patient's (this might not be the case, for example, if the patients are related to one another), the random variables X_1, X_2, X_3, and X_4 will have a multinomial distribution with a probability mass function

$$P(X_1 = x_1, X_2 = x_2, X_3 = x_3, X_4 = x_4)$$
$$= \frac{9!}{x_1!\, x_2!\, x_3!\, x_4!} \times 0.12^{x_1} \times 0.28^{x_2} \times 0.33^{x_3} \times 0.27^{x_4}$$

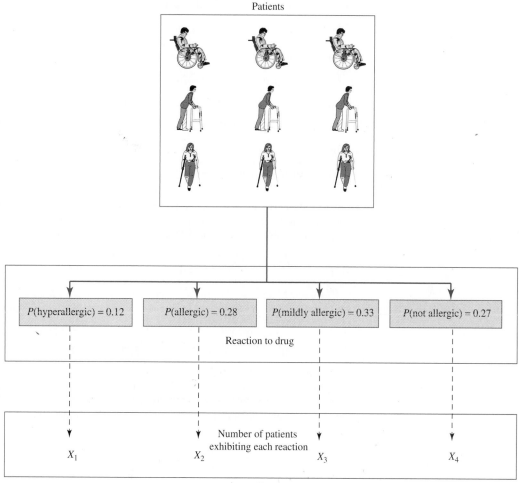

Patients

FIGURE 3.23

The number of patients exhibiting each kind of drug reaction has a *multinomial* distribution

The probability that *two* patients are hyperallergic, *one* patient is allergic, *four* patients are mildly allergic, and *two* patients exhibit no reaction is therefore

$$P(X_1 = 2, X_2 = 1, X_3 = 4, X_4 = 2)$$
$$= \frac{9!}{2!\,1!\,4!\,2!} \times 0.12^2 \times 0.28^1 \times 0.33^4 \times 0.27^2 = 0.013$$

The probability that *no* patients are hyperallergic, *one* patient is allergic, *four* patients are mildly allergic, and *four* patients exhibit no reaction is

$$P(X_1 = 0, X_2 = 1, X_3 = 4, X_4 = 4)$$
$$= \frac{9!}{0!\,1!\,4!\,4!} \times 0.12^0 \times 0.28^1 \times 0.33^4 \times 0.27^4 = 0.011$$

The expected number of hyperallergic reactions is

$$E(X_1) = np_1 = 9 \times 0.12 = 1.08$$

the expected number of allergic reactions is

$$E(X_2) = np_2 = 9 \times 0.28 = 2.52$$

the expected number of mildly allergic reactions is

$$E(X_3) = np_3 = 9 \times 0.33 = 2.97$$

and the expected number of patients not exhibiting a reaction is

$$E(X_4) = np_4 = 9 \times 0.27 = 2.43$$

GAMES OF CHANCE

Suppose that *eight* cards are chosen at random from a pack of cards with replacement. What is the distribution of the number of cards obtained from each of the four suits?

Since the drawings are made with replacement, each of the four suits is equally likely to be chosen at each of the drawings, so that the numbers of hearts X_1, clubs X_2, diamonds X_3, and spades X_4 have a multinomial distribution with probability mass function

$$P(X_1 = x_1, X_2 = x_2, X_3 = x_3, X_4 = x_4)$$

$$= \frac{8!}{x_1!\, x_2!\, x_3!\, x_4!} \times \left(\frac{1}{4}\right)^{x_1} \times \left(\frac{1}{4}\right)^{x_2} \times \left(\frac{1}{4}\right)^{x_3} \times \left(\frac{1}{4}\right)^{x_4}$$

However, since $X_1 + X_2 + X_3 + X_4 = 8$, this simplifies to

$$P(X_1 = x_1, X_2 = x_2, X_3 = x_3, X_4 = x_4) = \frac{8!}{x_1!\, x_2!\, x_3!\, x_4!} \times \left(\frac{1}{4}\right)^{8}$$

The probability of drawing *three* hearts, *two* clubs, *two* diamonds, and *one* spade is therefore

$$P(X_1 = 3, X_2 = 2, X_3 = 2, X_4 = 1) = \frac{8!}{3!\, 2!\, 2!\, 1!} \times \left(\frac{1}{4}\right)^{8} = \frac{105}{4096} = 0.026$$

and the expected number of cards from each suit is two. The probability of actually drawing two cards from each suit is

$$P(X_1 = 2, X_2 = 2, X_3 = 2, X_4 = 2) = \frac{8!}{2!\, 2!\, 2!\, 2!} \times \left(\frac{1}{4}\right)^{8} = \frac{315}{8192} = 0.038$$

■ 3.5.3 Problems

3.5.1 A garage sells three types of tires, type A, type B, and type C. A customer purchases type A with probability 0.23, type B with probability 0.48, and type C with probability 0.29.
(a) What is the probability that the next 11 customers purchase four sets of type A, five sets of type B, and two sets of type C?
(b) What is the probability that fewer than three sets of type A are sold to the next seven customers?

3.5.2 A fair die is rolled 15 times. Calculate the probability that there are:
(a) Exactly three 6s and three 5s
(b) Exactly three 6s, three 5s, and four 4s
(c) Exactly two 6s $\text{binomial} \quad \frac{1}{6} \quad \frac{5}{6}$
What is the expected number of 6s obtained?

3.5.3 Recall Problems 3.1.4 and 3.2.5, where an archer hits a bull's-eye with a probability of 0.09. Suppose also that the archer misses the target completely with a probability of

0.12. If the archer shoots *eight* arrows whose performances are independent of each other, calculate the probability of:

(a) Scoring exactly two bull's-eyes and missing the target exactly once

(b) Scoring exactly one bull's-eye and missing the target exactly twice

(c) Scoring at least two bull's-eyes

What is the expected number of times the archer misses the target?

3.5.4 A researcher plants 22 seedlings. After one month, independent of the other seedlings, each seedling has a probability of 0.08 of being dead, a probability of 0.19 of exhibiting slow growth, a probability of 0.42 of exhibiting medium growth, and a probability of 0.31 of exhibiting strong growth. What is the expected number of seedlings in each of these four categories after one month? Calculate

the probability that after one month:

(a) Exactly three seedlings are dead, exactly four exhibit slow growth, and exactly six exhibit medium growth.

(b) Exactly five seedlings are dead, exactly five exhibit slow growth, and exactly seven exhibit strong growth.

 (c) No more than two seedlings have died.

3.5.5 Recall Problems 3.1.9 and 3.2.9, where a company receives 60% of its orders over the Internet. Suppose that 30% of the orders received over the Internet are large orders, and 40% of the orders received by other means are large orders. Out of eight independently placed orders, what is the probability that two will be large orders received over the Internet, two will be small orders received over the Internet, two will be large orders not received over the Internet, and two will be small orders not received over the Internet?

3.6 Case Study: Microelectronic Solder Joints

 Recall that there is a probability of 0.12 that a solder joint will have an hourglass shape. If an assembly is comprised of 64 solder joints, and if their shapes can be taken to be independent of each other, then the total number of hourglass-shaped solder joints on the assembly is distributed as a $B(64, 0.12)$ random variable. This has an expectation of $64 \times 0.12 = 7.68$ with a standard deviation $\sqrt{64 \times 0.12 \times 0.88} = 2.60$. The probability that there are at most two hourglass-shaped solder joints on the assembly is

$$P(X = 0) + P(X = 1) + P(X = 2)$$
$$= 0.88^{64} + (64 \times 0.12 \times 0.88^{63}) + \left(\frac{64 \times 63}{2} \times 0.12^2 \times 0.88^{62}\right) = 0.0132$$

The probability that a solder joint is hourglass shaped and is cracked is

$$P(\text{hourglass}) \times P(\text{cracked}|\text{hourglass}) = 0.12 \times 0.005 = 0.0006$$

Consequently, if a researcher examines solder joints one at a time until a cracked hourglass-shaped solder joint is found, the number of solder joints that need to be examined has a geometric distribution with parameter $p = 0.0006$. The probability that more than 1000 solder joints will need to be examined is

$$1 - P(X \le 1000) = 1 - (1 - (1 - 0.0006)^{1000}) = 0.9994^{1000} = 0.549$$

If the researcher wishes to find two cracked hourglass-shaped joints, then the number of solder joints that need to be examined has a negative binomial distribution with parameters $r = 2$ and $p = 0.0006$. This has an expectation of $2/0.0006 = 3333.3$.

Finally, the number of barrel-shaped, cylinder-shaped, and hourglass-shaped solder joints on an assembly comprising of 16 solder joints has a multinomial distribution with $n = 16$ and probabilities $p_1 = 0.85$, $p_2 = 0.03$, and $p_3 = 0.12$ as given in Chapter 1. The probability that the assembly will have 12 barrel-shaped solder joints, 1 cylinder-shaped solder joint, and 3 hourglass-shaped solder joints is therefore

$$\frac{16!}{12! \, 1! \, 13!} \times 0.85^{12} \times 0.03^{1} \times 0.12^{3} = 0.054$$

3.7   Supplementary Problems

3.7.1 An integrated circuit manufacturer produces wafers that contain 18 chips. Each chip has a probability of 0.085 of not being placed quite correctly on the wafer.
  (a) Find the probability that a wafer contains at least three incorrectly placed chips.
  (b) What is the probability that a wafer contains no more than one incorrectly placed chip?
  (c) What is the expected number of incorrectly placed chips on a wafer?

3.7.2 A biologist has a culture consisting of 13 cells. In a period of one hour, independent of the other cells, there is a probability of 0.4 that each of these cells splits into 2 cells. What is the probability that after one hour the biologist has at least 16 cells? What is the expected number of cells after one hour?

3.7.3 A beverage company has three different formulas for its soft drink product. Tests reveal that 40% of consumers prefer formula I, 25% of consumers prefer formula II, and 35% of consumers prefer formula III. If eight consumers are chosen at random, calculate the probability that:
  (a) Exactly two prefer formula I and exactly three prefer formula II.
  (b) Exactly three prefer formula I and exactly four prefer formula III.
  (c) No more than two prefer formula III.

3.7.4 A company's toll-free complaints line receives an average of about 40 calls per hour. Use the Poisson distribution to estimate the probability that in *one minute* there are
  (a) no calls
  (b) exactly one call
  (c) three or more calls

3.7.5 The number of radioactive particles passing through a counter in one minute has a Poisson distribution with $\lambda = 3.3$. What is the probability that in one minute there are exactly two particles passing through the counter? What is the probability that in one minute there are at least six particles passing through the counter?

3.7.6 In a typical sports playoff series, two teams play a sequence of games until one team, the eventual winner, has won four games. Suppose that in each game team A beats team B with a probability of 0.55, and that the results of different games are independent.
  (a) Explain how the negative binomial distribution can be used to analyze this problem.
  (b) What is the probability that team A wins the series in game seven?
  (c) What is the probability that team A wins the series in game six?
  (d) What is the probability that the series is over after game five?
  (e) What is the probability that team A wins the series?

3.7.7 A golf shop sells both right-handed and left-handed sets of clubs, and 42% of customers purchase right-handed sets whereas 58% of customers purchase left-handed sets. The owner opens the shop one day and waits for customers to arrive.
  (a) What is the probability that the ninth set of clubs sold in the day is the third set of left-handed clubs sold that day?
  (b) What is the probability that four sets of right-handed clubs are sold before four sets of left-handed clubs are sold?

3.7.8 Box A contains six red balls and five blue balls. Box B contains five red balls and six blue balls. A fair coin is tossed, and if a head is obtained, three balls are taken at random from box A and placed in a jar. If a tail is obtained, three balls are taken at random from box B and placed in a jar. If the jar contains two red balls and one blue ball after the toss, what is the probability that a head had been obtained?

3.7.9 Suppose that a box contains 40 items of which 4 are defective. If a random sample of 5 items is chosen, what is the probability that it contains no more than 1 defective item? If a random sample of 5 items is chosen from a collection of 4,000,000 items of which 400,000 are defective, what is the probability that it contains no more than 1 defective item?

3.7.10 **(a)** If a fair die is rolled 22 times, what is the probability that a 6 is obtained exactly 3 times?

(b) If a fair die is rolled, what is the probability that the third time that a 6 is obtained is on the tenth roll?

(c) If a fair coin is tossed 11 times, what is the probability that three or fewer heads are obtained?

3.7.11 A box contains 11 red balls and 8 blue balls. Six balls are taken at random from the box without replacement. What is the probability that out of these 6 balls, exactly 3 are red and exactly 3 are blue?

3.7.12 Are the following statements true or false?

(a) An unfair coin, for which the probabilities of a head and a tail are different, is tossed seven times. The probability of getting three heads and four tails cannot be the same as the probability of getting four heads and three tails.

(b) A fair die is rolled. The probability that a 6 is not obtained until the eighth roll is 78,125/1,679,616.

(c) A fair coin is tossed. The probability that the fourth head occurs on the seventh toss is 5/32.

(d) The proportion of heads in 16 tosses of a fair coin has a standard deviation 1/8.

3.7.13 **(a)** A fair die is rolled 10 times. What is the probability of obtaining the outcome 6 exactly 3 times?

(b) A fair die is repeatedly rolled. What is the probability that the outcome 6 is obtained for the fourth time on the twentieth roll?

(c) A fair die is rolled nine times. What is the probability of obtaining the outcomes 5 and 6 both exactly two times each?

3.7.14 The number of imperfections in an object has a Poisson distribution with a mean $\lambda = 8.3$. If the number of imperfections is 4 or less, the object is called "top quality." If the number of imperfections is between 5 and 8 inclusive, the object is called "good quality." If the number of imperfections is between 9 and 12 inclusive, the object is called "normal quality." If the number of imperfections is 13 or more, the object is called "bad quality." The number of imperfections in different objects are independent of each other.

(a) A set of seven articles is taken. What is the probability that the set has exactly two top-quality, two good-quality, two normal-quality and one bad-quality objects?

(b) A set of 10 articles is taken. What are the expectation and the standard deviation of the number of normal quality objects in the set.

(c) A set of eight articles is taken. What is the probability that the sum of the number of top quality and good quality objects is three or less?

CHAPTER FOUR

Continuous Probability Distributions

Most of the common continuous probability distributions are presented in Chapters 4 and 5. The probability density functions of these distributions are described by formulas that depend on some **parameter** values. The expectations and variances of the distributions are specified in terms of these parameters. The probability values associated with these continuous distributions are sometimes straightforward to calculate, although some distributions require the use of a software package.

In this chapter the most simple continuous distribution, the **uniform distribution**, is considered first, followed by the **exponential distribution**, which is often appropriate for modeling failure rates or waiting times. The exponential distribution can be generalized to both the **gamma distribution** and the **Weibull distribution**. Finally, the **beta distribution** that can be useful for modeling proportions is discussed. The exponential and gamma distributions are used to discuss the **Poisson process**, which is a simple **stochastic process**.

However, the most important continuous probability distribution for statistical data analysis—the distribution that will be used the most frequently in the remainder of this book— is the **normal distribution**. This distribution is discussed in Chapter 5 together with same other continuous probability distributions that are related to the normal distribution.

4.1 The Uniform Distribution

4.1.1 Definition of the Uniform Distribution

The simplest continuous probability distribution has a **flat probability density function** between two points a and b, say, as illustrated in Figure 4.1. It is called a **uniform distribution** between a and b and can be written

$$X \sim U(a, b)$$

In order for the area under the probability density function to be equal to 1, it must have a height of $1/(b - a)$, so that

$$f(x) = \frac{1}{b - a}$$

for $a \leq x \leq b$ and $f(x) = 0$ elsewhere. The cumulative distribution function is

$$F(x) = \int_{y=a}^{x} \frac{1}{b - a} \, dy = \frac{x - a}{b - a}$$

for $a \leq x \leq b$.

A random variable $X \sim U(a, b)$ has the simple interpretation that it is "equally" likely to take values anywhere between a and b. More precisely, the random variable is equally

FIGURE 4.1

Probability density function of a
$U(a, b)$ distribution

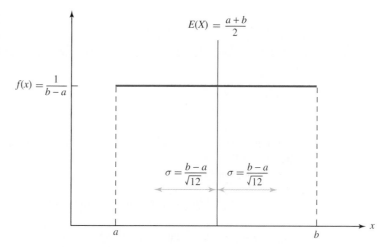

likely to take a value within any interval of length δ that is contained between a and b. Moreover, the probability that the random variable does fall within a given interval of length δ is $\delta/(b - a)$.

A $U(0, 1)$ distribution is often thought of as being a "standard" uniform distribution. A standard uniform distribution can be obtained from any uniform distribution using a linear transformation, since if

$$X \sim U(a, b)$$

then

$$Y = \frac{X - a}{b - a} \sim U(0, 1)$$

A uniform distribution is a symmetric distribution, so that its expectation is its middle value

$$E(X) = \frac{a + b}{2}$$

Also,

$$E(X^2) = \int_a^b x^2 \frac{1}{b - a} \, dx = \frac{b^3 - a^3}{3(b - a)} = \frac{a^2 + ab + b^2}{3}$$

and the variance of a $U(a, b)$ random variable is therefore

$$\text{Var}(X) = E(X^2) - (E(X))^2 = \frac{a^2 + ab + b^2}{3} - \frac{(a + b)^2}{4} = \frac{(b - a)^2}{12}$$

Notice that the variance increases as $b - a$ gets larger and the uniform distribution becomes more spread out.

The Uniform Distribution

A random variable X with a flat probability density function between two points a and b, so that

$$f(x) = \frac{1}{b-a}$$

for $a \leq x \leq b$ and $f(x) = 0$ elsewhere, is said to have a **uniform distribution**, which is written $X \sim U(a, b)$. The cumulative distribution function is

$$F(x) = \frac{x-a}{b-a}$$

for $a \leq x \leq b$, and the expectation and variance are

$$E(X) = \frac{a+b}{2} \quad \text{and} \quad \text{Var}(X) = \frac{(b-a)^2}{12}.$$

Since the distribution is symmetric, the median of a $U(a, b)$ distribution is, like the expectation, equal to the middle value $(a+b)/2$. In general, the pth quantile of the distribution is

$$(1-p)a + pb$$

and the interquartile range is $(b-a)/2$.

4.1.2   Examples of the Uniform Distribution

Example 30
Pearl Oyster Farming

When pearl oysters are opened, pearls of various sizes are found. Suppose that each oyster contains a pearl with a diameter in mm that has a $U(0, 10)$ distribution. The expected pearl diameter is therefore 5 mm, with a variance of

$$\text{Var}(X) = \frac{(10-0)^2}{12} = 8.33$$

and a standard deviation of $\sigma = \sqrt{8.33} = 2.89$ mm, as shown in Figure 4.2.

FIGURE 4.2

Distribution of pearl diameters

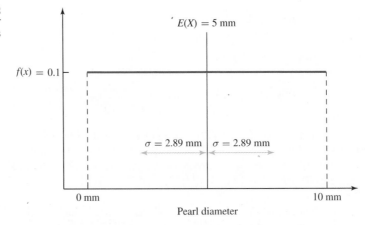

Pearls with a diameter of at least 4 mm have commercial value. The probability that an oyster contains a pearl of commercial value is therefore

$$P(X \geq 4) = 1 - F(4) = 1 - 0.4 = 0.6$$

Suppose that a farmer retrieves 10 oysters out of the water and that the random variable Y represents the number of them containing pearls of commercial value. If the oysters grow pearls independently of one another, Y has a binomial distribution with parameters $n = 10$ and $p = 0.6$, and the probability that at least 8 of the oysters contain pearls of commercial value is

$$P(Y \geq 8) = P(Y = 8) + P(Y = 9) + P(Y = 10)$$
$$= \binom{10}{8} \times 0.6^8 \times 0.4^2 + \binom{10}{9} \times 0.6^9 \times 0.4^1 + \binom{10}{10} \times 0.6^{10} \times 0.4^0$$
$$= 0.121 + 0.040 + 0.006 = 0.167$$

GAMES OF CHANCE

In the dial-spinning game discussed in Section 2.2, both the angle θ and the winnings X have uniform distributions. Specifically, $\theta \sim U(0, 180)$ and $X \sim U(0, 1000)$. Clearly,

$$E(\theta) = 90 \quad \text{and} \quad E(X) = 500$$

and the formula for the variance gives

$$\text{Var}(\theta) = \frac{(180 - 0)^2}{12} = 2700$$

with a standard deviation of $\sigma_\theta = \sqrt{2700} = 51.96$, and

$$\text{Var}(X) = \frac{(1000 - 0)^2}{12} = 83,333$$

with a standard deviation of $\sigma_X = \sqrt{83,333} = 288.7$.

■ 4.1.3 Problems

4.1.1 Suppose that $X \sim U(-3, 8)$. Find:
 (a) $E(X)$
 (b) The standard deviation of X
 (c) The upper quartile of the distribution
 (d) $P(0 \leq X \leq 4)$

4.1.2 A new battery supposedly with a charge of 1.5 volts actually has a voltage with a uniform distribution between 1.43 and 1.60 volts.
 (a) What is the expectation of the voltage?
 (b) What is the standard deviation of the voltage? $= \sqrt{V(X)}$
 (c) What is the cumulative distribution function of the voltage?
 (d) What is the probability that a battery has a voltage less than 1.48 volts?

 (e) If a box contains 50 batteries, what are the expectation and variance of the number of batteries in the box with a voltage less than 1.5 volts?

4.1.3 A computer random-number generator produces numbers that have a uniform distribution between 0 and 1.
 (a) If 20 random numbers are generated, what are the expectation and variance of the number of them that lie in each of the four intervals $[0.00, 0.30)$, $[0.30, 0.50)$, $[0.50, 0.75)$, and $[0.75, 1.00)$?
 (b) What is the probability that exactly five numbers lie in each of the four intervals?

4.1.4 The lengths in meters of pieces of scrap wood found on a building site are uniformly distributed between 0.0 and 2.5.

(a) What are the expectation and variance of the lengths?

(b) What is the probability that at least 20 out of 25 pieces of scrap wood are longer than 1 meter?

4.1.5 Suppose that a metal pin has a diameter that has a uniform distribution between 4.182 mm and 4.185 mm.

(a) What is the probability that a pin will fit into a hole that has a diameter of 4.184 mm?

(b) If a pin does fit into the hole, what is the probability that the difference between the diameter of the hole and the diameter of the pin is less than 0.0005 mm?

4.1.6 When employees undergo an evaluation, their scores are independent and uniformly distributed between 60 and 100.

(a) If six employees take the evaluation, what is the probability that half of them score more than 85 and half less?

(b) If six employees take the evaluation, what is the probability that two of them score less than 80, two of them score between 80 and 90, and the remaining two score more than 90?

(c) If the employees are tested sequentially, what is the expected number of employees who need to be tested before three are found with scores higher than 90?

4.2   The Exponential Distribution

4.2.1   Definition of the Exponential Distribution

The **exponential distribution** has a state space $x \geq 0$ and is often used to model failure or waiting times and interarrival times. It has a probability density function

$$f(x) = \lambda e^{-\lambda x}$$

for $x \geq 0$ and $f(x) = 0$ for $x < 0$, which depends upon a parameter $\lambda > 0$. The cumulative distribution function is

$$F(x) = \int_0^x \lambda e^{-\lambda y} \, dy = 1 - e^{-\lambda x}$$

for $x \geq 0$.

The expectation of an exponential distribution with parameter λ can be calculated using integration by parts as

$$E(X) = \int_0^\infty x \lambda e^{-\lambda x} \, dx = \frac{1}{\lambda}$$

Similarly, integration by parts reveals that

$$E(X^2) = \int_0^\infty x^2 \lambda e^{-\lambda x} \, dx = \frac{2}{\lambda^2}$$

so that

$$\text{Var}(X) = E(X^2) - E(X)^2 = \frac{2}{\lambda^2} - \frac{1}{\lambda^2} = \frac{1}{\lambda^2}$$

Notice that the standard deviation of the distribution is consequently $1/\lambda$, which is also the expectation of the distribution.

The Exponential Distribution

An **exponential distribution** with parameter $\lambda > 0$ has a probability density function

$$f(x) = \lambda e^{-\lambda x}$$

for $x \geq 0$ and $f(x) = 0$ for $x < 0$, and a cumulative distribution function

$$F(x) = 1 - e^{-\lambda x}$$

for $x \geq 0$. It is useful for modeling failure times and waiting times. Its expectation and variance are

$$E(X) = \frac{1}{\lambda} \quad \text{and} \quad \text{Var}(X) = \frac{1}{\lambda^2}$$

Figure 4.3 shows the probability density function of an exponential distribution with parameter $\lambda = 1$, and Figure 4.4 shows the probability density function of an exponential distribution with parameter $\lambda = 1/2$. The first distribution has a mean and standard deviation of 1, and the second distribution has a mean and standard deviation equal to 2. Notice that the shapes of the probability density functions are smooth exponential decays.

The pth quantile of an exponential distribution, that is, the value of x that satisfies $F(x) = p$, is

$$-\frac{\ln(1 - p)}{\lambda}$$

In particular, the median of the distribution is

$$-\frac{\ln(1.0 - 0.5)}{\lambda} = \frac{0.693}{\lambda} = 0.693 \times E(X)$$

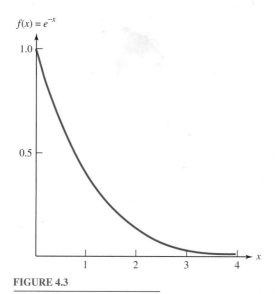

FIGURE 4.3

Probability density function of an exponential distribution with parameter $\lambda = 1$

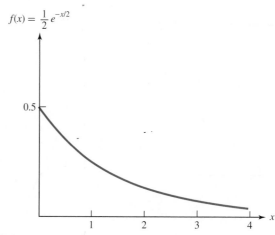

FIGURE 4.4

Probability density function of an exponential distribution with parameter $\lambda = 1/2$

The fact that the median is smaller than the expected value is because of the long right tail of the probability density function.

  The probability values of an exponential distribution are easily calculated because of the simple form of the cumulative distribution function. Nevertheless, you should also find that they are available on your computer package. Make sure that you specify the distribution correctly since some packages define the parameter of the exponential distribution to be $1/\lambda$ instead of λ.

4.2.2 The Memoryless Property of the Exponential Distribution

An important aspect of the exponential distribution is its **memoryless property**. This property states that if X has an exponential distribution with parameter λ, then conditional on $X \geq x_0$ for some fixed value x_0, the quantity $X - x_0$ also has an exponential distribution with parameter λ. In other words, if X measures the time until a certain event occurs and the event has not occurred by time x_0, the *additional* waiting time for the event to occur beyond x_0 has the same exponential distribution as X. The process seems to "forget" that a time x_0 has already elapsed and acts as though it is just starting afresh at time zero.

The memoryless property can be shown in the following manner. Notice that if X has an exponential distribution with parameter λ, then

$$P(X \geq x) = 1 - F(x) = e^{-\lambda x}$$

Then if the random variable Y represents the additional time beyond x_0 that elapses before the event occurs,

$$P(Y \geq y) = P(X \geq x_0 + y \mid X \geq x_0) = \frac{P(X \geq x_0 + y)}{P(X \geq x_0)} = \frac{e^{-\lambda(x_0+y)}}{e^{-\lambda x_0}} = e^{-\lambda y}$$

so that Y also has an exponential distribution with parameter λ. In graphical terms, the memoryless property follows from the fact that the section of the probability density function of an exponential distribution beyond a certain point x_0 is just a *scaled* version of the whole probability density function, as illustrated in Figure 4.5.

FIGURE 4.5

Illustration of the memoryless property of the exponential distribution. The part of the probability density function beyond x_0 is a **scaled** version of the whole probability density function

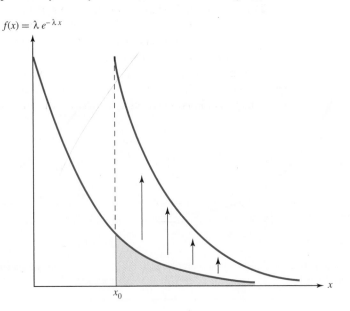

$f(x) = \lambda e^{-\lambda x}$

The implications of the memoryless property can be rather confusing when first encountered. Suppose that you are waiting at a bus stop and that the time in minutes until the arrival of the bus has an exponential distribution with $\lambda = 0.2$. The expected time that you will wait is consequently $1/\lambda = 5$ minutes. However, if after 1 minute the bus has not yet arrived, what is the expectation of the *additional* time that you must wait?

Unfortunately, it has not been reduced to 4 minutes but is still, as before, 5 minutes. This is because the additional waiting time until the bus arrives beyond the first minute during which you know the bus did not arrive still has an exponential distribution with $\lambda = 0.2$. In fact, as long as the bus has not arrived, no matter how long you have waited, you always have an expected additional waiting time of 5 minutes! This is true right up until the time you first spot the bus coming.

The memoryless property of the exponential distribution makes it attractive for modeling many processes. However, at the same time it reveals that the exponential distribution is unsuitable for modeling processes that clearly do not possess such a property. The gamma distribution and the Weibull distribution, which are discussed in subsequent sections, are generalizations of the exponential distribution that do not possess this memoryless property (in fact, the exponential distribution is the only continuous distribution with a memoryless property) and which may be more suitable than the exponential distribution for modeling certain waiting times.

In particular, if an electronic component fails due to a random voltage fluctuation, then it may be appropriate to model the failure time with an exponential distribution since the memoryless property is plausible in this case. However, if the failure is due to wearout, then the memoryless property is not plausible and the exponential distribution would not be appropriate to model the failure times. A further discussion of modeling failure times is presented in Section 17.2, and Example 76 discusses how the exponential distribution is often used to model the time until an electrical discharge occurs.

4.2.3 The Poisson Process

A **stochastic process** can be thought of as being a *series* of random events. In particular, a simple stochastic process may consist of a sequence of events occurring over time, in which case it can be defined by specifying the distributions of the time intervals between the occurrences of adjacent events. A **Poisson process** (with parameter λ) is one such process where these time intervals are independent random variables having exponential distributions with parameter λ, as shown in Figure 4.6.

A Poisson process may, for example, be used to model the arrival of calls at a switchboard, the addition of new elements to a queue, or the positions of deformities within a substance. Some examples of Poisson processes are discussed in this section. In all cases, the modeling

FIGURE 4.6

A Poisson process. The time intervals between events have independent exponential distributions with parameter λ

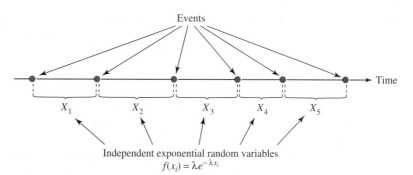

FIGURE 4.7

A Poisson process. The number of events occurring in a time interval of length t has a Poisson distribution with mean λt

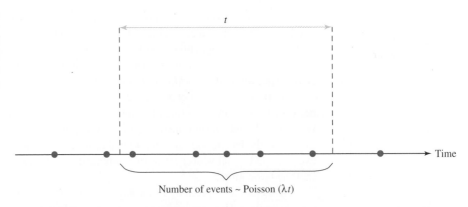

Number of events ~ Poisson (λt)

is based upon the assumption that the "distances" or times between events are independently distributed with identical exponential distributions.

The expected waiting time between two events in a Poisson process is $1/\lambda$ because it is simply the expected value of an exponential distribution with parameter λ. Furthermore, the expected number of events occurring within a fixed time interval of length t is λt. Moreover, the number of events occurring within such a time interval has a Poisson distribution with mean λt. In other words, if the random variable X counts the number of events occurring within a fixed time interval of length t, then

$$X \sim P(\lambda t)$$

as illustrated in Figure 4.7. This is why these processes are called Poisson processes.

4.2.4 Examples of the Exponential Distribution

Example 31
Shipwreck Hunts

A team of underwater salvage experts sets sail to search the ocean floor for the wreckage of a ship that is thought to have sunk within a certain area. Their boat is equipped with underwater sonar with which they hope to detect unusual objects lying on the ocean floor.

The captain's experience is that in similar situations it has taken an average of 20 days to locate a wreck. Consequently, the captain surmises that the time in days taken to locate the wreck can be modeled by an exponential distribution with parameter

$$\lambda = \frac{1}{E(X)} = \frac{1}{20} = 0.05$$

The captain considers the memoryless property of the exponential distribution to be suitable since, with such vast areas of the ocean floor to be searched, the unfruitful searching of certain areas does not appreciably alter the chance of finding the wreck in the future.

The captain's contractors have offered a sizeable bonus if it is possible to reduce searching costs by locating the wreck within the first week. The captain estimates the probability of this to be

$$P(X \leq 7) = F(7) = 1 - e^{-0.05 \times 7} = 0.30$$

On the other hand, the captain is only authorized to search for at most 4 weeks before calling off the search. The probability that the captain has to call off the search without success is

FIGURE 4.8

Probability density function for shipwreck hunt

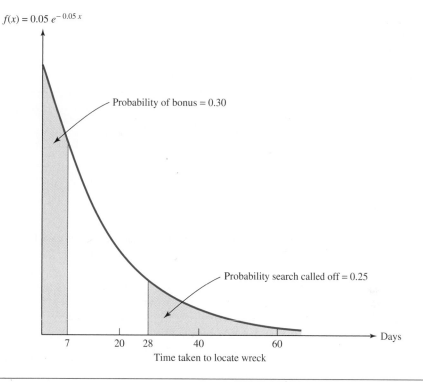

$f(x) = 0.05\, e^{-0.05\,x}$

Probability of bonus = 0.30

Probability search called off = 0.25

Days

Time taken to locate wreck

estimated to be

$$P(X \geq 28) = 1 - F(28) = e^{-0.05 \times 28} = 0.25$$

These probabilities are illustrated in Figure 4.8.

Example 32
Steel Girder Fractures

An engineer examines the edges of steel girders for hairline fractures. The girders are 10 m long, and it is discovered that they have an average of 42 fractures each. If a girder has 42 fractures, then there are 43 "gaps" between fractures or between the ends of the girder and the adjacent fractures. The average length of these gaps is therefore $10/43 = 0.23$ m. The fractures appear to be randomly spaced on the girders, so the engineer proposes that the location of fractures on a particular girder can be modeled by a Poisson process with

$$\lambda = \frac{1}{0.23} = 4.3$$

According to this model, the length of a gap between any two adjacent fractures has an exponential distribution with $\lambda = 4.3$, as illustrated in Figure 4.9. In this case, the probability that a gap is less than 10 cm long is

$$P(X \leq 0.10) = F(0.10) = 1 - e^{-4.3 \times 0.10} = 0.35$$

The probability that a gap is longer than 30 cm is

$$P(X \geq 0.30) = 1 - F(0.30) = e^{-4.3 \times 0.30} = 0.28$$

If a 25-cm segment of a girder is selected, the number of fractures it contains has a Poisson distribution with mean

$$\lambda \times 0.25 = 4.3 \times 0.25 = 1.075$$

FIGURE 4.9

Poisson process modeling fracture
locations on a steel girder

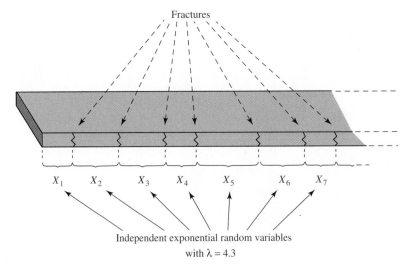

Independent exponential random variables
with $\lambda = 4.3$

FIGURE 4.10

The number of fractures in a 25-cm
segment of the steel girder has
a Poisson distribution with
mean 1.075

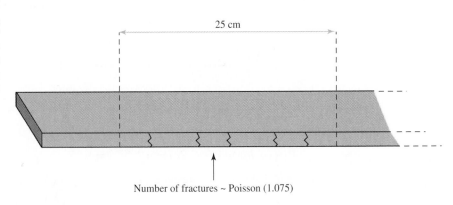

Number of fractures ~ Poisson (1.075)

as illustrated in Figure 4.10. The probability that the segment contains at least two fractures
is therefore

$$P(X \geq 2) = 1 - P(X = 0) - P(X = 1)$$

$$= 1 - \frac{e^{-1.075} \times 1.075^0}{0!} - \frac{e^{-1.075} \times 1.075^1}{1!}$$

$$= 1 - 0.341 - 0.367 = 0.292$$

Example 9

Car Body Assembly Line

The engineer in charge of the car panel manufacturing process pays particular attention to the
arrival of metal sheets at the beginning of the panel construction lines. These metal sheets are
brought one by one from other parts of the factory floor, where they have been cut into the
required sizes. On average, about 96 metal sheets are delivered to the panel construction lines
in 1 hour.

The engineer decides to model the arrival of the metal sheets with a Poisson process. The
average waiting time between arrivals is $60/96 = 0.625$ minute, so a value of

$$\lambda = \frac{1}{0.625} = 1.6$$

is used. This model assumes that the waiting times between arrivals of metal sheets are
independently distributed as exponential distributions with $\lambda = 1.6$, as shown in Figure 4.11.

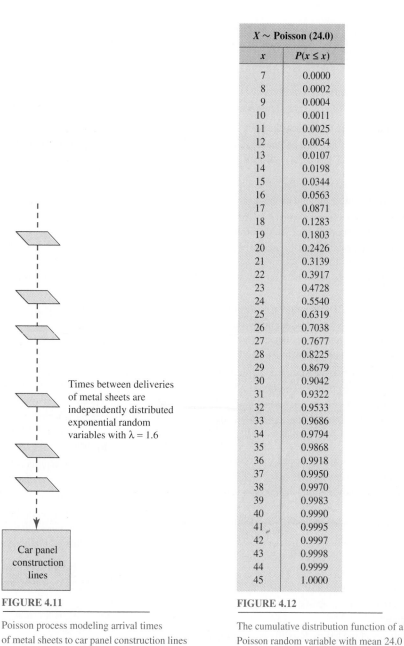

$X \sim$ **Poisson (24.0)**	
x	$P(x \leq x)$
7	0.0000
8	0.0002
9	0.0004
10	0.0011
11	0.0025
12	0.0054
13	0.0107
14	0.0198
15	0.0344
16	0.0563
17	0.0871
18	0.1283
19	0.1803
20	0.2426
21	0.3139
22	0.3917
23	0.4728
24	0.5540
25	0.6319
26	0.7038
27	0.7677
28	0.8225
29	0.8679
30	0.9042
31	0.9322
32	0.9533
33	0.9686
34	0.9794
35	0.9868
36	0.9918
37	0.9950
38	0.9970
39	0.9983
40	0.9990
41	0.9995
42	0.9997
43	0.9998
44	0.9999
45	1.0000

Times between deliveries of metal sheets are independently distributed exponential random variables with $\lambda = 1.6$

Car panel construction lines

FIGURE 4.11

Poisson process modeling arrival times of metal sheets to car panel construction lines

FIGURE 4.12

The cumulative distribution function of a Poisson random variable with mean 24.0

For example, the probability that there is a wait of more than 3 minutes between arrivals is

$$P(X \geq 3) = 1 - F(3) = e^{-1.6 \times 3} = 0.008$$

The number of metal sheets arriving at the panel construction lines during a specific 15-minute period has a Poisson distribution with mean

$$\lambda \times 15 = 1.6 \times 15 = 24.0$$

Figure 4.12 shows the cumulative distribution function of this Poisson distribution. The probability that no more than 16 sheets arrive during the 15-minute period, for example, is about

0.056. On the other hand, the engineer can be about 95% confident that no more than 32 metal sheets will arrive during the period under consideration.

■ 4.2.5 Problems

4.2.1 Use integration by parts to show that if X has an exponential distribution with parameter λ, then
(a) $E(X) = 1/\lambda$ (b) $E(X^2) = 2/\lambda^2$

4.2.2 Suppose that you are waiting for a friend to call you and that the time you wait in minutes has an exponential distribution with parameter $\lambda = 0.1$.
(a) What is the expectation of your waiting time?
(b) What is the probability that you will wait longer than 10 minutes?
(c) What is the probability that you will wait less than 5 minutes?
(d) Suppose that after 5 minutes you are still waiting for the call. What is the distribution of your *additional* waiting time? In this case, what is the probability that your total waiting time is longer than 15 minutes?
(e) Suppose now that the time you wait in minutes for the call has a $U(0, 20)$ distribution. What is the expectation of your waiting time? If after 5 minutes you are still waiting for the call, what is the distribution of your *additional* waiting time?

4.2.3 The time in days between breakdowns of a machine is exponentially distributed with $\lambda = 0.2$.
(a) What is the expected time between machine breakdowns?
(b) What is the standard deviation of the time between machine breakdowns?
(c) What is the median time between machine breakdowns?
(d) What is the probability that after the machine is repaired it lasts at least a week before failing again?
(e) If the machine has performed satisfactorily for six days, what is the probability that it lasts at least two more days before breaking down?

4.2.4 A researcher plants 12 seeds whose germination times in days are independent exponential distributions with $\lambda = 0.31$.
(a) What is the probability that a given seed germinates within five days?
(b) What are the expectation and variance of the number of seeds germinating within five days?

(c) What is the probability that no more than nine seeds have germinated within five days?

4.2.5 A double exponential distribution, often called a **Laplace distribution**, has a probability density function

$$f(x) = \frac{1}{2} \lambda \, e^{-\lambda|x-\theta|}$$

for $-\infty \le x \le \infty$, depending on two parameters λ and θ. Sketch the probability density function and cumulative distribution function of this distribution. What is the expectation of the distribution? If $\lambda = 3$ and $\theta = 2$, calculate:
(a) $P(X \le 0)$ (b) $P(X \ge 1)$

4.2.6 Imperfections in an optical fiber are distributed according to a Poisson process such that the distance between imperfections in meters has an exponential distribution with parameter $\lambda = 2\text{m}^{-1}$.
(a) What is the expected distance between imperfections?
(b) What is the probability that the distance between two imperfections is longer than 1 meter?
(c) What is the distribution of the number of imperfections in a 3-meter stretch of fiber?
(d) What is the probability that a 3-meter stretch of fiber has no more than four imperfections?
(This problem is continued in Problem 4.3.5.)

4.2.7 The arrival times of workers at a factory first-aid room satisfy a Poisson process with an average of 1.8 per hour.
(a) What is the value of the parameter λ of the Poisson process?
(b) What is the expectation of the time between two arrivals at the first-aid room?
(c) What is the probability that there is at least one hour between two arrivals at the first-aid room?
(d) What is the distribution of the number of workers visiting the first-aid room during a four-hour period?
(e) What is the probability that at least four workers visit the first-aid room during a four-hour period?
(This problem is continued in Problem 4.3.6.)

4.2.8 Engineers observe that about 90% of graphite samples fracture within five hours when subjected to a certain stress.
   (a) If the time to fracture is modeled with an exponential distribution, what would be a suitable value for the parameter λ?
   (b) Use the model to estimate the probability that a fracture occurs within three hours.

4.2.9 Consider a Poisson process with parameter $\lambda = 0.8$.
   (a) What is the probability that the time between two adjacent events is longer than 1.5?
   (b) What is the probability that there will be at least three events in a period of length 2?

4.2.10 The lengths of telephone calls can be modeled by an exponential distribution with parameter $\lambda = 0.3$ per minute, with the call lengths being independent. What is the probability that out of 10 telephone calls, 2 will be shorter than 1 minute, 4 will last between 1 minute and 3 minutes, and the other 4 will last longer than 3 minutes?

4.2.11 Customers arrive at a service window according to a Poisson process with parameter $\lambda = 0.2$ per minute.

   (a) What is the probability that the time between two successive arrivals is less than 6 minutes?
   (b) What is the probability that there will be exactly three arrivals during a given 10-minute period?

4.2.12 Suppose that components have failure times that are independent and that can be modeled with an exponential distribution with $\lambda = 0.0065$ per day. If a box contains 10 components, what is the probability that the box has at least 8 components that last longer than 150 days?

4.2.13 As a metal detector is passed over the ground, signals are received according to a Poisson process with $\lambda = 0.022$ per meter. What is the probability that there is no more than one signal in a 100-meter stretch?

4.2.14 Ninety identical electrical circuits are monitored at an extreme temperature to see how long they last before failing. The 50th failure occurs after 263 minutes. If the failure times are modeled with an exponential distribution, when would you predict that the 80th failure will occur?

4.3 The Gamma Distribution

4.3.1 Definition of the Gamma Distribution

The **gamma distribution** has many important applications in areas such as reliability theory, and it is also used in the analysis of a Poisson process. It has a state space $x \geq 0$ and a probability density function

$$f(x) = \frac{\lambda^k \, x^{k-1} \, e^{-\lambda x}}{\Gamma(k)}$$

for $x \geq 0$ and $f(x) = 0$ for $x < 0$, which depends upon two parameters $k > 0$ and $\lambda > 0$. The function $\Gamma(k)$ is known as the **gamma function**. It provides the correct scaling to ensure that the total area under the probability density function is equal to 1.

The Gamma Function

The **gamma function** is defined to be

$$\Gamma(k) = \int_0^\infty x^{k-1} e^{-x} \, dx$$

Some special cases are $\Gamma(1) = 1$ and $\Gamma(1/2) = \sqrt{\pi}$, and in general,

$$\Gamma(k) = (k-1) \, \Gamma(k-1)$$

The Gamma Function, continued

---

for $k > 1$. If n is a positive integer, then

$$\Gamma(n) = (n - 1)!$$

but except for these special cases there is in general no closed-form expression for the gamma function.

Notice that if $k = 1$, the gamma distribution simplifies to the exponential distribution with parameter λ. The expectation and variance of a gamma distribution are given in the following box.

The Gamma Distribution

A **gamma distribution** with parameters $k > 0$ and $\lambda > 0$ has a probability density function

$$f(x) = \frac{\lambda^k \, x^{k-1} \, e^{-\lambda x}}{\Gamma(k)}$$

for $x \geq 0$ and $f(x) = 0$ for $x < 0$, where $\Gamma(k)$ is the gamma function. It has an expectation and variance of

$$E(X) = \frac{k}{\lambda} \quad \text{and} \quad \text{Var}(X) = \frac{k}{\lambda^2}$$

The parameter k is often referred to as the *shape* parameter of the gamma distribution, and λ is referred to as the *scale* parameter. Figure 4.13 shows the probability density functions of gamma distributions with $\lambda = 1$ and $k = 1, 3$, and 5. As the shape parameter increases, the peak of the density function is seen to move farther to the right. Figure 4.14 shows the probability density functions of gamma distributions with $\lambda = 1, 2$, and 3 and $k = 3$. This illustrates how the parameter λ "scales" the distribution function.

One important property of a random variable that has a gamma distribution with an integer value of the parameter k is that it can be obtained as the sum of a set of independent exponential random variables. Specifically, if $X_1, \ldots, X_k$ are independent random variables each having an exponential distribution with parameter λ, then the random variable

$$X = X_1 + \cdots + X_k$$

has a gamma distribution with parameters k and λ. This fact implies that for a Poisson process with parameter λ, the time taken for k events to occur has a gamma distribution with parameters k and λ, since the time taken until the first event occurs, and the times between subsequent events, each have independent exponential distributions with parameter λ.

COMPUTER NOTE

The probability values of gamma distributions are generally intractable and are best obtained from software packages. However, when you do this it is important to ensure that you know what *parameterization* the package is using so that you define the distribution properly. In particular, many packages use parameters k and $1/\lambda$ instead of k and λ.

FIGURE 4.13

Probability density functions of gamma distributions

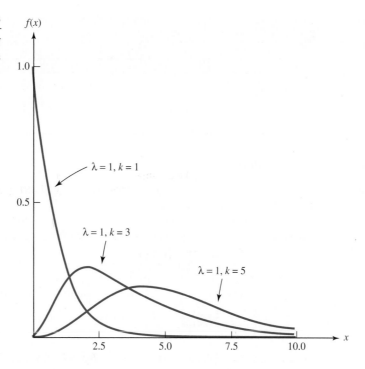

FIGURE 4.14

Probability density functions of gamma distributions

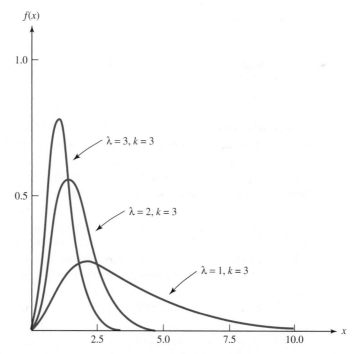

FIGURE 4.15

Distance to fifth fracture has a gamma distribution with parameters $k = 5$ and $\lambda = 4.3$

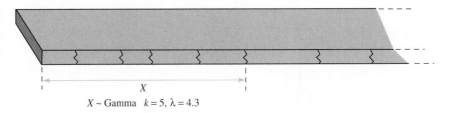

$$X \sim \text{Gamma} \quad k = 5, \lambda = 4.3$$

4.3.2   Examples of the Gamma Distribution

Example 32

Steel Girder Fractures

Suppose that the random variable X measures the length between one end of a girder and the *fifth* fracture along the girder, as shown in Figure 4.15. If the fracture locations are modeled by a Poisson process as discussed previously, X has a gamma distribution with parameters $k = 5$ and $\lambda = 4.3$. The expected distance to the fifth fracture is therefore

$$E(X) = \frac{k}{\lambda} = \frac{5}{4.3} = 1.16 \text{ m}$$

A software package can be used to show that the 0.05 quantile point of this distribution is $x = 0.458$ m, so that

$$F(0.458) = 0.05$$

Consequently, the engineer can be 95% sure that the fifth fracture is at least 46 cm away from the end of the girder. A software package can also be used to calculate the probability that the fifth fracture is within 1 m of the end of the girder, which is

$$F(1) = 0.4296$$

It is interesting to note that this latter probability can also be obtained using the Poisson distribution. The number of fractures within a 1-m section of the girder has a Poisson distribution with mean

$$\lambda \times 1 = 4.3$$

The probability that the fifth fracture is within 1 m of the end of the girder is the probability that there are at least five fractures within the first 1-m section, which is therefore

$$P(Y \geq 5) = 0.4296$$

where $Y \sim P(4.3)$.

Example 9

Car Body Assembly Line

Suppose that the engineer in charge of the car panel manufacturing process is interested in how long it will take for 20 metal sheets to be delivered to the panel construction lines. Under the Poisson process model, this time X has a gamma distribution with parameters $k = 20$ and $\lambda = 1.6$. The expected waiting time is consequently

$$E(X) = \frac{k}{\lambda} = \frac{20}{1.6} = 12.5 \text{ minutes}$$

The variance of the waiting time is

$$\text{Var}(X) = \frac{k}{\lambda^2} = \frac{20}{1.6^2} = 7.81$$

so that the standard deviation is $\sigma = \sqrt{7.81} = 2.80$ minutes, as illustrated in Figure 4.16.

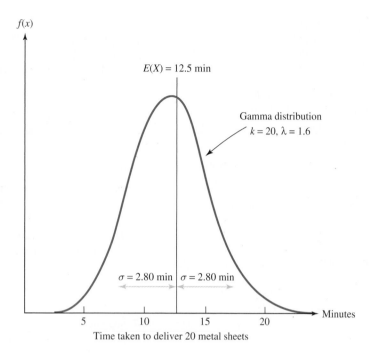

$f(x)$

$E(X) = 12.5$ min

Gamma distribution
$k = 20$, $\lambda = 1.6$

$\sigma = 2.80$ min | $\sigma = 2.80$ min

5   10   15   20   Minutes

Time taken to deliver 20 metal sheets

A software package can be used to show that for this distribution,

$$F(17.42) = 0.95 \quad \text{and} \quad F(15) = 0.8197$$

The engineer can therefore be 95% confident that 20 metal sheets will have arrived within 18 minutes, say. Furthermore, there is a probability of about 0.82 that they will all arrive within 15 minutes. This latter probability can also be obtained from the probabilities of a Poisson distribution with mean 24.0, which is shown in Figure 4.12. This Poisson distribution is the distribution of the number of sheets arriving during a 15-minute period, and $1 - 0.1803 = 0.8197$ is seen to be the probability that at least 20 sheets arrive during this time interval.

■ 4.3.3 Problems

4.3.1 Use integration by parts to show that

$$\Gamma(k) = (k - 1)\,\Gamma(k - 1)$$

for $k > 1$. Use the fact that $\Gamma(0.5) = \sqrt{\pi}$ to evaluate $\Gamma(5.5)$.

4.3.2 Recall that if $X_1, \ldots, X_k$ are independent random variables each having an exponential distribution with parameter λ, then the random variable

$$X = X_1 + \cdots + X_k$$

has a gamma distribution with parameters k and λ.

(a) Use this fact to verify the expectation and variance of a gamma distribution. Check that you get the same answer for the expectation of a gamma random variable using the formula

$$E(X) = \int_0^\infty x\, f(x)\, dx$$

where $f(x)$ is the probability density function of the gamma distribution.

(b) If X has a gamma distribution with parameters k_1 and λ, and Y has a gamma distribution with parameters k_2

and λ, where k_1 and k_2 are both positive integers and X and Y are independent random variables, explain why $Z = X + Y$ has a gamma distribution with parameters $k_1 + k_2$ and λ.

4.3.3 Use a computer package to find both the probability density function and cumulative distribution function at $x = 3$ and the median of gamma distributions with the following parameter values:
  (a) $k = 3.2, \lambda = 0.8$
  (b) $k = 7.5, \lambda = 5.3$
  (c) $k = 4.0, \lambda = 1.4$
  In part (c), check the value of the probability density function from its formula.

4.3.4 A day's sales in $1000 units at a gas station have a gamma distribution with parameters $k = 5$ and $\lambda = 0.9$.
  (a) What is the expectation of a day's sales?
  (b) What is the standard deviation of a day's sales?
  (c) What are the upper and lower quartiles of a day's sales?
  (d) What is the probability that a day's sales are more than $6000?
  (This problem is continued in Problem 5.3.9.)

4.3.5 Recall Problem 4.2.6 concerning imperfections in an optical fiber. Suppose that five adjacent imperfections are located on a fiber.
  (a) What is the distribution of the distance between the first imperfection and the fifth imperfection?

  (b) What is the expectation of the distance between the first imperfection and the fifth imperfection?
  (c) What is the standard deviation of the distance between the first imperfection and the fifth imperfection?
  (d) Consider the probability that the distance between the first imperfection and the fifth imperfection is longer than three meters. Show how this probability can be obtained using the gamma distribution and also by using the Poisson distribution.

4.3.6 Recall Problem 4.2.7 concerning the arrivals at a factory first-aid room.
  (a) What is the distribution of the time between the first arrival of the day and the fourth arrival?
  (b) What is the expectation of this time?
  (c) What is the variance of this time?
  (d) By using (i) the gamma distribution and (ii) the Poisson distribution, show how to calculate the probability that this time is longer than three hours.

4.3.7 Suppose that the time in minutes taken by a worker on an assembly line to complete a particular task has a gamma distribution with parameters $k = 44$ and $\lambda = 0.7$.
  (a) What are the expectation and standard deviation of the time taken to complete the task?
  (b) Use a software package to find the probability that the task is completed within an hour.

4.4   The Weibull Distribution

4.4.1   Definition of the Weibull Distribution

The **Weibull distribution** is often used to model failure and waiting times. It has a state space $x \geq 0$ and a probability density function

$$f(x) = a\,\lambda^a\,x^{a-1}\,e^{-(\lambda x)^a}$$

for $x \geq 0$ and $f(x) = 0$ for $x < 0$, which depends upon two parameters $a > 0$ and $\lambda > 0$. Notice that taking $a = 1$ gives the exponential distribution as a special case.

The cumulative distribution function of a Weibull distribution is

$$F(x) = \int_0^x a\,\lambda^a\,y^{a-1}\,e^{-(\lambda y)^a}\,dy = 1 - e^{-(\lambda x)^a}$$

for $x \geq 0$. The expectation and variance of a Weibull distribution depend upon the gamma function and are given in the following box.

The Weibull Distribution

A **Weibull distribution** with parameters $a > 0$ and $\lambda > 0$ has a probability density function

$$f(x) = a\,\lambda^a\,x^{a-1}\,e^{-(\lambda x)^a}$$

for $x \geq 0$ and $f(x) = 0$ for $x < 0$, and a cumulative distribution function

$$F(x) = 1 - e^{-(\lambda x)^a}$$

for $x \geq 0$. It has an expectation

$$E(X) = \frac{1}{\lambda}\,\Gamma\left(1 + \frac{1}{a}\right)$$

and a variance

$$\mathrm{Var}(X) = \frac{1}{\lambda^2}\left\{\Gamma\left(1 + \frac{2}{a}\right) - \Gamma\left(1 + \frac{1}{a}\right)^2\right\}$$

As with the gamma distribution, λ is called the *scale* parameter of the distribution, and a is called the *shape* parameter. A useful property of the Weibull distribution is that the probability density function can exhibit a wide variety of forms, depending on the choice of the two parameters. Figure 4.17 illustrates some probability density functions with $\lambda = 1$ and various values of the shape parameter a. Figure 4.18 illustrates some probability density functions with $a = 3$ and with various values of λ.

FIGURE 4.17

Probability density functions of the Weibull distribution

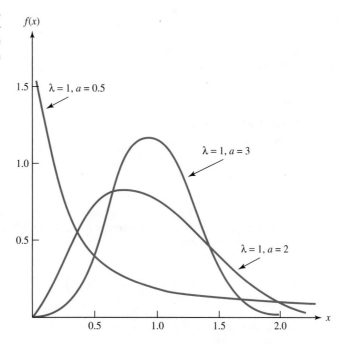

FIGURE 4.18

Probability density functions of the
Weibull distribution

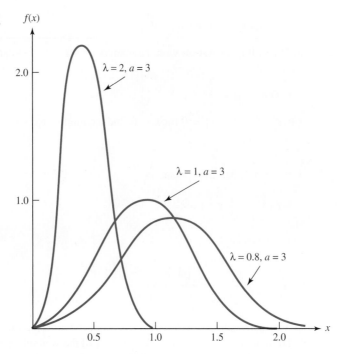

Notice that the pth quantile of the Weibull distribution is easily calculated to be

$$\frac{(-\ln(1-p))^{1/a}}{\lambda}$$

COMPUTER NOTE   The probability values of a Weibull distribution are easy to calculate because of the simple form of the cumulative distribution function. However, they should also be available on your computer package. As with the exponential and gamma distributions, make sure that you check on the parameterization used by your package which may be a and $1/\lambda$ instead of a and λ.

4.4.2   Examples of the Weibull Distribution

Example 33
Bacteria Lifetimes

Suppose that the random variable X measures the lifetime of a bacterium at a certain high temperature, and that it has a Weibull distribution with $a = 2$ and $\lambda = 0.1$. This distribution is illustrated in Figure 4.19.

The expected survival time of a bacterium is

$$E(X) = \frac{1}{0.1} \times \Gamma\left(1 + \frac{1}{2}\right) = 10 \times \frac{1}{2} \times \Gamma\left(\frac{1}{2}\right) = 10 \times \frac{1}{2} \times \sqrt{\pi} = 8.86 \text{ minutes}$$

The variance of the bacteria lifetimes is

$$\text{Var}(X) = \frac{1}{0.1^2} \times \left\{ \Gamma\left(1 + \frac{2}{2}\right) - \Gamma\left(1 + \frac{1}{2}\right)^2 \right\}$$

$$= 100 \times \left\{ 1 - \left(\frac{\sqrt{\pi}}{2}\right)^2 \right\} = 21.46$$

so that the standard deviation is $\sigma = \sqrt{21.46} = 4.63$ minutes.

FIGURE 4.19

Distribution of bacteria survival
times

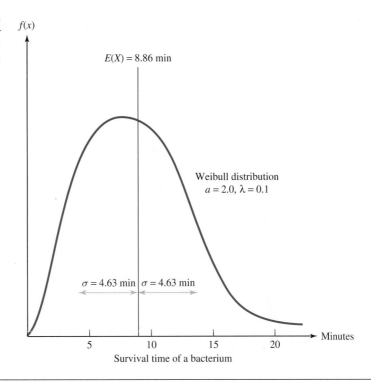

$f(x)$

$E(X) = 8.86$ min

Weibull distribution
$a = 2.0$, $\lambda = 0.1$

$\sigma = 4.63$ min | $\sigma = 4.63$ min

Minutes

5   10   15   20

Survival time of a bacterium

The probability that a bacterium dies within 5 minutes is

$$P(X \leq 5) = F(5) = 1 - e^{-(0.1 \times 5)^2} = 0.22$$

and the probability that a bacterium lives longer than 15 minutes is

$$P(X \geq 15) = 1 - F(15) = e^{-(0.1 \times 15)^2} = 0.11$$

Notice that if $F(x) = 0.95$, then

$$0.95 = 1 - e^{-(0.1 \times x)^2}$$

which can be solved to give $x = 17.31$ minutes. Consequently, within a large group of bacteria, it will take about 17.3 minutes for 95% of the bacteria to die.

**Example 34
Car Brake Pad Wear**

A brake pad made from a new compound is tested in cars that are driven in city traffic. The random variable X, which measures the mileage in 1000-mile units that the cars can be driven before the brake pads wear out, has a Weibull distribution with parameters $a = 3.5$ and $\lambda = 0.12$. This distribution is shown in Figure 4.20.

The median car mileage is the value x satisfying

$$0.5 = F(x) = 1 - e^{-(0.12 \times x)^{3.5}}$$

which can be solved to give $x = 7.50$. Consequently, it should be expected that about half of the brake pads will last longer than 7500 miles. The probability that a set of brake pads last longer than 10,000 miles is

$$P(X \geq 10) = 1 - F(10) = e^{-(0.12 \times 10)^{3.5}} = 0.15.$$

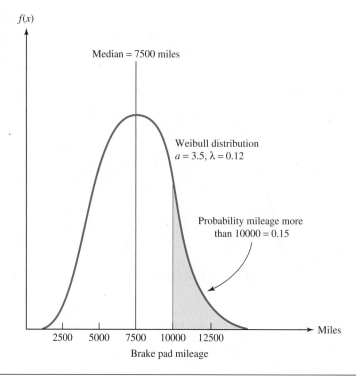

FIGURE 4.20

Distribution of brake pad mileage

4.4.3   Problems

4.4.1 Use the definition of the gamma function to derive the expectation and variance of a Weibull distribution.

4.4.2 Suppose that the random variable X has a Weibull distribution with parameters $a = 4.9$ and $\lambda = 0.22$. Find:
   (a) The median of the distribution
   (b) The upper and lower quartiles of the distribution
   (c) $P(2 \leq X \leq 7)$

4.4.3 Suppose that the random variable X has a Weibull distribution with parameters $a = 2.3$ and $\lambda = 1.7$. Find:
   (a) The median of the distribution
   (b) The upper and lower quartiles of the distribution
   (c) $P(0.5 \leq X \leq 1.5)$

4.4.4 The time to failure in hours of an electrical circuit subjected to a high temperature has a Weibull distribution with parameters $a = 3$ and $\lambda = 0.5$.
   (a) What is the median failure time of a circuit?
   (b) The circuit engineers can be 99% confident that a circuit will last as long as what time?
   (c) What are the expectation and variance of the circuit failure times?

   (d) If a circuit has three equivalent backup circuits that have independent failure times, what is the probability that at least one circuit is working after three hours?

4.4.5 A biologist models the time in minutes between the formation of a cell and the moment at which it splits into two new cells using a Weibull distribution with parameters $a = 0.4$ and $\lambda = 0.5$.
   (a) What is the median value of this distribution?
   (b) What are the upper and lower quartiles of this distribution?
   (c) What are the 95th and 99th percentiles of this distribution?
   (d) What is the probability that the cell "lifetime" is between three and five minutes?

4.4.6 The lifetime in minutes of a mechanical component has a Weibull distribution with parameters $a = 1.5$ and $\lambda = 0.03$.
   (a) What are the median, upper quartile, and 99th percentile of the lifetime of a component?

(b) If 500 independent components are considered, what are the expectation and variance of the number of components still operating after 30 minutes?

4.4.7 Suppose that the time in days taken for bacteria cultures to develop after they have been prepared can be modeled by a Weibull distribution with parameters $\lambda = 0.3$ and $a = 0.6$. A biologist prepares several sets of cultures at the same time, and after four days opens them one by one until five developed cultures have been found. What is the probability that the biologist opens exactly 10 cultures?

4.4.8 A physician conducts a study to investigate the time taken to recover from an ailment under a certain treatment. A group of 82 patients with the ailment are given the treatment, and when they are checked 7 days later, it is found that 9 of them have recovered. The remaining 73 patients are checked 14 days after receiving the treatment, and an additional 15 of them are found to have recovered. If the time to recovery is modeled with a Weibull distribution, estimate the median time to recovery.

4.5 The Beta Distribution

4.5.1 Definition of the Beta Distribution

The **beta distribution** has a state space $0 \le x \le 1$ and is often used to model proportions.

The Beta Distribution

A **beta distribution** with parameters $a > 0$ and $b > 0$ has a probability density function

$$f(x) = \frac{\Gamma(a+b)}{\Gamma(a)\Gamma(b)} \, x^{a-1} \, (1-x)^{b-1}$$

for $0 \le x \le 1$ and $f(x) = 0$ elsewhere. It is useful for modeling proportions. Its expectation and variance are

$$E(X) = \frac{a}{a+b} \quad \text{and} \quad \text{Var}(X) = \frac{ab}{(a+b)^2(a+b+1)}$$

Figure 4.21 illustrates the probability density functions of beta distributions with $a = b = 0.5$ and $a = b = 2$. While their shapes are quite different, they are both symmetric about $x = 0.5$. In fact, all beta distributions with $a = b$ are symmetric. Figure 4.22 illustrates the probability density functions of beta distributions with $a = 0.5, b = 2$ and with $a = 4, b = 2$.

COMPUTER NOTE

Unless the parameters a and b take integer values, the cumulative distribution function of the beta distribution is generally intractable so that a software package is essential to calculate the probability values of beta distributions. As before, check on the parameterization that your package employs.

4.5.2 Examples of the Beta Distribution

Example 35
Stock Prices

A Wall Street analyst has built a model for the performance of the stock market. In this model the *proportion* of listed stocks showing an increase in value on a particular day has a beta distribution with parameter values a and b, which depend upon various economic and political

FIGURE 4.21

Probability density functions of the beta distribution

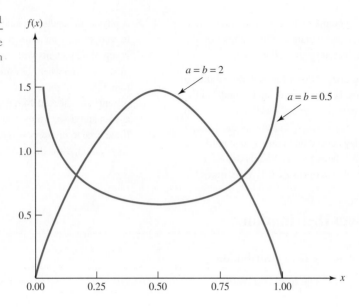

FIGURE 4.22

Probability density functions of the beta distribution

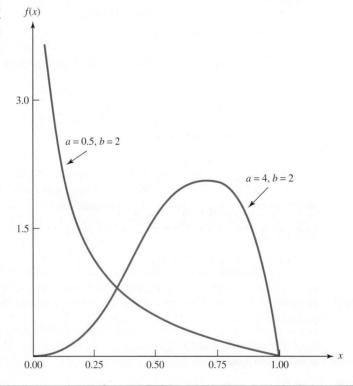

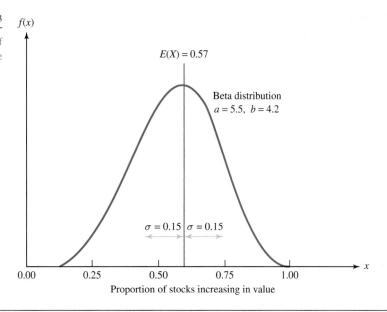

factors. On each day the analyst predicts suitable values of the parameters for modeling the
subsequent day's stock prices. Suppose that on Monday the analyst predicts that parameter
values $a = 5.5$ and $b = 4.2$ are suitable for the next day. What does this indicate about stock
prices on Tuesday?

The distribution of the proportion of stocks increasing in value on Tuesday is shown in
Figure 4.23. The expected proportion of stocks increasing in value on Tuesday is

$$E(X) = \frac{5.5}{5.5 + 4.2} = 0.57$$

The variance in the proportion is

$$\text{Var}(X) = \frac{5.5 \times 4.2}{(5.5 + 4.2)^2 \times (5.5 + 4.2 + 1)} = 0.0229$$

so that the standard deviation is $\sigma = \sqrt{0.0229} = 0.15$. A software package can be used to
calculate the probability that more than 75% of the stocks increase in value as

$$P(X \geq 0.75) = 1 - F(0.75) = 1 - 0.881 = 0.119$$

Example 36
Bee Colonies

When a queen bee leaves a bee colony to start a new hive, a certain proportion of the worker
bees take flight and follow her. An entomologist models the proportion X of the worker bees
that leave with the queen using a beta distribution with parameters $a = 2.0$ and $b = 4.8$. This
distribution is illustrated in Figure 4.24.

The expected proportion of bees leaving is

$$E(X) = \frac{2.0}{2.0 + 4.8} = 0.29$$

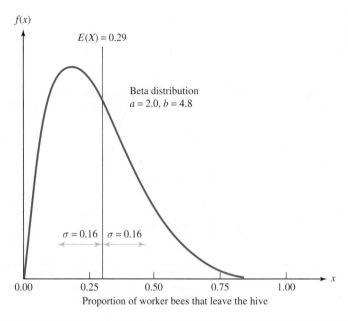

The variance in the proportion is

$$\text{Var}(X) = \frac{2.0 \times 4.8}{(2.0 + 4.8)^2 \times (2.0 + 4.8 + 1)} = 0.0266$$

so that the standard deviation is $\sigma = \sqrt{0.0266} = 0.16$. The probability that more than half of the bee colony leaves with the queen can be calculated from a software package to be

$$P(X \geq 0.5) = 1 - F(0.5) = 1 - 0.878 = 0.122$$

■ **4.5.3   Problems**

4.5.1 Consider the probability density function

$$f(x) = Ax^3 (1 - x)^2$$

for $0 \leq x \leq 1$ and $f(x) = 0$ elsewhere.
(a) Find the value of A by direct integration.
(b) Find by direct integration the expectation and variance of this distribution.
(c) What are the parameter values a and b of a beta distribution for which this is the probability density function? Check your answers to part (b) using the general formulas for beta distributions.

4.5.2 Consider the beta probability density function

$$f(x) = Ax^9 (1 - x)^3$$

for $0 \leq x \leq 1$ and $f(x) = 0$ elsewhere.

(a) What are the values of the parameters a and b?
(b) Use the answer to part (a) to calculate the value of A.
(c) What is the expectation of this distribution?
(d) What is the standard deviation of this distribution?
(e) Calculate the cumulative distribution function of this distribution.

4.5.3 Use a computer package to find the probability density function and cumulative distribution function at $x = 0.5$, and the upper quartile, of beta distributions with the following parameter values:
(a) $a = 3.3, b = 4.5$
(b) $a = 0.6, b = 1.5$
(c) $a = 2, b = 6$
In part (c), check the value of the probability density function using the general formula.

4.5.4 Suppose that the random variable X has a beta distribution with parameters $a = b = 2.1$, and consider the random variable

$$Y = 3 + 4X$$

(a) What is the state space of the random variable Y?
(b) What are the expectation and variance of the random variable Y?
(c) What is $P(Y \leq 5)$?

4.5.5 The purity of a chemical batch, expressed as a percentage, is equal to $100X$, where the random variable X has a beta distribution with parameters $a = 7.2$ and $b = 2.3$.

(a) What are the expectation and variance of the purity levels?
(b) What is the probability that a chemical batch has a purity of at least 90%?

4.5.6 The proportion of tin in a metal alloy has a beta distribution with parameters $a = 8.2$ and $b = 11.7$.

(a) What is the expected proportion of tin in the alloy?
(b) What is the standard deviation of the proportion of tin in the alloy?
(c) What is the median proportion of tin in the alloy?

4.6 Case Study: Microelectronic Solder Joints

A Weibull distribution can be used to model the number of temperature cycles that an assembly can be subjected to before it fails. In this case, experience dictates that it is best to define the cumulative distribution function of the failure time distribution in terms of the logarithm of the number of cycles, so that

$$P(\text{assembly fails within } t \text{ cycles}) = P(X \leq t) = 1 - e^{-(\lambda \ln(t))^a}$$

The values of the parameters a and λ will depend upon the specific design of the assembly. Suppose that if an epoxy of type I is used for the underfill, then $a = 25.31$ and $\lambda = 0.120$, whereas if an epoxy of type II is used, then $a = 27.42$ and $\lambda = 0.116$. The solution of

$$P(X \leq t) = 1 - e^{-(0.120 \ln(t))^{25.31}} = 0.01$$

is $t = 1041$, and

$$P(X \leq t) = 1 - e^{-(0.120 \ln(t))^{25.31}} = 0.5$$

is solved with $t = 3691$. Consequently, if epoxy of type I is used for the underfill, then 99% of the assemblies can survive 1041 temperature cycles, whereas half of them can survive 3691 temperature cycles. In addition, the solution of

$$P(X \leq t) = 1 - e^{-(0.116 \ln(t))^{27.42}} = 0.01$$

is $t = 1464$ and

$$P(X \leq t) = 1 - e^{-(0.116 \ln(t))^{27.42}} = 0.5$$

is solved with $t = 4945$, so that if epoxy of type II is used for the underfill, then 99% of the assemblies can survive 1464 temperature cycles, whereas half of them can survive 4945 temperature cycles. These calculations reveal that an underfill with epoxy of type II produces an assembly with better reliability.

4.7   Supplementary Problems

4.7.1 A dial is spun and an angle θ is measured, which can be taken to be uniformly distributed between $0°$ and $360°$. If $0 \le \theta \le 90$, a player wins nothing; if $90 \le \theta \le 270$, then a player wins $\$(2\theta - 180)$; and if $270 \le \theta \le 360$, then a player wins $\$(\theta^2 - 72{,}540)$. Draw the cumulative distribution function of the player's winnings.

4.7.2 A commercial bleach eventually becomes ineffective because the chlorine in it becomes attached to other molecules. The company that manufactures the bleach estimates that the *median* time for this to happen is about one and a half years.
  (a) If an exponential distribution is used to model the time taken for a sample of bleach to become ineffective, what is a suitable value for the parameter λ?
  (b) Estimate the probability that a sample of bleach is still effective after two years, and the probability that a sample of bleach becomes ineffective within one year.

4.7.3 A ship navigating through the southern regions of the North Atlantic ice flows encounters icebergs according to a Poisson process. The distances between icebergs in nautical miles are exponentially distributed with a parameter $\lambda = 0.7$.
  (a) What is the expected distance between iceberg encounters?
  (b) What is the probability that there is a distance of at least three nautical miles between iceberg encounters?
  (c) What is the median distance between icebergs?
  (d) What is the distribution of the number of icebergs encountered in a stretch of 10 nautical miles?
  (e) What is the probability that at least five icebergs are encountered in a 10-nautical-mile stretch?

  (f) What is the distribution of the distance traveled by the ship before encountering 10 icebergs? What are the expectation and variance of this distance?

4.7.4 Calls arriving at a switchboard follow a Poisson process with parameter $\lambda = 5.2$ per minute.
  (a) What is the expected waiting time between the arrivals of two calls?
  (b) What is the probability that the waiting time between the arrivals of 2 calls is less than 10 seconds?
  (c) What is the distribution of the time taken for 10 calls to arrive at the switchboard?
  (d) What is the expectation of the time taken for 10 calls to arrive at the switchboard?
  (e) What is the probability that more than 5 calls arrive at the switchboard during a 1-minute period?

4.7.5 Figure 4.25 shows the probability density function of a *triangle distribution* $T(a, b)$ with endpoints a and b.
  (a) What is the height of the probability density function at $(a + b)/2$?
  (b) If the random variable X has a $T(a, b)$ distribution, what is $P(X \le a/4 + 3b/4)$?
  (c) What is the variance of a $T(a, b)$ distribution?
  (d) Calculate the cumulative distribution function of a $T(a, b)$ distribution.

4.7.6 The fermentation time in weeks required by a brewery for a particular kind of beer has a Weibull distribution with parameters $a = 4$ and $\lambda = 0.2$.
  (a) What are the median, upper quartile, and 95th percentile of the fermentation times?
  (b) What are the expectation and variance of the fermentation times?

FIGURE 4.25

The probability density function of a triangle distribution, $T(a, b)$, with endpoints a and b

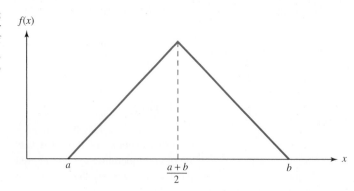

(c) What is the probability that a batch requires a fermentation time between five and six weeks?

4.7.7 The proportion of a day that a tiger spends hunting for food has a beta distribution with parameters $a = 2.7$ and $b = 2.9$.
- **(a)** What is the expected amount of time per day that the tiger spends hunting for food?
- **(b)** What is the standard deviation of the amount of time per day that the tiger spends hunting for food?
- **(c)** What is the probability that on a particular day the tiger spends more than half the day hunting for food?

4.7.8 The starting time of a class is uniformly distributed between 10:00 and 10:05. If a student arrives early and has to wait t minutes for the class to start, then the student incurs a penalty of $A_1 t$, which accounts for the waste in the student's time. On the other hand, if the student arrives t minutes after the class has started, then the student incurs a penalty of $A_2 t$, which accounts for the information the student has missed. If the student arrives at x minutes after 10:00, what is the expected penalty incurred by the student? What value of x minimizes the expected penalty?

4.7.9 An herbalist finds that about 25% of plants sprout within 35 days, and that about 75% of plants sprout within 65 days.
- **(a)** If the time of sprouting is modeled with a Weibull distribution, what parameter values would be appropriate?
- **(b)** Use this model to estimate the time by which 90% of plants sprout.

4.7.10 The strength of a chemical solution is measured on a scale between 0 and 1, with values smaller than 0.5 being too weak, values between 0.5 and 0.8 being satisfactory, and values larger than 0.8 being too strong. If chemical batches have strengths that are independently distributed according to a beta distribution with parameters $a = 18$ and $b = 11$, what is the probability that if 10 batches are produced, exactly 1 batch will be weak, 1 batch will be strong, and the other 8 batches will all be satisfactory? (This problem is continued in Problem 5.3.11.)

4.7.11 Suppose that visits to a website can be modeled by a Poisson process with parameter $\lambda = 4$ per hour.
- **(a)** What is the probability that there are exactly 10 visits within a given 2-hour interval?
- **(b)** A supervisor starts to monitor the website from the start of a new shift. What is the distribution of the time waited by the supervisor until the 10th visit to the website during that shift?

4.7.12 A hole is drilled into the Antarctic ice shelf and a core is extracted that provides information on the climate when the ice was formed at different times in the past. Suppose that a researcher is interested in high-temperature years, and that the places in the core corresponding to high-temperature years occur according to a Poisson process with parameter $\lambda = 0.48$ per cm.
- **(a)** What is the expected distance in cm between adjacent high-temperature years?
- **(b)** What is the expected distance in cm between one high-temperature year and the 10th high-temperature year that followed after it?
- **(c)** What is the probability that the distance between two adjacent high-temperature years is less that 0.5 cm?
- **(d)** Suppose that a 20-cm section of core is analyzed. What is the probability that the number of high-temperature years in this section of core is between 8 and 12 inclusive?

4.7.13 Are the following statements true or false?
- **(a)** If a Beta distribution has the parameter a larger than the parameter b, then its expectation is smaller than $1/2$.
- **(b)** The uniform distribution is a symmetric distribution.
- **(c)** In a Poisson process the distances between events are identically distributed.
- **(d)** The exponential distribution is a special case of the Weibull distribution.

4.7.14 Suppose that after operation, the electrical charge remaining on a circuit component has a uniform distribution between 50 and 100, and that these charges are independent of each other for different operations. If the machine is operated five times, what is the probability that the residual charge is between 50 and 70 exactly two times, between 70 and 90 exactly two times, and between 90 and 100 exactly one time?

4.7.15 Consider a Poisson process with parameter $\lambda = 8$.
- **(a)** Consider an interval of length 0.5. What is the probability of obtaining exactly four events within this interval?
- **(b)** What is the probability that the interval between two adjacent events is shorter than 0.2?

4.7.16 Suppose that customer waiting times are independent and can be modeled by a Weibull distribution with $a = 2.3$ and $\lambda = 0.09$ per minute. What is the probability that out of 10 customers, exactly 3 wait less than 8 minutes, exactly 4 wait between 8 and 12 minutes, and exactly 3 wait more than 12 minutes?

CHAPTER FIVE

The Normal Distribution

In this chapter the **normal** or **Gaussian distribution** is discussed. It is the most important of all continuous probability distributions and is used extensively as the basis for many statistical inference methods. Its importance stems from the fact that it is a natural probability distribution for directly modeling error distributions and many other naturally occurring phenomena. In addition, by virtue of the **central limit theorem**, which is discussed in Section 5.3, the normal distribution provides a useful, simple, and accurate approximation to the distribution of general sample averages.

5.1 Probability Calculations Using the Normal Distribution

5.1.1 Definition of the Normal Distribution

HISTORICAL NOTE

Carl Friedrich Gauss (1777–1855) ranks as one of the greatest mathematicians of all time. He studied mathematics at the University of Göttingen, Germany, between 1795 and 1798 and later in 1807 became professor of astronomy at the same university, where he remained until his death. His work on the normal distribution was performed around 1820. He is reported to have been deeply religious, aristocratic, and conservative. He did not enjoy teaching and consequently had only a few students.

The Normal Distribution

The **normal** or **Gaussian distribution** has a probability density function

$$f(x) = \frac{1}{\sigma\sqrt{2\pi}}\, e^{-(x-\mu)^2/2\sigma^2}$$

for $-\infty \le x \le \infty$, depending upon two parameters, the mean and the variance

$$E(X) = \mu \qquad \text{and} \qquad \text{Var}(X) = \sigma^2$$

of the distribution. The probability density function is a bell-shaped curve that is symmetric about μ. The notation

$$X \sim N(\mu, \sigma^2)$$

denotes that the random variable X has a normal distribution with mean μ and variance σ^2. In addition, the random variable X can be referred to as being "normally distributed."

The probability density function of a normal random variable is symmetric about the mean value μ and has what is known as a "bell-shaped" curve. Figure 5.1 shows the probability density functions of normal distributions with $\mu = 5$, $\sigma = 2$ and with $\mu = 10$, $\sigma = 2$ and illustrates the fact that as the mean value μ is changed, the shape of the density function remains unaltered while the location of the density function changes. On the other hand, Figure 5.2 shows the probability density functions of normal distributions with $\mu = 5$, $\sigma = 2$ and with $\mu = 5$, $\sigma = 0.5$. The central location of the density function has not changed, but the shape has. Large values of the variance σ^2 result in long, flat bell-shaped curves, whereas small values of the variance σ^2 result in thinner, sharper bell-shaped curves.

FIGURE 5.1

The effect of changing the mean of
a normal distribution

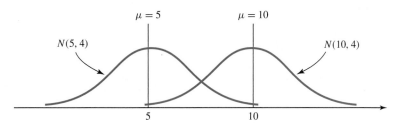

FIGURE 5.1

The effect of changing the mean of
a normal distribution

FIGURE 5.2

The effect of changing the variance
of a normal distribution

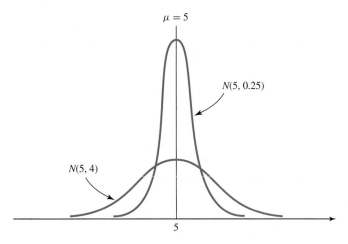

 COMPUTER NOTE There is no simple closed-form solution for the cumulative distribution function of a normal distribution. Nevertheless, in this section it is shown how probability values of normal random variables are easily calculated from tables of the standard normal distribution. In addition, it is worthwhile to discover how your computer software package can be used to obtain such values.

5.1.2 The Standard Normal Distribution

A normal distribution with mean $\mu = 0$ and variance $\sigma^2 = 1$ is known as the **standard normal distribution**. Its probability density function has the notation $\phi(x)$ and is given by

$$\phi(x) = \frac{1}{\sqrt{2\pi}} \, e^{-x^2/2}$$

for $-\infty \le x \le \infty$, as illustrated in Figure 5.3.

The notation $\Phi(x)$ is used for the cumulative distribution function of a standard normal distribution, which is calculated from the expression

$$\Phi(x) = \int_{-\infty}^{x} \phi(y) \, dy$$

as illustrated in Figure 5.4. The cumulative distribution function is shown in Figure 5.5 and is often referred to as an "S-shaped" curve. Notice that $\Phi(0) = 0.5$ because the standard normal distribution is symmetric about $x = 0$, and that the cumulative distribution function $\Phi(x)$ approaches 1 as x tends to ∞ and approaches 0 as x tends to $-\infty$.

FIGURE 5.3

The standard normal distribution

$$f(x) = \frac{1}{\sqrt{2\pi}} e^{-x^2/2}$$

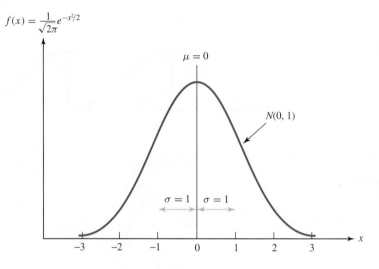

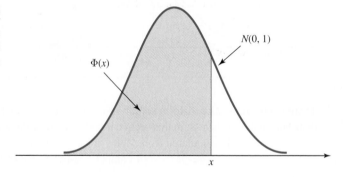

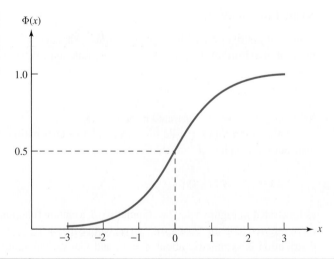

The symmetry of the standard normal distribution about 0 implies that if the random variable Z has a standard normal distribution, then

$$1 - \Phi(x) = P(Z \geq x) = P(Z \leq -x) = \Phi(-x)$$

as illustrated in Figure 5.6. This equation can be rearranged to provide the easily remembered relationship

$$\Phi(x) + \Phi(-x) = 1$$

The cumulative distribution function of the standard normal distribution $\Phi(x)$ is tabulated in Table I at the end of the book. This table provides values of $\Phi(x)$ to four decimal places for values of x between -3.49 and 3.49. For values of x less than -3.49, $\Phi(x)$ is very close to 0, and for values of x greater than 3.49, $\Phi(x)$ is very close to 1.

As an example of the use of Table I, suppose that the random variable Z has a standard normal distribution. Table I then indicates that

$$P(Z \leq 0.31) = \Phi(0.31) = 0.6217$$

as illustrated in Figure 5.7. Table I also reveals that

$$P(Z \geq 1.05) = 1 - \Phi(1.05) = 1 - 0.8531 = 0.1469$$

as illustrated in Figure 5.8, and that

$$P(-1.50 \leq Z \leq 1.18) = \Phi(1.18) - \Phi(-1.50) = 0.8810 - 0.0668 = 0.8142$$

as illustrated in Figure 5.9.

FIGURE 5.6

$\Phi(-x) = 1 - \Phi(x)$

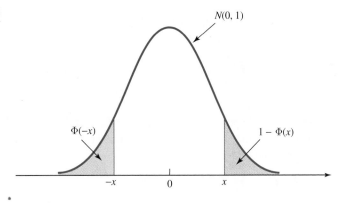

FIGURE 5.7

Probability calculations for a standard normal distribution

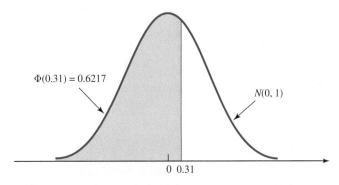

FIGURE 5.8

Probability calculations for a
standard normal distribution

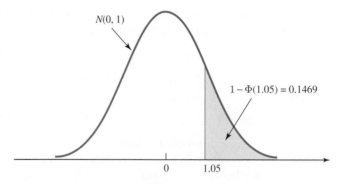

FIGURE 5.9

Probability calculations for a
standard normal distribution

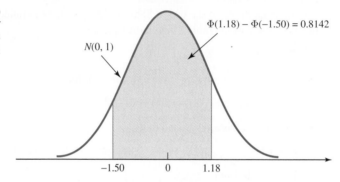

Table I can also be used to find percentiles of the standard normal distribution. For example, the 80th percentile satisfies

$$\Phi(x) = 0.8$$

and can be found by using the table "backward" by searching for the value 0.8 in the body of the table. It is found that $\Phi(0.84) = 0.7995$ and that $\Phi(0.85) = 0.8023$, so that the 80th percentile point is somewhere between 0.84 and 0.85. (If further accuracy is required, interpolation may be attempted or a computer software package may be utilized.) If the value x is required for which

$$P(|Z| \le x) = 0.7$$

as illustrated in Figure 5.10, notice that the symmetry of the standard normal distribution implies that

$$\Phi(-x) = 0.15$$

Table I then indicates that the required value of x lies between 1.03 and 1.04.

The percentiles of the standard normal distribution are used so frequently that they have their own notation. For $\alpha < 0.5$, the $(1 - \alpha) \times 100$th percentile of the distribution is denoted by z_α, so that

$$\Phi(z_\alpha) = 1 - \alpha$$

as illustrated in Figure 5.11, and some of these percentile points are given in Table I. The percentiles z_α are often referred to as the "critical points" of the standard normal distribution.

FIGURE 5.10

Symmetric tails of the normal distribution

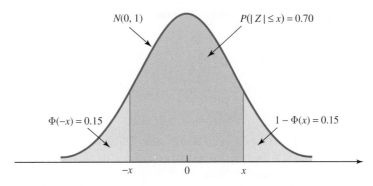

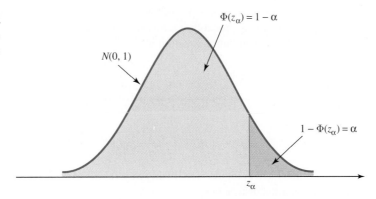

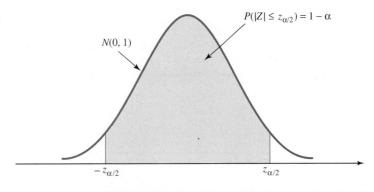

Notice that the symmetry of the standard normal distribution implies that if $Z \sim N(0, 1)$, then

$$P(|Z| \le z_{\alpha/2}) = P(-z_{\alpha/2} \le Z \le z_{\alpha/2})$$
$$= \Phi(z_{\alpha/2}) - \Phi(-z_{\alpha/2}) = (1 - \alpha/2) - \alpha/2 = 1 - \alpha$$

as illustrated in Figure 5.12.

5.1.3 Probability Calculations for General Normal Distributions

A very important general result is that if

$$X \sim N(\mu, \sigma^2)$$

then the transformed random variable

$$Z = \frac{X - \mu}{\sigma}$$

has a *standard normal distribution*. This result indicates that any normal distribution can be related to the standard normal distribution by appropriate scaling and location changes. Notice that the transformation operates by first subtracting the mean value μ and by then dividing by the standard deviation σ. The random variable Z is known as the "standardized" version of the random variable X.

A consequence of this result is that the probability values of any normal distribution can be related to the probability values of a standard normal distribution and, in particular, to the cumulative distribution function $\Phi(x)$. For example,

$$P(a \le X \le b) = P\left(\frac{a - \mu}{\sigma} \le \frac{X - \mu}{\sigma} \le \frac{b - \mu}{\sigma}\right)$$

$$= P\left(\frac{a - \mu}{\sigma} \le Z \le \frac{b - \mu}{\sigma}\right)$$

$$= \Phi\left(\frac{b - \mu}{\sigma}\right) - \Phi\left(\frac{a - \mu}{\sigma}\right)$$

as illustrated in Figure 5.13. In this way Table I can be used to calculate probability values for any normal distribution.

FIGURE 5.13

Relating normal probabilities to $\Phi(x)$

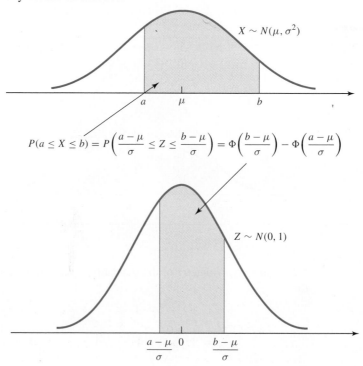

$$P(a \le X \le b) = P\left(\frac{a - \mu}{\sigma} \le Z \le \frac{b - \mu}{\sigma}\right) = \Phi\left(\frac{b - \mu}{\sigma}\right) - \Phi\left(\frac{a - \mu}{\sigma}\right)$$

Probability Calculations for Normal Distributions

If $X \sim N(\mu, \sigma^2)$, then

$$Z = \frac{X - \mu}{\sigma} \sim N(0, 1)$$

The random variable Z is known as the "standardized" version of the random variable X. This result implies that the probability values of a general normal distribution can be related to the cumulative distribution function of the standard normal distribution $\Phi(x)$ through the relationship

$$P(a \leq X \leq b) = \Phi\left(\frac{b - \mu}{\sigma}\right) - \Phi\left(\frac{a - \mu}{\sigma}\right)$$

As an illustration of this result, suppose that $X \sim N(3, 4)$. Then

$$P(X \leq 6) = P(-\infty \leq X \leq 6) = \Phi\left(\frac{6 - 3}{2}\right) - \Phi\left(\frac{-\infty - 3}{2}\right)$$

$$= \Phi(1.5) - \Phi(-\infty) = 0.9332 - 0 = 0.9332$$

as illustrated in Figure 5.14, and

$$P(2.0 \leq X \leq 5.4) = \Phi\left(\frac{5.4 - 3.0}{2.0}\right) - \Phi\left(\frac{2.0 - 3.0}{2.0}\right)$$

$$= \Phi(1.2) - \Phi(-0.5) = 0.8849 - 0.3085 = 0.5764$$

as illustrated in Figure 5.15.

FIGURE 5.14

Using the standard normal distribution to calculate the probabilities of a normal distribution

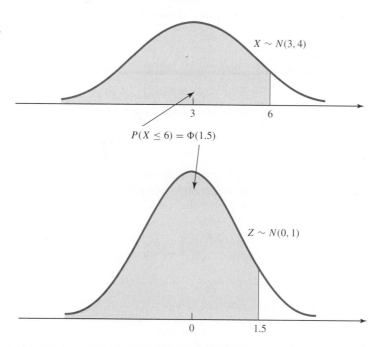

FIGURE 5.15

Using the standard normal
distribution to calculate the
probabilities of a normal
distribution

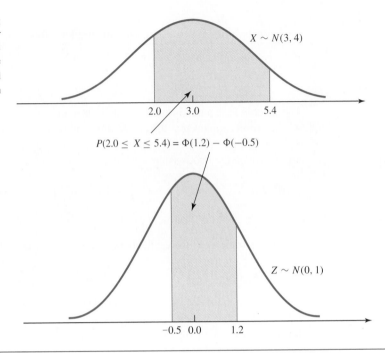

In general, if $X \sim N(\mu, \sigma^2)$, notice that

$$P(\mu - c\sigma \leq X \leq \mu + c\sigma) = P(-c \leq Z \leq c)$$

where $Z \sim N(0, 1)$. Table I reveals that when $c = 1$ this probability is about 68%, when $c = 2$ this probability is about 95%, and when $c = 3$ this probability is about 99.7%, as shown in Figure 5.16. These calculations can be summarized in the following general rules.

Normal Random Variables

- There is a probability of about 68% that a normal random variable takes a value within *one* standard deviation of its mean.

- There is a probability of about 95% that a normal random variable takes a value within *two* standard deviations of its mean.

- There is a probability of about 99.7% that a normal random variable takes a value within *three* standard deviations of its mean.

The percentiles of a $N(\mu, \sigma^2)$ distribution are related to the percentiles of a standard normal distribution through the relationship

$$P(X \leq \mu + \sigma z_\alpha) = P(Z \leq z_\alpha) = 1 - \alpha$$

For example, since the 95th percentile of the standard normal distribution is $z_{0.05} = 1.645$, the 95th percentile of a $N(3, 4)$ distribution is

$$\mu + \sigma z_{0.05} = 3 + (2 \times z_{0.05}) = 3 + (2 \times 1.645) = 6.29$$

FIGURE 5.16

The probability values of lying within one, two, and three standard deviations of the mean of a normal distribution

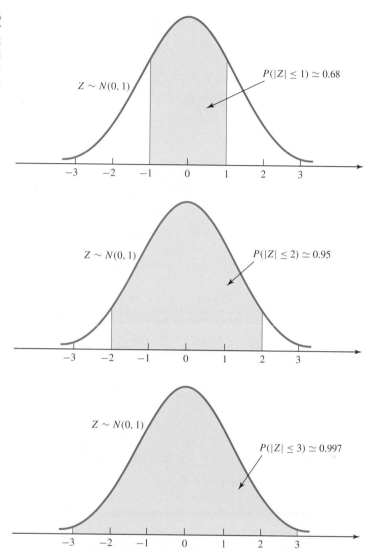

5.1.4   Examples of the Normal Distribution

Example 18
Tomato Plant Heights

Recall that three weeks after planting, the heights of tomato plants have a mean of 29.4 cm and a standard deviation of 2.1 cm. Chebyshev inequality was used to show that there is *at least* a 75% chance that a tomato plant has a height within two standard deviations of the mean, that is, within the interval [25.2, 33.6].

However, if the tomato plant heights are taken to be *normally* distributed, then this probability can be calculated much more precisely. In fact, the probability is about 95%. Moreover, there is a probability of about 99.7% that a tomato plant has a height within three standard deviations of the mean, that is, within the interval [23.1, 35.7].

More generally, there is a probability of $1 - \alpha$ that a tomato plant has a height within the interval

$$[\mu - \sigma z_{\alpha/2}, \mu + \sigma z_{\alpha/2}] = [29.4 - 2.1 z_{\alpha/2}, 29.4 + 2.1 z_{\alpha/2}]$$

FIGURE 5.17

Probability density function of
tomato plant heights

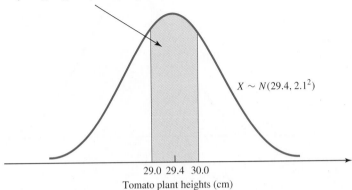

$$P(29.0 \leq X \leq 30.0) = \Phi(0.29) - \Phi(-0.19) = 0.1894$$

$X \sim N(29.4, 2.1^2)$

29.0 29.4 30.0
Tomato plant heights (cm)

Since $z_{0.05} = 1.645$, there is therefore a 90% chance that a tomato plant has a height within the interval

$$[29.4 - (2.1 \times 1.645), 29.4 + (2.1 \times 1.645)] = [25.95, 32.85]$$

Consequently, the researcher can predict that about 9 out of 10 tomato plants will have a height between 26 cm and 33 cm three weeks after planting.

The probability that a tomato plant height is between 29 cm and 30 cm can be calculated to be

$$P(29.0 \leq X \leq 30.0) = \Phi\left(\frac{30.0 - \mu}{\sigma}\right) - \Phi\left(\frac{29.0 - \mu}{\sigma}\right)$$

$$= \Phi\left(\frac{30.0 - 29.4}{2.1}\right) - \Phi\left(\frac{29.0 - 29.4}{2.1}\right)$$

$$= \Phi(0.29) - \Phi(-0.19) = 0.6141 - 0.4247 = 0.1894$$

as illustrated in Figure 5.17.

Example 37
Concrete Block
Weights

A company manufactures concrete blocks that are used for construction purposes. Suppose that the weights of the individual concrete blocks are normally distributed with a mean value of $\mu = 11.0$ kg and a standard deviation of $\sigma = 0.3$ kg.

Since $z_{0.005} = 2.576$ the company can be 99% confident that a randomly selected concrete block has a weight within the interval

$$[\mu - \sigma z_{0.005}, \mu + \sigma z_{0.005}] = [11.0 - (0.3 \times 2.576), 11.0 + (0.3 \times 2.576)]$$

$$= [10.23, 11.77]$$

The probability that a concrete block weighs less than 10.5 kg is

$$P(X \leq 10.5) = P(-\infty \leq X \leq 10.5) = \Phi\left(\frac{10.5 - \mu}{\sigma}\right) - \Phi\left(\frac{-\infty - \mu}{\sigma}\right)$$

$$= \Phi\left(\frac{10.5 - 11.0}{0.3}\right) - \Phi\left(\frac{-\infty - 11.0}{0.3}\right)$$

$$= \Phi(-1.67) - \Phi(-\infty) = 0.0475 - 0 = 0.0475$$

as illustrated in Figure 5.18. Consequently, only about 1 in 20 concrete blocks weighs less than 10.5 kg.

FIGURE 5.18

Probability density function of
concrete block weights

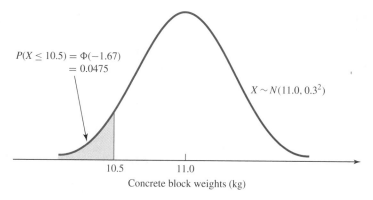

$P(X \leq 10.5) = \Phi(-1.67)$
$= 0.0475$

$X \sim N(11.0, 0.3^2)$

10.5   11.0

Concrete block weights (kg)

FIGURE 5.19

Probability density function of
annual return from stock of
company A

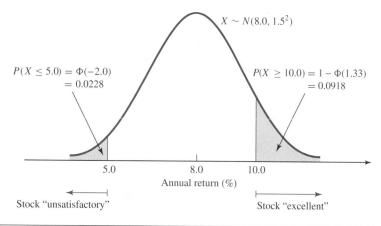

$X \sim N(8.0, 1.5^2)$

$P(X \leq 5.0) = \Phi(-2.0)$
$= 0.0228$

$P(X \geq 10.0) = 1 - \Phi(1.33)$
$= 0.0918$

5.0   8.0   10.0

Annual return (%)

Stock "unsatisfactory"

Stock "excellent"

Example 35
Stock Prices

A Wall Street analyst estimates that the annual return from the stock of company A can be
considered to be an observation from a normal distribution with mean $\mu = 8.0\%$ and standard
deviation $\sigma = 1.5\%$. The analyst's investment choices are based upon the considerations that
any return greater than 5% is "satisfactory" and a return greater than 10% is "excellent."

The probability that company A's stock will prove to be "unsatisfactory" is

$$P(X \leq 5.0) = P(-\infty \leq X \leq 5.0)$$
$$= \Phi\left(\frac{5.0 - \mu}{\sigma}\right) - \Phi\left(\frac{-\infty - \mu}{\sigma}\right)$$
$$= \Phi\left(\frac{5.0 - 8.0}{1.5}\right) - \Phi\left(\frac{-\infty - 8.0}{1.5}\right)$$
$$= \Phi(-2.00) - \Phi(-\infty) = 0.0228 - 0 = 0.0228$$

and the probability that Company A's stock will prove to be "excellent" is

$$P(10.0 \leq X) = P(10.0 \leq X \leq \infty)$$
$$= \Phi\left(\frac{\infty - \mu}{\sigma}\right) - \Phi\left(\frac{10.0 - \mu}{\sigma}\right)$$
$$= \Phi\left(\frac{\infty - 8.0}{1.5}\right) - \Phi\left(\frac{10.0 - 8.0}{1.5}\right)$$
$$= \Phi(\infty) - \Phi(1.33) = 1 - 0.9082 = 0.0918$$

These probabilities are illustrated in Figure 5.19.

■ **5.1.5 Problems**

5.1.1 Suppose that $Z \sim N(0, 1)$. Find:
(a) $P(Z \leq 1.34)$
(b) $P(Z \geq -0.22)$
(c) $P(-2.19 \leq Z \leq 0.43)$
(d) $P(0.09 \leq Z \leq 1.76)$
(e) $P(|Z| \leq 0.38)$
(f) The value of x for which $P(Z \leq x) = 0.55$
(g) The value of x for which $P(Z \geq x) = 0.72$
(h) The value of x for which $P(|Z| \leq x) = 0.31$

5.1.2 Suppose that $Z \sim N(0, 1)$. Find:
(a) $P(Z \leq -0.77)$
(b) $P(Z \geq 0.32)$
(c) $P(-3.09 \leq Z \leq -1.59)$
(d) $P(-0.82 \leq Z \leq 1.80)$
(e) $P(|Z| \geq 0.91)$
(f) The value of x for which $P(Z \leq x) = 0.23$
(g) The value of x for which $P(Z \geq x) = 0.51$
(h) The value of x for which $P(|Z| \geq x) = 0.42$

5.1.3 Suppose that $X \sim N(10, 2)$. Find:
(a) $P(X \leq 10.34)$
(b) $P(X \geq 11.98)$
(c) $P(7.67 \leq X \leq 9.90)$
(d) $P(10.88 \leq X \leq 13.22)$
(e) $P(|X - 10| \leq 3)$
(f) The value of x for which $P(X \leq x) = 0.81$
(g) The value of x for which $P(X \geq x) = 0.04$
(h) The value of x for which $P(|X - 10| \geq x) = 0.63$

5.1.4 Suppose that $X \sim N(-7, 14)$. Find:
(a) $P(X \leq 0)$
(b) $P(X \geq -10)$
(c) $P(-15 \leq X \leq -1)$
(d) $P(-5 \leq X \leq 2)$
(e) $P(|X + 7| \geq 8)$
(f) The value of x for which $P(X \leq x) = 0.75$
(g) The value of x for which $P(X \geq x) = 0.27$
(h) The value of x for which $P(|X + 7| \leq x) = 0.44$

5.1.5 Suppose that $X \sim N(\mu, \sigma^2)$ and that

$$P(X \leq 5) = 0.8 \quad \text{and} \quad P(X \geq 0) = 0.6$$

What are the values of μ and σ^2?

5.1.6 Suppose that $X \sim N(\mu, \sigma^2)$ and that

$$P(X \leq 10) = 0.55 \quad \text{and} \quad P(X \leq 0) = 0.40$$

What are the values of μ and σ^2?

5.1.7 Suppose that $X \sim N(\mu, \sigma^2)$. Show that

$$P(X \leq \mu + \sigma z_\alpha) = 1 - \alpha$$

and that

$$P(\mu - \sigma z_{\alpha/2} \leq X \leq \mu + \sigma z_{\alpha/2}) = 1 - \alpha$$

5.1.8 What are the upper and lower quartiles of a $N(0, 1)$ distribution? What is the interquartile range? What is the interquartile range of a $N(\mu, \sigma^2)$ distribution?

5.1.9 The thicknesses of glass sheets produced by a certain process are normally distributed with a mean of $\mu = 3.00$ mm and a standard deviation of $\sigma = 0.12$ mm.
(a) What is the probability that a glass sheet is thicker than 3.2 mm?
(b) What is the probability that a glass sheet is thinner than 2.7 mm?
(c) What is the value of c for which there is a 99% probability that a glass sheet has a thickness within the interval $[3.00 - c, 3.00 + c]$?
(This problem is continued in Problem 5.2.8.)

5.1.10 The amount of sugar contained in 1-kg packets is actually normally distributed with a mean of $\mu = 1.03$ kg and a standard deviation of $\sigma = 0.014$ kg.
(a) What proportion of sugar packets are underweight?
(b) If an alternative package-filling machine is used for which the weights of the packets are normally distributed with a mean of $\mu = 1.05$ kg and a standard deviation of $\sigma = 0.016$ kg, does this result in an increase or a decrease in the proportion of underweight packets?
(c) In each case, what is the expected value of the *excess* package weight above the advertised level of 1 kg?
(This problem is continued in Problem 5.2.9.)

5.1.11 The thicknesses of metal plates made by a particular machine are normally distributed with a mean of 4.3 mm and a standard deviation of 0.12 mm.
(a) What are the upper and lower quartiles of the metal plate thicknesses?
(b) What is the value of c for which there is 80% probability that a metal plate has a thickness within the interval $[4.3 - c, 4.3 + c]$?
(This problem is continued in Problem 5.2.4.)

5.1.12 The density of a chemical solution is normally distributed with mean 0.0046 and variance 9.6×10^{-8}.

(a) What is the probability that the density is less than 0.005?

(b) What is the probability that the density is between 0.004 and 0.005?

(c) What is the 10th percentile of the density level?

(d) What is the 99th percentile of the density level?

5.1.13 The resistance of one meter of copper cable at a certain temperature is normally distributed with mean $\mu = 23.8$ and variance $\sigma^2 = 1.28$.

(a) What is the probability that a one-meter segment of copper cable has a resistance less than 23.0?

(b) What is the probability that a one-meter segment of copper cable has a resistance greater than 24.0?

(c) What is the probability that a one-meter segment of copper cable has a resistance between 24.2 and 24.5?

(d) What is the upper quartile of the resistance level?

(e) What is the 95th percentile of the resistance level?

5.1.14 The weights of bags filled by a machine are normally distributed with a standard deviation of 0.05 kg and a mean that can be set by the operator. At what level should the mean be set if it is required that only 1% of the bags weigh less than 10 kg?

5.1.15 Suppose a certain mechanical component produced by a company has a width that is normally distributed with a mean $\mu = 2600$ and a standard deviation $\sigma = 0.6$.

(a) What proportion of the components have a width outside the range 2599 to 2601?

(b) If the company needs to be able to guarantee to its purchaser that no more than 1 in 1000 of the components have a width outside the range 2599 to 2601, by how much does the value of σ need to be reduced?

(This problem is continued in Problem 5.2.10.)

5.1.16 Bricks have weights that are independently distributed with a normal distribution that has a mean 1320 and a standard deviation of 15. A set of 10 bricks is chosen at random. What is the probability that exactly 3 bricks will weigh less than 1300, exactly 4 bricks will weigh between 1300 and 1330, and exactly 3 bricks will weigh more than 1330?

5.1.17 Manufactured items have a strength that has a normal distribution with a standard deviation of 4.2. The mean strength can be altered by the operator. At what value should the mean strength be set so that exactly 95% of the items have a strength less than 100?

5.2 Linear Combinations of Normal Random Variables

5.2.1 The Distribution of Linear Combinations of Normal Random Variables

In this section an attractive feature of the normal distribution is discussed, which is that linear combinations of normal random variables are also normally distributed. The means and variances of these linear combinations can be found from the general results presented in Section 2.6.

In the simplest case, Section 2.6.1 provides general results for the mean and variance of a linear function

$$Y = aX + b$$

of a random variable X. If the random variable X has a normal distribution, then an additional point is that the linear function Y also has a normal distribution. This result is summarized in the following box and is illustrated in Figure 5.20.

Linear Functions of a Normal Random Variable
If $X \sim N(\mu, \sigma^2)$ and a and b are constants, then $$Y = aX + b \sim N(a\mu + b, a^2\sigma^2)$$

Notice that if $a = 1/\sigma$ and $b = -\mu/\sigma$, the resulting linear function of X has a standard normal distribution, as discussed in Section 5.1.3.

FIGURE 5.20

A linear function of a normal
random variable

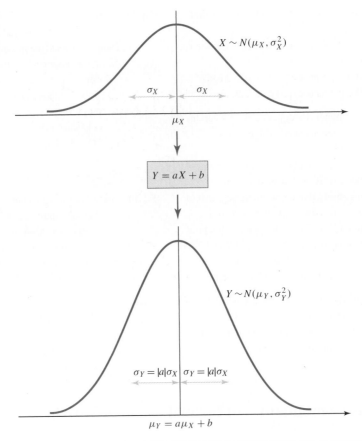

Section 2.6.2 provides some general results for the mean and variance of a linear combination of random variables. An additional point now is that a linear combination of normally
distributed random variables is also normally distributed. In the simple case involving the
summation of two independent random variables, the following result is obtained, which is
illustrated in Figure 5.21.

The Sum of Two Independent Normal Random Variables

If $X_1 \sim N(\mu_1, \sigma_1^2)$ and $X_2 \sim N(\mu_2, \sigma_2^2)$ are independent random variables, then

$$Y = X_1 + X_2 \sim N\left(\mu_1 + \mu_2, \sigma_1^2 + \sigma_2^2\right)$$

It is also worth noting that if two normal random variables are not independent, then their sum
is still normally distributed, but the variance of the sum depends on the covariance of the two
random variables.

The two results presented so far can be synthesized into the following general result.

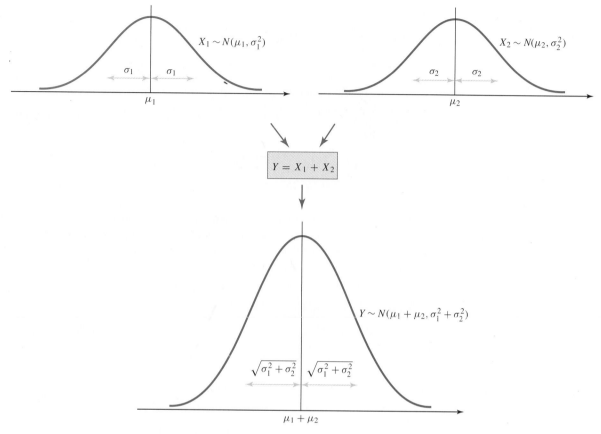

FIGURE 5.21

The sum of two independent normal random variables

Linear Combinations of Independent Normal Random Variables

If $X_i \sim N(\mu_i, \sigma_i^2)$, $1 \leq i \leq n$, are independent random variables and if a_i, $1 \leq i \leq n$, and b are constants, then

$$Y = a_1 X_1 + \cdots + a_n X_n + b \sim N(\mu, \sigma^2)$$

where

$$\mu = a_1 \mu_1 + \cdots + a_n \mu_n + b$$

and

$$\sigma^2 = a_1^2 \sigma_1^2 + \cdots + a_n^2 \sigma_n^2$$

A special case of this result concerns the situation in which interest is directed toward the average $\bar{X}$ of a set of independent identically distributed $N(\mu, \sigma^2)$ random variables. With $b = 0$, $a_i = 1/n$, $\mu_i = \mu$, and $\sigma_i^2 = \sigma^2$, the result above implies that $\bar{X}$ is normally distributed with

$$E(\bar{X}) = \mu \qquad \text{and} \qquad \text{Var}(\bar{X}) = \frac{\sigma^2}{n}$$

Averaging Independent Normal Random Variables

If $X_i \sim N(\mu, \sigma^2)$, $1 \leq i \leq n$, are independent random variables, then their average $\bar{X}$ is distributed

$$\bar{X} \sim N\left(\mu, \frac{\sigma^2}{n}\right)$$

Notice that averaging reduces the variance to σ^2/n, so that the average $\bar{X}$ has a tendency to be closer to the mean value μ than do the individual random variables X_i. This tendency increases as n increases and the average of more and more random variables X_i is taken.

As an illustration of this idea, recall that there is a probability of about 68% that a normal random variable takes a value within one standard deviation of its mean, so that

$$P(\mu - \sigma \leq X_i \leq \mu + \sigma) = 0.68$$

If $n = 10$ so that an average of 10 of these random variables is taken, then

$$\bar{X} \sim N\left(\mu, \frac{\sigma^2}{10}\right)$$

and

$$P(\mu - \sigma \leq \bar{X} \leq \mu + \sigma) = \Phi\left(\frac{\sigma}{\sqrt{\sigma^2/10}}\right) - \Phi\left(-\frac{\sigma}{\sqrt{\sigma^2/10}}\right)$$
$$= \Phi(3.16) - \Phi(-3.16) = 0.9992 - 0.0008 = 0.9984$$

In other words, while there is only a 68% chance that a $N(\mu, \sigma^2)$ random variable lies within the interval $[\mu - \sigma, \mu + \sigma]$, the *average* of 10 independent random variables of this kind has more than a 99.8% chance of taking a value within this interval.

5.2.2 Examples of Linear Combinations of Normal Random Variables

Example 23
Piston Head Construction

Recall that the radius of a piston head X_1 has a mean value of 30.00 mm and a standard deviation of 0.05 mm, and that the inside radius of a cylinder X_2 has a mean value of 30.25 mm and a standard deviation of 0.06 mm. The gap between the piston head and the cylinder $Y = X_2 - X_1$ therefore has a mean value of

$$\mu = 30.25 - 30.00 = 0.25$$

and a variance of

$$\sigma^2 = 0.05^2 + 0.06^2 = 0.0061$$

If the piston head radius and the cylinder radius are taken to be *normally distributed*, then the gap Y is also normally distributed since it is obtained as a linear combination of the normal random variables X_1 and X_2. Specifically,

$$Y \sim N(0.25, 0.0061)$$

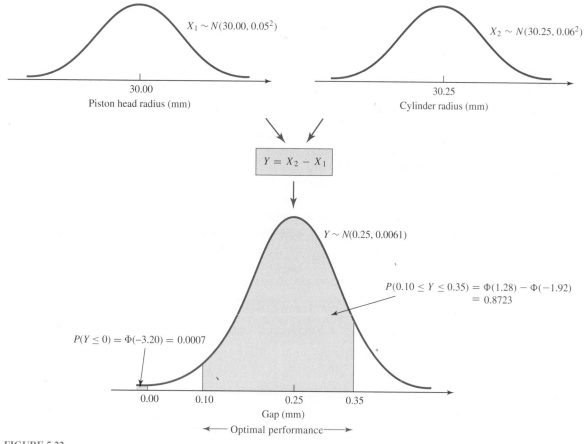

FIGURE 5.22

Probability density functions for piston head construction

The probability that a piston head will not fit within a cylinder can then be calculated to be

$$P(Y \leq 0) = \Phi\left(\frac{0 - 0.25}{\sqrt{0.0061}}\right) = \Phi(-3.20) = 0.0007$$

as illustrated in Figure 5.22.

Suppose that a piston performs optimally when the gap Y is between 0.10 mm and 0.35 mm. The probability that a piston performs optimally is then

$$P(0.10 \leq Y \leq 0.35) = \Phi\left(\frac{0.35 - 0.25}{\sqrt{0.0061}}\right) - \Phi\left(\frac{0.10 - 0.25}{\sqrt{0.0061}}\right)$$

$$= \Phi(1.28) - \Phi(-1.92) = 0.8997 - 0.0274 = 0.8723$$

Example 18
Tomato Plant Heights

Suppose that 20 tomato plants are planted. What is the distribution of the *average* tomato plant height after three weeks of growth?

The distribution of the individual tomato plant heights is

$$X_i \sim N(29.4, 2.1^2) = N(29.4, 4.41)$$

Consequently, the average of 20 of these heights is distributed

$$\bar{X} \sim N\left(29.4, \frac{4.41}{20}\right) = N(29.4, 0.2205)$$

Since $z_{0.025} = 1.96$, there is therefore a 95% chance that the average tomato plant height lies within the interval

$$[29.4 - (1.96 \times \sqrt{0.2205}), 29.4 + (1.96 \times \sqrt{0.2205})] = [28.48, 30.32]$$

Notice that the standard deviation of the average tomato plant height $\sigma = \sqrt{0.2205} = 0.47$ cm is considerably smaller than the standard deviation of the individual tomato plant heights, which is $\sigma = 2.10$ cm.

Example 37

Concrete Block Weights

Suppose that a wall is constructed from 24 concrete blocks as illustrated in Figure 5.23. What is the distribution of the total weight of the wall?

The weights of the individual concrete blocks X_i, $1 \le i \le 24$, are distributed

$$X_i \sim N(11.0, 0.3^2) = N(11.0, 0.09)$$

Consequently, the weight of the wall Y is distributed

$$Y = X_1 + \cdots + X_{24} \sim N(\mu, \sigma^2)$$

where

$$\mu = 11.0 + \cdots + 11.0 = 264.0$$

FIGURE 5.23

Distribution of the total weight of the wall

$$\mu = 11.0 + \cdots + 11.0 = 264.0$$
$$\sigma^2 = 0.09 + \cdots + 0.09 = 2.16$$

Total weight of wall is $N(264.0, 2.16)$.

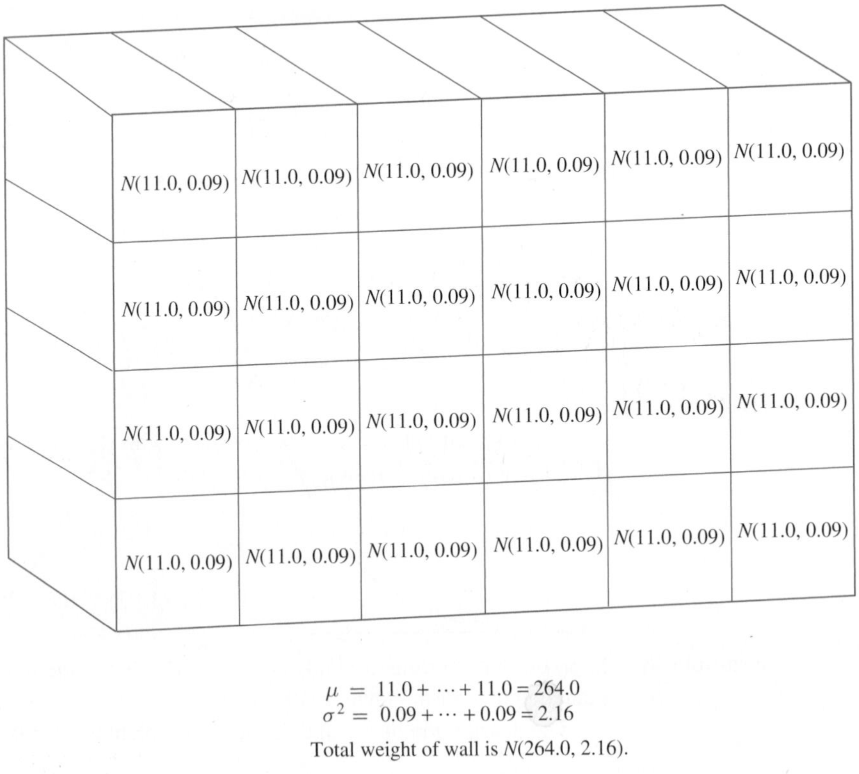

and

$$\sigma^2 = 0.09 + \cdots + 0.09 = 2.16$$

Thus the wall has an expected weight of 264.0 kg with a standard deviation of $\sqrt{2.16} = 1.47$ kg.

There is about a 99.7% chance that the wall has a weight within three standard deviations of its mean value, that is, within the interval

$$[264.0 - (3 \times 1.47), 264.0 + (3 \times 1.47)] = [259.59, 268.41]$$

Consequently, the builders can be confident that the wall weighs somewhere between 259 kg and 269 kg.

Example 35
Stock Prices

Recall that the annual return from the stock of company A, X_A say, is distributed

$$X_A \sim N(8.0, 1.5^2) = N(8.0, 2.25)$$

In addition, suppose that the annual return from the stock of company B, X_B say, is distributed

$$X_B \sim N(9.5, 4.00)$$

independent of the stock of company A.

The probability that company B's stock proves to be "unsatisfactory" is

$$P(X_B \leq 5.0) = \Phi\left(\frac{5.0 - 9.5}{2.0}\right) = \Phi(-2.25) = 0.0122$$

and the probability that company B's stock proves to be "excellent" is

$$P(10.0 \leq X) = 1 - \Phi\left(\frac{10.0 - 9.5}{2.0}\right) = 1 - \Phi(0.25) = 1 - 0.5987 = 0.4013$$

These probabilities are illustrated in Figure 5.24.

FIGURE 5.24

Probability density function of
annual return from stock of
company B

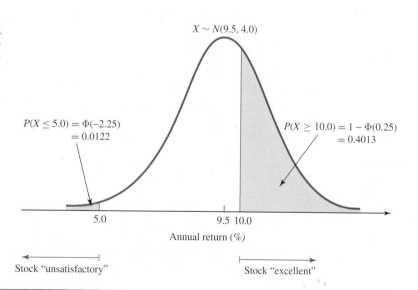

$X \sim N(9.5, 4.0)$

$P(X \leq 5.0) = \Phi(-2.25)$
$= 0.0122$

$P(X \geq 10.0) = 1 - \Phi(0.25)$
$= 0.4013$

5.0

9.5 10.0

Annual return (%)

Stock "unsatisfactory"

Stock "excellent"

What is the probability that company B's stock performs better than company A's stock? If $Y = X_B - X_A$, then

$$Y \sim N(9.5 - 8.0, 4.00 + 2.25) = N(1.5, 6.25)$$

The required probability is therefore

$$P(Y \geq 0) = 1 - \Phi\left(\frac{0 - 1.5}{\sqrt{6.25}}\right) = 1 - \Phi(-0.6) = 1 - 0.2743 = 0.7257$$

The probability that company B's stock performs at least two percentage points better than company A's stock is

$$P(Y \geq 2.0) = 1 - \Phi\left(\frac{2.0 - 1.5}{\sqrt{6.25}}\right) = 1 - \Phi(0.2) = 1 - 0.5793 = 0.4207$$

Example 38

Chemical Concentration Levels

A chemist has two different methods for measuring the concentration level C of a chemical solution. Method A produces a measurement X_A that is distributed

$$X_A \sim N(C, 2.97)$$

so that the chemist can be 99.7% certain that the measured value lies within the interval

$$[C - (3 \times \sqrt{2.97}), C + (3 \times \sqrt{2.97})] = [C - 5.17, C + 5.17]$$

Method B involves a different kind of analysis and produces a measurement X_B that is distributed

$$X_B \sim N(C, 1.62)$$

In this case the chemist can be 99.7% certain that the measured value lies within the interval

$$[C - (3 \times \sqrt{1.62}), C + (3 \times \sqrt{1.62})] = [C - 3.82, C + 3.82]$$

The variability in method B is smaller than the variability in method A, and so the chemist correctly feels that the measurement reading x_B obtained from method B is more "accurate" than the measurement reading x_A obtained from method A. Should the measurement reading x_A therefore be completely ignored?

In fact, the most sensible course for the chemist to take is to *combine* the two measurement values x_A and x_B into one value

$$y = px_A + (1 - p)x_B$$

for a suitable value of p between 0 and 1. The final measurement value y is thus a *weighted average* of the two measurement values x_A and x_B and has a distribution

$$Y = pX_A + (1 - p)X_B \sim N\left(\mu_Y, \sigma_Y^2\right)$$

where

$$\mu_Y = pE(X_A) + (1 - p)E(X_B) = pC + (1 - p)C = C$$

and

$$\sigma_Y^2 = p^2 \text{Var}(X_A) + (1 - p)^2 \text{Var}(X_B) = p^2 2.97 + (1 - p)^2 1.62$$

It is sensible to choose p in a way that *minimizes* the variance σ_Y^2. The derivative of σ_Y^2 with respect to p is

$$\frac{d\sigma_Y^2}{dp} = 5.94p - 3.24(1 - p)$$

Setting this expression equal to 0 gives an "optimal" value of $p = 0.35$, in which case

$$\sigma_Y^2 = (0.35^2 \times 2.97) + (0.65^2 \times 1.62) = 1.05$$

Notice that this variance is *smaller* than the variances of both method A and method B, as illustrated in Figure 5.25.

In conclusion, the chemist's best estimate of the concentration level is

$$y = 0.35x_A + 0.65x_B$$

and the chemist can be 99.7% certain that this value lies within the interval

$$[C - (3 \times \sqrt{1.05}), C + (3 \times \sqrt{1.05})] = [C - 3.07, C + 3.07]$$

FIGURE 5.25

Probability density functions for chemical concentration level measurements

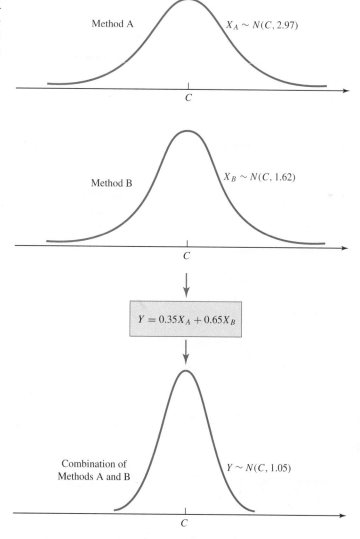

Method A $X_A \sim N(C, 2.97)$

C

Method B $X_B \sim N(C, 1.62)$

C

$Y = 0.35X_A + 0.65X_B$

Combination of
Methods A and B $Y \sim N(C, 1.05)$

C

■ 5.2.3 Problems

5.2.1 Suppose that $X \sim N(3.2, 6.5)$, $Y \sim N(-2.1, 3.5)$, and $Z \sim N(12.0, 7.5)$ are independent random variables. Find the probability that
(a) $X + Y \geq 0$
(b) $X + Y - 2Z \leq -20$
(c) $3X + 5Y \geq 1$
(d) $4X - 4Y + 2Z \leq 25$
(e) $|X + 6Y + Z| \geq 2$
(f) $|2X - Y - 6| \leq 1$

5.2.2 Suppose that $X \sim N(-1.9, 2.2)$, $Y \sim N(3.3, 1.7)$, and $Z \sim N(0.8, 0.2)$ are independent random variables. Find the probability that
(a) $X - Y \geq -3$
(b) $2X + 3Y + 4Z \leq 10$
(c) $3Y - Z \leq 8$
(d) $2X - 2Y + 3Z \leq -6$
(e) $|X + Y - Z| \geq 1.5$
(f) $|4X - Y + 10| \leq 0.5$

5.2.3 Consider a sequence of independent random variables X_i, each with a standard normal distribution.
(a) What is $P(|X_i| \leq 0.5)$?
(b) If $\bar{X}$ is the average of eight of these random variables, what is $P(|\bar{X}| \leq 0.5)$?
 (c) In general, if $\bar{X}$ is the average of n of these random variables, what is the smallest value of n for which $P(|\bar{X}| \leq 0.5) \geq 0.99$?

5.2.4 Recall Problem 5.1.11 where metal plate thicknesses are normally distributed with a mean of 4.3 mm and a standard deviation of 0.12 mm.
(a) If one metal plate is placed on top of another, what is the distribution of their combined thickness?
(b) What is the distribution of the average thickness of 12 metal plates?
(c) What is the smallest number of metal plates required in order for their average thickness to be between 4.25 and 4.35 mm with a probability of at least 99.7%?

5.2.5 A machine part is assembled by fastening *two* components of type A and *three* components of type B end to end. The lengths of components of type A in mm are independent $N(37.0, 0.49)$ random variables, and the lengths of components of type B in mm are independent $N(24.0, 0.09)$ random variables. What is the probability that a machine part has a length between 144 and 147 mm?

5.2.6 (a) Suppose that $X_1 \sim N(\mu_1, \sigma_1^2)$ and $X_2 \sim N(\mu_2, \sigma_2^2)$ are independently distributed. What is the variance of
$$Y = pX_1 + (1 - p)X_2?$$
Show that the variance is minimized when
$$p = \frac{\frac{1}{\sigma_1^2}}{\frac{1}{\sigma_1^2} + \frac{1}{\sigma_2^2}}$$
What is the variance of Y in this case?
(b) More generally suppose that $X_i \sim N(\mu_i, \sigma_i^2)$, $1 \leq i \leq n$, are independently distributed, and that
$$Y = p_1 X_1 + \cdots + p_n X_n$$
where $p_1 + \cdots + p_n = 1$. What values of the p_i minimize the variance of Y, and what is the minimum variance?

5.2.7 If \$$x$ is invested in mutual fund I, its worth after one year is distributed
$$X_I \sim N(1.05x, 0.0002x^2)$$
and if \$$x$ is invested in mutual fund II, its worth after one year is distributed
$$X_{II} \sim N(1.05x, 0.0003x^2)$$
Suppose that you have \$1000 to invest and that you place \$$y$ in mutual fund I and \$$(1000 - y)$ in mutual fund II.
(a) What is the expected value of the total worth of your investments after one year?
(b) What is the variance of the total worth of your investments after one year?
 (c) What value of y minimizes the variance of the total worth of your investments after one year? If you adopt this "conservative" strategy, what is the probability that after one year the total worth of your investments is more than \$1060?

5.2.8 Recall Problem 5.1.9 where glass sheets have a $N(3.00, 0.12^2)$ distribution.
(a) What is the probability that three glass sheets placed one on top of another have a total thickness greater than 9.50 mm?
(b) What is the probability that seven glass sheets have an average thickness less than 3.10 mm?

5.2.9 Recall Problem 5.1.10 where sugar packets have weights with $N(1.03, 0.014^2)$ distributions. A box contains 22 sugar packets.

(a) What is the distribution of the total weight of sugar in a box?

(b) What are the upper and lower quartiles of the total weight of sugar in a box?

5.2.10 Recall Problem 5.1.15, where mechanical components have a width that is normally distributed with a mean $\mu = 2600$ and a standard deviation $\sigma = 0.6$. In an assembly procedure, four of these components need to be fitted side by side into a slot in another part.

(a) Suppose that the slots have a width of 10,402.5. What proportion of the time will four randomly selected components be able to fit into a slot?

(b) More generally, suppose that the widths of the slots vary according to a normal distribution with mean $\mu = 10,402.5$ and standard deviation $\sigma = 0.4$. In this case, what proportion of the time will four randomly selected components be able to fit into a randomly selected slot?

5.2.11 Let $X_1, \ldots, X_{15}$ be independent identically distributed $N(4.5, 0.88)$ random variables, with an average $\bar{X}$.

(a) Calculate $P(4.2 \le \bar{X} \le 4.9)$.

(b) Find the value of c for which
$$P(4.5 - c \le \bar{X} \le 4.5 + c) = 0.99.$$

5.2.12 Five students are waiting to talk to the TA when office hours begin. The TA talks to the students one at a time, starting with the first student and ending with the fifth student, with no breaks between students. Suppose that the time taken by the TA to talk to a student has a normal distribution with a mean of 8 minutes and a standard deviation of 2 minutes, and suppose that the times taken by the students are independent of each other.

(a) What is the probability that the total time taken by the TA to talk to all five students is longer than 45 minutes?

(b) Suppose that the time that elapses between when the TA starts talking to the first student, and when the TA starts to have a headache, has a normal distribution with a mean of 28 minutes and a standard deviation of 5 minutes, which is independent of the times taken to talk to the students. What is the probability that the TA's headache starts at a time after the TA has finished talking to the third student?

5.2.13 Components of type A have heights that are independently distributed as a normal distribution with a mean 190 and a standard deviation of 10. Components of type B have heights that are independently distributed as a normal distribution with a mean 150 and a standard deviation of 8. What is the probability that a stack of four components of type A placed one on top of the other will be taller than a stack of five components of type B placed one on top of the other?

5.2.14 The times taken for worker 1 to perform a task are independently distributed as a normal distribution with mean 13 minutes and standard deviation 0.5 minutes. The times taken for worker 2 to perform a task are independently distributed as a normal distribution with mean 17 minutes and standard deviation 0.6 minutes, and they are independent of the times taken by worker 1. At the beginning of the day, both workers start their first task at the same time, and when they have finished a task, they immediately start another task. What is the probability that worker 1 will finish his fourth task before worker 2 has finished his third task?

5.2.15 Bricks' weights are independently distributed as a normal distribution with mean 110 and standard deviation 2. What is the smallest value of n such that there is a probability of at least 99% that the average weight of n randomly selected bricks is less than 111?

5.2.16 A piece of wire is cut, and the length of the wire has a normal distribution with a mean 7.2 m and a standard deviation 0.11 m. If the piece of wire is then cut exactly in half, what are the mean and the standard deviation of the lengths of the two pieces?

5.2.17 The amount of timber available from a certain type of fully grown tree has a mean of 63400 with a standard deviation of 2500.

(a) What are the mean and the standard deviation of the total amount of timber available from 20 trees?

(b) What are the mean and the standard deviation of the average amount of timber available from 30 trees?

5.2.18 A chemist can set the target value for the elasticity of a polymer compound. The resulting elasticity is normally distributed with a mean equal to the target value and a standard deviation of 47.

(a) What target value should be set if it is required that there is only a 10% probability that the elasticity is less than 800?

(b) Suppose that a target value of 850 is used. What is the probability that the average elasticity of 10 samples is smaller than 875?

5.3   Approximating Distributions with the Normal Distribution

A very useful property of the normal distribution is that it provides good approximations to the probability values of certain other distributions. In these special cases, the cumulative distribution function of a rather complicated distribution can sometimes be related to the cumulative distribution function of a normal distribution that is easily evaluated.

The most common example is when the normal distribution is used to approximate the binomial distribution. If the parameter n of the binomial distribution is large, then its cumulative distribution function is tedious to compute. However, these binomial probabilities are approximated very well by the probabilities of a corresponding normal distribution. Some more general theory provided by the **central limit theorem** indicates that the normal distribution is appropriate to approximate the distribution of an average of a set of identically distributed random variables, irrespective of the distribution of the individual random variables.

5.3.1   The Normal Approximation to the Binomial Distribution

An examination of the probability mass functions of the binomial distributions that are graphed in Section 3.1 reveals that they have a "bell-shaped" curve similar to the probability density function of a normal distribution. In fact, it turns out that the normal distribution is very good at providing an approximation to the probability values of a binomial distribution when the parameter n is reasonably large and when the success probability p is not too close to 0 or to 1.

Recall that a $B(n, p)$ distribution has an expected value of np and a variance of $np(1 - p)$. This distribution can be approximated by a normal distribution with the same mean and variance, that is, a

$$N(np, np(1 - p))$$

distribution.

For example, suppose that $X \sim B(16, 0.5)$, in which case X has a mean of 8 and a variance of 4. The probability mass function of this distribution is shown in Figure 5.26 together with the probability density function of the random variable $Y \sim N(8, 4)$. Even though the random variable X has a *discrete* distribution and the random variable Y has a *continuous* distribution, the shapes of their respective probability mass function and probability density function are

FIGURE 5.26

Comparison of the probability mass function of a $B(16, 0.5)$ random variable and the probability density function of a $N(8, 4)$ random variable

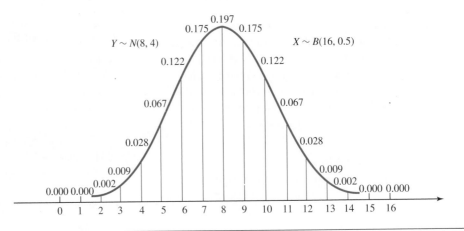

quite similar. The lines of the probability mass function of X are not exactly the same height as those of the probability density function of Y, but they are very close.

It is interesting to compute some probability values of the $B(16, 0.5)$ and $N(8, 4)$ distributions to see how well they compare. The probability that the binomial distribution takes a value no larger than 5 is

$$P(X \leq 5) = \sum_{x=0}^{5} \binom{16}{x} \times 0.5^x \times 0.5^{16-x} = 0.1051$$

The best way to approximate this probability using the normal distribution is to compute the probability that $Y \leq 5.5$. Notice that a "continuity correction" of 0.5 is added to the value 5 in order to improve the approximation of the discrete binomial distribution by a continuous distribution, as illustrated in Figure 5.27. The approximate probability value obtained from the normal distribution is

$$P(Y \leq 5.5) = \Phi\left(\frac{5.5 - \mu}{\sigma}\right) = \Phi\left(\frac{5.5 - 8.0}{2.0}\right) = \Phi(-1.25) = 0.1056$$

It is seen that the approximation provided by the normal distribution is remarkably good.

As another example, illustrated in Figure 5.28, the probability that the binomial distribution takes a value between 8 and 11 inclusive is

$$P(8 \leq X \leq 11) = \sum_{x=8}^{11} \binom{16}{x} \times 0.5^x \times 0.5^{16-x} = 0.5598$$

FIGURE 5.27

Approximating $P(X \leq 5)$ with a probability from a normal distribution

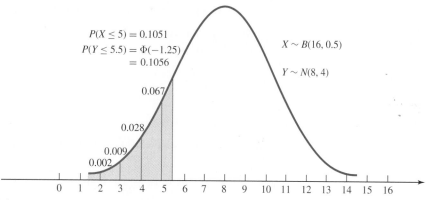

FIGURE 5.28

Approximating $P(8 \leq X \leq 11)$ with a probability from a normal distribution

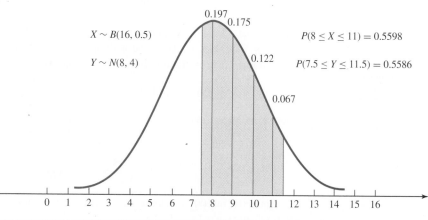

The normal approximation to this, with continuity corrections, is

$$P(7.5 \leq Y \leq 11.5) = \Phi\left(\frac{11.5 - \mu}{\sigma}\right) - \Phi\left(\frac{7.5 - \mu}{\sigma}\right)$$

$$= \Phi\left(\frac{11.5 - 8.0}{2.0}\right) - \Phi\left(\frac{7.5 - 8.0}{2.0}\right)$$

$$= \Phi(1.75) - \Phi(-0.25) = 0.9599 - 0.4013 = 0.5586$$

Again, the normal approximation is seen to be remarkably good.

The normal approximation to the binomial distribution improves in accuracy as the parameter n of the binomial distribution increases. However, for a given value of n, the approximation may not be very good if the success probability p is too close to 0 or to 1. How close is "too close" depends upon the value of n. A general rule is that the approximation is reasonable as long as both

$$np \geq 5 \quad \text{and} \quad n(1 - p) \geq 5$$

It should also be noted that the exact probability values of binomial distributions are easily obtainable from computer software packages, so that in practice there may be no need to approximate their probability values unless n is larger than, say, 100. In such cases the software package that you use may actually be using the normal approximation to calculate the binomial probabilities.

Normal Approximations to the Binomial Distribution

The probability values of a $B(n, p)$ distribution can be approximated by those of a $N(np, np(1 - p))$ distribution. If $X \sim B(n, p)$, then

$$P(X \leq x) \simeq \Phi\left(\frac{x + 0.5 - np}{\sqrt{np(1 - p)}}\right)$$

and

$$P(X \geq x) \simeq 1 - \Phi\left(\frac{x - 0.5 - np}{\sqrt{np(1 - p)}}\right)$$

These approximations work well as long as $np \geq 5$ and $n(1 - p) \geq 5$.

GAMES OF CHANCE

Suppose that a fair coin is tossed n times. The distribution of the number of heads obtained, X, is $B(n, 0.5)$. If $n = 100$, what is the probability of obtaining between 45 and 55 heads? With a normal approximation this probability is easily calculated as

$$P(45 \leq X \leq 55) \simeq P(44.5 \leq Y \leq 55.5)$$

where $Y \sim N(50, 25)$, which can be evaluated to be

$$\Phi\left(\frac{55.5 - 50.0}{\sqrt{25}}\right) - \Phi\left(\frac{44.5 - 50.0}{\sqrt{25}}\right) = \Phi(1.1) - \Phi(-1.1)$$

$$= 0.8643 - 0.1357 = 0.7286$$

Consequently, in 100 coin tosses there is a probability of about 0.73 that the *proportion* of heads is between 0.45 and 0.55.

FIGURE 5.29

The probabilities that in n tosses of
a fair coin the proportion of heads
lies between 0.45 and 0.55

Number of tosses	$P(0.45 \leq X/n \leq 0.55)$
$n = 100$	73%
$n = 150$	81%
$n = 200$	86%
$n = 500$	98%
$n = 750$	99.4%
$n = 1000$	99.9%

For a general value of n, the distribution of $X \sim B(n, 0.5)$ can be approximated by $Y \sim N(n/2, n/4)$. In general, the probability that the proportion of heads lies between 0.45 and 0.55 is therefore equal to

$$P(0.45n \leq X \leq 0.55n) \simeq P(0.45n - 0.5 \leq Y \leq 0.55n + 0.5)$$

$$= \Phi\left(\frac{0.55n + 0.5 - 0.5n}{\sqrt{n/4}}\right) - \Phi\left(\frac{0.45n - 0.5 - 0.5n}{\sqrt{n/4}}\right)$$

$$= \Phi\left(0.1\sqrt{n} + \frac{1}{\sqrt{n}}\right) - \Phi\left(-0.1\sqrt{n} - \frac{1}{\sqrt{n}}\right)$$

These probability values are given in Figure 5.29 for various values of n. Notice that they are increasing and tend toward a limiting value of one as the number of tosses n increases.

5.3.2   The Central Limit Theorem

Consider a sequence $X_1, \ldots, X_n$ of independent identically distributed random variables. Suppose that these random variables have an expectation and variance

$$E(X_i) = \mu \qquad \text{and} \qquad \text{Var}(X_i) = \sigma^2$$

If

$$\bar{X} = \frac{X_1 + \cdots + X_n}{n}$$

then it was shown in Section 2.6.2 that

$$E(\bar{X}) = \mu \qquad \text{and} \qquad \text{Var}(\bar{X}) = \frac{\sigma^2}{n}$$

Furthermore, it was also shown in Section 5.2.1 that if $X_i \sim N(\mu, \sigma^2)$, then

$$\bar{X} \sim N\left(\mu, \frac{\sigma^2}{n}\right)$$

In summary, it is known that if a set of independent random variables is obtained and each has the same distribution with mean μ and variance σ^2, then their average always has mean μ and variance σ^2/n, and their average is normally distributed if the individual random variables are normally distributed.

The **central limit theorem** provides an important extension to these results by stating that regardless of the actual distribution of the individual random variables X_i, the distribution of their average $\bar{X}$ is closely approximated by a $N(\mu, \sigma^2/n)$ distribution. In other words, the *average* of a set of *independent identically distributed* random variables is always approximately normally distributed. The accuracy of the approximation improves as n increases and the average is taken over more random variables.

A general rule is that the approximation is adequate as long as $n \geq 30$, although the approximation is often good for much smaller values of n, particularly if the distribution of the random variables X_i has a probability density function with a shape reasonably similar to the normal bell-shaped curve.

Notice that if

$$\bar{X} \sim N\left(\mu, \frac{\sigma^2}{n}\right)$$

then

$$X_1 + \cdots + X_n \sim N(n\mu, n\sigma^2)$$

and so the central limit theorem can also be used to show that the distribution of the sum $X_1 + \cdots + X_n$ can be approximated by a $N(n\mu, n\sigma^2)$ distribution.

The central limit theorem is regarded as one of the most important theorems in the whole of probability theory and, in fact, in the whole of mathematics. It may explain why many naturally occurring phenomena are observed to have distributions similar to the normal distribution, because they may be considered to be composed of the aggregate of many smaller random events.

The central limit theorem also explains why the normal distribution provides a good approximation to the binomial distribution, since if $X \sim B(n, p)$, then

$$X = X_1 + \cdots + X_n$$

where the random variables X_i have independent Bernoulli distributions with parameter p. Since $E(X_i) = p$ and $\text{Var}(X_i) = p(1 - p)$, the central limit theorem indicates that the distribution of X can be approximated by a $N(np, np(1 - p))$ distribution, as discussed in Section 5.3.1.

The Central Limit Theorem

If $X_1, \ldots, X_n$ is a sequence of independent identically distributed random variables with a mean μ and a variance σ^2, then the distribution of their average $\bar{X}$ can be approximated by a

$$N\left(\mu, \frac{\sigma^2}{n}\right)$$

distribution. Similarly, the distribution of the sum $X_1 + \cdots + X_n$ can be approximated by a

$$N(n\mu, n\sigma^2)$$

distribution.

In conclusion, the central limit theorem provides a very convenient way of approximating the probability values of an average of a set of identically distributed random variables. The exact distribution of this average will in general have a complicated distribution, whereas the approximate probability values are easily obtained from the appropriate normal distribution. The central limit theorem has important consequences for statistical analysis methods (which are discussed in later chapters) because it indicates that a sample average may be taken to be normally distributed regardless of the actual distribution of the individual sample observations.

5.3.3 Simulation Experiment 1: An Investigation of the Central Limit Theorem

It is instructive to check the central limit theorem by simulating sets of random numbers on a computer. Most computer packages allow you to simulate **random numbers** from a specified distribution. If the distribution is discrete, then these random numbers are the actual elements of the state space that are produced with frequencies corresponding to their probability values. For continuous distributions, the random numbers produced have the property that they fall into specific intervals with probability values governed by the probability density function of the distribution being used.

An idea of the *shape* of the probability mass function or probability density function of a distribution can be obtained by simulating a large number m of random variables from the distribution and then constructing a **histogram** of their values (histograms are discussed in more detail in Chapter 6). The shape of the histogram approximates the shape of the underlying density function of the random variables, and the accuracy of the approximation improves as the number of simulated observations m increases.

Suppose that we wish to investigate the probability density function of the random variable

$$Y = \frac{X_1 + X_2 + X_3 + X_4 + X_5 + X_6 + X_7 + X_8 + X_9 + X_{10}}{10}$$

where the random variables X_i are independent *exponential* random variables with parameter $\lambda = 0.1$. We can do this by simulating $m = 500$ values y_i, $1 \leq i \leq 500$, from this distribution. Each value y_i can be obtained by simulating ten values x_{ij}, $1 \leq j \leq 10$, having the specified exponential distribution, and then by taking their average. (In practice, a convenient way to do this is to form 10 columns of random numbers, each of length 500, and then to form a new column equal to their average.)

Notice that overall this requires the simulation of 5000 exponential random variables. The probability density function of the random variable Y is approximated by a histogram of the values y_i. Figure 5.30 presents a histogram obtained from this simulation experiment. Of course, if you try it yourself, you'll produce a slightly different histogram.

FIGURE 5.30

Histogram of the averages of simulated exponential random variables

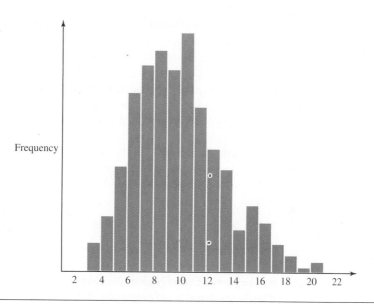

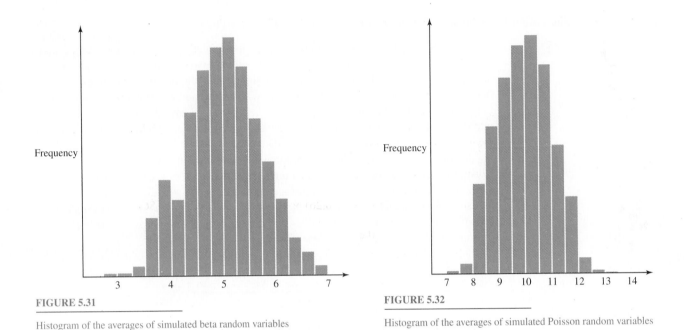

FIGURE 5.31

Histogram of the averages of simulated beta random variables

FIGURE 5.32

Histogram of the averages of simulated Poisson random variables

The random variables X_i have a mean of 10 and a variance of 100, so that the random variable Y has a mean of 10 and a variance of 10. Therefore, the central limit theorem indicates that the distribution of Y can be approximated by a $N(10, 10)$ distribution. Figure 5.30 clearly illustrates that the distribution of Y is starting to take on a normal bell-shaped curve, and that it is quite different from the exponential distribution.

Figure 5.31 presents a histogram obtained from a similar simulation experiment where the random variables X_i are taken to have beta distributions with parameter values $a = b = 2$ (this distribution is illustrated in Figure 4.21). This beta distribution has a mean of 0.5 and a variance of 0.05, so that the central limit theorem indicates that the distribution of Y can be approximated by a $N(0.5, 0.005)$ distribution, and this is confirmed by the histogram in Figure 5.31, which clearly has the shape of a normal distribution.

Finally, Figure 5.32 presents a histogram obtained from this simulation experiment when the random variables X_i are taken to have a Poisson distribution with parameter $\lambda = 10$. This Poisson distribution has a mean and a variance of 10, so that the central limit theorem indicates that the distribution of Y can be approximated by a $N(10, 1)$ distribution. Again, this is confirmed by the histogram in Figure 5.32, which clearly has the shape of a normal distribution.

When you perform simulations of this kind, it is useful to remember that the closeness of the histogram that you obtain to the shape of a normal distribution depends on three factors (m, n, and the distribution of X_i) in the following ways.

■ As the number m of simulated values y_i of the random variable Y increases, the histogram of the y_i becomes a more and more accurate representation of the *true probability density function* of Y.

■ As the number n increases, so that the random variable Y is obtained as the average of a larger number of random variables, then the *true probability density function* of Y becomes closer to the probability density function of a normal distribution.

■ In general, for a given value of n, the *true probability density function* of Y becomes closer to the probability density function of a normal distribution as the probability density function of the random variables X_i has a shape that looks more like a normal probability density function.

5.3.4 Examples of Employing Normal Approximations

Example 17
Milk Container Contents

Recall that there is a probability of 0.261 that a milk container is underweight. Consequently, the number of underweight containers X in a box of 20 containers has a $B(20, 0.261)$ distribution. This distribution may be approximated by

$$Y \sim N(20 \times 0.261, \ 20 \times 0.261 \times (1 - 0.261)) = N(5.22, 3.86)$$

Figure 5.33 shows the probability mass function of X together with the probability density function of Y, and they are seen to match well. The probability that a box contains no more than three underweight containers was calculated in Section 3.1.3 to be

$$P(X \leq 3) = 0.1935$$

The normal approximation to this probability is

$$P(Y \leq 3.5) = \Phi\left(\frac{3.5 - 5.22}{\sqrt{3.86}}\right) = \Phi(-0.87) = 0.1922$$

which is very close.

Suppose that 25 boxes of milk are delivered to a supermarket. What is the distribution of the number of underweight containers? Altogether there are now 500 milk containers so that the number of underweight containers X has a $B(500, 0.261)$ distribution. This distribution may be approximated by

$$Y \sim N(500 \times 0.261, \ 500 \times 0.261 \times (1 - 0.261)) = N(130.5, 96.44)$$

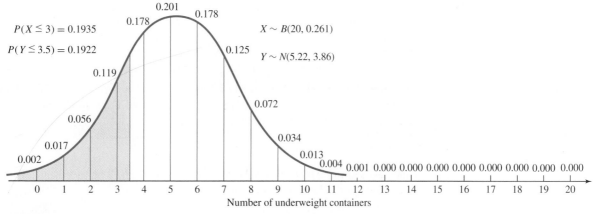

FIGURE 5.33

Approximating the distribution of the number of underweight containers with a normal distribution

The probability that at least 150 out of the 500 milk containers are underweight can then be calculated to be

$$P(X \geq 150) \simeq P(Y \geq 149.5) = 1 - \Phi\left(\frac{149.5 - 130.5}{\sqrt{96.44}}\right)$$

$$= 1 - \Phi(1.93) = 1 - 0.9732 = 0.0268$$

Example 39
Cattle Inoculations

Cattle are routinely inoculated to help prevent the spread of diseases among them. Suppose that a particular vaccine has a probability of 0.0005 of provoking a serious adverse reaction when administered to an animal. This probability value implies that, on average, only 1 in 2000 animals suffers this reaction.

Suppose that the vaccine is to be administered to 500,000 head of cattle. The number of animals X that will suffer an adverse reaction has a binomial distribution with parameters $n = 500,000$ and $p = 0.0005$. Since

$$E(X) = 500,000 \times 0.0005 = 250$$

and

$$\text{Var}(X) = 500,000 \times 0.0005 \times 0.9995 = 249.9$$

this distribution can be approximated by

$$Y \sim N(250, 249.9)$$

An interval of three standard deviations about the mean value for this normal distribution is

$$[250 - (3 \times \sqrt{249.9}), \ 250 + (3 \times \sqrt{249.9})] = [202.6, 297.4]$$

Consequently, the veterinarians can be confident (at least 99.7% certain) that if they inoculate 500,000 head of cattle, then the number of animals that suffer a serious adverse reaction will be somewhere between 200 and 300.

Example 27
Glass Sheet Flaws

Recall that the number of flaws in a glass sheet has a Poisson distribution with a parameter value $\lambda = 0.5$. This implies that both the expectation and variance of the number of flaws in a sheet are

$$\mu = \sigma^2 = 0.5 \ .$$

What is the distribution of the total number of flaws X in 100 sheets of glass?

The expected number of flaws in 100 sheets of glass is $100 \times \mu = 50$, and the variance of the number of flaws in 100 sheets of glass is $100 \times \sigma^2 = 50$. The central limit theorem then implies that the distribution of X can be approximated by a

$$N(50, 50)$$

distribution. The probability that there are fewer than 40 flaws in 100 sheets of glass can therefore be approximated as

$$P(X \leq 40) \simeq \Phi\left(\frac{40 - 50}{\sqrt{50}}\right) = \Phi(-1.41) = 0.0793$$

or about 8%.

The central limit theorem indicates that the average number of flaws per sheet in 100 sheets of glass, $\bar{X}$, has a distribution that can be approximated by a

$$N\left(\mu, \frac{\sigma^2}{100}\right) = N(0.5, 0.005)$$

distribution. The probability that this average is between 0.45 and 0.55 is therefore

$$P(0.45 \leq \bar{X} \leq 0.55) \simeq \Phi\left(\frac{0.55 - 0.50}{\sqrt{0.005}}\right) - \Phi\left(\frac{0.45 - 0.50}{\sqrt{0.005}}\right)$$

$$= \Phi(0.71) - \Phi(-0.71) = 0.7611 - 0.2389 = 0.5222$$

which is about 52%.

The key point in these probability calculations is that even though the number of flaws in an individual sheet of glass follows a Poisson distribution, the central limit theorem indicates that probability calculations concerning the total number or average number of flaws in 100 sheets of glass can be found using the normal distribution.

Example 30

Pearl Oyster Farming

Recall that there is a probability of 0.6 that an oyster produces a pearl with a diameter of at least 4 mm, which is consequently of commercial value. How many oysters does an oyster farmer need to farm in order to be 99% confident of having at least 1000 pearls of commercial value?

If the oyster farmer farms n oysters, then the distribution of the number of pearls of commercial value is

$$X \sim B(n, 0.6)$$

This distribution can be approximated by

$$Y \sim N(0.6n, 0.24n)$$

The probability of having at least 1000 pearls of commercial value is then

$$P(X \geq 1000) \simeq P(Y \geq 999.5) = 1 - \Phi\left(\frac{999.5 - 0.6n}{\sqrt{0.24n}}\right)$$

Now if

$$1 - \Phi(x) \geq 0.99$$

or equivalently

$$\Phi(x) \leq 0.01$$

it can be seen from Table I that $x \leq -2.33$. Therefore, it follows that the farmer's requirements can be met as long as

$$\frac{999.5 - 0.6n}{\sqrt{0.24n}} \leq -2.33$$

which is satisfied for $n \geq 1746$.

In conclusion, the farmer should farm about 1750 oysters in order to be 99% confident of having at least 1000 pearls of commercial value. In this case, the expected number of pearls of commercial value is $1750 \times 0.6 = 1050$, and the standard deviation is

$$\sqrt{1750 \times 0.6 \times 0.4} = 20.5$$

Recall that each pearl has an expected diameter and variance of

$$\mu = 5.0 \quad \text{and} \quad \sigma^2 = 8.33$$

If 1050 pearls are actually obtained, then the *average* diameter of the pearls will have an expectation and variance of

$$\mu = 5.0 \quad \text{and} \quad \frac{\sigma^2}{1050} = \frac{8.33}{1050} = 0.00793$$

The central limit theorem indicates that the average diameter has approximately a normal distribution (with such a large value of n the approximation will be very accurate), so that there is about a 99.7% chance that the average pearl diameter will lie within three standard deviations of the mean value, that is, within the interval

$$[5.0 - (3 \times \sqrt{0.00793}), 5.0 + (3 \times \sqrt{0.00793})] = [4.7, 5.3]$$

In other words, the farmer can be very confident of harvesting a collection of pearls with an average diameter between 4.7 mm and 5.3 mm.

■ 5.3.5 Problems

5.3.1 Calculate the following probabilities both exactly and by using a normal approximation:
(a) $P(X \geq 8)$ where $X \sim B(10, 0.7)$
(b) $P(2 \leq X \leq 7)$ where $X \sim B(15, 0.3)$
(c) $P(X \leq 4)$ where $X \sim B(9, 0.4)$
(d) $P(8 \leq X \leq 11)$ where $X \sim B(14, 0.6)$

5.3.2 Calculate the following probabilities both exactly and by using a normal approximation:
(a) $P(X \geq 7)$ where $X \sim B(10, 0.3)$
(b) $P(9 \leq X \leq 12)$ where $X \sim B(21, 0.5)$
(c) $P(X \leq 3)$ where $X \sim B(7, 0.2)$
(d) $P(9 \leq X \leq 11)$ where $X \sim B(12, 0.65)$

5.3.3 Suppose that a fair coin is tossed n times. Estimate the probability that the proportion of heads obtained lies between 0.49 and 0.51 for $n = 100, 200, 500, 1000$, and 2000.

5.3.4 Suppose that a fair die is rolled 1000 times.
(a) Estimate the probability that the number of 6s is between 150 and 180.
(b) What is the smallest value of n for which there is a probability of at least 99% of obtaining at least 50 6s in n rolls of a fair die?

5.3.5 The number of cracks in a ceramic tile has a Poisson distribution with parameter $\lambda = 2.4$.
(a) How would you approximate the distribution of the total number of cracks in 500 ceramic tiles?
(b) Estimate the probability that there are more than 1250 cracks in 500 ceramic tiles.

5.3.6 In a test for a particular illness, a *false-positive* result is obtained about 1 in 125 times the test is administered. If the test is administered to 15,000 people, estimate the probability of there being more than 135 false-positive results.

5.3.7 Despite a series of quality checks by a company that makes television sets, there is a probability of 0.0007 that when a purchaser unpacks a newly purchased television set it does not work properly. If the company sells 250,000 television sets a year, estimate the probability that there will be no more than 200 unhappy purchasers in a year.

5.3.8 A multiple-choice test consists of a series of questions, each with *four* possible answers.
(a) If there are 60 questions, estimate the probability that a student who guesses blindly at each question will get at least 30 questions right.
(b) How many questions are needed in order to be 99% confident that a student who guesses blindly at each question scores no more than 35% on the test?

5.3.9 Recall Problem 4.3.4 in which a day's sales in $1000 units at a gas station have a gamma distribution with $k = 5$ and $\lambda = 0.9$. If the sales on different days are distributed independently of each other, estimate the probability that in one year the gas station takes in more than $2 million.

5.3.10 Recall Problem 3.1.9, where a company receives 60% of its orders over the Internet. Estimate the probability that at least 925 of the company's next 1500 orders will be received over the Internet.

5.3.11 Consider again Supplementary Problem 4.7.10, where the strength of a chemical solution has a beta distribution with parameters $a = 18$ and $b = 11$. Estimate the probability that the average strength of 20 independently produced chemical solutions is between 0.60 and 0.65.

5.3.12 Suppose that a course has a capacity of at most 240 people, but that 1550 invitations are sent out. If each person who receives an invitation has a probability of 0.135 of attending the course, independently of everybody else, what is the probability that the number of people attending the course will exceed the capacity?

5.3.13 The lifetimes of batteries are independent with an exponential distribution with a mean of 84 days. Consider a random selection of 350 batteries. What is the probability that at least 55 of the batteries have lifetimes between 60 and 100 days?

5.3.14 The time to failure of an electrical component has a Weibull distribution with parameters $\lambda = 0.056$ and $a = 2.5$. A random collection of 500 components is obtained. Estimate the probability that at least 125 of the 500 components will have failure times larger than 20.

5.3.15 Suppose that components have weights that are independent and uniformally distributed between 890 and 892.

    (a) Suppose that components are weighed one by one. What is the probability that the sixth component weighed is the third component that weighs more than 891.2?

    (b) If a box contains 200 components, what is the probability that at least half of them weigh more than 890.7?

5.3.16 Suppose that the time taken for food to spoil using a certain packaging method has an exponential distribution with a mean of 8 days. If a random sample of 100 packets are tested after 10 days, what are the expectation and the standard deviation of the number of packets that will be found with spoiled food? What is the probability that at least 75 of the packets will contain spoiled food?

5.4 Distributions Related to the Normal Distribution

The normal distribution is the basis for the construction of various other important probability distributions. The **lognormal distribution** has a positive state space and can be used to model response times and failure times as well as many other phenomena. The **chi-square distribution**, the **t-distribution**, and the **F-distribution** are important tools in statistical data analysis, as will be seen in later chapters. Finally, the **multivariate normal distribution** is used to develop much of the theory behind statistical inference methods.

5.4.1 The Lognormal Distribution

A random variable X has a lognormal distribution with parameters μ and σ^2 if the transformed random variable $Y = \ln(X)$ has a normal distribution with mean μ and variance σ^2.

The Lognormal Distribution

A random variable X has a **lognormal distribution** with parameters μ and σ^2 if

$$Y = \ln(X) \sim N(\mu, \sigma^2)$$

The probability density function of X is

$$f(x) = \frac{1}{\sqrt{2\pi}\,\sigma x}\, e^{-(\ln(x)-\mu)^2/2\sigma^2}$$

for $x \geq 0$ and $f(x) = 0$ elsewhere, and the cumulative distribution function is

$$F(x) = \Phi\left(\frac{\ln(x) - \mu}{\sigma}\right)$$

A lognormal distribution has expectation and variance

$$E(X) = e^{\mu+\sigma^2/2} \quad \text{and} \quad \text{Var}(X) = e^{2\mu+\sigma^2}\left(e^{\sigma^2} - 1\right)$$

FIGURE 5.34

Probability density functions of the
lognormal distribution

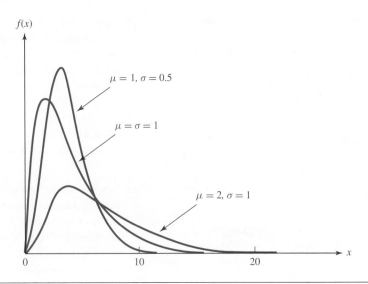

Notice that the cumulative distribution function of a lognormal distribution is easily calculated using the cumulative distribution function of a standard normal distribution $\Phi(x)$, since

$$P(X \leq x) = P(\ln(X) \leq \ln(x)) = P\left(\frac{Y - \mu}{\sigma} \leq \frac{\ln(x) - \mu}{\sigma}\right) = \Phi\left(\frac{\ln(x) - \mu}{\sigma}\right)$$

Figure 5.34 shows the probability density functions of lognormal distributions with parameter values $\mu = \sigma = 1$, $\mu = 2$ and $\sigma = 1$, and $\mu = 1$ and $\sigma = 0.5$. It can be seen that these distributions all have long, gradually decreasing right-hand tails, which is a general property of lognormal distributions. The cumulative distribution function of a lognormal distribution indicates that the *median* value is e^{μ}, which is always smaller than the mean value. This is a consequence of the long right-hand tail of the distribution.

Notice that, in general, $F(x) = 1 - \alpha$ implies that

$$\Phi\left(\frac{\ln(x) - \mu}{\sigma}\right) = 1 - \alpha$$

so that

$$\frac{\ln(x) - \mu}{\sigma} = z_\alpha$$

where z_α is the critical point of the standard normal distribution. This implies that the $(1 - \alpha)$th quantile of a lognormal distribution is

$$x = e^{\mu + \sigma z_\alpha}$$

COMPUTER NOTE

Probability values of the lognormal distribution are easily calculated due to the simple form of the cumulative distribution function, and they should also be available on your computer package.

Example 40

Testing Reaction Times

Suppose that the reaction time in seconds of a person, that is, the time elapsing between the arrival of a certain stimulus and a consequent action by the person, can be modeled by a

FIGURE 5.35

Probability density function of
reaction times

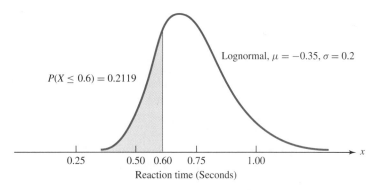

$P(X \leq 0.6) = 0.2119$

Lognormal, $\mu = -0.35$, $\sigma = 0.2$

0.25   0.50 0.60   0.75   1.00

Reaction time (Seconds)

lognormal distribution with parameter values $\mu = -0.35$ and $\sigma = 0.2$. This distribution is illustrated in Figure 5.35.

The mean reaction time is

$$E(X) = e^{-0.35 + 0.2^2/2} = 0.719$$

and the variance of the reaction times is

$$\text{Var}(X) = e^{-(2 \times 0.35) + 0.2^2} \left(e^{0.2^2} - 1 \right) = 0.021$$

Also, the median reaction time is $e^{-0.35} = 0.705$ seconds.

The probability that a reaction time is less than 0.6 seconds is

$$P(X \leq 0.6) = \Phi \left(\frac{\ln(0.6) + 0.35}{0.2} \right) = \Phi(-0.80) = 0.2119$$

The fifth percentile of the reaction times is the value x that satisfies

$$P(X \leq x) = \Phi \left(\frac{\ln(x) + 0.35}{0.2} \right) = 0.05$$

This equation is satisfied when

$$\frac{\ln(x) + 0.35}{0.2} = -1.645$$

so that $x = 0.507$ seconds. Consequently, only about 5% of the reaction times are less than 0.51 seconds.

5.4.2 The Chi-Square Distribution

If the random variable X has a standard normal distribution, then the random variable

$$Y = X^2$$

is said to have a **chi-square distribution** with *one degree of freedom*. More generally, if the random variables $X_i \sim N(0, 1)$, $1 \leq i \leq n$, are independent, then the random variable

$$Y = X_1^2 + \cdots + X_n^2$$

is said to have a chi-square distribution with n *degrees of freedom*.

The degrees of freedom of a chi-square distribution are usually denoted by the Greek letter ν and can take any positive integer value. The notation

$$X \sim \chi_\nu^2$$

is used to denote that the random variable X has a chi-square distribution with ν *degrees of freedom*. Notice that if the random variables

$$X_1 \sim \chi_{\nu_1}^2 \qquad \text{and} \qquad X_2 \sim \chi_{\nu_2}^2$$

are independently distributed, then it follows from their representation as the sum of squares of standard normal random variables that

$$Y = X_1 + X_2 \sim \chi_{\nu_1 + \nu_2}^2$$

The chi-square distribution is in fact a gamma distribution with parameter values $\lambda = 1/2$ and $k = \nu/2$. Its probability density function is

$$f(x) = \frac{1}{2^{\nu/2}\,\Gamma(\nu/2)}\, x^{\nu/2 - 1}\, e^{-x/2}$$

for $x \geq 0$ and $f(x) = 0$ elsewhere. Its expectation and variance are given in the following box. It is worth noting that a chi-square distribution with noninteger (but positive) degrees of freedom can also be defined from the gamma distribution.

The Chi-Square Distribution

A **chi-square** random variable with ν *degrees of freedom*, X, can be generated as

$$X = X_1^2 + \cdots + X_\nu^2$$

where the X_i are independent standard normal random variables. A chi-square distribution with ν degrees of freedom is a gamma distribution with parameter values $\lambda = 1/2$ and $k = \nu/2$, and it has an expectation of ν and a variance of 2ν.

Figure 5.36 illustrates chi-square distributions with degrees of freedom $\nu = 5, 10$, and 15. Notice that as the degrees of freedom increase, the distribution becomes more symmetric and more spread out. In fact, since a chi-square distribution with ν degrees of freedom is generated as the sum of ν independent, identically distributed random variables (i.e., X_i^2 where $X_i \sim N(0, 1)$), the central limit theorem implies that for large values of ν a chi-square distribution can be approximated by a $N(\nu, 2\nu)$ distribution.

The critical points of chi-square distributions are denoted by $\chi_{\alpha,\nu}^2$ and are defined by

$$P\left(X \geq \chi_{\alpha,\nu}^2\right) = \alpha$$

where X has a chi-square distribution with ν degrees of freedom, as illustrated in Figure 5.37. Table II contains the values of these critical points for various α levels and degrees of freedom ν. The chi-square distribution and these critical points will be used in the statistical inference methodologies discussed in Chapters 8, 10, 15, and 17.

COMPUTER NOTE

The critical points and other probability values of chi-square distributions should be available from your software package. Usually, there is a "chi-square" command for which you need to specify only the degrees of freedom ν, but in any case do not forget that the chi-square distribution is just a special case of a gamma distribution.

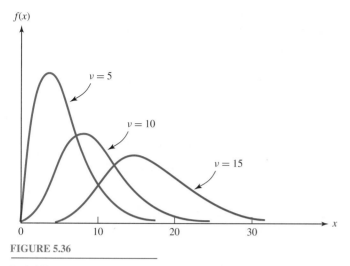

FIGURE 5.36

Probability density functions of the chi-square distribution

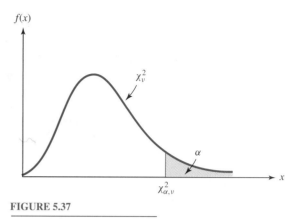

FIGURE 5.37

The critical points $\chi^2_{\alpha,v}$ of the chi-square distribution

5.4.3   The t-distribution

HISTORICAL NOTE

William Sealey Gosset (1876–1937) studied mathematics and chemistry at Oxford University in England. In 1899 he moved to Dublin, Ireland, and worked as a statistician for the Guinness brewery. During his work on the quality of barley and hops, Gosset proposed the use of the t-distribution, and in 1908 he published his ideas in an academic article using the pseudonym "Student" because Guinness forbade its employees from publishing their own research results. Consequently, the t-distribution is often referred to as Student's t-distribution.

If a standard normal random variable is divided by the square root of an independent χ^2_v/v random variable, then the resulting random variable is said to have a **t-distribution** with v degrees of freedom. This can be written

$$t_v \sim \frac{N(0,\,1)}{\sqrt{\chi^2_v/v}}$$

The t-distribution is often referred to as "Student's t-distribution" (see the Historical Note).

Figure 5.38 shows a t-distribution with 5 degrees of freedom superimposed upon a standard normal distribution. The t-distribution has a shape very similar to that of a standard normal distribution. It has a symmetric bell-shaped curve centered at 0, but it is actually a little "flatter" than the standard normal distribution. However, as the degrees of freedom v increase, the t-distribution becomes closer and closer to a standard normal distribution, and the standard normal distribution is in fact the limiting value of the t-distribution as $v \to \infty$.

The t-distribution

A **t-distribution** with v degrees of freedom is defined to be

$$t_v \sim \frac{N(0,\,1)}{\sqrt{\chi^2_v/v}}$$

where the $N(0,\,1)$ and χ^2_v random variables are independently distributed. The t-distribution has a shape similar to a standard normal distribution but is a little flatter. As $v \to \infty$, the t-distribution tends to a standard normal distribution.

The critical points of a t-distribution are denoted by $t_{\alpha,v}$ and are defined by

$$P(X \geq t_{\alpha,v}) = \alpha$$

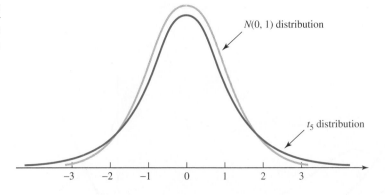

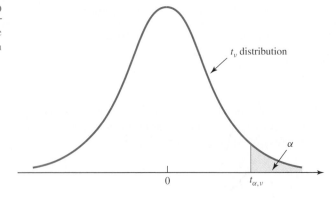

where the random variable X has a *t*-distribution with ν degrees of freedom, as illustrated in Figure 5.39. Some of these critical points are given in Table III for various values of ν and $\alpha \leq 0.1$.

Notice that the symmetry of the *t*-distribution implies that $t_{1-\alpha,\nu} = -t_{\alpha,\nu}$. Furthermore, notice that if X has a *t*-distribution with ν degrees of freedom, then

$$P(|X| \leq t_{\alpha/2,\nu}) = P(-t_{\alpha/2,\nu} \leq X \leq t_{\alpha/2,\nu}) = 1 - \alpha$$

as illustrated in Figure 5.40.

The last row in Table III with $\nu = \infty$ corresponds to the standard normal distribution, and it is seen that for a fixed value of α, the critical points $t_{\alpha,\nu}$ *decrease* to $t_{\alpha,\infty} = z_\alpha$ as the degrees of freedom ν increase. For example, with $\alpha = 0.05$, $t_{0.05,5} = 2.015$, $t_{0.05,10} = 1.812$, $t_{0.05,25} = 1.708$, and $t_{0.05,\infty} = z_{0.05} = 1.645$. The *t*-distribution and these critical points will be used in the statistical inference methodologies discussed in Chapters 8, 9, 12, and 13.

COMPUTER NOTE   Check to see how the critical values given in Table III and additional probability values of the *t*-distribution are obtained from your computer package.

FIGURE 5.40

The critical points $t_{\alpha/2,\nu}$ of the
t-distribution

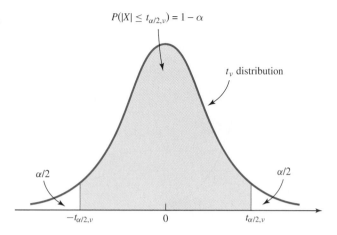

FIGURE 5.41

Probability density functions of the
F-distribution

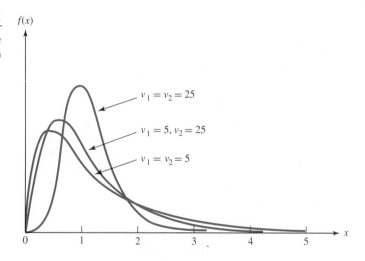

5.4.4 The F-distribution

The ratio of two independent chi-square random variables that have been divided by their respective degrees of freedom is defined to be an **F-distribution**. This ratio can be written

$$F_{\nu_1,\nu_2} \sim \frac{\chi^2_{\nu_1}/\nu_1}{\chi^2_{\nu_2}/\nu_2}$$

An F-distribution has degrees of freedom ν_1 and ν_2 that correspond to the degrees of freedom of first the numerator chi-square distribution and then the denominator chi-square distribution. Notice that in general an F_{ν_1,ν_2} distribution is not the same as an F_{ν_2,ν_1} distribution.

An F-distribution has a state space $x \geq 0$ and is unimodal with a long right-hand tail. Figure 5.41 shows the probability density functions of F-distributions with degrees of freedom $\nu_1 = \nu_2 = 5$, $\nu_1 = 5$ and $\nu_2 = 25$, and $\nu_1 = \nu_2 = 25$. The expectation of an F_{ν_1,ν_2} distribution is $\nu_2/(\nu_2 - 2)$ (for $\nu_2 \geq 3$), which is roughly equal to one for reasonably large values of ν_2. In addition, the variance of the F-distribution decreases as the degrees of freedom ν_1 and ν_2 become larger, in which case the probability density function becomes more and more sharply spiked about the value one.

FIGURE 5.42

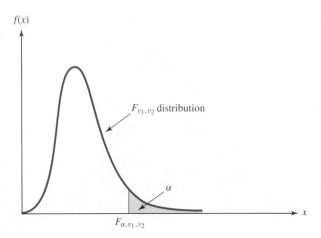

The critical point F_{α,ν_1,ν_2} of the F-distribution

The F-distribution

An **F-distribution** with degrees of freedom ν_1 and ν_2 is defined to be

$$F_{\nu_1,\nu_2} \sim \frac{\chi^2_{\nu_1}/\nu_1}{\chi^2_{\nu_2}/\nu_2}$$

where the two chi-square random variables are independently distributed. The F-distribution has a positive state space, an expectation close to one, and a variance that decreases as the degrees of freedom ν_1 and ν_2 increase.

The critical points of F-distributions are denoted by F_{α,ν_1,ν_2}, as illustrated in Figure 5.42. Table IV contains some of these critical points for $\alpha = 0.10, 0.05,$ and 0.01. In addition, it follows from the definition of the F-distribution that

$$F_{1-\alpha,\nu_1,\nu_2} = \frac{1}{F_{\alpha,\nu_2,\nu_1}}$$

so that Table IV can also be used to find the values F_{α,ν_1,ν_2} for $\alpha = 0.90, 0.95,$ and 0.99. The F-distribution and these critical points will be used in the statistical inference methodologies discussed in Chapters 11, 12, 13, and 14.

COMPUTER NOTE Check to see how the critical values given in Table IV and additional probability values of the F-distribution are obtained from your computer package.

5.4.5 The Multivariate Normal Distribution

A **bivariate normal distribution** for a pair of random variables (X, Y) has five parameters. These are the means μ_X and μ_Y and the variances σ_X^2 and σ_Y^2 of the marginal distributions of the random variables X and Y, together with the correlation ρ $(-1 \le \rho \le 1)$ between the two random variables. A useful property of the bivariate normal distribution is that both marginal distributions and any conditional distributions are all normal distributions.

FIGURE 5.43

A bivariate normal distribution
with correlation $\rho = 0$

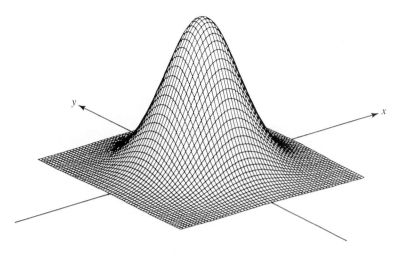

The random variables X and Y are independent when $\rho = 0$, in which case the joint probability density function of X and Y is the product of their normal marginal distributions. If the two marginal distributions are standard normal distributions, so that $\mu_X = \mu_Y = 0$ and $\sigma_X^2 = \sigma_Y^2 = 1$, then the joint probability density function of X and Y is

$$f(x, y) = \frac{1}{2\pi} e^{-(x^2+y^2)/2}$$

for $-\infty < x, y < \infty$, which is shown in Figure 5.43. This density function is rotationally symmetric, and any "slice" of it on a plane that is perpendicular to the (x, y)-plane produces a curve that is proportional to a normal probability density function.

If the two marginal distributions are standard normal distributions, so that $\mu_X = \mu_Y = 0$ and $\sigma_X^2 = \sigma_Y^2 = 1$, then the joint probability density function of X and Y for a general correlation value ρ is

$$f(x, y) = \frac{1}{2\pi \sqrt{1 - \rho^2}} e^{-(x^2+y^2-2\rho xy)/2(1-\rho^2)}$$

for $-\infty < x, y < \infty$. Two views of this probability density function are shown in Figure 5.44 when the correlation is $\rho = 0.8$.

Notice that this positive correlation tends to associate large values of X with large values of Y, and similarly small values of X with small values of Y, so that the probablity density function is concentrated close to the line $x = y$. Again, both marginal distributions and any conditional distributions are normally distributed, so that, as before, any "slice" of the joint probability density function on a plane that is perpendicular to the (x, y)-plane produces a curve that is proportional to a normal probability density function.

The ideas of a bivariate normal distribution can be extended to a more general *multivariate* normal distribution for any dimension. The notation

$$\mathbf{X} \sim N_k(\boldsymbol{\mu}, \Sigma)$$

indicates that the vector of random variables

$$\mathbf{X} = (X_1, \ldots, X_k)$$

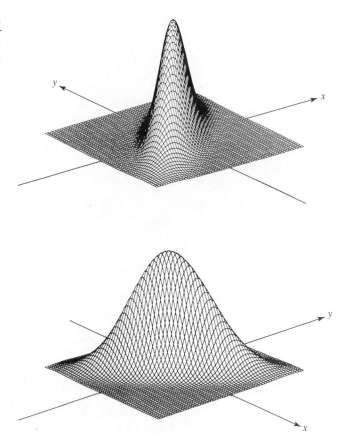

has a k-dimensional multivariate normal distribution with a mean vector

$$\boldsymbol{\mu} = (\mu_1, \ldots, \mu_k)$$

and a variance-covariance matrix Σ. The variance-covariance matrix is a symmetric $k \times k$ matrix with diagonal elements σ_i^2 equal to the variances of the marginal distributions of the random variables X_i, and off-diagonal terms equal to the covariances of the random variables.

The joint probability density function of $\mathbf{X}$ is

$$f(\mathbf{x}) = \left(\frac{1}{2\pi}\right)^{k/2} \left(\frac{1}{|\Sigma|}\right)^{1/2} e^{-(x-\mu)'\Sigma^{-1}(x-\mu)/2}$$

where $|\Sigma|$ is the determinant of the matrix Σ. The marginal distributions of the random variables X_i are

$$X_i \sim N\left(\mu_i, \sigma_i^2\right)$$

and all of the conditional distributions are also normally distributed. The multivariate normal distribution is important in the theoretical development of statistical methodologies.

■ **5.4.6 Problems**

5.4.1 Suppose that the random variable X has a lognormal distribution with parameter values $\mu = 1.2$ and $\sigma = 1.5$. Find:
   (a) $E(X)$
   (b) $\text{Var}(X)$
   (c) The upper quartile of X
   (d) The lower quartile of X
   (e) The interquartile range
   (f) $P(5 \leq X \leq 8)$

5.4.2 Suppose that the random variable X has a lognormal distribution with parameter values $\mu = -0.3$ and $\sigma = 1.1$. Find:
   (a) $E(X)$
   (b) $\text{Var}(X)$
   (c) The upper quartile of X
   (d) The lower quartile of X
   (e) The interquartile range
   (f) $P(0.1 \leq X \leq 7.0)$

5.4.3 Consider a sequence of random variables X_i that are independently identically distributed with a positive state space. Explain why the central limit theorem implies that the random variable

$$X = X_1 \times \cdots \times X_n$$

has approximately a lognormal distribution for large values of n.

5.4.4 A researcher grows cultures of bacteria. Suppose that after one day's growth, the size of the culture has a lognormal distribution with parameters $\mu = 2.3$ and $\sigma = 0.2$.
   (a) What is the expected size of the culture after one day?
   (b) What is the median size of the culture after one day?
   (c) What is the upper quartile of the size of the culture after one day?
   (d) What is the probability that the size of the culture after one day is greater than 15?
   (e) What is the probability that the size of the culture after one day is smaller than 6?

5.4.5 Use your computer package to find the following critical points, and check that they match the values given in Table II.
   (a) $\chi^2_{0.10,9}$ **(b)** $\chi^2_{0.05,20}$
   (c) $\chi^2_{0.01,26}$ **(d)** $\chi^2_{0.90,50}$
   (e) $\chi^2_{0.95,6}$

5.4.6 Use your computer package to find the following critical points:
   (a) $\chi^2_{0.12,8}$ **(b)** $\chi^2_{0.54,19}$ **(c)** $\chi^2_{0.023,32}$
   If the random variable X has a chi-square distribution with 12 degrees of freedom, use your computer package to find:
   (d) $P(X \leq 13.3)$
   (e) $P(9.6 \leq X \leq 15.3)$

5.4.7 Use your computer package to find the following critical points, and check that they match the values given in Table III.
   (a) $t_{0.10,7}$ **(b)** $t_{0.05,19}$
   (c) $t_{0.01,12}$ **(d)** $t_{0.025,30}$
   (e) $t_{0.005,4}$

5.4.8 Use your computer package to find the following critical points:
   (a) $t_{0.27,14}$ **(b)** $t_{0.09,22}$ **(c)** $t_{0.016,7}$
   If the random variable X has a t-distribution with 22 degrees of freedom, use your computer package to find:
   (d) $P(X \leq 1.78)$
   (e) $P(-0.65 \leq X \leq 2.98)$
   (f) $P(|X| \geq 3.02)$

5.4.9 Use your computer package to find the following critical points, and check that they match the values given in Table IV.
   (a) $F_{0.10,9,10}$ **(b)** $F_{0.05,6,20}$
   (c) $F_{0.01,15,30}$ **(d)** $F_{0.05,4,8}$
   (e) $F_{0.01,20,13}$

5.4.10 Use your computer package to find the following critical points:
   (a) $F_{0.04,7,37}$ **(b)** $F_{0.87,17,43}$ **(c)** $F_{0.035,3,8}$
   If the random variable X has an F-distribution with degrees of freedom $\nu_1 = 5$ and $\nu_2 = 33$, use your computer package to find:
   (d) $P(X \geq 2.35)$
   (e) $P(0.21 \leq X \leq 2.92)$

5.4.11 If the random variable X has a t-distribution with ν degrees of freedom, explain why the random variable $Y = X^2$ has an F-distribution with degrees of freedom 1 and ν.

5.4.12 (a) There is a probability of 0.90 that a t random variable with 23 degrees of freedom lies between $-x$ and x. Find the value of x.

(b) There is a probability of 0.975 that a t random variable with 60 degrees of freedom is larger than y. Find the value of y.

(c) What is the probability that a chi-square random variable with 29 degrees of freedom takes a value between 19.768 and 42.557?

5.4.13 The probability that an $F_{5,20}$ random variable takes a value greater than 4.00 is (a) greater than 10%, (b) between 5% and 10%, (c) between 1% and 5%, or (d) less than 1%?

5.4.14 The probability that a t_{35} random variable takes a value greater than 2.50 is (a) greater than 10%, (b) between 5% and 10%, (c) between 1% and 5%, or (d) less than 1%?

5.4.15 Use the tables to put bounds on these probabilities.
  (a) $P(F_{10,50} \geq 2.5)$
  (b) $P(\chi^2_{17} \leq 12)$
  (c) $P(t_{24} \geq 3)$
  (d) $P(t_{14} \geq -2)$

5.4.16 Use the tables to put bounds on these probabilities.
  (a) $P(t_{21} \leq 2.3)$
  (b) $P(\chi^2_6 \geq 13.0)$
  (c) $P(t_{10} \leq -1.9)$
  (d) $P(t_7 \geq -2.7)$

5.4.17 Use the tables to put bounds on these probabilities.
  (a) $P(t_{16} \leq 1.9)$
  (b) $P(\chi^2_{25} \geq 42.1)$
  (c) $P(F_{9,14} \leq 1.8)$
  (d) $P(-1.4 \leq t_{29} \leq 3.4)$

5.5   Case Study: Microelectronic Solder Joints

The thickness of the gold layer at the top of the bonding pad has an important effect on the quality of the conductive bond that is established between the solder joint and the substrate. Suppose that a certain manufacturing process produces a gold layer thickness that is normally distributed with a mean of 0.08 microns and a standard deviation of 0.01 microns. The probability that the gold layer thickness on a particular bond is within the range 0.075 to 0.085 microns can be calculated as

$$P(0.075 \leq N\left(0.08, 0.01^2\right) \leq 0.085)$$

$$= P\left(\frac{0.075 - 0.08}{0.01} \leq N(0,1) \leq \frac{0.085 - 0.08}{0.01}\right)$$

$$= \Phi(0.5) - \Phi(-0.5) = 0.6915 - 0.3085 = 0.3830$$

Furthermore, if an assembly consists of 16 solder joints and the thicknesses of the gold layers on the bond pads are independent of each other, the probability that the average gold layer thickness lies within the range 0.075 to 0.085 microns can be calculated to be

$$P\left(0.075 \leq N\left(0.08, \frac{0.01^2}{16}\right) \leq 0.085\right)$$

$$= P\left(\frac{0.075 - 0.08}{0.0025} \leq N(0,1) \leq \frac{0.085 - 0.08}{0.0025}\right)$$

$$= \Phi(2) - \Phi(2) = 0.9772 - 0.0228 = 0.9544$$

Recall that there is a probability of 0.12 that a solder joint has an hourglass shape, and suppose that an assembly consists of 512 solder joints. Then if the solder joint shapes are independent of each other, the number of hourglass-shaped solder joints on the assembly will have a $B(512, 0.12)$ distribution. This has a mean of $512 \times 0.12 = 61.44$ and a variance of $512 \times 0.12 \times 0.88 = 54.0672$, and so it can be approximated by a $N(61.44, 54.0672)$

distribution. The probability that there will be no more than 50 hourglass-shaped solder joints on the assembly can therefore be estimated to be

$$P(B(512, 0.12) \leq 50) \simeq P(N(61.44, 54.0672) \leq 50.5)$$
$$= P\left(N(0, 1) \leq \frac{50.5 - 61.44}{\sqrt{54.0672}}\right) = \Phi(-1.488) = 0.068$$

5.6 Supplementary Problems

5.6.1 The amount of sulfur dioxide escaping from the ground in a certain volcanic region in one day is normally distributed with a mean $\mu = 500$ tons and a standard deviation $\sigma = 50$ tons under ordinary conditions. However, if a volcanic eruption is imminent, there are much larger sulfur dioxide emissions.
(a) Under ordinary conditions, what is the probability of there being a daily sulfur dioxide emission larger than 625 tons?
(b) What is the 99th percentile of daily sulfur dioxide emissions under ordinary conditions?
(c) If your instruments indicate that 700 tons of sulfur dioxide have escaped from the ground on a particular day, would you advise that an eruption is imminent? Why? How sure would you be?

5.6.2 The breaking strengths of nylon fibers are normally distributed with a mean of 12,500 and a variance of 200,000.
(a) What is the probability that a fiber strength is more than 13,000?
(b) What is the probability that a fiber strength is less than 11,400?
(c) What is the probability that a fiber strength is between 12,200 and 14,000?
(d) What is the 95th percentile of the fiber strengths?

5.6.3 Adult salmon have lengths that are normally distributed with a mean of $\mu = 70$ cm and a standard deviation of $\sigma = 5.4$ cm.
(a) What is the probability that an adult salmon is longer than 80 cm?
(b) What is the probability that an adult salmon is shorter than 55 cm?
(c) What is the probability that an adult salmon is between 65 and 78 cm long?
(d) What is the value of c for which there is a 95% probability that an adult salmon has a length within the interval $[70 - c, 70 + c]$?

5.6.4 Consider again Problem 5.6.3 where the lengths of adult salmon have $N(70, 5.4^2)$ distributions.
(a) If you go fishing with a friend, what is the probability that the first adult salmon you catch is longer than the first adult salmon your friend catches?
(b) What is the probability that the first adult salmon you catch is at least 10 cm longer than the first adult salmon your friend catches?
(c) What is the probability that the average length of the first two adult salmon you catch is at least 10 cm longer than the first adult salmon your friend catches?

5.6.5 Suppose that the lengths of plastic rods produced by a machine are normally distributed with a mean of 2.30 m and a standard deviation of 2 cm. If two rods are placed side by side, what is the probability that the difference in their lengths is less than 3 cm?

5.6.6 A new 1.5-volt battery has an actual voltage that is uniformly distributed between 1.43 and 1.60 volts. Estimate the probability that the sum of the voltages from 120 new batteries lies between 180 and 182 volts.

5.6.7 The germination time in days of a newly planted seed is exponentially distributed with parameter $\lambda = 0.31$. If the germination times of different seeds are independent of one another, estimate the probability that the average germination time of 2000 seeds is between 3.10 and 3.25 days.

5.6.8 A publisher sends out advertisements in the mail asking people to subscribe to a magazine. Suppose that there is a probability of 0.06 that a recipient of the advertisement does subscribe to the magazine. If 350,000 advertisements are mailed out, estimate the probability that the magazine gains at least 20,800 new subscribers.

5.6.9 Suppose that if I invest $1000 today in a high-risk new-technology company, my return after 10 years has a lognormal distribution with parameters $\mu = 5.5$ and $\sigma = 2.0$.

(a) What are the median, upper, and lower quartiles of my 10-year return?

(b) What is the probability that my 10-year return is at least $75,000?

(c) What is the probability that my 10-year return is less than $1000?

5.6.10 Recall Problem 3.4.8, where the number of misrecorded pieces of information in a scanning process has a Poisson distribution with parameter $\lambda = 9.2$. Estimate the probability that there are fewer than 1000 total pieces of misrecorded information when 100 different scans are performed.

5.6.11 When making a connection at an airport, Jasmine arrives on a plane that is due to arrive at 2:15 P.M. However, the amount by which her plane arrives late has a normal distribution with a mean $\mu = 32$ minutes and a standard deviation $\sigma = 11$ minutes. Jasmine wants to transfer to a plane that is due to depart at 3:25 P.M., although the actual departure time is late by an amount that is normally distributed with a mean $\mu = 10$ minutes and a standard deviation $\sigma = 3$ minutes. If Jasmine needs 30 minutes at the airport to get from the arrival gate to the departure gate, what is the probability that she will be able to make her connection?

5.6.12 A clinic has four different physicians, A, B, C, and D, one of whom is selected by each new patient. If the new patients are equally likely to choose each of the four physicians independently of each other, estimate the probability that physician A will get at least 25 out of the next 80 new patients. If physician D leaves the clinic and the new patients are equally likely to select each of the remaining three physicians, what then is the probability that physician A will get at least 25 out of the next 80 new patients?

5.6.13 An aircraft can seat 220 passengers, and each of the passengers booked on the flight has a probability of 0.9 of actually arriving at the gate to board the plane, independent of the other passengers.

(a) Suppose the airline books 235 passengers on the flight. What is the probability that there will be insufficient seats to accommodate all of the passengers who wish to board the plane?

(b) If the airline wants to be 75% confident that there will be no more than 220 passengers who wish to board the plane, how many passengers can be booked on the flight?

5.6.14 (a) What is the probability that a random variable with a standard normal distribution takes a value between 0.6 and 2.2?

(b) What is the probability that a random variable with a normal distribution with $\mu = 4.1$ and $\sigma = 0.25$ takes a value between 3.5 and 4.5?

(c) What is the probability that a random variable with a chi-square distribution with 28 degrees of freedom takes a value between 16.928 and 18.939?

(d) What is the probability that a random variable with a t-distribution with 22 degrees of freedom takes a value between -1.717 and 2.819?

5.6.15 Components have lifetimes in minutes that are independent of each other with a lognormal distribution with parameters $\mu = 3.1$ and $\sigma = 0.1$. Suppose that a random sample of 200 components is taken. What is the probability that 30 or more of the components will have a lifetime of at least 25 minutes?

5.6.16 Are the following statements true or false?

(a) A t-distribution with 60 degrees of freedom has a larger variance than a standard normal distribution.

(b) The probability that a normal random variable with mean 10 and standard deviation 2 is less than 14 is equal to the probability that a normal random variable with mean 20 and standard deviation 3 is greater than 14.

(c) $P(\chi^2_{30} \leq 42) \geq 0.90$

(d) $P(-2 \leq t_9 \leq 2) \leq 0.95$

(e) $P(F_{10,15} \geq 6.5) \leq 0.01$

5.6.17 When an order is placed with a company, there is a probability of 0.2 that it is an express order. Estimate the probability that 90 or more of the next 400 orders will be express orders.

5.6.18 In genetic profiling, the expression of a gene is measured for a set of different samples. Suppose that the expressions are modeled as being independently normally distributed with a mean 0.768 and a standard deviation 0.083.

(a) If six samples are measured, what is the probability that at least half of them have expressions larger than 0.800?

(b) If six samples are measured, what is the probability that two of them have expressions smaller than 0.700, two of them have expressions between 0.700 and 0.750, and the remaining two have expressions larger than 0.780?

(c) If the samples are tested sequentially, what is the probability that the sixth sample tested is the third sample with an expression smaller than 0.760?

(d) If the samples are tested sequentially, what is the probability that the fifth sample tested is the first sample with an expression smaller than 0.680?

(e) Suppose that 10 samples are tested and exactly 5 of them have expressions smaller than 0.750. Furthermore, 6 samples are randomly selected from these 10 samples and are sent to another laboratory. What is the probability that exactly half of the samples sent to the other laboratory have an expression smaller than 0.750?

5.6.19 Suppose that electrical components have lifetimes that are independent and that come from a normal distribution with a mean of 8200 minutes and a standard deviation of 350 minutes.

(a) If three components are selected, what is the probability that 1 lasts for less than 8000 minutes, 1 lasts between 8000 and 8300 minutes, and 1 lasts for more than 8300 minutes?

(b) A consumer buys a box of 10 components. What is the probability that the sixth component that the consumer uses is the second one to last less than 7900 minutes?

(c) If 7 components are selected, what is the probability that exactly 3 of them last for more than 8500 minutes?

5.6.20 The time taken by operator A to finish a task has a normal distribution with a mean 220 minutes and a standard deviation 11 minutes. The time taken by operator B to finish a task has a normal distribution with a mean 185 minutes and a standard deviation 9 minutes, independent of operator A. Operator A began working at 9 A.M. The probability that operator A finishes before operator B is 0.90. What time did operator B start working?

5.6.21 When users connect to a server, the lengths of time in minutes that they are connected are independently distributed with a Weibull distribution with $\lambda = 0.03$ and $a = 0.8$.

(a) Suppose that 5 users connect to the server. What is the probability that 2 of the users are connected for a time less than 30 minutes and that 3 of the users are connected for a time greater than 30 minutes?

(b) Suppose that 500 users connect to the server. What is the probability that no more than 210 of the users are connected for a time greater than 30 minutes?

5.6.22 Tiles have weights that are independently normally distributed with a mean of 45.3 and a standard deviation of 0.02. What is the probability that the total weight of three tiles is no more than 135.975?

5.6.23 Components of type A have lengths that are independently normally distributed with a mean of 67.2 and a standard deviation of 1.9. Components of type B have lengths that are independently normally distributed with a mean of 33.2 and a standard deviation of 1.1. What is the probability that two components of type B will have a total length shorter than one component of type A?

5.6.24 Suppose that the failure time of a component is modeled with an exponential distribution with a mean of 32 days. A company acquires a batch of 240 components. If the failure times of these components are taken to be independent of each other, estimate the probability that at least half of the components will last longer than 25 days.

5.6.25 Machine A produces components with holes whose diameter is normally distributed with a mean 56,000 and a standard deviation 10. Machine B produces components with holes whose diameter is normally distributed with a mean 56,005 and a standard deviation 8. Machine C produces pins whose diameter is normally distributed with a mean 55,980 and a standard deviation 10. Machine D produces pins whose diameter is normally distributed with a mean 55,985 and a standard deviation 9.

(a) What is the probability that a pin from machine C will have a larger diameter than a pin from machine D?

(b) What is the probability that a pin from machine C will fit inside the hole of a component from machine A?

(c) If a component is taken from machine A and a component is taken from machine B, what is the probability that both holes will be smaller than 55,995?

5.6.26 (a) What is the probability that a t random variable with 40 degrees of freedom lies between -1.303 and 2.021?

(b) Use Table III to put bounds on the probability that a t random variable with 17 degrees of freedom is greater than 2.7.

5.6.27 Use the tables to put bounds on these probabilities.

(a) $P(F_{16,20} \leq 2)$ (b) $P(\chi^2_{28} \geq 47)$
(c) $P(t_{29} \geq 1.5)$ (d) $P(t_7 \leq -1.3)$
(e) $P(t_{10} \geq -2)$

5.6.28 Use the tables to put bounds on these probabilities.

(a) $P(\chi^2_{40} > 65.0)$ (b) $P(t_{20} < -1.2)$

(c) $P(t_{26} < 3.0)$ (d) $P(F_{8,14} > 4.8)$

5.6.29 A patient has a doctor's appointment that is scheduled for 9:40 A.M. However, the amount of time after the scheduled time that the doctor's consultation actually starts has a normal distribution with a mean of 22 minutes and a standard deviation of 4 minutes. The doctor's consultation lasts for a period that has a normal distribution with a mean of 17 minutes and a standard deviation of 5 minutes. After the doctor's consultation has finished, the patient visits the laboratory and then the pharmacy. It takes the patient 1 minute to go from the doctor's consultation to the laboratory, and 1 minute to go from the laboratory to the pharmacy. The amount of time spent at the laboratory has a normal distribution with a mean of 11 minutes and a standard deviation of 3 minutes, and the amount of time spent at the pharmacy has a normal distribution with a mean of 15 minutes and a standard deviation of 5 minutes. If the times taken by each component of the patient's visit are all independent, what is the probability that the patient will be finished by 11:00 A.M?

CHAPTER SIX

Descriptive Statistics

Now that probability theory has been presented, an important change occurs at this point in the book. The first five chapters on probability theory described how the properties of a random variable can be understood using the probability mass function or probability density function of the random variable. For these purposes the probability mass function or probability density function was taken to be *known*.

Of course, in most applications the probability mass function or probability density function of a random variable is not known by an experimenter, and one of the first tasks of the experimenter is to find out as much as possible about the probability distribution of the random variable under consideration. This is done through *experimentation* and the collection of a **data set** relating to the random variable. The science of deducing properties of an underlying probability distribution from such a data set is known as the science of **statistical inference**.

In this chapter the collection of a data **sample** from an overall **population** is discussed, together with various basic data investigations. These include initial graphical presentations of the data set and the calculation of useful summary statistics of the data set.

6.1 Experimentation

6.1.1 Samples

Consider Example 1 concerning machine breakdowns, which are classified as due either to electrical causes, to mechanical causes, or to operator misuse. The probability mass function of the breakdowns, that is, the probability values of each of the three causes, summarizes the breakdown characteristics of the machine. In other words, the probability mass function can be thought of as summarizing the "true state of nature."

In practice, however, this underlying probability mass function is unknown. Consequently, an obvious task of an experimenter or manager who wishes to understand as fully as possible the breakdowns of the machine is to estimate the probability mass function. This can be done by collecting a **data set** relating to the machine breakdowns, which in this case would just be a record of how many machine breakdowns are actually attributable to each of the three causes.

In general, suppose that a (continuous) random variable X of interest to an experimenter has a probability density function $f(x)$. With reference to Example 17, X may represent the amount of milk in a milk container. The probability density function $f(x)$ provides complete information about the probabilistic properties of the random variable X and is *unknown* to the experimenter. Again, it represents the "true state of nature" that the experimenter wishes to find out about.

The experimenter proceeds by obtaining a **sample** of observations of the random variable X, which may be written

$$x_1, x_2, \ldots, x_n$$

For the milk container example, this sample or data set is obtained by weighing the contents of n milk containers. Since these data observations are *governed* by the unknown underlying

FIGURE 6.1

The relationship between probability theory and statistical inference

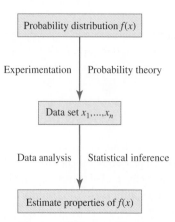

probability density function $f(x)$, an appropriate analysis of the data affords the experimenter a glimpse of $f(x)$. Such an analysis is known as **statistical inference**, as illustrated in Figure 6.1.

A great deal of care needs to be taken to ensure that the data set is obtained in an appropriate manner. The expression "garbage in, garbage out" is often used by statisticians to make the point that any statistical analysis based upon inaccurate or poor quality data is necessarily misleading. To judge the quality of a sample of data observations it is often useful to envision a **population** of potential observations from which the sample should be drawn in a *representative* manner.

For example, in the milk container example the population can be thought of as *all* of the milk containers. A representative sample can then be obtained by taking a **random sample**, whereby milk containers are chosen at random and weighed. If the milk company has three machines that fill milk containers, an example of a sample that is potentially *not* representative is one in which only containers from two of the three machines are selected and weighed, since it is possible that the amount of milk in a container may depend upon which machine it comes from. Indeed, one purpose of the statistical analysis may be to investigate whether there is any difference between the three machines, using the techniques described in Chapter 11.

The notion of a population is a fairly flexible one, and its definition depends upon the particular question being investigated by the experimenter. For example, if the experimenter specifically wishes to investigate whether the three filling machines are actually operating differently from one another, then conceptually it is appropriate to envision *three* populations and *three* samples. The milk containers filled by each of the three machines constitute three different populations, and a random sample can be obtained from each of the three populations by in turn selecting at random milk containers from each of the three machines.

Populations and Samples

A **population** consists of all possible observations available from a particular probability distribution. A **sample** is a particular subset of the population that an experimenter measures and uses to investigate the unknown probability distribution. A **random sample** is one in which the elements of the sample are chosen at random from the population, and this procedure is often used to ensure that the sample is *representative* of the population.

The data observations $x_1, \ldots, x_n$ can be of several general types. **Categorical** or **nominal data** record which of several categories or types an observation takes. A machine breakdown classified as either mechanical, electrical, or misuse is an example of a categorical data observation. **Numerical data** may be either integers or real numbers. It is important to be aware of the type of data that one is dealing with since it affects the choice of an appropriate analysis, as illustrated in the following examples.

6.1.2   Examples

Example 1
Machine Breakdowns

Breakdown cause	Frequency
Electrical	9
Mechanical	24
Misuse	13
Total	46

FIGURE 6.2

Data set of machine breakdowns

The engineer in charge of the maintenance of the machine keeps records on the breakdown causes over a period of a year. Altogether there are 46 breakdowns, of which 9 are attributable to electrical causes, 24 are attributable to mechanical causes, and 13 are attributable to operator misuse. This data set is shown in Figure 6.2 and can be used to estimate the probabilities of the breakdowns being attributable to each of the three causes.

Notice that this data set actually consists of 46 *categorical* observations, $x_1, \ldots, x_{46}$, with each observation taking one of the values

{electrical, mechanical, misuse}

However, the data set can be summarized by simply recording the *frequencies* of occurrence of the three categories, as in Figure 6.2.

The 46 breakdowns that occurred over the year constitute the sample of observations available to the engineer. What is the population from which this sample is drawn? This is a rather confusing question, but it may be helpful to envisage the population as consisting of all breakdowns during the year under consideration together with breakdowns from previous years and future years.

In practice, the most sensible thing for the engineer to do here is to concentrate directly on whether the sample obtained is a representative one. This judgment can be made only after the purpose of the data analysis has been decided. If the engineer is conducting the data analysis in order to predict the types and frequencies of machine breakdowns that will occur in the future, then the appropriate question is

How representative is this year's data set of future years?

This question is most easily answered by noticing whether there are any factors that suggest that the data may not be representative. For example, if the machine was operated this year by a skilled operator but will be operated next year by an inexperienced trainee, then it is probably reasonable to anticipate a greater proportion of breakdowns due to operator misuse next year. Similarly, if it is anticipated that next year the machine must be operated at higher speeds than this year in order to meet larger production targets, then a greater proportion of breakdowns due to mechanical reasons may be expected. These are the kinds of issues that a good statistician will investigate in order to assess the quality and representativeness of the data set being dealt with.

Example 2
Defective Computer Chips

Recall that a company sells computer chips in boxes of 500 chips. How can the company investigate the probability distribution of the number of defective chips per box? Figure 6.3 shows a data set of 80 observations corresponding to the number of defective chips found in a random sample of 80 boxes. An appropriate analysis of this data set will reveal

FIGURE 6.3

Data set of the number of defective computer chips in a box

1	3	4	7	2	7	5	5	2	2	4	2	4	3	2
2	7	1	3	3	2	5	0	0	1	2	5	5	4	1
3	2	6	3	8	2	2	3	1	6	3	4	1	2	5
1	3	3	3	2	1	2	5	5	4	1	4	3	1	0
2	1	2	4	4	5	3	3	4	0	5	2	5	6	2
5	3	3	3	1										

properties of the underlying unknown probability distribution of the number of defective chips per box.

The population of interest here can be thought of as *all* the boxes produced by the company within a certain time period. The representativeness of the sample of 80 boxes examined can be justified on the basis that it is a *random* sample, that is, the 80 boxes have been selected on some random basis. There are various ways in which this might have been done. For example, if each box has a code number assigned to it, a table of random numbers or a random-number generator on a computer may be used to identify the 80 boxes that can be chosen to make up the random sample.

In contrast, if the boxes are all stored in a warehouse, then boxes may be selected by randomly choosing aisles and shelves and then randomly selecting a particular box on a shelf. Alternatively, a random sample may be obtained by selecting every 100th box, say, to come off a production line.

For each box selected into the sample, all 500 chips must be tested to determine whether or not they are defective. The data observations $x_1, \ldots, x_{80}$ will then take integer values between 0 and 500. A useful way to summarize the data set is to record the *frequencies* of the number of defective chips found, as illustrated in Figure 6.4.

Number of defective chips	Frequency
0	4
1	12
2	18
3	17
4	10
5	12
6	3
7	3
8	1
≥ 9	0
Total	80

FIGURE 6.4

Data set of defective computer chips

Example 14

Metal Cylinder Production

In order to investigate the actual probability distribution of the diameters of the cylinders that it produces, the company selects a random sample of 60 cylinders and measures their diameters. This data set is shown in Figure 6.5. The population may be the set of all cylinders produced by the company (within a certain time period) or, if attention is directed at only one production line, say, the population may be all cylinders produced from that production line. The random selection of the 60 cylinders that constitute the sample should ensure that it is a representative sample. Notice that in this case the data observations $x_1, \ldots, x_{60}$, which represent the diameters in mm of the cylinders, are numbers recorded to two decimal places.

Example 17

Milk Container Contents

A random sample of 50 milk containers is selected and their milk contents are weighed. This data set is shown in Figure 6.6 and it can be used to investigate the unknown underlying probability distribution of the milk container weights. The population in this experiment is the collection of all the milk containers produced and, again, the random selection of the sample should ensure that it is a representative one. Notice that for this experiment, the data observations $x_1, \ldots, x_{50}$, which represent the milk contents in liters, are numbers recorded to three decimal places.

COMPUTER NOTE

Remember that all of the data sets discussed in the examples and problems are available on the book's CD and website.

FIGURE 6.5	50.08	49.81	50.00
Data set of metal cylinder diameters in mm	49.78	50.02	50.03
	50.02	50.26	49.92
	50.02	49.90	50.07
	50.13	50.01	49.89
	49.74	50.04	49.99
	49.84	50.01	50.01
	49.97	49.79	50.09
	49.93	50.36	49.90
	50.02	50.21	50.05
	50.05	50.17	49.95
	49.94	50.12	50.20
	50.19	50.00	50.03
	49.86	50.01	49.92
	50.03	49.85	50.02
	50.04	49.93	49.97
	49.96	49.84	50.27
	49.90	50.20	49.77
	49.87	49.94	50.07
	50.13	49.74	50.07

FIGURE 6.6	1.958	1.951	2.107	2.092	1.955	2.162	2.168	2.134	1.971
Data set of milk weights in liters	2.072	2.049	2.017	2.117	1.977	2.034	2.062	2.110	1.974
	1.992	2.018	2.135	2.107	2.084	2.169	2.085	2.018	1.977
	2.116	1.988	2.066	2.126	2.167	1.969	2.198	2.078	2.119
	2.088	2.172	2.133	2.112	2.066	2.128	2.142	2.042	2.050
	2.102	2.000	2.188	1.960	2.128				

■ **6.1.3 Problems**

Answer these questions for each data set. The data sets labeled DS are on the accompanying CD.

(a) Define the population from which the sample is taken. Do you think that it is a representative sample?

(b) Are there any other factors that should be taken into account in interpreting the data set? Are there any issues pertaining to the way in which the sample has been collected that you would want to investigate before interpreting the data set?

6.1.1 DS 6.1.1 shows the outcomes obtained from a series of rolls of a six-sided die. (This problem is continued in Problems 6.2.5, 6.3.4, and 7.3.10.)

6.1.2 Television Set Quality
One Friday morning at a television manufacturing company the quality inspector recorded the grades assigned to the pictures on the television sets that were ready to be shipped. The grades, presented in DS 6.1.2, are "perfect," "good," "satisfactory," or "fail." (This problem is continued in Problems 6.2.6 and 7.3.11.)

6.1.3 Eye Colors
DS 6.1.3 presents the eye colors of a group of students who are registered for a course on computer programming. (This problem is continued in Problems 6.2.7 and 7.3.12.)

6.1.4 Restaurant Service Times
One Saturday a researcher recorded the times taken to serve customers at a fast-food restaurant. DS 6.1.4 shows the service times in seconds for all the customers who were served between 2:00 and 3:00 in the afternoon. (This problem is continued in Problems 6.2.8, 6.3.5, 7.3.13, 8.1.18, 8.2.16, and 9.3.21.)

6.1.5 Fruit Spoilage

Every day in the summer months a supermarket receives a shipment of peaches. The supermarket's quality inspector arranges to have one box randomly selected from each shipment for which the number of "spoiled" peaches (out of 48 peaches in the box), which cannot be put out on the supermarket shelves, is recorded. DS 6.1.5 shows the data set obtained after 55 days. (This problem is continued in Problems 6.2.9, 6.3.6, and 7.3.14.)

6.1.6 Telephone Switchboard Activity

A researcher records the number of calls received by a switchboard during a one-minute period. These one-minute intervals are chosen at evenly spaced times during a working week. The data set obtained by the researcher is shown in DS 6.1.6. (This problem is continued in Problems 6.2.10, 6.3.7, 7.3.15, 8.1.19, and 8.2.17.)

6.1.7 Paving Slab Weights

A builder orders a large shipment of paving slabs from a particular company. The weights of a sample of randomly selected slabs are given in DS 6.1.7. (This problem is continued in Problems 6.2.11, 6.3.8, 7.3.16, 8.1.20, 8.2.18, and 9.3.17.)

6.1.8 Spray Painting Procedure

Car panels are spray painted by a machine. An inspector selects 1 in every 20 panels coming off a production line and measures the paint thickness at a specified point on the panels. The resulting data set is given in DS 6.1.8. (This problem is continued in Problems 6.2.12, 6.3.9, 7.3.17, 8.1.21, 8.2.19, and 9.3.18.)

6.1.9 Plastic Panel Bending Capabilities

The bending capabilities of plastic panels are investigated by measuring the angle of bend at which a deformity first appears in the panel. A researcher collects 80 plastic panels made by a machine and measures their deformity angles. The resulting data set is shown in DS 6.1.9. (This problem is continued in Problems 6.2.13, 6.3.10, 7.3.18, 8.1.22, and 8.2.20.)

6.2 Data Presentation

Once a data set has been collected, the experimenter's next task is to find an informative way of presenting it. In this chapter various graphical techniques for presenting data sets are discussed. In general, a table of numbers is not very informative, whereas a picture or graphical representation of the data set can be quite informative. If "a picture is worth a thousand words," then it is worth at least a million numbers.

6.2.1 Bar Charts and Pareto Charts

A **bar chart** is a simple graphical technique for illustrating a **categorical** data set. Each category has a bar whose length is proportional to the frequency associated with that category. A **Pareto chart** is a bar chart that is popular in quality control (see Chapter 16) where the categories are arranged in order of decreasing frequency.

Example 1

Machine Breakdowns

Figure 6.7 shows a bar chart for the data set of 46 machine breakdowns.

Example 41

Internet Commerce

A manager in an Internet-based company that sells a certain range of products from its website is interested in the causes of customer dissatisfaction. The complaints that the company received over a certain period of time are classsified as being due to the late delivery of an order, to the delivery of a damaged product, to the delivery of a wrong order, to errors in the billing procedure, or to any other type of complaint, as shown in Figure 6.8. A Pareto chart

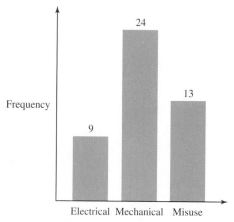

FIGURE 6.7

Bar chart of machine breakdown data set

Cause of complaint	Frequency
Late delivery	481
Damaged product	134
Wrong product	83
Billing error	44
Other	21
Total	763

FIGURE 6.8

Data set of customer complaints for Internet company

FIGURE 6.9

Pareto chart of customer complaints for Internet company

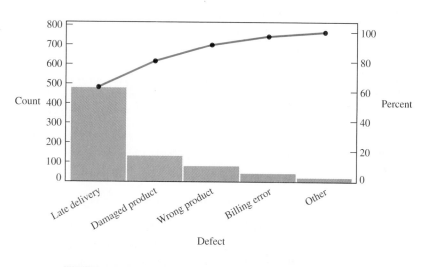

Count	481	134	83	44	21
Percent	63.0	17.6	10.9	5.8	2.8
Cumulative %	63.0	80.6	91.5	97.2	100.0

for this data set is shown in Figure 6.9. The Pareto chart arranges the causes of the complaints in order of decreasing frequency and presents a bar chart together with a line representing the cumulative count. For example, the two most common causes of complaint are the late delivery of a product and the delivery of a damaged product, which together account for 80.6% of all of the complaints.

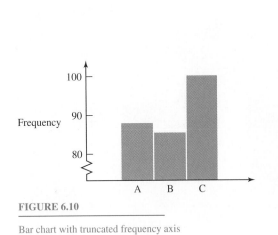

Frequency

FIGURE 6.10

Bar chart with truncated frequency axis

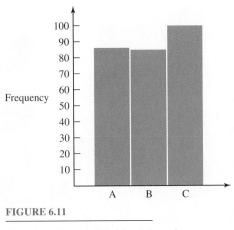

FIGURE 6.11

Bar chart without truncated frequency axis

When one looks at bar charts it is always prudent to check the frequency axis to make sure that it has not been truncated. The bar chart shown in Figure 6.10 conveys the visual impression that category C has a frequency at least twice as large as the frequencies of the other categories. However, the frequency axis has been truncated and does not start at zero. Figure 6.11 shows the bar chart without truncation of the axis.

COMPUTER NOTE   Find out how to obtain bar charts on your software package. Do not confuse bar charts with histograms, which are discussed in Section 6.2.3 and which look similar. Most spreadsheet and data management packages will produce bar charts for you.

6.2.2   Pie Charts

Pie charts are an alternative way of presenting the frequencies of categorical data in a graphical manner. A pie chart emphasizes the *proportion* of the total data set that is taken up by each of the categories. If a data set of n observations has r observations in a specific category, then that category receives a "slice" of the pie with an angle of

$$\frac{r}{n} \times 360°$$

This means that the angles of the pie slices are proportional to the frequencies of the various categories. Even though pie charts are a very simple graphical tool, their effectiveness should not be underestimated.

Example 1
Machine Breakdowns   Figure 6.12 shows a pie chart for the data set of 46 machine breakdowns. Notice that this chart immediately conveys the information that more than half of the breakdowns were attributable to mechanical causes.

Example 41
Internet Commerce   Figure 6.13 shows a pie chart of the customer complaints for the Internet company. The late delivery of orders is clearly seen to generate the most customer complaints.

COMPUTER NOTE   Find out how to obtain pie charts on your software package. Again, most spreadsheet and data management packages will produce pie charts for you.

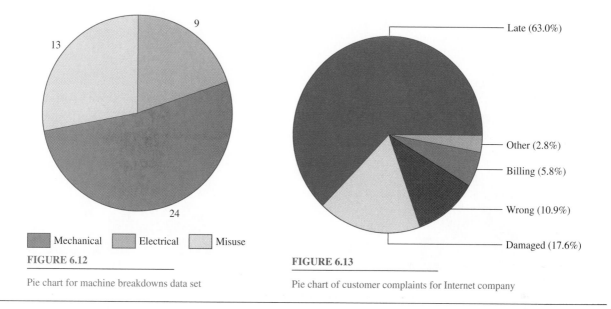

FIGURE 6.12

Pie chart for machine breakdowns data set

FIGURE 6.13

Pie chart of customer complaints for Internet company

6.2.3 Histograms

Histograms look similar to bar charts, but they are used to present *numerical* data rather than categorical data. In bar charts the "x-axis" lists the various categories under consideration, whereas in histograms the "x-axis" is a numerical scale. A histogram consists of a number of bands whose length is proportional to the number of data observations that take a value within that band. An important consideration in the construction of a histogram is an appropriate choice of the bandwidth.

Example 2
Defective Computer Chips

Figure 6.14 shows a histogram of the data set in Figure 6.4, which consists of the number of defective chips found in a sample of 80 boxes. Since the data observations are the integers $0, 1, 2, \ldots, 8$, the bands of the histogram are chosen to be

$$(-0.5, 0.5], (0.5, 1.5], \ldots, (7.5, 8.5]$$

The histogram shows that no defectives were found in 4 of the boxes, exactly one defective was found in 12 of the boxes, and so on.

The histogram in Figure 6.14 provides a graphical indication of the shape of the probability mass function of the number of defective chips in a box. It suggests that the probability mass function increases to a peak value at either 2 or 3 and then decreases rapidly. In other words, the actual probability mass function can be thought of as being a *smoothed* version of the histogram. An important question to ask is

> *How close is the shape of the histogram to the actual shape of the probability mass function?*

This question is addressed in later chapters using some technical statistical inference tools.

Example 14
Metal Cylinder Production

Figure 6.15 shows a histogram of the data set of metal cylinder diameters given in Figure 6.5. The bandwidth is chosen to be 0.04 mm. Figure 6.16 shows how the histogram changes when bandwidths of 0.10 mm and 0.02 mm are employed.

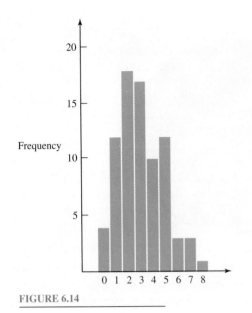

FIGURE 6.14

Histogram of computer chips data set

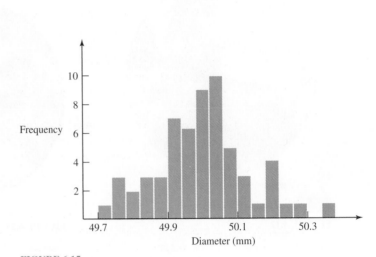

FIGURE 6.15

Histogram of metal cylinder diameter data set

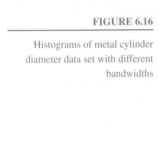

FIGURE 6.16

Histograms of metal cylinder
diameter data set with different
bandwidths

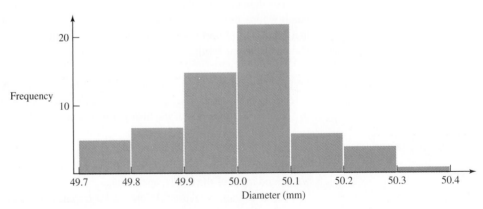

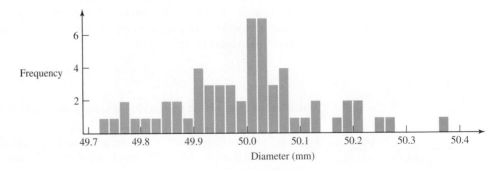

Common sense should be used to decide what is the best bandwidth when constructing a histogram. If the bandwidth is too large, then the histogram fails to convey all of the "structure" within the data set. On the other hand, if the bandwidth is too small, then the histogram becomes too "spiky" and may have gaps in it. For this data set, a bandwidth of 0.04 mm seems to be about right.

The shape of the histogram in Figure 6.15 provides an indication of the shape of the unknown underlying probability density function of the cylinder diameters. It is to be expected that the actual probability density function is some smooth curve that mimics the shape of the histogram.

Would you guess that the true probability density function is symmetric? This seems to be possible, since the histogram appears to be fairly symmetric about a value close to 50.00 mm. In Section 2.2.2, for illustrative purposes the probability density function of the metal cylinders was taken to be

$$f(x) = 1.5 - 6(x - 50.0)^2$$

for $49.5 \le x \le 50.5$, which is drawn in Figure 2.21. In view of the data that are now available, does this look like a plausible probability density function? The answer is probably not, since the histogram appears to have longer, flatter tails than the probability density function drawn in Figure 2.21. In fact, the histogram in Figure 6.15 appears to have a shape fairly similar to that of a *normal* distribution.

Example 17
Milk Container
Contents

Figure 6.17 shows a histogram of the data set of milk weights given in Figure 6.6. The band intervals employed are

$$(1.95, 1.97], (1.97, 1.99], \ldots, (2.19, 2.21]$$

In Section 2.2.2 a probability density function of

$$f(x) = 40.976 - 16x - 30e^{-x}$$

for $1.95 \le x \le 2.20$ was assumed for the milk content weights, which is drawn in Figure 2.24. In retrospect, given the data set at hand, does this density function appear reasonable? It's difficult to say with this sample size. The histogram in Figure 6.17 appears to have a shape

FIGURE 6.17

Histogram of milk weights

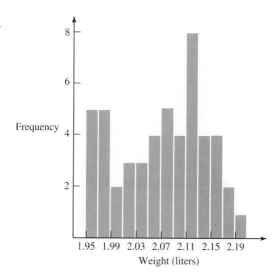

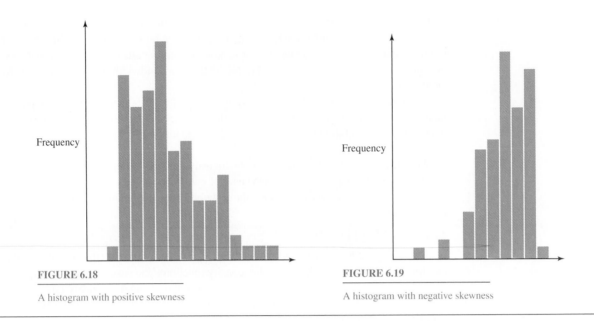

FIGURE 6.18

A histogram with positive skewness

FIGURE 6.19

A histogram with negative skewness

fairly similar to the probability density function drawn in Figure 2.24, except for the "spike" at about 2.12 liters and the "dip" between 2.00 and 2.04 liters. The shape of the probability density function will become clearer if a histogram is constructed from a larger sample size.

When one looks at a histogram, one of the first considerations is usually to determine whether or not it appears to be *symmetric*. For histograms that are not symmetric, it is often useful to talk about **skewness**. The histogram in Figure 6.18 is said to be **right-skewed** or **positively skewed** because its right-hand tail is much longer and flatter than its left-hand tail. Similarly, the histogram in Figure 6.19 is said to be **left-skewed** or **negatively skewed** because its left-hand tail is much longer and flatter than its right-hand tail. Skewness can also be used to describe probability density functions. For example, the Weibull distribution shown in Figure 4.19 is slightly positively skewed, whereas the beta distribution ($a = 4$, $b = 2$) shown in Figure 4.22 is negatively skewed.

Of course, not all histograms are *unimodal*. Figure 6.20 shows a histogram that is *bimodal* since it has two separate peaks. How should such a histogram be interpreted? It may be that the data set is actually a combination of two data sets corresponding to two different probability distributions. For example, a data set measuring some attribute of "people" may more usefully be separated into one data set for men and one for women.

COMPUTER NOTE

Find out how to draw histograms using your software package. Again, most spreadsheet and data management packages will also produce histograms for you. Find out how to change the bandwidths and band center points of histograms. You'll probably find that your package chooses these for you automatically unless you specify them.

6.2.4   Outliers

Graphical presentations of a data set sometimes indicate odd-looking data points that don't seem to fit in with the rest of the data set. For example, consider the histogram shown in Figure 6.21. This indicates a fairly symmetric distribution centered close to zero, except for one strange data point lying at about 4.5. Such a data point may be considered to be an **outlier**,

FIGURE 6.20

A histogram for a bimodal
distribution

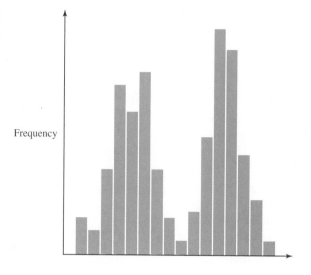

FIGURE 6.21

Histogram of a data set with a
possible outlier

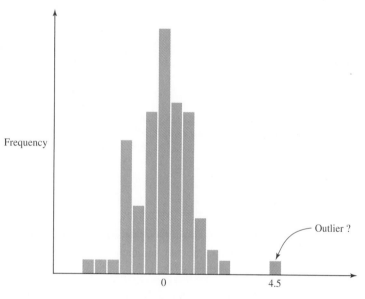

and in general, outliers can be defined to be data points that appear to be separate from the rest of the data set.

Should outliers be removed from the data set? It is usually sensible to remove outliers from the data set before any statistical inference techniques (discussed in subsequent chapters) are applied. The experimenter should certainly notice the outlier and investigate the data observation to see whether there is anything special about it. In many cases an outlier will be discovered to be a *misrecorded* data observation and can be corrected. In other cases it may be discovered that the data point corresponds to some special conditions that were not in effect when the other data points were collected.

The important lesson here is that an experimenter should be aware of outliers in a data set, as identified through a graphical presentation of the data set, and should take steps to deal with them in an appropriate manner. The basic issue is whether the outlier represents true variation in the population under consideration or whether it is caused by an "outside" influence.

■ 6.2.5   Problems

6.2.1 Fabric Types

DS 6.2.1 shows a data set of fabric types. Construct a bar chart and a pie chart for the data set.

6.2.2 Software Evaluations

DS 6.2.2 shows the evaluations of a new piece of software from a group of 60 trial users. Construct a bar chart and a pie chart for the data set.

6.2.3 Piston Rod Lengths

DS 6.2.3 shows the lengths of 30 piston rods. Construct a histogram of the data set with appropriate band widths. Do you think that there are any outliers in the data set? (This problem is continued in Problem 6.3.2.)

6.2.4 Physical Training Course Completion Times

DS 6.2.4 shows the times taken by 25 students to finish a physical training course. Construct a histogram of the data set with appropriate band widths. Do you think that there are any outliers in the data set? (This problem is continued in Problem 6.3.3.)

Use a statistical software package to obtain appropriate graphical presentations of each of the following data sets. Obtain more than one graphical presentation where appropriate. Indicate any data observations that might be considered to be outliers. What do your pictures tell you about the data sets?

6.2.5 The data set of die rolls given in DS 6.1.1.

6.2.6 Television Set Quality

The data set of television picture grades given in DS 6.1.2.

6.2.7 Eye Colors

The data set of eye colors given in DS 6.1.3.

6.2.8 Restaurant Service Times

The data set of service times given in DS 6.1.4.

6.2.9 Fruit Spoilage

The data set of spoiled peaches given in DS 6.1.5.

6.2.10 Telephone Switchboard Activity

The data set of calls received by a switchboard given in DS 6.1.6.

6.2.11 Paving Slab Weights

The data set of paving slab weights given in DS 6.1.7.

6.2.12 Spray Painting Procedure

The data set of paint thicknesses given in DS 6.1.8.

6.2.13 Plastic Panel Bending Capabilities

The data set of plastic panel bending capabilities given in DS 6.1.9.

6.2.14 Explain the difference between a bar chart and a histogram.

6.3   Sample Statistics

Sample statistics such as the sample mean, the sample median, and the sample standard deviation provide numerical summary measures of a data set in the same way that the expectation, median, and standard deviation provide numerical summary measures of a probability distribution.

6.3.1   Sample Mean

The **sample mean** of a data set $\bar{x}$ is simply the arithmetic average of the data observations. Specifically, if a data set consists of the n observations $x_1, \ldots, x_n$, then the sample mean is

$$\bar{x} = \frac{\sum_{i=1}^{n} x_i}{n}$$

The sample mean can be thought of as indicating a "middle value" of the data set in the same way that the expectation $E(X)$ of a random variable X indicates a "middle value" of the probability distribution of X. Moreover, the sample mean $\bar{x}$ can be thought of as being an *estimate* of the expectation of the unknown underlying probability distribution of the observations in the data set. Statistical estimation is discussed in more detail in Chapter 7.

Figure 6.22 shows a data set of 20 observations that have a sample mean $\bar{x} = 3.725$.

FIGURE 6.22	0.9	1.3	1.8	2.5	2.6	2.8	3.6	4.0	4.1	4.2
	4.3	4.3	4.6	4.6	4.6	4.7	4.8	4.9	4.9	5.0

Illustrative data set

Sample mean $\bar{x} = \frac{0.9 + \cdots + 5.0}{20} = 3.725$

Sample median $\frac{4.2 + 4.3}{2} = 4.25$

Sample trimmed mean $\frac{1.8 + \cdots + 4.9}{16} = 3.90$

6.3.2 Sample Median

The **sample median** is the value of the "middle" of the ordered data points. For example, if a data set consists of 31 observations, the sample median is the 16th largest data point, so that there are at least 15 data points no larger than the sample median and at least 15 data points no smaller than the sample median. If a data set consists of 30 observations, say, then the sample median is usually taken to be the average of the 15th and 16th largest data points.

The sample median can be considered to be an estimate of the median value of the unknown underlying probability distribution of the observations in the data set. The relationship between a sample mean and a sample median is similar to the relationship between the expectation and median of a probability distribution. A symmetric sample has a sample mean and a sample median roughly equal. However, a sample with positive skewness has a sample mean larger than the sample median, and a sample with negative skewness has a sample mean smaller than the sample median.

For example, consider the sample of salaries of professional athletes in a major sport. Typically, a small group of athletes earn salaries vastly higher than their fellow athletes so that the sample of salaries has positive skewness. What then is an "average salary"? The mean salary is influenced by the few very large salaries, so that considerably fewer than half of the athletes earn more than the mean salary. However, the median salary, which is smaller than the mean salary, may be more appropriate as an "average salary" since half of the athletes earn less than the median amount and half of the athletes earn more than the median amount.

In Figure 6.22, the 10th and the 11th largest data observations are 4.2 and 4.3, so that the sample median is 4.25. Notice that this is larger than the sample mean 3.725 due to the negative skewness of the sample.

6.3.3 Sample Trimmed Mean

A **trimmed mean** is obtained by deleting some of the largest and some of the smallest data observations, and by then taking the mean of the remaining observations. Usually a 10% trimmed mean is employed, whereby the top 10% of the data observations are removed together with the bottom 10% of the data points. For example, if there are $n = 50$ data observations, then the largest 5 and the smallest 5 are removed, and the mean is taken of the remaining 40 data points.

The advantage of a trimmed mean compared with a general sample mean is that the trimmed mean is not as sensitive to the tails of the data set as the overall mean. In particular, if the data set contains an outlier, then this affects the sample mean but does not affect the trimmed mean since the outlier will be one of the points removed. On the other hand, the tails of the sample may consist of valid data points, in which case the trimmed mean is "wasteful" because it does not use these data points. The trimmed mean is often referred to as a **robust** estimator of the expectation of the unknown underlying probability distribution of the observations in the data set since it is not sensitive to the largest and smallest elements of the data set.

The trimmed mean in Figure 6.22 is calculated as the average of the 16 data points when the 2 largest and 2 smallest data points have been removed. It has a value of 3.90, which is larger than the overall sample mean 3.725 but smaller than the sample median 4.25. Again, this is due to the negative skewness of the data set. In general, a trimmed mean can be considered to be a compromise between the sample mean and the sample median, as shown in Figure 6.23 for positively and negatively skewed data sets.

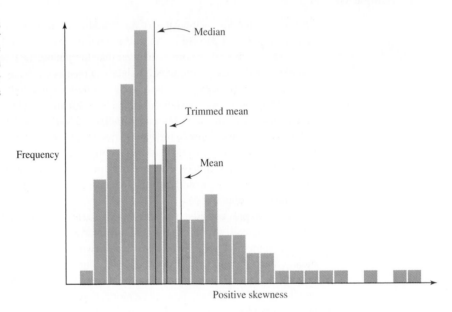

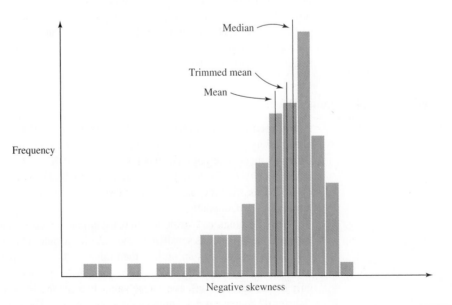

6.3.4 Sample Mode

For categorical or discrete data, the **sample mode** may be used to denote the category or data value that contains the largest number of data observations. In other words, the sample mode is the value with the highest frequency and can be thought of as estimating the category or value that has the highest probability. Figure 6.14, which shows the histogram of the number of defective chips in a box, reveals that the sample mode for this data set is two defective chips per box.

6.3.5 Sample Variance

The **sample variance** of a set of data observations $x_1, \ldots, x_n$ is defined to be

$$s^2 = \frac{\sum_{i=1}^{n} (x_i - \bar{x})^2}{n - 1}$$

and the sample standard deviation is s. Notice that the numerator of the formula for s^2 is composed of the sum of the squares of the deviances of the data observations x_i about the sample average $\bar{x}$. Also, notice that the denominator of the formula for s^2 is $n - 1$ and not n, although for reasonably large data sets there is very little difference between using $n - 1$ and n (an explanation of the use of $n - 1$ rather than n is provided in Chapter 7).

The sample variance s^2 can be thought of as an estimate of the variance σ^2 of the unknown underlying probability distribution of the observations in the data set. It provides an indication of the variability in the sample in the same way that the variance σ^2 provides an indication of the variability of a probability distribution.

Alternative computational formulas for the sample variance s^2 are

$$s^2 = \frac{\left(\sum_{i=1}^{n} x_i^2\right) - n\bar{x}^2}{n - 1}$$

and

$$s^2 = \frac{\left(\sum_{i=1}^{n} x_i^2\right) - \left(\sum_{i=1}^{n} x_i\right)^2 / n}{n - 1}$$

These are usually the easiest way to calculate s^2 by hand, since they require knowledge of only $\sum_{i=1}^{n} x_i^2$ and either $\bar{x}$ or $\sum_{i=1}^{n} x_i$.

For the data set given in Figure 6.22

$$\bar{x} = 3.725$$

and

$$\sum_{i=1}^{20} x_i^2 = 0.9^2 + \cdots + 5.0^2 = 308.61$$

so that the sample variance is

$$s^2 = \frac{308.61 - (20 \times 3.725^2)}{19} = 1.637$$

The sample standard deviation is therefore $s = \sqrt{1.637} = 1.279$.

6.3.6 Sample Quantiles

The pth **sample quantile** is a value that has a proportion p of the sample taking values smaller than it and a proportion $1 - p$ taking values larger than it. Clearly, it is an estimate of the pth quantile of the unknown underlying probability distribution of the sample observations. The terminology **sample percentile** is often used in place of sample quantile in the obvious manner.

The sample median is the 50th percentile of the sample, and the upper and lower **sample quartiles** are respectively the 75th percentile and 25th percentile of the sample. The **sample interquartile range** denotes the difference between the upper and lower sample quartiles.

Sample quantiles usually take a value between two data observations and are usually presented as an appropriate weighted average of the two data observations. For example, what is the upper sample quartile of the data set given in Figure 6.22? The 15th largest data observation is 4.6, and the 16th largest data observation is 4.7. The upper sample quartile is then usually given as

$$\left(\frac{1}{4} \times 4.6 \right) + \left(\frac{3}{4} \times 4.7 \right) = 4.675$$

On the other hand, the fifth largest data observation is 2.6, and the sixth largest data observation is 2.8, so that the lower sample quartile is usually given as

$$\left(\frac{3}{4} \times 2.6 \right) + \left(\frac{1}{4} \times 2.8 \right) = 2.65$$

Notice that the weighting of the two data observations may be performed in accordance with the proportion p of the quantile being calculated, although different software packages may do the weighting in different ways. The important point is that the sample quartile is between the appropriate two adjacent data values.

Finally, it is worth remarking that the *empirical cumulative distribution function* discussed in Section 15.1.1 is a simple graphical representation of a data set from which the sample quantiles can easily be found.

6.3.7   Boxplots

A **boxplot** is a schematic presentation of the sample median, the upper and lower sample quartiles, and the largest and smallest data observations. As Figure 6.24 shows, a box is constructed whose ends are the lower and upper sample quartiles, and a vertical line in the middle of the box represents the sample median. Horizontal lines stretch out from the ends of the box to the largest and smallest data observations.

A boxplot provides a simple and immediate graphical representation of the shape of a data set. Notice that half of the data observations lie within the box and half lie outside. If the sample histogram is fairly symmetric, then the two lines on the ends of the box should be

FIGURE 6.24

Boxplot of a data set

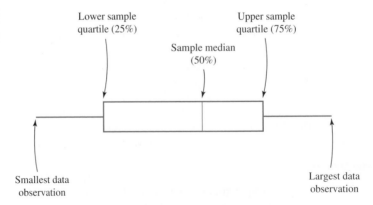

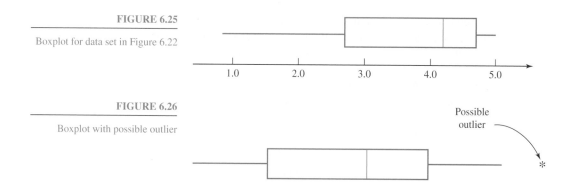

FIGURE 6.25

Boxplot for data set in Figure 6.22

FIGURE 6.26

Boxplot with possible outlier

roughly the same length, and the median should lie roughly in the center of the box. If the data are skewed, then the two lines are not the same length, and the median does not lie in the center of the box.

Figure 6.25 shows a boxplot for the data set given in Figure 6.22. Notice that the line on the left of the box is much longer than the line on the right of the box, and the median lies to the right of the center of the box. This illustrates the negative skewness of the data set.

Various additions to a boxplot are often employed to convey more information. If very large or very small data observations are considered to be possible outliers, then they may be represented by an asterisk and the line does not extend all the way to them, as shown in Figure 6.26. Some statistical packages allow you to custom design boxplots by, for example, adding various notches on the lines to indicate sample percentiles such as the 10th percentile and the 90th percentile.

6.3.8 Coefficient of Variation

Recall that the sample mean $\bar{x}$ and the sample standard deviation s are both measured in the same units as the data observations, and that they provide information on the middle value and the spread of the data set respectively. Sometimes it may be useful to consider the spread of the data *relative* to the middle value, which can be measured by the **coefficient of variation** defined by

$$CV = \frac{s}{\bar{x}}$$

This is a positive unitless quantity that can be useful to make comparisons between different data sets in terms of their variabilities expressed relative to their sample averages. Large values of the coefficient of variation imply that the variability is large relative to the sample average, while small values indicate that the variability is small relative to the sample average.

It is also worth noting that the coefficient of variation can be applied to probability distributions where it is calculated as $CV = \sigma/\mu$ and measures the standard deviation relative to the mean. For example, it takes a value of $\sqrt{(1-p)/np}$ for a binomial distribution, $1/\sqrt{\lambda}$ for a Poisson distribution, 1 for an exponential distribution, and $1/\sqrt{k}$ for a gamma distribution.

Example 42

Elephants and Mice

A zoologist is interested in the variations in the weights of different kinds of animals. A data set of adult male African elephants provided weights with a sample average of $\bar{x}_e = 4550$ kg and a sample standard deviation of $s_e = 150$ kg, while a data set concerning a certain kind of mouse provided weights with a sample average of $\bar{x}_m = 30$ g and a sample standard deviation of $s_m = 1.67$ g.

Obviously, the variation in the elephant weights is larger than the variation in the mice weights when compared directly because the elephant weights are so much larger. However, the coefficient of variation for the elephant weights is

$$CV_e = \frac{s_e}{\bar{x}_e} = \frac{150}{4550} = 0.033$$

while the coefficient of variation for the mice weights is

$$CV_m = \frac{s_m}{\bar{x}_m} = \frac{1.67}{30} = 0.056$$

Consequently, it can be seen that the mice have more variability in their weights than the elephants relative to their respective average weights.

6.3.9   Examples

Example 17
Milk Container Contents

Figure 6.27 shows a boxplot of the data set of milk container weights together with a set of sample statistics. The sample mean of 2.0727 liters, the sample median of 2.0845 liters, and the sample 10% trimmed mean of 2.0730 liters are all close together, which confirms that the data set is fairly symmetric, as suggested by the histogram in Figure 6.17. Furthermore, the boxplot is also fairly symmetric.

Example 2
Defective Computer Chips

Figure 6.28 shows a dotplot and the summary statistics of the computer chips data set. The dotplot of the data set simply records the data observations on a linear scale, and for this data set it provides a visual representation similar to the histogram in Figure 6.14. The sample median is 3, and the lower and upper sample quartiles are 2 and 4, which indicates that at least half of the boxes examined contained between two and four defective chips. The sample mean is 3.075, and the sample standard deviation is 1.813.

FIGURE 6.27

Boxplot and summary statistics for milk weights data set

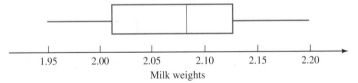

Sample size = 50   Sample standard deviation = 0.0711   Sample median = 2.0845
Sample mean = 2.0727   Sample maximum = 2.1980   Sample lower quartile = 2.0127
Sample trimmed mean = 2.0730   Sample upper quartile = 2.1280   Sample minimum = 1.9510

FIGURE 6.28

Dotplot and summary statistics for computer chips data set

Sample size = 80   Sample standard deviation = 1.813   Sample median = 3
Sample mean = 3.075   Sample maximum = 8   Sample lower quartile = 2
Sample trimmed mean = 3.014   Sample upper quartile = 4   Sample minimum = 0

FIGURE 6.29

Boxplot and summary statistics for
metal cylinder diameters data set

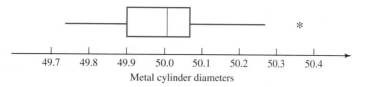

49.7   49.8   49.9   50.0   50.1   50.2   50.3   50.4

Metal cylinder diameters

Sample size = 60	Sample standard deviation = 0.134	Sample median = 50.010
Sample mean = 49.999	Sample maximum = 50.360	Sample lower quartile = 49.905
Sample trimmed mean = 49.996	Sample upper quartile = 50.070	Sample minimum = 49.740

Example 14
Metal Cylinder
Production

Figure 6.29 presents sample statistics and a boxplot for the data set of metal cylinder diameters. This boxplot has been drawn by the statistical software package Minitab that has indicated that the largest data observation may be an outlier by representing it with an asterisk and by curtailing the top line. Notice also that the sample mean of 49.999 mm and the sample median of 50.01 mm are very close, which confirms the suggestion from the histogram in Figure 6.15 that the data set is fairly symmetric.

Finally, it worth remarking that boxplots are useful tools for providing a graphical comparison of samples from two or more populations. Figure 9.25 shows boxplots of two samples drawn to the same scale, and Figure 11.12 presents a comparison of six samples using boxplots.

COMPUTER NOTE

Find out how to obtain sample statistics and boxplots on your software package.

■ 6.3.10   Problems

6.3.1 Consider the data set given in DS 6.3.1. Calculate by hand the sample mean, sample median, sample trimmed mean, and sample standard deviation. Calculate the upper and lower sample quartiles, and draw a boxplot of the data set.

6.3.2 Piston Rod Lengths
Consider the data set of 30 piston rod lengths given in DS 6.2.3. Calculate the sample mean, sample median, sample trimmed mean, and sample standard deviation. Calculate the upper and lower sample quartiles, and draw a boxplot of the data set.

6.3.3 Physical Training Course Completion Times
Consider the data set of physical training course completion times given in DS 6.2.4. Calculate the sample mean, sample median, sample trimmed mean, and sample standard deviation. Calculate the upper and lower sample quartiles, and draw a boxplot of the data set.

Use a statistical software package to obtain sample statistics and boxplots for the following data sets. What do the sample statistics and boxplots tell you about the data set?

6.3.4 The data set of die rolls given in DS 6.1.1.

6.3.5 Restaurant Service Times
The data set of service times given in DS 6.1.4.

6.3.6 Fruit Spoilage
The data set of spoiled peaches given in DS 6.1.5.

6.3.7 Telephone Switchboard Activity
The data set of calls received by a switchboard given in DS 6.1.6.

6.3.8 Paving Slab Weights
The data set of paving slab weights given in DS 6.1.7.

6.3.9 Spray Painting Procedure
The data set of paint thicknesses given in DS 6.1.8.

6.3.10 Plastic Panel Bending Capabilities
The data set of plastic panel bending capabilities given in DS 6.1.9.

6.3.11 Consider the data set

6   7   12   18   22

together with a sixth value x. What value of x minimizes the sample standard deviation of all six data points?

6.4   Examples

Example 43
Rolling Mill Scrap

In a rolling mill process, illustrated in Figure 6.30, ingots of bronze metals such as brass or copper are repeatedly heated, rolled, and cooled until a desired thickness and hardness of metal plate are obtained. After each pass through the rolling machines, the metal plates are trimmed on the sides and ends to remove material that has cracked or is otherwise damaged. Much of this scrap material can be recycled, although some is lost.

It is useful for the company to be able to predict the amount of scrap obtained from each order. Figure 6.31 shows a data set of 95 observations which are the % scrap for 95 orders that required only one pass through the rolling machines. The variable % scrap is defined to be

$$\% \text{ scrap} = \left(1 - \frac{\text{finished weight of plate}}{\text{input ingot weight}} \right) \times 100\%$$

Figures 6.32 and 6.33 show a histogram and a boxplot of this data set together with summary statistics. The histogram and the boxplot suggest that the data set has a *slight* negative skewness. This is consistent with the sample mean 20.81% being slightly smaller than the sample median 21.05%. As expected, the trimmed mean 20.91% lies between these two values. However, the smallest observation 7.69% might be considered to be an outlier. If it is removed from the data set, then the sample looks much more symmetric.

Finally, notice that the sample standard deviation is 4.878% and that half the data observations lie between the lower sample quartile 16.67% and the upper sample quartile 24.44%.

Example 44
Army Physical Fitness Test

The Army Physical Fitness Test, illustrated in Figure 6.34, consists of two minutes of pushups followed by two minutes of situps, and is completed with a two-mile run. Figure 6.35 presents

FIGURE 6.30

Rolling mill process

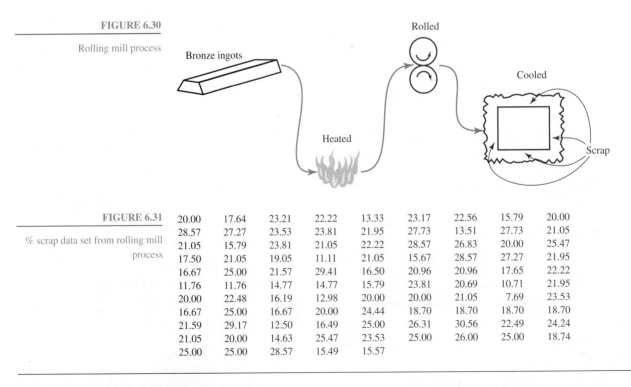

FIGURE 6.31

% scrap data set from rolling mill process

20.00	17.64	23.21	22.22	13.33	23.17	22.56	15.79	20.00
28.57	27.27	23.53	23.81	21.95	27.73	13.51	27.73	21.05
21.05	15.79	23.81	21.05	22.22	28.57	26.83	20.00	25.47
17.50	21.05	19.05	11.11	21.05	15.67	28.57	27.27	21.95
16.67	25.00	21.57	29.41	16.50	20.96	20.96	17.65	22.22
11.76	11.76	14.77	14.77	15.79	23.81	20.69	10.71	21.95
20.00	22.48	16.19	12.98	20.00	20.00	21.05	7.69	23.53
16.67	25.00	16.67	20.00	24.44	18.70	18.70	18.70	18.70
21.59	29.17	12.50	16.49	25.00	26.31	30.56	22.49	24.24
21.05	20.00	14.63	25.47	23.53	25.00	26.00	25.00	18.74
25.00	25.00	28.57	15.49	15.57				

FIGURE 6.32

Histogram of rolling mill scrap
data set

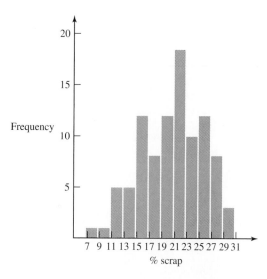

FIGURE 6.33

Boxplot and summary statistics for
rolling mill scrap data set

Sample size = 95
Sample mean = 20.810
Sample trimmed mean = 20.913

Sample standard deviation = 4.878
Sample maximum = 30.560
Sample upper quartile = 24.440

Sample median = 21.050
Sample lower quartile = 16.670
Sample minimum = 7.690

2 minutes of pushups

2 minutes of situps

2-mile run

847	880	870
887	905	931
879	895	930
919	720	808
816	712	828
814	703	719
814	741	707
855	792	934
980	808	939
954	761	977
1078	785	896
1001	801	921
766	810	815
916	1013	838
798	882	854
782	861	1063
836	845	1024
837	865	780
791	883	813
838	881	850
853	921	902
840	816	906
740	837	865
763	1056	886
778	1034	881
855	774	825
875	821	821
868	850	832

FIGURE 6.34

The Army Physical Fitness Test

FIGURE 6.35

Data set of run times in seconds

287

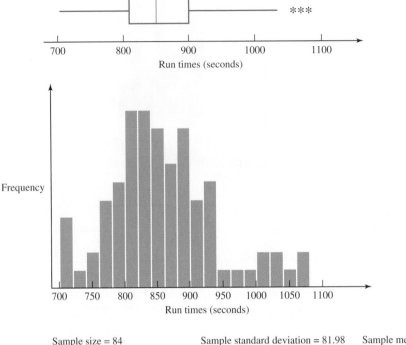

Sample size = 84   Sample standard deviation = 81.98   Sample median = 850.00
Sample mean = 857.70   Sample maximum = 1078.00   Sample lower quartile = 808.50
Sample trimmed mean = 854.93   Sample upper quartile = 900.50   Sample minimum = 703.00

FIGURE 6.36

Boxplot, histogram, and summary statistics of run times data set

84 run times in seconds for a group of male army officers. Figure 6.36 shows a boxplot, histogram, and summary statistics for this data set.

The histogram and boxplot reveal that the data set has a slight positive skewness due to a long right tail made up of some relatively slow runners, which the boxplot has indicated might be considered as outliers. Correspondingly, the mean run time, which is 857.7 seconds = 14 minutes 17.7 seconds, is larger than the median run time, which is 850.0 seconds = 14 minutes 10.0 seconds.

The sample standard deviation is 81.98 seconds, and half of the run times are between the lower sample quartile of 808.5 seconds = 13 minutes 28.5 seconds and the upper sample quartile of 900.5 seconds = 15 minutes 0.5 second. The quickest run recorded is 703.0 seconds = 11 minutes 43.0 seconds, and the slowest is 1078.0 seconds = 17 minutes 58.0 seconds.

Example 45
Fabric Water
Absorption Properties

An experiment is conducted to investigate the water absorption properties of cotton fabric. This absorption level is important in understanding the dyeing behavior of the fabric. A diagram of the experimental apparatus employed is shown in Figure 6.37. The cotton fabric is scoured and bleached and then run vertically between two rollers containing a bath of water. The water pickup of the fabric is defined to be

$$\% \text{ pickup} = \left(\frac{\text{final fabric weight}}{\text{initial fabric weight}} - 1 \right) \times 100\%$$

FIGURE 6.37

Apparatus for fabric water
absorption experiment

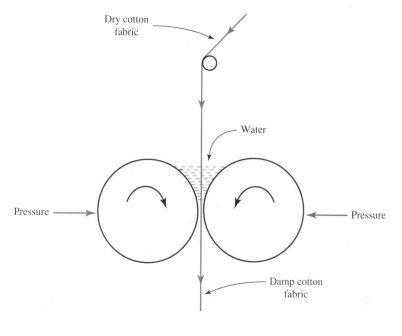

FIGURE 6.37

Apparatus for fabric water
absorption experiment

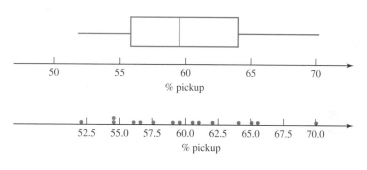

51.8	61.8	57.3	54.5	64.0
59.5	61.2	64.9	54.5	70.2
59.1	55.8	65.4	60.4	56.7

FIGURE 6.38

% pickup data set

Sample size = 15

Sample mean = 59.81

Sample trimmed mean = 59.62

Sample standard deviation = 4.94

Sample maximum = 70.20

Sample upper quartile = 64.00

Sample median = 59.50

Sample lower quartile = 55.80

Sample minimum = 51.80

FIGURE 6.39

Boxplot, dotplot, and summary statistics for fabric water absorption data set

Figure 6.38 contains 15 data observations of % pickup obtained when the two rollers rotated at 24 revolutions per minute with a pressure of 10 pounds per square inch between them. Figure 6.39 shows a boxplot of the data set, a dotplot, and summary statistics.

Again, the dotplot of the data set simply records the data observations on a linear scale. With so few data observations, it provides a better representation of the data set than a histogram. The dotplot indicates that the largest observation 70.2% appears to be far away from the rest of the data points and may perhaps be considered an outlier. Finally, notice that the sample mean, median, and trimmed mean are all roughly equal.

| 2.72 | 2.79 | 2.81 | 2.75 | 2.77 | 2.76 | 2.75 | 2.75 |
| 2.81 | 2.75 | 2.74 | 2.77 | 2.79 | 2.78 | 2.80 | 2.76 |

FIGURE 6.40

Data set of nickel layer thicknesses on substrate bond pads (microns)

Barrel shape	Cylinder shape	Hourglass shape	Total
451	8	53	512

FIGURE 6.41

Data set of solder joint shape frequencies from an assembly with 512 solder joints

6.5   Case Study: Microelectronic Solder Joints

 A researcher is investigating a new method for applying the nickel layer onto the bond pads in the substrate, and the thickness of the nickel layer is of particular interest. An assembly with 16 bond pads is examined and the nickel layer thickness is measured for each pad, resulting in the data set shown in Figure 6.40.

This data set has a sample size $n = 16$, a sample mean $\bar{x} = 2.7688$ microns, and a sample standard deviation $s = 0.0260$ microns. The sample median is the average of the eighth largest data point 2.76 and the ninth largest 2.77, and so it is 2.765 microns. The minimum data point is 2.72 microns and the largest is 2.81 microns.

In a separate experiment, the researcher examines each of 512 solder joints on an assembly to determine their shapes, and the categorical data set shown in Figure 6.41 is obtained.

6.6   Supplementary Problems

The following data sets can be used to practice the generation and interpretation of summary statistics and graphical representations.

6.6.1   Bird Species Identification

Three species of bird inhabit an island and they are classified as having either brown, grey, or black markings. DS 6.6.1 shows the types of birds observed by an ornithologist during a stay on the island. (This problem is continued in Problem 7.6.6.)

6.6.2   Oil Rig Accidents

DS 6.6.2 presents the number of accidents occurring on a collection of oil rigs for each month during a two-year span. (This problem is continued in Problem 7.6.7.)

6.6.3   Programming Errors

A software development company keeps track of the number of errors found in the programs written by the company employees. DS 6.6.3 shows the number of errors found in the 30 programs that were written during a particular month. (This problem is continued in Problem 7.6.8.)

6.6.4   Osteoporosis Patient Heights

DS 6.6.4 shows the heights in inches of 60 adult males with osteoporosis who visit a medical clinic during a particular week. (This problem is continued in Problems 7.6.9 and 8.5.5.)

6.6.5   Bamboo Cultivation

A researcher grows bamboo under controlled conditions in a greenhouse. DS 6.6.5 presents the heights of a set of bamboo shoots 40 days after planting. (This problem is continued in Problems 7.6.10, 8.5.6, and 9.6.5.)

6.6.6   Soil Compressibility Tests

The knowledge of soil behavior is an important issue in civil engineering. When soil is subjected to a load, there is a change in the volume of the soil due to drainage of water. A consolidation test can be performed to evaluate the compressibility of soil, so that the amount of settlement of buildings and other structures can be estimated. DS 6.6.6 contains the measurements of compressibility of 44 soil samples taken from a construction site. (This problem is continued in Problems 7.6.14 and 8.5.9.)

6.6.7 Glass Fiber Reinforced Polymer Tensile Strengths
Specimens of a glass fiber reinforced polymer were placed in a tension testing machine. Increasing amounts of tensile stress were applied until failure, and the maximum loads are shown in DS 6.6.7.

6.6.8 Infant Blood Levels of Hydrogen Peroxide
High blood levels of hydrogen peroxide in infants can be indicative of a dangerous infection. In order to understand what levels of hydrogen peroxide are unusually high, the data set in DS 6.6.8 was collected of hydrogen peroxide levels in the blood of infants who were known to be free of infection.

6.6.9 Paper Mill Operation of a Lime Kiln
A lime kiln is a large cylinder made of metal that is used at a paper mill to extract lime from calcium carbonate by heating it to a high temperature. An engineer was interested in variations in the temperature of the kiln, and DS 6.6.9 shows the temperature of the kiln every ten minutes during a five-hour period.

6.6.10 River Salinity Levels
DS 6.6.10 shows the salinity levels in parts per trillion (ppt) at various points along a river.

6.6.11 Dew Point Readings from Coastal Buoys
Buoys floating in the ocean provide important information on weather conditions, including the dew point measurement which is defined to be the temperature at which water will condense in the air. DS 6.6.11 shows the dew point measurements from a set of buoys at a certain time.

6.6.12 Brain pH Levels
Psychiatrists are interested in how the pH levels of brains may change for patients with mental illnesses. DS 6.6.12 shows the pH levels of brains of 20 healthy individuals which the psychiatrists hope to use as a reference point.

6.6.13 Silicon Dioxide Percentages in Ocean Floor Volcanic Glass
DS 6.6.13 shows the silicon dioxide percentages for samples of volcanic glass found in the Atlantic Ocean.

6.6.14 Network Server Response Times
DS 6.6.14 shows the times in milliseconds taken by a server to fulfill a standard task.

6.6.15 Are the following statements true or false?
- **(a)** The shape of a boxplot provides some information on the amount of skewness in the data set.
- **(b)** Histograms are not a good way to detect skewness in a data set.
- **(c)** Outliers may be misrecorded data points.
- **(d)** A boxplot indicates the value of the sample mean.

CHAPTER SEVEN

Statistical Estimation and Sampling Distributions

Estimators provide the basis for the first technical discussion of statistical inference, which is presented in this chapter. Based on the ideas discussed in the previous chapter on descriptive statistics, an important distinction is made between population properties (**parameters**) and sample properties (**statistics**). The basic statistical inference problem of *estimating* parameters is formulated, and various desirable properties of estimators are considered. Finally, the sampling distributions of common estimators are discussed, and general techniques for constructing good estimators are described.

7.1 Point Estimates

7.1.1 Parameters

It is very important to have a clear understanding of the difference between a **parameter** and a **statistic**. A parameter, which can be generically denoted by θ, is a property of an underlying probability distribution governing a particular observation. Parameters of obvious interest are the mean μ and variance σ^2 of the probability distribution. For continuous probability distributions other parameters of interest may be the various quantiles of the probability distribution, and for discrete probability distributions the probability values of particular categories may be parameters of interest.

Parameters can be thought of as representing a quantity of interest about a general population. In Chapters 1–5 probability calculations were made based on given values of the parameters of the probability distributions, but in practice the parameters are *unknown* since the probability distribution that characterizes observations from the population is unknown. An experimenter's goal is to find out as much as possible about these parameters since they provide an understanding of the underlying probability distribution that characterizes the population.

Parameters

In statistical inference, the term **parameter** is used to denote a quantity θ, say, that is a property of an unknown probability distribution. For example, it may be the mean, variance, or a particular quantile of the probability distribution. Parameters are unknown, and one of the goals of statistical inference is to estimate them.

Example 1
Machine Breakdowns

Let p_o be the probability that a machine breakdown is due to operator misuse. This is a parameter because it depends upon the probability distribution that governs the causes of the machine breakdowns. In practice p_o is an unknown quantity, but it may be estimated from the records of machine breakdown causes.

Example 43 **Rolling Mill Scrap**	Let μ and σ^2 be the mean and variance of the probability distribution of % scrap when an ingot is passed once through the rollers. These are unknown parameters that are properties of the unknown underlying probability distribution governing the % scrap obtained from the rolling process. Other parameters of interest may be the upper quartile and the lower quartile of the % scrap distribution.

Example 45 **Fabric Water** **Absorption Properties**	In this example, suppose that the parameters μ and σ^2 are the mean and variance of the unknown probability distribution governing % pickup. In other words, a particular observation of % pickup is considered to be an observation from a probability distribution with mean μ and variance σ^2. These parameters are unknown but may be estimated from the sample of observations of % pickup.

7.1.2 Statistics

Whereas a parameter is a property of a population or a probability distribution, a **statistic** is a property of a sample from the population. Specifically, a statistic is defined to be any function of a set of data observations. In contrast to parameters, statistics take observed values and consequently can be thought of as being *known*. However, in the discussion of statistical estimation it is useful to remember that statistics are actually observations of random variables with their own probability distributions.

For example, suppose that a sample of size n is collected of observations from a particular probability distribution $f(x)$. The data values recorded, $x_1, \ldots, x_n$, are the observed values of a set of n random variables $X_1, \ldots, X_n$, and each has the probability distribution $f(x)$. In general, a statistic is any function

$$A(X_1, \ldots, X_n)$$

of these random variables. The observed value of the statistic

$$A(x_1, \ldots, x_n)$$

can be calculated from the observed data values $x_1, \ldots, x_n$.

Common statistics are, of course, the sample mean

$$\bar{X} = \frac{X_1 + \cdots + X_n}{n}$$

and the sample variance

$$S^2 = \frac{\sum_{i=1}^{n}(X_i - \bar{X})^2}{n - 1}$$

For a given data set $x_1, \ldots, x_n$, these statistics take the observed values

$$\bar{x} = \frac{x_1 + \cdots + x_n}{n} \qquad \text{and} \qquad s^2 = \frac{\sum_{i=1}^{n}(x_i - \bar{x})^2}{n - 1}$$

as discussed in Chapter 6. For continuous data, other useful statistics are the sample quantiles and perhaps the sample trimmed mean. For discrete data, the cell frequencies are statistics of obvious interest.

Statistics

In statistical inference, the term **statistic** is used to denote a quantity that is a property of a sample. For example, it may be a sample mean, a sample variance, or a particular sample quantile. Statistics are random variables whose observed values can be calculated from a set of observed data observations. Statistics can be used to estimate unknown parameters.

7.1.3 Estimation

Estimation is a procedure by which the information contained within a sample is used to investigate properties of the population from which the sample is drawn. In particular, a **point estimate** of an unknown *parameter* θ is a *statistic* $\hat{\theta}$ that is in some sense a "best guess" of the value of θ. The relationship between a point estimate $\hat{\theta}$ calculated from a sample, and the unknown parameter θ is illustrated in Figure 7.1. Notice that a caret or "hat" placed over a parameter signifies a statistic used as an estimate of the parameter.

Of course, an experimenter does not in general believe that a point estimate $\hat{\theta}$ is exactly equal to the unknown parameter θ. Nevertheless, good point estimates are chosen to be good indicators of the actual values of the unknown parameter θ. In certain situations, however, there may be two or more good point estimates of a certain parameter which could yield slightly different numerical values.

Remember that point estimates can only be as good as the data set from which they are calculated. Again, this is a question of how representative the sample is of the population relating to the parameter that is being estimated. In addition, if a data set has some obvious outliers, then these observations should be removed from the data set before the point estimates are calculated.

Point Estimates of Parameters

A **point estimate** of an unknown parameter θ is a statistic $\hat{\theta}$ that represents a "best guess" at the value of θ. There may be more than one sensible point estimate of a parameter.

FIGURE 7.1

The relationship between a point estimate $\hat{\theta}$ and an unknown parameter θ

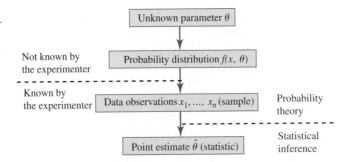

The statistic $\hat{\theta}$ is the "best guess" of the parameter θ.

FIGURE 7.2

Estimation of the population mean by the sample mean

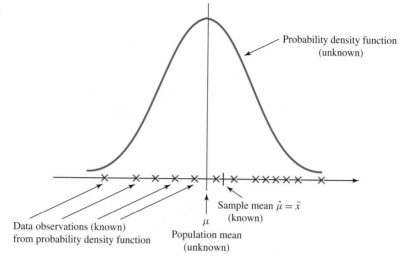

Probability density function (unknown)

Data observations (known) from probability density function

Population mean (unknown)

Sample mean $\hat{\mu} = \bar{x}$ (known)

μ

FIGURE 7.3

Estimating the probability that a machine breakdown is due to operator misuse

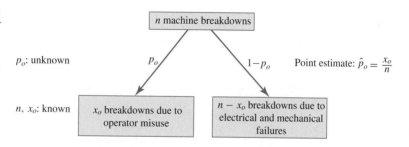

n machine breakdowns

p_o: unknown

p_o

$1 - p_o$

Point estimate: $\hat{p}_o = \frac{x_o}{n}$

n, x_o: known

x_o breakdowns due to operator misuse

$n - x_o$ breakdowns due to electrical and mechanical failures

As a simple example of a point estimate, notice that an obvious point estimate of the mean μ of a probability distribution is the sample mean $\bar{x}$ of data observations obtained from the probability distribution. In this case $\hat{\mu} = \bar{x}$, as illustrated in Figure 7.2.

Example 1
Machine Breakdowns

Consider the unknown parameter p_o, which represents the probability that a machine breakdown is due to operator misuse. Suppose that a representative sample of n machine breakdowns is recorded, of which x_o are due to operator misuse. As illustrated in Figure 7.3, the statistic x_o/n is an obvious point estimate of the unknown parameter p_o, and this may be written

$$\hat{p}_o = \frac{x_o}{n}$$

For the data set shown in Figure 6.2, $n = 46$ and $x_o = 13$. Consequently, based upon this data set a point estimate

$$\hat{p}_o = \frac{13}{46} = 0.28$$

is obtained.

Example 43
Rolling Mill Scrap

Given a representative sample $x_1, \ldots, x_n$ of % scrap values, obvious point estimates of the unknown parameters μ and σ^2, the mean and variance of the probability distribution of %

FIGURE 7.4

Estimating the population mean
and variance of the rolling mill
scrap

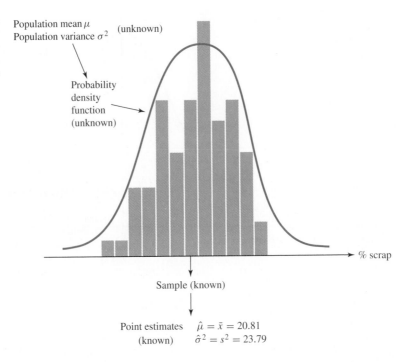

Population mean μ (unknown)
Population variance σ^2

Probability
density
function
(unknown)

% scrap

Sample (known)

Point estimates $\hat{\mu} = \bar{x} = 20.81$
(known) $\hat{\sigma}^2 = s^2 = 23.79$

scrap when an ingot is passed once through the rollers, are

$$\hat{\mu} = \bar{x} = \frac{x_1 + \cdots + x_n}{n} \qquad \text{and} \qquad \hat{\sigma}^2 = s^2 = \frac{\sum_{i=1}^{n}(x_i - \bar{x})^2}{n-1}$$

In other words, the sample mean and sample variance can be used as point estimates of the
population mean and population variance.

For the data set given in Figure 6.31, these point estimates take the values

$$\hat{\mu} = 20.81 \qquad \text{and} \qquad \hat{\sigma}^2 = 4.878^2 = 23.79$$

as shown in Figure 7.4. In addition, the upper quartile $\theta_{0.75}$ and the lower quartile $\theta_{0.25}$ of the
% scrap distribution may be estimated by the upper and lower sample quartiles, so that

$$\hat{\theta}_{0.75} = 24.44 \qquad \text{and} \qquad \hat{\theta}_{0.25} = 16.67$$

Is it sensible to use the sample trimmed mean as a point estimate of μ instead of the sample
mean $\bar{x}$? In fact, this is a situation where there is more than one sensible point estimate of the
parameter μ, since both the sample mean and the trimmed sample mean provide a good point
estimate of μ. Actually, the sample mean usually has a smaller variance than the trimmed
sample mean, but, as discussed in the previous chapter, the trimmed sample mean is a more
robust estimator and is not sensitive to data observations that may be outliers.

Example 45
Fabric Water
Absorption Properties

Consider the data set of % pickup observations given in Figure 6.38. Should any outliers be
removed before point estimates of the mean μ and variance σ^2 of the fabric absorption are
calculated? The dotplot in Figure 6.39 suggests that the largest data observation is suspect.
However, since the data set has only 15 observations, it is not clear whether this data point is
really "unusual" or not.

The experimenter checked this data observation and could not find anything unusual about it, so it is probably best to leave it in the data set. In this case the point estimates are

$$\hat{\mu} = 59.81 \qquad \text{and} \qquad \hat{\sigma}^2 = 4.94^2 = 24.40$$

7.2 Properties of Point Estimates

This section considers two basic criteria for determining good point estimates of a particular parameter, namely, unbiased estimates and minimum variance estimates. These criteria help us decide which statistics to use as point estimates. In general, when there is more than one obvious point estimate for a parameter, these criteria can be used to compare the possible choices of point estimate.

7.2.1 Unbiased Estimates

A point estimate $\hat{\theta}$ for a parameter θ is said to be **unbiased** if

$$E(\hat{\theta}) = \theta$$

Remember that a point estimate is an observation of a random variable with a probability distribution. The property of unbiasedness requires a point estimate $\hat{\theta}$ to have a probability distribution with a mean equal to θ, the value of the parameter being estimated. If a point estimate has a symmetric probability distribution, then as Figure 7.5 illustrates, it is unbiased if the probability distribution is centered at the parameter value θ.

Unbiasedness is clearly a nice property for a point estimate to possess. If a point estimate is not unbiased, then its **bias** can be defined to be

$$\text{bias} = E(\hat{\theta}) - \theta$$

Figure 7.6 illustrates the bias of a point estimate with a symmetric distribution. If two different point estimates are being compared, then the one with the *smaller* absolute bias is usually preferable (although their variances may also affect the choice between them).

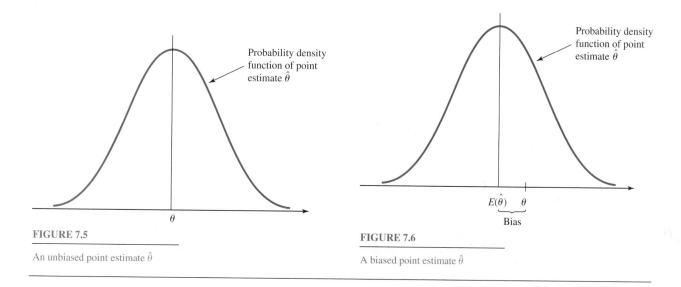

FIGURE 7.5

An unbiased point estimate $\hat{\theta}$

FIGURE 7.6

A biased point estimate $\hat{\theta}$

Unbiased and Biased Point Estimates

A point estimate $\hat{\theta}$ for a parameter θ is said to be **unbiased** if

$$E(\hat{\theta}) = \theta$$

Unbiasedness is a good property for a point estimate to possess. If a point estimate is not unbiased, then its **bias** can be defined to be

$$\text{bias} = E(\hat{\theta}) - \theta$$

All other things being equal, the smaller the absolute value of the bias of a point estimate, the better.

The common point estimates discussed in the previous section can now be investigated to determine whether or not they are unbiased. Consider first a sequence of Bernoulli trials with a constant unknown success probability p. This unknown parameter p can be estimated by conducting a sequence of trials and by observing how many of them result in a success. Suppose that n trials are conducted and that the random variable X counts the number of successes observed. The obvious point estimate of p is

$$\hat{p} = \frac{X}{n}$$

Is it an unbiased point estimate?

Notice that the number of successes X has a binomial distribution

$$X \sim B(n, p)$$

Therefore, the expected value of X is

$$E(X) = np$$

Consequently,

$$E(\hat{p}) = E\left(\frac{X}{n}\right) = \frac{1}{n}E(X) = \frac{1}{n}np = p$$

so that $\hat{p} = X/n$ is indeed an unbiased point estimate of the success probability p.

Point Estimate of a Success Probability

Suppose that $X \sim B(n, p)$. Then

$$\hat{p} = \frac{X}{n}$$

is an unbiased point estimate of the success probability p. This result implies that the proportion of successes in a sequence of Bernoulli trials with a constant success probability p is an unbiased point estimate of the success probability.

Example 1
Machine Breakdowns

Notice that the number of machine breakdowns due to operator misuse, X_o, has the binomial distribution

$$X_o \sim B(n, p_o)$$

Consequently, the point estimate

$$\hat{p}_o = \frac{X_o}{n}$$

is an unbiased point estimate of p_o.

Now suppose that $X_1, \ldots, X_n$ is a sample of observations from a probability distribution with a mean μ and a variance σ^2. Is the sample mean

$$\hat{\mu} = \bar{X}$$

an unbiased point estimate of the population mean μ? Clearly it is since

$$E(X_i) = \mu, \quad 1 \le i \le n$$

so that

$$E(\hat{\mu}) = E(\bar{X}) = \frac{1}{n} E\left(\sum_{i=1}^{n} X_i\right) = \frac{1}{n} \sum_{i=1}^{n} E(X_i) = \frac{1}{n} n\mu = \mu$$

Point Estimate of a Population Mean

If $X_1, \ldots, X_n$ is a sample of observations from a probability distribution with a mean μ, then the sample mean

$$\hat{\mu} = \bar{X}$$

is an unbiased point estimate of the population mean μ.

Is a trimmed sample mean an unbiased estimate of the population mean μ? If the probability distribution of the data observations is *symmetric*, then the answer is yes. However, a trimmed sample mean is in general *not* an unbiased point estimate of the population mean when the probability distribution is not symmetric. Furthermore, a sample median is in general *not* an unbiased point estimate of the population median when the probability distribution is not symmetric.

However, this does not necessarily imply that the sample median should not be used as a point estimate of the population median. Whether it should or not depends on whether there are "better" point estimates than the sample median. For example, other point estimates may have a smaller bias than the bias of the sample median. The important point to notice here is that sometimes obvious point estimates may not be unbiased, although this does not imply that better point estimates are available.

The sample variance S^2 is an unbiased estimate of the population variance σ^2. This is because

$$E(S^2) = \frac{1}{n-1} E\left(\sum_{i=1}^{n} (X_i - \bar{X})^2\right)$$

$$= \frac{1}{n-1} E\left(\sum_{i=1}^{n} ((X_i - \mu) - (\bar{X} - \mu))^2\right)$$

$$= \frac{1}{n-1} E\left(\sum_{i=1}^{n} (X_i - \mu)^2 - 2(\bar{X} - \mu) \sum_{i=1}^{n}(X_i - \mu) + n(\bar{X} - \mu)^2\right)$$

$$= \frac{1}{n-1} E\left(\sum_{i=1}^{n} (X_i - \mu)^2 - n(\bar{X} - \mu)^2\right)$$

$$= \frac{1}{n-1} \left(\sum_{i=1}^{n} E((X_i - \mu)^2) - nE((\bar{X} - \mu)^2)\right)$$

Now notice that

$$E(X_i) = \mu$$

so that

$$E((X_i - \mu)^2) = \text{Var}(X_i) = \sigma^2$$

Furthermore,

$$E(\bar{X}) = \mu$$

so that

$$E((\bar{X} - \mu)^2) = \text{Var}(\bar{X}) = \frac{\sigma^2}{n}$$

Putting this all together gives

$$E(S^2) = \frac{1}{n-1} \left(\sum_{i=1}^{n} \sigma^2 - n\left(\frac{\sigma^2}{n}\right)\right) = \sigma^2$$

so that S^2 is indeed an unbiased estimate of σ^2.

Point Estimate of a Population Variance

If $X_1, \ldots, X_n$ is a sample of observations from a probability distribution with a variance σ^2, then the sample variance

$$\hat{\sigma}^2 = S^2 = \frac{\sum_{i=1}^{n}(X_i - \bar{X})^2}{n-1}$$

is an unbiased point estimate of the population variance σ^2.

In fact, unbiasedness is the reason the denominator of S^2 is chosen to be $n-1$ rather than the perhaps more obvious choice of n. If the denominator is chosen to be n, so that the point estimate is

$$\hat{\sigma}^2 = \frac{\sum_{i=1}^{n}(X_i - \bar{X})^2}{n}$$

then this estimate has an expectation of

$$E(\hat{\sigma}^2) = \left(\frac{n-1}{n}\right)\sigma^2$$

so that it is not unbiased. It has a bias of

$$E(\hat{\sigma}^2) - \sigma^2 = \left(\frac{n-1}{n}\right)\sigma^2 - \sigma^2 = -\frac{\sigma^2}{n}$$

Notice that as the sample size n increases, the bias becomes increasingly small, and clearly for large sample sizes it is unimportant whether $n-1$ or n is used in the calculation of the sample variance. However, in general the unbiasedness criterion dictates the use of $n-1$ in the denominator of S^2.

| Example 43 | The point estimates |

Rolling Mill Scrap

$$\hat{\mu} = 20.81 \quad \text{and} \quad \hat{\sigma}^2 = 23.79$$

which are the sample mean and sample variance, are the observed values of unbiased point estimates. They are good and sensible estimates of the true mean and variance of the % scrap amounts.

7.2.2 Minimum Variance Estimates

As well as looking at the expectation $E(\hat{\theta})$ of a point estimate $\hat{\theta}$, it is important to consider the variance $\text{Var}(\hat{\theta})$ of the point estimate. It is generally desirable to have unbiased point estimates with as small a variance as possible.

For example, suppose that two point estimates $\hat{\theta}_1$ and $\hat{\theta}_2$ have symmetric distributions as shown in Figure 7.7. Moreover, suppose that their distributions are both centered at θ so that they are both unbiased point estimates of θ. Which is the better point estimate? Since

$$\text{Var}(\hat{\theta}_1) > \text{Var}(\hat{\theta}_2)$$

FIGURE 7.7

The unbiased point estimate $\hat{\theta}_2$ is better than the unbiased point estimate $\hat{\theta}_1$ because it has a smaller variance

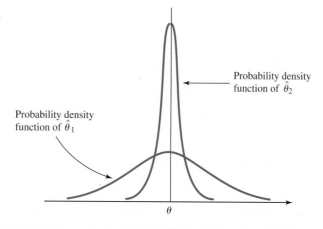

Probability density function of $\hat{\theta}_2$

Probability density function of $\hat{\theta}_1$

θ

FIGURE 7.8

$\hat{\theta}_2$ is a better point estimate than $\hat{\theta}_1$

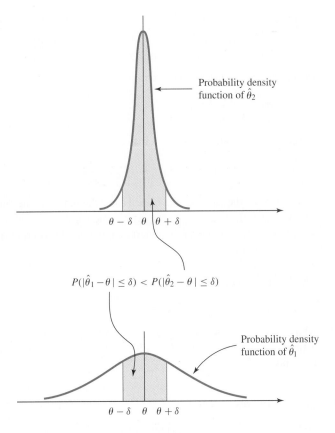

Probability density function of $\hat{\theta}_2$

$\theta - \delta \quad \theta \quad \theta + \delta$

$P(|\hat{\theta}_1 - \theta| \leq \delta) < P(|\hat{\theta}_2 - \theta| \leq \delta)$

Probability density function of $\hat{\theta}_1$

$\theta - \delta \quad \theta \quad \theta + \delta$

$\hat{\theta}_2$ is clearly a better point estimate than $\hat{\theta}_1$. It is better in the sense that it is likely to provide an estimate closer to the true value θ than the estimate provided by $\hat{\theta}_1$.

In mathematical terms, this can be written

$$P(|\hat{\theta}_1 - \theta| \leq \delta) < P(|\hat{\theta}_2 - \theta| \leq \delta)$$

for any value of $\delta > 0$, as illustrated in Figure 7.8. This inequality says that the probability that the point estimate $\hat{\theta}_2$ provides an estimate no more than an amount δ away from the real value of θ is larger than the corresponding probability for the point estimate $\hat{\theta}_1$.

The best possible situation is to be able to construct a point estimate that is unbiased and that also has the smallest possible variance. An unbiased point estimate that has a smaller variance than any other unbiased point estimate is called a **minimum variance unbiased estimate** (MVUE). Such estimates are clearly good ones. A great deal of mathematical theory has been developed to detect and investigate MVUEs for various problems. For our purpose it is sufficient to note that if $X_1, \ldots, X_n$ is a sample of observations that are independently *normally* distributed with a mean μ and a variance σ^2, then the sample mean $\bar{X}$ is a minimum variance unbiased estimate of the mean μ.

The *efficiency* of an unbiased point estimate is calculated as the ratio of the variance of the point estimate and the variance of the MVUE. In this sense the MVUE can be described as the "most efficient" point estimate. More generally, the **relative efficiency** of two unbiased point estimates is defined to be the ratio of their variances.

Relative Efficiency

The **relative efficiency** of an unbiased point estimate $\hat{\theta}_1$ to an unbiased point estimate $\hat{\theta}_2$ is

$$\frac{\mathrm{Var}(\hat{\theta}_2)}{\mathrm{Var}(\hat{\theta}_1)}$$

As a simple example of the calculation and interpretation of relative efficiency, suppose that $X_1, \ldots, X_{20}$ are independent, identically distributed random variables with an unknown mean μ and a variance σ^2. If a point estimate of μ is required, then the sample mean

$$\bar{X} = \frac{X_1 + \cdots + X_{20}}{20}$$

is known to provide an unbiased point estimate. However, suppose that it is suggested that the point estimate

$$\bar{X}_{10} = \frac{X_1 + \cdots + X_{10}}{10}$$

should be used to estimate μ. Is this a sensible point estimate?

Intuitively we feel uncomfortable with the point estimate $\bar{X}_{10}$ because it uses only half the data set. Consequently, it is not utilizing all the "information" available for estimating μ. Nevertheless, $\bar{X}_{10}$ is an unbiased point estimate of μ, so the criterion of unbiasedness does not allow us to distinguish between the point estimates $\bar{X}$ and $\bar{X}_{10}$.

Of course, the reason that $\bar{X}$ is a better point estimate than $\bar{X}_{10}$ is that it has a smaller variance, since

$$\mathrm{Var}(\bar{X}) = \frac{\sigma^2}{20}$$

while

$$\mathrm{Var}(\bar{X}_{10}) = \frac{\sigma^2}{10}$$

In fact, the relative efficiency of the point estimate $\bar{X}_{10}$ to the point estimate $\bar{X}$ is

$$\frac{\mathrm{Var}(\bar{X})}{\mathrm{Var}(\bar{X}_{10})} = \frac{1}{2}$$

In conclusion, in this example our intuitive desire to use all of the data available to estimate μ corresponds in mathematical terms to obtaining a point estimate with as small a variance as possible.

Example 38

Chemical Concentration Levels

Recall that a chemist has two independent measurements X_A and X_B available to estimate a concentration level C, and that

$$X_A \sim N(C, 2.97) \qquad \text{and} \qquad X_B \sim N(C, 1.62)$$

With our present knowledge of estimation theory, we are in a position to understand more fully how the chemist arrives at an optimum point estimate of the concentration level C.

Notice that $\hat{C}_A = X_A$, say, and $\hat{C}_B = X_B$ are both point estimates of the unknown concentration level C. Moreover, they are both unbiased point estimates since

$$E(\hat{C}_A) = C \quad \text{and} \quad E(\hat{C}_B) = C$$

However, since

$$\text{Var}(\hat{C}_A) = 2.97 \quad \text{and} \quad \text{Var}(\hat{C}_B) = 1.62$$

$\hat{C}_B$ is a more efficient estimate than $\hat{C}_A$. In fact the relative efficiency of $\hat{C}_A$ to $\hat{C}_B$ is

$$\frac{\text{Var}(\hat{C}_B)}{\text{Var}(\hat{C}_A)} = \frac{1.62}{2.97} = 0.55$$

In summary, $\text{Var}(\hat{C}_B)$ is a better point estimate than $\text{Var}(\hat{C}_A)$.

Consider now the point estimate

$$\hat{C} = a\hat{C}_A + b\hat{C}_B$$

where a and b are two constants. Can a and b be chosen to make this a better point estimate than $\hat{C}_B$? First, notice that

$$E(\hat{C}) = aE(\hat{C}_A) + bE(\hat{C}_B) = (a + b)C$$

Therefore, in order for $\hat{C}$ to be an unbiased point estimate of C, it is necessary to choose

$$a + b = 1$$

Setting $a = p$ and $b = 1 - p$ gives the point estimate

$$\hat{C} = p\hat{C}_A + (1 - p)\hat{C}_B$$

Finally, how should the value of p be chosen? Now the objective is to minimize the variance of $\hat{C}$. The calculations made before showed that this goal is met by taking $p = 0.35$, so that

$$\hat{C} = 0.35\hat{C}_A + 0.65\hat{C}_B$$

which has a variance

$$\text{Var}(\hat{C}) = 1.05$$

This is a better point estimate than $\hat{C}_B$, and the relative efficiency of $\hat{C}_B$ to $\hat{C}$ is

$$\frac{\text{Var}(\hat{C})}{\text{Var}(\hat{C}_B)} = \frac{1.05}{1.62} = 0.65$$

In some circumstances it may be useful to compare two point estimates that have different expectations and different variances. For example, in Figure 7.9, the point estimate $\hat{\theta}_1$ has a smaller bias than the point estimate $\hat{\theta}_2$, but it also has a larger variance. In such cases, it is usual to prefer the point estimate that minimizes the value of **mean square error (MSE)**, which is defined to be

$$\text{MSE}(\hat{\theta}) = E((\hat{\theta} - \theta)^2)$$

Notice that the mean square error is simply the expectation of the squared deviation of the point estimate about the value of the parameter of interest. Moreover, notice that

$$
\begin{aligned}
\text{MSE}(\hat{\theta}) &= E((\hat{\theta} - \theta)^2) \\
&= E(((\hat{\theta} - E(\hat{\theta})) + (E(\hat{\theta}) - \theta))^2) \\
&= E((\hat{\theta} - E(\hat{\theta}))^2) + 2(E(\hat{\theta}) - \theta)E(\hat{\theta} - E(\hat{\theta})) + (E(\hat{\theta}) - \theta)^2
\end{aligned}
$$

FIGURE 7.9

Comparing point estimates with
different biases and different
variances

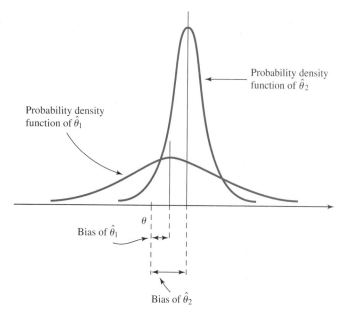

However,

$$E(\hat{\theta} - E(\hat{\theta})) = E(\hat{\theta}) - E(\hat{\theta}) = 0$$

so that

$$\text{MSE}(\hat{\theta}) = E((\hat{\theta} - E(\hat{\theta}))^2) + (E(\hat{\theta}) - \theta)^2 = \text{Var}(\hat{\theta}) + \text{bias}^2$$

Thus, the mean square error of a point estimate is the sum of its variance and the square of its bias. For unbiased point estimates, the mean square error is simply equal to the variance of the point estimate.

For example, suppose that in Figure 7.9

$$\hat{\theta}_1 \sim N(1.1\theta, 0.04\theta^2) \qquad \text{and} \qquad \hat{\theta}_2 \sim N(1.2\theta, 0.02\theta^2)$$

Then $\hat{\theta}_1$ has a bias of 0.1θ and a variance of $0.04\theta^2$, so that its mean square error is

$$\text{MSE}(\hat{\theta}_1) = 0.04\theta^2 + (0.1\theta)^2 = 0.05\theta^2$$

Similarly, $\hat{\theta}_2$ has a mean square error

$$\text{MSE}(\hat{\theta}_2) = 0.02\theta^2 + (0.2\theta)^2 = 0.06\theta^2$$

so that, based upon this criterion, the point estimate $\hat{\theta}_1$ is preferable to $\hat{\theta}_2$.

Finally, it is worth remarking that the properties of a point estimate generally depend on the size n of the sample from which they are constructed. In particular, the variances of sensible point estimates decrease as the sample size n increases. Notice that it is reassuring if the variance of a point estimate tends to 0 as the sample size becomes larger and larger, and if the point estimate is either unbiased or has a bias that also tends to 0 as the sample size becomes larger and larger (such point estimates are said to be *consistent*), since in this case the point estimate can be made to be as accurate as required by taking a sufficiently large sample size.

■ 7.2.3 Problems

7.2.1 Suppose that $E(X_1) = \mu$, $\mathrm{Var}(X_1) = 10$, $E(X_2) = \mu$, and $\mathrm{Var}(X_2) = 15$, and consider the point estimates

$$\hat{\mu}_1 = \frac{X_1}{2} + \frac{X_2}{2}$$

$$\hat{\mu}_2 = \frac{X_1}{4} + \frac{3X_2}{4}$$

$$\hat{\mu}_3 = \frac{X_1}{6} + \frac{X_2}{3} + 9$$

(a) Calculate the bias of each point estimate. Is any one of them unbiased?

(b) Calculate the variance of each point estimate. Which one has the smallest variance?

(c) Calculate the mean square error of each point estimate. Which point estimate has the smallest mean square error when $\mu = 8$?

7.2.2 Suppose that $E(X_1) = \mu$, $\mathrm{Var}(X_1) = 7$, $E(X_2) = \mu$, $\mathrm{Var}(X_2) = 13$, $E(X_3) = \mu$, and $\mathrm{Var}(X_3) = 20$, and consider the point estimates

$$\hat{\mu}_1 = \frac{X_1}{3} + \frac{X_2}{3} + \frac{X_3}{3}$$

$$\hat{\mu}_2 = \frac{X_1}{4} + \frac{X_2}{3} + \frac{X_3}{5}$$

$$\hat{\mu}_3 = \frac{X_1}{6} + \frac{X_2}{3} + \frac{X_3}{4} + 2$$

(a) Calculate the bias of each point estimate. Is any one of them unbiased?

(b) Calculate the variance of each point estimate. Which one has the smallest variance?

(c) Calculate the mean square error of each point estimate. Which point estimate has the smallest mean square error when $\mu = 3$?

7.2.3 Suppose that $E(X_1) = \mu$, $\mathrm{Var}(X_1) = 4$, $E(X_2) = \mu$, and $\mathrm{Var}(X_2) = 6$.

(a) What is the variance of

$$\hat{\mu}_1 = \frac{X_1}{2} + \frac{X_2}{2}$$

(b) What value of p minimizes the variance of

$$\hat{\mu} = pX_1 + (1 - p)X_2?$$

(c) What is the relative efficiency of $\hat{\mu}_1$ to the point estimate with the smallest variance that you have found?

7.2.4 Repeat Problem 7.2.3 with $\mathrm{Var}(X_1) = 1$ and $\mathrm{Var}(X_2) = 7$.

7.2.5 Suppose that a sequence of independent random variables $X_1, \ldots, X_n$ each has an expectation μ and variance σ^2, and consider the point estimate

$$\hat{\mu} = a_1 X_1 + \cdots + a_n X_n$$

for some constants $a_1, \ldots, a_n$.

(a) What is the condition on the constants a_i for this to be an unbiased point estimate of μ?

(b) Subject to this condition, what value of the constants a_i minimizes the variance of the point estimate?

7.2.6 If

$$\hat{\theta}_1 \sim N(1.13\theta, 0.02\theta^2)$$
$$\hat{\theta}_2 \sim N(1.05\theta, 0.07\theta^2)$$
$$\hat{\theta}_3 \sim N(1.24\theta, 0.005\theta^2)$$

which point estimate would you prefer to estimate θ? Why?

7.2.7 Suppose that $X \sim N(\mu, \sigma^2)$ and consider the point estimate

$$\hat{\mu} = \frac{X + \mu_0}{2}$$

for some fixed value μ_0. Show that this point estimate has a smaller mean square error than X when

$$|\mu - \mu_0| \leq \sqrt{3}\sigma$$

Explain why it is not surprising that $\hat{\mu}$ has a smaller mean square error than X when μ is close to μ_0.

7.2.8 Suppose that $X \sim B(10, p)$ and consider the point estimate

$$\hat{p} = \frac{X}{11}$$

(a) What is the bias of this point estimate?

(b) What is the variance of this point estimate?

(c) Show that this point estimate has a mean square error of

$$\frac{10p - 9p^2}{121}$$

(d) Show that this mean square error is smaller than the mean square error of $X/10$ when $p \leq 21/31$.

7.2.9 Suppose that X_1 is an estimate of a parameter θ with a standard deviation 5.39, and that X_2 is an estimate of θ with a standard deviation 9.43. If the estimates X_1 and X_2 are independent, what is the standard deviation of the estimate $(X_1 + X_2)/2$?

7.3 Sampling Distributions

The probability distributions or **sampling distributions** of the sample proportion $\hat{p}$, the sample mean $\bar{X}$, and the sample variance S^2 are now considered in more detail.

7.3.1 Sample Proportion

If $X \sim B(n, p)$, then an unbiased estimate of the success probability p is

$$\hat{p} = \frac{X}{n}$$

This estimate can be referred to as a sample proportion since it represents the proportion of successes observed in a sample of n trials. For large enough values of n, the normal approximation to the binomial distribution (discussed in Section 5.3.1) implies that X, and similarly $\hat{p}$, may be taken to have normal distributions. Notice that because $\text{Var}(X) = np(1 - p)$, it follows that

$$\text{Var}(\hat{p}) = \frac{p(1 - p)}{n}$$

Sample Proportion

If $X \sim B(n, p)$, then the sample proportion $\hat{p} = X/n$ has the approximate distribution

$$\hat{p} \sim N\left(p, \frac{p(1 - p)}{n}\right)$$

The standard deviation of $\hat{p}$ is referred to as its **standard error** and is

$$\text{s.e.}(\hat{p}) = \sqrt{\frac{p(1 - p)}{n}}$$

The standard error provides an indication of the "accuracy" of the point estimate $\hat{p}$. Smaller values of the standard error indicate that the point estimate is likely to be more accurate because its variability about the true value of p is smaller. Notice that the standard error is inversely proportional to the square root of the sample size n, so that as the sample size increases, the standard error decreases and $\hat{p}$ becomes a more accurate estimate of the success probability p.

Of course, since the success probability p is unknown, the standard error is really also unknown since it depends upon p. However, it is customary to estimate the standard error by replacing p by the observed value $\hat{p} = x/n$, so that

$$\text{s.e.}(\hat{p}) = \sqrt{\frac{\hat{p}(1 - \hat{p})}{n}} = \frac{1}{n}\sqrt{\frac{x(n - x)}{n}}$$

Example 1
Machine Breakdowns

Recall that 13 out of 46 machine breakdowns are attributable to operator misuse. The point estimate of the probability of a breakdown being attributable to operator misuse is

$$\hat{p}_o = \frac{13}{46} = 0.28$$

which has a standard error of

$$\text{s.e.}(\hat{p}_o) = \frac{1}{n}\sqrt{\frac{x_o(n - x_o)}{n}} = \frac{1}{46}\sqrt{\frac{13 \times (46 - 13)}{46}} = 0.066$$

Example 39

Cattle Inoculations

Suppose that the probability p that a vaccine provokes a serious adverse reaction is unknown. If the vaccine is administered to $n = 500{,}000$ head of cattle and then $x = 372$ are observed to suffer the reaction, the point estimate of p is

$$\hat{p} = \frac{372}{500{,}000} = 7.44 \times 10^{-4}$$

with a standard error of

$$\text{s.e.}(\hat{p}) = \frac{1}{500{,}000}\sqrt{\frac{372 \times (500{,}000 - 372)}{500{,}000}} = 3.86 \times 10^{-5}$$

A comparison of this calculation with the previous discussion of this example in Section 5.3.3 provides a distinct contrast between the different uses of probability theory and statistical inference. In Section 5.3.3 the probability of an adverse reaction p is taken to be *known*, and probability theory then allows the number of cattle suffering a reaction to be *predicted*. However, the situation is now reversed. In this discussion, the number of cattle suffering a reaction is observed, and hence is *known*, and statistical inference is used to *estimate* the probability of an adverse reaction p.

GAMES OF CHANCE

A coin that is suspected of being biased is tossed many times in order to investigate the possible bias. Consider the following two scenarios:

■ Scenario I : The coin is tossed 100 times and 40 heads are obtained.

■ Scenario II : The coin is tossed 1000 times and 400 heads are obtained.

What is the difference, if any, between the interpretations of these two sets of experimental results?

In either case, the probability of obtaining a head is estimated to be $\hat{p} = 0.4$. However, in scenario II the total number of coin tosses is larger than in scenario I, and so we feel that the point estimate obtained from scenario II is more "accurate" than the point estimate obtained from scenario I.

Mathematically, this is reflected in the point estimate having a smaller standard error in scenario II than in scenario I. The standard error in scenario I is

$$\text{s.e.}(\hat{p}) = \frac{1}{100}\sqrt{\frac{40 \times (100 - 40)}{100}} = 0.0490$$

whereas the standard error in scenario II is

$$\text{s.e.}(\hat{p}) = \frac{1}{1000}\sqrt{\frac{400 \times (1000 - 400)}{1000}} = 0.0155$$

As a result of increasing the sample size by a factor of 10, the standard error has been reduced by a factor of $\sqrt{10} = 3.16$. This problem is analyzed further in Section 10.1.

7.3.2   Sample Mean

Consider a set of independent, identically distributed random variables $X_1, \ldots, X_n$ with a mean μ and a variance σ^2. The central limit theorem (discussed in Section 5.3.2) indicates that the sample mean $\bar{X}$ has the approximate distribution

$$\bar{X} \sim N\left(\mu, \frac{\sigma^2}{n}\right)$$

This distribution is exact if the random variables X_i are normally distributed.

Sample Mean

If $X_1, \ldots, X_n$ are observations from a population with a mean μ and a variance σ^2, then the central limit theorem indicates that the sample mean $\hat{\mu} = \bar{X}$ has the approximate distribution

$$\hat{\mu} = \bar{X} \sim N\left(\mu, \frac{\sigma^2}{n}\right)$$

The standard error of the sample mean is

$$\text{s.e.}(\bar{X}) = \frac{\sigma}{\sqrt{n}}$$

which again is inversely proportional to the square root of the sample size. Thus, if the sample size is doubled, the standard error is reduced by a factor of $1/\sqrt{2} = 0.71$. Similarly, in order to *halve* the standard error, the sample size needs to be multiplied by *four*.

If a sample size $n = 20$ is used, what is the probability that the value of $\hat{\mu} = \bar{X}$ lies within $\sigma/4$ of the true mean μ? From the properties of the normal distribution, this probability can be calculated to be

$$P\left(\mu - \frac{\sigma}{4} \le \bar{X} \le \mu + \frac{\sigma}{4}\right) = P\left(\mu - \frac{\sigma}{4} \le N\left(\mu, \frac{\sigma^2}{20}\right) \le \mu + \frac{\sigma}{4}\right)$$

$$= P\left(-\frac{\sqrt{20}}{4} \le N(0, 1) \le \frac{\sqrt{20}}{4}\right)$$

$$= \Phi(1.12) - \Phi(-1.12)$$

$$= 0.8686 - 0.1314 = 0.7372$$

However, if a sample size of $n = 40$ is used, this probability increases to

$$P\left(-\frac{\sqrt{40}}{4} \le N(0, 1) \le \frac{\sqrt{40}}{4}\right) = \Phi(1.58) - \Phi(-1.58)$$

$$= 0.9429 - 0.0571 = 0.8858$$

These probability values, which are illustrated in Figure 7.10, demonstrate the increase in accuracy obtained with a larger sample size.

Since the standard deviation σ is usually unknown, it can be replaced by the observed value s, so that in practice the standard error of an observed sample mean $\bar{x}$ is calculated as

$$\text{s.e.}(\bar{x}) = \frac{s}{\sqrt{n}}$$

Example 43

Rolling Mill Scrap

The standard error of the sample mean $\hat{\mu} = \bar{x} = 20.81$ is

$$\text{s.e.}(\bar{x}) = \frac{s}{\sqrt{n}} = \frac{4.878}{\sqrt{95}} = 0.500$$

FIGURE 7.10

The increase in accuracy of $\hat{\mu} = \bar{X}$
as the sample size increases

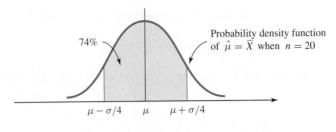

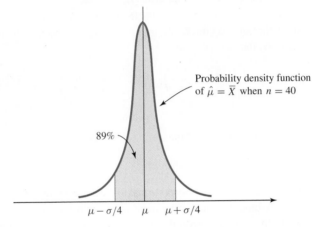

Example 44

**Army Physical Fitness
Test**

The standard error of the sample mean $\hat{\mu} = \bar{x} = 857.7$ is

$$\text{s.e.}(\bar{x}) = \frac{s}{\sqrt{n}} = \frac{81.98}{\sqrt{84}} = 8.94$$

Example 45

**Fabric Water
Absorption Properties**

The standard error of the sample mean $\hat{\mu} = \bar{x} = 59.81$ is

$$\text{s.e.}(\bar{x}) = \frac{s}{\sqrt{n}} = \frac{4.94}{\sqrt{15}} = 1.28$$

7.3.3   Sample Variance

For a sample $X_1, \ldots, X_n$ obtained from a population with a mean μ and a variance σ^2, consider the variance estimate

$$\hat{\sigma}^2 = S^2 = \frac{\sum_{i=1}^{n}(X_i - \bar{X})^2}{n - 1}$$

Sample Variance

If $X_1, \ldots, X_n$ are normally distributed with a mean μ and a variance σ^2, then the sample variance S^2 has the distribution

$$S^2 \sim \sigma^2 \frac{\chi^2_{n-1}}{(n - 1)}$$

Thus, S^2 is distributed as a scaled chi-square random variable with $n - 1$ degrees of freedom, where the scaling factor is $\sigma^2/(n-1)$. Notice that for a sample of size n, the degrees of freedom of the chi-square random variable are $n - 1$.

This distributional result turns out to be very important for the problem of estimating a normal population *mean*. This is because, as shown above, the standard error of the sample mean $\hat{\mu} = \bar{X}$ is $\sigma/\sqrt{n}$. The dependence of the standard error on the unknown variance σ^2 is rather awkward, but the sample variance S^2 can be used to overcome the problem.

The elimination of the unknown variance σ^2 is accomplished as follows using the t-distribution. The distribution of the sample mean

$$\bar{X} \sim N\left(\mu, \frac{\sigma^2}{n}\right)$$

can be rearranged as

$$\frac{\sqrt{n}}{\sigma}(\bar{X} - \mu) \sim N(0, 1)$$

Also, notice that

$$\frac{S}{\sigma} \sim \sqrt{\frac{\chi^2_{n-1}}{(n-1)}}$$

so that

$$\frac{\sqrt{n}(\bar{X} - \mu)}{S} = \frac{\frac{\sqrt{n}}{\sigma}(\bar{X} - \mu)}{\left(\frac{S}{\sigma}\right)} \sim \frac{N(0, 1)}{\sqrt{\frac{\chi^2_{n-1}}{(n-1)}}} \sim t_{n-1}$$

Again, notice that the degrees of freedom of the t-distribution are one fewer than the sample size n. For a given value of μ, the quantity

$$\frac{\sqrt{n}(\bar{X} - \mu)}{S}$$

is known as a **t-statistic**.

---

t-statistic

If $X_1, \ldots, X_n$ are normally distributed with a mean μ, then

$$\frac{\sqrt{n}(\bar{X} - \mu)}{S} \sim t_{n-1}$$

This result is very important since in practice an experimenter knows the values of n and the observed sample mean $\bar{x}$ and sample variance s^2, and so knows everything in the quantity

$$\frac{\sqrt{n}(\bar{x} - \mu)}{s}$$

except for μ. This allows the experimenter to make useful inferences about μ, as described in Chapter 8.

7.3.4 Simulation Experiment 2: An Investigation of Sampling Distributions

Suppose that an experimenter can measure some variables that are taken to be normally distributed with an unknown mean and variance. Using simulation methods, for specific values of the mean and variance we can *simulate* the data values that the experimenter might obtain. More interestingly, we can simulate lots of possible samples of which, in reality, the experimenter would observe only one. Performing this simulation experiment allows us to check on the sampling distributions of the parameter estimates that we have discussed in this section.

Let us suppose that $\mu = 10$ and $\sigma^2 = 3$. This is something that the experimenter does not know, and indeed the experimenter is conducting the experiment in order to find out what these parameter values are. Suppose that the experimenter decides to take a sample of $n = 30$ observations. We can use the computer to simulate the data values that the experimenter might observe. This simply involves obtaining 30 random observations from a $N(10, 3)$ distribution.

When this is done, a data set of 30 observations is obtained with $\bar{x} = 10.589$ and $s^2 = 3.4622$, as illustrated in Figure 7.11. An experimenter who obtained these data values would therefore estimate the parameters as $\hat{\mu} = 10.589$ and $\hat{\sigma}^2 = 3.4622$. With our knowledge of the true parameter values, we can see that the experimenter is not doing too badly.

Suppose that we now simulate lots of different samples of data observations. Specifically, let's simulate 500 samples. Since each sample contains 30 data observations, notice that this requires a total of 15,000 random observations from a $N(10, 3)$ distribution. For each sample we can calculate a value of $\bar{x}$ and s^2.

Figure 7.12 shows a histogram of the 500 values of the sample mean $\bar{x}$ obtained from the simulation. The sampling distribution theory discussed in this section tells us that the sample means $\bar{x}$ are observations from a normal distribution with a mean of $\mu = 10$ and a variance of $\sigma^2/n = 3/30 = 0.1$. The histogram in Figure 7.12 is seen to have a shape similar to a normal distribution, and in fact the average of the 500 $\bar{x}$ values is 10.006 and they have a (sample) variance of 0.091, so that the simulation results agree well with the theory.

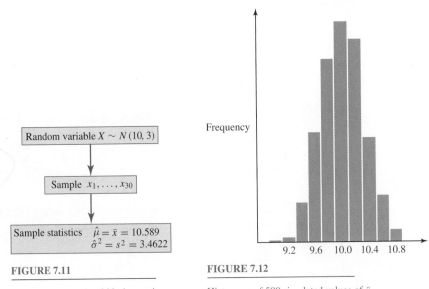

FIGURE 7.11	**FIGURE 7.12**
A simulated sample of 30 observations	Histogram of 500 simulated values of $\bar{x}$

In reality, the experimenter obtains just one sample, and exactly how close the estimate $\hat{\mu} = \bar{x}$ is to the true value of μ is a matter of luck. The point estimate obtained by the experimenter is a random observation from the normal distribution indicated by the histogram in Figure 7.12. In our 500 simulated samples, it turns out that the largest value of $\bar{x}$ obtained is 10.876 and the smallest is 9.073. However, about 250 of the simulations produced a value of $\bar{x}$ between 9.8 and 10.2.

Figure 7.13 shows a histogram of the 500 values of the sample variance s^2 obtained from the simulation. The sampling distribution theory presented in this section tells us that the distribution of the sample variance is

$$\sigma^2 \frac{\chi^2_{n-1}}{(n-1)} = 3 \times \frac{\chi^2_{29}}{29}$$

The histogram in Figure 7.13 exhibits the positive skewness that a chi-square distribution possesses. The average of the 500 simulated values of s^2 is 3.068 and half of them take values between 2.54 and 3.56. The largest simulated value is 6.75 and the smallest is 1.33.

How close will the experimenter's value of $\hat{\sigma}^2$ be to the true value $\sigma^2 = 3$? Again, it's a matter of luck. The point estimate obtained by the experimenter is a random observation from the scaled chi-square distribution indicated by the histogram in Figure 7.13.

Finally, let's look at the values of the t-statistics

$$\frac{\sqrt{n}(\bar{x} - \mu)}{s} = \frac{\sqrt{30}(\bar{x} - 10)}{s}$$

Figure 7.14 shows a histogram of the 500 simulated values of these t-statistics. The theory presented in this section tells us that the t-statistics should have a t-distribution with $n - 1 = 29$ degrees of freedom, which is very similar to a standard normal distribution but with a slightly larger variance. In fact, the histogram in Figure 7.14 is seen to have a shape similar to a normal distribution, and the 500 t-statistics have an average of 0.0223 and a variance of 0.96, which is in general agreement with the theory. This simulation experiment is continued in Section 8.1.4.

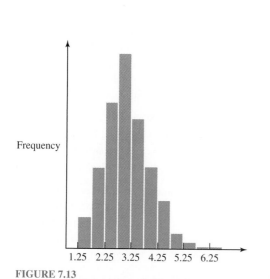

FIGURE 7.13

Histogram of 500 simulated values of s^2

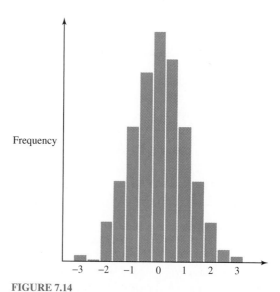

FIGURE 7.14

Histogram of 500 simulated t-statistics

■ 7.3.5   Problems

7.3.1 Suppose that $X_1 \sim B(n_1, p)$ and $X_2 \sim B(n_2, p)$. What is the relative efficiency of the point estimate

$$\frac{X_1}{n_1}$$

to the point estimate

$$\frac{X_2}{n_2}$$

for estimating the success probability p?

7.3.2 Consider a sample $X_1, \ldots, X_n$ of normally distributed random variables with mean μ and variance $\sigma^2 = 1$.
(a) If $n = 10$, what is the probability that $|\mu - \bar{X}| \leq 0.3$?
(b) What is this probability when $n = 30$?

7.3.3 Consider a sample $X_1, \ldots, X_n$ of normally distributed random variables with mean μ and variance $\sigma^2 = 7$.
(a) If $n = 15$, what is the probability that $|\mu - \bar{X}| \leq 0.4$?
(b) What is this probability when $n = 50$?

7.3.4 Consider a sample $X_1, \ldots, X_n$ of normally distributed random variables with variance $\sigma^2 = 5$. Suppose that $n = 31$.
(a) What is the value of c for which $P(S^2 \leq c) = 0.90$?
(b) What is the value of c for which $P(S^2 \leq c) = 0.95$?

7.3.5 Repeat Problem 7.3.4 with $n = 21$ and $\sigma^2 = 32$.

7.3.6 Consider a sample $X_1, \ldots, X_n$ of normally distributed random variables with mean μ. Suppose that $n = 16$.
(a) What is the value of c for which
$P(|4(\bar{X} - \mu)/S| \leq c) = 0.95$?
(b) What is the value of c for which
$P(|4(\bar{X} - \mu)/S| \leq c) = 0.99$?

7.3.7 Consider a sample $X_1, \ldots, X_n$ of normally distributed random variables with mean μ. Suppose that $n = 21$.
(a) What is the value of c for which
$P(|(\bar{X} - \mu)/S| \leq c) = 0.95$?
(b) What is the value of c for which
$P(|(\bar{X} - \mu)/S| \leq c) = 0.99$?

7.3.8 In a consumer survey, 234 people out of a representative sample of 450 people say that they prefer product A to product B. Let p be the proportion of all consumers who prefer product A to product B. Construct a point estimate of p. What is the standard error of your point estimate?

7.3.9 The breaking strengths of 35 pieces of cotton thread are measured. The sample mean is $\bar{x} = 974.3$ and the sample variance is $s^2 = 452.1$. Construct a point estimate of the average breaking strength of this type of cotton thread. What is the standard error of your point estimate?

7.3.10 Consider the data set of die rolls given in DS 6.1.1. Construct a point estimate of the probability of scoring a 6. What is the standard error of your point estimate?

7.3.11 Television Set Quality
Consider the data set of television picture grades given in DS 6.1.2. Construct a point estimate of the probability that a television picture is satisfactory. What is the standard error of your point estimate?

7.3.12 Eye Colors
Consider the data set of eye colors given in DS 6.1.3. Construct a point estimate of the probability that a student has blue eyes. What is the standard error of your point estimate?

7.3.13 Restaurant Serving Times
Consider the data set of service times given in DS 6.1.4. Construct a point estimate of the average service time. What is the standard error of your point estimate?

7.3.14 Fruit Spoilage
Consider the data set of spoiled peaches given in DS 6.1.5. Construct a point estimate of the average number of spoiled peaches per box. What is the standard error of your point estimate?

7.3.15 Telephone Switchboard Activity
Consider the data set of calls received by a switchboard given in DS 6.1.6. Construct a point estimate of the average number of calls per minute. What is the standard error of your point estimate?

7.3.16 Paving Slab Weights
Consider the data set of paving slab weights given in DS 6.1.7. Construct a point estimate of the average slab weight. What is the standard error of your point estimate?

7.3.17 Spray Painting Procedure
Consider the data set of paint thicknesses given in DS 6.1.8. Construct a point estimate of the average paint thickness. What is the standard error of your point estimate?

7.3.18 Plastic Panel Bending Capabilities
Consider the data set of plastic panel bending capabilities given in DS 6.1.9. Construct a point estimate of the

average deformity angle. What is the standard error of your point estimate?

7.3.19 Unknown to an experimenter, the probability of a prototype etching procedure producing a defective part is $p = 0.24$. The experimenter examines 100 randomly selected parts and finds out whether or not each one is defective. What is the probability that the experimenter's point estimate of p is within 0.05 of the true value? How does this probability change if the experimenter examines 200 randomly selected parts?

7.3.20 The capacitances of certain electronic components have a normal distribution with a mean $\mu = 174$ and a standard deviation $\sigma = 2.8$. If an engineer randomly selects a sample of $n = 30$ components and measures their capacitances, what is the probability that the engineer's point estimate of the mean μ will be within the interval $(173, 175)$?

7.3.21 Unknown to an experimenter, when a coin is tossed there is a probability of $p = 0.63$ of obtaining a head. The experimenter tosses the coin 300 times in order to estimate the probability p. What is the probability that the experimenter's point estimate of p will be within the interval $(0.62, 0.64)$?

7.3.22 The weights of bricks are normally distributed with $\mu = 110.0$ and $\sigma = 0.4$. If the weights of 22 randomly selected bricks are measured, what is the probability that the resulting point estimate of μ will be in the interval $(109.9, 110.1)$?

7.3.23 A scientist reports that the proportion of defective items from a process is 12.6%. If the scientist's estimate is based on the examination of a random sample of 360 items from the process, what is the standard error of the scientist's estimate?

7.3.24 Suppose that components have weights that are normally distributed with $\mu = 341$ and $\sigma = 2$. An experimenter measures the weights of a random sample of 20 components in order to estimate μ. What is the probability that the experimenter's estimate of μ will be less than 341.5?

7.3.25 Unknown to an experimenter, the corrosion rate of a certain type of chilled cast iron has a standard deviation of 5.2. The experimenter measures the corrosion rates of 18 random samples of the chilled cast iron and estimates the mean corrosion rate. What is the probability that the

experimenter's estimate will be more than 2 away from the correct value?

7.3.26 In a poll a random sample of 1400 respondents are asked whether they are in support or against a proposal. What is the largest possible value of the standard error of the estimate of the overall proportion in favor of the proposal?

7.3.27 Unknown to an experimenter, the failure time of a component has an exponential distribution with parameter $\lambda = 0.02$ per minute. The experimenter takes 110 components, and finds out how many of them last longer than one hour. This allows the experimenter to estimate the probability that a component will last longer than one hour. What is the probability that the experimenter's estimate is within 0.05 of the correct answer?

7.3.28 The pH levels of food items prepared in a certain way are normally distributed with a standard deviation of $\sigma = 0.82$. An experimenter estimates the mean pH level by averaging the pH levels of a random sample of n items.

(a) If $n = 5$, what is the probability that the experimenter's estimate is within 0.5 of the true mean value?

(b) If $n = 10$, what is the probability that the experimenter's estimate is within 0.5 of the true mean value?

(c) What sample size n is needed to ensure that there is a probability of at least 99% that the experimenter's estimate is within 0.5 of the true mean value?

7.3.29 A company has installed 3288 flow meters throughout an extensive sewer system. Unknown to the company, 592 of these meters are operating outside acceptable tolerance limits, whereas the other 2696 meters are operating satisfactorily. The company decides to estimate the unknown proportion p of the meters that are operating outside acceptable tolerance limits based on the inspection of a random sample of 20 meters.

(a) What is the probability that the company's estimate of p will be within 0.1 of the correct value?

(b) Suppose that 2012 of the meters are easily accessible, whereas the other 1276 meters are not easily accessible. In addition, suppose that only 184 of the easily accessible meters are operating outside acceptable tolerance limits. If the company's sample of 20 meters is biased due to the fact that the meters were randomly chosen from the subset of easily accessible meters, what is the probability that the company's estimate of p will be within 0.1 of the correct value?

7.4   Constructing Parameter Estimates

In this chapter the obvious point estimates for a success probability, a population mean, and a population variance have been considered in detail. However, it is often of interest to estimate parameters that require less obvious point estimates. For example, if an experimenter observes a data set that is taken to consist of observations from a beta distribution, how should the parameters of the beta distribution be estimated?

Two general methods of estimation can be used to solve questions of this kind. They are the **method of moments** and **maximum likelihood estimation**. They are described next and are illustrated on the standard problems of estimating a success probability and a normal mean and variance. The two methods are then applied to some more complicated examples.

7.4.1   The Method of Moments

Method of Moments Point Estimate for One Parameter

If a data set consists of observations $x_1, \ldots, x_n$ from a probability distribution that depends upon one unknown parameter θ, the **method of moments** point estimate $\hat{\theta}$ of the parameter is found by solving the equation

$$\bar{x} = E(X)$$

In other words, the point estimate is found by setting the sample mean equal to the population mean.

As an example of the implementation of this estimation method, suppose that $x_1, \ldots, x_n$ are a set of Bernoulli observations, with each taking the value 1 with probability p and the value 0 with probability $1 - p$. The expectation of the Bernoulli distribution is $E(X) = p$, so that the method of moments point estimate of p is found from the equation

$$\bar{x} = p$$

This simply provides the usual point estimate $\hat{p} = x/n$, where x is the number of data observations that take the value 1.

Method of Moments Point Estimates for Two Parameters

If a data set consists of observations $x_1, \ldots, x_n$ from a probability distribution that depends upon two unknown parameters, the method of moments point estimates of the parameters are found by solving the equations

$$\bar{x} = E(X) \qquad \text{and} \qquad s^2 = \text{Var}(X)$$

This is an intuitively reasonable method of estimation since it simply sets the population mean and variance equal to the sample mean and variance. Some practitioners may use n instead of $n - 1$ in the denominator of s^2 here, but it generally makes little difference to the point estimates.

Normally distributed data provide a simple example of estimating two parameters by the method of moments. Since $E(X) = \mu$ and $\mathrm{Var}(X) = \sigma^2$ for a $N(\mu, \sigma^2)$ distribution, the method of moments immediately gives the usual point estimates

$$\bar{x} = \hat{\mu} \qquad \text{and} \qquad s^2 = \hat{\sigma}^2$$

In general, the method of moments is a simple, easy-to-use method for obtaining sensible point estimates. However, it is not foolproof. Suppose that the data observations

$$2.0 \quad 2.4 \quad 3.1 \quad 3.9 \quad 4.5 \quad 4.8 \quad 5.7 \quad 9.9$$

are obtained from a $U(0, \theta)$ distribution. In this case the upper endpoint of the uniform distribution is the unknown parameter to be estimated. Since the expectation of a $U(0, \theta)$ distribution is

$$E(X) = \frac{\theta}{2}$$

and the sample mean is $\bar{x} = 4.5375$, the method of moments point estimate of θ is obtained from the equation

$$4.5375 = \frac{\theta}{2}$$

This gives

$$\hat{\theta} = 2 \times 4.5375 = 9.075$$

The problem with this point estimate is that it is clearly impossible! One of the data observations 9.9 exceeds the value $\hat{\theta}$, whereas the true value of θ must necessarily be larger than all the data values. Nevertheless, even though this example shows that point estimation using the method of moments may be unsuitable in certain cases, in general it is a simple and sensible method.

The method of moments can be generalized to problems with three or more unknown parameters by equating additional population moments

$$E(X - \mu)^k$$

where $k \geq 3$, with the corresponding sample moments. However, examples of this kind are rare.

7.4.2 Maximum Likelihood Estimates

Maximum likelihood estimation is a more technical method of obtaining point estimates, yet it is a very powerful method with a great deal of theoretical justification behind its use. Consider a set of data values $x_1, \ldots, x_n$ that are taken to be observations with a probability density function $f(x, \theta)$ depending on one unknown parameter θ. The joint density function of the data observations is therefore

$$f(x_1, \ldots, x_n, \theta) = f(x_1, \theta) \times \cdots \times f(x_n, \theta)$$

which can be thought of as the "likelihood" of observing the data values $x_1, \ldots, x_n$ for a given value of θ.

Maximum Likelihood Estimate for One Parameter

If a data set consists of observations $x_1, \ldots, x_n$ from a probability distribution $f(x, \theta)$ depending upon one unknown parameter θ, the **maximum likelihood estimate** $\hat{\theta}$ of the parameter is found by maximizing the likelihood function

$$L(x_1, \ldots, x_n, \theta) = f(x_1, \theta) \times \cdots \times f(x_n, \theta)$$

This method of estimation has an intuitive appeal to it, since it asks the question

For what parameter value is the observed data "most likely" to have arisen?

In practice, the maximization of the likelihood function is usually performed by taking the derivative of the likelihood function with respect to the parameter value. Often, however, it is convenient to take the *natural log* of the likelihood function before differentiating. Since the *natural log* is a monotonic function, maximizing the log-likelihood is equivalent to maximizing the likelihood.

To illustrate this estimation method, suppose again that $x_1, \ldots, x_n$ are a set of Bernoulli observations, with each taking the value 1 with probability p and the value 0 with probability $1 - p$. In this case, the probability distribution (actually a probability mass function) is

$$f(1, p) = p \quad \text{and} \quad f(0, p) = 1 - p$$

A succinct way of writing this is

$$f(x_i, p) = p^{x_i}(1 - p)^{1 - x_i}$$

The likelihood function is therefore

$$L(x_1, \ldots, x_n, p) = \prod_{i=1}^{n} p^{x_i}(1 - p)^{1 - x_i} = p^x(1 - p)^{n - x}$$

where $x = x_1 + \cdots + x_n$, and the maximum likelihood estimate $\hat{p}$ is the value that maximizes this. The log-likelihood is

$$\ln(L) = x \ln(p) + (n - x) \ln(1 - p)$$

and

$$\frac{d \ln(L)}{dp} = \frac{x}{p} - \frac{n - x}{1 - p}$$

Setting this expression equal to 0 and solving for p produce

$$\hat{p} = \frac{x}{n}$$

which can be checked to be a true maximum of the likelihood function. Consequently, the method of maximum likelihood estimation is seen to produce the usual estimate of the success probability p, which is the proportion of the sample that are successes.

> ### Maximum Likelihood Estimate for Two Parameters
>
> If a data set consists of observations $x_1, \ldots, x_n$ from a probability distribution $f(x, \theta_1, \theta_2)$ depending upon two unknown parameters, the maximum likelihood estimates $\hat{\theta}_1$ and $\hat{\theta}_2$ are the values of the parameters that jointly maximize the likelihood function
>
> $$L(x_1, \ldots, x_n, \theta_1, \theta_2) = f(x_1, \theta_1, \theta_2) \times \cdots \times f(x_n, \theta_1, \theta_2)$$

Again, the best way to perform the joint maximization is usually to take derivatives of the log-likelihood with respect to θ_1 and θ_2 and to set the two resulting expressions equal to 0.

The normal distribution is an example of a distribution with two parameters, with a probability density function

$$f(x, \mu, \sigma^2) = \frac{1}{\sqrt{2\pi}\sigma} e^{-(x-\mu)^2/2\sigma^2}$$

The likelihood of a set of normal observations is therefore

$$L(x_1, \ldots, x_n, \mu, \sigma^2) = \prod_{i=1}^{n} f(x_i, \mu, \sigma^2)$$

$$= \left(\frac{1}{2\pi\sigma^2}\right)^{n/2} \exp\left\{-\sum_{i=1}^{n}(x_i - \mu)^2/2\sigma^2\right\}$$

so that the log-likelihood is

$$\ln(L) = -\frac{n}{2}\ln(2\pi\sigma^2) - \frac{\sum_{i=1}^{n}(x_i - \mu)^2}{2\sigma^2}$$

Taking derivatives with respect to the parameter values μ and σ^2 gives

$$\frac{d\ln(L)}{d\mu} = \frac{\sum_{i=1}^{n}(x_i - \mu)}{\sigma^2}$$

and

$$\frac{d\ln(L)}{d\sigma^2} = -\frac{n}{2\sigma^2} + \frac{\sum_{i=1}^{n}(x_i - \mu)^2}{2\sigma^4}$$

Setting $d\ln(L)/d\mu = 0$ gives

$$\hat{\mu} = \bar{x}$$

and setting $d\ln(L)/d\sigma^2 = 0$ then gives

$$\hat{\sigma}^2 = \frac{\sum_{i=1}^{n}(x_i - \hat{\mu})^2}{n} = \frac{\sum_{i=1}^{n}(x_i - \bar{x})^2}{n}$$

which are consequently the maximum likelihood estimates of the parameters. It is interesting to notice that these point estimates come out to be the usual estimates that have been discussed in this chapter, except that the variance estimate uses n rather than $n - 1$ in the denominator.

As with point estimates produced by the method of moments, maximum likelihood estimates are generally sensible point estimates, and theoretical results show that they have very good properties when the sample size n is reasonably large. If there are three or more unknown parameters to be estimated, then the method of maximum likelihood estimation can be generalized in the obvious manner.

In most cases the two methods of estimation produce identical point estimates, although in certain cases the estimates may differ slightly. In certain cases the point estimates obtained from these methods may not be unbiased, as was seen with the maximum likelihood estimate of the normal variance, but any bias is usually small and decreases as the sample size n increases.

7.4.3 Examples

Example 27
Glass Sheet Flaws

Suppose that the quality inspector at the glass manufacturing company inspects 30 randomly selected sheets of glass and records the number of flaws found in each sheet. These data values are shown in Figure 7.15. If the distribution of the number of flaws per sheet is taken to have a Poisson distribution, how should the parameter λ of the Poisson distribution be estimated?

If the random variable X has a Poisson distribution with parameter λ, then

$$E(X) = \lambda$$

Consequently, the method of moments immediately suggests that the parameter estimate should be

$$\hat{\lambda} = \bar{x}$$

This is also the maximum likelihood estimate, which can be shown as follows. The probability mass function of a data observation x_i is

$$f(x_i, \lambda) = \frac{e^{-\lambda} \lambda^{x_i}}{x_i!}$$

so that the likelihood is

$$L(x_1, \ldots, x_n, \lambda) = \prod_{i=1}^{n} f(x_i, \lambda) = \frac{e^{-n\lambda} \lambda^{(x_1 + \cdots + x_n)}}{(x_1! \times \cdots \times x_n!)}$$

The log-likelihood is therefore

$$\ln(L) = -n\lambda + (x_1 + \cdots + x_n) \ln(\lambda) - \ln(x_1! \times \cdots \times x_n!)$$

so that

$$\frac{d \ln(L)}{d\lambda} = -n + \frac{(x_1 + \cdots + x_n)}{\lambda}$$

Setting this expression equal to 0 gives $\hat{\lambda} = \bar{x}$.

The sample average of the 30 data observations in Figure 7.15 is 0.567, so that the quality inspector should use the point estimate

$$\hat{\lambda} = 0.567$$

In addition, since each data observation has a variance of λ,

$$\mathrm{Var}(\bar{X}) = \frac{\lambda}{n}$$

FIGURE 7.15

Glass sheet flaws data set

| 0 | 1 | 1 | 1 | 0 | 0 | 0 | 2 | 0 | 1 | 0 | 1 | 0 | 0 | 0 |
| 0 | 0 | 1 | 0 | 2 | 0 | 0 | 3 | 1 | 2 | 0 | 0 | 1 | 0 | 0 |

so that the standard error of the estimate of a Poisson parameter can be calculated as

$$\text{s.e.}(\hat{\lambda}) = \sqrt{\frac{\hat{\lambda}}{n}}$$

The quality inspector's point estimate $\hat{\lambda} = 0.567$ consequently has a standard error of

$$\text{s.e.}(\hat{\lambda}) = \sqrt{\frac{0.567}{30}} = 0.137$$

Example 26

Fish Tagging and Recapture

Fish tagging and recapture present a way to estimate the size of a fish population. Suppose that a fisherman wants to estimate the fish stock N of a lake and that 34 fish have been tagged and released back into the lake. If, over a period of time, the fisherman catches 50 fish (without release) and 9 of them are tagged, an intuitive point estimate of the total number of fish in the lake is

$$\hat{N} = \frac{34 \times 50}{9} \simeq 189$$

This point estimate is based upon the reasoning that the proportion of fish in the lake that are tagged should be roughly equal to the proportion of the fisherman's catch that is tagged.

This point estimate is also the method of moments point estimate. Under the assumption that all the fish are equally likely to be caught, the distribution of the number of *tagged* fish X in the fisherman's catch of 50 fish is a hypergeometric distribution with $r = 34$, $n = 50$, and N unknown. The expectation of X is therefore

$$E(X) = \frac{nr}{N} = \frac{50 \times 34}{N}$$

and the method of moments point estimate of N is found by equating this to the observed value $x = 9$. Notice that here there is only one data observation x, which is therefore $\bar{x}$.

A similar point estimate is arrived at if the binomial approximation to the hypergeometric distribution is employed. In this case the success probability $p = r/N$ is estimated to be

$$\hat{p} = \frac{x}{n} = \frac{9}{50}$$

with

$$\hat{N} = \frac{r}{\hat{p}}$$

Example 36

Bee Colonies

An entomologist collects data on the proportion of worker bees that leave a colony with a queen bee. Calculations from 14 colonies provide the data values given in Figure 7.16. If the entomologist wishes to model this proportion with a beta distribution, how should the parameters be estimated?

The simplest way to answer this question is to use the method of moments. Recall that a beta distribution with parameters a and b has an expectation and variance

$$E(X) = \frac{a}{a+b} \quad \text{and} \quad \text{Var}(X) = \frac{ab}{(a+b)^2(a+b+1)}$$

The 14 data observations have a mean of 0.3007 and a variance of 0.01966.

FIGURE 7.16

Bee colony data set

| 0.28 | 0.32 | 0.09 | 0.35 | 0.45 | 0.41 | 0.06 | 0.16 | 0.16 | 0.46 | 0.35 |
| 0.52 | 0.29 | 0.31 | | | | | | | | |

The point estimates $\hat{a}$ and $\hat{b}$ are consequently the solutions to the equations

$$\frac{a}{a+b} = 0.3007$$

and

$$\frac{ab}{(a+b)^2(a+b+1)} = 0.01966$$

which are $\hat{a} = 2.92$ and $\hat{b} = 6.78$.

■ 7.4.4 Problems

7.4.1 Suppose that 23 observations are collected from a Poisson distribution, and the sample average is $\bar{x} = 5.63$. Construct a point estimate of the parameter of the Poisson distribution and calculate its standard error.

7.4.2 Suppose that a set of observations is collected from a beta distribution, with an average of $\bar{x} = 0.782$ and a variance of $s^2 = 0.0083$. Obtain point estimates of the parameters of the beta distribution.

7.4.3 Consider a set of independent data observations $x_1, \ldots, x_n$ that have an exponential distribution with an unknown parameter λ. Show that the method of moments and maximum likelihood estimation both produce the point estimate

$$\hat{\lambda} = \frac{1}{\bar{x}}$$

7.4.4 If the random variables $X_1, \ldots, X_k$ have a multinomial distribution with parameters n and $p_1, \ldots, p_k$, the likelihood function is

$$L(x_1, \ldots, x_k, p_1, \ldots, p_k) = \frac{n!}{x_1! \cdots x_k!} p_1^{x_1} \cdots p_k^{x_k}$$

Maximize this likelihood subject to the condition that

$$p_1 + \cdots + p_k = 1$$

in order to find the maximum likelihood estimates $\hat{p}_i$, $1 \le i \le k$.

7.4.5 Consider a set of independent data observations $x_1, \ldots, x_n$ that have a gamma distribution with $k = 5$ and an unknown parameter λ. Show that the method of moments and maximum likelihood estimation both produce the point estimate

$$\hat{\lambda} = \frac{5}{\bar{x}}$$

7.5   Case Study: Microelectronic Solder Joints

Recall the data set in Figure 6.40 of the nickel layer thicknesses on the substrate bond pads produced by a new method. If μ represents the average amount of nickel deposited by this new method, then it can be estimated by

$$\hat{\mu} = \bar{x} = 2.7688$$

with a standard error

$$s.e.(\hat{\mu}) = \frac{s}{\sqrt{n}} = \frac{0.0260}{\sqrt{16}} = 0.0065$$

The data set in Figure 6.41 can be used to estimate p_b, the probability that a solder joint will have a barrel shape for that production method, as

$$\hat{p}_b = \frac{451}{512} = 0.881$$

which has a standard error

$$s.e.(\hat{p}_b) = \sqrt{\frac{\hat{p}_b(1 - \hat{p}_b)}{n}} = \sqrt{\frac{0.881(1 - 0.881)}{512}} = 0.014$$

7.6 Supplementary Problems

7.6.1 Suppose that X_1 and X_2 are independent random variables with

$$E(X_1) = E(X_2) = \mu$$

and

$$\text{Var}(X_1) = \text{Var}(X_2) = 1$$

Show that the point estimate

$$\hat{\mu}_1 = \frac{X_1 + X_2}{4} + 5$$

has a smaller mean square error than the point estimate

$$\hat{\mu}_2 = \frac{X_1 + X_2}{2}$$

when

$$|\mu - 10| \leq \frac{\sqrt{6}}{2}$$

Why would you expect $\hat{\mu}_1$ to have a smaller mean square error than $\hat{\mu}_2$ when μ is close to 10?

7.6.2 Suppose that $X \sim B(12, p)$ and consider the point estimate

$$\hat{p} = \frac{X}{14}$$

(a) What is the bias of this point estimate?
(b) What is the variance of this point estimate?
(c) Show that this point estimate has a mean square error of

$$\frac{3p - 2p^2}{49}$$

(d) Show that this mean square error is smaller than the mean square error of $X/12$ when $p \leq 0.52$.

7.6.3 Let $X_1, \ldots, X_n$ be a set of independent random variables with a $U(0, \theta)$ distribution, and let

$$T = \max\{X_1, \ldots, X_n\}$$

(a) Explain why the cumulative distribution function of T is

$$F(t) = \left(\frac{t}{\theta}\right)^n$$

for $0 \leq t \leq \theta$.

(b) Show that the probability density function of T is

$$f(t) = n\frac{t^{n-1}}{\theta^n}$$

for $0 \leq t \leq \theta$.

(c) Show that

$$\hat{\theta} = \frac{n + 1}{n}T$$

is an unbiased point estimate of θ.

(d) What is the standard error of $\hat{\theta}$?
(e) Suppose that $n = 10$ and that the following data values are obtained:

1.2 6.3 7.3 6.4 3.5 0.2 4.6 7.1 5.0 1.8

What are the values of $\hat{\theta}$ and the standard error of $\hat{\theta}$?

7.6.4 As in Problem 7.6.3, let $X_1, \ldots, X_n$ be a set of independent random variables with a $U(0, \theta)$ distribution, and let

$$T = \max\{X_1, \ldots, X_n\}$$

Explain why the likelihood function $L(x_1, \ldots, x_n, \theta)$ is equal to

$$\frac{1}{\theta^n}$$

if $\theta \geq t = \max\{x_1, \ldots, x_n\}$, and is equal to 0 otherwise. Sketch the likelihood function against θ, and deduce that the maximum likelihood estimate of θ is $\hat{\theta} = t$. What is the bias of this point estimate?

7.6.5 Consider a set of independent data observations $x_1, \ldots, x_n$ that have a geometric distribution with an unknown parameter p. Show that the method of moments and maximum likelihood estimation both produce the point estimate

$$\hat{p} = \frac{1}{\bar{x}}$$

7.6.6 Bird Species Identification
Consider the data set of bird species given in DS 6.6.1. Construct a point estimate of the probability that a bird has black markings. What is the standard error of your point estimate?

7.6.7 Oil Rig Accidents

Consider the data set of monthly accidents given in DS 6.6.2. Construct a point estimate of the average number of accidents per month. What is the standard error of your point estimate?

7.6.8 Programming Errors

Consider the data set of programming errors given in DS 6.6.3. Construct a point estimate of the average number of errors per month. What is the standard error of your point estimate?

7.6.9 Osteoporosis Patient Heights

Consider the data set of osteoporosis patient heights given in DS 6.6.4. Construct a point estimate of the average height. What is the standard error of your point estimate?

7.6.10 Bamboo Cultivation

Consider the data set of bamboo shoot heights given in DS 6.6.5. Construct a point estimate of the average height. What is the standard error of your point estimate?

7.6.11 Consider the usual point estimates s_1^2 and s_2^2 of the variance σ^2 of a normal distribution based on sample sizes n_1 and n_2, respectively. What is the relative efficiency of the point estimate s_1^2 to the point estimate s_2^2?

7.6.12 Suppose that among 24,839 customers of a certain company, exactly 11,842 feel "very satisfied" with the service they received. In order to estimate the satisfaction levels of the customers, a manager contacts a random sample of 80 of these customers and finds out how many of them were "very satisfied." What is the probability that the manager's estimate of the proportion of "very satisfied" customers in this group is within 0.10 of the true value?

7.6.13 The viscosities of chemical infusions obtained from a specific production technique are normally distributed with a standard deviation $\sigma = 3.9$. If a chemist is able to measure the viscosities of 15 independent samples of the infusions, what is the probability that the resulting point estimate of the mean μ will be within 0.5 of the true value? How does this probability change if a sample of 40 independent infusions is obtained?

7.6.14 Soil Compressibility Tests

Recall the data set of soil compressibility measurements given in DS 6.6.6. Construct a point estimate of the average soil compressibility, and find its standard error. What is a point estimate of the upper quartile of the distribution of soil compressibilities?

7.6.15 An engineer assumes that the distribution of the breaking strengths of fibers is $N(280, 2.5)$ and uses this distribution to perform an analysis of whether the average breaking strength of a collection of 20 fibers will exceed a specified value. Is the engineer doing probability theory or statistical inference?

7.6.16 An experimenter assumes a probability distribution for the lengths of telephone calls arriving at a hotline, and predicts the lengths of calls that will be obtained in a random sample of calls. Is the experimenter using probability theory or statistical inference?

7.6.17 Suppose that an engineer wishes to estimate the proportion of defective products from a production line. A random sample of 220 products are tested, of which 39 are found to be defective. What is the standard error of the engineer's estimate of the proportion of defective products?

7.6.18 The probability that a medical treatment is effective is 0.68, unknown to a researcher. In an experiment to investigate the effectiveness of the treatment, the researcher applies the treatment in 140 cases and measures whether the treatment is effective or not. What is the probability that the researcher's estimate of the probability that the medical treatment is effective is within 0.05 of the correct answer?

7.6.19 The biomass of 12 samples was measured, and the following values were obtained:

78   67   58   93   63   70   59   82   88   66   50   73

(a) What is the estimate of the mean biomass?
(b) What is the standard error of the estimate of the mean biomass?
(c) What is the sample median?

7.6.20 An experimenter measures the weights of a random sample of 20 items and uses the information to estimate the overall population mean. Is the experimenter using probability theory or statistical inference?

7.6.21 A random sample of components from a supplier is tested in order to estimate the probability that a component from that supplier satisfies the design requirements. Is this probability theory or statistical inference?

7.6.22 Are the following statements true or false?

(a) Statistical inference uses the results of an experiment to make inferences on some properties of an unknown underlying probability distribution.

(b) The margin of error in a political poll is based on the standard error of the estimate obtained.

(c) An experimenter collects some data from a process and uses it to estimate some properties of the process. The experimenter is using statistical inference, not probability theory, because the known data is used to make inferences about the unknown parameters of the process.

(d) The standard error of a point estimate provides an indication of its accuracy.

7.6.23 Components have lengths that are independently distributed as a normal distribution with $\mu = 723$ and $\sigma = 3$. If an experimenter measures the lengths of a random sample of 11 components, what is the probability that the experimenter's estimate of μ will be between 722 and 724?

7.6.24 An experimenter wishes to estimate the mean weight of some components where the weights have a normal distribution with a standard deviation of 40.0.

(a) If the experimenter has a sample size of 10, what is the probability that the estimate is within 20.0 of the correct value?

(b) What is the probability if the sample size is 20?

7.6.25 In a political poll, responses were obtained from a sample of 1962 people about which candidate they preferred.

There were 852 people who reported that they preferred candidate A. What is the estimate of the proportion of the overall electorate who prefer candidate A? What is the standard error of this estimate?

For Problems 7.6.26–7.6.33 use the data sets to practice finding parameter point estimates and their standard errors.

7.6.26 Glass Fiber Reinforced Polymer Tensile Strengths
The data set in DS 6.6.7.

7.6.27 Infant Blood Levels of Hydrogen Peroxide
The data set in DS 6.6.8.

7.6.28 Paper Mill Operation of a Lime Kiln
The data set in DS 6.6.9.

7.6.29 River Salinity Levels
The data set in DS 6.6.10.

7.6.30 Dew Point Readings from Coastal Buoys
The data set in DS 6.6.11.

7.6.31 Brain pH levels
The data set in DS 6.6.12.

7.6.32 Silicon Dioxide Percentages in Ocean Floor Volcanic Glass
The data set in DS 6.6.13.

7.6.33 Network Server Response Times
The data set in DS 6.6.14.

CHAPTER EIGHT

Inferences on a Population Mean

Random variable theory and estimation methods are combined in this chapter to provide an analysis of a single sample of continuous data observations taken from a particular population. Inference procedures designed to investigate the population mean μ are described. These inference procedures are **confidence interval construction** and **hypothesis testing**, which are two fundamental techniques of statistical inference. The methodologies discussed are commonly referred to as "t-intervals" and "t-tests," and these are among the most basic and widely employed of all statistical inference methods.

8.1 Confidence Intervals

8.1.1 Confidence Interval Construction

The discussions in this chapter concern the analysis of a sample of data observations $x_1, \ldots, x_n$ that are independent observations from some unknown continuous probability distribution. Statistical methodologies for investigating the unknown population mean are described. The data set of metal cylinder diameters given in Figure 6.5 is a typical data set of this kind for which μ is the average diameter of cylinders produced in this manner.

A **confidence interval** for μ is an interval that contains "plausible" values of the parameter μ (the notion of plausibility is given a rigorous definition in Section 8.2). It is a simple combination of the point estimate $\hat{\mu} = \bar{x}$ together with its estimated standard error $s/\sqrt{n}$. A confidence interval is associated with a **confidence level**, which is usually written as $1 - \alpha$, and which indicates the confidence that the experimenter has that the parameter μ actually lies within the given confidence interval. Confidence levels of 90%, 95%, and 99% are typically used, which correspond to α values of 0.10, 0.05, and 0.01, respectively.

Confidence Intervals

A **confidence interval** for an unknown parameter θ is an interval that contains a set of plausible values of the parameter. It is associated with a confidence level $1 - \alpha$, which measures the probability that the confidence interval actually contains the unknown parameter value.

The t-intervals discussed in this section, and more generally any t-procedure such as these t-intervals or the hypothesis tests discussed in Section 8.2, are appropriate for making inferences on a population mean in a wide variety of settings. Technically, the implementation of these procedures requires that the sample mean be an observation from a normal distribution, and for sample sizes $n \geq 30$ the central limit theorem ensures that this will be a reasonable assumption. For smaller sample sizes the requirement is met if the data are normally distributed, and in fact the t-intervals provide a sensible analysis unless the data observations are clearly not normally distributed. In this latter case the general nonparametric inference methods discussed in Chapter 15 may be employed, or alternative procedures for specific distributions may be used such as the procedure described in Section 17.3.1 for data from an exponential distribution.

Inferences on a Population Mean

Inference methods on a population mean based upon the t-procedure are appropriate for large sample sizes $n \geq 30$ and also for small sample sizes as long as the data can reasonably be taken to be approximately normally distributed. Nonparametric techniques can be employed for small sample sizes with data that are clearly not normally distributed.

The most commonly used confidence interval for a population mean μ based on a sample of n continuous data observations with a sample mean $\bar{x}$ and a sample standard deviation s is a **two-sided t-interval**, which is constructed as

$$\mu \in \left(\bar{x} - \frac{t_{\alpha/2,n-1}s}{\sqrt{n}}, \bar{x} + \frac{t_{\alpha/2,n-1}s}{\sqrt{n}} \right)$$

As illustrated in Figure 8.1, the interval is centered at the "best guess" $\hat{\mu} = \bar{x}$ and extends on either side by an amount equal to a *critical point* $t_{\alpha/2,n-1}$ (as defined in Section 5.4.3) multiplied by the standard error of $\hat{\mu}$. Thus, it is useful to understand that the confidence interval is constructed as

$$\mu \in (\hat{\mu} - \text{critical point} \times \text{s.e.}(\hat{\mu}), \hat{\mu} + \text{critical point} \times \text{s.e.}(\hat{\mu}))$$

FIGURE 8.1

A two-sided t-interval

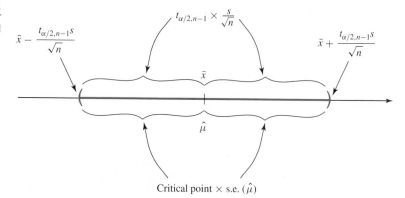

Two-Sided t-Interval

A confidence interval with confidence level $1 - \alpha$ for a population mean μ based upon a sample of n continuous data observations with a sample mean $\bar{x}$ and a sample standard deviation s is

$$\mu \in \left(\bar{x} - \frac{t_{\alpha/2, n-1} s}{\sqrt{n}}, \bar{x} + \frac{t_{\alpha/2, n-1} s}{\sqrt{n}} \right)$$

The interval is known as a **two-sided t-interval** or variance unknown confidence interval.

The length of the confidence interval is

$$L = \frac{2\, t_{\alpha/2, n-1} s}{\sqrt{n}} = 2 \times \text{critical point} \times \text{s.e.}(\hat{\mu})$$

which is proportional to the standard error of $\hat{\mu}$. As the standard error of $\hat{\mu}$ decreases, so that $\hat{\mu} = \bar{x}$ becomes a more "accurate" estimate of μ, the length of the confidence interval decreases so that there are fewer plausible values for μ. In other words, a more accurate estimate of μ allows the experimenter to eliminate more values of μ from contention.

The length of the confidence interval L also depends upon the critical point $t_{\alpha/2, n-1}$. Recall that this critical point is defined by

$$P(X \geq t_{\alpha/2, n-1}) = \alpha/2$$

where the random variable X has a t-distribution with degrees of freedom $n - 1$. The confidence interval depends upon the confidence level $1 - \alpha$ through this critical point. As the confidence level increases, so that α decreases, the critical point $t_{\alpha/2, n-1}$ also increases so that the confidence interval becomes longer. This relationship is illustrated in Figure 8.2 and may be summarized as

Higher confidence levels require longer confidence intervals.

Finally, notice that the degrees of freedom of the critical point are *one fewer* than the sample size n.

FIGURE 8.2

Higher confidence levels require longer confidence intervals

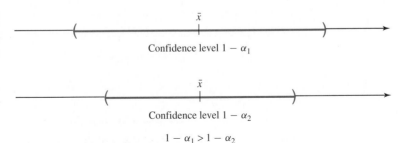

Effect of the Confidence Level on the Confidence Interval Length

The length of a confidence interval depends upon the confidence level $1 - \alpha$ through the critical point. As the confidence level $1 - \alpha$ increases, the length of the confidence interval also increases.

Example 17

Milk Container Contents

The data set of milk container weights is given in Figure 6.6, and summary statistics are given in Figure 6.27.

Suppose that a confidence interval is required with confidence level 95%. In this case $\alpha = 0.05$, so that the relevant critical point is $t_{\alpha/2,n-1} = t_{0.025,49} = 2.0096$ (which can be obtained exactly from the computer or approximately from Table III). Consequently, the confidence interval is

$$
\left(\bar{x} - \frac{t_{\alpha/2,n-1}s}{\sqrt{n}}, \ \bar{x} + \frac{t_{\alpha/2,n-1}s}{\sqrt{n}} \right)
$$

$$
= \left(2.0727 - \frac{2.0096 \times 0.0711}{\sqrt{50}}, \ 2.0727 + \frac{2.0096 \times 0.0711}{\sqrt{50}} \right)
$$

$$
= (2.0525, 2.0929)
$$

This result has the interpretation that the experimenter is 95% confident that the average milk container content is between about 2.053 and 2.093 liters.

Since $t_{0.005,49} = 2.680$, a confidence interval with confidence level 99% is

$$
\left(2.0727 - \frac{2.680 \times 0.0711}{\sqrt{50}}, \ 2.0727 + \frac{2.680 \times 0.0711}{\sqrt{50}} \right) = (2.0457, \ 2.0996)
$$

Similarly, $t_{0.05,49} = 1.6766$, so that a confidence interval with confidence level 90% is

$$
\left(2.0727 - \frac{1.6766 \times 0.0711}{\sqrt{50}}, \ 2.0727 + \frac{1.6766 \times 0.0711}{\sqrt{50}} \right) = (2.0558, \ 2.0895)
$$

The three confidence intervals are shown in Figure 8.3, and clearly the confidence interval length increases as the confidence level rises.

FIGURE 8.3

Confidence intervals for the mean milk container weight

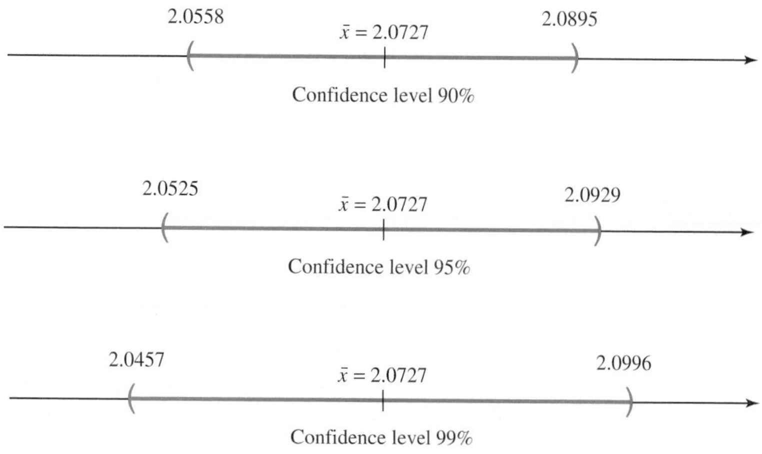

Example 14

Metal Cylinder Production

A data set of 60 metal cylinder diameters is given in Figure 6.5, and summary statistics are given in Figure 6.29. The critical points required for confidence interval construction are given in Figure 8.4.

With $\alpha = 0.10$, $t_{0.05,59} = 1.671$ so that a confidence interval with confidence level 90% is

$$\left(49.999 - \frac{1.671 \times 0.134}{\sqrt{60}}, 49.999 + \frac{1.671 \times 0.134}{\sqrt{60}}\right) = (49.970, 50.028)$$

Similarly, $t_{0.025,59} = 2.001$ so that a confidence interval with confidence level 95% is

$$\left(49.999 - \frac{2.001 \times 0.134}{\sqrt{60}}, 49.999 + \frac{2.001 \times 0.134}{\sqrt{60}}\right) = (49.964, 50.034)$$

and $t_{0.005,59} = 2.662$ so that a confidence interval with confidence level 99% is

$$\left(49.999 - \frac{2.662 \times 0.134}{\sqrt{60}}, 49.999 + \frac{2.662 \times 0.134}{\sqrt{60}}\right) = (49.953, 50.045)$$

These confidence intervals are illustrated in Figure 8.5.

A sensible way to summarize these results might be to notice that based upon this sample of 60 randomly selected cylinders, the experimenter can conclude (with over 99% certainty) that the average cylinder diameter lies within 0.05 mm of 50.00 mm, that is, within the interval (49.95, 50.05). Of course, it is important to remember that this confidence interval is for the *mean* cylinder diameter, and not for the actual diameter of a randomly selected cylinder. In fact, the sample contains a cylinder as thin as 49.737 mm and as thick as 50.362 mm.

The justification of the two-sided t-intervals introduced in this section is straightforward. The result at the end of Section 7.3.3 states that

$$\frac{\sqrt{n}(\bar{X} - \mu)}{S} \sim t_{n-1}$$

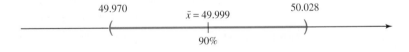

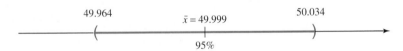

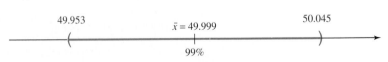

Sample size $n = 60$	
Confidence level 90%:	$t_{0.05,59} = 1.671$
Confidence level 95%:	$t_{0.025,59} = 2.001$
Confidence level 99%:	$t_{0.005,59} = 2.662$

FIGURE 8.4

Critical points for the construction of two-sided confidence intervals for the mean metal cylinder diameter

FIGURE 8.5

Confidence intervals for the mean metal cylinder diameter

and so the definition of the critical points of the t-distribution ensures that

$$P\left(-t_{\alpha/2,n-1} \leq \frac{\sqrt{n}(\bar{X} - \mu)}{S} \leq t_{\alpha/2,n-1}\right) = 1 - \alpha$$

However, the inequality

$$-t_{\alpha/2,n-1} \leq \frac{\sqrt{n}(\bar{X} - \mu)}{S}$$

can be rewritten

$$\mu \leq \bar{X} + \frac{t_{\alpha/2,n-1}S}{\sqrt{n}}$$

and the inequality

$$\frac{\sqrt{n}(\bar{X} - \mu)}{S} \leq t_{\alpha/2,n-1}$$

can be rewritten

$$\bar{X} - \frac{t_{\alpha/2,n-1}S}{\sqrt{n}} \leq \mu$$

so that

$$P\left(\bar{X} - \frac{t_{\alpha/2,n-1}S}{\sqrt{n}} \leq \mu \leq \bar{X} + \frac{t_{\alpha/2,n-1}S}{\sqrt{n}}\right) = 1 - \alpha$$

This probability expression indicates that there is a probability of $1 - \alpha$ that the parameter value μ lies within the two-sided t-interval. A subtle but important point to remember is that μ is a fixed value but that the confidence interval limits are random quantities. Thus, the probability statement should be interpreted as saying that there is a probability of $1 - \alpha$ that the random confidence interval limits take values that "straddle" the fixed value μ. This interpretation of a confidence interval is further clarified by the simulation experiment in Section 8.1.4.

Technically speaking, $\sqrt{n}(\bar{X} - \mu)/S$ has a t-distribution only when the random variables X_i are normally distributed. Nevertheless, as discussed at the beginning of this section, the central limit theorem ensures that the distribution of $\bar{X}$ is approximately normal for reasonably large sample sizes, and in such cases it is sensible to construct t-intervals regardless of the actual distribution of the data observations. Alternative *nonparametric* confidence intervals are discussed in Chapter 15 for situations where the sample size is small (less than 30, say) and the data observations are evidently not normally distributed.

8.1.2 Effect of the Sample Size on Confidence Intervals

The sample size n has an important effect on the confidence interval length

$$L = \frac{2\,t_{\alpha/2,n-1}\,s}{\sqrt{n}}$$

Effect of the Sample Size on the Confidence Interval Length

For a fixed critical point, a confidence interval length L is inversely proportional to the square root of the sample size n

$$L \propto \frac{1}{\sqrt{n}}$$

Thus a *fourfold* increase in the sample size reduces the confidence interval length by half.

The critical point $t_{\alpha/2,n-1}$ also depends upon the sample size n, although this dependence is generally minimal. Recall that as the sample size n increases, the critical point $t_{\alpha/2,n-1}$ decreases to the standard normal critical point $z_{\alpha/2}$. For example, with $\alpha = 0.05$, it can be seen from Table III that

$$t_{0.025,10} = 2.228$$
$$t_{0.025,20} = 2.086$$
$$t_{0.025,30} = 2.042$$
$$t_{0.025,\infty} = z_{0.025} = 1.960$$

Notice that this dependence of the critical point on the sample size also serves to produce smaller confidence intervals with larger sample sizes.

Some simple calculations can be made to determine what sample size n is required to obtain a confidence interval of a certain length. Specifically, if a confidence interval with a length no larger than L_0 is required, then a sample size

$$n \geq 4 \times \left(\frac{t_{\alpha/2,n-1}s}{L_0} \right)^2$$

must be used. This inequality can be used to find a suitable sample size n if approximate values or upper bounds are used for $t_{\alpha/2,n-1}$ and s.

For example, suppose that an experimenter wishes to construct a 95% confidence interval with a length no larger than $L_0 = 2.0$ mm for the mean thickness of plastic sheets produced by a particular process. Previous experience with the process enables the experimenter to be certain that the standard deviation of the sheet thicknesses cannot be larger than 4.0 mm, and a large enough sample size is expected so that the critical point $t_{0.025,n-1}$ will be less than 2.1, say. Consequently, the experimenter can expect that a sample size

$$n \geq 4 \times \left(\frac{t_{\alpha/2,n-1}s}{L_0} \right)^2 = 4 \times \left(\frac{2.1 \times 4.0}{2.0} \right)^2 = 70.56$$

is sufficient. A random sample of at least 71 plastic sheets should then meet the experimenter's requirement.

Similar calculations can also be employed to ascertain what *additional* sampling is required to reduce the length of a confidence interval that has been constructed from an initial sample. In this case, the values of $t_{\alpha/2,n-1}$ and s employed in the initial confidence interval can be used as approximate values.

For example, if an initial sample of n_1 observations is obtained that has a sample standard deviation s, then a confidence interval of length

$$L = \frac{2\,t_{\alpha/2,n_1-1}s}{\sqrt{n_1}}$$

can be constructed. If the experimenter decides that additional sampling is required in order to reduce the confidence interval length to $L_0 < L$, the experimenter can expect that a *total* sample size n will be sufficient as long as

$$n \geq 4 \times \left(\frac{t_{\alpha/2,n_1-1}s}{L_0} \right)^2$$

The difference $n - n_1$ is the size of the additional sample required, which can then be combined with the initial sample of size n_1.

Example 17
Milk Container Contents

With a sample of $n = 50$ milk containers, a confidence interval for the mean container content with confidence level 99% is constructed to be

$$(2.0457, 2.0996)$$

This interval has a length of $2.0996 - 2.0457 = 0.0539$ liters. Suppose that the engineers decide that they need a 99% confidence interval that has a length no larger than 0.04 liters. How much additional sampling is required?

Using the values $t_{0.005,49} = 2.680$ and $s = 0.0711$ employed in the initial analysis, it appears that a total sample size

$$n \geq 4 \times \left(\frac{t_{\alpha/2,n-1}s}{L_0} \right)^2 = 4 \times \left(\frac{2.680 \times 0.0711}{0.04} \right)^2 = 90.77$$

is required. The engineers can therefore predict that if an additional random sample of at least $91 - 50 = 41$ milk containers is obtained, a confidence interval based upon the combination of the two samples will have a length no larger than 0.04 liters.

Example 14
Metal Cylinder Production

With a sample of $n = 60$ metal cylinders, a 99% confidence interval

$$(49.953, 50.045)$$

has been obtained with a length of $50.045 - 49.953 = 0.092$ mm. How much additional sampling is required to provide the increased precision of a confidence interval with a length of 0.08 mm at the same confidence level?

Using $t_{0.005,59} = 2.662$ and $s = 0.134$, a total sample size of

$$n \geq 4 \times \left(\frac{t_{\alpha/2,n-1}s}{L_0} \right)^2 = 4 \times \left(\frac{2.662 \times 0.134}{0.08} \right)^2 = 79.53$$

is required. Therefore, the engineers can anticipate that an additional sample of at least $80 - 60 = 20$ cylinders is needed to meet the specified goal.

8.1.3 Further Examples

Example 43
Rolling Mill Scrap

Recall that a random sample of $n = 95$ ingots that were passed through the rolling machines provided % scrap observations with a sample mean of $\bar{x} = 20.810$ and a sample standard deviation of $s = 4.878$. Since $t_{0.05,94} = 1.6612$, a confidence interval for the mean % scrap with a confidence level of 90% is

$$\left(\bar{x} - \frac{t_{\alpha/2,n-1}s}{\sqrt{n}}, \ \bar{x} + \frac{t_{\alpha/2,n-1}s}{\sqrt{n}} \right)$$

$$= \left(20.810 - \frac{1.6612 \times 4.878}{\sqrt{95}}, \ 20.810 + \frac{1.6612 \times 4.878}{\sqrt{95}} \right) = (19.978, \ 21.641)$$

With a confidence level of 99%, the confidence interval increases in length to (19.494, 22.126).

Perhaps a good summary of these results is to say that there is a high degree of confidence that the mean value of % scrap is somewhere between about 19.5% and 22%. This is very useful information for the rolling mill managers because it indicates the amount of scrap that they should expect over a certain period of time. Even though the amount of scrap obtained from each ingot varies considerably (in the sample of 95 ingots % scrap varied from about 7% to 31%), if a large number of ingots are to be rolled over a reasonably long period of time, then the managers can be fairly confident that the amount of scrap obtained during the period will be about 19.5%–22% of the total weight of the ingots used.

Example 44
Army Physical Fitness Test

The sample of $n = 84$ run times has a sample mean $\bar{x} = 857.70$ and a sample standard deviation $s = 81.98$. Since $t_{0.025,83} = 1.9890$, a confidence interval for the mean run time with a confidence level of 95% is

$$\left(\bar{x} - \frac{t_{\alpha/2,n-1}s}{\sqrt{n}}, \ \bar{x} + \frac{t_{\alpha/2,n-1}s}{\sqrt{n}} \right)$$

$$= \left(857.70 - \frac{1.9890 \times 81.98}{\sqrt{84}}, \ 857.70 + \frac{1.9890 \times 81.98}{\sqrt{84}} \right) = (839.91, \ 875.49)$$

With $t_{0.005,83} = 2.6364$, a confidence interval with confidence level 99% is

$$\left(857.70 - \frac{2.6364 \times 81.98}{\sqrt{84}}, \ 857.70 + \frac{2.6364 \times 81.98}{\sqrt{84}} \right) = (834.12, 881.28)$$

Consequently, with 99% confidence, the mean run time is found to lie between 834 and 882 seconds, which is between 13 minutes 54 seconds and 14 minutes 42 seconds. With 95% confidence this interval can be reduced to 839 and 876 seconds, that is, between 13 minutes 59 seconds and 14 minutes 36 seconds.

Example 45
Fabric Water Absorption Properties

The sample of $n = 15$ fabric % pickup observations has a sample mean $\bar{x} = 59.81$ and a sample standard deviation $s = 4.94$. Since $t_{0.005,14} = 2.9769$, a confidence interval for the mean % pickup value with a confidence level of 99% is

$$\left(\bar{x} - \frac{t_{\alpha/2,n-1}s}{\sqrt{n}}, \ \bar{x} + \frac{t_{\alpha/2,n-1}s}{\sqrt{n}} \right)$$

$$= \left(59.81 - \frac{2.9769 \times 4.94}{\sqrt{15}}, \ 59.81 + \frac{2.9769 \times 4.94}{\sqrt{15}} \right) = (56.01, \ 63.61)$$

This analysis reveals that the mean water pickup of the cotton fabric under examination lies between about 56% and 64%. Suppose that the textile engineers decide that they need more precision, and that specifically they require a 99% confidence interval with a length no larger than $L_0 = 5\%$. Using the values $t_{0.005,14} = 2.9769$ and $s = 4.94$, the engineers can predict that a total sample size of

$$n \geq 4 \times \left(\frac{t_{\alpha/2,n-1}s}{L_0} \right)^2 = 4 \times \left(\frac{2.9769 \times 4.94}{5} \right)^2 = 34.6$$

will suffice. Therefore, a second sample of at least $35 - 15 = 20$ observations is required.

8.1.4   Simulation Experiment 3: An Investigation of Confidence Intervals

The simulation experiment described in Section 7.3.4 can be extended to illustrate the probabilistic properties of confidence intervals. Recall that when an initial sample of 30 observations was simulated, a sample mean of $\bar{x} = 10.589$ and a sample standard deviation of $s = \sqrt{3.4622} = 1.861$ were obtained. With $t_{0.025,29} = 2.0452$, these values provide a 95% confidence interval

$$\left(10.589 - \frac{2.0452 \times 1.861}{\sqrt{30}}, \ 10.589 + \frac{2.0452 \times 1.861}{\sqrt{30}} \right) = (9.89, \ 11.28)$$

We know that the 30 observations were simulated with a mean of $\mu = 10$, so we know that this confidence interval does indeed contain the true value of μ. In fact, any confidence interval calculated in this fashion using simulated observations has a probability of 0.95 of containing the value $\mu = 10$. Notice that the value of the mean $\mu = 10$ is fixed, and that the upper and lower endpoints of the confidence interval, in this case 9.89 and 11.28, are random variables that depend upon the simulated data set.

Figure 8.6 shows 95% confidence intervals for some of the 500 samples of simulated data observations. For example, in simulation 1 the sample statistics are $\bar{x} = 10.3096$ and $s = 1.25211$, so that the 95% confidence interval is

$$\left(10.3096 - \frac{2.0452 \times 1.25211}{\sqrt{30}}, \ 10.3096 + \frac{2.0452 \times 1.25211}{\sqrt{30}} \right)$$
$$= (9.8421, 10.7772)$$

which again does indeed contain the true value $\mu = 10$. However, notice that in simulation 24 the confidence interval

$$(10.2444, 11.2548)$$

does not include the correct value $\mu = 10$, as is the case with simulation 37 where the confidence interval is

$$(8.6925, 9.8505)$$

Remember that each simulation provides a 95% confidence interval, which has a probability of 0.05 of not containing the value $\mu = 10$. Since the simulations are independent of each other, the number of simulations out of 500 for which the confidence interval does not contain $\mu = 10$ has a binomial distribution with $n = 500$ and $p = 0.05$. The expected number of simulations where this happens is therefore $np = 500 \times 0.05 = 25$.

Figure 8.7 presents a graphical illustration of some of the simulated confidence intervals showing how they generally straddle the value $\mu = 10$, although simulations 24 and 37 are exceptions, for example. Notice that the lengths of the confidence intervals vary from one simulation to another due to changes in the value of the sample standard deviation s. Remember that, in practice, an experimenter observes just one data set, and it has a probability of 0.95 of providing a 95% confidence interval that does indeed straddle the true value μ.

8.1.5 One-Sided Confidence Intervals

One-sided confidence intervals can be useful if only an *upper bound* or only a *lower bound* on the population mean μ is of interest. Since

$$\frac{\sqrt{n}(\bar{X} - \mu)}{S} \sim t_{n-1}$$

the definition of the critical point $t_{\alpha,n-1}$ implies that

$$P\left(-t_{\alpha,n-1} \le \frac{\sqrt{n}(\bar{X} - \mu)}{S} \right) = 1 - \alpha$$

This may be rewritten

$$P\left(\mu \le \bar{X} + \frac{t_{\alpha,n-1} S}{\sqrt{n}} \right) = 1 - \alpha$$

FIGURE 8.6

Confidence interval construction
from simulation results

Simulation	$\bar{x}$	s	Confidence interval Lower bound	Upper bound
1	10.3096	1.25211	9.8421	10.7772
2	10.0380	1.85805	9.3442	10.7318
3	9.4313	1.52798	8.8608	10.0019
4	9.5237	1.99115	8.7802	10.2671
5	9.8644	1.78138	9.1993	10.5296
6	9.9980	1.75209	9.3438	10.6523
7	10.0151	1.67648	9.3891	10.6410
8	9.5657	2.02599	8.8092	10.3222
9	9.9897	1.96631	9.2555	10.7239
10	9.8570	1.73049	9.2108	10.5032
11	10.2514	2.16112	9.4444	11.0584
12	10.1104	1.80382	9.4368	10.7839
13	10.0157	1.41898	9.4859	10.5455
14	10.3813	1.38312	9.8648	10.8977
15	9.4689	1.94572	8.7432	10.1954
16	9.7135	1.62459	9.1069	10.3201
17	10.2732	1.82360	9.5923	10.9542
18	9.6372	1.74860	8.9842	10.2901
19	10.1828	1.83197	9.4987	10.8668
20	10.2726	1.45160	9.7305	10.8146
21	9.9432	1.74806	9.2905	10.5959
22	10.1797	1.84898	9.4893	10.8701
23	10.2311	1.59639	9.6351	10.8272
24	10.7496	1.35291	10.2444	11.2548 ←
25	10.2216	1.85089	9.5305	10.9127
26	10.3936	1.80703	9.7189	11.0684
27	10.1002	1.79356	9.4305	10.7699
28	10.0762	1.53233	9.5040	10.6484
29	10.2444	1.87772	9.5433	10.9456
30	10.2307	2.22136	9.4012	11.0601
31	10.3165	1.95971	9.5848	11.0483
32	9.9352	2.15672	9.1298	10.7405
33	10.1056	1.36460	9.5961	10.6152
34	10.3469	1.53834	9.7725	10.9213
35	10.0949	1.62171	9.4893	10.7004
36	10.0132	1.81789	9.3344	10.6920
37	9.2715	1.55059	8.6925	9.8505 ←
38	10.0483	1.40464	9.5238	10.5728
39	9.8531	2.07755	9.0773	10.6289
40	9.8680	1.18785	9.4244	10.3115
41	10.0369	1.64264	9.4236	10.6503
42	9.8179	1.91046	9.1046	10.5313
43	10.4725	1.57621	9.8839	11.0610
44	9.8642	1.68885	9.2336	10.4949
45	10.2753	2.02283	9.5199	11.0306
46	9.8033	1.95644	9.0728	10.5338
47	10.0100	1.57821	9.4207	10.5993
48	9.9486	1.75946	9.2916	10.6056
49	9.9372	1.55260	9.3574	10.5169
50	9.7926	1.52264	9.2128	10.3724
⋮	⋮	⋮	⋮	⋮
500	9.6115	1.59441	9.0161	10.2068

Confidence interval
does not contain
$\mu = 10$

FIGURE 8.7

Confidence intervals from simulation experiment

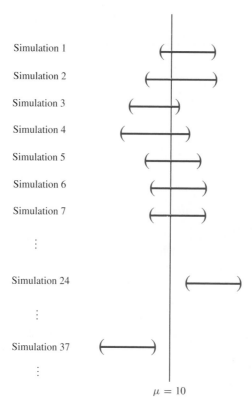

so that

$$\mu \in \left(-\infty, \bar{x} + \frac{t_{\alpha,n-1}s}{\sqrt{n}} \right)$$

is a *one-sided* confidence interval for μ with a confidence level of $1 - \alpha$. This confidence interval provides an upper bound on the population mean μ.

Similarly, the result

$$P \left(\frac{\sqrt{n}(\bar{X} - \mu)}{S} \leq t_{\alpha,n-1} \right) = 1 - \alpha$$

implies that

$$P \left(\bar{X} - \frac{t_{\alpha,n-1}S}{\sqrt{n}} \leq \mu \right) = 1 - \alpha$$

so that

$$\mu \in \left(\bar{x} - \frac{t_{\alpha,n-1}s}{\sqrt{n}}, \infty \right)$$

is also a *one-sided* confidence interval for μ with a confidence level of $1 - \alpha$. This confidence interval provides a lower bound on the population mean μ. These confidence intervals are known as **one-sided t-intervals**.

FIGURE 8.8

Comparison of two-sided and
one-sided confidence intervals

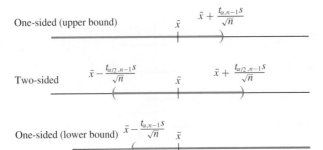

One-Sided t-Interval

One-sided confidence intervals with confidence levels $1 - \alpha$ for a population mean μ based on a sample of n continuous data observations with a sample mean $\bar{x}$ and a sample standard deviation s are

$$\mu \in \left(-\infty, \bar{x} + \frac{t_{\alpha,n-1}s}{\sqrt{n}} \right)$$

which provides an upper bound on the population mean μ, and

$$\mu \in \left(\bar{x} - \frac{t_{\alpha,n-1}s}{\sqrt{n}}, \infty \right)$$

which provides a lower bound on the population mean μ. These confidence intervals are known as **one-sided t-intervals**.

Figure 8.8 compares the one-sided t-intervals with a two-sided t-interval. Notice that $t_{\alpha,n-1} < t_{\alpha/2,n-1}$, so that the one-sided t-intervals provide a lower or an upper bound that is closer to $\hat{\mu} = \bar{x}$ than the limits of the two-sided t-interval.

Example 46

Hospital Worker Radiation Exposures

Hospital workers who are routinely involved in administering radioactive tracers to patients are subject to a radiation exposure emanating from the skin of the patient. In an experiment to assess the amount of this exposure, radiation levels were measured at a distance of 50 cm from $n = 28$ patients who had been injected with a radioactive tracer, and a sample mean $\bar{x} = 5.145$ and sample standard deviation $s = 0.7524$ are obtained. With a critical point $t_{0.01,27} = 2.473$, a 99% one-sided confidence interval providing an upper bound for μ is

$$\mu \in \left(-\infty, \bar{x} + \frac{t_{\alpha,n-1}s}{\sqrt{n}} \right) = \left(-\infty, 5.145 + \frac{2.473 \times 0.7524}{\sqrt{28}} \right) = (-\infty, 5.496)$$

Consequently, with a confidence level of 0.99 the experimenter can conclude that the average radiation level at a 50-cm distance from a patient is *no more* than about 5.5.

8.1.6 z-Intervals

In some circumstances an experimenter may wish to use a "known" value of the population standard deviation σ in a confidence interval in place of the sample standard deviation s. In this case, the standard normal critical point $z_{\alpha/2}$ is used in place of $t_{\alpha/2,n-1}$ for a two-sided confidence interval.

Two-Sided z-Interval

If an experimenter wishes to construct a confidence interval for a population mean μ based on a sample of size n with a sample mean $\bar{x}$ and using an assumed *known* value for the population standard deviation σ, then the appropriate confidence interval is

$$\mu \in \left(\bar{x} - \frac{z_{\alpha/2}\sigma}{\sqrt{n}}, \bar{x} + \frac{z_{\alpha/2}\sigma}{\sqrt{n}} \right)$$

which is known as a **two-sided z-interval** or variance known confidence interval.

If a one-sided confidence interval using a "known" value of the population standard deviation σ is required, then a critical point z_α should be used in place of $t_{\alpha,n-1}$.

One-Sided z-Interval

One-sided $1 - \alpha$ level confidence intervals for a population mean μ based on a sample of n observations with a sample mean $\bar{x}$ and using a "known" value of the population standard deviation σ are

$$\mu \in \left(-\infty, \bar{x} + \frac{z_\alpha\sigma}{\sqrt{n}} \right) \quad \text{and} \quad \mu \in \left(\bar{x} - \frac{z_\alpha\sigma}{\sqrt{n}}, \infty \right)$$

These confidence intervals are known as **one-sided z-intervals**.

When a confidence interval is required for a population mean μ, as discussed in this chapter, it is almost always the case that a t-interval should be used rather than a z-interval. A z-interval might be appropriate if an experimenter has "prior" information from previous experimentation on the population standard deviation σ and wishes to use this value in the confidence interval. However, in most cases, whereas an experimenter may have some idea as to the value of the population standard deviation σ, it is proper to estimate it with the sample standard deviation s and to construct a t-interval.

Nevertheless, for a reasonably large sample size n there is little difference between the critical points $t_{\alpha,n-1}$ and z_α, and so with $s = \sigma$ there is little difference between z-intervals and t-intervals. Consequently, it may be helpful to think of the z-intervals as *large-sample* confidence intervals and the t-intervals as *small-sample* confidence intervals.

As with t-intervals, the z-intervals require that the sample mean $\bar{x}$ is an observation from a normal distribution, and for small sample sizes with data observations that are obviously not normally distributed it is best to use the nonparametric procedures discussed in Chapter 15.

COMPUTER NOTE   Find out how to obtain confidence intervals using your computer package. Software packages usually allow you to choose whether you want a "t-interval" or a "z-interval." The intervals may be alternatively described as "variance unknown" or "variance known" intervals. If you select a z-interval, you also have to specify the assumed value of σ. You should also have the option of specifying whether you want a "two-sided" or a "one-sided" confidence interval. Finally, do not forget that you will need to specify a confidence level $1 - \alpha$.

■ **8.1.7 Problems**

8.1.1 A sample of 31 data observations has a sample mean $\bar{x} = 53.42$ and a sample standard deviation $s = 3.05$. Construct a 95% two-sided t-interval for the population mean. (This problem is continued in Problem 8.1.9.)

8.1.2 A random sample of 41 glass sheets is obtained and their thicknesses are measured. The sample mean is $\bar{x} = 3.04$ mm and the sample standard deviation is $s = 0.124$ mm. Construct a 99% two-sided t-interval for the mean glass thickness. Do you think it is plausible that the mean glass thickness is 2.90 mm? (This problem is continued in Problems 8.1.10 and 8.5.12.)

8.1.3 The breaking strengths of a random sample of 20 bundles of wool fibers have a sample mean $\bar{x} = 436.5$ and a sample standard deviation $s = 11.90$. Construct 90%, 95%, and 99% two-sided t-intervals for the average breaking strength μ. Compare the lengths of the confidence intervals. Do you think it is plausible that the average breaking strength is equal to 450.0? (This problem is continued in Problems 8.1.11 and 8.5.13.)

8.1.4 A random sample of 16 one-kilogram sugar packets is obtained and the actual weights of the packets are measured. The sample mean is $\bar{x} = 1.053$ kg and the sample standard deviation is $s = 0.058$ kg. Construct a 99% two-sided t-interval for the average sugar packet weight. Do you think it is plausible that the average weight is 1.025 kg? (This problem is continued in Problem 8.5.14.)

8.1.5 A sample of 28 data observations has a sample mean $\bar{x} = 0.0328$. If an experimenter wishes to use a "known" value $\sigma = 0.015$ for the population standard deviation, construct an appropriate 95% two-sided confidence interval for the population mean μ.

8.1.6 The resilient moduli of 10 samples of a clay mixture are measured and the sample mean is $\bar{x} = 19.50$. If an experimenter wishes to use a "known" value $\sigma = 1.0$ for the standard deviation of the resilient modulus measurements based upon prior experience, construct appropriate 90%, 95%, and 99% two-sided confidence intervals for the average resilient modulus μ. Compare the lengths of the confidence intervals. Do you think it is plausible that the average resilient modulus is equal to 20.0?

8.1.7 An experimenter feels that a population standard deviation is no larger than 10.0 and would like to construct a 95% two-sided t-interval for the population mean that has a length at most 5.0. What sample size would you recommend?

8.1.8 An experimenter would like to construct a 99% two-sided t-interval, with a length at most 0.2 ohms, for the average resistance of a segment of copper cable of a certain length. If the experimenter feels that the standard deviation of such resistances is no larger than 0.15 ohms, what sample size would you recommend?

8.1.9 Consider the sample of 31 data observations discussed in Problem 8.1.1. How many additional data observations should be obtained to construct a 95% two-sided t-interval for the population mean μ with a length no larger than $L_0 = 2.0$?

8.1.10 Consider the sample of 41 glass sheets discussed in Problem 8.1.2. How many additional glass sheets should be sampled to construct a 99% two-sided t-interval for the average sheet thickness with a length no larger than $L_0 = 0.05$ mm?

8.1.11 Consider the sample of 20 breaking strength measurements discussed in Problem 8.1.3. How many additional data observations should be obtained to construct a 99% two-sided t-interval for the average breaking strength with a length no larger than $L_0 = 10.0$?

8.1.12 A sample of 30 data observations has a sample mean $\bar{x} = 14.62$ and a sample standard deviation $s = 2.98$. Find the value of c for which $\mu \in (-\infty, c)$ is a one-sided 95% t-interval for the population mean μ. Is it plausible that $\mu \geq 16$?

8.1.13 A sample of 61 bottles of chemical solution is obtained and the solution densities are measured. The sample mean is $\bar{x} = 0.768$ and the sample standard deviation is $s = 0.0231$. Find the value of c for which $\mu \in (c, \infty)$ is a one-sided 99% t-interval for the average solution density μ. Is it plausible that the average solution density is less than 0.765?

8.1.14 A sample of 19 data observations has a sample mean of $\bar{x} = 11.80$. If an experimenter wishes to use a "known" value $\sigma = 2.0$ for the population standard deviation, find the value of c for which $\mu \in (c, \infty)$ is a one-sided 95% confidence interval for the population mean μ.

8.1.15 A sample of 29 measurements of radiation levels in a research laboratory taken at random times has a sample mean of $\bar{x} = 415.7$. If an experimenter wishes to use a "known" value $\sigma = 10.0$ for the standard deviation of these radiation levels based upon prior experience, find the value of c for which $\mu \in (-\infty, c)$ is a one-sided 99% confidence interval for the mean radiation level μ. Is it plausible that the mean radiation level is larger than 418.0?

8.1.16 The pH levels of a random sample of 16 chemical mixtures from a process were measured, and a sample mean $\bar{x} = 6.861$ and a sample standard deviation $s = 0.440$ were obtained. The scientists presented a confidence interval (6.668, 7.054) for the average pH level of chemical mixtures from the process. What is the confidence level of this confidence interval?

8.1.17 Chilled cast iron is used for mechanical components that need particularly high levels of hardness and durability. In an experiment to investigate the corrosion properties of a particular type of chilled cast iron, a collection of $n = 10$ samples of this chilled cast iron provided corrosion rates with a sample mean of $\bar{x} = 2.752$ and a sample standard deviation of $s = 0.280$. Construct a two-sided 99% confidence interval for the average corrosion rate for chilled cast iron of this type. Is 3.1 a plausible value for the average corrosion rate? (This problem is continued in Problem 8.2.15.)

For Problems 8.1.18–8.1.22 use the summary statistics that you calculated for the data sets to construct by hand confidence intervals for the appropriate population mean. Use your statistical software package to obtain confidence intervals and check your answers. Show how you would describe what you have learned from your analysis.

8.1.18 Restaurant Service Times
The data set of service times given in DS 6.1.4.

8.1.19 Telephone Switchboard Activity
The data set of calls received by a switchboard given in DS 6.1.6.

8.1.20 Paving Slab Weights
The data set of paving slab weights given in DS 6.1.7.

8.1.21 Spray Painting Procedure
The data set of paint thicknesses given in DS 6.1.8.

8.1.22 Plastic Panel Bending Capabilities
The data set of plastic panel bending capabilities given in DS 6.1.9.

8.1.23 The yields of nine batches of a chemical process were measured and a sample mean of 2.843 and a sample standard deviation of 0.150 were obtained. The experimenter presented a confidence interval of $(2.773, \infty)$ for the average yield of the process. What is the confidence level of this confidence interval?

8.1.24 Consider the data set

$$34 \quad 45 \quad 27 \quad 33 \quad 38 \quad 41 \quad 45 \quad 29 \quad 30 \quad 39$$
$$34 \quad 40 \quad 28 \quad 33 \quad 36$$

(a) What is the sample median?
(b) Construct a 99% two-sided confidence interval for the population mean.

8.1.25 A random sample of 14 chemical solutions is obtained, and their strengths are measured. The sample mean is 5437.2 and the sample standard deviation is 376.9.
(a) Construct a two-sided 95% confidence interval for the average strength.
(b) Estimate how many additional chemical solutions need to be measured in order to obtain a two-sided 95% confidence interval for the average strength with a length no larger than 300.

8.1.26 A boot manufacturer is testing the quality of leather provided by a potential supplier. The manufacturer wants to construct a two-sided confidence interval with a confidence level of 95% that has a length no larger than 0.1, and from previous experience it is believed that the variability in the leather is such that the standard deviation is no larger than 0.2031. What sample size would you recommend?

8.2 Hypothesis Testing

8.2.1 Hypotheses

So far statistical inferences about an unknown population mean μ have been based upon the calculation of a point estimate and the construction of a confidence interval. An additional methodology discussed in this section is **hypothesis testing**, which allows an experimenter to assess the plausibility or credibility of a specific statement or hypothesis.

For example, an experimenter may be interested in the plausibility of the statement $\mu = 20$, say. In other words, an experimenter may be interested in the plausibility that the population mean is equal to a specific fixed value. If this fixed value is denoted by μ_0, then the experimenter's statement may formally be described by a **null hypothesis**

$$H_0 : \mu = \mu_0$$

The word *hypothesis* indicates that this statement will be tested with an appropriate data set.

It is useful to associate a null hypothesis with an **alternative hypothesis**, which is defined to be the "opposite" of the null hypothesis. The null hypothesis above has an alternative hypothesis

$$H_A : \mu \neq \mu_0$$

This is known as a *two-sided* problem since the alternative hypothesis concerns values of μ both larger and smaller than μ_0. In a *one-sided* problem the experimenter allows the null hypotheses to be broader so as to indicate that the specified value μ_0 provides either an upper or a lower bound for the population mean μ.

Hypothesis Tests of a Population Mean

A **null hypothesis** H_0 for a population mean μ is a statement that designates possible values for the population mean. It is associated with an **alternative hypothesis** H_A, which is the "opposite" of the null hypothesis. A *two-sided* set of hypotheses is

$$H_0 : \mu = \mu_0 \quad \text{versus} \quad H_A : \mu \neq \mu_0$$

for a specified value of μ_0, and a *one-sided* set of hypotheses is either

$$H_0 : \mu \leq \mu_0 \quad \text{versus} \quad H_A : \mu > \mu_0$$

or

$$H_0 : \mu \geq \mu_0 \quad \text{versus} \quad H_A : \mu < \mu_0$$

Example 47
Graphite-Epoxy Composites

A supplier claims that its products made from a graphite-epoxy composite material have a tensile strength of 40. An experimenter may test this claim by collecting a random sample of products and measuring their tensile strengths. The experimenter is interested in testing the hypothesis

$$H_0 : \mu = 40$$

where μ is the actual mean of the tensile strengths, against the two-sided alternative hypothesis

$$H_A : \mu \neq 40$$

In this case, the null hypothesis states that the supplier's claim concerning the tensile strength is correct.

Example 14
Metal Cylinder Production

The machine that produces metal cylinders is set to make cylinders with a diameter of 50 mm. Is it calibrated correctly? Regardless of the machine setting there is always some variation in the cylinders produced, so it makes sense to conclude that the machine is calibrated correctly if the mean cylinder diameter μ is equal to the set amount. Consequently, the two-sided hypotheses of interest are

$$H_0 : \mu = 50 \quad \text{versus} \quad H_A : \mu \neq 50$$

where the null hypothesis states that the machine is calibrated correctly.

Example 48
Car Fuel Efficiency

A manufacturer claims that its cars achieve an average of at least 35 miles per gallon in highway driving. A consumer interest group tests this claim by driving a random selection of the cars in highway conditions and measuring their fuel efficiency. If μ denotes the true average miles per gallon achieved by the cars, then the consumer interest group is interested in testing the one-sided hypotheses

$$H_0 : \mu \geq 35 \quad \text{versus} \quad H_A : \mu < 35$$

For this experiment, the null hypothesis states that the manufacturer's claim regarding the fuel efficiency of its cars is correct.

Example 45
Fabric Water Absorption Properties

Suppose that a fabric is unsuitable for dyeing if its water pickup is less than 55%. Is the cotton fabric under consideration suitable for dyeing? This question can be formulated as a set of one-sided hypotheses

$$H_0 : \mu \leq 55\% \quad \text{versus} \quad H_A : \mu > 55\%$$

where μ is the mean water pickup of the cotton fabric. These hypotheses have been chosen so that the null hypothesis corresponds to the fabric being unsuitable for dyeing and the alternative hypothesis corresponds to the fabric being suitable for dyeing.

With one-sided sets of hypotheses, considerable care needs to be directed toward deciding which should be the *null* hypothesis and which should be the *alternative* hypothesis. For instance, in the fabric absorption example, why not take the null hypothesis to be that the cotton fabric is suitable for dyeing? This matter is addressed below with the discussion of *p*-values.

8.2.2 Interpretation of *p*-Values

The plausibility of a null hypothesis is measured with a ***p*-value**, which is a probability that takes a value between 0 and 1. The *p*-value is sometimes referred to as the *observed level of significance*. A *p*-value is constructed from a data set as illustrated in Figure 8.9. A useful way of interpreting a *p*-value is to consider it as the *plausibility* or credibility of the null hypothesis. The *p*-value is directly proportional to the plausibility of the null hypothesis, so that

The smaller the p-value, the less plausible is the null hypothesis.

_p_-Values

A data set can be used to measure the plausibility of a null hypothesis H_0 through the construction of a ***p*-value**. The smaller the *p*-value, the less plausible is the null hypothesis.

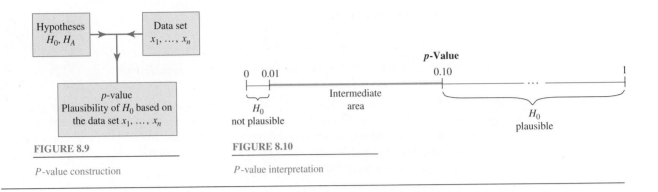

FIGURE 8.9

P-value construction

FIGURE 8.10

P-value interpretation

Figure 8.10 shows how an experimenter can interpret different levels of a *p*-value. If the *p*-value is very small, less than 1% say, then an experimenter can conclude that the null hypothesis is *not* a plausible statement. In other words, a *p*-value less than 0.01 indicates to the experimenter that the null hypothesis H_0 is not a credible statement. The experimenter can then consider the alternative hypothesis H_A to be true. In such situations, the null hypothesis is said to have been *rejected* in favor of the alternative hypothesis.

Rejection of the Null Hypothesis

A *p*-value smaller than 0.01 is generally taken to indicate that the null hypothesis H_0 is not a plausible statement. The null hypothesis H_0 can then be *rejected* in favor of the alternative hypothesis H_A.

If a *p*-value larger than 10% is obtained, then an experimenter should conclude that there is no substantial evidence that the null hypothesis is not a plausible statement. In other words, a *p*-value larger than 0.10 implies that there is no substantial evidence that the null hypothesis H_0 is false. The experimenter has learned that the null hypothesis is a credible statement based upon the fact that there is no strong "inconsistency" between the data set and the null hypothesis. In these situations, the null hypothesis is said to have been *accepted*.

It is important to realize that when a *p*-value larger than 0.10 is obtained, the experimenter should *not* conclude that the null hypothesis has been *proven*. If a null hypothesis is accepted, then this simply means that the null hypothesis is a plausible statement. However, there will be many other plausible statements and consequently many other different null hypotheses that can also be accepted. The acceptance of a null hypothesis therefore indicates that the data set does not provide enough evidence to reject the null hypothesis, but it does not indicate that the null hypothesis has been proven to be true.

Acceptance of the Null Hypothesis

A *p*-value larger than 0.10 is generally taken to indicate that the null hypothesis H_0 is a plausible statement. The null hypothesis H_0 is therefore *accepted*. However, this does not mean that the null hypothesis H_0 has been proven to be true.

A p-value in the range 1%–10% is in an intermediate area. There is some evidence that the null hypothesis is not plausible, but the evidence is not overwhelming. In a sense the experiment is inconclusive but suggests that perhaps a further look at the problem is warranted. If it is possible, the experimenter may wish to collect more information, that is, a larger data set, to help clarify the matter. Sometimes a cutoff value of 0.05 is employed (see Section 8.2.4 on *significance levels*) and the null hypothesis is accepted if the p-value is larger than 0.05 and is rejected if the p-value is smaller than 0.05.

Intermediate p-Values

A p-value in the range 1%–10% is generally taken to indicate that the data analysis is inconclusive. There is some evidence that the null hypothesis is not plausible, but the evidence is not overwhelming.

With a two-sided hypothesis testing problem

$$H_0 : \mu = \mu_0 \quad \text{versus} \quad H_A : \mu \neq \mu_0$$

rejection of the null hypothesis allows the experimenter to conclude that $\mu \neq \mu_0$. Acceptance of the null hypothesis indicates that μ_0 is a plausible value of μ, together with many other plausible values. Acceptance of the null hypothesis does not prove that μ is equal to μ_0.

With the one-sided hypothesis testing problem

$$H_0 : \mu \leq \mu_0 \quad \text{versus} \quad H_A : \mu > \mu_0$$

rejection of the null hypothesis allows the experimenter to conclude that $\mu > \mu_0$. Acceptance of the null hypothesis, however, indicates that it is plausible that $\mu \leq \mu_0$, but that this has not been proven. Consequently, it is seen that the "strongest" inference is available when the null hypothesis is rejected.

The preceding consideration is important when an experimenter decides which should be the null hypothesis and which should be the alternative hypothesis for one-sided problems. In order to "prove" or establish the statement $\mu > \mu_0$ it is necessary to take it as the *alternative* hypothesis. It can then be established by demonstrating that its opposite $\mu \leq \mu_0$ is implausible. Remember that

A null hypothesis cannot be proven to be true; it can only be shown to be implausible.

Example 47
Graphite-Epoxy Composites

For this problem, the onus is on the experimenter to *disprove* the supplier's claim that $\mu = 40$. That is why it is appropriate to take the null hypothesis as

$$H_0 : \mu = 40$$

A small p-value (less than 0.01) will demonstrate that this null hypothesis is not plausible and consequently will establish that the supplier's claim is not credible. If the p-value is not small, then the experimenter must conclude that there is not enough evidence to disprove the supplier's claim.

It may be helpful to realize that the supplier is being given the benefit of the doubt or, putting it in legal terms, the supplier is "innocent" until proven "guilty." In this sense, "guilt" (the alternative hypothesis $H_A : \mu \neq 40$) is established by showing that the supplier's "innocence" (the null hypothesis) is implausible. If "innocence" is plausible (a large p-value), then the null hypothesis is accepted and the supplier is acquitted. The important point is that the acquittal is as a result of the failure to prove guilt, and not as a result of a proof of innocence.

<table>
<tr><td>

Example 14

Metal Cylinder Production

</td><td>

For this problem, the question is whether the machine can be shown to be calibrated incorrectly. It is therefore appropriate to take the alternative hypothesis to be

$$H_A : \mu \neq 50$$

which corresponds to a miscalibration. With a small p-value, the null hypothesis $H_0 : \mu = 50$ is rejected and the machine is demonstrated to be miscalibrated. With a large p-value, the null hypothesis is accepted and the experimenter concludes that there is no evidence that the machine is calibrated incorrectly.

</td></tr>
</table>

<table>
<tr><td>

Example 48

Car Fuel Efficiency

</td><td>

The onus is on the consumer interest group to demonstrate that the car manufacturer's claim is incorrect. The manufacturer's claim is incorrect if $\mu < 35$, and so this should be taken as the alternative hypothesis. The one-sided hypotheses that should be tested are therefore

$$H_0 : \mu \geq 35 \quad \text{versus} \quad H_A : \mu < 35$$

If a small p-value is obtained, the null hypothesis is rejected and the consumer interest group has demonstrated that the manufacturer's claim is incorrect. A large p-value indicates that there is insufficient evidence to establish that the manufacturer's claim is incorrect.

</td></tr>
</table>

<table>
<tr><td>

Example 45

Fabric Water Absorption Properties

</td><td>

How can the experimenter establish that the cotton fabric is suitable for dyeing? In other words, how can the experimenter establish that $\mu > 55\%$? With this question in mind, it is appropriate to use the one-sided hypotheses

$$H_0 : \mu \leq 55\% \quad \text{versus} \quad H_A : \mu > 55\%$$

If a small p-value is obtained, the null hypothesis is rejected and the cotton fabric is demonstrated to be fit for dyeing. A large p-value indicates that it is plausible that the cotton fabric is unfit for dyeing.

</td></tr>
</table>

8.2.3 Calculation of p-Values

A p-value for a particular null hypothesis based on an observed data set is defined in the following way:

> *The p-value is the probability of obtaining this data set or worse when the null hypothesis is true.*

There are two important components of this definition:

> *(i) this data set or worse*

and

> *(ii) when the null hypothesis is true.*

In the first component, *worse* is interpreted as meaning to have less affinity with the null hypothesis. In other words, a "worse" data set is one for which the null hypothesis is *less* plausible than it is for the actual observed data set. The second component of the definition indicates that the probability calculation is made under the assumption that the null hypothesis is true, which in practice means calculating a probability under the assumption that $\mu = \mu_0$.

Definition of a p-Value

A p-value for a particular null hypothesis H_0 based on an observed data set is defined to be "the probability of obtaining the data set or worse when the null hypothesis is true." A "worse" data set is one that has less affinity with the null hypothesis.

This definition of a p-value explains the interpretation of p-values discussed in the previous section. A p-value smaller than 0.01 reveals that if the null hypothesis H_0 is true, then the chance of observing the kind of data observed (or "worse") is less than 1 in 100. If the null hypothesis is true, then it is unlikely that the experimenter obtains the kind of data set that has been obtained. It is this argument that leads the experimenter to conclude that the null hypothesis is implausible.

On the other hand, a p-value larger than 0.10 reveals that if the null hypothesis H_0 is true, then the chance of observing the kind of data observed is at least 1 in 10. In other words, if the null hypothesis is true, then it is not at all unlikely that the experimenter obtains the kind of data set that has been obtained. Consequently, the null hypothesis is a plausible statement and should be accepted.

Two-Sided Problems Consider the two-sided hypothesis testing problem

$$H_0 : \mu = \mu_0 \quad \text{versus} \quad H_A : \mu \neq \mu_0$$

Suppose that a data set of n observations is obtained, and the observed sample mean and standard deviation are $\bar{x}$ and s, respectively. The "discrepancy" between the data set and the null hypothesis is measured through a **t-statistic**

$$t = \frac{\sqrt{n}(\bar{x} - \mu_0)}{s}$$

The discrepancy is smallest when $\bar{x} = \mu_0$, which gives $t = 0$, because this indicates that the sample mean coincides exactly with the hypothesized value μ_0 of the population mean. The discrepancy between the data set and the null hypothesis increases as the absolute value of the t-statistic $|t|$ increases, as illustrated in Figure 8.11. Consequently, a data set is considered to be "worse" (to have less affinity with the null hypothesis) than the observed data set if it has a t-statistic with an absolute value larger than $|t|$, as shown in Figure 8.11.

The p-value is therefore calculated as the probability that a data set generated with $\mu = \mu_0$ (that is, under the null hypothesis H_0) has a t-statistic with an absolute value larger than $|t|$. However, if $\bar{x}$ and s are the sample mean and sample standard deviation of a data set generated with $\mu = \mu_0$, the t-statistic

$$\frac{\sqrt{n}(\bar{x} - \mu_0)}{s}$$

is known to be an observation from a t-distribution with $n - 1$ degrees of freedom. Therefore, the p-value is

$$p\text{-value} = P(X \geq |t|) + P(X \leq -|t|)$$

FIGURE 8.11

Measuring discrepancy between a
data set and H_0 for a two-sided
problem

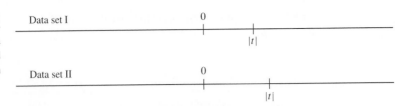

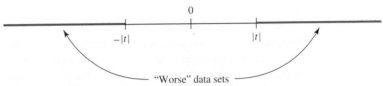

There is more discrepancy between data set II and H_0 than between data set I and H_0.

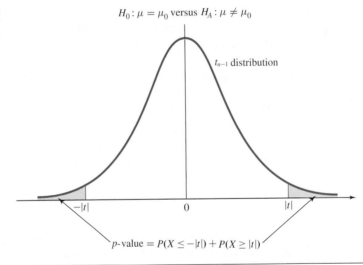

"Worse" data sets
than data set II have values of the t-statistic in these regions.

FIGURE 8.12

P-value for two-sided t-test

$H_0 : \mu = \mu_0$ versus $H_A : \mu \neq \mu_0$

t_{n-1} distribution

$p\text{-value} = P(X \leq -|t|) + P(X \geq |t|)$

where the random variable X has a t-distribution with $n - 1$ degrees of freedom, as illustrated in Figure 8.12. However, the symmetry of the t-distribution ensures that

$$P(X \geq |t|) = P(X \leq -|t|)$$

and so the p-value may be calculated as

$$p\text{-value} = 2 \times P(X \geq |t|)$$

as illustrated in Figure 8.13. This testing procedure is known as a **two-sided t-test**.

FIGURE 8.13

P-value for two-sided *t*-test

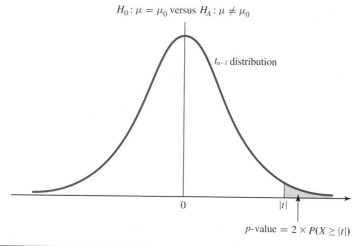

$H_0 : \mu = \mu_0$ versus $H_A : \mu \neq \mu_0$

t_{n-1} distribution

0

$|t|$

p-value $= 2 \times P(X \geq |t|)$

Two-Sided *t*-Test

The *p*-value for the two-sided hypothesis testing problem

$$H_0 : \mu = \mu_0 \quad \text{versus} \quad H_A : \mu \neq \mu_0$$

based on a data set of n observations with a sample mean $\bar{x}$ and a sample standard deviation s, is

$$p\text{-value} = 2 \times P(X \geq |t|)$$

where the random variable X has a *t*-distribution with $n - 1$ degrees of freedom, and

$$t = \frac{\sqrt{n}(\bar{x} - \mu_0)}{s}$$

which is known as the ***t*-statistic**. This testing procedure is called a **two-sided *t*-test**.

As an illustration of the calculation of a *p*-value for a two-sided hypothesis testing problem, consider the hypotheses

$$H_0 : \mu = 10.0 \quad \text{versus} \quad H_A : \mu \neq 10.0$$

Suppose that a data set is obtained with $n = 15$, $\bar{x} = 10.6$, and $s = 1.61$. The *t*-statistic is

$$t = \frac{\sqrt{n}(\bar{x} - \mu_0)}{s} = \frac{\sqrt{15}(10.6 - 10.0)}{1.61} = 1.44$$

Therefore any data set with a *t*-statistic larger than 1.44 or smaller than -1.44 is "worse" than the observed data set. The *p*-value is

$$p\text{-value} = 2 \times P(X \geq 1.44)$$

where the random variable X has a *t*-distribution with $n - 1 = 14$ degrees of freedom. A computer package can be used to show that this value is

$$p\text{-value} = 2 \times 0.086 = 0.172$$

as illustrated in Figure 8.14.

FIGURE 8.14

Two-sided p-value calculation

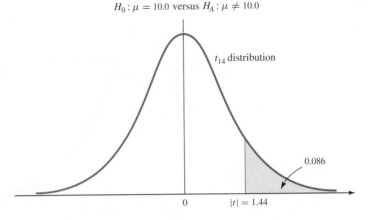

$$p\text{-value} = 2 \times P(X \geq 1.44) = 2 \times 0.086 = 0.172$$

FIGURE 8.15

Two-sided p-value calculation

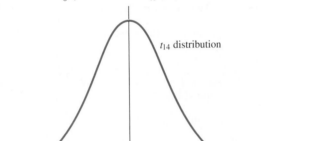

$$p\text{-value} = 2 \times P(X \geq 3.61) = 2 \times 0.0014 = 0.0028$$

This large p-value (greater than 0.10) indicates that the null hypothesis should be accepted. There is not enough evidence to conclude that the null hypothesis is implausible. In other words, based upon the data set observed, it is plausible that $\mu = 10.0$. More specifically, if $\mu = 10.0$, there is a probability of over 17% of observing a data set with a t-statistic larger than 1.44 or smaller than -1.44, and so this data set does not cast any doubt on the plausibility of the null hypothesis.

Suppose instead that the data set has $\bar{x} = 11.5$. In this case the t-statistic is

$$t = \frac{\sqrt{15}(11.5 - 10.0)}{1.61} = 3.61$$

and the p-value is

$$p\text{-value} = 2 \times P(X \geq 3.61) = 2 \times 0.0014 = 0.0028$$

as illustrated in Figure 8.15. Since this p-value is smaller than 0.01, the experimenter now concludes that the null hypothesis is not a credible statement. In this case, the data set provides

FIGURE 8.16

P-value calculation for graphite-epoxy composites

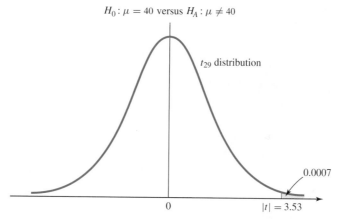

$$H_0 : \mu = 40 \text{ versus } H_A : \mu \neq 40$$

t_{29} distribution

0.0007

0

$|t| = 3.53$

$p\text{-value} = 2 \times P(X \geq 3.53) = 2 \times 0.0007 = 0.0014$

enough evidence to conclude that the population mean μ cannot be equal to 10.0, because if the population mean were equal to 10.0, the probability of getting these data or "worse" is only 0.0028.

Example 47
Graphite-Epoxy Composites

When the tensile strengths of 30 randomly selected products are measured, a sample mean of $\bar{x} = 38.518$ and a sample standard deviation of $s = 2.299$ are obtained. Since $\mu_0 = 40.0$, the *t*-statistic is

$$t = \frac{\sqrt{30}(38.518 - 40.0)}{2.299} = -3.53$$

Since this is a two-sided problem, the *p*-value is

$$p\text{-value} = 2 \times P(X \geq |-3.53|) = 2 \times P(X \geq 3.53)$$

where X has a *t*-distribution with $n - 1 = 29$ degrees of freedom. This can be shown to be

$$p\text{-value} = 2 \times 0.0007 = 0.0014$$

as illustrated in Figure 8.16.

Since the *p*-value is so small, the null hypothesis can be rejected and there is sufficient evidence to conclude that the mean tensile strength cannot be equal to the claimed value of 40. In fact, since $\bar{x} = 38.518 < \mu_0 = 40.0$, it is clear that the actual mean tensile strength is smaller than the claimed value.

Example 14
Metal Cylinder Production

The data set of metal cylinder diameters has $n = 60$, $\bar{x} = 49.99856$, and $s = 0.1334$, so that with $\mu_0 = 50.0$ the *t*-statistic is

$$t = \frac{\sqrt{60}(49.99856 - 50.0)}{0.1334} = -0.0836$$

Since this is a two-sided problem, the *p*-value is

$$p\text{-value} = 2 \times P(X \geq 0.0836)$$

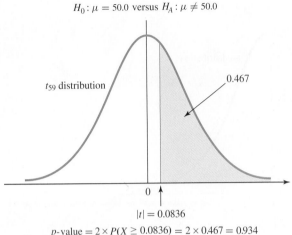

$H_0 : \mu = 50.0$ versus $H_A : \mu \neq 50.0$

t_{59} distribution

0.467

0

$|t| = 0.0836$

p-value $= 2 \times P(X \geq 0.0836) = 2 \times 0.467 = 0.934$

FIGURE 8.17

P-value calculation for metal cylinder diameters

$H_0 : \mu \leq \mu_0$ versus $H_A : \mu > \mu_0$

$t = \dfrac{\sqrt{n}(\bar{x} - \mu_0)}{s}$

"Worse" data sets

FIGURE 8.18

Worse data sets for one-sided problems

where X has a t-distribution with $n - 1 = 59$ degrees of freedom, which can be shown to be

$$p\text{-value} = 2 \times 0.467 = 0.934$$

as illustrated in Figure 8.17. With such a large p-value the null hypothesis is accepted and the experimenter can conclude that there is not sufficient evidence to establish that the machine that produces the metal cylinders is calibrated incorrectly.

One-Sided Problems   The calculation of p-values for one-sided hypothesis testing problems involves defining "worse" data sets in one direction rather than in two directions. For example, with the t-statistic

$$t = \frac{\sqrt{n}(\bar{x} - \mu_0)}{s}$$

and the one-sided hypotheses

$$H_0 : \mu \leq \mu_0 \quad \text{versus} \quad H_A : \mu > \mu_0$$

"worse" data sets are those that have a t-statistic *greater* than t, as illustrated in Figure 8.18. This is because for this one-sided problem, the discrepancy between the data set and the null hypothesis is measured by how much *larger* the sample mean $\bar{x}$ is than μ_0. Figure 8.19 shows that in this case the p-value is calculated as

$$p\text{-value} = P(X \geq t)$$

where again, the random variable X has a t-distribution with $n - 1$ degrees of freedom. This inference method is known as a **one-sided t-test**.

Notice that if $\bar{x} \leq \mu_0$, then $t \leq 0$ and the p-value is larger than 0.5. The null hypothesis is therefore accepted, which is clearly the right decision since the sample mean actually takes a value that is consistent with the null hypothesis. This situation is illustrated in Figure 8.20. There really isn't any point in calculating a p-value in this case because the null hypothesis

FIGURE 8.19

P-value calculation for a
one-sided problem

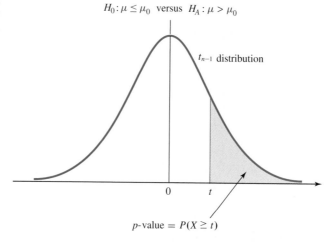

FIGURE 8.20

P-value larger than 0.50 for a
one-sided t-test

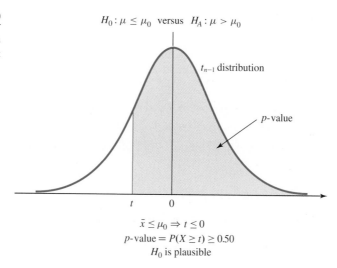

obviously cannot be shown to be an implausible statement. However, if $\bar{x} > \mu_0$, the calculation
of a p-value is useful because it indicates whether $\bar{x}$ is close enough to μ_0 for the null
hypothesis to be considered a plausible statement or whether $\bar{x}$ is so far away from μ_0 that
the null hypothesis is not credible.

For the one-sided hypotheses

$$H_0 : \mu \geq \mu_0 \quad \text{versus} \quad H_A : \mu < \mu_0$$

worse data sets are those that have a t-statistic *smaller* than t, as illustrated in Figure 8.21. In
this case the p-value is calculated as

$$p\text{-value} = P(X \leq t)$$

If $\bar{x} \geq \mu_0$, then the null hypothesis is clearly a plausible statement and the p-value calculation
(which will result in a p-value of at least 0.5) is really not necessary.

FIGURE 8.21

P-value calculation for a one-sided
problem

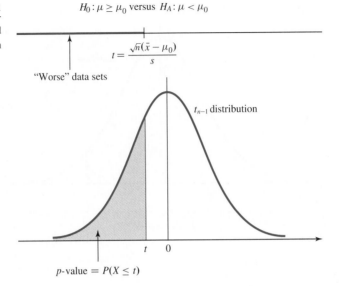

$$H_0 : \mu \geq \mu_0 \text{ versus } H_A : \mu < \mu_0$$

$$t = \frac{\sqrt{n}(\bar{x} - \mu_0)}{s}$$

"Worse" data sets

t_{n-1} distribution

t 0

p-value $= P(X \leq t)$

One-Sided *t*-Test

Based upon a data set of n observations with a sample mean $\bar{x}$ and a sample standard deviation s, the p-value for the one-sided hypothesis testing problem,

$$H_0 : \mu \leq \mu_0 \quad \text{versus} \quad H_A : \mu > \mu_0$$

is

$$p\text{-value} = P(X \geq t)$$

and the p-value for the one-sided hypothesis testing problem

$$H_0 : \mu \geq \mu_0 \quad \text{versus} \quad H_A : \mu < \mu_0$$

is

$$p\text{-value} = P(X \leq t)$$

where the random variable X has a t-distribution with $n - 1$ degrees of freedom, and

$$t = \frac{\sqrt{n}(\bar{x} - \mu_0)}{s}$$

These testing procedures are called **one-sided *t*-tests**.

As an illustration of p-value calculations for one-sided problems, consider the one-sided hypotheses

$$H_0 : \mu \leq 125.0 \quad \text{versus} \quad H_A : \mu > 125.0$$

Suppose that a sample mean of $\bar{x} = 122.3$ is observed, as illustrated in Figure 8.22. What is the p-value? Since the sample mean takes a value that corresponds to a population mean

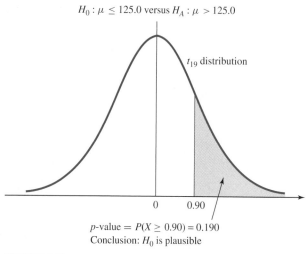

$H_0 : \mu \leq 125.0$ versus $H_A : \mu > 125.0$

t_{19} distribution

0   0.90

p-value $= P(X \geq 0.90) = 0.190$
Conclusion: H_0 is plausible

FIGURE 8.23

P-value calculation for a one-sided *t*-test

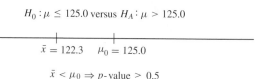

$H_0 : \mu \leq 125.0$ versus $H_A : \mu > 125.0$

$\bar{x} = 122.3$ $\mu_0 = 125.0$

$\bar{x} < \mu_0 \Rightarrow p\text{-value} > 0.5$
Conclusion: H_0 is plausible

FIGURE 8.22

P-value larger than 0.50 for a one-sided *t*-test

μ contained within the null hypothesis, the *p*-value is immediately known to be at least 0.5, and its exact value is immaterial. The data obviously do not indicate that the null hypothesis is implausible, and the null hypothesis should be accepted.

Suppose instead that a sample mean of $\bar{x} = 128.4$ is observed, with $n = 20$ and $s = 16.9$. Since $\bar{x} = 128.4 > \mu_0 = 125.0$, the data *suggest* that the null hypothesis is false. How plausible is the null hypothesis? The *t*-statistic is

$$t = \frac{\sqrt{20}(128.4 - 125.0)}{16.9} = 0.90$$

so that, as illustrated in Figure 8.23, the *p*-value is

$$p\text{-value} = P(X \geq 0.90)$$

where the random variable X has a *t*-distribution with $n - 1 = 19$ degrees of freedom. A computer package can be used to show that this is

$$p\text{-value} = 0.190$$

so that the null hypothesis is accepted and the experimenter concludes that the data set does not provide sufficient evidence to establish that the population mean is larger than $\mu_0 = 125.0$.

However, if a sample mean of $\bar{x} = 137.8$ is observed instead, then the *t*-statistic is

$$t = \frac{\sqrt{20}(137.8 - 125.0)}{16.9} = 3.39$$

and the *p*-value is

$$p\text{-value} = P(X \geq 3.39) = 0.0015$$

as illustrated in Figure 8.24. In this case the null hypothesis is rejected and the experimenter has established that the population mean is larger than $\mu_0 = 125.0$.

$H_0 : \mu \leq 125.0$ versus $H_A : \mu > 125.0$

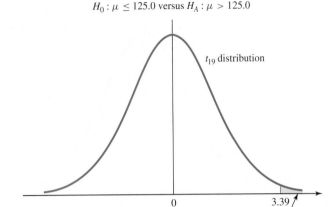

t_{19} distribution

0   3.39

p-value $= P(X \geq 3.39) = 0.0015$
Conclusion: H_0 is not plausible

FIGURE 8.24

P-value calculation for a one-sided *t*-test

$H_0 : \mu \geq 35.0$ versus $H_A : \mu < 35.0$

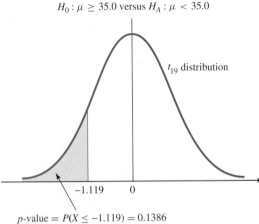

t_{19} distribution

−1.119   0

p-value $= P(X \leq -1.119) = 0.1386$
Conclusion: H_0 is plausible

FIGURE 8.25

P-value calculation for car fuel efficiency

Example 48
Car Fuel Efficiency

A sample of $n = 20$ cars driven under varying highway conditions achieved fuel efficiencies with a sample mean of $\bar{x} = 34.271$ miles per gallon and a sample standard deviation of $s = 2.915$ miles per gallon. With $\mu_0 = 35.0$ the *t*-statistic is therefore

$$t = \frac{\sqrt{20}(34.271 - 35.0)}{2.915} = -1.119$$

The alternative hypothesis is $H_A : \mu < 35$, so that the *p*-value is

$$p\text{-value} = P(X \leq -1.119)$$

where X has a *t*-distribution with $n - 1 = 19$ degrees of freedom. This value can be shown to be

$$p\text{-value} = 0.1386$$

as illustrated in Figure 8.25. This *p*-value is larger than 0.10 and so the null hypothesis $H_0 : \mu \geq 35.0$ should be accepted. Even though $\bar{x} = 34.271 < \mu_0 = 35.0$ this data set does *not* provide sufficient evidence for the consumer interest group to conclude that the average miles per gallon achieved in highway driving is any less than 35.

Example 45
Fabric Water
Absorption Properties

The data set of % pickup values has $n = 15$, $\bar{x} = 59.81\%$, and $s = 4.94\%$. With $\mu_0 = 55\%$ the *t*-statistic is

$$t = \frac{\sqrt{15}(59.81 - 55.0)}{4.94} = 3.77$$

The alternative hypothesis is $H_A : \mu > 55\%$, and so the *p*-value is calculated as

$$p\text{-value} = P(X \geq 3.77)$$

where X has a *t*-distribution with $n - 1 = 14$ degrees of freedom, which can be shown to be

$$p\text{-value} = 0.0010$$

as illustrated in Figure 8.26. This small p-value indicates that the null hypothesis can be rejected and that there is sufficient evidence to conclude that $\mu > 55\%$. Therefore the cotton fabric under consideration has been shown to be suitable for dyeing.

FIGURE 8.26

P-value calculation for fabric water absorption data set

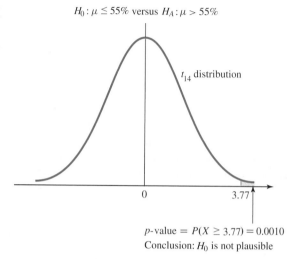

$H_0 : \mu \leq 55\%$ versus $H_A : \mu > 55\%$

t_{14} distribution

0   3.77

p-value $= P(X \geq 3.77) = 0.0010$
Conclusion: H_0 is not plausible

Example 49

Sand Blast Paint Removal

Sand blasting can be a convenient way for removing paint from items without damaging their surfaces. The efficiency of the procedure depends on various factors such as the particle size of the sand or medium that is used, the blasting pressure, the distance of the blaster from the item, and the blasting angle. The data set shown in Figure 8.27 is the times in minutes taken to

FIGURE 8.27

Illustration of the stages of a hypothesis test for the sand blast paint removal example

Data and Question   Data set of blast times in minutes: 10.3 9.3 11.2 8.8 9.5 9.0   Question: What evidence is there that the average blast time is less than 10 minutes?
Stage I: Data Summary   Sample average $n = 6$, sample mean $\bar{x} = 9.683$, sample standard deviation $s = 0.906$.
Stage II: Determination of Suitable Hypotheses   Since the objective is to assess whether there is sufficient evidence to conclude that $\mu < 10$, this should be the alternative hypothesis. $H_0 : \mu \geq 10$ versus $H_A : \mu < 10$.
Stage III: Calculation of the Test Statistic $t = \frac{\sqrt{n}(\bar{x}-\mu_0)}{s} = \frac{\sqrt{6}(9.683-10.000)}{0.906} = -0.857$
Stage IV: Expression for the p-value p-value $= P(X \leq -0.857)$ where the random variable X has a t-distribution with $n - 1 = 5$ degrees of freedom.
Stage V: Evaluation of the p-value   Table III gives $t_{0.10,5} = 1.476$, and consequently it is known that the p-value is larger than 0.10. Alternatively, exact computer calculation gives the p-value as 0.216.
Stage VI: Decision   Since the p-value is larger than 0.10, the null hypothesis is accepted.
Stage VII: Conclusion   This data set does not provide sufficient evidence to establish that the average blast time is less than 10 minutes.

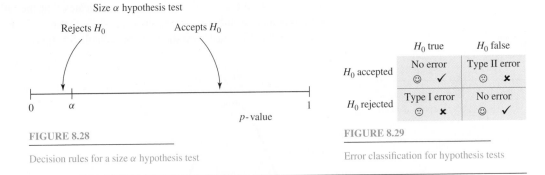

FIGURE 8.28

Decision rules for a size α hypothesis test

FIGURE 8.29

Error classification for hypothesis tests

remove paint from a sample of items with equivalent paint thicknesses for a certain blasting method. It was of interest to assess the evidence that the average blast time was less than 10 minutes, and each stage in the hypothesis test is identified to clarify the process.

8.2.4   Significance Levels

Hypothesis tests may be defined formally in terms of a **significance level** or **size** α. As Figure 8.28 shows, a hypothesis test at size α *rejects* the null hypothesis H_0 if a p-value *smaller* than α is obtained and *accepts* the null hypothesis H_0 if a p-value *larger* than α is obtained.

The significance level α is also referred to as the probability of a **Type I error**. As Figure 8.29 shows, a Type I error occurs when the null hypothesis is rejected when it is really true. A Type II error occurs when the null hypothesis is accepted when it is really false. Having a small probability of a Type I error, that is, having a small significance level α, is consistent with the "protection" of the null hypothesis discussed in Section 8.2.2. This means that the null hypothesis is rejected only when there is sufficient evidence that it is false.

It is common to use significance levels of $\alpha = 0.10$, $\alpha = 0.05$, and $\alpha = 0.01$, which tie in with the p-value interpretations given in Section 8.2.2. A p-value larger than 0.10 implies that hypothesis tests with $\alpha = 0.10$, $\alpha = 0.05$, and $\alpha = 0.01$ all accept the null hypothesis. Similarly, a p-value smaller than 0.01 implies that hypothesis tests with $\alpha = 0.10$, $\alpha = 0.05$, and $\alpha = 0.01$ all reject the null hypothesis.

A p-value in the range 0.01 to 0.10 is in the intermediate area. A hypothesis test with size $\alpha = 0.10$ rejects the null hypothesis, whereas a hypothesis test with size $\alpha = 0.01$ accepts the null hypothesis. A hypothesis test with size $\alpha = 0.05$ may accept or reject the null hypothesis depending on whether the p-value is larger or smaller than 0.05.

Significance Level of a Hypothesis Test

A hypothesis test with a **significance level** or **size** α *rejects* the null hypothesis H_0 if a p-value *smaller* than α is obtained and *accepts* the null hypothesis H_0 if a p-value *larger* than α is obtained. In this case, the probability of a **Type I error**, that is, the probability of rejecting the null hypothesis when it is true, is no larger than α.

An important point to remember is that

p-values are more informative than knowing whether a size α hypothesis test accepts or rejects the null hypothesis.

This is because if the p-value is known, then the outcome of a hypothesis test at any significance level α can be deduced by comparing α with the p-value. However, the acceptance or rejection of a size α hypothesis test provides only a lower bound (p-value $\geq \alpha$) or an upper bound (p-value $< \alpha$) on the p-value. Nevertheless, it will be seen that hypothesis tests at a fixed size level are easy to perform by hand because the test statistics need only be compared with a tabulated critical point, whereas a p-value calculation requires the determination of the cumulative distribution function of the appropriate t-distribution. Of course, computer packages generally indicate the exact p-value.

Two-Sided Problems   For the t-statistic

$$t = \frac{\sqrt{n}(\bar{x} - \mu_0)}{s}$$

the decision as to whether a size α two-sided hypothesis test rejects or accepts can be made by determining whether the **test statistic $|t|$** falls in the

>   **rejection region** $|t| > t_{\alpha/2, n-1}$

or in the

>   **acceptance region** $|t| \leq t_{\alpha/2, n-1}$

as illustrated in Figure 8.30.

This is because the p-value for a two-sided problem is

$$2 \times P(X \geq |t|)$$

where the random variable X has a t-distribution with $n - 1$ degrees of freedom. Since the critical point $t_{\alpha/2, n-1}$ has the property that

$$P(X \geq t_{\alpha/2, n-1}) = \frac{\alpha}{2}$$

it is clear that the p-value is greater than α if $|t| \leq t_{\alpha/2, n-1}$ and is smaller than α if $|t| > t_{\alpha/2, n-1}$. In other words, comparing the test statistic $|t|$ with the critical point $t_{\alpha/2, n-1}$ indicates whether the p-value is smaller or greater than α.

FIGURE 8.30

Size α two-sided t-test

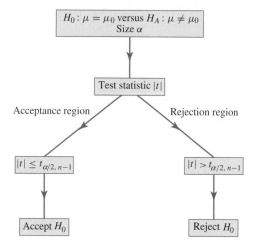

Two-Sided Hypothesis Test for a Population Mean

A size α test for the two-sided hypotheses

$$H_0 : \mu = \mu_0 \quad \text{versus} \quad H_A : \mu \neq \mu_0$$

rejects the null hypothesis H_0 if the **test statistic** $|t|$ falls in the **rejection region**

$$|t| > t_{\alpha/2,n-1}$$

and accepts the null hypothesis H_0 if the test statistic $|t|$ falls in the **acceptance region**

$$|t| \leq t_{\alpha/2,n-1}$$

As an example of a two-sided hypothesis testing problem with fixed significance levels, suppose that a sample of $n = 18$ observations is obtained. Then Table III provides the critical points

$$t_{0.05,17} = 1.740$$

for $\alpha = 0.10$,

$$t_{0.025,17} = 2.110$$

for $\alpha = 0.05$, and

$$t_{0.005,17} = 2.898$$

for $\alpha = 0.01$, which are illustrated in Figure 8.31. If the test statistic is $|t| = 3.24$, then hypothesis tests with $\alpha = 0.10$, $\alpha = 0.05$, and $\alpha = 0.01$ all reject the null hypothesis, since the test statistic is larger than the respective critical points. The actual p-value is smaller than 0.01.

If the test statistic is $|t| = 1.625$, then hypothesis tests with $\alpha = 0.10$, $\alpha = 0.05$, and $\alpha = 0.01$ all accept the null hypothesis, because the test statistic is smaller than the respective critical points. The actual p-value is larger than 0.10. A test statistic in the region $1.740 \leq |t| \leq 2.898$ falls in the intermediate region with a p-value between 0.01 and 0.10. For example, if $|t| = 2.021$, then a hypothesis test with $\alpha = 0.10$ rejects the null hypothesis, whereas a hypothesis test with $\alpha = 0.05$ or $\alpha = 0.01$ accepts the null hypothesis. In this case

FIGURE 8.31

Hypothesis tests at fixed significance levels

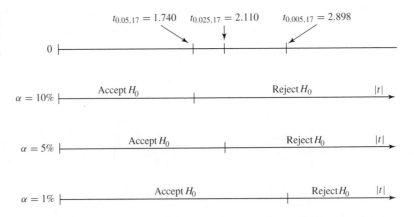

it is sensible to summarize the situation by concluding that there is some evidence that the null hypothesis is false, but that the evidence is not overwhelming.

Example 47
Graphite-Epoxy
Composites

For this problem the test statistic is

$$|t| = 3.53$$

It can be seen from Table III that the critical points are $t_{0.05,29} = 1.699$ for a size $\alpha = 0.10$ hypothesis test, $t_{0.025,29} = 2.045$ for a size $\alpha = 0.05$ hypothesis test, and $t_{0.005,29} = 2.756$ for a size $\alpha = 0.01$ hypothesis test. The test statistic exceeds each of these critical points, and so the hypothesis tests all reject the null hypothesis. This indicates that the p-value is smaller than 0.01, which is consistent with the previous analysis, which found the p-value to be 0.0014.

Example 14
Metal Cylinder
Production

The data set of metal cylinder diameters gives a test statistic of

$$|t| = 0.0836$$

The critical points are $t_{0.05,59} = 1.671$ for a size $\alpha = 0.10$ hypothesis test, $t_{0.025,59} = 2.001$ for a size $\alpha = 0.05$ hypothesis test, and $t_{0.005,59} = 2.662$ for a size $\alpha = 0.01$ hypothesis test. The test statistic is smaller than each of these critical points, and so the hypothesis tests all accept the null hypothesis. The p-value is therefore known to be larger than 0.10, and in fact the previous analysis found the p-value to be 0.934.

Example 50
Engine Oil Viscosity

An engine oil is supposed to have a mean viscosity of $\mu_0 = 85.0$. A sample of $n = 25$ viscosity measurements resulted in a sample mean of $\bar{x} = 88.3$ and a sample standard deviation of $s = 7.49$. What is the evidence that the mean viscosity is not as stated?

It is appropriate to test the two-sided set of hypotheses

$$H_0 : \mu = 85.0 \quad \text{versus} \quad H_A : \mu \neq 85.0$$

With $\mu_0 = 85.0$ the t-statistic is

$$t = \frac{\sqrt{25}(88.3 - 85.0)}{7.49} = 2.203$$

It can be seen from Table III that the critical points are $t_{0.05,24} = 1.711$ for a size $\alpha = 0.10$ hypothesis test, $t_{0.025,24} = 2.064$ for a size $\alpha = 0.05$ hypothesis test, and $t_{0.005,24} = 2.797$ for a size $\alpha = 0.01$ hypothesis test. The test statistic $|t| = 2.203$ exceeds the first two of these values but not the third, so that the null hypothesis is rejected with $\alpha = 0.10$ and $\alpha = 0.05$ but not with $\alpha = 0.01$.

This result indicates that the p-value lies somewhere between 0.01 and 0.05, and in fact it can be calculated to be

$$p\text{-value} = 2 \times P(X \geq 2.203) = 2 \times 0.0187 = 0.0374$$

where X has t-distribution with 24 degrees of freedom. In summary, there is some evidence that the mean viscosity is not equal to 85.0, but the evidence is not overwhelming. The experimenter may wish to obtain a larger sample size to clarify the matter.

There is an important connection between confidence intervals and hypothesis testing that provides additional insights into the interpretation of a confidence interval. A two-sided confidence interval for μ with a confidence level of $1 - \alpha$ actually consists of the values μ_0

FIGURE 8.32

Relationship between hypothesis
testing and confidence intervals for
two-sided problems

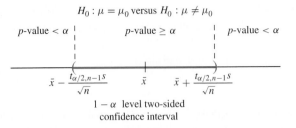

for which the hypothesis testing problem

$$H_0 : \mu = \mu_0 \quad \text{versus} \quad H_A : \mu \neq \mu_0$$

with size α accepts the null hypothesis. In other words, the value μ_0 is contained within a $1 - \alpha$ level two-sided confidence interval for μ if the p-value for this two-sided hypothesis test is larger than α. Thus a confidence interval for μ with a confidence level of $1 - \alpha$ consists of "plausible" values for μ, where plausibility is defined in terms of having a p-value larger than α. This relationship between two-sided confidence intervals and hypothesis tests is illustrated in Figure 8.32.

Relationship Between Confidence Intervals and Hypothesis Tests

The value μ_0 is contained within a $1 - \alpha$ level two-sided confidence interval

$$\left(\bar{x} - \frac{t_{\alpha/2, n-1} s}{\sqrt{n}}, \bar{x} + \frac{t_{\alpha/2, n-1} s}{\sqrt{n}} \right)$$

if the p-value for the two-sided hypothesis test

$$H_0 : \mu = \mu_0 \quad \text{versus} \quad H_A : \mu \neq \mu_0$$

is larger than α. Therefore if μ_0 is contained within the $1 - \alpha$ level confidence interval, the hypothesis test with size α accepts the null hypothesis, and if μ_0 is not contained within the $1 - \alpha$ level confidence interval, the hypothesis test with size α rejects the null hypothesis.

It is useful to remember that

Constructing a confidence interval with a confidence level of $1 - \alpha$ for μ is more informative than performing a size α hypothesis test.

This is because the decision made by a size α hypothesis test of $H_0 : \mu = \mu_0$ can be deduced from a $1 - \alpha$ level confidence interval for μ by noticing whether or not μ_0 is inside the confidence interval. Thus the confidence interval portrays the decisions made by hypothesis tests for all possible values of μ_0.

It is important to have a clear understanding of the relationships between *confidence intervals*, *p-values*, and hypothesis tests at fixed *significance levels*. Confidence intervals and p-values for specific hypotheses of interest generally provide the most useful statistical inferences.

<table>
<tr><td>**Example 47**</td></tr>
<tr><td>**Graphite-Epoxy**
Composites</td></tr>
</table>

With $t_{0.005,29} = 2.756$, a 99% two-sided t-interval for the mean tensile stength is

$$\left(\bar{x} - \frac{t_{\alpha/2,n-1}s}{\sqrt{n}}, \ \bar{x} + \frac{t_{\alpha/2,n-1}s}{\sqrt{n}} \right)$$

$$= \left(38.518 - \frac{2.756 \times 2.299}{\sqrt{30}}, \ 38.518 + \frac{2.756 \times 2.299}{\sqrt{30}} \right) = (37.36, \ 39.67)$$

Notice that the value $\mu_0 = 40.0$ is not contained within this confidence interval, which is consistent with the hypothesis testing problem

$$H_0 : \mu = 40.0 \quad \text{versus} \quad H_A : \mu \neq 40.0$$

having a p-value of 0.0014, so that the null hypothesis is rejected at size $\alpha = 0.01$.
   In fact, the 99% confidence interval implies that the hypothesis testing problem

$$H_0 : \mu = \mu_0 \quad \text{versus} \quad H_A : \mu \neq \mu_0$$

has a p-value larger than 0.01 for $37.36 \leq \mu_0 \leq 39.67$ and a p-value smaller than 0.01 otherwise.

Example 14
Metal Cylinder
Production

In Section 8.1.1 a 90% two-sided t-interval for the mean cylinder diameter was found to be

$$(49.970, 50.028)$$

This contains the value $\mu_0 = 50.0$ and so is consistent with the hypothesis testing problem

$$H_0 : \mu = 50.0 \quad \text{versus} \quad H_A : \mu \neq 50.0$$

having a p-value of 0.934, so that the null hypothesis is accepted at size $\alpha = 0.10$.
   Moreover, the 90% confidence interval implies that the hypothesis testing problem

$$H_0 : \mu = \mu_0 \quad \text{versus} \quad H_A : \mu \neq \mu_0$$

has a p-value larger than 0.10 for $49.970 \leq \mu_0 \leq 50.028$ and a p-value smaller than 0.10 otherwise.

Example 50
Engine Oil Viscosity

With $t_{0.025,24} = 2.064$, a 95% two-sided t-interval for the mean oil viscosity is

$$\left(88.3 - \frac{2.064 \times 7.49}{\sqrt{25}}, 88.3 + \frac{2.064 \times 7.49}{\sqrt{25}} \right) = (85.21, 91.39)$$

and with $t_{0.005,24} = 2.797$, a 99% two-sided t-interval for the mean oil viscosity is

$$\left(88.3 - \frac{2.797 \times 7.49}{\sqrt{25}}, 88.3 + \frac{2.797 \times 7.49}{\sqrt{25}} \right) = (84.11, 92.49)$$

Notice that the value $\mu_0 = 85.0$ is contained within the 99% confidence interval but is not contained within the 95% confidence interval, which is consistent with the hypothesis testing problem

$$H_0 : \mu = 85.0 \quad \text{versus} \quad H_A : \mu \neq 85.0$$

having a p-value of 0.0374, which lies between 0.01 and 0.05.

One-Sided Problems   The relationships between confidence intervals, p-values, and significance levels are the same for one-sided problems as they are for two-sided problems. For the one-sided hypotheses

$$H_0 : \mu \leq \mu_0 \quad \text{versus} \quad H_A : \mu > \mu_0$$

a size α hypothesis test rejects the null hypothesis if the test statistic

$$t = \frac{\sqrt{n}(\bar{x} - \mu_0)}{s}$$

is greater than the critical point $t_{\alpha,n-1}$ and accepts the null hypothesis if t is smaller than $t_{\alpha,n-1}$. In other words, the *rejection* region is

$$t > t_{\alpha,n-1}$$

and the *acceptance* region is

$$t \leq t_{\alpha,n-1}$$

as illustrated in Figure 8.33.

A size α test for the one-sided hypotheses

$$H_0 : \mu \geq \mu_0 \quad \text{versus} \quad H_A : \mu < \mu_0$$

has a *rejection* region

$$t < -t_{\alpha,n-1}$$

and an *acceptance* region

$$t \geq -t_{\alpha,n-1}$$

as illustrated in Figure 8.34. For both of these one-sided problems, the null hypothesis is rejected when the p-value is smaller than α and is accepted when the p-value is larger than α.

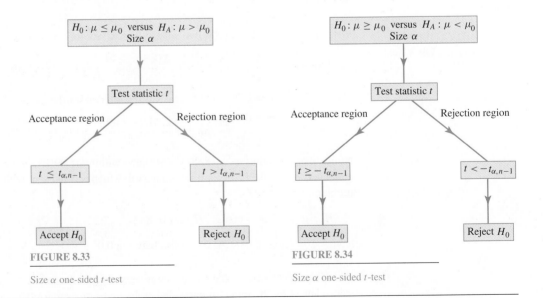

FIGURE 8.33

Size α one-sided t-test

FIGURE 8.34

Size α one-sided t-test

The relationship between one-sided confidence intervals and one-sided hypothesis testing problems is as follows. The $1 - \alpha$ level one-sided confidence interval

$$\mu \in \left(-\infty, \bar{x} + \frac{t_{\alpha,n-1}s}{\sqrt{n}} \right)$$

consists of the values μ_0 for which the hypothesis testing problem

$$H_0 : \mu \geq \mu_0 \quad \text{versus} \quad H_A : \mu < \mu_0$$

has a p-value larger than α, as illustrated in Figure 8.35. Similarly, the $1 - \alpha$ level one-sided confidence interval

$$\mu \in \left(\bar{x} - \frac{t_{\alpha,n-1}s}{\sqrt{n}}, \infty \right)$$

consists of the values μ_0 for which the hypothesis testing problem

$$H_0 : \mu \leq \mu_0 \quad \text{versus} \quad H_A : \mu > \mu_0$$

has a p-value larger than α, as illustrated in Figure 8.36. Figure 8.37 summarizes the relationships between confidence intervals, p-values, and significance levels for two-sided problems and one-sided problems.

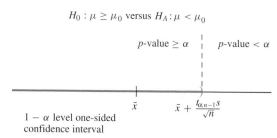

$H_0 : \mu \geq \mu_0$ versus $H_A : \mu < \mu_0$

FIGURE 8.35

Relationship between hypothesis testing and confidence intervals for one-sided problems

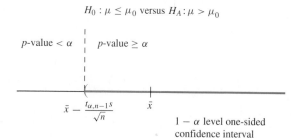

$H_0 : \mu \leq \mu_0$ versus $H_A : \mu > \mu_0$

FIGURE 8.36

Relationship between hypothesis testing and confidence intervals for one-sided problems

FIGURE 8.37

Relationship between hypothesis testing and confidence intervals for two-sided and one-sided problems

Hypothesis tests $H_0: \mu = \mu_0$ versus $H_A: \mu \neq \mu_0$ and confidence intervals $\left(\bar{x} - \frac{t_{\alpha/2,n-1}s}{\sqrt{n}}, \bar{x} + \frac{t_{\alpha/2,n-1}s}{\sqrt{n}} \right)$

Hypothesis tests $H_0: \mu \geq \mu_0$ versus $H_A: \mu < \mu_0$ and confidence intervals $\left(-\infty, \bar{x} + \frac{t_{\alpha,n-1}s}{\sqrt{n}} \right)$

Hypothesis tests $H_0: \mu \leq \mu_0$ versus $H_A: \mu > \mu_0$ and confidence intervals $\left(\bar{x} - \frac{t_{\alpha,n-1}s}{\sqrt{n}}, \infty \right)$

| | Significance Levels | | | Confidence Levels | | |
p-Value	$\alpha = 0.10$	$\alpha = 0.05$	$\alpha = 0.01$	$1 - \alpha = 0.90$	$1 - \alpha = 0.95$	$1 - \alpha = 0.99$
≥ 0.10	accept H_0	accept H_0	accept H_0	contains μ_0	contains μ_0	contains μ_0
0.05–0.10	reject H_0	accept H_0	accept H_0	does not contain μ_0	contains μ_0	contains μ_0
0.01–0.05	reject H_0	reject H_0	accept H_0	does not contain μ_0	does not contain μ_0	contains μ_0
< 0.01	reject H_0	reject H_0	reject H_0	does not contain μ_0	does not contain μ_0	does not contain μ_0

One-Sided Inferences on a Population Mean (H_0: $\mu \leq \mu_0$)

A size α test for the one-sided hypotheses

$$H_0 : \mu \leq \mu_0 \quad \text{versus} \quad H_A : \mu > \mu_0$$

rejects the null hypothesis when

$$t > t_{\alpha,n-1}$$

and accepts the null hypothesis when

$$t \leq t_{\alpha,n-1}$$

The $1 - \alpha$ level one-sided confidence interval

$$\mu \in \left(\bar{x} - \frac{t_{\alpha,n-1}s}{\sqrt{n}}, \infty \right)$$

consists of the values μ_0 for which this hypothesis testing problem has a p-value larger than α, that is, the values μ_0 for which the size α hypothesis test accepts the null hypothesis.

One-Sided Inferences on a Population Mean (H_0: $\mu \geq \mu_0$)

A size α test for the one-sided hypotheses

$$H_0 : \mu \geq \mu_0 \quad \text{versus} \quad H_A : \mu < \mu_0$$

rejects the null hypothesis when

$$t < -t_{\alpha,n-1}$$

and accepts the null hypothesis when

$$t \geq -t_{\alpha,n-1}$$

The $1 - \alpha$ level one-sided confidence interval

$$\mu \in \left(-\infty, \bar{x} + \frac{t_{\alpha,n-1}s}{\sqrt{n}} \right)$$

consists of the values μ_0 for which this hypothesis testing problem has a p-value larger than α, that is, the values μ_0 for which the size α hypothesis test accepts the null hypothesis.

Example 48

Car Fuel Efficiency

The one-sided hypotheses of interest here are

$$H_0 : \mu \geq 35.0 \quad \text{versus} \quad H_A : \mu < 35.0$$

and since the test statistic $t = -1.119$ is larger than the critical point $-t_{0.10,19} = -1.328$, a size $\alpha = 0.10$ hypothesis test *accepts* the null hypothesis. This conclusion is consistent with the previous analysis where the p-value was found to be 0.1386, which is larger than $\alpha = 0.10$.

Furthermore, the one-sided 90% t-interval

$$\mu \in \left(-\infty, \bar{x} + \frac{t_{\alpha,n-1}s}{\sqrt{n}} \right) = \left(-\infty, 34.271 + \frac{1.328 \times 2.915}{\sqrt{20}} \right) = (-\infty, 35.14)$$

contains the value $\mu_0 = 35.0$, as expected. In fact, this confidence interval indicates that the hypothesis testing problem

$$H_0 : \mu \geq \mu_0 \quad \text{versus} \quad H_A : \mu < \mu_0$$

has a p-value larger than 0.10 for any value of $\mu_0 \leq 35.14$.

| Example 45 | For the hypotheses |

Example 45

Fabric Water Absorption Properties

For the hypotheses

$$H_0 : \mu \leq 55\% \quad \text{versus} \quad H_A : \mu > 55\%$$

the t-statistic $t = 3.77$ is larger than the critical point $t_{0.01,14} = 2.624$, so the null hypothesis is rejected at size $\alpha = 0.01$. This conclusion is consistent with the previous analysis where the p-value was shown to be 0.0010, which is smaller than $\alpha = 0.01$.

A one-sided 99% t-interval for the mean water pickup μ is

$$\mu \in \left(\bar{x} - \frac{t_{\alpha,n-1}s}{\sqrt{n}}, \infty \right) = \left(59.81 - \frac{2.624 \times 4.94}{\sqrt{15}}, \infty \right) = (56.46, \infty)$$

which, as expected, does not contain the value $\mu_0 = 55.0$. Furthermore, this confidence interval indicates that the hypothesis testing problem

$$H_0 : \mu \leq \mu_0 \quad \text{versus} \quad H_A : \mu > \mu_0$$

has a p-value smaller than 0.01 for any value of $\mu_0 \leq 56.46$

Example 46

Hospital Worker Radiation Exposures

A 99% one-sided t-interval for the mean radiation level μ was found to be

$$(-\infty, 5.496)$$

This implies that the one-sided hypothesis testing problem

$$H_0 : \mu \geq \mu_0 \quad \text{versus} \quad H_A : \mu < \mu_0$$

has a p-value smaller than 0.01 for $\mu_0 > 5.496$ and a p-value larger than 0.01 for $\mu_0 \leq 5.496$.

Power Levels The significance level α of a hypothesis test designates the probability of a Type I error, that is, the probability that the null hypothesis is rejected when it is true (see Figure 8.29). Small significance levels are employed in hypothesis tests so that this probability is small. However, it is also useful to consider the probability of a Type II error, which is the probability that the null hypothesis is accepted when the alternative hypothesis is true.

Power of a Hypothesis Test

The *power* of a hypothesis test is defined to be

power = 1 − (probability of Type II error)

which is the probability that the null hypothesis is rejected when it is false.

FIGURE 8.38

The specification of two quantities
determines the third quantity

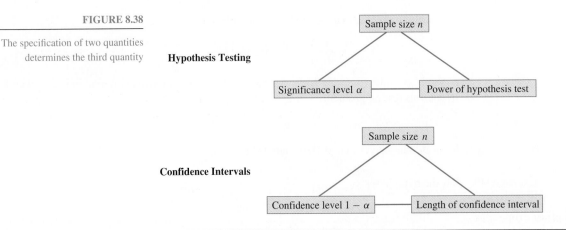

Obviously, large values of power are good. Larger power levels and shorter confidence intervals are both indications of an increase in the "precision" of an experiment. However, as shown in Figure 8.38, an experimenter can choose *only two* quantities out of

- the sample size n, the significance level α, and the power of a hypothesis test

or similarly *only two* out of

- the sample size n, the confidence level $1 - \alpha$, and the length of a confidence interval

Once two of these quantities have been specified, the third is automatically determined.

Usually, the sample size n obtained in the experiment and the choice of a significance level α determine the power of the hypothesis test, just as the sample size and a confidence level determine a confidence interval length. Sometimes, an experimenter may investigate what sample size n is required to achieve a specified significance level and power level. However, it is generally more convenient to base sample size determination on confidence interval lengths, as described in Section 8.1.2. It is important to realize that for a fixed significance level α, the power of a hypothesis test increases as the sample size n increases.

Relationship Between Power and Sample Size

For a fixed significance level α, the *power* of a hypothesis test increases as the sample size n increases.

8.2.5   z-Tests

Testing procedures similar to the t-tests can be employed when an experimenter wishes to use an assumed "known" value of the population standard deviation σ rather than the sample standard deviation s. These "variance known" tests are called z-tests and are based upon the z-statistic

$$z = \frac{\sqrt{n}(\bar{x} - \mu_0)}{\sigma}$$

which has a standard normal distribution when $\mu = \mu_0$.

For a two-sided hypothesis testing problem, the p-value is calculated as

$$p\text{-value} = 2 \times P(X \geq |z|)$$

where the random variable X has a standard normal distribution. With the standard normal cumulative distribution function $\Phi(x)$, this value can be written

$$p\text{-value} = 2 \times (1 - \Phi(|z|)) = 2 \times \Phi(-|z|)$$

For one-sided hypothesis testing problems the p-value is either

$$p\text{-value} = P(X \geq z) = 1 - \Phi(z) \qquad \text{or} \qquad p\text{-value} = P(X \leq z) = \Phi(z)$$

If a fixed significance level α is employed, then a critical point $z_{\alpha/2}$ is appropriate for a two-sided hypothesis testing problem and a critical point z_α is appropriate for a one-sided hypothesis testing problem.

Two-Sided z-Test

The p-value for the two-sided hypothesis testing problem

$$H_0 : \mu = \mu_0 \quad \text{versus} \quad H_A : \mu \neq \mu_0$$

based upon a data set of n observations with a sample mean $\bar{x}$ and an *assumed* "*known*" population standard deviation σ, is

$$p\text{-value} = 2 \times \Phi(-|z|)$$

where $\Phi(x)$ is the standard normal cumulative distribution function and

$$z = \frac{\sqrt{n}(\bar{x} - \mu_0)}{\sigma}$$

which is known as the **z-statistic**. This testing procedure is called a **two-sided z-test**.

A size α test rejects the null hypothesis H_0 if the test statistic $|z|$ falls in the *rejection region*

$$|z| > z_{\alpha/2}$$

and accepts the null hypothesis H_0 if the test statistic $|z|$ falls in the *acceptance region*

$$|z| \leq z_{\alpha/2}$$

The $1 - \alpha$ level two-sided confidence interval

$$\mu \in \left(\bar{x} - \frac{z_{\alpha/2}\sigma}{\sqrt{n}}, \bar{x} + \frac{z_{\alpha/2}\sigma}{\sqrt{n}} \right)$$

consists of the values μ_0 for which this hypothesis testing problem has a p-value larger than α, that is, the values μ_0 for which the size α hypothesis test accepts the null hypothesis.

One-Sided z-Test ($H_0: \mu \leq \mu_0$)

The p-value for the one-sided hypothesis testing problem

$$H_0 : \mu \leq \mu_0 \quad \text{versus} \quad H_A : \mu > \mu_0$$

based upon a data set of n observations with a sample mean $\bar{x}$ and an assumed known population standard deviation σ, is

$$p\text{-value} = 1 - \Phi(z)$$

This testing procedure is called a **one-sided z-test**.

A size α test rejects the null hypothesis when

$$z > z_\alpha$$

and accepts the null hypothesis when

$$z \leq z_\alpha$$

The $1 - \alpha$ level one-sided confidence interval

$$\mu \in \left(\bar{x} - \frac{z_\alpha \sigma}{\sqrt{n}}, \infty \right)$$

consists of the values μ_0 for which this hypothesis testing problem has a p-value larger than α, that is, the values μ_0 for which the size α hypothesis test accepts the null hypothesis.

One-Sided z-Test ($H_0: \mu \geq \mu_0$)

The p-value for the one-sided hypothesis testing problem

$$H_0 : \mu \geq \mu_0 \quad \text{versus} \quad H_A : \mu < \mu_0$$

based upon a data set of n observations with a sample mean $\bar{x}$ and an assumed known population standard deviation σ, is

$$p\text{-value} = \Phi(z)$$

This testing procedure is called a **one-sided z-test**.

A size α test rejects the null hypothesis when

$$z < -z_\alpha$$

and accepts the null hypothesis when

$$z \geq -z_\alpha$$

The $1 - \alpha$ level one-sided confidence interval

$$\mu \in \left(-\infty, \bar{x} + \frac{z_\alpha \sigma}{\sqrt{n}} \right)$$

consists of the values μ_0 for which this hypothesis testing problem has a p-value larger than α, that is, the values μ_0 for which the size α hypothesis test accepts the null hypothesis.

COMPUTER NOTE Find out how to do t-tests (variance unknown) and z-tests (variance known) with your software package. You will need to specify the value of μ_0 and also whether you want an alternative hypothesis of $H_A : \mu < \mu_0$, $H_A : \mu \neq \mu_0$, or $H_A : \mu > \mu_0$. If you wish to use a z-test, you will also need to specify the value of the "known" population standard deviation σ. Usually, the computer will provide you with an exact p-value and you will not need to specify a significance level α.

■ 8.2.6 Problems

8.2.1 A sample of $n = 18$ observations has a sample mean of $\bar{x} = 57.74$ and a sample standard deviation of $s = 11.20$. Consider the hypothesis testing problems
(**a**) $H_0 : \mu = 55.0$ versus $H_A : \mu \neq 55.0$
(**b**) $H_0 : \mu \geq 65.0$ versus $H_A : \mu < 65.0$
In each case, write down an expression for the p-value. What do the critical points in Table III tell you about the p-values? Use a computer package to evaluate the p-values exactly.

8.2.2 A sample of $n = 39$ observations has a sample mean of $\bar{x} = 5532$ and a sample standard deviation of $s = 287.8$. Consider the hypothesis testing problems
(**a**) $H_0 : \mu = 5680$ versus $H_A : \mu \neq 5680$
(**b**) $H_0 : \mu \leq 5450$ versus $H_A : \mu > 5450$
In each case, write down an expression for the p-value. What do the critical points in Table III tell you about the p-values? Use a computer package to evaluate the p-values exactly.

8.2.3 A sample of $n = 13$ observations has a sample mean of $\bar{x} = 2.879$. If an assumed known standard deviation of $\sigma = 0.325$ is used, calculate the p-values for the hypothesis testing problems
(**a**) $H_0 : \mu = 3.0$ versus $H_A : \mu \neq 3.0$
(**b**) $H_0 : \mu \geq 3.1$ versus $H_A : \mu < 3.1$

8.2.4 A sample of $n = 44$ observations has a sample mean of $\bar{x} = 87.90$. If an assumed known standard deviation of $\sigma = 5.90$ is used, calculate the p-values for the hypothesis testing problems
(**a**) $H_0 : \mu = 90.0$ versus $H_A : \mu \neq 90.0$
(**b**) $H_0 : \mu \leq 86.0$ versus $H_A : \mu > 86.0$

8.2.5 An experimenter is interested in the hypothesis testing problem

$$H_0 : \mu = 3.0 \text{ mm} \quad \text{versus} \quad H_A : \mu \neq 3.0 \text{ mm}$$

where μ is the average thickness of a set of glass sheets. Suppose that a sample of $n = 41$ glass sheets is obtained and their thicknesses are measured.

(**a**) For what values of the t-statistic does the experimenter *accept* the null hypothesis with a size $\alpha = 0.10$?
(**b**) For what values of the t-statistic does the experimenter *reject* the null hypothesis with a size $\alpha = 0.01$?
Suppose that the sample mean is $\bar{x} = 3.04$ mm and the sample standard deviation is $s = 0.124$ mm.
(**c**) Is the null hypothesis accepted or rejected with $\alpha = 0.10$? With $\alpha = 0.01$?
(**d**) Write down an expression for the p-value and evaluate it using a computer package.

8.2.6 An experimenter is interested in the hypothesis testing problem

$$H_0 : \mu = 430.0 \quad \text{versus} \quad H_A : \mu \neq 430.0$$

where μ is the average breaking strength of a bundle of wool fibers. Suppose that a sample of $n = 20$ wool fiber bundles is obtained and their breaking strengths are measured.

(**a**) For what values of the t-statistic does the experimenter *accept* the null hypothesis with a size $\alpha = 0.10$?
(**b**) For what values of the t-statistic does the experimenter *reject* the null hypothesis with a size $\alpha = 0.01$?
Suppose that the sample mean is $\bar{x} = 436.5$ and the sample standard deviation is $s = 11.90$.
(**c**) Is the null hypothesis accepted or rejected with $\alpha = 0.10$? With $\alpha = 0.01$?
(**d**) Write down an expression for the p-value and evaluate it using a computer package.

8.2.7 An experimenter is interested in the hypothesis testing problem

$$H_0 : \mu = 1.025 \text{ kg} \quad \text{versus} \quad H_A : \mu \neq 1.025 \text{ kg}$$

where μ is the average weight of a 1-kilogram sugar packet. Suppose that a sample of $n = 16$ sugar packets is obtained and their weights are measured.

(a) For what values of the *t*-statistic does the experimenter *accept* the null hypothesis with a size $\alpha = 0.10$?

(b) For what values of the *t*-statistic does the experimenter *reject* the null hypothesis with a size $\alpha = 0.01$?

Suppose that the sample mean is $\bar{x} = 1.053$ kg and the sample standard deviation is $s = 0.058$ kg.

(c) Is the null hypothesis accepted or rejected with $\alpha = 0.10$? With $\alpha = 0.01$?

(d) Write down an expression for the *p*-value and evaluate it using a computer package.

8.2.8 An experimenter is interested in the hypothesis testing problem

$$H_0 : \mu = 20.0 \quad \text{versus} \quad H_A : \mu \neq 20.0$$

where μ is the average resilient modulus of a clay mixture. Suppose that a sample of $n = 10$ resilient modulus measurements is obtained and that the experimenter wishes to use a value of $\sigma = 1.0$ for the resilient modulus standard deviation.

(a) For what values of the *z*-statistic does the experimenter *accept* the null hypothesis with a size $\alpha = 0.10$?

(b) For what values of the *z*-statistic does the experimenter *reject* the null hypothesis with a size $\alpha = 0.01$?

Suppose that the sample mean is $\bar{x} = 19.50$.

(c) Is the null hypothesis accepted or rejected with $\alpha = 0.10$? With $\alpha = 0.01$?

(d) Calculate the exact *p*-value.

8.2.9 An experimenter is interested in the hypothesis testing problem

$$H_0 : \mu \leq 0.065 \quad \text{versus} \quad H_A : \mu > 0.065$$

where μ is the average density of a chemical solution. Suppose that a sample of $n = 61$ bottles of the chemical solution is obtained and their densities are measured.

(a) For what values of the *t*-statistic does the experimenter *accept* the null hypothesis with a size $\alpha = 0.10$?

(b) For what values of the *t*-statistic does the experimenter *reject* the null hypothesis with a size $\alpha = 0.01$?

Suppose that the sample mean is $\bar{x} = 0.0768$ and the sample standard deviation is $s = 0.0231$.

(c) Is the null hypothesis accepted or rejected with $\alpha = 0.10$? With $\alpha = 0.01$?

(d) Write down an expression for the *p*-value and evaluate it using a computer package.

8.2.10 An experimenter is interested in the hypothesis testing problem

$$H_0 : \mu \geq 420.0 \quad \text{versus} \quad H_A : \mu < 420.0$$

where μ is the average radiation level in a research

laboratory. Suppose that a sample of $n = 29$ radiation level measurements is obtained and that the experimenter wishes to use a value of $\sigma = 10.0$ for the standard deviation of the radiation levels.

(a) For what values of the *z*-statistic does the experimenter *accept* the null hypothesis with a size $\alpha = 0.10$?

(b) For what values of the *z*-statistic does the experimenter *reject* the null hypothesis with a size $\alpha = 0.01$?

Suppose that the sample mean is $\bar{x} = 415.7$.

(c) Is the null hypothesis accepted or rejected with $\alpha = 0.10$? With $\alpha = 0.01$?

(d) Calculate the exact *p*-value.

8.2.11 A machine is set to cut metal plates to a length of 44.350 mm. The lengths of a random sample of 24 metal plates have a sample mean of $\bar{x} = 44.364$ mm and a sample standard deviation of $s = 0.019$ mm. Is there any evidence that the machine is miscalibrated?

8.2.12 A food manufacturer claims that at the time of purchase by a consumer the average age of its product is no more than 120 days. In an experiment to test this claim a random sample of 36 items are found to have ages at the time of purchase with a sample mean of $\bar{x} = 122.5$ days and a sample standard deviation of $s = 13.4$ days. With this information how do you feel about the manufacturer's claim?

8.2.13 A chemical plant is required to maintain ambient sulfur levels in the working environment atmosphere at an average level of no more than 12.50. The results of 15 randomly timed measurements of the sulfur level produced a sample mean of $\bar{x} = 14.82$ and a sample standard deviation of $s = 2.91$. What is the evidence that the chemical plant is in violation of the working code?

8.2.14 A company advertises that its electric motors provide an efficiency that is at least 25% higher than the industry norm. A consumer interest group ran an experiment with a sample of 23 machines for which the increases in efficiency over the industry norm had a sample mean of $\bar{x} = 22.8\%$ and a sample standard deviation of $s = 8.72\%$. What evidence does the consumer interest group have that the advertised claim is false?

8.2.15 Recall Problem 8.1.17 where a collection of $n = 10$ samples of chilled cast iron provided corrosion rates with a sample mean of $\bar{x} = 2.752$ and a sample standard deviation of $s = 0.280$. Is there sufficient evidence to conclude that the average corrosion rate of chilled cast iron of this type is larger than 2.5?

8.2.16 Restaurant Service Times

Consider the data set of service times given in DS 6.1.4. The manager of the fast-food restaurant claims that at the time the survey was conducted, the average service time was less than 65 seconds. What is the evidence that this claim is false?

8.2.17 Telephone Switchboard Activity

Consider the data set of calls received by a switchboard given in DS 6.1.6. A manager claims that the switchboard needs additional staffing because the average number of calls taken per minute is at least 13. How do you feel about this claim?

8.2.18 Paving Slab Weights

Consider the data set of paving slab weights given in DS 6.1.7. The slabs are supposed to have an average weight of 1.1 kg. Is there any evidence that the manufacturing process needs adjusting?

8.2.19 Spray Painting Procedure

Consider the data set of paint thicknesses given in DS 6.1.8. The spray painting machine is supposed to spray paint to a mean thickness of 0.225 mm. What is the evidence that the spray painting machine is not performing properly?

8.2.20 Plastic Panel Bending Capabilities

Consider the data set of plastic panel bending capabilities given in DS 6.1.9. The plastic panels are designed to be able to bend on average to at least $9.5°$ without deforming. Is there any evidence that this design criterion has not been met?

8.2.21 An experimenter randomly selects $n = 16$ batteries from a production line and measures their voltages. An average $\bar{x} = 239.13$ is obtained, with a sample standard deviation $s = 2.80$. Does this experiment provide sufficient evidence for the experimenter to conclude that the average voltage of the batteries from the production line is at least 238.5?

8.2.22 A two-sided t-procedure is performed. Use Table III to put bounds on the p-value if:

(a) $n = 12, t = 3.21$
(b) $n = 24, t = 1.96$
(c) $n = 30, t = 3.88$

8.2.23 A company claims that its components have an average length of 82.50 mm. An experimenter tested this claim by measuring the lengths of a random sample of 25 components. It was found that $\bar{x} = 82.40$ and $s = 0.14$. Use a hypothesis test to assess whether the

experimenter has sufficient evidence to conclude that the average length of the components is different from 82.50.

8.2.24 A random sample of 25 components is obtained, and their weights are measured. The sample mean is 71.97 g and the sample standard deviation is 7.44 g. Conduct a hypothesis test to assess whether there is sufficient evidence to establish that the components have an average weight larger than 70 g.

8.2.25 A random sample of 28 plastic items is obtained, and their breaking strengths are measured. The sample mean is 7.442 and the sample standard deviation is 0.672. Conduct a hypothesis test to assess whether there is any evidence that the average breaking strength is not 7.000.

8.2.26 An experimenter measures the failure times of a random sample of 25 components. The sample average is 53.43 hours and the sample standard deviation is 3.93 hours. Use a hypothesis test to determine whether there is sufficient evidence for the experimenter to conclude that the average failure time of the components is at least 50 hours.

8.2.27 An experimenter is planning an experiment to assess whether it can be established that an unknown failure rate μ is smaller than 25. Write down the null hypothesis and the alternative hypothesis that the experimenter should use for the analysis.

8.2.28 Use Table III to indicate whether the p-values for the following t-tests are less than 1%, between 1% and 10%, or more than 10%.

(a) $H_0 : \mu = 10, H_A : \mu \neq 10, n = 20, \bar{x} = 12.49, s = 1.32$
(b) $H_0 : \mu \leq 3.2, H_A : \mu > 3.2, n = 43, \bar{x} = 3.03, s = 0.11$
(c) $H_0 : \mu \geq 85, H_A : \mu < 85, n = 16, \bar{x} = 73.43, s = 16.44$

8.2.29 Toxicity of Salmon Fillets

An experiment is conducted to investigate the time taken for salmon fillets to become toxic under certain storage conditions. Eight samples are prepared, and the times to toxicity in days are given in DS 8.2.1.

(a) Does this experiment provide sufficient evidence to conclude that the average time to toxicity of salmon fillets under these storage conditions is more than 11 days?

(b) Construct a two-sided 99% confidence interval for the average time to toxicity of salmon fillets under these storage conditions.

8.3   Summary

Figure 8.39 shows a summary of the process an experimenter goes through in order to make suitable inferences on a population mean μ. The process consists of two questions followed by a choice of inference methods.

Question I relates to whether a t-procedure or a z-procedure is appropriate. It is almost always the case that the experimenter can employ a t-procedure. However, a z-procedure is appropriate if the experimenter wishes to use an assumed known value for the population standard deviation σ, presumably obtained from prior experience. With $s = \sigma$, the t-procedure and z-procedure are identical for large sample sizes.

Remember that for sample sizes smaller than about 30, the test procedures require that the distribution of the sample observations should be approximately normally distributed. For sample sizes larger than about 30, the test procedures are appropriate regardless of the actual distribution of the sample observations because of the central limit theorem.

If the sample size is small, the t-procedure generally provides fairly sensible results unless the data observations are clearly not normally distributed. The nonparametric inference methods discussed in Chapter 15 offer an alternative approach in these situations.

Question II relates to whether a two-sided or a one-sided procedure is appropriate. Generally, a two-sided inference method is appropriate, and if in doubt it is always appropriate to employ a two-sided inference method. However, in certain situations where the experimenter is interested in obtaining only upper or lower bounds on the population mean, a one-sided approach provides a more efficient analysis.

Figures 8.40 and 8.41 summarize how t-procedures and z-procedures are employed. A confidence interval for the population mean is usually the best way to summarize the results of an experiment. It provides a range of plausible values for the population mean, and most people find it easy to interpret. Nevertheless, if there is a value of the population mean μ_0 that is of particular interest to the experimenter, then it can also be useful to calculate a p-value to assess how plausible that particular value is.

Hypothesis testing at a fixed significance level α, so that the result is reported as either acceptance or rejection of the null hypothesis, can be employed, but it is not as informative an inference method as confidence interval construction and p-value calculation. Remember that the result of a size α hypothesis test can be inferred either from a $1 - \alpha$ level confidence interval or from the exact p-value of a hypothesis testing problem.

A final matter is how an experimenter may determine an appropriate sample size n when this option is available. Obviously, larger sample sizes allow a more precise statistical analysis

FIGURE 8.39

Decision process for inferences on a population mean

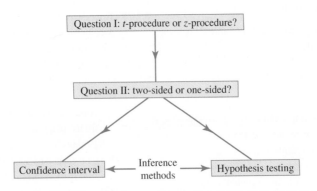

FIGURE 8.40

Summary of the t-procedure

t-procedure		
(sample size $n \geq 30$ or a small sample size with normally distributed data; variance unknown)		
One-sided	**Two-sided**	**One-sided**
$1 - \alpha$ level Confidence Intervals		
$\left(-\infty, \bar{x} + \frac{t_{\alpha,n-1}s}{\sqrt{n}}\right)$	$\left(\bar{x} - \frac{t_{\alpha/2,n-1}s}{\sqrt{n}}, \bar{x} + \frac{t_{\alpha/2,n-1}s}{\sqrt{n}}\right)$	$\left(\bar{x} - \frac{t_{\alpha,n-1}s}{\sqrt{n}}, \infty\right)$
Hypothesis Testing: test statistic $t = \frac{\sqrt{n}(\bar{x}-\mu_0)}{s}$; $X \sim t_{n-1}$		
$H_0 : \mu \geq \mu_0, H_A : \mu < \mu_0$	$H_0 : \mu = \mu_0, H_A : \mu \neq \mu_0$	$H_0 : \mu \leq \mu_0, H_A : \mu > \mu_0$
p-value $= P(X \leq t)$	p-value $= 2 \times P(X \geq |t|)$	p-value $= P(X \geq t)$
Size α hypothesis tests		
accept H_0 reject H_0 $t \geq -t_{\alpha,n-1}$ $t < -t_{\alpha,n-1}$	accept H_0 reject H_0 $|t| \leq t_{\alpha/2,n-1}$ $|t| > t_{\alpha/2,n-1}$	accept H_0 reject H_0 $t \leq t_{\alpha,n-1}$ $t > t_{\alpha,n-1}$

FIGURE 8.41

Summary of the z-procedure

z-procedure		
(sample size $n \geq 30$ or a small sample size with normally distributed data; variance known)		
One-sided	**Two-sided**	**One-sided**
$1 - \alpha$ level Confidence Intervals		
$\left(-\infty, \bar{x} + \frac{z_{\alpha}\sigma}{\sqrt{n}}\right)$	$\left(\bar{x} - \frac{z_{\alpha/2}\sigma}{\sqrt{n}}, \bar{x} + \frac{z_{\alpha/2}\sigma}{\sqrt{n}}\right)$	$\left(\bar{x} - \frac{z_{\alpha}\sigma}{\sqrt{n}}, \infty\right)$
Hypothesis Testing: test statistic $z = \frac{\sqrt{n}(\bar{x}-\mu_0)}{\sigma}$		
$H_0 : \mu \geq \mu_0, H_A : \mu < \mu_0$	$H_0 : \mu = \mu_0, H_A : \mu \neq \mu_0$	$H_0 : \mu \leq \mu_0, H_A : \mu > \mu_0$
p-value $= \Phi(z)$	p-value $= 2 \times \Phi(-|z|)$	p-value $= 1 - \Phi(z)$
Size α hypothesis tests		
accept H_0 reject H_0 $z \geq -z_{\alpha}$ $z < -z_{\alpha}$	accept H_0 reject H_0 $|z| \leq z_{\alpha/2}$ $|z| > z_{\alpha/2}$	accept H_0 reject H_0 $z \leq z_{\alpha}$ $z > z_{\alpha}$

but also incur a greater cost. The most convenient way to assess the precision afforded by a certain sample size is to estimate the length of the resulting two-sided confidence interval for the population mean.

8.4 Case Study: Microelectronic Solder Joints

The new method that the researcher is investigating is supposed to deposit a nickel layer with an average thickness of 2.775 microns on the substrate bond pad. The researcher's data set provided a sample average of $\bar{x} = 2.7688$ microns, which suggests that the method may

FIGURE 8.42

Two-sided hypothesis test of
whether the average nickel layer
thickness is 2.775 microns

Data and Question
Data set of nickel layer thicknesses in microns (given in Figure 6.40).
Question: What evidence is there that the average thickness is not 2.775 microns?

Stage I: Data Summary
Sample average $n = 16$, sample mean $\bar{x} = 2.7688$, sample standard deviation $s = 0.0260$.

Stage II: Determination of Suitable Hypotheses
Since this is a two-sided problem concerning whether $\mu \neq 2.775$, this should be the alternative hypothesis.
$H_0 : \mu = 2.775$ versus $H_A : \mu \neq 2.775$.

Stage III: Calculation of the Test Statistic
$t = \frac{\sqrt{n}(\bar{x} - \mu_0)}{s} = \frac{\sqrt{16}(2.7688 - 2.775)}{0.0260} = -0.954$

Stage IV: Expression for the p-value
p-value $= 2 \times P(X \geq 0.954)$
where the random variable X has a t-distribution with $n - 1 = 15$ degrees of freedom.

Stage V: Evaluation of the p-value
Table III gives $t_{0.10,15} = 1.341$, and consequently it is known that the p-value is larger than $2 \times 0.10 = 0.20$.
Alternatively, exact computer calculation gives the p-value as 0.352.

Stage VI: Decision
Since the p-value is larger than 0.10, the null hypothesis is accepted.

Stage VII: Conclusion
This data set does not provide sufficient evidence to establish that the average nickel layer thickness
is not 2.775 microns.

not be depositing enough nickel. But is the difference between the sample average and the target value statistically significant? Figure 8.42 shows how a two-sided hypothesis test can be employed to show that, in fact, the difference is not statistically significant.

With $t_{0.005,15} = 2.947$, a 99% confidence level two-sided confidence interval for the average thickness is

$$\mu \in \left(\bar{x} - \frac{t_{0.005,15}s}{\sqrt{n}}, \bar{x} - \frac{t_{0.005,15}s}{\sqrt{n}} \right)$$

$$= \left(2.7688 - \frac{2.947 \times 0.0260}{\sqrt{16}}, 2.7688 + \frac{2.947 \times 0.0260}{\sqrt{16}} \right) = (2.750, 2.788)$$

This confidence interval contains the target value of 2.775 microns, so just as the hypothesis test indicated, it is plausible that the average thickness really is 2.775 microns. However, it is important to remember that the hypothesis test has not proved that the average thickness is 2.775 microns, and the confidence interval indicates that it could be as small as 2.750 microns, or as large as 2.788 microns.

The researcher decides that it is worthwhile to commit further resources toward investigating the thicknesses of the nickel layers and decides that it would be useful to be able to have a 99% confidence level two-sided confidence interval for the average thickness that has a length no longer than 0.02 microns. It can be estimated that this would require a total sample size of

$$n \geq 4 \times \left(\frac{t_{0.005,15}s}{L_0} \right)^2 = 4 \times \left(\frac{2.947 \times 0.0260}{0.02} \right)^2 = 58.7$$

Consequently, it can be estimated that an additional $59 - 16 = 43$ nickel layer thicknesses need to be measured.

8.5   Supplementary Problems

8.5.1 In an experiment to investigate when a radar picks up a certain kind of target, a total of $n = 15$ trials are conducted in which the distance of the target from the radar is measured when the target is detected. A sample mean of $\bar{x} = 67.42$ miles is obtained, with a sample standard deviation of $s = 4.947$ miles.

    (a) Is there enough evidence for the scientists to conclude that the average distance at which the target is detected is at least 65 miles?

    (b) Construct a 99% one-sided t-interval that provides a lower bound on the average distance at which the target can be detected.

8.5.2 A company is planning a large telephone survey and is interested in assessing how long it will take. In a short pilot study, 40 people are contacted by telephone and are asked the specified set of questions. The times of these 40 telephone surveys have a sample mean of $\bar{x} = 9.39$ minutes, with a sample standard deviation of $s = 1.041$ minutes.

    (a) Can the company safely conclude that the telephone surveys will last on average no more than 10 minutes each?

    (b) Construct a 99% one-sided t-interval that provides an upper bound on the average time of each telephone call.

8.5.3 A paper company sells paper that is supposed to have a weight of 75.0 g/m^2. In a quality inspection, the weights of 30 random samples of paper are measured. The sample mean of these weights is $\bar{x} = 74.63$ g/m^2, with a sample standard deviation of $s = 2.095$ g/m^2.

    (a) Is there any evidence that the paper does not have an average weight of 75.0 g/m^2?

    (b) Construct a 99% two-sided t-interval for the average weight of the paper.

    (c) If a 99% two-sided t-interval for the average weight of the paper is required with a length no longer than 1.5 g/m^2, how many additional paper samples would you recommend need to be weighed?

8.5.4 A group of medical researchers is investigating how artery disease affects the rigidity of the arteries. Deformity measurements are made on a sample of 14 diseased arteries, and a sample mean of $\bar{x} = 0.497$ is obtained, with a sample standard deviation of $s = 0.0764$.

    (a) What is the evidence that the average deformity value of diseased arteries is less than 0.50?

    (b) Construct a 99% two-sided t-interval for the average deformity value of diseased arteries.

    (c) If a 99% two-sided t-interval for the average deformity value of diseased arteries is required with a length no larger than 0.10, how many additional arteries would you recommend be analyzed?

8.5.5 Osteoporosis Patient Heights

Consider the data set of osteoporosis patient heights given in DS 6.6.4. Use a computer package to construct 90%, 95%, and 99% two-sided t-intervals for the mean height. Is 70 inches a plausible value for the mean height?

8.5.6 Bamboo Cultivation

Consider the data set in DS 6.6.5 of bamboo shoot heights 40 days after planting. Use a computer package to construct 90%, 95%, and 99% two-sided t-intervals for the mean shoot height. A previous study reported that under similar growing conditions the mean shoot height after 40 days was more than 35 cm. Does the new data set confirm the previous study? Does it contradict the previous study?

8.5.7 The breaking strengths of a random sample of 26 molded plastic housings were measured, and a sample mean of $\bar{x} = 479.42$ and a sample standard deviation of $s = 12.55$ were obtained. A confidence interval (472.56, 486.28) for the average strength of molded plastic housings was constructed from these results. What is the confidence level of this confidence interval?

8.5.8 Composites are materials that are made by embedding a fiber, such as glass or carbon, inside a matrix, such as a metal or a ceramic. Composites are used in civil engineering structures, and their degradation when subjected to weather conditions is an important issue. In an experiment to investigate the effect of moisture on a certain kind of composite, the weight gains of a collection of 18 samples of composite subjected to water diffusion were obtained. The sample mean was $\bar{x} = 0.337\%$, with a sample standard deviation of $s = 0.025\%$.

    (a) Is it safe to conclude from the results of this experiment that the average weight gain for composites of this kind is smaller than 0.36%?

    (b) Construct a 99% confidence interval that provides an upper bound for the average weight gain for composites of this kind.

8.5.9 Soil Compressibility Tests

Recall the data set of soil compressibility measurements given in DS 6.6.6. Construct a 99% one-sided confidence interval that provides an upper bound on the average soil compressibility. Can the engineers conclude that the average soil compressibility is no larger than 25.5?

8.5.10 Confidence Interval for a Population Variance

For use with Problems 8.5.11–8.5.14.

Recall that if the data observations are *normally distributed*, then the sample variance S^2 has the distribution

$$S^2 \sim \sigma^2 \frac{\chi^2_{n-1}}{n-1}$$

(a) Show that this result implies that

$$P\left(\chi^2_{1-\alpha/2,n-1} \leq \frac{(n-1)S^2}{\sigma^2} \leq \chi^2_{\alpha/2,n-1}\right) = 1-\alpha$$

(b) Deduce that

$$P\left(\frac{(n-1)S^2}{\chi^2_{\alpha/2,n-1}} \leq \sigma^2 \leq \frac{(n-1)S^2}{\chi^2_{1-\alpha/2,n-1}}\right) = 1-\alpha$$

so that

$$\left(\frac{(n-1)s^2}{\chi^2_{\alpha/2,n-1}}, \frac{(n-1)s^2}{\chi^2_{1-\alpha/2,n-1}}\right)$$

is a $1-\alpha$ level two-sided confidence interval for the population variance σ^2.

(c) Explain why

$$\left(\sqrt{\frac{(n-1)s^2}{\chi^2_{\alpha/2,n-1}}}, \sqrt{\frac{(n-1)s^2}{\chi^2_{1-\alpha/2,n-1}}}\right)$$

is a $1-\alpha$ level two-sided confidence interval for the population standard deviation σ.

Even though the population mean μ is usually the parameter of primary interest to an experimenter, in certain situations it may be helpful to use this method to construct a confidence interval for the population variance σ^2 or population standard deviation σ. An unfortunate aspect of these confidence intervals is that they depend heavily on the data being normally distributed, and they should be used only when that is a fair assumption. You may be able to obtain these confidence intervals on your computer package.

8.5.11
A sample of $n = 18$ observations has a sample standard deviation of $s = 6.48$. Use the method above to construct 99% and 95% two-sided confidence intervals for the population variance σ^2.

8.5.12
Consider the data set of 41 glass sheet thicknesses described in Problem 8.1.2. Construct a 99% two-sided confidence interval for the standard deviation σ of the sheet thicknesses.

8.5.13
Consider the data set of breaking strengths of wool fiber bundles described in Problem 8.1.3. Construct a 95% two-sided confidence interval for the variance σ^2 of the breaking strengths.

8.5.14
Consider the data set of sugar packet weights described in Problem 8.1.4. Construct 90%, 95%, and 99% two-sided confidence intervals for the standard deviation σ of the packet weights.

8.5.15
A two-sided t-test is performed. Use Table III to put bounds on the p-value if:
(a) $n = 8, t = 1.31$
(b) $n = 30, t = -2.82$
(c) $n = 25, t = 1.92$

8.5.16
An experimenter measures the compressibility of 16 samples of clay randomly selected from a particular location, and they have a sample mean of 76.99 and a sample standard deviation of 5.37. Does this provide sufficient evidence for the experimenter to conclude that the average clay compressibility at the location is less than 81?

8.5.17
A sample of 14 fibers was tested. Their strengths had a sample average of 266.5 and a sample standard deviation of 18.6. Use a hypothesis test to assess whether it is safe to conclude that the average strength of fibers of this type is at least 260.0.

8.5.18
Consider the data set

> 34 54 73 38 89 52 75 33 50 39 42 42
> 40 66 72 85 28 71

which is a random sample from a distribution with an unknown mean μ. Calculate the following.
(a) The sample size.
(b) The sample median.
(c) The sample mean.
(d) The sample standard deviation.
(e) The sample variance.
(f) The standard error of the sample mean.
(g) A one-sided 99% confidence interval that provides a lower bound for μ.
(h) Consider the hypothesis test $H_0 : \mu = 50$ versus $H_A : \mu \neq 50$. What bounds can you put on the p-value using Table III?

8.5.19 Are the following statements true or false?

(a) In hypothesis testing the null hypothesis can never be proved to be correct.

(b) For a given data set a two-sided confidence interval for a parameter with a confidence level 99% is shorter than a two-sided confidence interval for the parameter with a confidence level 95%.

(c) A statistical proof that a statement is true is achieved when the null hypothesis that the statement is false is rejected.

(d) A hypothesis test addresses the question of whether or not there is sufficient evidence to establish that the null hypothesis is false.

(e) A p-value between 1% and 10% should be interpreted as implying that there is some evidence that the null hypothesis is false, but that the evidence is not overwhelming.

(f) z-intervals are sometimes referred to as large sample intervals.

(g) If the p-value is 0.39 the null hypothesis is accepted at size $\alpha = 0.05$.

8.5.20 A sample of 22 wires was tested. Their resistances had a sample average of 193.7 and a sample standard deviation of 11.2. It is claimed that the average resistance of wires of this type is 200.0. Use an appropriate hypothesis test to investigate this claim.

8.5.21 An engineer selects 10 components at random and measures their strengths. It is reported that the average strength of the components is between 72.3 and 74.5 with 99% confidence.

(a) What is the sample standard deviation of the 10 component strengths?

(b) If a 99% two-sided confidence interval is desired with a length no longer than 1.0, about how many additional components would you recommend be tested?

8.5.22 A random sample of 10 items gives $\bar{x} = 614.5$ and $s = 42.9$.

(a) Use a hypothesis test to determine whether there is sufficient evidence for the experimenter to conclude that the population average is not 600.

(b) Construct a 99% two-sided confidence interval for the population average.

(c) If a 99% two-sided confidence interval for the population average is required with a total length no larger than 30, approximately how many additional items do you think need to be sampled?

8.5.23 Twelve samples of a metal alloy are tested. The flexibility measurements had a sample average of 732.9 and a sample standard deviation of 12.5.

(a) Is there sufficient evidence to conclude that the flexibility of this kind of metal alloy is smaller than 750? Use an appropriate hypothesis test to investigate this question.

(b) Construct a 99% confidence interval that provides an upper bound on the flexibility of this kind of metal alloy.

8.5.24 Flowrates in Urban Sewer Systems

Flow meters are installed in urban sewer systems to measure the flows through the pipes. In dry weather conditions (no rain) the flows are generated by waste water from households and industries, together with some possible drainage from water stored in the topsoil from previous rainfalls. In a study of an urban sewer system, the values given in DS 8.5.1 were obtained for flowrates during dry weather conditions.

(a) What is the sample mean?

(b) What is the sample median?

(c) What is the sample standard deviation?

(d) Construct a 99% two-sided confidence interval for the average flowrate under dry weather conditions.

(e) Construct a 95% one-sided confidence interval that provides an upper bound for the average flowrate under dry weather conditions.

(f) If a 99% two-sided confidence interval for the average flowrate under dry weather conditions is required with a length no larger than 50, how much additional sampling would you recommend?

(g) Show how to test $H_0 : \mu = 440$ against $H_A : \mu \neq 440$.

(h) Show how to test $H_0 : \mu \geq 480$ against $H_A : \mu < 480$.

8.5.25 Polymer Compound Densities

Eight samples of a polymer compound were obtained and their densities were measured as given in DS 8.5.2.

(a) Use an appropriate hypothesis test to assess whether there is sufficient evidence to establish that the average density of these kind of compounds is larger than 3.50.

(b) Construct a 99% confidence interval that provides a lower bound on the average density of these kind of compounds.

8.5.26 In a sample of size 33 a sample mean of 382.97 and a sample standard deviation of 3.81 are obtained.

(a) Use an appropriate hypothesis test to assess whether there is sufficient evidence to establish that the population mean is different from 385.

(b) Construct a 99% two-sided confidence interval for the population mean.

8.5.27 Show how Table III can be used to put bounds on the p-values for these hypothesis tests.

(a) $n = 24$, $\bar{x} = 2.39$, $s = 0.21$, $H_0 : \mu = 2.5$, $H_A : \mu \neq 2.5$

(b) $n = 30$, $\bar{x} = 0.538$, $s = 0.026$, $H_0 : \mu \geq 0.540$, $H_A : \mu < 0.540$

(c) $n = 10$, $\bar{x} = 143.6$, $s = 4.8$, $H_0 : \mu \leq 135.0$, $H_A : \mu > 135.0$

You can use the data sets referred to in Problems 8.5.28–8.5.35 to practice confidence interval construction and hypothesis testing for an unknown population mean.

8.5.28 Glass Fiber Reinforced Polymer Tensile Strengths
The data set in DS 6.6.7.

8.5.29 Infant Blood Levels of Hydrogen Peroxide
The data set in DS 6.6.8.

8.5.30 Paper Mill Operation of a Lime Kiln
The data set in DS 6.6.9.

8.5.31 River Salinity Levels
The data set in DS 6.6.10.

8.5.32 Dew Point Readings from Coastal Buoys
The data set in DS 6.6.11.

8.5.33 Brain pH levels
The data set in DS 6.6.12.

8.5.34 Silicon Dioxide Percentages in Ocean Floor Volcanic Glass
The data set in DS 6.6.13.

8.5.35 Network Server Response Times
The data set in DS 6.6.14.

CHAPTER NINE

Comparing Two Population Means

9.1 Introduction

9.1.1 Two-Sample Problems

One of the most important statistical problems is making comparisons between two probability distributions, which is considered in this chapter. This issue is often referred to as a **two-sample problem** since an experimenter typically has a set of data observations

$$x_1, \ldots, x_n$$

from one population, population A say, and an additional set of data observations

$$y_1, \ldots, y_m$$

from another population, population B say. The sample of data observations x_i are taken to be a set of independent observations from the unknown probability distribution governing population A, with a cumulative distribution function $F_A(x)$. Similarly, the sample of data observations y_i are taken to be a set of independent observations from the unknown probability distribution governing population B, with a cumulative distribution function $F_B(x)$. The sample sizes n and m of the two data sets need not be equal, although experiments are often designed to have equal sample sizes.

In general, an experimenter is interested in assessing the evidence that there is a difference between the two probability distributions $F_A(x)$ and $F_B(x)$. One important aspect of this assessment is a comparison between the means of the two probability distributions, μ_A and μ_B, as illustrated in Figure 9.1. Thus if $\mu_A = \mu_B$, the two populations have equal means and this may be sufficient for the experimenter to conclude that for practical purposes, the populations are "identical" (although, in addition, a comparison of the variances of the two populations may be informative, as illustrated in Figure 9.2). If the data analysis provides evidence that $\mu_A \neq \mu_B$, this indicates that the population probability distributions are different.

| Example 51 |
| Acrophobia Treatments |

The standard treatment of acrophobia, the fear of heights, involves desensitizing patients' fear of heights by asking them to imagine being in high places and by taking them to high places, as illustrated in Figure 9.3. A proposed new treatment using virtual reality provides a patient with a head-mounted display that simulates the appearance of being in a building and allows the patient to "travel" around the building. This device allows the patient to "explore" high buildings while actually remaining in a safe place.

In an experiment to investigate whether the new treatment is effective or not, a group of 30 patients suffering from acrophobia are randomly assigned to one of the two treatment methods. Thus, 15 patients undergo the standard treatment, treatment A say, and 15 patients undergo the proposed new treatment, treatment B. At the conclusion of the treatments the patients are given a score that measures how much their condition has improved. The scores

FIGURE 9.1

Comparison of the means of two
probability distributions

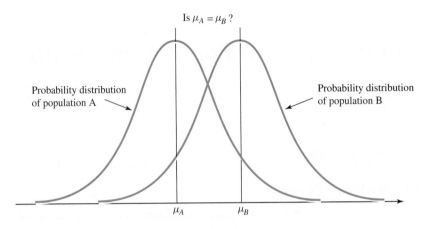

FIGURE 9.2

Comparison of the variances of two
probability distributions

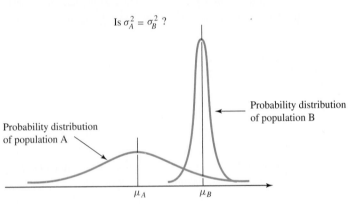

of the patients undergoing the standard treatment provide the data observations

$$x_1, \ldots, x_{15}$$

and the scores of the patients undergoing the new treatment provide the data observations

$$y_1, \ldots, y_{15}$$

For this example, a comparison of the population means μ_A and μ_B provides an indication of whether the new treatment is any better or any worse than the standard treatment.

Example 51 provides a good example of the use of a **control group**, which in this case is the group of patients who undergo the standard treatment. In general, a control group provides a standard against which a new procedure or treatment can be measured. Notice that it is good experimental practice to **randomize** the allocation of subjects or experimental objects between the standard treatment and the new treatment, as shown in Figure 9.4. Randomization helps to eliminate any bias that may otherwise arise if certain kinds of subject are "favored" and given a particular treatment.

Control groups are particularly important in medical or clinical trials where the efficacy of a new treatment or drug is under investigation. In these experiments the patients in the control group may actually be administered a *placebo*, so that in effect they have no treatment at all. For example, half the group may be given the new pills that are being tested and the other half may be given identical looking pills that actually contain no medicine at all. Good

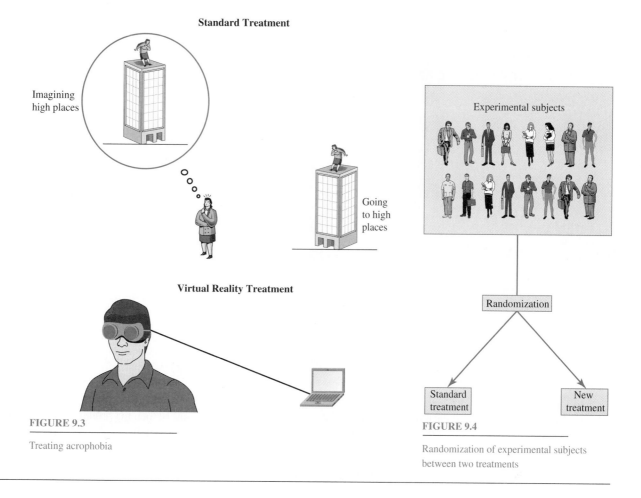

Standard Treatment

Imagining high places

Going to high places

Experimental subjects

Randomization

Standard treatment

New treatment

Virtual Reality Treatment

FIGURE 9.3

Treating acrophobia

FIGURE 9.4

Randomization of experimental subjects between two treatments

experimental practice dictates that it is usually appropriate to run **blind** experiments where the patients do not know which treatment they are receiving (Figure 9.5). In addition, these experiments are often run in a **double-blind** manner whereby the person taking measurements also does not know which treatment each patient received. These practices help to alleviate any bias that may arise from patients or experimenters inadvertently allowing their perceptions or hopes of what should happen to influence the results.

Of course, some experiments cannot be run blind. In Example 51 concerning acrophobia treatments the patients obviously know if they are receiving the new virtual reality treatment. However, it still may be advisable to arrange for the person measuring the progress made by the patients to be unaware of which patients received which treatment.

Example 52
Kaolin Processing

Kaolin, a white clay material, is processed in a calciner to remove impurities. An important characteristic of the processed kaolin is its "brightness" since this determines its suitability for use in such things as paper products, ceramics, paints, medicines, and cosmetics.

A processing company has two calciners and the manager is interested in investigating whether they are equally effective in processing the kaolin. A batch of kaolin is fed into the two calciners and 12 randomly selected samples of the processed material are collected from

FIGURE 9.5

In a blind experiment the experimental subjects do not know which treatment they receive

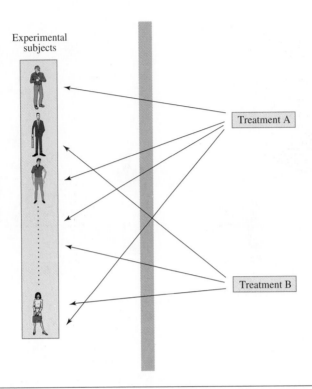

each of the calciners. If the calciners are labeled A and B, then the brightness measurements of the 12 samples from calciner A provide the data observations

$$x_1, \ldots, x_{12}$$

and the brightness measurements of the 12 samples from calciner B provide the data observations

$$y_1, \ldots, y_{12}$$

A comparison of the population means μ_A and μ_B provides an indication of whether the two calciners are equally effective.

Example 53
Kudzu Pulping

Chemical engineers are interested in finding nonwood fibers that can be used as an alternative to wood pulp in paper manufacture. In an experiment to investigate the utility of kudzu, a fast-growing vine that covers much of the southeastern United States, kudzu batches are pulped with and without the addition of anthraquinone. One question of interest is whether the addition of anthraquinone increases pulp yield.

A set of 20 experiments performed without anthraquinone (the control group) provide pulp yield measurements

$$x_1, \ldots, x_{20}$$

and a set of 25 experiments performed with anthraquinone provide pulp yield measurements

$$y_1, \ldots, y_{25}$$

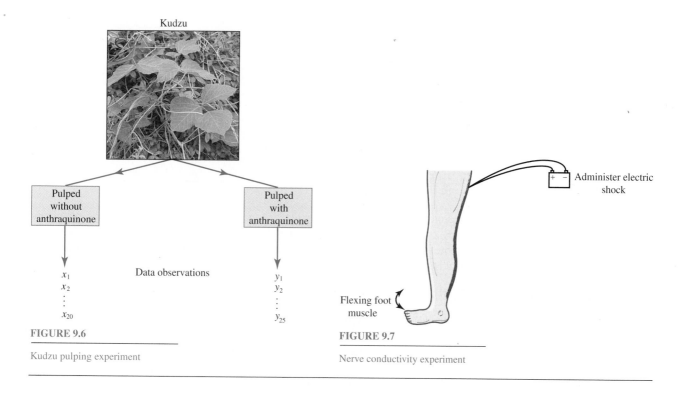

Kudzu

Pulped without anthraquinone

Pulped with anthraquinone

x_1
x_2
$\vdots$
x_{20}

Data observations

y_1
y_2
$\vdots$
y_{25}

FIGURE 9.6

Kudzu pulping experiment

Administer electric shock

Flexing foot muscle

FIGURE 9.7

Nerve conductivity experiment

as illustrated in Figure 9.6. In this experiment, a comparison of the population means μ_A and μ_B indicates whether the addition of anthraquinone increases pulp yield.

Example 54
Nerve Conductivity
Speeds

A neurologist is investigating how diseases of the periphery nerves in humans influence the conductivity speed of the nervous system. As Figure 9.7 shows, the conductivity speed of nerves is determined by administering an electric shock to a patient's leg and measuring the time it takes to flex a muscle in the patient's foot.

Nerve conductivity speed measurements are made on $n = 32$ healthy patients and on $m = 27$ patients who are known to have a periphery nerve disorder. The comparison of the population means μ_A and μ_B provides an indication of whether diseases of the periphery nerves affect the conductivity speed of the nervous system.

Example 45
Fabric Water
Absorption Properties

With the experimental apparatus shown in Figure 6.37 an experimenter can alter the revolutions per minute of the rollers and the pressure between them. If the rollers rotate at 24 revolutions per minute, how does changing the pressure from 10 pounds per square inch to 20 pounds per square inch influence the water pickup of the fabric?

This question can be investigated by collecting some data observations x_i of the fabric water pickup with a pressure of 10 pounds per square inch and some data observations y_i of the fabric water pickup with a pressure of 20 pounds per square inch. A comparison of the population means μ_A and μ_B shows how the average fabric water pickup is influenced by the change in pressure.

The comparison between the unknown parameters μ_A and μ_B may involve the construction of a confidence interval for the difference $\mu_A - \mu_B$. This confidence interval is centered at the point estimate $\bar{x} - \bar{y}$. It is particularly interesting to discover whether or not the confidence

FIGURE 9.8

Interpretation of confidence
intervals for $\mu_A - \mu_B$

Two-sided confidence interval for $\mu_A - \mu_B$

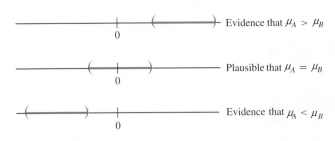

interval contains 0, because this provides information on the plausibility of the population means μ_A and μ_B being equal, as shown in Figure 9.8.

A more direct approach to assessing the plausibility that the population means μ_A and μ_B are equal is to calculate a *p*-value for the hypotheses

$$H_0 : \mu_A = \mu_B \quad \text{versus} \quad H_A : \mu_A \neq \mu_B$$

Small *p*-values (less than 0.01) indicate that the null hypothesis is not a plausible statement, and the experimenter can conclude that there is sufficient evidence to establish that the two population means are different. A large *p*-value (greater than 0.10) indicates that there is *not* sufficient evidence to establish that the two population means are different. Notice that the hypothesis testing problem is formulated so that the equality of the population means is considered to be plausible unless the data present sufficient evidence to prove that this cannot be the case. One-sided versions of this hypothesis test can also be used.

9.1.2 Paired Samples Versus Independent Samples

When collecting and analysing data for the comparison of two populations, it is important to pay some attention to the **experimental design**. This term refers to the manner in which the data are collected. In this chapter a distinction will be made between **paired samples** and **independent samples**, and the appropriate analysis method depends upon which of these two experimental designs is employed. The advantage of paired samples is that they can alleviate the effect of variabilities in a factor other than the difference between the two populations. This concept is illustrated in the following example.

Example 55

Heart Rate Reductions

A new drug for inducing a temporary reduction in a patient's heart rate is to be compared with a standard drug. The drugs are to be administered to a patient at rest, and the percentage reduction in the heart rate is to be measured after five minutes.

Since the drug efficacy is expected to depend heavily on the particular patient involved, a **paired experiment** is run whereby each of 40 patients is administered one drug on one day and the other drug on the following day. The spacing of the two experiments over two days ensures that there is no "carryover" effect since the drugs are only temporarily effective. Nevertheless, as Figure 9.9 illustrates, the *order* in which the two drugs are administered is decided in a random manner so that one patient may have the standard drug followed by the new drug and another patient may have the new drug followed by the standard drug. The comparison between the two drugs is based upon the *differences* for each patient in the percentage heart rate reductions achieved by the two drugs.

FIGURE 9.9

Heart rate reduction experiment

	Day 1	Day 2	Difference
patient 1	standard drug x_1	new drug y_1	$z_1 = x_1 - y_1$
patient 2	new drug y_2	standard drug x_2	$z_2 = x_2 - y_2$
patient 3	standard drug x_3	new drug y_3	$z_3 = x_3 - y_3$
patient 4	new drug y_4	standard drug x_4	$z_4 = x_4 - y_4$
$\vdots$	$\vdots$	$\vdots$	$\vdots$
patient 39	standard drug x_{39}	new drug y_{39}	$z_{39} = x_{39} - y_{39}$
patient 40	new drug y_{40}	standard drug x_{40}	$z_{40} = x_{40} - y_{40}$

FIGURE 9.10

The distinction between paired
and independent samples

Paired Samples

Independent Samples

As this example illustrates, data from paired samples are of the form

$$(x_1, y_1), (x_2, y_2), \ldots, (x_n, y_n)$$

which arise from each of n experimental subjects being subjected to both "treatments." The data observation x_i represents the measurement of treatment A applied to the ith experimental subject, and the data observation y_i represents the measurement of treatment B applied to the same subject. The comparison between the two treatments is then based upon the pairwise differences

$$z_i = x_i - y_i \qquad 1 \le i \le n$$

Notice that with paired samples the sample sizes n and m from populations A and B obviously have to be equal. The distinction between paired samples and independent (unpaired) samples is illustrated in Figure 9.10.

Conducting an experiment in a paired manner, when it is possible to do so, is a specific example of a more general experimental design concept called **blocking**. The experimenter attempts to "block out" unwanted sources of variation that otherwise might cloud the comparisons of real interest. Additional experimental designs and analyses that use blocking are presented in Section 11.2.

In Example 55, the medical researchers know that the efficacies of the two drugs vary considerably from one patient to another. Suppose that the experiment had been run on a group of 80 patients with 40 patients being assigned randomly to each of the two drugs, as illustrated in Figure 9.11. If the new treatment then appears to be better than the standard

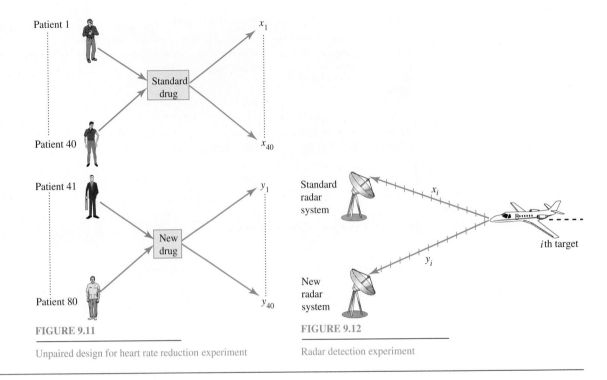

FIGURE 9.11

Unpaired design for heart rate reduction experiment

FIGURE 9.12

Radar detection experiment

treatment, could it be because the new treatment happened to be administered to patients who are more receptive to drug treatment? In statistical terms, the variability in the patients creates more "noisy" data, which make it more difficult to detect a difference in the efficacies of the two drugs. Conducting a paired experiment and looking at the differences in the two measurements for each patient neutralizes the variability among the patients. Thus, the paired experiment is more *efficient* in that it provides more information for the given amount of data collection.

Example 56

Radar Detection Systems

A new radar system for detecting airborne objects is being tested against a standard system. Different types of targets are flown in different atmospheric conditions, and the distance of the target from the radar system location is measured at the time when the target is first detected by the radar. It is obviously sensible to conduct this experiment with a paired design whereby both radar systems attempt to detect the same target at the same time, assuming that the two systems do not interfere with each other while operating simultaneously. The data observations then consist of pairs (x_i, y_i), where x_i is the distance of the ith target when detected by the standard radar system and y_i is the distance of the ith target when detected by the new radar system, as shown in Figure 9.12.

An unpaired design for this experiment would consist of a series of targets being tested against one radar system, with a rerun of the targets (or additional targets) for the other radar system. The comparison between the two radar systems would then be clouded by possible variations in the "detectability" of the targets due to possible changes in the experimental conditions such as atmospheric conditions.

In conclusion, when an extraneous source of variation can be identified, such as variations in a patient's receptiveness to a drug or variations in a target's detectability, it is best to employ a paired experimental design where possible. Unfortunately, in many cases it is impossible to employ a paired experimental design due to the nature of the problem.

For instance, in Example 51 where two treatments for acrophobia are compared, it is to be expected that a given treatment works better on some patients than on others. In other words, there is likely to be some patient variability. However, the experiment cannot be run in a paired fashion since this would require patients to undergo one treatment and then to "revert" to their previous state before undergoing the second treatment! Undergoing one treatment changes a subject so that an equivalent assessment of the other treatment on the same subject cannot be undertaken. Similarly, if in Example 56 the two radar detection systems interfere with each other if they are operated at the same time from the same place, then the paired design as described is not feasible.

Section 9.2 discusses the analysis of paired samples based upon their reduction to a one-sample problem and the employment of the one-sample techniques discussed in Chapter 8. In Section 9.3 new techniques for analyzing two independent (unpaired) samples are discussed.

9.2 Analysis of Paired Samples

9.2.1 Methodology

The analysis of paired samples with data observations

$$(x_1, y_1), (x_2, y_2), \ldots, (x_n, y_n)$$

is performed by reducing the problem to a one-sample problem. This is achieved by calculating the differences

$$z_i = x_i - y_i \qquad 1 \leq i \leq n$$

The data observations z_i can be taken to be independent, identically distributed observations from some probability distribution with mean μ. The one-sample techniques discussed in Chapter 8 can be applied to the data set

$$z_1, \ldots, z_n$$

in order to make inferences about the unknown mean μ. The parameter μ can be interpreted as being the average difference between the "treatments" A and B.

Positive values of μ indicate that the random variables X_i tend to be larger than the random variables Y_i, so that the mean of population A, μ_A, is larger than the mean of population B, μ_B. Similarly, negative values of μ indicate that $\mu_A < \mu_B$. It is usually particularly interesting to test the hypotheses

$$H_0 : \mu = 0 \quad \text{versus} \quad H_A : \mu \neq 0$$

If the null hypothesis is a plausible statement, then it implies that there is not sufficient evidence of a difference between the mean values of the probability distributions of population A and population B.

It can be instructive to build a simple **model** for the data observations. The observation x_i, that is, the observation obtained when treatment A is applied to the ith experimental subject, can be thought of as a treatment A effect μ_A, together with a subject i effect γ_i say, and with some random error ϵ_i^A. Thus

$$x_i = \mu_A + \gamma_i + \epsilon_i^A$$

Similarly, the observation y_i, that is, the observation obtained when treatment B is applied to the ith experimental subject, can be thought of as being formed as a treatment B effect μ_B, together with the same subject i effect γ_i, and with a random error ϵ_i^B, so that

$$y_i = \mu_B + \gamma_i + \epsilon_i^B$$

Notice that in these models the pairing of the data implies that the subject effects γ_i are the same in the two equations. Also, it is important to notice that μ_A, μ_B, and γ_i are fixed unknown parameters, while the error terms ϵ_i^A and ϵ_i^B are observations of random variables with expectations equal to 0.

With these model representations, it is clear that the differences z_i can be represented as

$$z_i = \mu_A - \mu_B + \epsilon_i^{AB}$$

where the error term is

$$\epsilon_i^{AB} = \epsilon_i^A - \epsilon_i^B$$

Since this error term is an observation from a distribution with a zero expectation, the differences z_i are consequently observations from a distribution with expectation

$$\mu = \mu_A - \mu_B$$

which does not depend on the subject effect γ_i.

9.2.2   Examples

Example 55

Heart Rate Reductions

Figure 9.13 contains the percentage reductions in heart rate for the standard drug x_i and the new drug y_i, together with the differences z_i, for the 40 experimental subjects. First of all, notice that the patients exhibit a wide variability in their response to the drugs. For some patients the heart rate reductions are close to 20%, while for others they are over 40%. This variability confirms the appropriateness of a paired experiment.

An initial investigation of the data observations z_i reveals that 30 out of 40 are negative. This result suggests that $\mu = \mu_A - \mu_B < 0$, so that the new drug has a stronger effect on average. This suggestion is reinforced by a negative value for the sample average $\bar{z} = -2.655$. The sample standard deviation of the differences z_i is $s = 3.730$, so that with $\mu = 0$ the t-statistic is

$$t = \frac{\sqrt{n}(\bar{z} - \mu)}{s} = \frac{\sqrt{40} \times (-2.655)}{3.730} = -4.50$$

The p-value for the two-sided hypothesis testing problem

$$H_0 : \mu = 0 \quad \text{versus} \quad H_A : \mu \neq 0$$

is therefore

$$p\text{-value} = 2 \times P(X > 4.50) \simeq 0.0001$$

where the random variable X has a t-distribution with 39 degrees of freedom.

This analysis reveals that it is *not* plausible that $\mu = 0$, and so the experimenter can conclude that there is evidence that the new drug has a different effect from the standard drug. From the critical point $t_{0.005,39} = 2.7079$, a 99% two-sided confidence interval for the

FIGURE 9.13

Heart rate reductions data set
(% reduction in heart rate)

Patient	Standard drug x_i	New drug y_i	$z_i = x_i - y_i$
1	28.5	34.8	−6.3
2	26.6	37.3	−10.7
3	28.6	31.3	−2.7
4	22.1	24.4	−2.3
5	32.4	39.5	−7.1
6	33.2	34.0	−0.8
7	32.9	33.4	−0.5
8	27.9	27.4	0.5
9	26.8	35.4	−8.6
10	30.7	35.7	−5.0
11	39.6	40.4	−0.8
12	34.9	41.6	−6.7
13	31.1	30.8	0.3
14	21.6	30.5	−8.9
15	40.2	40.7	−0.5
16	38.9	39.9	−1.0
17	31.6	30.2	1.4
18	36.0	34.5	1.5
19	25.4	31.2	−5.8
20	35.6	35.5	0.1
21	27.0	25.3	1.7
22	33.1	34.5	−1.4
23	28.7	30.9	−2.2
24	33.7	31.9	1.8
25	33.7	36.9	−3.2
26	34.3	27.8	6.5
27	32.6	35.7	−3.1
28	34.5	38.4	−3.9
29	32.9	36.7	−3.8
30	29.3	36.3	−7.0
31	35.2	38.1	−2.9
32	29.8	32.1	−2.3
33	26.1	29.1	−3.0
34	25.6	33.5	−7.9
35	27.6	28.7	−1.1
36	25.1	31.4	−6.3
37	23.7	22.4	1.3
38	36.3	43.7	−7.4
39	33.4	30.8	2.6
40	40.1	40.8	−0.7

difference between the average effects of the drugs is

$$
\mu = \mu_A - \mu_B \in \left(\bar{z} - \frac{t_{0.005,39}s}{\sqrt{40}}, \bar{z} + \frac{t_{0.005,39}s}{\sqrt{40}} \right)
$$

$$
= \left(-2.655 - \frac{2.7079 \times 3.730}{\sqrt{40}}, -2.655 + \frac{2.7079 \times 3.730}{\sqrt{40}} \right)
$$

$$
= (-4.252, -1.058)
$$

Consequently, based upon this data set the experimenter can conclude that the new drug provides a reduction in a patient's heart rate of somewhere between 1% and 4.25% more on average than the standard drug.

FIGURE 9.14

Radar detection systems data set
(distance of target in miles when
detected)

Target	Standard radar system x_i	New radar system y_i	$z_i = x_i - y_i$
1	48.40	51.14	−2.74
2	47.73	46.48	1.25
3	51.30	50.90	0.40
4	50.49	49.82	0.67
5	47.06	47.99	−0.93
6	53.02	53.20	−0.18
7	48.96	46.76	2.20
8	52.03	54.44	−2.41
9	51.09	49.85	1.24
10	47.35	47.45	−0.10
11	50.15	50.66	−0.51
12	46.59	47.92	−1.33
13	52.03	52.37	−0.34
14	51.96	52.90	−0.94
15	49.15	50.67	−1.52
16	48.12	49.50	−1.38
17	51.97	51.29	0.68
18	53.24	51.60	1.64
19	55.87	54.48	1.39
20	45.60	45.62	−0.02
21	51.80	52.24	−0.44
22	47.64	47.33	0.31
23	49.90	51.13	−1.23
24	55.89	57.86	−1.97

Example 56

Radar Detection Systems

Figure 9.14 shows the radar detection distances in miles for 24 targets. The observations x_i are for the standard system and the observations y_i are for the new system. An initial look at the data indicates that the detectability of the targets varies from about 45 miles in some cases to over 55 miles in other cases, and this confirms the advisability of a paired experiment.

The differences z_i have a sample mean $\bar{z} = -0.261$ and a sample standard deviation $s = 1.305$. With $\mu = 0$ the t-statistic is therefore

$$t = \frac{\sqrt{n}(\bar{z} - \mu)}{s} = \frac{\sqrt{24} \times (-0.261)}{1.305} = -0.980$$

If the experimenter is interested in ascertaining whether or not the new radar system can detect targets at a greater distance than the standard system, it is appropriate to consider the one-sided hypothesis testing problem

$$H_0 : \mu \geq 0 \quad \text{versus} \quad H_A : \mu < 0$$

This is because the experimenter is asking whether there is sufficient evidence to establish that $\mu < 0$. In this case the p-value is

$$P(X \leq -0.980) = 0.170$$

where the random variable X has a t-distribution with 23 degrees of freedom. With such a large p-value the analysis indicates that this data set does not provide sufficient evidence to establish that the new radar system is any better than the standard radar system.

■ 9.2.3   Problems

9.2.1 Production Line Assembly Methods
DS 9.2.1 shows the data obtained from a paired experiment performed to examine which of two assembly methods is quicker on average. A random sample of 35 workers on an assembly line were selected and all were timed while they assembled an item in the standard manner (method A) and while they assembled an item in the new manner (method B). The times in seconds are recorded. Analyze the data set and present your conclusions on how the new assembly method differs from the standard assembly method. Why are the two data samples paired? Why did the experimenter decide to perform a paired experiment rather than an unpaired experiment?

9.2.2 Red Blood Cell Adherence to Endothelial Cells
Researchers into the genetic disease sickle cell anemia are interested in how red blood cells adhere to endothelial cells, which form the innermost lining of blood vessels. A set of 14 blood samples are obtained, and each sample is split in half. One half of the blood sample is profused over an endothelial monolayer of type A, and the other half of the blood sample is profused over an endothelial monolayer of type B. The two types of monolayer differ in respect to the stimulation conditions of the endothelial cells. The data recorded in DS 9.2.2 are the number of adherent red blood cells per mm^2. Is there any evidence that the different stimulation conditions affect the adhesion of red blood cells?

9.2.3 Tire Tread Wear
An experiment is performed to assess whether a new tire wears more slowly than a standard tire. A set of 20 trucks is chosen. A new tire is placed on one of the front wheels of each truck, and a standard tire is placed on the other front wheel. Right and left positions of the two kinds of tire are randomized over the 20 trucks. The trucks are driven over varying road conditions, and then the *reductions* in the tread depths of the tires are measured. DS 9.2.3 contains these data values in mm. Analyze the data and present your conclusions. Why is this a paired experiment? Why did the experimenter decide to perform a paired experiment rather than an unpaired experiment?

9.2.4 Calculus Teaching Methods
A new teaching method for a calculus class is being evaluated. A set of 80 students is formed into 40 pairs, where the two-pair members have roughly equal mathematics test scores. The pairs are then randomly split,

with one member being assigned to section A where the standard teaching method is used and with one member being assigned to section B where the new teaching method is tried. At the end of the course all the students take the same exam and their scores are shown in DS 9.2.4. Analyze the data and present your conclusions regarding how the new teaching method compares with the standard approach. Why is this a paired experiment? Why was it decided to perform a paired experiment rather than an unpaired experiment?

9.2.5 Radioactive Carbon Dating
Two independently operated laboratories provide historical dating services using radioactive carbon dating methods. A researcher suspects that one laboratory tends to provide older datings than the other laboratory. To investigate this supposition 18 samples of old material are split in half. One half is sent to laboratory A for dating, and the other half is sent to laboratory B for dating. The laboratories are asked to submit their answers to the nearest decade, and the results obtained are presented in DS 9.2.5. Is there any evidence that one laboratory tends to provide older datings than the other laboratory?

9.2.6 Golf Ball Design
A sports manufacturer has developed a new golf ball with special dimples that it hopes will cause the ball to travel farther than standard golf balls. This question is examined by asking 24 golfers to hit 10 new balls and 10 standard balls. Each ball is hit off a tee, and randomization techniques are employed to account for any fatigue or wind effects. The distances traveled by the balls are measured, and DS 9.2.6 presents the average distances in yards for the 10 shots for each of the 24 golfers and the two types of ball. What should the company conclude from this experiment? Why was it a good idea to use a paired design for this experiment?

9.2.7 Stimulus Reaction Times
An experiment was conducted to compare two procedures (A and B) for measuring a person's reaction time to a stimulus. Ten volunteers participated in the experiment, and each volunteer was given the stimulus twice. For each person the reaction time was measured once with procedure A and once with procedure B, as shown in DS 9.2.7. A reviewer comments that the differences between the reaction times obtained for procedures A and B can be explained by the fact that each time a person is given the

stimulus the actual reaction time varies. Do you agree with the reviewer, or do you think that there is evidence that procedures A and B do give different readings on average?

9.2.8 Antibiotic Efficacies

Eight cultures of a bacterium are split in half. One half is tested using a standard antibiotic and the other half is tested using a new antibiotic. The data values in DS 9.2.8 are the times taken to kill the bacterium. Use an appropriate hypothesis test to assess whether there is any evidence that the new antibiotic is quicker than the standard antibiotic.

9.2.9 Uranium-Oxide Removal from Water

An experiment is conducted to investigate how the addition of a surfactant affects the ability of magnetized steel wool

to remove uranium-oxide particles from water. Six batches of uranium-oxide contaminated water are obtained that are each split in half, and the surfactant is added to one of the two halves for each batch. The uranium-oxide levels are measured for each of the resulting 12 samples of water both before and after they are passed through some magnetized steel wool, and the reductions in the uranium-oxide levels are calculated. The resulting data set is shown in DS 9.2.9. Perform a hypothesis test to investigate whether this experiment provides sufficient evidence for the experimenter to conclude that the addition of the surfactant has an effect on the ability of magnetized steel wool to remove uranium-oxide particles from water.

9.3 Analysis of Independent Samples

The analysis of two independent (unpaired) samples is now considered. The data consist of a sample of n observations x_i from population A with a sample mean $\bar{x}$ and a sample standard deviation s_x, together with a sample of m observations y_i from population B with a sample mean $\bar{y}$ and a sample standard deviation s_y, as shown in Figure 9.15.

The point estimate of the difference in the population means $\mu_A - \mu_B$ is $\bar{x} - \bar{y}$. Since $\text{Var}(\bar{x}) = \sigma_A^2/n$ and $\text{Var}(\bar{y}) = \sigma_B^2/m$, where σ_A^2 and σ_B^2 are the two population variances, this point estimate has a standard error

$$\text{s.e.}(\bar{x} - \bar{y}) = \sqrt{\frac{\sigma_A^2}{n} + \frac{\sigma_B^2}{m}}$$

Three procedures for making inferences about the difference of the population means $\mu_A - \mu_B$ are outlined below, and they differ with respect to how the standard error of $\bar{x} - \bar{y}$ is estimated.

The first "general procedure" estimates the standard error as

$$\text{s.e.}(\bar{x} - \bar{y}) = \sqrt{\frac{s_x^2}{n} + \frac{s_y^2}{m}}$$

A second "pooled variance procedure" is based on the assumption that the population variances σ_A^2 and σ_B^2 are equal, and estimates the standard error as

$$\text{s.e.}(\bar{x} - \bar{y}) = s_p \sqrt{\frac{1}{n} + \frac{1}{m}}$$

where s_p^2 is a pooled estimate of the common population variance. These two procedures are referred to as **two-sample t-tests**. Finally, a **two-sample z-test** can be used when the population variances σ_A^2 and σ_B^2 are assumed to take "known" values. In each case, the choice

FIGURE 9.15

Summary statistics for analysis of two independent samples

	Sample size	Sample mean	Sample standard deviation
Population A	n	$\bar{x}$	s_x
Population B	m	$\bar{y}$	s_y

of the estimate of the standard error affects the probability distribution used to calculate p-values and critical points. In the first two cases t-distributions are appropriate, though with different degrees of freedom, and in the final case the standard normal distribution is used. Out of these three procedures the general procedure can always be used, although in certain cases an experimenter may prefer one of the other two procedures.

As with one-sample t-tests and z-tests, these two-sample tests are based on the assumption that the data are normally distributed. For large sample sizes the central limit theorem implies that the sample means are normally distributed and this is sufficient to ensure that the tests are appropriate. For small sample sizes the tests also behave satisfactorily unless the data observations are clearly not normally distributed, in which case it is wise to employ one of the nonparametric procedures described in Chapter 15.

9.3.1 General Procedure

A general method for making inferences about the difference of the population means $\mu_A - \mu_B$ uses a point estimate $\bar{x} - \bar{y}$ whose standard error is estimated by

$$\text{s.e.}(\bar{x} - \bar{y}) = \sqrt{\frac{s_x^2}{n} + \frac{s_y^2}{m}}$$

In this case p-values and critical points are calculated from a t-distribution. The degrees of freedom ν of the t-distribution are usually calculated to be

$$\nu = \frac{\left(\frac{s_x^2}{n} + \frac{s_y^2}{m}\right)^2}{\frac{s_x^4}{n^2(n-1)} + \frac{s_y^4}{m^2(m-1)}}$$

rounded down to the nearest integer. The simpler choice $\nu = \min\{n, m\} - 1$ can also be used, although it is a little less powerful.

A two-sided $1 - \alpha$ level confidence interval for $\mu_A - \mu_B$ is therefore

$$\mu_A - \mu_B \in \left(\bar{x} - \bar{y} - t_{\alpha/2,\nu} \sqrt{\frac{s_x^2}{n} + \frac{s_y^2}{m}}, \; \bar{x} - \bar{y} + t_{\alpha/2,\nu} \sqrt{\frac{s_x^2}{n} + \frac{s_y^2}{m}} \right)$$

which, as shown in Figure 9.16, is constructed using the standard format

$$\mu \in (\hat{\mu} - \text{critical point} \times \text{s.e.}(\hat{\mu}), \; \hat{\mu} + \text{critical point} \times \text{s.e.}(\hat{\mu}))$$

FIGURE 9.16

A two-sample two-sided t-interval

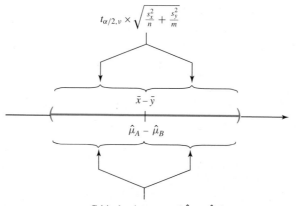

Critical point $\times$ s.e. $(\hat{\mu}_A - \hat{\mu}_B)$

with $\mu = \mu_A - \mu_B$ in this case. One-sided confidence intervals are

$$\mu_A - \mu_B \in \left(-\infty, \bar{x} - \bar{y} + t_{\alpha,\nu} \sqrt{\frac{s_x^2}{n} + \frac{s_y^2}{m}} \right)$$

and

$$\mu_A - \mu_B \in \left(\bar{x} - \bar{y} - t_{\alpha,\nu} \sqrt{\frac{s_x^2}{n} + \frac{s_y^2}{m}}, \infty \right)$$

For the two-sided hypothesis testing problem

$$H_0 : \mu_A - \mu_B = \delta \quad \text{versus} \quad H_A : \mu_A - \mu_B \neq \delta$$

for some fixed value δ of interest (usually $\delta = 0$), the appropriate t-statistic is

$$t = \frac{\bar{x} - \bar{y} - \delta}{\sqrt{\frac{s_x^2}{n} + \frac{s_y^2}{m}}}$$

The two-sided p-value is calculated as

$$p\text{-value} = 2 \times P(X > |t|)$$

where the random variable X has a t-distribution with ν degrees of freedom, and a size α hypothesis test accepts the null hypothesis if

$$|t| \leq t_{\alpha/2,\nu}$$

and rejects the null hypothesis when

$$|t| > t_{\alpha/2,\nu}$$

A one-sided hypothesis testing problem

$$H_0 : \mu_A - \mu_B \leq \delta \quad \text{versus} \quad H_A : \mu_A - \mu_B > \delta$$

has a p-value

$$p\text{-value} = P(X > t)$$

and a size α hypothesis test accepts the null hypothesis if

$$t \leq t_{\alpha,\nu}$$

and rejects the null hypothesis if

$$t > t_{\alpha,\nu}$$

Similarly, the one-sided hypothesis testing problem

$$H_0 : \mu_A - \mu_B \geq \delta \quad \text{versus} \quad H_A : \mu_A - \mu_B < \delta$$

has a p-value

$$p\text{-value} = P(X < t)$$

and a size α hypothesis test accepts the null hypothesis if

$$t \geq -t_{\alpha,\nu}$$

and rejects the null hypothesis if

$$t < -t_{\alpha,\nu}$$

As an illustration of these inference procedures, suppose that data are obtained with $n = 24$, $\bar{x} = 9.005$, $s_x = 3.438$ and $m = 34$, $\bar{y} = 11.864$, $s_y = 3.305$. The hypotheses

$$H_0 : \mu_A = \mu_B \quad \text{versus} \quad H_A : \mu_A \neq \mu_B$$

are tested with the t-statistic

$$t = \frac{\bar{x} - \bar{y}}{\sqrt{\frac{s_x^2}{n} + \frac{s_y^2}{m}}} = \frac{9.005 - 11.864}{\sqrt{\frac{3.438^2}{24} + \frac{3.305^2}{34}}} = -3.169$$

Two-Sample t-Procedure (Unequal Variances)

Consider a sample of size n from population A with a sample mean $\bar{x}$ and a sample standard deviation s_x, and a sample of size m from population B with a sample mean $\bar{y}$ and a sample standard deviation s_y.

A two-sided $1 - \alpha$ level confidence interval for the difference in population means $\mu_A - \mu_B$ is

$$\mu_A - \mu_B \in \left(\bar{x} - \bar{y} - t_{\alpha/2,\nu} \sqrt{\frac{s_x^2}{n} + \frac{s_y^2}{m}}, \bar{x} - \bar{y} + t_{\alpha/2,\nu} \sqrt{\frac{s_x^2}{n} + \frac{s_y^2}{m}} \right)$$

where the degrees of freedom of the critical point are

$$\nu = \frac{\left(\frac{s_x^2}{n} + \frac{s_y^2}{m} \right)^2}{\frac{s_x^4}{n^2(n-1)} + \frac{s_y^4}{m^2(m-1)}}$$

One-sided confidence intervals are

$$\mu_A - \mu_B \in \left(-\infty, \bar{x} - \bar{y} + t_{\alpha,\nu} \sqrt{\frac{s_x^2}{n} + \frac{s_y^2}{m}} \right)$$

and

$$\mu_A - \mu_B \in \left(\bar{x} - \bar{y} - t_{\alpha,\nu} \sqrt{\frac{s_x^2}{n} + \frac{s_y^2}{m}}, \infty \right)$$

The appropriate t-statistic for the null hypothesis $H_0 : \mu_A - \mu_B = \delta$ is

$$t = \frac{\bar{x} - \bar{y} - \delta}{\sqrt{\frac{s_x^2}{n} + \frac{s_y^2}{m}}}$$

A two-sided p-value is calculated as $2 \times P(X > |t|)$, where the random variable X has a t-distribution with ν degrees of freedom, and one-sided p-values are $P(X > t)$ and $P(X < t)$. A size α two-sided hypothesis test accepts the null hypothesis if

$$|t| \leq t_{\alpha/2,\nu}$$

and rejects the null hypothesis when

$$|t| > t_{\alpha/2,\nu}$$

and size α one-sided hypothesis tests have rejection regions $t > t_{\alpha,\nu}$ or $t < -t_{\alpha,\nu}$.

These procedures are known as **two-sample t-tests** *without* a pooled variance estimate.

FIGURE 9.17

Calculation of a two-sided p-value

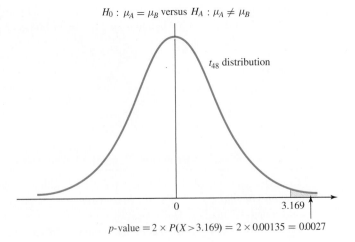

$H_0 : \mu_A = \mu_B$ versus $H_A : \mu_A \neq \mu_B$

t_{48} distribution

p-value $= 2 \times P(X > 3.169) = 2 \times 0.00135 = 0.0027$

The two-sided p-value is therefore

$$p\text{-value} = 2 \times P(X > 3.169)$$

where the random variable X has a t-distribution with degrees of freedom

$$\nu = \frac{\left(\frac{3.438^2}{24} + \frac{3.305^2}{34} \right)^2}{\frac{3.438^4}{24^2 \times 23} + \frac{3.305^4}{34^2 \times 33}} = 48.43$$

Using the integer value $\nu = 48$ gives

$$p\text{-value} \simeq 2 \times 0.00135 = 0.0027$$

as illustrated in Figure 9.17, so that there is very strong evidence that the null hypothesis is not a plausible statement, and the experimenter can conclude that $\mu_A \neq \mu_B$.

With a critical point $t_{0.005,48} = 2.6822$, a 99% two-sided confidence interval for the difference in population means can be calculated as

$$\mu_A - \mu_B \in \left(9.005 - 11.864 - 2.6822 \sqrt{\frac{3.438^2}{24} + \frac{3.305^2}{34}}, \right.$$
$$\left. 9.005 - 11.864 + 2.6822 \sqrt{\frac{3.438^2}{24} + \frac{3.305^2}{34}} \right)$$
$$= (-5.28, -0.44)$$

The fact that 0 is not contained within this confidence interval implies that the null hypothesis $H_0 : \mu_A = \mu_B$ has a two-sided p-value smaller than 0.01, which is consistent with the result of the hypothesis test.

Remember that the relationships between confidence intervals, p-values, and size α hypothesis tests for two-sample problems are exactly the same as they are for one-sample problems. A size α hypothesis test rejects when the p-value is less than α and accepts when it is larger than α. Furthermore, a $1 - \alpha$ level confidence interval for $\mu_A - \mu_B$ contains the values of δ for which the null hypothesis $H_0 : \mu_A - \mu_B = \delta$ has a p-value larger than α. These relationships are illustrated in Figure 9.18 for two-sample two-sided problems.

	Hypothesis Testing			Confidence Intervals		
	$H_0 : \mu_A - \mu_B = \delta$ versus $H_A : \mu_A - \mu_B \neq \delta$			$\left(\bar{x} - \bar{y} - t_{\alpha/2,\nu} \sqrt{\frac{s_x^2}{n} + \frac{s_y^2}{m}} , \ \bar{x} - \bar{y} + t_{\alpha/2,\nu} \sqrt{\frac{s_x^2}{n} + \frac{s_y^2}{m}} \right)$		
	Significance Levels			Confidence Levels		
p-Value	$\alpha = 0.10$	$\alpha = 0.05$	$\alpha = 0.01$	$1 - \alpha = 0.90$	$1 - \alpha = 0.95$	$1 - \alpha = 0.99$
≥ 0.10	accept H_0	accept H_0	accept H_0	contains δ	contains δ	contains δ
0.05–0.10	reject H_0	accept H_0	accept H_0	does not contain δ	contains δ	contains δ
0.01–0.05	reject H_0	reject H_0	accept H_0	does not contain δ	does not contain δ	contains δ
< 0.01	reject H_0	reject H_0	reject H_0	does not contain δ	does not contain δ	does not contain δ

FIGURE 9.18

Relationship between hypothesis testing and confidence intervals for two-sample two-sided problems

9.3.2 Pooled Variance Procedure

The general procedures described above can always be used, except with small sample sizes when the data are obviously not normally distributed, and in particular they are appropriate when the population variances σ_A^2 and σ_B^2 are unequal. However, in certain circumstances an experimenter may be willing to make the assumption that $\sigma_A^2 = \sigma_B^2$, and this allows a slightly more powerful analysis based on a **pooled variance estimate**.

If the population variances σ_A^2 and σ_B^2 are assumed to be equal to a common value σ^2, then this can be estimated by

$$\hat{\sigma}^2 = s_p^2 = \frac{(n-1)s_x^2 + (m-1)s_y^2}{n + m - 2}$$

which is known as the pooled variance estimator. In this case the standard error of $\bar{x} - \bar{y}$ is

$$\text{s.e.}(\bar{x} - \bar{y}) = \sqrt{\frac{\sigma_A^2}{n} + \frac{\sigma_B^2}{m}} = \sigma \sqrt{\frac{1}{n} + \frac{1}{m}}$$

which can be estimated by

$$\text{s.e.}(\bar{x} - \bar{y}) = s_p \sqrt{\frac{1}{n} + \frac{1}{m}}$$

When a pooled variance estimate is employed, p-values and critical points are calculated from a t-distribution with $n + m - 2$ degrees of freedom. For example, a two-sided $1 - \alpha$ level confidence interval for $\mu_A - \mu_B$ is therefore

$$\mu_A - \mu_B \in \left(\bar{x} - \bar{y} - t_{\alpha/2,n+m-2} \, s_p \sqrt{\frac{1}{n} + \frac{1}{m}}, \ \bar{x} - \bar{y} + t_{\alpha/2,n+m-2} \, s_p \sqrt{\frac{1}{n} + \frac{1}{m}} \right)$$

which again is constructed using the standard format

$$\mu \in (\hat{\mu} - \text{critical point} \times \text{s.e.}(\hat{\mu}), \ \hat{\mu} + \text{critical point} \times \text{s.e.}(\hat{\mu}))$$

with $\mu = \mu_A - \mu_B$. The box on the two-sample t-procedure (equal variances) shows the other applications of this methodology.

Consider again the data obtained with $n = 24$, $\bar{x} = 9.005$, $s_x = 3.438$ and $m = 34$, $\bar{y} = 11.864$, $s_y = 3.305$. The sample standard deviations are similar and so it may be reasonable to assume that the population variances are equal. In this case, the estimate of the common standard deviation is

$$s_p = \sqrt{\frac{(n-1)s_x^2 + (m-1)s_y^2}{n+m-2}} = \sqrt{\frac{(23 \times 3.438^2) + (33 \times 3.305^2)}{24 + 34 - 2}} = 3.360$$

The hypotheses

$$H_0 : \mu_A = \mu_B \quad \text{versus} \quad H_A : \mu_A \neq \mu_B$$

are now tested with the t-statistic

$$t = \frac{\bar{x} - \bar{y}}{s_p \sqrt{\frac{1}{n} + \frac{1}{m}}} = \frac{9.005 - 11.864}{3.360 \sqrt{\frac{1}{24} + \frac{1}{34}}} = -3.192$$

The two-sided p-value is therefore

$$p\text{-value} = 2 \times P(X > 3.192) \simeq 2 \times 0.00115 = 0.0023$$

where the random variable X has a t-distribution with degrees of freedom $n + m - 2 = 56$, as illustrated in Figure 9.19.

With a critical point $t_{0.005,56} = 2.6665$, a 99% two-sided confidence interval for the difference in population means can be calculated as

$$\mu_A - \mu_B \in \left(9.005 - 11.864 - 2.6665 \times 3.360 \times \sqrt{\frac{1}{24} + \frac{1}{34}}, \right.$$

$$\left. 9.005 - 11.864 + 2.6665 \times 3.360 \times \sqrt{\frac{1}{24} + \frac{1}{34}} \right)$$

$$= (-5.25, -0.47)$$

FIGURE 9.19

Calculation of a two-sided p-value

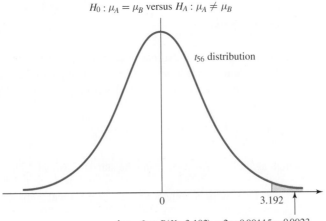

$H_0 : \mu_A = \mu_B$ versus $H_A : \mu_A \neq \mu_B$

t_{56} distribution

0

3.192

$p\text{-value} = 2 \times P(X > 3.192) = 2 \times 0.00115 = 0.0023$

Two-Sample t-Procedure (Equal Variances)

Consider a sample of size n from population A with a sample mean $\bar{x}$ and a sample standard deviation s_x, and a sample of size m from population B with a sample mean $\bar{y}$ and a sample standard deviation s_y. If an experimenter assumes that the population variances σ_A^2 and σ_B^2 are equal, then the common variance can be estimated by

$$s_p^2 = \frac{(n-1)s_x^2 + (m-1)s_y^2}{n+m-2}$$

which is known as the **pooled variance estimate**.

A two-sided $1 - \alpha$ level confidence interval for the difference in population means $\mu_A - \mu_B$ is

$$\mu_A - \mu_B \in \left(\bar{x} - \bar{y} - t_{\alpha/2,n+m-2}\, s_p \sqrt{\frac{1}{n} + \frac{1}{m}}, \bar{x} - \bar{y} + t_{\alpha/2,n+m-2}\, s_p \sqrt{\frac{1}{n} + \frac{1}{m}} \right)$$

One-sided confidence intervals are

$$\mu_A - \mu_B \in \left(-\infty, \bar{x} - \bar{y} + t_{\alpha,n+m-2}\, s_p \sqrt{\frac{1}{n} + \frac{1}{m}} \right)$$

and

$$\mu_A - \mu_B \in \left(\bar{x} - \bar{y} - t_{\alpha,n+m-2}\, s_p \sqrt{\frac{1}{n} + \frac{1}{m}}, \infty \right)$$

The appropriate t-statistic for the null hypothesis $H_0 : \mu_A - \mu_B = \delta$ is

$$t = \frac{\bar{x} - \bar{y} - \delta}{s_p \sqrt{\frac{1}{n} + \frac{1}{m}}}$$

A two-sided p-value is calculated as $2 \times P(X > |t|)$, where the random variable X has a t-distribution with $n + m - 2$ degrees of freedom, and one-sided p-values are $P(X > t)$ and $P(X < t)$. A size α two-sided hypothesis test accepts the null hypothesis if

$$|t| \le t_{\alpha/2,n+m-2}$$

and rejects the null hypothesis when

$$|t| > t_{\alpha/2,n+m-2}$$

and size α one-sided hypothesis tests have rejection regions $t > t_{\alpha,n+m-2}$ or $t < -t_{\alpha,n+m-2}$.

These procedures are known as two-sample t-tests *with* a pooled variance estimate.

These results are seen to match well with the corresponding analyses conducted previously using the general procedure, which does not require the equality of the population variances.

When should an experimenter use this method with a pooled variance estimate, and when should the general procedure be employed that does not require the equality of the population variances? The safest answer is to always use the general procedure since it provides a valid analysis even when the population variances are equal. However, if the population variances are equal or quite similar, then the pooled variance procedure generally provides a slightly more powerful analysis with slightly shorter confidence intervals and slightly smaller *p*-values.

An experimenter's assessment of whether or not the population variances can be taken to be equal may be based on prior experience with the kind of data under consideration, or may be based on a comparison of the sample standard deviations s_x and s_y. A formal test for the equality of the population variances σ_A^2 and σ_B^2 is described in the Supplementary Problems section at the end of this chapter. However, this test has some drawbacks and its employment is not widely recommended.

In general, analyses performed with and without the use of a pooled variance estimate will often provide similar results. In fact, if the results are quite different, then this will be because the sample standard deviations s_x and s_y are quite different, in which case it is proper to use the general procedure that does not require the equality of the population variances.

9.3.3 *z*-Procedure

Two-Sample *z*-Procedure

Consider a sample of size n from population A with a sample mean $\bar{x}$, and a sample of size m from population B with a sample mean $\bar{y}$. Suppose that the population variances are assumed to take known values σ_A^2 and σ_B^2.

A two-sided $1 - \alpha$ level confidence interval for the difference in population means $\mu_A - \mu_B$ is

$$\mu_A - \mu_B \in \left(\bar{x} - \bar{y} - z_{\alpha/2} \sqrt{\frac{\sigma_A^2}{n} + \frac{\sigma_B^2}{m}}, \bar{x} - \bar{y} + z_{\alpha/2} \sqrt{\frac{\sigma_A^2}{n} + \frac{\sigma_B^2}{m}} \right)$$

One-sided confidence intervals are

$$\mu_A - \mu_B \in \left(-\infty, \bar{x} - \bar{y} + z_{\alpha} \sqrt{\frac{\sigma_A^2}{n} + \frac{\sigma_B^2}{m}} \right)$$

and

$$\mu_A - \mu_B \in \left(\bar{x} - \bar{y} - z_{\alpha} \sqrt{\frac{\sigma_A^2}{n} + \frac{\sigma_B^2}{m}}, \infty \right)$$

The appropriate *z*-statistic for the null hypothesis $H_0 : \mu_A - \mu_B = \delta$ is

$$z = \frac{\bar{x} - \bar{y} - \delta}{\sqrt{\frac{\sigma_A^2}{n} + \frac{\sigma_B^2}{m}}}$$

A two-sided *p*-value is calculated as $2 \times \Phi(-|z|)$, and one-sided *p*-values are $1 - \Phi(z)$ and $\Phi(z)$. A size α two-sided hypothesis test accepts the null hypothesis if

$$|z| \leq z_{\alpha/2}$$

and rejects the null hypothesis when

$$|z| > z_{\alpha/2}$$

and size α one-sided hypothesis tests have rejection regions $z > z_{\alpha}$ or $z < -z_{\alpha}$.

These procedures are known as **two-sample *z*-tests**.

Two-sample z-tests are used when an experimenter wishes to use "known" values of the population standard deviations σ_A and σ_B in place of the sample standard deviations s_x and s_y. In this case p-values and critical points are calculated from the standard normal distribution.

As with one-sample procedures, two-sample t-tests with large sample sizes are essentially equivalent to the two-sample z-test. Consequently, the two-sample z-test can be thought of as a large-sample procedure and the two-sample t-tests can be thought of as small-sample procedures.

COMPUTER NOTE Find out how to perform two-sample procedures for independent samples on your computer package. You should anticipate being able to make the following choices:

- t-procedure or z-procedure

- two-sided procedure or one-sided procedure

- general procedure or pooled variance procedure

- confidence interval or hypothesis test

9.3.4 Examples

Example 51

Acrophobia Treatments

Figure 9.20 shows the scores of the 15 patients who underwent the standard treatment and the scores of the 15 patients who underwent the new treatment. Higher scores correspond to greater improvements in the patient's condition. There are two different sets of patients undergoing the two therapies, so these are independent (unpaired) samples.

Figure 9.21 shows descriptive statistics and boxplots of the two samples. The boxplots, which are drawn with the same scale, clearly indicate that the scores with the new treatment appear to be slightly higher than the scores with the standard treatment. In fact, the average of the scores with the new treatment is $\bar{y} = 51.20$, whereas the average of the scores with the standard treatment is $\bar{x} = 47.47$. Is this difference statistically significant?

Standard treatment x_i	New treatment y_i
33	65
54	61
62	37
46	47
52	45
42	53
34	53
51	69
26	49
68	42
47	40
40	67
46	46
51	43
60	51

FIGURE 9.20

Acrophobia treatments data set (improvement scores)

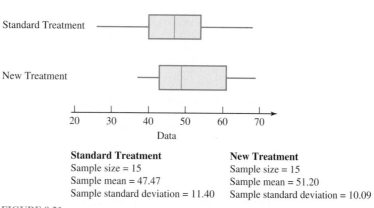

FIGURE 9.21

Descriptive statistics and boxplots for acrophobia treatments data set

Standard Treatment
Sample size = 15
Sample mean = 47.47
Sample standard deviation = 11.40

New Treatment
Sample size = 15
Sample mean = 51.20
Sample standard deviation = 10.09

The sample standard deviations are $s_x = 11.40$ and $s_y = 10.09$, which are similar. The boxplots also indicate that the variabilities of the two samples are roughly the same, so that an experimenter may decide to use a pooled variance analysis, although the general procedure is also appropriate.

In order to assess the evidence that the new treatment is *better* than the standard treatment (that is, $\mu_A < \mu_B$), the one-sided hypothesis testing problem

$$H_0 : \mu_A \geq \mu_B \quad \text{versus} \quad H_A : \mu_A < \mu_B$$

is considered. For the general (unpooled) procedure, the appropriate degrees of freedom are

$$\nu = \frac{\left(\frac{11.40^2}{15} + \frac{10.09^2}{15} \right)^2}{\frac{11.40^4}{15^2 \times 14} + \frac{10.09^4}{15^2 \times 14}} = 27.59$$

which can be rounded down to $\nu = 27$. The t-statistic is

$$t = \frac{47.47 - 51.20}{\sqrt{\frac{11.40^2}{15} + \frac{10.09^2}{15}}} = -0.949$$

so that the one-sided p-value is

$$p\text{-value} = P(X < -0.949) = 0.175$$

where the random variable X has a t-distribution with $\nu = 27$ degrees of freedom.

Table III shows that $t_{0.01, 27} = 2.473$, so that the 99% one-sided confidence interval is

$$\mu_A - \mu_B \in \left(-\infty, 47.47 - 51.20 + 2.473\sqrt{\frac{11.40^2}{15} + \frac{10.09^2}{15}} \right)$$
$$= (-\infty, 5.99)$$

For the pooled variance analysis, the appropriate degrees of freedom are $\nu = n + m - 2 = 28$. The pooled variance estimate is

$$s_p^2 = \frac{(14 \times 11.40^2) + (14 \times 10.09^2)}{28} = 115.88$$

so that the pooled standard deviation is $s_p = \sqrt{115.88} = 10.76$. In this case the t-statistic is

$$t = \frac{47.47 - 51.20}{10.76\sqrt{\frac{1}{15} + \frac{1}{15}}} = -0.946$$

which is almost the same as in the unpooled case. The p-value is

$$p\text{-value} = P(X < -0.946) = 0.175$$

where the random variable X has a t-distribution with $v = 28$ degrees of freedom. Also, the pooled variance analysis provides a 99% one-sided confidence interval $\mu_A - \mu_B \in (-\infty, 5, 97)$.

Both the unpooled and pooled analyses report a p-value of 0.175, so that the experimenter must conclude that there is *insufficient* evidence to reject the null hypothesis or, in other words, that there is *insufficient* evidence to establish that the new treatment is any better on average than the standard treatment. The one-sided 99% confidence intervals indicate that the standard treatment may be up to six points better than the new treatment, which again confirms that it is impossible to conclude that the new treatment is any better than the standard treatment.

It is worthwhile to reflect a moment on what the statistical analysis has achieved. An experimenter unversed in statistical inference matters may observe that $\bar{y} > \bar{x}$ and claim that the new treatment has been shown to be better than the standard treatment. However, with the statistical knowledge that we have gained so far we can see that this claim is not justifiable. Our comparison of the difference in the sample averages with the sample variabilities, which results in a large p-value, leads us to conclude that the difference observed can be attributed to randomness rather than to a real difference in the treatments, and so we realize that there is no real evidence of a treatment difference.

Example 52

Kaolin Processing

Figure 9.22 contains the brightness measurements of the processed kaolin from the two calciners. The two samples are independent, not paired.

Figure 9.23 contains summary statistics of the two samples and boxplots drawn to the same scale. The boxplots indicate that the brightness measurements appear to be slightly higher on average for calciner B and that the variability in the measurements may be slightly smaller for calciner B than for calciner A. The sample averages are $\bar{x} = 91.558$ and $\bar{y} = 92.500$, and the sample standard deviations are $s_x = 2.323$ and $s_y = 1.563$.

Whether the difference in population means suggested by the data is really statistically significant is investigated by the following statistical analysis for the two-sided hypotheses

$$H_0 : \mu_A = \mu_B \quad \text{versus} \quad H_A : \mu_A \neq \mu_B$$

For this problem it would appear to be safest to use the general procedure with degrees of freedom

$$\nu = \frac{\left(\frac{2.323^2}{12} + \frac{1.563^2}{12} \right)^2}{\frac{2.323^4}{12^2 \times 11} + \frac{1.563^4}{12^2 \times 11}} \simeq 19.3$$

Calciner A	Calciner B
x_i	y_i
88.4	92.6
93.2	93.2
87.4	89.2
94.3	94.8
93.0	93.3
94.3	94.0
89.0	93.2
90.5	91.7
90.8	91.5
93.1	92.0
92.8	90.7
91.9	93.8

FIGURE 9.22

Kaolin processing data set (brightness measurements)

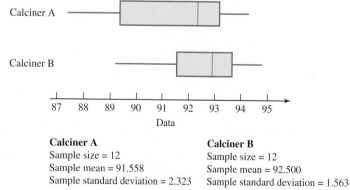

FIGURE 9.23

Descriptive statistics and boxplots for kaolin processing data set

Calciner A
Sample size = 12
Sample mean = 91.558
Sample standard deviation = 2.323

Calciner B
Sample size = 12
Sample mean = 92.500
Sample standard deviation = 1.563

which can be rounded down to $\nu = 19$. The t-statistic is

$$t = \frac{91.558 - 92.500}{\sqrt{\frac{2.323^2}{12} + \frac{1.563^2}{12}}} = -1.165$$

so that the two-sided p-value is

$$p\text{-value} = 2 \times P(X > 1.165) = 0.258$$

where the random variable X has a t-distribution with 19 degrees of freedom. Consequently, the experimenter concludes that there is not sufficient evidence to establish a difference in the average brightness of the kaolin processed by the two calciners.

Example 53	Figure 9.24 contains the percentage yield measurements x_i for the $n = 20$ kudzu pulpings

Example 53

Kudzu Pulping

Figure 9.24 contains the percentage yield measurements x_i for the $n = 20$ kudzu pulpings without the addition of anthraquinone and the percentage yield measurements y_i for the $m = 25$ kudzu pulpings with the addition of anthraquinone. The boxplots and summary statistics shown in Figure 9.25 clearly suggest that the addition of anthraquinone increases average yield. The sample averages are $\bar{x} = 38.55$ and $\bar{y} = 44.17$, and the sample standard deviations are $s_x = 3.627$ and $s_y = 3.994$. The closeness of the two-sample standard deviations suggests that a pooled variance analysis may be acceptable.

Without anthraquinone x_i	With anthraquinone y_i
39.7	43.5
42.4	41.6
34.6	47.9
35.6	39.0
40.6	48.9
41.0	49.2
37.9	46.2
30.2	49.5
44.5	50.3
43.0	37.6
36.0	41.0
35.7	40.4
38.9	47.4
38.2	48.3
39.8	49.4
40.3	44.4
35.7	42.0
41.3	41.0
42.2	38.5
33.5	39.4
	42.6
	46.9
	46.0
	42.3
	41.2

FIGURE 9.24

Kudzu pulping data set (percentage yield measurements)

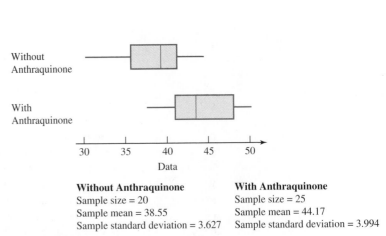

FIGURE 9.25

Summary statistics and boxplots for kudzu pulping experiment

Without Anthraquinone
Sample size = 20
Sample mean = 38.55
Sample standard deviation = 3.627

With Anthraquinone
Sample size = 25
Sample mean = 44.17
Sample standard deviation = 3.994

Healthy subjects x_i	Nerve disorder subjects y_i
52.20	50.68
53.81	47.49
53.68	51.47
54.47	48.47
54.65	52.50
52.43	48.55
54.43	45.96
54.06	50.40
52.85	45.07
54.12	48.21
54.17	50.06
55.09	50.63
53.91	44.99
52.95	47.22
54.41	48.71
54.14	49.64
55.12	47.09
53.35	48.73
54.40	45.08
53.49	45.73
52.52	44.86
54.39	50.18
55.14	52.65
54.64	48.50
53.05	47.93
54.31	47.25
55.90	53.98
52.23	
54.90	
55.64	
54.48	
52.89	

FIGURE 9.26

Nerve conductivity speeds data set (conductivity speeds in m/s)

In order to assess whether or not there is sufficient evidence to establish that the addition of anthraquinone *increases* yield, the one-sided hypothesis testing problem

$$H_0 : \mu_A \geq \mu_B \quad \text{versus} \quad H_A : \mu_A < \mu_B$$

is considered, where μ_A is the average yield without anthraquinone and μ_B is the average yield with anthraquinone.

With a pooled variance estimate

$$s_p^2 = \frac{(19 \times 3.627^2) + (24 \times 3.994^2)}{43} = 14.72$$

so that $s_p = \sqrt{14.72} = 3.836$, the t-statistic is

$$t = \frac{38.55 - 44.17}{3.836\sqrt{\frac{1}{20} + \frac{1}{25}}} = -4.884$$

The p-value is therefore

$$p\text{-value} = P(X < -4.884)$$

where the random variable X has a t-distribution with $\nu = n + m - 2 = 43$ degrees of freedom, which is less than 0.0001. The experimenter can therefore conclude that there is sufficient evidence to establish that the addition of anthraquinone does result in an increase in average yield.

With a critical point $t_{0.01,43} = 2.4163$, a one-sided 99% confidence interval for the difference in the average yields is

$$\mu_A - \mu_B \in \left(-\infty, \bar{x} - \bar{y} + t_{\alpha,n+m-2} s_p \sqrt{\frac{1}{n} + \frac{1}{m}} \right)$$

$$= \left(-\infty, 38.55 - 44.17 + 2.4163 \times 3.836 \times \sqrt{\frac{1}{20} + \frac{1}{25}} \right)$$

$$= (-\infty, -2.84)$$

Thus, the experimenter can conclude that the addition of anthraquinone increases the average yield by at least 2.8%.

Example 54

Nerve Conductivity Speeds

Figure 9.26 shows the data observations of nerve conductivity speeds, and Figure 9.27 shows boxplots and summary statistics of the data set. The boxplots suggest that the patients with a nerve disorder have slower conductivity speeds on average and more variability in their conductivity speeds. The sample averages are $\bar{x} = 53.994$ and $\bar{y} = 48.594$, and the sample standard deviations are $s_x = 0.974$ and $s_y = 2.490$.

A pooled variance analysis is clearly *not* appropriate here, and so the general procedure must be employed. Figure 9.28 shows the details of the statistical analysis, and the very small p-value indicates that the data set provides sufficient evidence to establish that the nerve conductivity speeds are different for the two groups of subjects. The 99% two-sided confidence interval for the difference in population means is

$$\mu_A - \mu_B \in (4.01, 6.79)$$

and consequently the experimenter can conclude that a periphery nerve disorder *reduces* the average nerve conductivity by somewhere between 4.0 and 6.8 m/s. The experimenter should

FIGURE 9.27

Descriptive statistics and boxplots
for the nerve conductivity
speeds data set

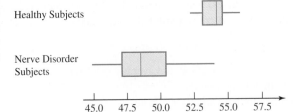

Healthy Subjects

Nerve Disorder
Subjects

Healthy Subjects
Sample size = 32
Sample mean = 53.994
Sample standard deviation = 0.974

Nerve Disorder Subjects
Sample size = 27
Sample mean = 48.594
Sample standard deviation = 2.490

FIGURE 9.28

Nerve conductivity speeds analysis

Stage I: Data Summary
Healthy subjects:
sample average $n = 32$, sample mean $\bar{x} = 53.994$, sample standard deviation $s_x = 0.974$
Nerve disorder subjects:
sample average $m = 27$, sample mean $\bar{y} = 48.594$, sample standard deviation $s_y = 2.490$

Stage II: Determination of Suitable Hypotheses
Question: Is there sufficient evidence to establish that the nerve conductivity speeds
are different for the two groups of subjects?
Use a two-sided procedure: $H_0 : \mu_A = \mu_B$ versus $H_A : \mu_A \neq \mu_B$.

Stage III: Determination of a Suitable Test Procedure
The sample standard deviations $s_x = 0.974$ and $s_y = 2.490$ are very different
so use the general (unequal variances) procedure.

Stage IV: Degrees of freedom
$$\frac{\left(\frac{0.974^2}{32} + \frac{2.490^2}{27} \right)^2}{\frac{0.974^4}{32^2 \times 31} + \frac{2.490^4}{27^2 \times 26}} = 32.69$$
Use $\nu = 32$ degrees of freedom.

Stage IV: Calculation of the Test Statistic
$$t = \frac{53.994 - 48.594}{\sqrt{\frac{0.974^2}{32} + \frac{2.490^2}{27}}} = 10.61$$

Stage VI: Expression for the p-value
p-value $= 2 \times P(X \geq 10.61) \simeq 0$
where the random variable X has a t-distribution with 32 degrees of freedom.

Stage VII: Decision
Since the p-value is less than 0.01, the null hypothesis is rejected.

Stage VIII: Conclusion
This data set provides sufficient evidence to establish that the average nerve conductivity
speeds are different for healthy subjects and for nerve disorder subjects.

Stage IX: Confidence Interval
Question: How different are the two groups?
Critical point: $t_{0.005,32} = 2.738$
99% confidence interval:
$$\mu_A - \mu_B \in 53.994 - 48.594 \pm 2.738 \sqrt{\frac{0.974^2}{32} + \frac{2.490^2}{27}} = (4.01, 6.79)$$

also take note of the fact that the variability in conductivity speeds is greater for the nerve disorder subjects than for the healthy subjects.

Example 45

Fabric Water Absorption Properties

The % pickup values for the $n = 15$ cases with a 10 pounds per square inch pressure and the $m = 15$ cases with a 20 pounds per square inch pressure are given in Figure 9.29. Figure 9.30 shows boxplots and summary statistics for this data set. The boxplots indicate a similarity in the variability of the two samples, but with higher water contents at the lower pressure. The sample averages are $\bar{x} = 59.807$ and $\bar{y} = 48.560$, and the sample standard deviations are $s_x = 4.943$ and $s_y = 4.991$.

A pooled variance analysis seems reasonable here, and the pooled variance estimate is

$$s_p^2 = \frac{(14 \times 4.943^2) + (14 \times 4.991^2)}{28} = 24.67$$

so that $s_p = \sqrt{24.67} = 4.967$. This gives a t-statistic of

$$t = \frac{59.807 - 48.560}{4.967\sqrt{\frac{1}{15} + \frac{1}{15}}} = 6.201$$

A two-sided p-value for the null hypothesis that the population means are equal is therefore

$$p\text{-value} = 2 \times P(X > 6.201) \simeq 0$$

where the random variable X has a t-distribution with degrees of freedom $n + m - 2 = 28$. Hence the experiment has established that changing the pressure does affect the average water absorption.

With a critical point $t_{0.005,28} = 2.763$ obtained from Table III, a 99% two-sided confidence interval for the difference in the population means can be calculated as

$$\mu_A - \mu_B \in \left(59.807 - 48.560 - 2.763 \times 4.967 \times \sqrt{\frac{1}{15} + \frac{1}{15}}, \right.$$

$$\left. 59.807 - 48.560 + 2.763 \times 4.967 \times \sqrt{\frac{1}{15} + \frac{1}{15}} \right)$$

$$= (6.24, 16.26)$$

The experimenter can therefore conclude that increasing the pressure between the rollers from 10 pounds per square inch to 20 pounds per square inch *decreases* the average water pickup by somewhere between about 6.2% and 16.3%.

FIGURE 9.29

10 lb/in²	20 lb/in²
x_i	y_i
51.8	55.6
61.8	44.6
57.3	46.7
54.5	45.8
64.0	49.9
59.5	51.9
61.2	44.1
64.9	52.3
54.5	51.0
70.2	39.9
59.1	51.6
55.8	42.5
65.4	45.5
60.4	58.0
56.7	49.0

Fabric water absorption properties data set (% water pickup)

FIGURE 9.30

Boxplots and summary statistics for the fabric water absorption properties data set

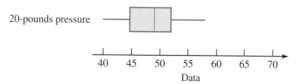

10-pounds pressure

20-pounds pressure

10-Pounds Pressure	20-Pounds Pressure
Sample size = 15	Sample size = 15
Sample mean = 59.807	Sample mean = 48.560
Sample standard deviation = 4.943	Sample standard deviation = 4.991

9.3.5 Sample Size Calculations

The determination of appropriate sample sizes n and m, or an assessment of the precision afforded by given sample sizes, is most easily performed by considering the length of a two-sided confidence interval for the difference in population means $\mu_A - \mu_B$. For the general procedure, this confidence interval length is

$$L = 2\, t_{\alpha/2,\nu} \sqrt{\frac{s_x^2}{n} + \frac{s_y^2}{m}}$$

To estimate the sample sizes that are required to obtain a confidence interval of length no larger than L_0, an experimenter needs to use estimated values of the population variances σ_A^2 and σ_B^2 (or at least upper bounds on these variances). The experimenter can then estimate that the sample sizes n and m are adequate as long as

$$L_0 \geq 2\, t_{\alpha/2,\nu} \sqrt{\frac{\sigma_A^2}{n} + \frac{\sigma_B^2}{m}}$$

where a suitable value of the critical point $t_{\alpha/2,\nu}$ can be used, depending on the confidence level $1 - \alpha$ required. If an experimenter wishes to have equal sample sizes $n = m$, then this inequality can be rewritten

$$n = m \geq \frac{4\, t_{\alpha/2,\nu}^2 \left(\sigma_A^2 + \sigma_B^2 \right)}{L_0^2}$$

For example, suppose that an experimenter wishes to construct a 99% confidence interval with a length no larger than $L_0 = 5.0$ mm for the difference between the mean thicknesses of two types of plastic sheets. Previous experience suggests that the thicknesses of sheets of type A have a standard deviation of no more than $\sigma_A = 4.0$ mm and that the thicknesses of sheets of type B have a standard deviation of no more than $\sigma_B = 2.0$ mm. It can be seen from Table III that as long as $\nu \geq 30$, the critical point $t_{0.005,\nu}$ is no larger than 2.75, so that it can be estimated that equal sample sizes

$$n = m \geq \frac{4 \times 2.75^2 \times (4.0^2 + 2.0^2)}{5.0^2} = 24.2$$

will suffice. Therefore, a random sample of 25 sheets of each type would appear to be sufficient to meet the experimenter's goal.

Similar calculations can also be employed to ascertain the *additional* sampling required to reduce the length of a confidence interval calculated from initial samples from the two populations. In this case the critical point $t_{\alpha/2,\nu}$ used in the initial analysis and the sample standard deviations s_x and s_y of the initial samples can be used in the sample size formula. This method is illustrated in the following example.

Example 45	Recall that a 99% confidence interval for the difference in the average percentage water pickup
Fabric Water	at the two pressure levels was calculated to be
Absorption Properties	

$$\mu_A - \mu_B \in (6.24,\ 16.26)$$

based upon samples with $n = m = 15$. This interval has a length of over 10%. How much additional sampling is required if the experimenter wants a 99% confidence interval with a length no larger than $L_0 = 4\%$?

The initial samples have sample standard deviations of $s_x = 4.943$ and $s_y = 4.991$, and the (pooled) analysis of the initial samples uses a critical point $t_{0.005,28} = 2.763$. Using these values in the formula

$$n = m \geq \frac{4 \, t_{\alpha/2, \nu}^2 \left(\sigma_A^2 + \sigma_B^2\right)}{L_0^2}$$

in place of the critical point and the population variances gives

$$n = m \geq \frac{4 \times 2.763^2 \times (4.943^2 + 4.991^2)}{4.0^2} = 94.2$$

Consequently, to meet the specified goal the experimenter can estimate that total sample sizes of $n = m = 95$ will suffice, so that additional sampling of 80 observations from each of the two pressure levels is required.

■ 9.3.6   Problems

9.3.1 An experimenter wishes to compare two treatments A and B and obtains some data observations x_i using treatment A and some data observations y_i using treatment B. It turns out that $\bar{x} > \bar{y}$ and so the experimenter concludes that treatment A results in larger data values on average than treatment B. How do you feel about the experimenter's conclusion? What other information would you like to know?

9.3.2 In an unpaired two-sample problem an experimenter observes $n = 14$, $\bar{x} = 32.45$, $s_x = 4.30$ from population A and $m = 14$, $\bar{y} = 41.45$, $s_y = 5.23$ from population B.
(a) Use the pooled variance method to construct a 99% two-sided confidence interval for $\mu_A - \mu_B$.
(b) Construct a 99% two-sided confidence interval for $\mu_A - \mu_B$ without assuming equal population variances.
(c) Consider a two-sided hypothesis test of $H_0 : \mu_A = \mu_B$ without assuming equal population variances. Does a size $\alpha = 0.01$ test accept or reject the null hypothesis? Write down an expression for the exact p-value.
(This problem is continued in Problem 9.3.14.)

9.3.3 In an unpaired two-sample problem an experimenter observes $n = 8$, $\bar{x} = 675.1$, $s_x = 44.76$ from population A and $m = 17$, $\bar{y} = 702.4$, $s_y = 38.94$ from population B.
(a) Use the pooled variance method to construct a 99% two-sided confidence interval for $\mu_A - \mu_B$.
(b) Construct a 99% two-sided confidence interval for $\mu_A - \mu_B$ without assuming equal population variances.
(c) Consider a two-sided hypothesis test of $H_0 : \mu_A = \mu_B$ using the pooled variance method. Does a size

$\alpha = 0.01$ test accept or reject the null hypothesis? Write down an expression for the exact p-value.

9.3.4 In an unpaired two-sample problem an experimenter observes $n = 10$, $\bar{x} = 7.76$, $s_x = 1.07$ from population A and $m = 9$, $\bar{y} = 6.88$, $s_y = 0.62$ from population B.
(a) Construct a 99% one-sided confidence interval $\mu_A - \mu_B \in (c, \infty)$ without assuming equal population variances.
(b) Does the value of c increase or decrease if a confidence level 95% is used?
(c) Consider a one-sided hypothesis test of $H_0 : \mu_A \leq \mu_B$ versus $H_A : \mu_A > \mu_B$ without assuming equal population variances. Does a size $\alpha = 0.01$ test accept or reject the null hypothesis? Write down an expression for the exact p-value.

9.3.5 In an unpaired two-sample problem an experimenter observes $n = 13$, $\bar{x} = 0.0548$, $s_x = 0.00128$ from population A and $m = 15$, $\bar{y} = 0.0569$, $s_y = 0.00096$ from population B.
(a) Construct a 95% one-sided confidence interval $\mu_A - \mu_B \in (-\infty, c)$ using the pooled variance method.
(b) Consider a one-sided hypothesis test of $H_0 : \mu_A \geq \mu_B$ versus $H_A : \mu_A < \mu_B$ using the pooled variance method. Does a size $\alpha = 0.05$ test accept or reject the null hypothesis? What about a size $\alpha = 0.01$ test? Write down an expression for the exact p-value.

9.3.6 The thicknesses of $n = 41$ glass sheets made using process A are measured and the statistics $\bar{x} = 3.04$ mm and $s_x = 0.124$ mm are obtained. In addition, the thicknesses of $m = 41$ glass sheets made using process B are measured and the statistics $\bar{y} = 3.12$ mm and $s_y = 0.137$ mm are obtained. Use a pooled variance procedure to answer the following questions.

(a) Does a two-sided hypothesis test with size $\alpha = 0.01$ accept or reject the null hypothesis that the two processes produce glass sheets with equal thicknesses on average?

(b) What is a two-sided 99% confidence interval for the difference in the average thicknesses of sheets produced by the two processes?

(c) Is there enough evidence to conclude that the average thicknesses of sheets produced by the two processes are different?

(This problem is continued in Problems 9.3.15 and 9.6.11.)

9.3.7 The breaking strengths of $n = 20$ bundles of wool fibers have a sample mean $\bar{x} = 436.5$ and a sample standard deviation $s_x = 11.90$. In addition, the breaking strengths of $m = 25$ bundles of synthetic fibers have a sample mean $\bar{y} = 452.8$ and a sample standard deviation $s_y = 4.61$. Answer the following questions without assuming that the two population variances are equal.

(a) Does a one-sided hypothesis test with size $\alpha = 0.01$ accept or reject the null hypothesis that the synthetic fiber bundles have an average breaking strength no larger than the wool fiber bundles?

(b) What is a one-sided 99% confidence interval that provides an *upper bound* on $\mu_A - \mu_B$, where μ_A is the average breaking strength of wool fiber bundles and μ_B is the average breaking strength of synthetic fiber bundles?

(c) Is there enough evidence to conclude that the average breaking strength of synthetic fiber bundles is larger than the average breaking strength of wool fiber bundles?

(This problem is continued in Problem 9.6.12.)

9.3.8 A random sample of $n = 16$ one-kilogram sugar packets of brand A have weights with a sample mean $\bar{x} = 1.053$ kg and a sample standard deviation $s_x = 0.058$ kg. In addition, a random sample of $m = 16$ one-kilogram sugar packets of brand B have weights with a sample mean $\bar{y} = 1.071$ kg and a sample standard deviation $s_y = 0.062$ kg. Is it safe to conclude that brand B sugar packets weigh slightly more on average than brand A sugar packets?

9.3.9 In an unpaired two-sample problem, an experimenter observes $n = 47$, $\bar{x} = 100.85$ from population A and

$m = 62$, $\bar{y} = 89.32$ from population B. Suppose that the experimenter wishes to use values $\sigma_A = 25$ and $\sigma_B = 20$ for the population standard deviations.

(a) What is the exact p-value for the hypothesis testing problem $H_0 : \mu_A = \mu_B + 3.0$ versus $H_A : \mu_A \neq \mu_B + 3.0$?

(b) Construct a 90% two-sided confidence interval for $\mu_A - \mu_B$.

9.3.10 In an unpaired two-sample problem, an experimenter observes $n = 38$, $\bar{x} = 5.782$ from population A and $m = 40$, $\bar{y} = 6.443$ from population B. Suppose that the experimenter wishes to use values $\sigma_A = \sigma_B = 2.0$ for the population standard deviations.

(a) What is the exact p-value for the hypothesis testing problem $H_0 : \mu_A \geq \mu_B$ versus $H_A : \mu_A < \mu_B$?

(b) Construct a 99% one-sided confidence interval that provides an *upper bound* for $\mu_A - \mu_B$.

9.3.11 The resilient moduli of $n = 10$ samples of a clay mixture of type A are measured and the sample mean is $\bar{x} = 19.50$. In addition, the resilient moduli of $m = 12$ samples of a clay mixture of type B are measured and the sample mean is $\bar{y} = 18.64$. Suppose that the experimenter wishes to use values $\sigma_A = \sigma_B = 1.0$ for the standard deviations of the resilient modulus of the two types of clay.

(a) What is the exact two-sided p-value for the null hypothesis that the two types of clay have equal average values of resilient modulus?

(b) Construct 90%, 95%, and 99% two-sided confidence intervals for the difference between the average resilient modulus of the two types of clay.

9.3.12 An experimenter feels that observations from population A have a standard deviation no larger than 10.0 and that observations from population B have a standard deviation no larger than 15.0. If the experimenter wants a two-sided 99% confidence interval for the difference in population means with a length no larger than $L_0 = 10.0$, what (equal) sample sizes would you recommend be obtained from the two populations?

9.3.13 An experimenter would like to construct a two-sided 95% confidence interval for the difference between the average resistance of two types of copper cable with a length no larger than $L_0 = 1.0$ ohms. If the experimenter feels that the standard deviations of the resistances of either type of cable are no larger than 1.2 ohms, what (equal) sample sizes would you recommend be obtained from the two types of copper cable?

9.3.14 Consider again the data set in Problem 9.3.2 with sample sizes $n = m = 14$. If a two-sided 99% confidence interval for the difference in population means is required with a length no larger than $L_0 = 5.0$, what *additional* sample sizes would you recommend be obtained from the two populations?

9.3.15 Consider again the data set of glass sheet thicknesses in Problem 9.3.6 with sample sizes $n = m = 41$. If a two-sided 99% confidence interval for the difference in the average thickness of glass sheets produced by the two processes is required with a length no larger than $L_0 = 0.1$ mm, how many *additional* glass sheets from the two processes do you think need to be sampled?

9.3.16 An experiment was conducted to investigate how the corrosion properties of chilled cast iron depend upon the chromium content of the alloy. A collection of $n = 12$ samples of chilled cast iron with 0.1% chromium content provided corrosion rates with a sample mean of $\bar{x} = 2.462$ and a sample standard deviation of $s_x = 0.315$, while a collection of $m = 13$ samples of chilled cast iron with 0.2% chromium content provided corrosion rates with a sample mean of $\bar{y} = 2.296$ and a sample standard deviation of $s_y = 0.297$.

  (a) Conduct a hypothesis test to investigate whether there is any evidence that the chromium content has an effect on the corrosion rate of chilled cast iron.

  (b) Construct a 99% two-sided confidence interval for the difference between the average corrosion rates of chilled cast iron at the two chromium contents.

9.3.17 Paving Slab Weights
Recall that DS 6.1.7 shows the weights of a sample of paving slabs from a certain company, manufacturer A say. In addition, DS 9.3.1 shows the weights of a sample of paving slabs from another company, manufacturer B. Is there evidence of any difference in the paving slab weights between the two manufacturers?

9.3.18 Spray Painting Procedure
An engineer compares the sample of paint thicknesses in DS 6.1.8 from production line A with a sample of paint thicknesses in DS 9.3.2 from production line B. What conclusions should the engineer draw?

9.3.19 Heel-Strike Force on a Treadmill
Physical disorders commonly experienced by long-distance runners are often related to large vertical reaction ground forces, and the minimization of such forces is the goal of much of the current research in sports biomechanics. DS 9.3.3 contains measurements of the heel-strike force, in newtons, of a particular runner on a standard treadmill and on a treadmill with a damped feature activated. Is the damped feature effective in reducing the heel-strike force?

9.3.20 Bleaching Agents
In the garment industry, bleaching is an important component of the manufacturing process. Chlorine bleach is very effective, but it can be unsatisfactory for environmental reasons. Consequently, various alternative bleaches such as hydrogen peroxide have been investigated. DS 9.3.4 contains the results of an experiment to compare the bleaching effectiveness of two levels of hydrogen peroxide, a low level and a high level. The data values are the whiteness levels, calculated from color measurements readings, for various samples of garments bleached with the hydrogen peroxide. What conclusions would you draw from this data set?

9.3.21 Restaurant Service Times
Recall that DS 6.1.4 shows the service times of customers at a fast-food restaurant who were served between 2:00 and 3:00 on a Saturday afternoon. In addition, DS 9.3.5 shows the service times of customers at the fast-food restaurant who were served between 9:00 and 10:00 in the morning on the same day. What do these data sets tell us about the difference between the service times at these two times of day?

9.3.22 The breaking strengths of 14 randomly selected objects produced from a standard procedure had a mean of 56.43 and a standard deviation of 6.30. In addition, the breaking strengths of 20 randomly selected objects produced from a new procedure had a mean of 62.11 and a standard deviation of 7.15. Perform a hypothesis test to investigate whether there is sufficient evidence to conclude that the new procedure has a larger breaking strength on average than the standard procedure.

9.3.23 Clinical Trial
A simple clinical trial was performed to compare two medicines. A total of 20 patients were obtained, and they were randomly split into two groups of 10 patients each. One group received medicine A, and the other group received medicine B. The patients' responses are given in DS 9.3.6. Does this experiment provide sufficient evidence to conclude that on average medicine A provides a higher response than medicine B?

9.3.24 An athlete recorded her practice times running a course. She had 8 times recorded when she ran in the morning and these had a sample average of 132.52 minutes with a standard deviation of 1.31 minutes. In addition, she had 10 times recorded when she ran in the afternoon and these had a sample average of 133.87 minutes with a standard deviation of 1.72 minutes.

(a) Is there sufficient evidence to conclude that the morning and afternoon are any different on average? Use an appropriate hypothesis test to investigate this question.

(b) Construct a two-sided 99% confidence interval for the difference between the average run time in the morning and the average run time in the afternoon.

9.3.25 A random sample of 10 observations from population A has a sample mean of 152.30 and a sample standard deviation of 1.83. A random sample of 8 observations from population B has a sample standard deviation of 1.94. If the p-value for the one-sided hypothesis test with an alternative hypothesis $H_A : \mu_A > \mu_B$ is less than 1%, what can you say about the sample mean of the observations from population B?

9.4 Summary

The first question in the design and analysis of a two-sample problem is whether it should be a paired problem or an unpaired problem. If an extraneous source of variability can be identified, it is appropriate to design a paired experiment when it is possible to do so. The analysis of a paired two-sample problem simplifies to a one-sample problem when differences are taken in the data observations within each pairing.

In many situations either there is no reason to design a paired experiment or a paired design is not feasible for the specific problem at hand. In these cases the experimenter has two independent or unpaired samples that may have unequal sample sizes. It is still very important to employ good experimental practices in the collection of these data sets, such as the random allocation of experimental subjects between the two treatments and the employment of blind or double-blind conditions where possible.

Two independent samples can be analyzed with two-sample t-tests or two-sample z-tests. The two-sample t-tests can be used without pooling the variances or with a pooled variance estimate. These two t-procedures are summarized in Figures 9.31 and 9.32. The pooled variance procedure assumes that the population variances σ_A^2 and σ_B^2 are equal, whereas the general procedure makes no assumptions about the population variances. It is usually safest to use the general procedure, although if the population variances are close, then the pooled variance approach may allow a slightly more accurate analysis.

The two-sample z-test is summarized in Figure 9.33, and it employs "known" values of the population variances σ_A^2 and σ_B^2. As with one-sample problems, it can also be thought of as the limiting value of the two-sample t-tests as the sample sizes n and m increase. In this sense it is often referred to as a large-sample procedure.

The two-sample procedures can be applied to either two-sided or one-sided problems and can be used to construct confidence intervals, to calculate p-values, or to perform hypothesis tests at a fixed size. As with one-sample procedures, two-sample procedures are based on the assumption that the data are normally distributed, although if the sample sizes are large enough, the central limit theorem ensures that the procedures are applicable. For small sample sizes and distributions that are evidently not normally distributed, the nonparametric procedures discussed in Chapter 15 provide a method of analysis.

Sample size calculations are most easily performed by considering the length that the sample sizes afford for a two-sided confidence interval for the difference in population means $\mu_A - \mu_B$. This assessment requires estimates of (or upper bounds on) the population variances σ_A^2 and σ_B^2. If follow-up studies are being planned, then the sample variances from the initial samples can be used as these estimates.

Two-sample t-procedure – general procedure		
(sample sizes $n, m \geq 30$ or small sample sizes with normally distributed data)		
$$\nu = \frac{\left(\frac{s_x^2}{n} + \frac{s_y^2}{m}\right)^2}{\frac{s_x^4}{n^2(n-1)} + \frac{s_y^4}{m^2(m-1)}}$$		
One-sided	Two-sided	One-sided
1 − α level Confidence Intervals		
$\left(-\infty, \bar{x} - \bar{y} + t_{\alpha,\nu}\sqrt{\frac{s_x^2}{n} + \frac{s_y^2}{m}}\right)$	$\left(\bar{x} - \bar{y} - t_{\alpha/2,\nu}\sqrt{\frac{s_x^2}{n} + \frac{s_y^2}{m}}, \bar{x} - \bar{y} + t_{\alpha/2,\nu}\sqrt{\frac{s_x^2}{n} + \frac{s_y^2}{m}}\right)$	$\left(\bar{x} - \bar{y} - t_{\alpha,\nu}\sqrt{\frac{s_x^2}{n} + \frac{s_y^2}{m}}, \infty\right)$
Hypothesis Testing: test statistic $t = \frac{\bar{x} - \bar{y} - \delta}{\sqrt{\frac{s_x^2}{n} + \frac{s_y^2}{m}}}$; $X \sim t_\nu$		
$H_0: \mu_A - \mu_B \geq \delta, H_A: \mu_A - \mu_B < \delta$	$H_0: \mu_A - \mu_B = \delta, H_A: \mu_A - \mu_B \neq \delta$	$H_0: \mu_A - \mu_B \leq \delta, H_A: \mu_A - \mu_B > \delta$
p-value $= P(X < t)$	p-value $= 2 \times P(X > \lvert t \rvert)$	p-value $= P(X > t)$
Size α hypothesis tests		
accept H_0 $\quad$ reject H_0 $t \geq -t_{\alpha,\nu}$ $\quad$ $t < -t_{\alpha,\nu}$	accept H_0 $\quad$ reject H_0 $\lvert t \rvert \leq t_{\alpha/2,\nu}$ $\quad$ $\lvert t \rvert > t_{\alpha/2,\nu}$	accept H_0 $\quad$ reject H_0 $t \leq t_{\alpha,\nu}$ $\quad$ $t > t_{\alpha,\nu}$

FIGURE 9.31

Summary of the general two-sample t-procedure

Finally, it should be remembered that the inference procedures discussed above are designed to investigate location differences between the two population probability distributions, that is, differences between the population means μ_A and μ_B. However, it is also important to notice whether there appears to be a difference between the two population variances σ_A^2 and σ_B^2. A test procedure for comparing two variances is described in the Supplementary Problems section, but it relies heavily on the normality of the population distributions and its widespread use is not recommended. In practice, it is sensible and often adequate for the experimenter to compare boxplots or histograms of the two samples, and to notice whether or not there is any obvious difference between the sample variabilities.

Two-sample t-procedure – pooled variance procedure		
(sample sizes $n, m \geq 30$ or small sample sizes with normally distributed data)		
Assumption: $\sigma_A^2 = \sigma_B^2$ $\quad\quad$ $s_p^2 = \frac{(n-1)s_x^2 + (m-1)s_y^2}{n+m-2}$		
One-sided	Two-sided	One-sided
$1 - \alpha$ level Confidence Intervals		
$\left(-\infty, \bar{x} - \bar{y} + t_{\alpha, n+m-2} s_p \sqrt{\frac{1}{n} + \frac{1}{m}}\right)$	$\left(\bar{x} - \bar{y} - t_{\alpha/2, n+m-2} s_p \sqrt{\frac{1}{n} + \frac{1}{m}}, \bar{x} - \bar{y} + t_{\alpha/2, n+m-2} s_p \sqrt{\frac{1}{n} + \frac{1}{m}}\right)$	$\left(\bar{x} - \bar{y} - t_{\alpha, n+m-2} s_p \sqrt{\frac{1}{n} + \frac{1}{m}}, \infty\right)$
Hypothesis Testing: test statistic $t = \dfrac{\bar{x} - \bar{y} - \delta}{s_p \sqrt{\frac{1}{n} + \frac{1}{m}}}$; $X \sim t_{n+m-2}$		
$H_0: \mu_A - \mu_B \geq \delta,\ H_A: \mu_A - \mu_B < \delta$	$H_0: \mu_A - \mu_B = \delta,\ H_A: \mu_A - \mu_B \neq \delta$	$H_0: \mu_A - \mu_B \leq \delta,\ H_A: \mu_A - \mu_B > \delta$
p-value $= P(X < t)$	p-value $= 2 \times P(X > |t|)$	p-value $= P(X > t)$
Size α hypothesis tests		
accept H_0 $\quad$ reject H_0 $t \geq -t_{\alpha, n+m-2}$ $\quad$ $t < -t_{\alpha, n+m-2}$	accept H_0 $\quad$ reject H_0 $|t| \leq t_{\alpha/2, n+m-2}$ $\quad$ $|t| > t_{\alpha/2, n+m-2}$	accept H_0 $\quad$ reject H_0 $t \leq t_{\alpha, n+m-2}$ $\quad$ $t > t_{\alpha, n+m-2}$

FIGURE 9.32

Summary of the pooled variance two-sample t-procedure

9.5   Case Study: Microelectronic Solder Joints

The researcher examines an assembly with 16 solder joints that was made using the original method for depositing nickel on the bond pads, and measures the nickel layer thicknesses. This new data set is shown in Figure 9.34, and the sample size is $m = 16$, the sample average is $\bar{y} = 2.7981$ microns, and the sample standard deviation is $s_y = 0.0256$ microns.

The sample average $\bar{y} = 2.7981$ microns of the nickel layer thicknesses for the assembly prepared using the original method (assembly 2) is larger than the sample average $\bar{x} = 2.7688$ microns of the nickel layer thicknesses for the assembly prepared using the new method (assembly 1) that is given in Figure 6.40, and the boxplots in Figure 9.35 confirm that the

Two-sample z-procedure		
(sample sizes $n, m \geq 30$ or small sample sizes with normally distributed data; variances known)		
One-sided	Two-sided	One-sided
$1 - \alpha$ level Confidence Intervals		
$\left(-\infty, \bar{x} - \bar{y} + z_\alpha \sqrt{\frac{\sigma_A^2}{n} + \frac{\sigma_B^2}{m}}\right)$	$\left(\bar{x} - \bar{y} - z_{\alpha/2} \sqrt{\frac{\sigma_A^2}{n} + \frac{\sigma_B^2}{m}}, \bar{x} - \bar{y} + z_{\alpha/2} \sqrt{\frac{\sigma_A^2}{n} + \frac{\sigma_B^2}{m}}\right)$	$\left(\bar{x} - \bar{y} - z_\alpha \sqrt{\frac{\sigma_A^2}{n} + \frac{\sigma_B^2}{m}}, \infty\right)$
Hypothesis Testing: test statistic $z = \dfrac{\bar{x} - \bar{y} - \delta}{\sqrt{\frac{\sigma_A^2}{n} + \frac{\sigma_B^2}{m}}}$		
$H_0: \mu_A - \mu_B \geq \delta, H_A: \mu_A - \mu_B < \delta$	$H_0: \mu_A - \mu_B = \delta, H_A: \mu_A - \mu_B \neq \delta$	$H_0: \mu_A - \mu_B \leq \delta, H_A: \mu_A - \mu_B > \delta$
p-value $= \Phi(z)$	p-value $= 2 \times \Phi(-\lvert z \rvert)$	p-value $= 1 - \Phi(z)$
Size α hypothesis tests		
accept H_0 reject H_0 $z \geq -z_\alpha$ $z < -z_\alpha$	accept H_0 reject H_0 $\lvert z \rvert \leq z_{\alpha/2}$ $\lvert z \rvert > z_{\alpha/2}$	accept H_0 reject H_0 $z \leq z_\alpha$ $z > z_\alpha$

FIGURE 9.33

Summary of the two-sample z-procedure

FIGURE 9.34

Data set of nickel layer thicknesses on substrate bond pads for assembly 2.

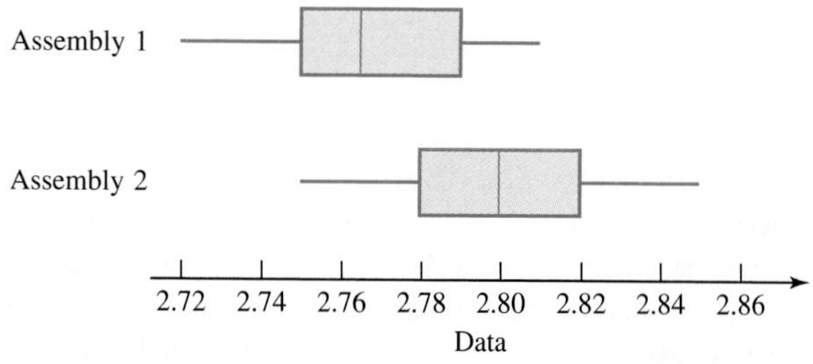

2.78	2.77	2.79	2.78	2.81	2.79	2.83	2.75
2.82	2.82	2.85	2.80	2.77	2.80	2.82	2.80

FIGURE 9.35

Boxplots of the nickel layer thickness for the new method (assembly 1) and the original method (assembly 2).

nickel layer thicknesses tend to be larger in assembly 2. The researcher decides to test whether this difference is statistically significant using the two-sided hypothesis test

$$H_0 : \mu_{new} = \mu_{original} \quad \text{versus} \quad H_A : \mu_{new} \neq \mu_{original}$$

Using the general procedure, the appropriate degrees of freedom are found to be $\nu = 29$ and the test statistic is

$$t = \frac{\bar{x} - \bar{y}}{\sqrt{\frac{s_x^2}{n} + \frac{s_y^2}{m}}} = \frac{2.7688 - 2.7981}{\sqrt{\frac{0.0260^2}{16} + \frac{0.0256^2}{16}}} = -3.22$$

The p-value is therefore $2 \times P(X \geq 3.22)$ where the random variable X has a t-distribution with 29 degrees of freedom, which is 0.003. Since the p-value is less than 1%, the null hypothesis is rejected and the researcher can conclude that these data sets provide sufficient evidence to establish that the new method is providing a nickel layer with a smaller average thickness than the original method.

Furthermore, with $t_{0.005,29} = 2.756$ a 99% two-sided confidence interval for the difference between the two means is

$$\mu_{new} - \mu_{original} \in \left(2.7688 - 2.7981 - 2.756\sqrt{\frac{0.0260^2}{16} + \frac{0.0256^2}{16}}, \right.$$

$$\left. 2.7688 - 2.7981 + 2.756\sqrt{\frac{0.0260^2}{16} + \frac{0.0256^2}{16}} \right)$$

$$= (-0.055, -0.004)$$

and so the researcher can conclude that the difference between the average nickel layer thicknesses of the two methods is somewhere between 0.004 microns and 0.055 microns.

9.6   Supplementary Problems

9.6.1 Video Display Designs

A researcher is interested in how a color video display rather than a black-and-white video display can help a person assimilate the information provided on a screen. A set of 22 experimental subjects are used, and each person undergoes five trials with a color display and five trials with a black-and-white display. In each trial the subject has to perform a task based upon information provided on the screen, and the time taken to perform the task is measured. For each of the 22 subjects, DS 9.6.1 shows the average time in seconds taken to perform the five color trials and the five black-and-white trials. Do the data show that color displays are more effective than black-and-white displays?

9.6.2 Fabric Water Absorption Properties

In assessing how the water absorption properties of cotton fabric differ with roller pressures of 10 pounds per square inch and 20 pounds per square inch, an experimenter suspects that different fabric samples may have different absorption properties. Therefore a paired experimental design is adopted whereby 14 samples are split in half, with one half being examined at one pressure and the other half being examined at the other pressure. DS 9.6.2 contains the % pickup values obtained. Do the water absorption properties of cotton fabric depend upon the roller pressure?

9.6.3 A researcher in the petroleum industry is interested in the sizes of wax crystals produced when wax dissolved in a supercritical fluid is sprayed through a capillary nozzle. In the first experiment, a pre-expansion temperature of $80°C$ is employed and the diameters of $n = 35$ crystals are measured with an electron microscope. A sample mean of $\bar{x} = 22.73 \ \mu m$ and a sample standard deviation of $s_x = 5.20 \ \mu m$ are obtained. In the second experiment, a pre-expansion temperature of $150°C$ is employed and the diameters of $m = 35$ crystals are measured, which have a sample mean of $\bar{y} = 12.66 \ \mu m$ and a sample standard deviation of $s_y = 3.06 \ \mu m$. The researcher decides that it is not appropriate to assume that the variances of the crystal diameters are the same under both sets of experimental conditions.

(a) Write down an expression for the p-value of the statement that the average crystal size does not depend upon the pre-expansion temperature. Do you think that this is a plausible statement?

(b) Construct a 99% two-sided confidence interval for the difference between the average crystal diameters at the two pre-expansion temperatures.

(c) If a 99% two-sided confidence interval for the difference between the average crystal diameters at the two pre-expansion temperatures is required with a length no larger than $L_0 = 4.0 \ \mu$m, how much additional sampling would you recommend?

9.6.4 A company is investigating how long it takes its drivers to deliver goods from its factory to a nearby port for export. Records reveal that with a standard specified driving route, the last $n = 48$ delivery times have a sample mean of $\bar{x} = 432.7$ minutes and a sample standard deviation of $s_x = 20.39$ minutes. A new driving route is proposed, and this has been tried $m = 10$ times with a sample mean of $\bar{y} = 403.5$ minutes and a sample standard deviation of $s_y = 15.62$ minutes. What is the evidence that the new route is quicker on average than the standard route?

9.6.5 Bamboo Cultivation
A researcher compares the bamboo shoot heights in DS 6.6.5 obtained under growing conditions A with the bamboo shoot heights in DS 9.6.3 obtained under growing conditions B. In each case the bamboo shoot heights are measured 40 days after planting, but growing conditions B allow 10% more sunlight than growing conditions A. Does the extra sunlight tend to increase the bamboo shoot heights?

9.6.6 Consumer Complaints Division Reorganization
In a quality drive a food manufacturer reorganizes its consumer complaints division. Before the reorganization, a study was conducted of the time that a consumer calling the toll-free complaints line had to wait before speaking to a company employee. After reorganization, a similar follow-up study was conducted. DS 9.6.4 contains the two samples of waiting times in seconds that were recorded. Does the reorganization appear to have been successful in affecting the time taken to answer calls?

9.6.7 Ocular Motor Measurements
Ocular motor measurements are designed to assess the amount of contraction in the muscles around the eyes.

High ocular motor measurements may be indicative of eyestrain, which may lead to spasms and headaches. A group of 10 subjects had their ocular motor measurements recorded after they had been reading a book for an hour and also after they had been reading a computer screen for an hour. The results are listed in DS 9.6.5. What conclusions can you draw from this data set?

9.6.8 Engine Oil Viscosity
The viscosity of oil after it has been used in an engine over a period of time may change from its initial value because the high temperature inside the engine can cause the oil to break down. An experiment was conducted to compare the effect on oil viscosity of two different engines. Various samples of the same type of oil with a constant viscosity were used, some in engine 1 and some in engine 2, and the engines were run under identical operating conditions. The resulting values of the oil viscosities after having been used in the engines are given in DS 9.6.6. Is there any evidence that the engines have different effects on the oil viscosity?

9.6.9 Comparing Two Population Variances
For use with Problems 9.6.10–9.6.12.

Recall that if S_x^2 is the sample variance of a set of n observations from a normal distribution with variance σ_A^2, then

$$S_x^2 \sim \sigma_A^2 \frac{\chi_{n-1}^2}{n-1}$$

and that if S_y^2 is the sample variance of a set of m observations from a normal distribution with variance σ_B^2, then

$$S_y^2 \sim \sigma_B^2 \frac{\chi_{m-1}^2}{m-1}$$

(a) Explain why

$$\frac{\sigma_A^2 S_y^2}{\sigma_B^2 S_x^2} \sim F_{m-1,n-1}$$

(b) Show that part (a) implies that

$$P\left(F_{1-\alpha/2,m-1,n-1} \le \frac{\sigma_A^2 S_y^2}{\sigma_B^2 S_x^2} \le F_{\alpha/2,m-1,n-1}\right)$$
$$= 1 - \alpha$$

or alternatively that

$$P\left(\frac{1}{F_{\alpha/2,n-1,m-1}} \le \frac{\sigma_A^2 S_y^2}{\sigma_B^2 S_x^2} \le F_{\alpha/2,m-1,n-1}\right)$$
$$= 1 - \alpha$$

(c) Deduce that

$$\frac{\sigma_A^2}{\sigma_B^2} \in \left(\frac{s_x^2}{s_y^2 \, F_{\alpha/2,n-1,m-1}} , \frac{s_x^2 \, F_{\alpha/2,m-1,n-1}}{s_y^2} \right)$$

is a $1 - \alpha$ level two-sided confidence interval for the ratio of the population variances.

If such a confidence interval contains the value 1, then this indicates that it is plausible that the population variances are equal. An unfortunate aspect of these confidence intervals, however, is that they depend heavily on the data being normally distributed, and they should be used only when that is a fair assumption. You may be able to obtain these confidence intervals on your computer package.

9.6.10 A sample of $n = 18$ observations from population A has a sample standard deviation of $s_x = 6.48$, and a sample of $m = 21$ observations from population B has a sample standard deviation of $s_y = 9.62$. Obtain a 90% confidence interval for the ratio of the population variances.

9.6.11 Consider again the data set of glass sheet thicknesses in Problem 9.3.6 with sample sizes $n = m = 41$. Construct a 90% confidence interval for the ratio of the variances of the thicknesses of glass sheets produced by the two processes.

9.6.12 Consider again the data set of the breaking strengths of $n = 20$ bundles of wool fibers and $m = 25$ bundles of synthetic fibers in Problem 9.3.7. Construct 90%, 95%, and 99% confidence intervals for the ratio of the variances of the breaking strengths of the two types of fiber.

9.6.13 The strengths of two types of canvas were compared in an experiment. Fourteen samples of type A gave an average strength of 327,433 with a standard deviation of 9,832. Twelve samples of type B gave an average strength of 335,537 with a standard deviation of 10,463. Use a hypothesis test to evaluate whether there is sufficient evidence to conclude that there is a difference between the strengths of the two canvas types.

9.6.14 Reinforced Cement Strengths
The strengths of nine reinforced cement samples were tested using two procedures. Each sample was split into two parts, with one part being tested with procedure 1 and the other part being tested with procedure 2. The resulting data set is given in DS 9.6.7. Use an appropriate hypothesis test to assess whether there is any evidence that the two testing procedures provide different results on average.

9.6.15 Are the following statements true or false?
(a) The advantage of paired experiments is that any carry-over effects from one treatment to the other treatment do not affect the analysis.
(b) In an experiment to compare two treatments with an independent samples design, the random allocation of the experimental units between the two treatments is a good tool to help eliminate any bias in the data collection.
(c) The experimental design refers to the manner in which the data are collected.
(d) An experiment is performed to compare two medical treatments. If one treatment is a placebo, then an unpaired analysis should always be used.
(e) In a double-blind experiment the researcher does not know the true values of the variable being measured.
(f) A paired experimental design can be conducted as a blind experiment.
(g) For the analysis of two independent samples, the unequal variances procedure can still be used even if the experimenter suspects that the two population variances may be equal.
(h) The manner in which the data are collected indicates whether a two-sample data set is paired or unpaired.
(i) For the analysis of two independent samples using the unequal variances procedure, the value obtained from the formula for the degrees of freedom should be rounded down to the nearest integer.

9.6.16 Comparisons of Experimental Drug Therapies
Eight people participated in an experiment to compare two experimental drug therapies. Each person was administered both therapies, but in a random order. The data values given in DS 9.6.8 were obtained. Perform a hypothesis test to investigate whether there is any evidence of a difference between the two therapies.

9.6.17 A sample of 20 items from manufacturer A were measured and a sample mean of 2376.3 and a sample standard deviation of 24.1 were obtained. Also, a sample of 24 items from manufacturer B were measured and a sample mean of 2402.0 and a sample standard deviation of 26.4 were obtained.
(a) Use a one-sided hypothesis test to assess whether there is sufficient evidence to conclude that the items from manufacturer B provide larger measurements on average than these items from manufacturer A.

(b) Construct a 95% one-sided confidence interval that provides an upper bound on how much larger on average the measurements of the items from manufacturer B are compared with the items from manufacturer A.

The data sets given in problems 9.6.18–9.6.22 can be used to practice the two-sample methodologies presented in this chapter.

9.6.18 Rubber Seal Curing Methods

An engineer is interested in whether the standard curing time for the rubber seal on a radial assembly can be replaced by a new rapid curing method that would reduce manufacturing costs. However, it is important to investigate whether the rapid curing method has any effect on the dimensions of the seal. DS 9.6.9 contains data on the inside diameter measurements of some seals prepared with the standard curing method and with the new rapid curing method.

9.6.19 Light and Dark Regimens for Plant Growth

In an experiment, plants were grown under controlled conditions whereby the light they received was from a sun lamp. In one regimen the plants were subjected to alternate 12 hour periods of light and dark, while in another regimen the plants were subjected to alternate 6 hour periods of light and dark. DS 9.6.10 contains the heights of the plants after a certain time period.

9.6.20 Joystick Design for Spinal Cord Injury Patients

Patients with spinal cord injuries can lose mobility in their arms and hands, and it is important to find the optimal design of a joystick that will enable them to perform tasks in the most efficient manner. An experiment was designed to compare two joystick designs. As a target moved across a computer screen, the patients were asked to use a joystick to follow the target with a cursor. Nine spinal cord injury patients participated in the study, and each patient tried out both joystick designs. DS 9.6.11 contains data on the mean error measurements that were calculated by aggregating the distances between the target and the cursor at a series of time points.

9.6.21 Ambient Air Carbon Monoxide Pollution Levels

A researcher hypothesizes that ambient air carbon monoxide pollution levels at a certain location should be higher in the winter than they are in the summer. The reasoning behind this hypothesis is that a major source of carbon monoxide in the air is from the incomplete combustion of fuels, and fuels tend to burn less efficiently at low temperatures. Moreover, it is felt that the stagnant winter air is more likely to trap the pollution. In order to investigate this hypothesis, the data set in DS 9.6.12 is collected which shows the ambient air carbon monoxide pollution levels (parts per million) for 10 Sunday mornings in the middle of winter and 10 Sunday mornings in the middle of summer.

9.6.22 Sphygmomanometer and Finger Monitor Systolic Blood Pressure Measurements

A sphygmomanometer is a standard instrument for measuring blood pressure in the arteries consisting of a pressure gauge and a rubber cuff that wraps around the upper arm. DS 9.6.13 compares blood pressure readings for 15 patients using this standard method and a new method based upon a simple finger monitor.

CHAPTER TEN

Discrete Data Analysis

Subsequent to Chapters 8 and 9, which showed how inferences can be made on the population means of continuous random variables, in this chapter the problem of making inferences on the population probabilities of *discrete random variables* is considered. Recall that discrete random variables may take only discrete values. For example, a random variable measuring the number of errors in a software product may take the discrete values

$$0, 1, 2, 3, 4, \ldots$$

A product's quality level may be *categorized* as

high, medium, or low,

or a machine breakdown may be characterized as being due to either

mechanical failure, electrical failure, or operator misuse.

Since discrete random variables often arise from assigning an event to one of several categories, discrete data analysis is also often referred to as *categorical data* analysis.

A discrete data set consists of the *frequencies* or *counts* of the observations found at each of the possible levels or *cells*. Discrete data analysis then consists of making inferences on the *cell probabilities*. The simplest example of this kind is the problem of making inferences about the success probability p of a binomial distribution. In this case there are two cells with probabilities p and $1 - p$. The first two sections of this chapter discuss inference procedures on the success probability of a binomial distribution and the comparison of two success probabilities. The next two sections of the chapter consider more complex data sets in which there may be three or more different categories or in which observations may be simultaneously subjected to more than one categorization process. These data sets can be represented in a tabular form known as a *contingency table*, which is often analyzed with a *chi-square* goodness of fit test.

10.1 Inferences on a Population Proportion

Suppose that a parameter p represents the unknown proportion of a population that possesses a particular characteristic. For instance, the population may represent all the items produced by a particular machine, a proportion p of which are defective. If an observation is taken at random from the population, it can then be thought of as having a probability p of exhibiting the characteristic. If a random sample of n observations is obtained from the population, each of the observations is then a realization of a Bernoulli random variable with "success probability" p, and so the number of successes x, or in other words the number of observations that possess the particular characteristic of interest, is a realization of a random variable X that has a binomial distribution with parameters n and p

$$X \sim B(n, p)$$

FIGURE 10.1

Generation of binary data

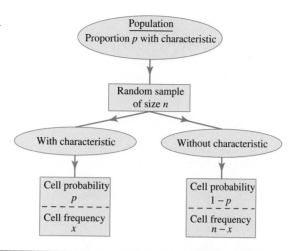

This framework is illustrated in Figure 10.1. Notice that each observation from the population is a discrete random variable with *two* values, namely, *possessing* the characteristic or *not possessing* the characteristic of interest. The probabilities of these two categories, that is, the cell probabilities, are

$$p \quad \text{and} \quad 1 - p$$

The cell frequencies are the number of observations that are observed to fall in each of the two categories, and in this case they are

$$x \quad \text{and} \quad n - x$$

respectively.

Recall from Section 7.3.1 that the cell probability or success probability p can be estimated by the sample proportion

$$\hat{p} = \frac{x}{n}$$

and that

$$E(\hat{p}) = p \quad \text{and} \quad \text{Var}(\hat{p}) = \frac{p(1-p)}{n}$$

Furthermore, for large enough values of n the sample proportion can be taken to have approximately the normal distribution

$$\hat{p} \sim N\left(p, \frac{p(1-p)}{n}\right)$$

This expression may be rewritten in terms of a standard normal distribution as

$$\frac{\hat{p} - p}{\sqrt{\frac{p(1-p)}{n}}} \sim N(0, 1)$$

The normal approximation is appropriate as long as both np and $n(1 - p)$ are **larger than 5**, which can be taken to be the case as long as both x and $n - x$ are larger than 5.

10.1.1   Confidence Intervals for Population Proportions

One of the most useful inferences on a population proportion p is a confidence interval that summarizes the values that it can plausibly take. Either two-sided or one-sided confidence intervals can be constructed as shown below.

Two-Sided Confidence Intervals   If $Z \sim N(0, 1)$, then

$$P(-z_{\alpha/2} \leq Z \leq z_{\alpha/2}) = 1 - \alpha$$

Consequently, it is approximately the case that

$$P\left(-z_{\alpha/2} \leq \frac{\hat{p} - p}{\sqrt{\frac{p(1-p)}{n}}} \leq z_{\alpha/2}\right) = 1 - \alpha$$

which can be rewritten

$$P\left(\hat{p} - z_{\alpha/2}\sqrt{\frac{p(1-p)}{n}} \leq p \leq \hat{p} + z_{\alpha/2}\sqrt{\frac{p(1-p)}{n}}\right) = 1 - \alpha$$

This expression implies that

$$p \in \left(\hat{p} - z_{\alpha/2}\sqrt{\frac{p(1-p)}{n}}, \ \hat{p} + z_{\alpha/2}\sqrt{\frac{p(1-p)}{n}}\right)$$

is a two-sided $1 - \alpha$ confidence level confidence interval for p. Notice that this confidence interval is of the form

$$p \in (\hat{p} - \text{critical point} \times \text{s.e.}(\hat{p}), \ \hat{p} + \text{critical point} \times \text{s.e.}(\hat{p}))$$

where the standard error (s.e.) of the estimate $\hat{p}$ is (see Section 7.3.1)

$$\text{s.e.}(\hat{p}) = \sqrt{\frac{p(1-p)}{n}}$$

However, as it stands, this confidence interval cannot be constructed because the standard error s.e.$(\hat{p})$ depends on the unknown probability p. It is therefore customary to estimate this standard error by replacing p with its estimated value $\hat{p}$, so that the confidence interval is constructed using

$$\text{s.e.}(\hat{p}) = \sqrt{\frac{\hat{p}(1-\hat{p})}{n}} = \frac{1}{n}\sqrt{\frac{x(n-x)}{n}}$$

where x is the observed number of "successes" in the random sample of size n.

Two-Sided Confidence Intervals for a Population Proportion

If the random variable X has a $B(n, p)$ distribution, then an approximate two-sided $1 - \alpha$ confidence level confidence interval for the success probability p based on an observed value of the random variable x is

$$p \in \left(\hat{p} - z_{\alpha/2} \sqrt{\frac{\hat{p}(1 - \hat{p})}{n}}, \hat{p} + z_{\alpha/2} \sqrt{\frac{\hat{p}(1 - \hat{p})}{n}} \right)$$

where the estimated success probability is $\hat{p} = x/n$. This confidence interval can also be written as

$$p \in \left(\hat{p} - \frac{z_{\alpha/2}}{n} \sqrt{\frac{x(n - x)}{n}}, \hat{p} + \frac{z_{\alpha/2}}{n} \sqrt{\frac{x(n - x)}{n}} \right)$$

If a random sample of n observations is taken from a population and x of the observations are of a certain type, then this expression provides a confidence interval for the proportion p of the population of that type.

The approximation is reasonable as long as both x and $n - x$ are larger than 5. If it turns out that the upper end of this confidence interval is larger than one, then it can of course be truncated at 1, and if the lower end of the confidence interval is smaller than 0, then it can be truncated at 0.

Example 57

Building Tile Cracks

One construction method employed in many public utility buildings and office blocks is to have the exterior building walls composed of a large number of small tiles. These tiles are generally cemented into place on the building wall using some type of resin mixture as the cement. Over time, the resin mixture may contract and expand, resulting in the building tiles becoming cracked.

A construction engineer is faced with the problem of assessing the tile damage in a certain group of downtown buildings. The total number of tiles on these buildings is around five million and it is far too costly to examine each tile in detail for cracks. Therefore the engineer constructs a sample of 1250 tiles chosen randomly from the blueprints of the building, as illustrated in Figure 10.2, and examines each tile in the sample for cracking.

The engineer is interested in the true overall proportion p of all the tiles that are cracked. Out of the sample of $n = 1250$ tiles, $x = 98$ are found to be cracked, and so the overall proportion p can be estimated as

$$\hat{p} = \frac{x}{n} = \frac{98}{1250} = 0.0784$$

With a critical point $z_{0.005} = 2.576$, a 99% two-sided confidence interval for the overall proportion of cracked tiles is

$$p \in \left(\hat{p} - \frac{z_{\alpha/2}}{n} \sqrt{\frac{x(n - x)}{n}}, \hat{p} + \frac{z_{\alpha/2}}{n} \sqrt{\frac{x(n - x)}{n}} \right)$$

$$= \left(0.0784 - \frac{2.576}{1250} \sqrt{\frac{98 \times (1250 - 98)}{1250}}, 0.0784 + \frac{2.576}{1250} \sqrt{\frac{98 \times (1250 - 98)}{1250}} \right)$$

$$= (0.0784 - 0.0196, 0.0784 + 0.0196) = (0.0588, 0.0980)$$

Consequently, based upon this sample of 1250 tiles, with 99% confidence the true proportion of cracked tiles is determined to lie somewhere between 5.88% and 9.8%, or somewhere between about 6% and 10%. If the total number of tiles on the buildings is about five million,

FIGURE 10.2

Random selection of tiles from all facades of the buildings

then the construction engineer can infer that the total number of cracked tiles is between about

$$0.06 \times 5{,}000{,}000 = 300{,}000 \qquad \text{and} \qquad 0.10 \times 5{,}000{,}000 = 500{,}000$$

Example 58

Overage Weedkiller Product

A chemical company produces a weedkiller that is sold in containers for consumers to apply in their yards and on their lawns. The company knows that after the time of manufacture there is an optimum time period in which the weedkiller should be used for maximum effect, and that applications of the weedkiller after this time period are not so effective. Products with an age older than this optimum time period are considered to be "overage" products.

In order to ensure the effectiveness of its product when it reaches the consumer's hands, the chemical company is interested is assessing how much of its product on the shelf waiting to be sold is in fact overage. A nationwide sampling scheme is developed whereby auditors visit randomly selected stores and determine whether the shelf product is overage or not from codings on the weedkiller containers indicating the date of manufacture.

The chemical company is interested in the true overall proportion p of the product on the shelf that is overage. The auditors examined $n = 54{,}965$ weedkiller containers and found that $x = 2779$ of them were overage. The overall proportion of overage product can then be estimated as

$$\hat{p} = \frac{2779}{54{,}965} = 0.0506$$

With a critical point $z_{0.005} = 2.576$, a 99% two-sided confidence interval for the overall proportion of overage product is

$$p \in \left(0.0506 - \frac{2.576}{54{,}965} \sqrt{\frac{2779 \times (54{,}965 - 2779)}{54{,}965}}, \right.$$

$$\left. 0.0506 + \frac{2.576}{54{,}965} \sqrt{\frac{2779 \times (54{,}965 - 2779)}{54{,}965}} \right)$$

$$= (0.0506 - 0.0024, 0.0506 + 0.0024) = (0.0482, 0.0530)$$

Therefore, the chemical company has discovered that somewhere between about 4.8% and 5.3% of its weedkiller product on the shelf waiting to be sold is overage.

GAMES OF CHANCE

Recall the problem discussed in Section 7.3.1 concerning the investigation of the bias of a coin. Two scenarios were considered.

- Scenario I : The coin is tossed 100 times and 40 heads are obtained.

- Scenario II : The coin is tossed 1000 times and 400 heads are obtained.

In either case, the probability of obtaining a head is estimated to be $\hat{p} = 0.4$, but the standard error of the estimate was shown to be much smaller in scenario II than in scenario I.

The smaller standard error in scenario II results in a smaller confidence interval for the probability p. For example, in scenario I with $z_{0.025} = 1.96$, a 95% two-sided confidence interval for p is calculated to be

$$p \in \left(\hat{p} - z_{\alpha/2}\sqrt{\frac{\hat{p}(1-\hat{p})}{n}}, \ \hat{p} + z_{\alpha/2}\sqrt{\frac{\hat{p}(1-\hat{p})}{n}} \right)$$

$$= \left(0.4 - 1.96\sqrt{\frac{0.4 \times (1-0.4)}{100}}, \ 0.4 + 1.96\sqrt{\frac{0.4 \times (1-0.4)}{100}} \right)$$

$$= (0.4 - 0.096, 0.4 + 0.096) = (0.304, 0.496)$$

However, in scenario II the confidence interval is calculated to be

$$p \in \left(0.4 - 1.96\sqrt{\frac{0.4 \times (1-0.4)}{1000}}, \ 0.4 + 1.96\sqrt{\frac{0.4 \times (1-0.4)}{1000}} \right)$$

$$= (0.4 - 0.030, 0.4 + 0.030) = (0.370, 0.430)$$

These two confidence intervals are illustrated in Figure 10.3. Notice that the larger value of n in scenario II results in a shorter confidence interval for p, reflecting the increase in precision of the estimate $\hat{p}$. In fact, for a fixed value of $\hat{p}$ and a fixed confidence level $1 - \alpha$, the confidence interval length is seen to be *inversely* proportional to the square root of the sample size n.

One-Sided Confidence Intervals   One-sided confidence intervals for a population proportion p can be used instead of a two-sided confidence interval if an experimenter is interested

FIGURE 10.3

Confidence intervals for the probability p of obtaining a head in a toss of a biased coin

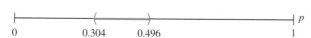

40 heads from 100 coin tosses

400 heads from 1000 coin tosses

in obtaining only an *upper bound* or a *lower bound* on the population proportion. Their format is similar to the two-sided confidence interval except that a (smaller) critical point z_α is employed in place of $z_{\alpha/2}$.

The one-sided confidence intervals can be constructed as follows. If $Z \sim N(0, 1)$, then

$$P(Z \le z_\alpha) = 1 - \alpha$$

so that it is approximately the case that

$$P \left(\frac{\hat{p} - p}{\sqrt{\frac{p(1-p)}{n}}} \le z_\alpha \right) = 1 - \alpha$$

which can be rewritten

$$P \left(\hat{p} - z_\alpha \sqrt{\frac{p(1 - p)}{n}} \le p \right) = 1 - \alpha$$

This expression implies that

$$p \in \left(\hat{p} - z_\alpha \sqrt{\frac{p(1 - p)}{n}}, 1 \right)$$

is a one-sided $1 - \alpha$ confidence level confidence interval that provides a *lower bound* for p.

Similarly,

$$P(-z_\alpha \le Z) = 1 - \alpha$$

so that it is approximately the case that

$$P \left(-z_\alpha \le \frac{\hat{p} - p}{\sqrt{\frac{p(1-p)}{n}}} \right) = 1 - \alpha$$

which can be rewritten

$$P \left(p \le \hat{p} + z_\alpha \sqrt{\frac{p(1 - p)}{n}} \right) = 1 - \alpha$$

Therefore

$$p \in \left(0, \hat{p} + z_\alpha \sqrt{\frac{p(1 - p)}{n}} \right)$$

is a one-sided $1 - \alpha$ confidence level confidence interval that provides an *upper bound* for p.

As in the two-sided confidence intervals, the unknown value of p is replaced by the estimate $\hat{p} = x/n$ in the expression

$$\sqrt{\frac{p(1 - p)}{n}}$$

One-Sided Confidence Intervals for a Population Proportion

If the random variable X has a $B(n, p)$ distribution, then approximate one-sided $1 - \alpha$ confidence level confidence intervals for the success probability p based upon an observed value x of the random variable are

$$p \in \left(\hat{p} - z_\alpha \sqrt{\frac{\hat{p}(1 - \hat{p})}{n}}, 1 \right) \qquad \text{and} \qquad p \in \left(0, \hat{p} + z_\alpha \sqrt{\frac{\hat{p}(1 - \hat{p})}{n}} \right)$$

where the estimated success probability is $\hat{p} = x/n$. These confidence intervals respectively provide a *lower bound* and an *upper bound* on the probability p, and can also be written as

$$p \in \left(\hat{p} - \frac{z_\alpha}{n} \sqrt{\frac{x(n - x)}{n}}, 1 \right) \qquad \text{and} \qquad p \in \left(0, \hat{p} + \frac{z_\alpha}{n} \sqrt{\frac{x(n - x)}{n}} \right)$$

The approximation is reasonable as long as both x and $n - x$ are larger than 5.

Example 39

Cattle Inoculations

Recall that when a vaccine was administered to $n = 500{,}000$ head of cattle, $x = 372$ were observed to suffer a serious adverse reaction. The estimate of the probability p of an animal suffering such a reaction is thus

$$\hat{p} = \frac{372}{500{,}000} = 7.44 \times 10^{-4}$$

In order to satisfy government safety regulations, the manufacturers of the vaccine must provide an upper bound on the probability of an animal suffering such a reaction. With the critical point $z_{0.01} = 2.326$, a one-sided 99% confidence interval for the probability p of an animal suffering a reaction can be calculated to be

$$p \in \left(0, \hat{p} + \frac{z_\alpha}{n} \sqrt{\frac{x(n - x)}{n}} \right)$$

$$= \left(0, 0.000744 + \frac{2.326}{500{,}000} \sqrt{\frac{372 \times (500{,}000 - 372)}{500{,}000}} \right)$$

$$= (0, 0.000744 + 0.000090) = (0, 0.000834)$$

as illustrated in Figure 10.4. Thus, with 99% confidence, the manufacturer can claim that the probability of an adverse reaction to the vaccine is no larger than 8.34×10^{-4}.

As a final point, notice that an *upper bound* on a probability p can be used to obtain a *lower bound* on the *complementary* probability $1 - p$. Thus since $1 - 8.34 \times 10^{-4} = 0.999166$, this result can be rephrased as "the probability that an animal does not suffer an adverse reaction is *at least* 0.999166."

10.1.2 Hypothesis Tests on a Population Proportion

An observation x from a random variable X with a $B(n, p)$ distribution can be used to test a hypothesis concerning the success probability p. A *two-sided* hypothesis testing problem would be

$$H_0 : p = p_0 \quad \text{versus} \quad H_A : p \neq p_0$$

FIGURE 10.4

Upper confidence bound on the
probability of an adverse reaction
from the cattle vaccine

Vaccine administered to $n = 500{,}000$ cattle
$x = 372$ suffer an adverse reaction
$p =$ probability of an adverse reaction

$\hat{p} = x/n = 0.000744$

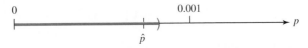

99% upper confidence bound
$p \in (0, 0.000834)$

---

for a particular fixed value p_0. This is appropriate if an experimenter wishes to determine whether there is significant evidence that the success probability is different from p_0. *One-sided* sets of hypotheses

$$H_0 : p \geq p_0 \quad \text{versus} \quad H_A : p < p_0$$

and

$$H_0 : p \leq p_0 \quad \text{versus} \quad H_A : p > p_0$$

can also be used.

The p-values for these hypothesis tests can be calculated using the cumulative distribution function of the binomial distribution, which for reasonably large values of n can be approximated by a normal distribution. If the normal approximation is employed, then

$$\frac{\hat{p} - p}{\sqrt{\frac{p(1-p)}{n}}}$$

is taken to have approximately a standard normal distribution, so that if $p = p_0$, the "z-statistic"

$$z = \frac{\hat{p} - p_0}{\sqrt{\frac{p_0(1-p_0)}{n}}}$$

can be taken to be an observation from a standard normal distribution. Notice that the hypothesized value p_0 is used inside the square root term

$$\sqrt{\frac{p_0(1 - p_0)}{n}}$$

of this expression, and that when the top and the bottom of the expression are multiplied by n, the z-statistic can be rewritten as

$$z = \frac{x - np_0}{\sqrt{np_0(1 - p_0)}}$$

The normal approximation can be improved with a continuity correction whereby the numerator of the z-statistic $x - np_0$ is replaced by either $x - np_0 - 0.5$ or $x - np_0 + 0.5$ as described in the next section.

Two-Sided Hypothesis Tests The exact p-value for the two-sided hypothesis testing problem

$$H_0 : p = p_0 \quad \text{versus} \quad H_A : p \neq p_0$$

is usually calculated as

$$p\text{-value} = 2 \times P(X \geq x)$$

if $\hat{p} = x/n > p_0$, and as

$$p\text{-value} = 2 \times P(X \leq x)$$

if $\hat{p} = x/n < p_0$, where the random variable X has a $B(n, p_0)$ distribution. This can be deduced from the definition of the p-value:

> *The p-value is the probability of obtaining this data set or worse*
> *when the null hypothesis is true.*

Notice that under the null hypothesis H_0, the expected value of the number of successes is np_0. Consequently, as Figure 10.5 shows, "worse" in the definition of the p-value means values of the random variable X farther away from np_0 than is observed. This is values larger than x when $x > np_0$ ($\hat{p} > p_0$), and values smaller than x when $x < np_0$ ($\hat{p} < p_0$). The tail probabilities of the binomial distribution

$$P(X \geq x) \qquad \text{and} \qquad P(X \leq x)$$

are then multiplied by 2 since it is a two-sided problem with the alternative hypothesis $H_A : p \neq p_0$ allowing values of p both smaller and larger than p_0. Of course if $\hat{p} = p_0$, then the p-value can be taken to be equal to 1, and there is clearly no evidence that the null hypothesis is not plausible.

FIGURE 10.5

Constructing two-sided
hypothesis tests

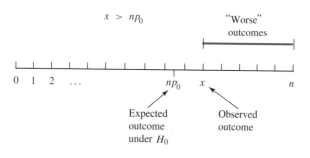

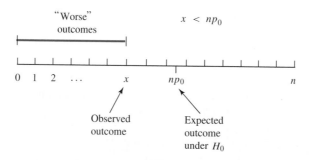

FIGURE 10.6

p-value calculations for two-sided
hypothesis tests

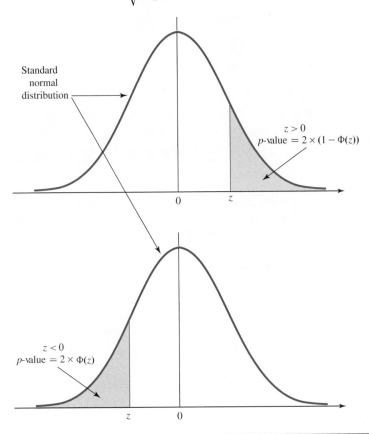

When the normal approximation to the distribution of $\hat{p}$ is appropriate (in this case it can be considered to be acceptable as long as np_0 and $n(1 - p_0)$ are both larger than 5), the statistic

$$z = \frac{\hat{p} - p_0}{\sqrt{\frac{p_0(1-p_0)}{n}}} = \frac{x - np_0}{\sqrt{np_0(1 - p_0)}}$$

is calculated and, as Figure 10.6 illustrates, the *p*-value is

$$p\text{-value} = 2 \times P(Z \geq z)$$

if $z > 0$, and

$$p\text{-value} = 2 \times P(Z \leq z)$$

if $z < 0$, where the random variable Z has a standard normal distribution. In either case, the *p*-value can be written as

$$p\text{-value} = 2 \times \Phi(-|z|)$$

where $\Phi(\cdot)$ is the standard normal cumulative distribution function.

The normal approximation can be improved by employing a continuity correction of 0.5 in the numerator of the z-statistic. If $x - np_0 > 0.5$, a z-statistic

$$z = \frac{x - np_0 - 0.5}{\sqrt{np_0(1 - p_0)}}$$

can be used, and if $x - np_0 < -0.5$, a z-statistic

$$z = \frac{x - np_0 + 0.5}{\sqrt{np_0(1 - p_0)}}$$

can be used. Notice that the continuity correction serves to bring the value of the z-statistic closer to 0. The effect of employing the continuity correction becomes less important as the sample size n gets larger.

Hypothesis tests at a fixed size α can be performed by comparing the p-value with the value of α. The null hypothesis is accepted if the p-value is larger than α, so that the *acceptance region* is

$$|z| \le z_{\alpha/2}$$

and the null hypothesis is rejected if the p-value is smaller than α, so that the *rejection region* is

$$|z| > z_{\alpha/2}$$

Two-Sided Hypothesis Tests for a Population Proportion

If the random variable X has a $B(n, p)$ distribution, then the exact p-value for the two-sided hypothesis testing problem

$$H_0 : p = p_0 \quad \text{versus} \quad H_A : p \ne p_0$$

based upon an observed value of the random variable x is

$$p\text{-value} = 2 \times P(X \ge x)$$

if $\hat{p} = x/n > p_0$, and

$$p\text{-value} = 2 \times P(X \le x)$$

if $\hat{p} = x/n < p_0$, where the random variable X has a $B(n, p_0)$ distribution. When np_0 and $n(1 - p_0)$ are both larger than 5, a normal approximation can be used to give a p-value of

$$p\text{-value} = 2 \times \Phi(-|z|)$$

where $\Phi(\cdot)$ is the standard normal cumulative distribution function and

$$z = \frac{\hat{p} - p_0}{\sqrt{\frac{p_0(1-p_0)}{n}}} = \frac{x - np_0}{\sqrt{np_0(1 - p_0)}}$$

In order to improve the normal approximation the value $x - np_0 - 0.5$ may be used in the numerator of the z-statistic when $x - np_0 > 0.5$, and the value $x - np_0 + 0.5$ may be used in the numerator of the z-statistic when $x - np_0 < -0.5$. A size α hypothesis test *accepts* the null hypothesis when

$$|z| \le z_{\alpha/2}$$

and *rejects* the null hypothesis when

$$|z| > z_{\alpha/2}$$

Example 59

Opossum Progeny Genders

A biologist is interested in whether opossums give birth to male and female progeny with equal probabilities. A group of opossums is observed, and out of 23 births, 14 are male and 9 are female.

Suppose that each opossum offspring has a probability p of being male, independent of any other births. The number of male births out of 23 births is then a random variable with a $B(23, p)$ distribution, and the hypotheses of interest are

$$H_0 : p = 0.5 \quad \text{versus} \quad H_A : p \neq 0.5$$

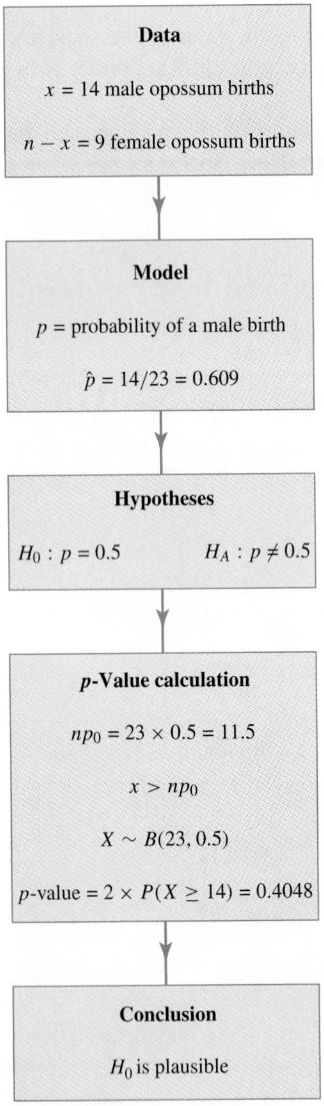

With $x = 14$ male births out of $n = 23$ total births, the estimated probability of a male birth is

$$\hat{p} = \frac{14}{23} = 0.609$$

which is larger than the hypothesized value of $p_0 = 0.5$. As Figure 10.7 shows, the exact p-value is therefore

$$p\text{-value} = 2 \times P(X \geq 14)$$

where the random variable X has a $B(23, 0.5)$ distribution. This value can be calculated to be

$$p\text{-value} = 2 \times 0.2024 = 0.4048$$

Since $np_0 = n(1 - p_0) = 23 \times 0.5 = 11.5 > 5$, a normal approximation to the distribution of X should be reasonable. The value of the z-statistic with continuity correction is

$$z = \frac{x - np_0 - 0.5}{\sqrt{np_0(1 - p_0)}} = \frac{14 - (23 \times 0.5) - 0.5}{\sqrt{23 \times 0.5 \times (1.0 - 0.5)}} = 0.83$$

which gives a p-value (calculated from Table I) of

$$p\text{-value} = 2 \times \Phi(-0.83) = 2 \times 0.2033 = 0.4066$$

It can be seen that the normal approximation is quite accurate, and that with such large p-values there is no reason to doubt the validity of the null hypothesis. Based on this data set the biologist realizes that there is not sufficient evidence to conclude that male and female births are not equally likely.

However, with only 23 births observed, it should be remembered that there is a wide range of other plausible values for the probability of a male birth p. In fact, with a critical point $z_{0.025} = 1.96$, a 95% two-sided confidence interval for the probability of a male birth is

$$p \in \left(\hat{p} - \frac{z_{\alpha/2}}{n}\sqrt{\frac{x(n - x)}{n}}, \hat{p} + \frac{z_{\alpha/2}}{n}\sqrt{\frac{x(n - x)}{n}} \right)$$

$$= \left(0.609 - \frac{1.96}{23}\sqrt{\frac{14 \times (23 - 14)}{23}}, 0.609 + \frac{1.96}{23}\sqrt{\frac{14 \times (23 - 14)}{23}} \right)$$

$$= (0.609 - 0.199, 0.609 + 0.199) = (0.410, 0.808)$$

Thus, the probability of a male birth could in fact be anywhere between about 0.4 and 0.8.

Data

$x = 14$ male opossum births

$n - x = 9$ female opossum births

Model

p = probability of a male birth

$\hat{p} = 14/23 = 0.609$

Hypotheses

$H_0 : p = 0.5$ $H_A : p \neq 0.5$

p-Value calculation

$np_0 = 23 \times 0.5 = 11.5$

$x > np_0$

$X \sim B(23, 0.5)$

$p\text{-value} = 2 \times P(X \geq 14) = 0.4048$

Conclusion

H_0 is plausible

FIGURE 10.7

Exact p-value calculation for opossum progeny genders example

Example 60	A mathematician is investigating various algorithms for simulating random variable observa-
Random Variable	tions on a computer. One algorithm is supposed to produce (independent) observations from
Simulations	a standard normal distribution. The mathematician obtains 10,000 simulations from the algo-

rithm and notices that 6702 of them have an absolute value no larger than 1. Does this cast any doubt on the validity of the algorithm?

Suppose that the algorithm produces a value between -1 and 1 with a probability p, so that the number of values out of 10,000 lying in this range has a $B(10,000, p)$ distribution. If the values really are observations from a standard normal distribution, then the success probability is

$$p = \Phi(1) - \Phi(-1) = 0.8413 - 0.1587 = 0.6826$$

so that the two-sided hypotheses of interest are

$$H_0 : p = 0.6826 \quad \text{versus} \quad H_A : p \neq 0.6826$$

With $n = 10,000$ a normal approximation to the p-value is appropriate and the z-statistic is

$$z = \frac{6702 - (10,000 \times 0.6826) + 0.5}{\sqrt{10,000 \times 0.6826 \times (1.0000 - 0.6826)}} = -2.65$$

although it can be seen that the continuity correction of 0.5 is not important here. The p-value (calculated from Table I) is therefore

$$p\text{-value} = 2 \times \Phi(-2.65) = 2 \times 0.0040 = 0.0080$$

and such a small value leads the mathematician to conclude that the null hypothesis is not plausible and that the algorithm is *not* doing a very good job of simulating standard normal random variables.

One-Sided Hypothesis Tests   The p-values for one-sided hypothesis tests are calculated in the obvious way as shown in the accompanying box. Exact p-values are calculated from the tail probabilities of the appropriate binomial distribution, and these can be approximated by a normal distribution in the usual circumstances. A continuity correction of either $+0.5$ or -0.5 should be used, depending on the direction of the one-sided problem.

Example 57	Legal agreements have been reached whereby if 10% or more of the building tiles are cracked,
Building Tile Cracks	then the construction company that originally installed the tiles must help pay for the building

repair costs. Do the survey results of 98 cracked tiles out of 1250 tiles indicate that the construction company should be required to contribute to the building repair costs?

The construction engineers approach this problem in the following way. If p is the probability that a tile is cracked, then the one-sided hypotheses

$$H_0 : p \geq 0.1 \quad \text{versus} \quad H_A : p < 0.1$$

should be considered. The null hypothesis corresponds to situations in which the construction company must contribute to the repair costs, and the alternative hypothesis corresponds to situations where it is not liable for any costs. A rejection of the null hypothesis would therefore establish that the construction company has no financial responsibilities.

One-Sided Hypothesis Tests for a Population Proportion

If the random variable X has a $B(n, p)$ distribution, then the exact p-value for the one-sided hypothesis testing problem

$$H_0 : p \geq p_0 \quad \text{versus} \quad H_A : p < p_0$$

based upon an observed value of the random variable x is

$$p\text{-value} = P(X \leq x)$$

where the random variable X has a $B(n, p_0)$ distribution. The normal approximation to this is

$$p\text{-value} = \Phi(z)$$

where $\Phi(\cdot)$ is the standard normal cumulative distribution function and

$$z = \frac{x - np_0 + 0.5}{\sqrt{np_0(1 - p_0)}}$$

A size α hypothesis test *accepts* the null hypothesis when

$$z \geq -z_\alpha$$

and *rejects* the null hypothesis when

$$z < -z_\alpha$$

For the one-sided hypothesis testing problem

$$H_0 : p \leq p_0 \quad \text{versus} \quad H_A : p > p_0$$

the p-value is

$$p\text{-value} = P(X \geq x)$$

where the random variable X has a $B(n, p_0)$ distribution. The normal approximation to this is

$$p\text{-value} = 1 - \Phi(z)$$

where

$$z = \frac{x - np_0 - 0.5}{\sqrt{np_0(1 - p_0)}}$$

A size α hypothesis test *accepts* the null hypothesis when

$$z \leq z_\alpha$$

and *rejects* the null hypothesis when

$$z > z_\alpha$$

A normal approximation is appropriate, and with $n = 1250$ and $x = 98$ the z-statistic is

$$z = \frac{x - np_0 + 0.5}{\sqrt{np_0(1 - p_0)}} = \frac{98 - (1250 \times 0.1) + 0.5}{\sqrt{1250 \times 0.1 \times (1.0 - 0.1)}} = -2.50$$

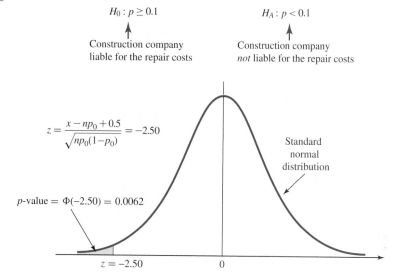

FIGURE 10.8

Hypothesis test analysis for building tile cracks example

$x = 98$ cracked tiles in a sample of $n = 1250$ tiles
$p =$ probability that a tile is cracked

$H_0 : p \geq 0.1$

Construction company
liable for the repair costs

$H_A : p < 0.1$

Construction company
not liable for the repair costs

$$z = \frac{x - np_0 + 0.5}{\sqrt{np_0(1-p_0)}} = -2.50$$

Standard
normal
distribution

p-value $= \Phi(-2.50) = 0.0062$

$z = -2.50$

0

Conclusion: The null hypothesis is *not* plausible and it has been established that the construction company is *not* liable for the repair costs.

The p-value is therefore

$$p\text{-value} = \Phi(-2.50) = 0.0062$$

which, as Figure 10.8 illustrates, indicates that the null hypothesis is not plausible and can be rejected. This establishes that the construction company is not required to contribute to the repair costs.

Example 39
Cattle Inoculations

Suppose that the vaccine can be approved for widespread use if it can be established that on average no more than one in a thousand cattle will suffer a serious adverse reaction. In other words, the probability of a serious adverse reaction must be *no larger* than 0.001. If the one-sided hypotheses

$$H_0 : p \geq 0.001 \quad \text{versus} \quad H_A : p < 0.001$$

are used, then the rejection of the null hypothesis will establish that the vaccine can be approved for widespread use.

With $x = 372$ reactions observed in a sample of $n = 500,000$ cattle, a 99% one-sided confidence interval for the probability of a reaction was calculated to be

$$p \in (0, 0.000834)$$

Since the upper bound of this confidence interval is smaller than $p_0 = 0.001$, this implies that the p-value for these one-sided hypotheses will be smaller than 1%. In fact, with a z-statistic

$$z = \frac{372 - (500,000 \times 0.001) + 0.5}{\sqrt{500,000 \times 0.001 \times (1.000 - 0.001)}} = -5.70$$

the p-value is calculated to be

$$p\text{-value} = \Phi(-5.70) \simeq 0$$

In conclusion, the null hypothesis has been shown not to be plausible, so that the probability of an adverse reaction is known to be less than one in a thousand and the vaccine can be approved for widespread use.

GAMES OF CHANCE

If I take a six-sided die, roll it 10 times, and score a 6 only once, should I have a reasonable suspicion that the die is weighted to *reduce* the chance of scoring a 6? If p is the probability of scoring a 6 on a roll of the die, the one-sided hypotheses of interest are

$$H_0 : p \geq \frac{1}{6} \quad \text{versus} \quad H_A : p < \frac{1}{6}$$

If the null hypothesis is plausible, then there is no reasonable suspicion that the die is weighted to reduce the chance of scoring a 6. However, a small p-value will indicate that the null hypothesis is not plausible and will imply that the die is weighted to reduce the chance of scoring a 6.

With $n = 10$ and $x = 1$, the exact p-value for these hypotheses is

$$p\text{-value} = P(X \leq 1)$$

where the random variable X has a $B(10, 1/6)$ distribution, which is

$$p\text{-value} = 0.4845$$

This large p-value indicates that the null hypothesis is quite plausible and that there is no reason to suspect that the chance of scoring a 6 is any smaller than $1/6$.

What if only one 6 is scored in 20 rolls of the die? In this case the p-value is

$$p\text{-value} = P(X \leq 1)$$

where the random variable X has a $B(20, 1/6)$ distribution, which is

$$p\text{-value} = 0.1304$$

Again, there is still not sufficient evidence to conclude that the chance of scoring a 6 is any smaller than $1/6$.

Figure 10.9 shows the p-values for additional values of n. If only one 6 is obtained in 30 rolls of the die, the situation is starting to look suspicious because the p-value is about 3%. With 40 rolls of the die the p-value falls below 1% and the null hypothesis that the probability of scoring a 6 is at least $1/6$ becomes unbelievable.

Figure 10.10 provides a summary chart of confidence interval construction and hypothesis testing for a population proportion p. As is always the case, the decisions of hypothesis tests at a fixed size α can be deduced from the corresponding p-value. In general, the inclusion of p_0 in a $1 - \alpha$ confidence level confidence interval for p implies a p-value greater than α for a hypothesized value p_0 for the population proportion p. Similarly, exclusion generally implies a p-value smaller than α. In some rare cases, however, there may be a slight discrepancy in these general rules due to the way in which the quantity $\sqrt{p(1-p)/n}$ is handled, with p being replaced either by $\hat{p}$ in confidence interval construction or by p_0 in hypothesis testing.

FIGURE 10.9

p-values for die example

One 6 is obtained in n rolls of a die. Is the die weighted to *reduce* the chance of scoring a 6?

p = probability of scoring a 6

$$H_0 : p \geq 1/6 \qquad\qquad H_A : p < 1/6$$

$$\uparrow$$

Die weighted to reduce chance
of scoring a 6

$$X \sim B(n, 1/6)$$

$$p\text{-value} = P(X \leq 1)$$

n	p-value
10	0.4845
20	0.1304
30	0.0295
40	0.0061
50	0.0012

FIGURE 10.10

Summary of inferences on a
population proportion

Inferences on a Population Proportion

Observe x successes from n trials $\qquad p$ = probability of success $\qquad \hat{p} = x/n$

Two-sided $1 - \alpha$ level confidence interval (both x and $n - x$ larger than 5)

$$p \in \left(\hat{p} - z_{\alpha/2} \sqrt{\tfrac{\hat{p}(1-\hat{p})}{n}}, \, \hat{p} + z_{\alpha/2} \sqrt{\tfrac{\hat{p}(1-\hat{p})}{n}} \right) = \left(\hat{p} - \tfrac{z_{\alpha/2}}{n} \sqrt{\tfrac{x(n-x)}{n}}, \, \hat{p} + \tfrac{z_{\alpha/2}}{n} \sqrt{\tfrac{x(n-x)}{n}} \right)$$

One-sided $1 - \alpha$ level confidence intervals (both x and $n - x$ larger than 5)

$$p \in \left(\hat{p} - z_{\alpha} \sqrt{\tfrac{\hat{p}(1-\hat{p})}{n}}, \, 1 \right) = \left(\hat{p} - \tfrac{z_{\alpha}}{n} \sqrt{\tfrac{x(n-x)}{n}}, \, 1 \right)$$

$$p \in \left(0, \, \hat{p} + z_{\alpha} \sqrt{\tfrac{\hat{p}(1-\hat{p})}{n}} \right) = \left(0, \, \hat{p} + \tfrac{z_{\alpha}}{n} \sqrt{\tfrac{x(n-x)}{n}} \right)$$

Hypothesis testing $\qquad X \sim B(n, p_0)$

$$H_0: p = p_0, \quad H_A: p \neq p_0$$

$$x - np_0 > 0.5 \qquad z = \tfrac{x-np_0-0.5}{\sqrt{np_0(1-p_0)}} \qquad p\text{-value} = 2 \times P(X \geq x) \simeq 2 \times (1 - \Phi(z))$$

$$x - np_0 < -0.5 \qquad z = \tfrac{x-np_0+0.5}{\sqrt{np_0(1-p_0)}} \qquad p\text{-value} = 2 \times P(X \leq x) \simeq 2 \times \Phi(z)$$

$$H_0: p \geq p_0, \quad H_A: p < p_0$$

$$z = \tfrac{x-np_0+0.5}{\sqrt{np_0(1-p_0)}} \qquad p\text{-value} = P(X \leq x) \simeq \Phi(z)$$

$$H_0: p \leq p_0, \quad H_A: p > p_0$$

$$z = \tfrac{x-np_0-0.5}{\sqrt{np_0(1-p_0)}} \qquad p\text{-value} = P(X \geq x) \simeq 1 - \Phi(z)$$

10.1.3 Sample Size Calculations

The sample size n affects the precision of the inference that can be made about a population proportion p. It is often useful to gauge the amount of precision afforded by a certain sample size before any sampling is performed. Furthermore, after the results of an initial sample are observed, it may be useful to determine how much additional sampling is required to attain a specified precision.

Within a hypothesis testing framework, increased sample sizes result in increased power for tests at a fixed significance level α. This means that when the null hypothesis is false, there is a greater chance that there will be enough evidence to reject it. However, as in previous chapters, the most convenient way to assess the amount of precision afforded by a sample size n is to consider the length L of a two-sided confidence interval for the population proportion p. If a confidence level $1 - \alpha$ is used, then the confidence interval length is

$$L = 2z_{\alpha/2}\sqrt{\frac{\hat{p}(1 - \hat{p})}{n}}$$

so that the sample size n required to achieve a confidence interval length L is

$$n = \frac{4z_{\alpha/2}^2\hat{p}(1 - \hat{p})}{L^2}$$

Notice that the required sample size n increases either as the confidence interval length L decreases or as the specified confidence level $1 - \alpha$ increases (so that α decreases and $z_{\alpha/2}$ increases).

A problem in using this formula to find the required sample size is that the term $\hat{p}(1 - \hat{p})$ is unknown. However, Figure 10.11 shows how the value of $p(1 - p)$ varies for p between 0 and 1, and it can be seen that the *largest* value taken is $1/4$ when $p = 1/2$. Consequently, a *worst case* scenario is to take

$$\hat{p}(1 - \hat{p}) = \frac{1}{4}$$

in which case the required sample size is

$$n = \frac{z_{\alpha/2}^2}{L^2}$$

Nevertheless, if the value of $\hat{p}$ is far from 0.5, this worst case scenario is wasteful and the requirement on the confidence interval length L can be met with a smaller sample size. If prior information or knowledge of the problem allows the experimenter to *bound* the value of $\hat{p}$ away from 0.5, then a smaller required sample size can be determined. Specifically, if the experimenter can reasonably expect that $\hat{p}$ will be *less* than some value $p^* < 0.5$, or alternatively if the experimenter can reasonably expect that $\hat{p}$ will be *greater* than a value

FIGURE 10.11

Value of $p(1 - p)$

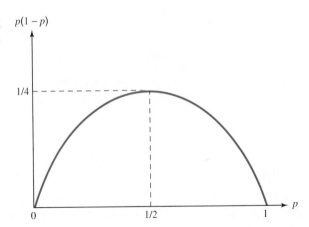

$p^* > 0.5$, then a sample size

$$n \simeq \frac{4z_{\alpha/2}^2 p^*(1 - p^*)}{L^2}$$

will suffice.

Example 61
Political Polling

A local newspaper wishes to poll the population of its readership area to determine the proportion p of them who agree with the statement

The city mayor is doing a good job.

The newspaper wishes to present the results as a percentage with a footnote reading "accurate to within ±3%." How many people do they need to poll?

The first point to notice here is that the newspaper decides to discard anybody who does not express an opinion. Thus, the sample results will consist of x people who agree with the statement and $n - x$ people who do not agree with the statement and, as Figure 10.12 shows, the paper will publish the estimate

$$\hat{p} = \frac{x}{n}$$

The pollsters then decide that if they can construct a 99% confidence interval for p with a length no larger than $L = 6\%$, they will have achieved the desired accuracy (because this confidence interval is $\hat{p} \pm 3\%$). In addition, the pollsters feel that the population may be fairly evenly spread on their agreement with the statement. Therefore a worst case scenario with $\hat{p} = 0.5$ is considered, and with $z_{0.005} = 2.576$ the required sample size (of people with an

FIGURE 10.12

Political polling example

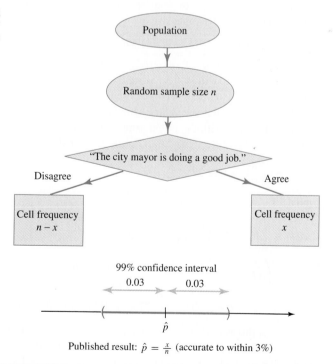

Published result: $\hat{p} = \frac{x}{n}$ (accurate to within 3%)

opinion on the statement) is calculated to be

$$n = \frac{z_{\alpha/2}^2}{L^2} = \frac{2.576^2}{0.06^2} = 1843.3$$

Consequently, the paper decides to obtain the opinion of a *representative* sample of at least 1844 people.

In fact, you will often see polls reported with a footnote "based on a sample of 1000 respondents, ±3% sampling error." How should this statement be interpreted? Under the worst case scenario with $\hat{p} = 0.5$, this sample size results in a confidence interval $\hat{p} \pm 3\%$ with a confidence level $1 - \alpha$ where

$$z_{\alpha/2} = \sqrt{n}L = \sqrt{1000} \times 0.06 = 1.90$$

This equation is satisfied with $\alpha \simeq 0.057$, so that the reader should interpret the sensitivity of the poll as implying that with about 95% confidence, the true proportion p is within three percentage points of the reported value. However, if $\hat{p}$ is far from 0.5, the confidence level may in fact be much larger than 95%.

Example 57

Building Tile Cracks

Recall that a 99% confidence interval for the overall proportion of cracked tiles was calculated to be

$$p \in (0.0588, 0.0980)$$

However, the lawyers handling this case decided that they needed to know the overall proportion of cracked tiles to within 1% with 99% confidence. How much additional sampling is necessary?

The lawyers' demands can be met with a 99% confidence interval with a length L of 2%. It is reasonable to expect, based on the confidence interval above, that the estimated proportion $\hat{p}$ after the second stage of sampling will be less than $p^* = 0.1$. Consequently, it can be estimated that a total sample size of

$$n \simeq \frac{4z_{\alpha/2}^2 p^*(1 - p^*)}{L^2} = \frac{4 \times 2.576^2 \times 0.1 \times (1.0 - 0.1)}{0.02^2} = 5972.2$$

or about 6000 tiles, will suffice. Since the initial sample consisted of 1250 tiles, the engineers therefore decided to take a secondary representative sample of 4750 tiles.

The results of the secondary sample reveal that 308 of the 4750 tiles examined were cracked. Together with the initial sample of 98 cracked tiles out of 1250, there are therefore $x = 308 + 98 = 406$ cracked tiles out of $n = 6000$. This gives

$$\hat{p} = \frac{406}{6000} = 0.0677$$

and a 99% confidence interval

$$p \in \left(\hat{p} - \frac{z_{\alpha/2}}{n} \sqrt{\frac{x(n - x)}{n}}, \hat{p} + \frac{z_{\alpha/2}}{n} \sqrt{\frac{x(n - x)}{n}} \right)$$

$$= \left(0.0677 - \frac{2.576}{6000} \sqrt{\frac{406 \times (6000 - 406)}{6000}}, 0.0677 + \frac{2.576}{6000} \sqrt{\frac{406 \times (6000 - 406)}{6000}} \right)$$

$$= (0.0677 - 0.0084, 0.0677 + 0.0084) = (0.0593, 0.0761)$$

Notice that this confidence interval has a length of about 1.7%, which is smaller than the required length of 2% (since $\hat{p}$ turned out to be smaller than $p^* = 0.1$). In conclusion, the overall proportion of cracked tiles can be reported to be 6.8% ± 1%, with at least 99% confidence.

■ 10.1.4   Problems

10.1.1 Suppose that $x = 11$ is an observation from a $B(32, p)$ random variable.

(a) Compute a two-sided 99% confidence interval for p.

(b) Compute a two-sided 95% confidence interval for p.

(c) Compute a one-sided 99% confidence interval that provides an upper bound for p.

(d) Consider the hypotheses

$$H_0 : p = 0.5 \quad \text{versus} \quad H_A : p \neq 0.5$$

Calculate an exact p-value using the tail probability of the binomial distribution and compare it with the corresponding p-value calculated using a normal approximation.

10.1.2 Suppose that $x = 21$ is an observation from a $B(27, p)$ random variable.

(a) Compute a two-sided 99% confidence interval for p.

(b) Compute a two-sided 95% confidence interval for p.

(c) Compute a one-sided 95% confidence interval that provides a lower bound for p.

(d) Consider the hypotheses

$$H_0 : p \leq 0.6 \quad \text{versus} \quad H_A : p > 0.6.$$

Calculate an exact p-value using the tail probability of the binomial distribution and compare it with the corresponding p-value calculated using a normal approximation.

10.1.3 A random-number generator is supposed to produce a sequence of 0s and 1s with each value being equally likely to be a 0 or a 1 and with all values being independent. In an examination of the random-number generator, a sequence of 50,000 values is obtained of which 25,264 are 0s.

(a) Formulate a set of hypotheses to test whether there is any evidence that the random-number generator is producing 0s and 1s with unequal probabilities, and calculate the corresponding p-value.

(b) Compute a two-sided 99% confidence interval for the probability p that a value produced by the random-number generator is a 0.

(c) If a two-sided 99% confidence interval for this probability is required with a total length no larger than 0.005, how many additional values need to be investigated?

10.1.4 A new radar system is being developed to detect packages dropped by airplane. In a series of trials, the radar detected the packages being dropped 35 times out of 44. Construct a 95% lower confidence bound on the probability that the radar successfully detects dropped packages. (This problem is continued in Problem 10.2.3.)

10.1.5 Two experiments are performed. In the first experiment a six-sided die is rolled 50 times and a 6 is scored twice. In the second experiment the die is rolled 100 times and a 6 is scored four times. Which of the two experiments provides the most support for the claim that the die has been weighted to reduce the chance of scoring a 6? (*Hint*: Form a suitable set of hypotheses and compare the p-values obtained from the two experiments.)

10.1.6 If 21 6s are obtained from 100 rolls of a die, should the null hypothesis that the probability of scoring a 6 is $1/6$ be rejected at the size $\alpha = 0.05$ level?

10.1.7 A court holds jurisdiction over five counties, and the juries are required to be made up in a representative manner from the eligible populations of these five counties. An investigator notices that the county where she lives has 14% of the total population of the five counties eligible for jury duty, yet records reveal that over the past five years only 122 out of the 1386 jurors used by the court reside in her county. Do you feel that this constitutes reasonable evidence that the jurors are not being randomly selected from the total population?

10.1.8 In trials of a medical screening test for a particular illness, 23 cases out of 324 positive results turned out to be "false-positive" results. The screening test is acceptable as long as p, the probability of a positive result being incorrect, is no larger than 10%. Calculate a p-value for the hypotheses

$$H_0 : p \geq 0.1 \quad \text{versus} \quad H_A : p < 0.1$$

Construct a 99% upper confidence bound on p. Do you think that the screening test is acceptable?

10.1.9 Suppose that you wish to find a population proportion p with accuracy $\pm 1\%$ with 95% confidence. What sample size n would you recommend if p could be 0.5? What if the population proportion p can be assumed to be larger than 0.75?

10.1.10 Suppose that you wish to find a population proportion p with accuracy $\pm 2\%$ with 99% confidence. What sample size n would you recommend if p could be 0.5? What if the population proportion p can be assumed to be no larger than 0.40?

10.1.11 In experimental bioengineering trials, a successful outcome was achieved 73 times out of 120 attempts. Construct a 99% two-sided confidence interval for the probability of a success under these conditions. If this probability is required to be known to a precision of ±5%, how many additional trials would you recommend be run? (This problem is continued in Problem 10.2.8.)

10.1.12 A manufacturer receives a shipment of 100,000 computer chips. A random sample of 200 chips is examined, and 8 of these are found to be defective. Construct a 95% confidence level upper bound on the total number of defective chips in the shipment. (This problem is continued in Problem 10.2.9.)

10.1.13 A glass tube is designed to withstand a pressure differential of 1.1 atmospheres. In testing, it was found that 12 out of 20 tubes could in fact withstand a pressure differential of 1.5 atmospheres. Calculate a two-sided 95% confidence interval for the probability that a tube can withstand a pressure differential of 1.5 atmospheres.

10.1.14 An audit of a federal assistance program implemented after a major regional disaster discovered that out of 85 randomly selected applications, 17 contained errors due to either applicant fraud or processing mistakes. If there were 7607 applications made to the federal assistance program, calculate a 95% lower bound on the total number of applications that contained errors. (This problem is continued in Problem 10.2.10.)

10.1.15 A city councilor asks your advice on how many householders should be polled in order to gauge the support for a tax increase to build more schools. The councilor wants to assess the support to within ±5% with 95% confidence. What sample size would you recommend if the councilor advises you that the householders may be evenly split on the issue? What if the councilor advises you that fewer than one in three householders are likely to support the tax increase?

10.1.16 In a particular day, 22 out of 542 visitors to a website followed a link provided by one of the advertisers. Calculate a 99% two-sided confidence interval for the probability that a user of the website will follow a link provided by an advertiser. (This problem is continued in Problem 10.2.12.)

10.1.17 Sometimes the following alternative way of constructing a two-sided confidence interval on a population proportion p is employed. Recall that there is a probability of $1 - \alpha$ that

$$\frac{|\hat{p} - p|}{\sqrt{\frac{p(1-p)}{n}}} \leq z_{\alpha/2}$$

By squaring both sides of this inequality and solving the resulting quadratic expression for p, show that this can be rewritten

$$l \leq p \leq u$$

where

$$l = \frac{2x + z_{\alpha/2}^2 - z_{\alpha/2}\sqrt{4x(1 - x/n) + z_{\alpha/2}^2}}{2(n + z_{\alpha/2}^2)}$$

and

$$u = \frac{2x + z_{\alpha/2}^2 + z_{\alpha/2}\sqrt{4x(1 - x/n) + z_{\alpha/2}^2}}{2(n + z_{\alpha/2}^2)}$$

This result implies that

$$p \in (l, u)$$

is a two-sided $1 - \alpha$ confidence level confidence interval for p. Notice that this confidence interval is *not* centered at $\hat{p}$.

  If $x = 14$ is an observation from a $B(39, p)$ distribution, compare the 99% two-sided confidence interval obtained from this method with the standard 99% two-sided confidence interval for p.

10.1.18 The dielectric breakdown strength of an electrical insulator is defined to be the voltage at which the insulator starts to leak detectable amounts of electrical current, and it is an important safety consideration. In an experiment, 62 insulators of a certain type were tested at 180°C, and it was found that 13 had a dielectric breakdown strength below a specified threshold level.

(a) Conduct a hypothesis test to investigate whether this experiment provides sufficient evidence to conclude that the probability of an insulator of this type having a dielectric breakdown strength below the specified threshold level is larger than 5%.

(b) Construct a one-sided 95% confidence interval that provides a lower bound on the probability of an insulator of this type having a dielectric breakdown strength below the specified threshold level.

(This problem is continued in Problem 10.2.13.)

10.1.19 Out of a random sample of 210 parts produced on a production line, 31 fail a quality inspection. Obtain a 99% two-sided confidence interval for the proportion of parts from the production line that will fail the quality inspection.

10.1.20 A random sample of 38 wheelchair users were asked whether they preferred cushion type A or B, and 28 of them preferred type A whereas only 10 of them preferred type B. Use a hypothesis test to assess whether it is fair to conclude that cushion type A is at least twice as popular as cushion type B.

10.1.21 A newspaper reported the results of a political poll about which candidate likely voters preferred, together with a note that the margin of error was plus or minus 3.5 percentage points and that the numbers were based on answers from 793 likely voters. How was this margin of error calculated? Do you agree with it?

10.2 Comparing Two Population Proportions

The problem of comparing two population proportions is now considered. Suppose that observations from population A have a success probability p_A and that observations from population B have a success probability p_B. If the random variable X measures the number of successes observed in a sample of size n from population A, then

$$X \sim B(n, p_A)$$

and similarly, if the random variable Y measures the number of successes observed in a sample of size m from population B, then

$$Y \sim B(m, p_B)$$

The experimenter's goal is to make inferences on the difference between the two population proportions

$$p_A - p_B$$

based on observed values x and y of the random variables X and Y.

A good way to do this is to calculate a two-sided confidence interval for $p_A - p_B$. Notice that, as Figure 10.13 illustrates, if the confidence interval contains 0, then it is plausible that $p_A = p_B$ and so there is no evidence that the two population proportions are different. However, if the confidence interval contains only *positive* values, then this implies that all the plausible values of the success probabilities satisfy $p_A > p_B$, and so it can be concluded that there is evidence that population A has a *larger* proportion or success probability than population B. Similarly, if the confidence interval contains only *negative* values, then this implies that all the plausible values of the success probabilities satisfy $p_A < p_B$, and so it

FIGURE 10.13

Interpretation of confidence intervals for $p_A - p_B$

Two-sided confidence interval for $p_A - p_B$

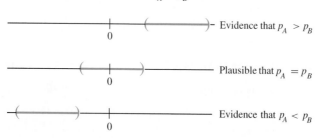

can be concluded that there is evidence that population A has a *smaller* proportion or success probability than population B.

If the experimenter wants to concentrate on assessing the evidence that the two population proportions are different, then it is useful to consider the hypotheses

$$H_0 : p_A = p_B \quad \text{versus} \quad H_A : p_A \neq p_B$$

One-sided hypothesis tests can also be considered, and one-sided confidence intervals for the difference $p_A - p_B$ can also be constructed.

The unbiased point estimates of the two population proportions are

$$\hat{p}_A = \frac{x}{n} \quad \text{and} \quad \hat{p}_B = \frac{y}{m}$$

and so an unbiased point estimate of the difference $p_A - p_B$ can be taken to be the difference in these two point estimates

$$\hat{p}_A - \hat{p}_B = \frac{x}{n} - \frac{y}{m}$$

Furthermore, since

$$\text{Var}(\hat{p}_A) = \frac{p_A(1 - p_A)}{n} \quad \text{and} \quad \text{Var}(\hat{p}_B) = \frac{p_B(1 - p_B)}{m}$$

and since the random variables X and Y are independent, it follows that

$$\text{Var}(\hat{p}_A - \hat{p}_B) = \text{Var}(\hat{p}_A) + \text{Var}(\hat{p}_B) = \frac{p_A(1 - p_A)}{n} + \frac{p_B(1 - p_B)}{m}$$

Both large-sample confidence interval construction and hypothesis testing are based on the formation of appropriate z-statistics with standard normal distributions, but as Figure 10.14 illustrates, they differ in the manner in which the variance term is estimated. In confidence interval construction the unknown population proportions p_A and p_B in the variance term are replaced by their point estimates, and the z-statistic

$$z = \frac{(\hat{p}_A - \hat{p}_B) - (p_A - p_B)}{\sqrt{\frac{\hat{p}_A(1-\hat{p}_A)}{n} + \frac{\hat{p}_B(1-\hat{p}_B)}{m}}} = \frac{(\hat{p}_A - \hat{p}_B) - (p_A - p_B)}{\sqrt{\frac{x(n-x)}{n^3} + \frac{y(m-y)}{m^3}}}$$

is employed.

For hypothesis tests with a null hypothesis $H_0 : p_A = p_B$, a *pooled* estimate of the common success probability

$$\hat{p} = \frac{x + y}{n + m}$$

is used, which is employed in place of both of the unknown population proportions p_A and p_B in the variance term. This results in a z-statistic

$$z = \frac{\hat{p}_A - \hat{p}_B}{\sqrt{\hat{p}(1 - \hat{p})\left(\frac{1}{n} + \frac{1}{m}\right)}}$$

Confidence Interval Construction

$$\hat{p}_A = \frac{x}{n} \qquad \hat{p}_B = \frac{y}{m}$$

Use

$$z = \frac{(\hat{p}_A - \hat{p}_B) - (p_A - p_B)}{\sqrt{\frac{\hat{p}_A(1-\hat{p}_A)}{n} + \frac{\hat{p}_B(1-\hat{p}_B)}{m}}}$$

Hypothesis Testing

$$H_0 : p_A = p_B$$

Pooled estimate $\hat{p} = \frac{x+y}{n+m}$

Use

$$z = \frac{\hat{p}_A - \hat{p}_B}{\sqrt{\hat{p}(1 - \hat{p})(\frac{1}{n} + \frac{1}{m})}}$$

FIGURE 10.14

Comparing two proportions p_A and p_B

10.2.1 Confidence Intervals for the Difference Between Two Population Proportions

For large enough sample sizes n and m, the z-statistic

$$z = \frac{(\hat{p}_A - \hat{p}_B) - (p_A - p_B)}{\sqrt{\frac{\hat{p}_A(1-\hat{p}_A)}{n} + \frac{\hat{p}_B(1-\hat{p}_B)}{m}}} = \frac{(\hat{p}_A - \hat{p}_B) - (p_A - p_B)}{\sqrt{\frac{x(n-x)}{n^3} + \frac{y(m-y)}{m^3}}}$$

can be taken to be an observation from a standard normal distribution. Roughly speaking, sample sizes for which $np_A, n(1 - p_A), mp_B$, and $m(1 - p_B)$ are all larger than 5 are adequate for the normal approximation to be appropriate, and these conditions can be taken to be satisfied as long as $x, n - x, y$, and $m - y$ are all larger than 5. Two-sided and one-sided confidence intervals for the difference $p_A - p_B$, given in the accompanying box, are obtained in the usual way by bounding the z-statistic with critical points from the standard normal distribution.

Confidence Intervals for the Difference of Two Population Proportions

If a random variable X has a $B(n, p_A)$ distribution and a random variable Y has a $B(m, p_B)$ distribution, then an approximate two-sided $1 - \alpha$ confidence level confidence interval for the difference between the success probabilities $p_A - p_B$ based upon observed values x and y of the random variables is

$$p_A - p_B \in \left(\hat{p}_A - \hat{p}_B - z_{\alpha/2}\sqrt{\frac{\hat{p}_A(1 - \hat{p}_A)}{n} + \frac{\hat{p}_B(1 - \hat{p}_B)}{m}}, \right.$$
$$\left. \hat{p}_A - \hat{p}_B + z_{\alpha/2}\sqrt{\frac{\hat{p}_A(1 - \hat{p}_A)}{n} + \frac{\hat{p}_B(1 - \hat{p}_B)}{m}} \right)$$

where the estimated success probabilities are $\hat{p}_A = x/n$ and $\hat{p}_B = y/m$. This confidence interval can also be written

$$p_A - p_B \in \left(\hat{p}_A - \hat{p}_B - z_{\alpha/2}\sqrt{\frac{x(n - x)}{n^3} + \frac{y(m - y)}{m^3}}, \right.$$
$$\left. \hat{p}_A - \hat{p}_B + z_{\alpha/2}\sqrt{\frac{x(n - x)}{n^3} + \frac{y(m - y)}{m^3}} \right)$$

One-sided approximate $1 - \alpha$ confidence level confidence intervals are

$$p_A - p_B \in \left(\hat{p}_A - \hat{p}_B - z_{\alpha}\sqrt{\frac{\hat{p}_A(1 - \hat{p}_A)}{n} + \frac{\hat{p}_B(1 - \hat{p}_B)}{m}}, 1 \right)$$
$$= \left(\hat{p}_A - \hat{p}_B - z_{\alpha}\sqrt{\frac{x(n - x)}{n^3} + \frac{y(m - y)}{m^3}}, 1 \right)$$

and

$$p_A - p_B \in \left(-1, \hat{p}_A - \hat{p}_B + z_{\alpha}\sqrt{\frac{\hat{p}_A(1 - \hat{p}_A)}{n} + \frac{\hat{p}_B(1 - \hat{p}_B)}{m}} \right)$$
$$= \left(-1, \hat{p}_A - \hat{p}_B + z_{\alpha}\sqrt{\frac{x(n - x)}{n^3} + \frac{y(m - y)}{m^3}} \right)$$

The approximations are reasonable as long as $x, n - x, y$, and $m - y$ are all larger than 5.

Example 57
Building Tile Cracks

Recall that a combination of two surveys of the tiles on a group of buildings, buildings A say, revealed a total of $x = 406$ cracked tiles out of $n = 6000$. Suppose that another group of buildings in another part of town, buildings B, were constructed about the same time as

FIGURE 10.15

Analysis of building tile cracks

Buildings A

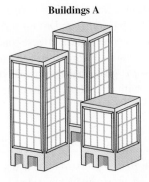

Buildings B

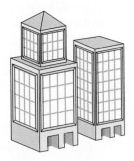

$x = 406$ cracked tiles out of $n = 6000$

$$\hat{p}_A = \frac{406}{6000} = 0.0677$$

$y = 83$ cracked tiles out of $m = 2000$

$$\hat{p}_B = \frac{83}{2000} = 0.0415$$

99% two-sided confidence interval

$$p_A - p_B \in (0.0120, 0.0404)$$

Conclusion: evidence that $p_A > p_B$

buildings A and have exterior walls composed of the same type of tiles. However, the tiles on buildings B were cemented into place with a *different* resin mixture than that used on buildings A. The construction engineers are interested in investigating whether the two types of resin mixtures have different expansion and contraction properties which affect the chances of the tiles becoming cracked.

As Figure 10.15 shows, a representative sample of $m = 2000$ tiles on buildings B is examined, and $y = 83$ are found to be cracked. If p_A represents the probability that a tile on buildings A becomes cracked, and if p_B represents the probability that a tile on buildings B becomes cracked, then

$$\hat{p}_A = \frac{x}{n} = \frac{406}{6000} = 0.0677 \qquad \text{and} \qquad \hat{p}_B = \frac{y}{m} = \frac{83}{2000} = 0.0415$$

With $z_{0.005} = 2.576$, a two-sided 99% confidence interval for the difference in these probabilities is

$$p_A - p_B \in \left(\hat{p}_A - \hat{p}_B - z_{\alpha/2} \sqrt{\frac{x(n-x)}{n^3} + \frac{y(m-y)}{m^3}}, \right.$$

$$\left. \hat{p}_A - \hat{p}_B + z_{\alpha/2} \sqrt{\frac{x(n-x)}{n^3} + \frac{y(m-y)}{m^3}} \right)$$

$$= \left(0.0677 - 0.0415 - 2.576 \sqrt{\frac{406 \times (6000 - 406)}{6000^3} + \frac{83 \times (2000 - 83)}{2000^3}}, \right.$$

$$\left. 0.0677 - 0.0415 + 2.576 \sqrt{\frac{406 \times (6000 - 406)}{6000^3} + \frac{83 \times (2000 - 83)}{2000^3}} \right)$$

$$= (0.0262 - 0.0142, 0.0262 + 0.0142) = (0.0120, 0.0404)$$

The fact that this confidence interval contains only positive values indicates that $p_A > p_B$, so that the resin mixture employed on buildings B is *better* than the resin mixture employed on buildings A. More specifically, the confidence interval indicates that the resin mixture employed on buildings A has a probability of causing a tile to crack somewhere between about 1.2% and 4.0% *larger* than the resin mixture employed on buildings B.

Example 58
Overage Weedkiller Product

The chemical company, company A, not only is interested in the proportion of its own weed-killer product that is overage, but in addition is interested in the overage proportion of its main competitor's weedkiller brand, produced by company B. Therefore, the auditors in the nationwide sampling scheme are also instructed to investigate the shelf product of company B to determine whether or not it is overage.

Recall that the auditors examined $n = 54{,}965$ weedkiller containers of company A's product and found that $x = 2779$ of them were overage. This finding results in an estimate of company A's overage proportion p_A of

$$\hat{p}_A = \frac{x}{n} = \frac{2779}{54{,}965} = 0.0506$$

In addition, the auditors examined $m = 47{,}892$ weedkiller containers of company B's product and found that $x = 3298$ of them were overage, which provides an estimate of company B's overage proportion p_B of

$$\hat{p}_B = \frac{y}{m} = \frac{3298}{47{,}892} = 0.0689$$

A two-sided 99% confidence interval for the difference in these probabilities is

$$p_A - p_B \in \left(0.0506 - 0.0689 - \right.$$

$$2.576 \times \sqrt{\frac{2779 \times (54{,}965 - 2779)}{54{,}965^3} + \frac{3298 \times (47{,}892 - 3298)}{47{,}892^3}},$$

$$0.0506 - 0.0689 +$$

$$\left. 2.576 \times \sqrt{\frac{2779 \times (54{,}965 - 2779)}{54{,}965^3} + \frac{3298 \times (47{,}892 - 3298)}{47{,}892^3}} \right)$$

$$= (-0.0183 - 0.0038, -0.0183 + 0.0038) = (-0.0221, -0.0145)$$

This confidence interval contains only negative values, which indicates that $p_A < p_B$. Thus, the sampling has provided evidence that company B has proportionally *more* overage product on sale than company A. However, the difference in the proportions is quite small, lying somewhere between about 1.4% and 2.2%.

Example 59
Opossum Progeny Genders

In the study of evolutionary behavior, the Trivers-Willard hypothesis indicates that healthy parents should tend to have more male offspring than female, and that weaker parents should tend to have more female offspring than male. This tendency may maximize the number of each parent's grandchildren (and thus help ensure that its genetic code is preserved) since a healthy male offspring can win many mates, but a relatively unhealthy offspring has the best chance of mating if it is a female.

In an experiment to examine this hypothesis, a group of 40 opossums were monitored and 20 of them were given an enhanced diet. Suppose that after a certain period of time, the opossums with the enhanced diet had raised 19 male offspring and 14 female offspring, and the opossums without the enhanced diet had raised 15 male offspring and 15 female offspring. Does this finding provide any evidence in support of the Trivers-Willard hypothesis?

Let p_A be the probability that opossums on the enhanced diet have a male offspring, and let p_B be the probability that opossums without the enhanced diet have a male offspring. Then

$$\hat{p}_A = \frac{19}{19 + 14} = 0.576 \quad \text{and} \quad \hat{p}_B = \frac{15}{15 + 15} = 0.500$$

The Trivers-Willard hypothesis suggests that the difference $p_A - p_B$ should be positive, and this can be examined by obtaining a one-sided confidence interval providing a lower bound on $p_A - p_B$. With $z_{0.05} = 1.645$, a 95% confidence interval of this kind is

$$p_A - p_B \in \left(\hat{p}_A - \hat{p}_B - z_\alpha \sqrt{\frac{x(n - x)}{n^3} + \frac{y(m - y)}{m^3}}, \ 1 \right)$$

$$= \left(0.576 - 0.500 - 1.645 \sqrt{\frac{19 \times (33 - 19)}{33^3} + \frac{15 \times (30 - 15)}{30^3}}, \ 1 \right)$$

$$= (0.076 - 0.206, \ 1) = (-0.130, \ 1)$$

This confidence interval contains some negative values and so clearly it is plausible that $p_A \leq p_B$. Consequently, this experiment does not provide any significant evidence to substantiate the Trivers-Willard hypothesis. Of course, this experiment does not *disprove* the Trivers-Willard hypothesis, and the collection of more data may demonstrate it to be valid in this situation.

10.2.2 Hypothesis Tests on the Difference Between Two Population Proportions

If an experimenter wants to concentrate on assessing the evidence that two population proportions p_A and p_B are different, then it is useful to consider the two-sided hypotheses

$$H_0 : p_A = p_B \quad \text{versus} \quad H_A : p_A \neq p_B$$

or the associated one-sided hypotheses. Since the null hypothesis specifies that the two proportions are identical, it is appropriate to employ a *pooled* estimate of the common success probability

$$\hat{p} = \frac{x + y}{n + m}$$

in the estimation of the variance of $\hat{p}_A - \hat{p}_B$. This estimate results in a z-statistic

$$z = \frac{\hat{p}_A - \hat{p}_B}{\sqrt{\hat{p}(1 - \hat{p}) \left(\frac{1}{n} + \frac{1}{m} \right)}}$$

which with sufficiently large sample sizes, can be taken to be an observation from a standard normal distribution when the null hypothesis is true. The calculation of p-values is performed in the usual manner by comparing the z-statistic with a standard normal distribution, as outlined in the accompanying box. The determination of whether a fixed size hypothesis test accepts or rejects the null hypothesis is similarly made in the usual manner.

Hypothesis Tests of the Equality of Two Population Proportions

---

Suppose that x is an observation from a $B(n, p_A)$ distribution and that y is an observation from a $B(m, p_B)$ distribution. Then the two-sided hypothesis testing problem

$$H_0 : p_A = p_B \quad \text{versus} \quad H_A : p_A \neq p_B$$

has a p-value

$$p\text{-value} = 2 \times \Phi(-|z|)$$

where

$$z = \frac{\hat{p}_A - \hat{p}_B}{\sqrt{\hat{p}(1 - \hat{p})\left(\frac{1}{n} + \frac{1}{m}\right)}} \quad \text{and} \quad \hat{p} = \frac{x + y}{n + m}$$

A size α hypothesis test *accepts* the null hypothesis when

$$|z| \leq z_{\alpha/2}$$

and *rejects* the null hypothesis when

$$|z| > z_{\alpha/2}$$

The one-sided hypothesis testing problem

$$H_0 : p_A - p_B \geq 0 \quad \text{versus} \quad H_A : p_A - p_B < 0$$

has a p-value

$$p\text{-value} = \Phi(z)$$

and a size α hypothesis test *accepts* the null hypothesis when

$$z \geq -z_\alpha$$

and *rejects* the null hypothesis when

$$z < -z_\alpha$$

The one-sided hypothesis testing problem

$$H_0 : p_A - p_B \leq 0 \quad \text{versus} \quad H_A : p_A - p_B > 0$$

has a p-value

$$p\text{-value} = 1 - \Phi(z)$$

and a size α hypothesis test *accepts* the null hypothesis when

$$z \leq z_\alpha$$

and *rejects* the null hypothesis when

$$z > z_\alpha$$

Example 61

Political Polling

When polling the agreement with the statement

The city mayor is doing a good job.

the local newspaper is also interested in how a person's support for this statement may depend upon his or her age. Therefore the pollsters also gather information on the ages of the respondents in their random sample.

As Figure 10.16 shows, the polling results consist of $n = 952$ people aged 18 to 39 of whom $x = 627$ agree with the statement, and $m = 1043$ people aged at least 40 of whom $y = 421$ agree with the statement. The estimate of p_A, the proportion of the younger group who agree with the statement, is therefore

$$\hat{p}_A = \frac{627}{952} = 0.659$$

and the estimate of p_B, the proportion of the older group who agree with the statement, is therefore

$$\hat{p}_B = \frac{421}{1043} = 0.404$$

Does the strength of support for the statement differ between the two age groups? This question can be examined with the two-sided hypotheses

$$H_0 : p_A = p_B \quad \text{versus} \quad H_A : p_A \neq p_B$$

Acceptance of the null hypothesis implies that there is no evidence of a difference in the proportions of the two age groups who agree with the statement, whereas rejection of the null hypothesis indicates that there is evidence of a difference between the two age groups.

FIGURE 10.16

Political polling example

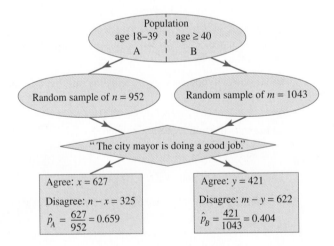

$$H_0 : p_A = p_B, \, H_A : p_A \neq p_B$$

$$\text{Pooled estimate } \hat{p} = \frac{627 + 421}{952 + 1043} = 0.525$$

$$z = 11.39, \, p\text{-value} = 2 \times \Phi(-11.39) \simeq 0$$

Conclusion: The null hypothesis is not plausible. There is evidence that the proportion agreeing with the statement differs between the two age groups.

The pooled estimate of a common proportion is

$$\hat{p} = \frac{x + y}{n + m} = \frac{627 + 421}{952 + 1043} = 0.525$$

and the z-statistic is

$$z = \frac{\hat{p}_A - \hat{p}_B}{\sqrt{\hat{p}(1 - \hat{p})\left(\frac{1}{n} + \frac{1}{m}\right)}}$$

$$= \frac{0.659 - 0.404}{\sqrt{0.525 \times (1.000 - 0.525) \times \left(\frac{1}{952} + \frac{1}{1043}\right)}} = 11.39$$

The p-value is therefore

$$p\text{-value} = 2 \times \Phi(-11.39) \simeq 0$$

Consequently, the null hypothesis has been shown to be not at all plausible, and the poll has demonstrated a difference in agreement with the statement between the two age groups.

In fact, a two-sided 99% confidence interval for the difference between the proportions who agree with the statement is

$$p_A - p_B \in \left(0.659 - 0.404 - 2.576\sqrt{\frac{627 \times (952 - 627)}{952^3} + \frac{421 \times (1043 - 421)}{1043^3}}, \right.$$

$$\left. 0.659 - 0.404 + 2.576\sqrt{\frac{627 \times (952 - 627)}{952^3} + \frac{421 \times (1043 - 421)}{1043^3}} \right)$$

$$= (0.255 - 0.056, 0.255 + 0.056) = (0.199, 0.311)$$

Therefore the poll shows that the proportion of the younger group who agree with the statement is somewhere between about 20% to 30% larger than the proportion of the older group who agree.

Example 59

Opossum Progeny Genders

Recall that $x = 19$ of the $n = 33$ offspring raised by opossums *with* the enhanced diet are male, and that $y = 15$ of the $m = 30$ offspring raised by opossums *without* the enhanced diet are male. The Trivers-Willard hypothesis suggests that $p_A > p_B$, and it can be tested with the one-sided hypotheses

$$H_0 : p_A - p_B \leq 0 \quad \text{versus} \quad H_A : p_A - p_B > 0$$

Acceptance of the null hypothesis indicates that there is not sufficient evidence to establish that the Trivers-Willard hypothesis is true, whereas rejection of the null hypothesis indicates that there is sufficient evidence to substantiate the Trivers-Willard hypothesis.

The pooled estimate of the probability of a male offspring is

$$\hat{p} = \frac{x + y}{n + m} = \frac{19 + 15}{33 + 30} = 0.540$$

and the z-statistic is

$$z = \frac{0.576 - 0.500}{\sqrt{0.540 \times (1.000 - 0.540) \times \left(\frac{1}{33} + \frac{1}{30}\right)}} = 0.60$$

The p-value is therefore

$$p\text{-value} = 1 - \Phi(0.60) = 1 - 0.7257 \simeq 0.27$$

Such a large p-value implies that there is no reason to conclude that the null hypothesis is not plausible, and so, as was found from the previous construction of a one-sided confidence interval for $p_A - p_B$, this experiment does not provide any substantiation of the Trivers-Willard hypothesis.

Figure 10.17 provides a summary chart of confidence interval construction and hypothesis testing procedures for comparing two population proportions.

Population A	**Population B**
x successes out of n trials	y successes out of m trials
p_A = probability of success	p_B = probability of success
$\hat{p}_A = \dfrac{x}{n}$	$\hat{p}_B = \dfrac{y}{m}$

Two-sided $1-\alpha$ level confidence interval

$$p_A - p_B \in \left(\hat{p}_A - \hat{p}_B - z_{\alpha/2} \sqrt{\frac{\hat{p}_A(1-\hat{p}_A)}{n} + \frac{\hat{p}_B(1-\hat{p}_B)}{m}}, \hat{p}_A - \hat{p}_B + z_{\alpha/2} \sqrt{\frac{\hat{p}_A(1-\hat{p}_A)}{n} + \frac{\hat{p}_B(1-\hat{p}_B)}{m}} \right)$$

$$= \left(\hat{p}_A - \hat{p}_B - z_{\alpha/2} \sqrt{\frac{x(n-x)}{n^3} + \frac{y(m-y)}{m^3}}, \hat{p}_A - \hat{p}_B + z_{\alpha/2} \sqrt{\frac{x(n-x)}{n^3} + \frac{y(m-y)}{m^3}} \right)$$

One-sided $1-\alpha$ level confidence interval

$$p_A - p_B \in \left(\hat{p}_A - \hat{p}_B - z_{\alpha} \sqrt{\frac{\hat{p}_A(1-\hat{p}_A)}{n} + \frac{\hat{p}_B(1-\hat{p}_B)}{m}}, 1 \right) = \left(\hat{p}_A - \hat{p}_B - z_{\alpha} \sqrt{\frac{x(n-x)}{n^3} + \frac{y(m-y)}{m^3}}, 1 \right)$$

$$p_A - p_B \in \left(-1, \hat{p}_A - \hat{p}_B + z_{\alpha} \sqrt{\frac{\hat{p}_A(1-\hat{p}_A)}{n} + \frac{\hat{p}_B(1-\hat{p}_B)}{m}} \right) = \left(-1, \hat{p}_A - \hat{p}_B + z_{\alpha} \sqrt{\frac{x(n-x)}{n^3} + \frac{y(m-y)}{m^3}} \right)$$

Hypothesis Testing

$$\hat{p} = \frac{x+y}{n+m} \qquad z = \frac{\hat{p}_A - \hat{p}_B}{\sqrt{\hat{p}(1-\hat{p})\left(\frac{1}{n}+\frac{1}{m}\right)}}$$

$H_0 : p_A - p_B \geq 0, H_A : p_A - p_B < 0$	$H_0 : p_A = p_B, H_A : p_A \neq p_B$	$H_0 : p_A - p_B \leq 0, H_A : p_A - p_B > 0$
$p\text{-value} = \Phi(z)$	$p\text{-value} = 2 \times \Phi(-\|z\|)$	$p\text{-value} = 1 - \Phi(z)$
Size α test	Size α test	Size α test
accept $H_0 : z \geq -z_\alpha$	accept $H_0 : \|z\| \leq z_{\alpha/2}$	accept $H_0 : z \leq z_\alpha$
reject $H_0 : z < -z_\alpha$	reject $H_0 : \|z\| > z_{\alpha/2}$	reject $H_0 : z > z_\alpha$

FIGURE 10.17

Summary of inference procedures for comparing two population proportions (valid when x, $n - x$, m, and $m - y$ are all larger than 5)

■ **10.2.3 Problems**

10.2.1 Suppose that $x = 14$ is an observation from a $B(37, p_A)$ random variable, and that $y = 7$ is an observation from a $B(26, p_B)$ random variable.
  (a) Compute a two-sided 99% confidence interval for $p_A - p_B$.
  (b) Compute a two-sided 95% confidence interval for $p_A - p_B$.
  (c) Compute a one-sided 99% confidence interval that provides a lower bound for $p_A - p_B$.
  (d) Calculate the p-value for the test of the hypotheses
$$H_0 : p_A = p_B \quad \text{versus} \quad H_A : p_A \neq p_B$$

10.2.2 Suppose that $x = 261$ is an observation from a $B(302, p_A)$ random variable, and that $y = 401$ is an observation from a $B(454, p_B)$ random variable.
  (a) Compute a two-sided 99% confidence interval for $p_A - p_B$.
  (b) Compute a two-sided 90% confidence interval for $p_A - p_B$.
  (c) Compute a one-sided 95% confidence interval that provides an upper bound for $p_A - p_B$.
  (d) Calculate the p-value for the test of the hypotheses
$$H_0 : p_A = p_B \quad \text{versus} \quad H_A : p_A \neq p_B$$

10.2.3 Suppose that the abilities of two new radar systems to detect packages dropped by airplane are being compared. In a series of trials, radar system A detected the packages being dropped 35 times out of 44, while radar system B detected the packages being dropped 36 times out of 52.
  (a) Construct a 99% two-sided confidence interval for the differences between the probabilities that the radar systems successfully detect dropped packages.
  (b) Calculate the p-value for the test of the two-sided null hypothesis that the two radar systems are equally effective.
  Interpret your answers.

10.2.4 Die A is rolled 50 times and a 6 is scored 4 times, while a 6 is obtained 10 times when die B is rolled 50 times.
  (a) Construct a two-sided 99% confidence interval for the difference in the probabilities of scoring a 6 on the two dice.
  (b) Calculate a p-value for the two-sided null hypothesis that the two dice have equal probabilities of scoring a 6.

  (c) What would your answers be if die A produced a 6 40 times in 500 rolls and die B produced a 6 100 times in 500 rolls?

10.2.5 In an experiment to determine the best conditions to produce suitable crystals for the recovery and purification of biological molecules such as enzymes and proteins, crystals had appeared within 24 hours in 27 out of 60 trials of a particular solution *without* seed crystals, and had appeared within 24 hours in 36 out of 60 trials of a particular solution *with* seed crystals. Construct a one-sided confidence interval and calculate a one-sided p-value to investigate the evidence that the presence of seed crystals increases the probability of crystallization within 24 hours using this method? (This problem is continued in Problem 10.6.1.)

10.2.6 A new drug is being compared with a standard drug for treating a particular illness. In the clinical trials, a group of 200 patients was randomly split into two groups, with one group being given the standard drug and one group the new drug. Altogether, 83 out of the 100 patients given the new drug improved their condition, while only 72 out of the 100 patients given the standard drug improved their condition. Construct a one-sided confidence interval and calculate a one-sided p-value to investigate the evidence that the new drug is better than the standard drug.

10.2.7 A company has two production lines for constructing television sets. Over a certain period of time, 23 out of 1128 television sets from production line A are found not to meet the company's quality standards, while 24 out of 962 television sets from production line B are found not to meet the company's quality standards. Use a two-sided confidence interval and a two-sided p-value to assess the evidence of a difference in operating standards between the two production lines.

10.2.8 In experimental bioengineering trials, a successful outcome was achieved 73 times out of 120 attempts using a standard procedure, whereas with a new procedure a successful outcome was achieved 101 times out of 120. What is the evidence that the new procedure is better than the standard procedure?

10.2.9 A manufacturing company has to choose between two potential suppliers of computer chips. A random sample of 200 chips from supplier A is examined and 8 are found to be defective, while 13 chips out of a random sample of 250 chips from supplier B are found to be defective. Use two-sided inference procedures to assess whether this finding should influence the company's choice of supplier.

10.2.10 An audit of a federal assistance program implemented after a major regional disaster discovered that out of 85 randomly selected applications processed during the first two weeks after the disaster, 17 contained errors due to either applicant fraud or processing mistakes. However, out of 132 randomly selected applications processed later than the first two weeks after the disaster, only 16 contained errors. Does this information substantiate the contention that errors in the assistance applications are more likely in the initial aftermath of the disaster?

10.2.11 Two scanning machines are compared. When 185 items were scanned with machine A, 159 of the the scans were free of errors. When the same 185 items were scanned with machine B, only 138 of the scans were free of errors. Is this sufficient evidence to conclude that in general the probability of an error-free scan is higher for machine A than for machine B?

10.2.12 Recall from Problem 10.1.16 that in a particular day, 22 out of 542 visitors to a website followed a link provided by an advertiser. After the advertisements were modified, it was found that 64 out of 601 visitors to the website on a day followed the link. Is there any evidence that the modifications to the advertisements attracted more customers?

10.2.13 Consider again Problem 10.1.18 where 62 insulators of a certain type were tested at 180°C, and it was found that

13 had a dielectric breakdown strength below a specified threshold level. In addition, 70 insulators of the same type were tested at 250°C, and it was found that 20 had a dielectric breakdown strength below the specified threshold level.

(a) Conduct a one-sided hypothesis test to investigate whether this experiment provides sufficient evidence to conclude that the probability of an insulator of this type having a dielectric breakdown strength below the specified threshold level is larger at 250°C than it is at 180°C.

(b) Construct a two-sided 99% confidence interval for the difference between the probabilities of an insulator of this type having a dielectric breakdown strength below the specified threshold level at 180°C and at 250°C.

10.2.14 A group of 250 patients was randomly split into two groups of 125 patients. The first group of 125 patients was given treatment A, and 72 of them improved their condition. The second group of 125 patients was given treatment B, and 60 of them improved their condition. Perform a hypothesis test to investigate whether there is evidence of a difference between the two treatments.

10.2.15 A company is performing a failure analysis for two of its products. It found that for the first product 76 out of 243 failures were due to operator misuse, while for the second product 122 out of 320 failures were due to operator misuse. Construct a 99% two-sided confidence interval for the difference between the two products of the probabilities that a failure is due to operator misuse. Based on this confidence interval, is there evidence that these probabilities are different for the two products?

10.3 Goodness of Fit Tests for One-Way Contingency Tables

In this section the analysis of classifications with more than two levels is considered. Thus, data sets are considered where each unit is assigned to one of three or more different categories. Whereas the binomial distribution was appropriate to analyze classifications with two levels, the *multinomial* distribution is appropriate when there are three or more classification levels.

Hypothesis tests are described for assessing whether the probability vector of the multinomial distribution takes a specified value. In particular, the hypothesis of **homogeneity** (which states that every classification is equally likely) is commonly of interest. Goodness of fit tests, often referred to as chi-square tests, are used to test the hypothesis. These methods can also be used to test the distributional assumptions of a data set.

10.3.1 One-Way Classifications

Consider the data set illustrated in Figure 10.18. Each of n observations is classified into one (and only one) of k categories or cells. The resulting **cell frequencies** are

$$x_1, x_2, \ldots, x_k$$

with

$$x_1 + x_2 + \cdots + x_k = n$$

For a fixed total sample size n, data sets of this kind can be modeled with a multinomial distribution that depends upon a set of **cell probabilities**

$$p_1, p_2, \ldots, p_k$$

with

$$p_1 + p_2 + \cdots + p_k = 1$$

The cell frequency x_i is an observation from a $B(n, p_i)$ distribution, so a particular cell probability p_i can be estimated by

$$\hat{p}_i = \frac{x_i}{n}$$

Furthermore, the methods described in Section 10.1 can be used to make inferences on a particular cell probability p_i either through confidence interval construction or with hypothesis testing.

FIGURE 10.18

One-way classification

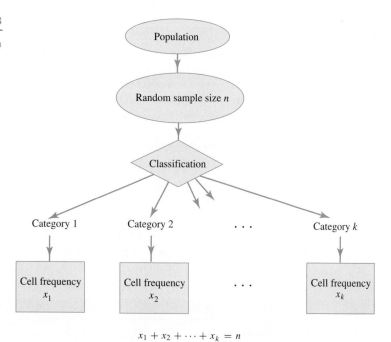

$$x_1 + x_2 + \cdots + x_k = n$$

However, it can be useful to assess the plausibility that the cell probabilities, taken all together rather than individually, take a set of specified values

$$p_1^*, p_2^*, \ldots, p_k^*$$

with

$$p_1^* + \cdots + p_k^* = 1$$

In other words, it can be useful to examine the null hypothesis

$$H_0 : p_i = p_i^* \qquad 1 \leq i \leq k$$

for specified values $p_1^*, \ldots, p_k^*$. This can be accomplished with a chi-square goodness of fit test. The null hypothesis of homogeneity is often of interest and states that the k cell probabilities are all equal, so that

$$p_i^* = \frac{1}{k} \qquad 1 \leq i \leq k$$

Notice that this hypothesis testing problem is intrinsically a *two-sided* problem. There are no one-sided versions of it. The implied alternative hypothesis, which is usually not stated, is that the null hypothesis is *false*. Thus, it consists of all the sets of probability values $p_1, \ldots, p_k$ except for the specific set of values $p_1^*, \ldots, p_k^*$.

Example 1
Machine Breakdowns

Recall that out of $n = 46$ machine breakdowns, $x_1 = 9$ are attributable to electrical problems, $x_2 = 24$ are attributable to mechanical problems, and $x_3 = 13$ are attributable to operator misuse. It is suggested that the probabilities of these three kinds of breakdown are respectively

$$p_1^* = 0.2, \quad p_2^* = 0.5, \quad p_3^* = 0.3$$

The plausibility of this suggestion can be examined with a chi-square goodness of fit test.

Example 13
Factory Floor Accidents

A factory is embarking on a new safety drive in an attempt to reduce the number of accidents occurring on the factory floor. A manager checks back through the records of factory floor accidents and finds the day of the week on which each of the last $n = 270$ accidents occurred, as shown in Figure 10.19. It appears that accidents are more likely on Mondays and Fridays than on other days. If this is really the case, then it is sensible to be particularly vigilant on these days or to make some changes to reduce the chances of accidents occurring on these days.

Does the data set really provide evidence that accidents are more likely on Mondays and Fridays than on other days? The null hypothesis of homogeneity

$$H_0 : p_i = \frac{1}{5} \qquad 1 \leq i \leq 5$$

states that accidents are equally likely to occur on any day of the week. If this hypothesis is plausible, then there is no evidence that accidents are more likely on any one day than on another. However, if the hypothesis is rejected, then it can be taken as evidence that accidents are more likely on Mondays and Fridays than on the other days.

FIGURE 10.19

Day of the week of factory floor accidents

Day of week	Monday	Tuesday	Wednesday	Thursday	Friday	
Number of accidents	65	43	48	41	73	$n = 270$

FIGURE 10.20

Observed and expected cell
frequencies

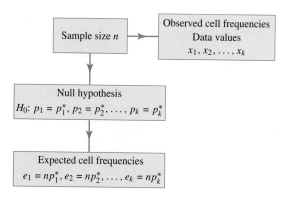

The null hypothesis

$$H_0 : p_i = p_i^* \qquad 1 \le i \le k$$

is tested by comparing a set of *observed cell frequencies*

$$x_1, x_2, \ldots, x_k$$

with a set of *expected cell frequencies*

$$e_1, e_2, \ldots, e_k$$

As Figure 10.20 illustrates, the observed cell frequencies x_i are simply the data values observed in each of the cells (more technically they are the realizations of the random variables $X_1, \ldots, X_k$, which have a multinomial distribution), and the expected cell frequencies e_i are given by

$$e_i = np_i^* \qquad 1 \le i \le k$$

Thus the expected cell frequencies e_i are the expected values of the multinomial random variables $X_1, \ldots, X_k$ when the null hypothesis is true. Notice that in contrast to the observed cell frequencies x_i, the expected cell frequencies e_i need not take integer values, but that

$$e_1 + e_2 + \cdots + e_k = n$$

A goodness of fit test operates by measuring the discrepancy between the observed cell frequencies x_i and the expected cell frequencies e_i. The closer these sets of frequencies are to each other, the *more* plausible is the null hypothesis. Conversely, the farther apart these sets of frequencies are, the *less* plausible is the null hypothesis.

Statistics that measure the discrepancy between the two sets of cell frequencies are usually called **chi-square statistics**, since a p-value is obtained by comparing them with a chi-square distribution. Two common ones are

$$X^2 = \sum_{i=1}^{k} \frac{(x_i - e_i)^2}{e_i} \qquad \text{and} \qquad G^2 = 2 \sum_{i=1}^{k} x_i \ln\left(\frac{x_i}{e_i}\right)$$

The former is known as the **Pearson chi-square statistic**, and the latter is known as the **likelihood ratio chi-square statistic**. These statistics both take positive values, and larger values of the statistics indicate a greater discrepancy between the two sets of cell frequencies. In the unlikely circumstance that the observed cell frequencies are all exactly equal to the

HISTORICAL NOTE

Karl Pearson (1857–1936) was one of the founders of modern statistics. He was born in London, England, and practiced law for three years and published two literary works before he was appointed professor of applied mathematics and mechanics at University College, London, in 1884, where he taught until his retirement in 1933. His work on the chi-square statistic began in 1893 through his attention to the problem of applying statistical techniques to the biological problems of heredity and evolution. He was a socialist and a self-described "free-thinker."

FIGURE 10.21

P-value calculation for chi-square goodness of fit tests

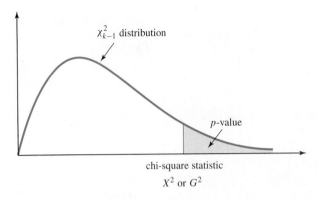

$$\text{\textit{p}-value} = P(\chi^2_{k-1} \geq X^2)$$
or
$$\text{\textit{p}-value} = P(\chi^2_{k-1} \geq G^2)$$

expected cell frequencies so that

$$x_i = e_i \qquad 1 \leq i \leq k$$

then

$$X^2 = G^2 = 0$$

indicating a "perfect fit." The two chi-square statistics arise from two different mathematical approaches to the testing problem, and there is generally no reason to prefer one statistic to the other (many statistical software packages provide both statistics).

A *p*-value is obtained by comparing the chi-square statistic, X^2 or G^2, with a chi-square distribution with $k - 1$ degrees of freedom. Specifically, the *p*-value is given by

$$\text{\textit{p}-value} = P\left(\chi^2_{k-1} \geq X^2\right) \qquad \text{or} \qquad \text{\textit{p}-value} = P\left(\chi^2_{k-1} \geq G^2\right)$$

as shown in Figure 10.21. Consequently, a size α hypothesis test accepts the null hypothesis if the chi-square statistic is no larger than a critical point $\chi^2_{\alpha,k-1}$ and rejects the null hypothesis if the chi-square statistic is larger than the critical point.

The values of the Pearson chi-square statistic X^2 and the likelihood ratio chi-square statistic G^2 are generally very close together, and so it usually makes little difference which one is employed. The *p*-value calculations are based upon an asymptotic (large expected cell frequencies) chi-square distribution for the test statistics, and this can be considered to be appropriate as long as each cell frequency e_i is no smaller than 5. If a cell has an expected frequency e_i less than 5, then standard practice is to *group* it with another cell, thereby reducing the number of cells k, so that the new grouped cell has an expected frequency of at least 5. Sometimes, three or more cells need to be grouped together for this purpose.

Notice that the *p*-value is obtained by comparing the chi-square statistic with a chi-square distribution with degrees of freedom $k - 1$, which is one less than the total number of cells. In situations where the hypothesized cell probabilities $p_1^*, \ldots, p_k^*$ depend upon the data values $x_1, \ldots, x_k$ in some manner, smaller degrees of freedom are appropriate, as illustrated in the following section on testing distributional assumptions.

Example 1

Machine Breakdowns

Consider the null hypothesis

$$H_0 : p_1 = 0.2, \ p_2 = 0.5, \ p_3 = 0.3$$

Under this null hypothesis the expected cell frequencies are

$$e_1 = np_1^* = 46 \times 0.2 = 9.2$$
$$e_2 = np_2^* = 46 \times 0.5 = 23.0$$
$$e_3 = np_3^* = 46 \times 0.3 = 13.8$$

As Figure 10.22 shows, the chi-square statistics are

$$X^2 = 0.0942 \qquad \text{and} \qquad G^2 = 0.0945$$

which, compared with a chi-square distribution with $k - 1 = 3 - 1 = 2$ degrees of freedom (which is an exponential distribution with mean 2), give a p-value of

$$p\text{-value} \simeq P\left(\chi_2^2 \geq 0.094\right) = 0.95$$

Goodness of Fit Tests for One-Way Contingency Tables

Consider a multinomial distribution with k cells and a set of unknown cell probabilities $p_1, \ldots, p_k$. Based upon a set of observed cell frequencies $x_1, \ldots, x_k$ with $x_1 + \cdots + x_k = n$, the null hypothesis

$$H_0 : p_i = p_i^* \qquad 1 \leq i \leq k$$

which states that the cell probabilities take the specific set of values $p_1^*, \ldots, p_k^*$, has a p-value that can be calculated as either

$$p\text{-value} = P\left(\chi_{k-1}^2 \geq X^2\right) \qquad \text{or} \qquad p\text{-value} = P\left(\chi_{k-1}^2 \geq G^2\right)$$

where the chi-square test statistics are

$$X^2 = \sum_{i=1}^{k} \frac{(x_i - e_i)^2}{e_i} \qquad \text{and} \qquad G^2 = 2 \sum_{i=1}^{k} x_i \ln\left(\frac{x_i}{e_i}\right)$$

with expected cell frequencies

$$e_i = np_i^* \qquad 1 \leq i \leq k$$

The two p-values calculated in this manner are usually similar, although they may differ slightly, and they are appropriate as long as the expected cell frequencies e_i are each no smaller than 5.

At size α, the null hypothesis is accepted if

$$X^2 \leq \chi_{\alpha,k-1}^2$$

(or if $G^2 \leq \chi_{\alpha,k-1}^2$), and the null hypothesis is rejected if

$$X^2 > \chi_{\alpha,k-1}^2$$

(or if $G^2 > \chi_{\alpha,k-1}^2$).

$$H_0 : p_1 = 0.2, p_2 = 0.5, p_3 = 0.3$$

	Electrical	Mechanical	Operator misuse	
Observed cell frequencies	$x_1 = 9$	$x_2 = 24$	$x_3 = 13$	$n = 46$
Expected cell frequencies	$e_1 = 46 \times 0.2 = 9.2$	$e_2 = 46 \times 0.5 = 23.0$	$e_3 = 46 \times 0.3 = 13.8$	$n = 46$

Pearson chi-square statistic: $X^2 = \dfrac{(9.0 - 9.2)^2}{9.2} + \dfrac{(24.0 - 23.0)^2}{23.0} + \dfrac{(13.0 - 13.8)^2}{13.8} = 0.0942$

Likelihood ratio chi-square statistic: $G^2 = 2 \times \left(9.0 \times \ln\left(\dfrac{9.0}{9.2}\right) + 24.0 \times \ln\left(\dfrac{24.0}{23.0}\right) + 13.0 \times \ln\left(\dfrac{13.0}{13.8}\right) \right) = 0.0945$

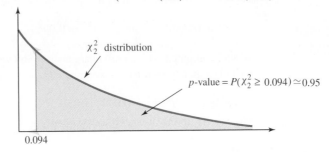

χ_2^2 distribution

p-value $= P(\chi_2^2 \geq 0.094) \simeq 0.95$

0.094

Conclusion: Null hypothesis is plausible

FIGURE 10.22

Goodness of fit test for the machine breakdown example

Clearly the null hypothesis is plausible, and such a large p-value indicates that there is a very close fit between the observed cell frequencies x_1, x_2, x_3 and the expected cell frequencies e_1, e_2, e_3, as might be observed from a visual comparison of their values.

Of course, the fact that the null hypothesis is plausible does not mean that it has been proven to be true. There are sets of plausible cell probabilities p_1, p_2, p_3 other than the hypothesized values 0.2, 0.5, and 0.3. In fact, with $z_{0.025} = 1.96$, the method described in Section 10.1 can be used to obtain a 95% confidence interval for p_2, the probability that a machine breakdown can be attributed to a mechanical failure, as

$$p_2 \in \left(\frac{24}{46} - \frac{1.96}{46} \sqrt{\frac{24 \times (46 - 24)}{46}}, \frac{24}{46} + \frac{1.96}{46} \sqrt{\frac{24 \times (46 - 24)}{46}} \right)$$
$$= (0.522 - 0.144, 0.522 + 0.144) = (0.378, 0.666)$$

Is the hypothesis of homogeneity plausible here? In other words, is it plausible that the three kinds of machine breakdown are equally likely? Under this hypothesis, the expected cell frequencies are

$$e_1 = e_2 = e_3 = \frac{46}{3} = 15.33$$

and the Pearson chi-square statistic is

$$X^2 = \frac{(9.00 - 15.33)^2}{15.33} + \frac{(24.00 - 15.33)^2}{15.33} + \frac{(13.00 - 15.33)^2}{15.33} = 7.87$$

This value is much larger than the previous value of 0.0942, which indicates, as expected, that the hypothesis of homogeneity does not provide as good a fit as the previous hypothesis. In fact, the p-value for the hypothesis of homogeneity is

$$p\text{-value} = P\left(X_2^2 \geq 7.87\right) \simeq 0.02$$

which casts serious doubts on the plausibility of the hypothesis. Notice also that the 95% confidence interval for p_2 does not contain the value $p_2 = 1/3$.

Example 13
Factory Floor Accidents

Under the null hypothesis

$$H_0 : p_i = \frac{1}{5} \qquad 1 \leq i \leq 5$$

the expected cell frequencies are

$$e_1 = e_2 = e_3 = e_4 = e_5 = \frac{270}{5} = 54$$

As Figure 10.23 shows, the chi-square statistics are

$$X^2 = 14.95 \qquad \text{and} \qquad G^2 = 14.64$$

which, compared with a chi-square distribution with $k - 1 = 5 - 1 = 4$ degrees of freedom, give p-values of

$$p\text{-value} \simeq P\left(X_4^2 \geq 14.95\right) = 0.0048 \qquad \text{and} \qquad p\text{-value} \simeq P\left(X_4^2 \geq 14.64\right) = 0.0055$$

$$H_0 : p_1 = p_2 = p_3 = p_4 = p_5 = \frac{1}{5}$$

	Monday	**Tuesday**	**Wednesday**	**Thursday**	**Friday**	
Observed cell frequencies	$x_1 = 65$	$x_2 = 43$	$x_3 = 48$	$x_4 = 41$	$x_5 = 73$	$n = 270$
Expected cell frequencies	$e_1 = 270 \times \frac{1}{5} = 54$	$e_2 = 270 \times \frac{1}{5} = 54$	$e_3 = 270 \times \frac{1}{5} = 54$	$e_4 = 270 \times \frac{1}{5} = 54$	$e_5 = 270 \times \frac{1}{5} = 54$	$n = 270$

Pearson chi-square statistic: $X^2 = \dfrac{(65-54)^2}{54} + \dfrac{(43-54)^2}{54} + \dfrac{(48-54)^2}{54} + \dfrac{(41-54)^2}{54} + \dfrac{(73-54)^2}{54} = 14.95$

Likelihood ratio chi-square statistic: $G^2 = 2 \times \left(65 \times \ln\left(\dfrac{65}{54}\right) + 43 \times \ln\left(\dfrac{43}{54}\right) + 48 \times \ln\left(\dfrac{48}{54}\right) + 41 \times \ln\left(\dfrac{41}{54}\right) + 73 \times \ln\left(\dfrac{73}{54}\right) \right) = 14.64$

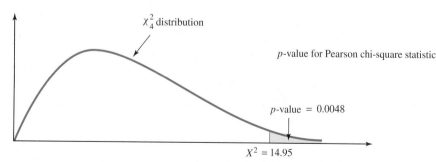

X_4^2 distribution

p-value for Pearson chi-square statistic

p-value = 0.0048

$X^2 = 14.95$

Conclusion: Null hypothesis is *not* plausible.

FIGURE 10.23

Goodness of fit test for the factory floor accidents example

Such small p-values lead to the conclusion that the hypothesis of homogeneity is not plausible. Thus, this data set provides sufficient evidence to conclude that factory floor accidents are *not* equally likely to occur on any day of the week, and one issue that ought to be considered by the safety drive is why Mondays and Fridays are particularly dangerous days.

10.3.2 Testing Distributional Assumptions

Goodness of fit tests for one-way layouts can be used to test the plausibility that a data set consists of independent observations from a particular distribution. The observed cell frequencies x_i are the number of data observations falling within the cells, and the expected cell frequencies e_i are the expected frequencies of the cells under the specific probability distribution of interest. For discrete distributions the cells can be taken to be the discrete levels of the distribution, although these may need to be grouped in some manner. For continuous distributions, cells are formed by assigning (usually arbitrarily) a set of intervals that cover the range of the random variable.

The following examples illustrate how to test whether software errors have a Poisson distribution, whether brake pad mileages before wearout have an exponential distribution, and whether concrete block weights can be taken to be normally distributed. In each case the null hypothesis is that the distribution is as specified, and so rejection of the null hypothesis indicates that the specified distribution is not plausible. Acceptance of the null hypothesis implies that the specified distribution is plausible, although this does not *prove* that the distribution is as specified. There will be other plausible distributions as well. It should be remembered that unless the sample size involved is very large, goodness of fit tests of this kind may not be very powerful, so that a wide range of different distributional assumptions may be plausible.

Finally, it is important to distinguish between two kinds of null hypotheses, which can be typified by the hypotheses

H_0 : the software errors have a Poisson distribution with mean $\lambda = 3$

and

H_0 : the software errors have a Poisson distribution

The first null hypothesis specifies precisely the distribution against which the data are to be tested. The second null hypothesis is more general, requiring only that the distribution be some Poisson distribution with any parameter value. In the second case, the data can be tested against a Poisson distribution with parameter $\lambda = \bar{x}$, the average of the data, because this is likely to give the best fit. The only difference in this latter case is that the degrees of freedom of the chi-square distribution used to calculate a p-value should be $k - 2$, where k is the number of cells, rather than the usual $k - 1$, which would be appropriate for testing the first null hypothesis.

A general rule is that the appropriate degrees of freedom are

number of cells $- 1 -$ number of parameters estimated from the data set

For example, if the null hypothesis is that the data are normally distributed with an unspecified mean and variance, then these two parameters can be estimated from the data set as $\bar{x}$ and s^2, and if there are k cells, then the appropriate degrees of freedom are $k - 3$.

It should also be pointed out that various other techniques are available to investigate the probability distribution governing a data set. The normal probability plots discussed in Section 12.7 are an example of more general **probability plots**, which provide a graphical

FIGURE 10.24

Distributional goodness of fit test
for the software errors example

Number of errors found in a software product	0	1	2	3	4	5	6	7	8	
Frequency	3	14	20	25	14	6	2	0	1	$n = 85$

H_0 : number of errors X has a Poisson distribution with mean $\lambda = 3.0$

Cell	Expected cell frequency	
$X = 0$	$e_1 = 85 \times P(X = 0) = 85 \times \dfrac{e^{-3} \times 3^0}{0!} = 4.23$	} Group
$X = 1$	$e_2 = 85 \times P(X = 1) = 85 \times \dfrac{e^{-3} \times 3^1}{1!} = 12.70$	
$X = 2$	$e_3 = 85 \times P(X = 2) = 85 \times \dfrac{e^{-3} \times 3^2}{2!} = 19.04$	
$X = 3$	$e_4 = 85 \times P(X = 3) = 85 \times \dfrac{e^{-3} \times 3^3}{3!} = 19.04$	
$X = 4$	$e_4 = 85 \times P(X = 4) = 85 \times \dfrac{e^{-3} \times 3^4}{4!} = 14.28$	
$X = 5$	$e_5 = 85 \times P(X = 5) = 85 \times \dfrac{e^{-3} \times 3^5}{5!} = 8.57$	
$X = 6$	$e_6 = 85 \times P(X = 6) = 85 \times \dfrac{e^{-3} \times 3^6}{6!} = 4.28$	} Group
$X = 7$	$e_7 = 85 \times P(X = 7) = 85 \times \dfrac{e^{-3} \times 3^7}{7!} = 1.84$	
$X = 8$	$e_8 = 85 \times P(X = 8) = 85 \times \dfrac{e^{-3} \times 3^8}{8!} = 0.69$	
$X \geq 9$	$e_9 = 85 \times P(X \geq 9)$ $\qquad\qquad\qquad = 0.33$	
	$n = 85.0$	

After grouping

Number of errors	0–1	2	3	4	5	≥ 6	
Observed cell frequency	$x_1 = 17$	$x_2 = 20$	$x_3 = 25$	$x_4 = 14$	$x_5 = 6$	$x_6 = 3$	$n = 85$
Expected cell frequency	$e_1 = 16.93$	$e_2 = 19.04$	$e_3 = 19.04$	$e_4 = 14.28$	$e_5 = 8.57$	$e_6 = 7.14$	$n = 85$

means of assessing the distribution governing a data set. The empirical cumulative distribution function discussed in Section 15.1 is an additional method.

Example 3
Software Errors

Figure 10.24 shows a data set of the number of errors found in a total of $n = 85$ software products. For example, 3 of the products had no errors, 14 had one error, and so on. Is it plausible that the number of errors has a Poisson distribution with mean $\lambda = 3$?

If the cells are taken to be

no errors, 1 error, . . . , 8 errors, at least 9 errors

then the expected cell frequencies are shown in Figure 10.24. For example, if the random variable X has a Poisson distribution with mean $\lambda = 3$, then

$$e_1 = n \times P(X = 0) = 85 \times \frac{e^{-3} \times 3^0}{0!} = 4.23$$

and

$$e_2 = n \times P(X = 1) = 85 \times \frac{e^{-3} \times 3^1}{1!} = 12.70$$

Since some of these expected values are smaller than 5, it is appropriate to group the cells as shown, so that there are eventually $k = 6$ cells, each with an expected cell frequency larger than 5.

The Pearson chi-square statistic is

$$X^2 = \frac{(17.00 - 16.93)^2}{16.93} + \frac{(20.00 - 19.04)^2}{19.04} + \frac{(25.00 - 19.04)^2}{19.04}$$
$$+ \frac{(14.00 - 14.28)^2}{14.28} + \frac{(6.00 - 8.57)^2}{8.57} + \frac{(3.00 - 7.14)^2}{7.14}$$
$$= 5.12$$

Comparison with a chi-square distribution with degrees of freedom $k - 1 = 6 - 1 = 5$ gives a p-value of

$$p\text{-value} = P\left(\chi_5^2 \geq 5.12\right) = 0.40$$

which indicates that it is quite plausible that the software errors have a Poisson distribution with mean $\lambda = 3$.

Notice that if the more general null hypothesis

$$H_0 : \text{the software errors have a Poisson distribution}$$

had been considered, then the expected cell frequencies e_i would have been calculated using a Poisson distribution with mean $\lambda = \bar{x} = 2.76$, and a p-value would have been calculated from a chi-square distribution with $k - 1 - 1 = 4$ degrees of freedom.

Example 34

Car Brake Pad Wear

Figure 10.25 shows the data collected by an engineer on how long a set of car brake pads lasted before they needed to be changed. Information on a total of $n = 232$ sets of brake pads is available, and their mileage is classified into one of six categories. The engineer would like to know whether it is plausible that the brake pad mileages to wearout can be modeled with an exponential distribution.

FIGURE 10.25

Distributional goodness of fit test for the car brake pad wear example

Brake pad wearout mileage in 1000-mile units	0.0–5.0	5.0–7.5	7.5–10.0	10.0–12.5	12.5–15.0	>15 0	
Frequency	81	48	41	36	22	4	$n = 232$

H_0 : wearout mileage X has an exponential distribution with parameter $\lambda = 1/\bar{x} = 1/7.6 = 0.1316$

Cell	Expected cell frequency	
$0.0 < X \leq 5.0$	$e_1 = 232 \times P(X \leq 5.0) = 232 \times (1 - e^{-0.1316 \times 5})$	$= 111.8$
$5.0 < X \leq 7.5$	$e_2 = 232 \times P(5.0 < X \leq 7.5) = 232 \times (e^{-0.1316 \times 5} - e^{-0.1316 \times 7.5})$	$= 33.6$
$7.5 < X \leq 10.0$	$e_3 = 232 \times P(7.5 < X \leq 10.0) = 232 \times (e^{-0.1316 \times 7.5} - e^{-0.1316 \times 10})$	$= 24.4$
$10.0 < X \leq 12.5$	$e_4 = 232 \times P(10.0 < X \leq 12.5) = 232 \times (e^{-0.1316 \times 10} - e^{-0.1316 \times 12.5})$	$= 17.4$
$12.5 < X \leq 15.0$	$e_5 = 232 \times P(12.5 < X \leq 15.0) = 232 \times (e^{-0.1316 \times 12.5} - e^{-0.1316 \times 15})$	$= 12.5$
$15.0 < X$	$e_6 = 232 \times P(15.0 < X) = 232 \times e^{-0.1316 \times 15}$	$= 32.3$
		$n = 232$

The engineer calculates that the average brake pad mileage, in 1000-mile units, is about $\bar{x} = 7.6$. Therefore the data can be tested against an exponential distribution with parameter

$$\lambda = \frac{1}{7.6} = 0.1316$$

The expected cell frequencies are shown in Figure 10.25 with, for example,

$$e_1 = n \times P(X \le 5) = 232 \times (1 - e^{-0.1316 \times 5}) = 111.8$$

where the random variable X has an exponential distribution with parameter $\lambda = 0.1316$, and

$$e_2 = n \times P(5 \le X \le 7.5) = 232 \times (e^{-0.1316 \times 5} - e^{-0.1316 \times 7.5}) = 33.6$$

A visual inspection of the observed and expected cell frequencies reveals that they appear to be quite different, and the Pearson chi-square statistic is

$$X^2 = \frac{(81.0 - 111.8)^2}{111.8} + \frac{(48.0 - 33.6)^2}{33.6} + \frac{(41.0 - 24.4)^2}{24.4}$$
$$+ \frac{(36.0 - 17.4)^2}{17.4} + \frac{(22.0 - 12.5)^2}{12.5} + \frac{(4.0 - 32.3)^2}{32.3}$$
$$= 77.78$$

Comparison with a chi-square distribution with *four* degrees of freedom, since there are six cells and the mean of the exponential distribution has been estimated from the data set, gives a *p*-value of

$$p\text{-value} = P\left(\chi_4^2 \ge 77.78\right) \simeq 0$$

Clearly the exponential distribution is *not* plausible for modeling the brake pad mileages, and actually it can be seen that the exponential distribution severely overpredicts both the number of brake pads lasting less than 5000 miles and the number of brake pads lasting more than 15,000 miles. This latter problem is related to the memoryless property of the exponential distribution, which results in a large upper tail of the exponential distribution.

Example 37
Concrete Block
Weights

Figure 10.26 shows the weights, in ascending order, of $n = 125$ randomly selected concrete blocks. Is it reasonable to model the concrete block weights as being normally distributed?
The data set has a sample mean of $\bar{x} = 10.94$ and a sample standard deviation of $s = 0.286$, and so it can be tested against a normal distribution with mean $\mu = 10.94$ and standard deviation $\sigma = 0.286$. If six cells are used based upon the intervals

$$(-\infty, 10.54], (10.54, 10.74], (10.74, 10.94], (10.94, 11.14], (11.14, 11.34], (11.34, \infty)$$

then the observed cell frequencies x_i and the expected cell frequencies e_i are shown in Figure 10.26. For example, there are 10 bricks that weigh less than or equal to 10.54 kg so that $x_1 = 10$, and if X is a normal random variable with mean $\mu = 10.94$ and standard deviation $\sigma = 0.286$, then

$$e_1 = n \times P(X \le 10.54) = 125 \times \Phi\left(\frac{10.54 - 10.94}{0.286}\right) = 125 \times 0.0810 = 10.12$$

10.10	10.34	10.39	10.42	10.48	10.48	10.49	10.52	10.52
10.54	10.55	10.57	10.59	10.59	10.59	10.61	10.61	10.62
10.64	10.65	10.67	10.68	10.69	10.69	10.70	10.70	10.70
10.71	10.72	10.72	10.73	10.74	10.74	10.74	10.74	10.75
10.75	10.75	10.76	10.78	10.79	10.79	10.79	10.79	10.80
10.80	10.80	10.81	10.82	10.82	10.84	10.85	10.86	10.87
10.89	10.89	10.91	10.91	10.92	10.93	10.93	10.93	10.93
10.95	10.97	10.99	11.00	11.01	11.01	11.02	11.02	11.03
11.03	11.04	11.04	11.04	11.05	11.05	11.06	11.06	11.07
11.08	11.08	11.08	11.08	11.08	11.08	11.08	11.09	11.09
11.10	11.10	11.11	11.12	11.13	11.13	11.14	11.14	11.15
11.16	11.16	11.17	11.17	11.17	11.18	11.18	11.20	11.21
11.22	11.26	11.27	11.30	11.31	11.32	11.33	11.34	11.35
11.40	11.41	11.43	11.55	11.57	11.58	11.58	11.61	

Summary statistics: $\bar{x} = 10.94$, $s = 0.286$

H_0 : block weight, X, has a $N(10.94, 0.286^2)$ distribution

Cell	Observed Cell Frequency	Expected Cell Frequency
$-\infty < X \leq 10.54$	$x_1 = 10$	$e_1 = 125 \times P(X \leq 10.54) = 125 \times \Phi\left(\frac{10.54-10.94}{0.286}\right)$ $= 10.12$
$10.54 < X \leq 10.74$	$x_2 = 25$	$e_2 = 125 \times P(10.54 < X \leq 10.74) = 125 \times \left(\Phi\left(\frac{10.74-10.94}{0.286}\right) - \Phi\left(\frac{10.54-10.94}{0.286}\right)\right) = 20.15$
$10.74 < X \leq 10.94$	$x_3 = 28$	$e_3 = 125 \times P(10.74 < X \leq 10.94) = 125 \times \left(\Phi\left(\frac{10.94-10.94}{0.286}\right) - \Phi\left(\frac{10.74-10.94}{0.286}\right)\right) = 32.23$
$10.94 < X \leq 11.14$	$x_4 = 35$	$e_4 = 125 \times P(10.94 < X \leq 11.14) = 125 \times \left(\Phi\left(\frac{11.14-10.94}{0.286}\right) - \Phi\left(\frac{10.94-10.94}{0.286}\right)\right) = 32.23$
$11.14 < X \leq 11.34$	$x_5 = 18$	$e_5 = 125 \times P(11.14 < X \leq 11.34) = 125 \times \left(\Phi\left(\frac{11.34-10.94}{0.286}\right) - \Phi\left(\frac{11.14-10.94}{0.286}\right)\right) = 20.15$
$11.34 < X$	$x_6 = 9$	$e_6 = 125 \times P(11.34 < X) = 125 \times \left(1 - \Phi\left(\frac{11.34-10.94}{0.286}\right)\right)$ $= 10.12$
	$n = 125$	$n = 125$

Similarly, there are 25 bricks that weigh more than 10.54 kg but less than or equal to 10.74 kg so that $x_2 = 25$, and

$$e_2 = n \times P(10.54 < X \leq 10.74)$$

$$= 125 \times \left(\Phi\left(\frac{10.74 - 10.94}{0.286}\right) - \Phi\left(\frac{10.54 - 10.94}{0.286}\right)\right)$$

$$= 125 \times (0.2422 - 0.0810)$$

$$= 20.15$$

The Pearson chi-square statistic is

$$X^2 = \frac{(10.00 - 10.12)^2}{10.12} + \frac{(25.00 - 20.15)^2}{20.15} + \frac{(28.00 - 32.23)^2}{32.23}$$

$$+ \frac{(35.00 - 32.23)^2}{32.23} + \frac{(18.00 - 20.15)^2}{20.15} + \frac{(9.00 - 10.12)^2}{10.12}$$

$$= 2.32$$

Comparison with a chi-square distribution with degrees of freedom $k - 1 - 2 = 3$, because the two parameters of the normal distribution are estimated from the data set, gives a

p-value of

$$p\text{-value} = P\left(\chi_3^2 \geq 2.32\right) = 0.51$$

which indicates that it is plausible that the concrete block weights are normally distributed.

It can be seen that a rather arbitrary aspect of this test is the manner in which the intervals have been chosen to define the cells. In this example, the intervals were chosen to be of equal length, except for the tail intervals. Another common practice is to choose the intervals (of different lengths) so that the expected cell frequencies are all equal.

Finally, there are other ways to test for normality that are often more sensitive than this application of goodness of fit tests. Simple graphical methods such as normal probability plots discussed in Section 12.7 can be effective. More generally, the empirical cumulative distribution function discussed in Section 15.1 can also be used to investigate the distribution from which a sample is drawn.

■ 10.3.3   Problems

10.3.1 DS 10.3.1 gives the results of $n = 500$ die rolls.
- **(a)** What are the expected cell frequencies if the die is a fair one?
- **(b)** Calculate the Pearson chi-square statistic X^2 for testing whether the die is fair.
- **(c)** Calculate the likelihood ratio chi-square statistic G^2 for testing whether the die is fair.
- **(d)** What p-values do these chi-square statistics give? Does a size $\alpha = 0.01$ test of the null hypothesis that the die is fair reject or accept the null hypothesis?
- **(e)** Calculate a 90% two-sided confidence interval for the probability of obtaining a 6.

10.3.2 DS 10.3.2 presents a data set on the number of rolls of a die required before a 6 is obtained. If the probability of scoring a 6 is $1/6$, then the distribution of the number of rolls required until a 6 is scored is a geometric distribution with parameter $p = 1/6$. Test whether this distribution is plausible.

10.3.3 Tire Sales
A garage sells tires of types A, B, and C, and the owner surmises that a customer is twice as likely to choose type A as type B, and twice as likely to choose type B as type C.
- **(a)** Is this supposition plausible based upon the data set in DS 10.3.3 of this year's sales?
- **(b)** Calculate a 99% two-sided confidence interval for the probability that a customer chooses type A tires.

(This problem is continued in Problem 10.6.10.)

10.3.4 Jury Selection
A court has jurisdiction over five counties, and of the people eligible for jury duty, 14% reside in county A, 22% reside in county B, 35% reside in county C, 16% reside in county D, and 13% reside in county E. DS 10.3.4 gives the residential locations of the jurors selected over a five-year period. Is there any evidence that the jurors have not been selected at random from the eligible population?

10.3.5 Infection Recovery
DS 10.3.5 presents a data set compiled from a series of clinical trials to investigate the effectiveness of a certain medication in healing an infection.
- **(a)** Is it appropriate to say that there is a 50% chance that the infection is completely healed and a 10% chance that there is no change in the infection (calculate the G^2 statistic)?
- **(b)** Calculate a 95% two-sided confidence interval for the probability that the infection is completely healed.

10.3.6 Taste Tests for Soft Drink Formulations
A beverage company has three formulations of a soft drink product. DS 10.3.6 gives the results of some taste tests where participants are asked to declare which formulation they like best. Is it plausible that the three formulations are equally popular?

10.3.7 Hospital Emergency Room Operation
DS 10.3.7 gives a data set of the number of arrivals at a hospital emergency room during one-hour periods. Is

there any evidence that it is not reasonable that the number of arrivals are modeled with a Poisson distribution with mean $\lambda = 7$?

10.3.8   Radioactive Particle Emissions

DS 10.3.8 gives a data set of the number of radioactive particles emitted from a substance and passing through a counter in one-minute periods. Is there any evidence that it is not reasonable to model these with a Poisson distribution?

10.3.9   Oyster Pearl Grades

Oyster pearls are graded A, B, C, and D depending on whether their diameter is 9–10 mm, 7–9 mm, 5–7 mm, or 0–5 mm, respectively. DS 10.3.9 shows how a farmer's pearl crop was categorized. Is it plausible that the pearl oyster diameters have a uniform distribution between 0 and 10 mm?

10.3.10   Photocopy Machine Breakdowns

DS 10.3.10 gives a data set on the time intervals between breakdowns of a photocopy machine. Is there any evidence that it is not reasonable to model these time intervals with an exponential distribution with a mean of five days?

10.3.11   Optical Cable Imperfections

DS 10.3.11 gives a data set on the distances between imperfections on an optical cable. Is there any evidence that it is not reasonable to model these distances with an exponential distribution with a mean of two meters?

10.3.12   Production Line Variability

DS 10.3.12 gives a data set on the actual weights of a set of 1-kg sugar packets. Is there any evidence that it is not reasonable to model these weights with a normal distribution with a mean of $\mu = 1.03$ kg and a standard deviation of $\sigma = 0.01$ kg?

10.3.13   Genetic Variations in Plants

In a biological experiment a large quantity of plants were grown. For each plant the stem length was classified as being either tall or dwarf, and the position of the flowers was classified as being either axial or terminal. All together there were 412 tall axial plants, 121 tall terminal plants, 148 dwarf axial plants, and 46 dwarf terminal plants. According to the proposed genetic theory, the probabilities of the plants displaying each of these four characteristics should have relative magnitudes of 9:3:3:1, respectively. Is the data set consistent with the proposed genetic theory?

10.3.14

Each of 205 consumers was asked to choose which of three products they preferred. Product A was chosen by 83 of the consumers, product B was chosen by 75 of the consumers, and product C was chosen by 47 of the consumers. Is there sufficient evidence to conclude that the three products do not have equal probabilities of being chosen?

10.3.15   Chemical Solution Acidities

The acidity of a chemical solution can be classified as "very high," "high," "normal," "low," and "very low." A total of 630 samples of the solution from a production process were obtained, and the acidities were classified as given in DS 10.3.13.

(a) Perform a two-sided hypothesis test of the null hypothesis that the probability that a solution has normal acidity is 0.80.

(b) It is claimed that the probability that a solution has very high acidity is 0.04, the probability that a solution has high acidity is 0.06, the probability that a solution has normal acidity is 0.80, the probability that a solution has low acidity is 0.06 and the probability that a solution has very low acidity is 0.04. Are the data consistent with this claim?

10.3.16

An experiment was performed to investigate how long batteries remain charged under certain storage conditions. A total of 125 batteries were charged to the same level and stored in the designated conditions. After 24 hours all 125 batteries were tested and it was found that 12 of them had charges that had dropped below the threshold level. After an additional 24 hours the remaining 113 batteries were tested and it was found that 53 of them had charges that had dropped below the threshold level. Finally, after an additional 24 hours the remaining 60 batteries were tested and it was found that 39 of them had charges that had dropped below the threshold level. It is claimed that for these batteries under these storage conditions the time in hours until the charge drops below the threshold level has a Weibull distribution with parameters $\lambda = 0.065$ and $a = 0.45$. Are the results of this experiment consistent with that claim?

10.3.17   Shark Attacks

The data set in DS 10.3.14 shows the number of shark attacks along a popular stretch of coastline for each of the past 76 years. Is it plausible that the number of shark attacks per year follows a Poisson distribution with mean 2.5?

10.4 Testing for Independence in Two-Way Contingency Tables

10.4.1 Two-Way Classifications

A two-way contingency table is a set of frequencies that summarize how a set of objects is *simultaneously* classified under two different categorizations. If the first categorization has r levels and the second categorization has c levels, then as Figure 10.27 shows, the data can be presented in tabular form as a set of frequencies

$$x_{ij} \qquad 1 \le i \le r \qquad 1 \le j \le c$$

Thus, the **cell frequency** x_{ij} is the number of objects that fall into the ith level of the first categorization and into the jth level of the second categorization. Data sets of this form are referred to as $r \times c$ **contingency tables**.

The **row marginal frequencies** $x_1., \ldots, x_r.$ are defined to be

$$x_{i.} = \sum_{j=1}^{c} x_{ij} \qquad 1 \le i \le r$$

so that they are the sum of the frequencies in each of the r levels of the first categorization. Similarly, the **column marginal frequencies** $x_{.1}, \ldots, x_{.c}$ are defined to be

$$x_{.j} = \sum_{i=1}^{r} x_{ij} \qquad 1 \le j \le c$$

so that they are the sum of the frequencies in each of the c levels of the second categorization. A subscript "·" is therefore taken to imply that the replaced subscript has been summed over. The sample size n may be written

$$n = x_{..} = \sum_{j=1}^{c} x_{.j} = \sum_{i=1}^{r} x_{i.}$$

FIGURE 10.27

A two-way ($r \times c$) contingency table

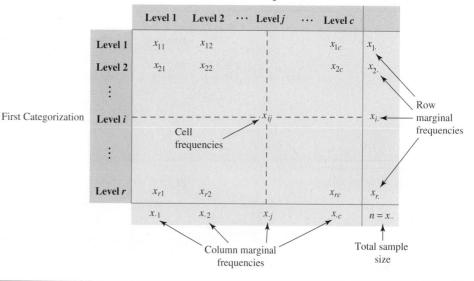

In certain data sets some of the marginal frequencies may be fixed due to the manner in which the data are collected. For example, certain fixed amounts of each of the levels of one categorization may be investigated to determine which level of the other categorization they fall into. In other cases all of the marginal frequencies may be random. These differences are illustrated in the following examples.

Example 29
Drug Allergies

Three drugs are compared with respect to the types of allergic reaction that they cause to patients. A group of $n = 300$ patients is randomly split into three groups of 100 patients, each of which is given one of the three drugs. The patients are then categorized as being *hyperallergic, allergic, mildly allergic,* or as having *no allergy*.

Figure 10.28 shows a 3×4 contingency table that presents the results of this experiment. Notice that the row marginal frequencies $x_{i.}$ are all equal to 100 since each drug is administered to exactly 100 patients. In contrast, the column marginal frequencies $x_{.j}$, which represent the total number of patients with each of the different allergy levels, are not fixed in advance of the experiment.

Example 58
Overage Weedkiller Product

Recall that in the nationwide survey the auditors examined $n = 54,965$ weedkiller containers and found 2779 of them to be overage. However, the weedkiller containers had three sizes (*small, medium,* and *large*), and the auditors were also required to record the size of the containers that they examined. Therefore, the full data set, shown in Figure 10.29, takes the form of a 2×3 contingency table. None of the marginal frequencies of this contingency table is fixed before the survey. The total number of containers examined, n, is also not fixed in advance, although it could have been estimated from knowledge of the number of stores that were to be visited in the survey.

Example 57
Building Tile Cracks

One of the most common forms of a two-way contingency table is a 2×2 table in which each of the two categorizations has only two levels. Such a data set has much in common with the problem of comparing two population proportions discussed in Section 10.2. Figure 10.30 shows how the survey results of cracked tiles on buildings A and on buildings B can be

FIGURE 10.28

Drug allergies data set

Reaction

Type of Drug		Hyperallergic	Allergic	Mildly allergic	No allergy	
	Drug A	$x_{11} = 11$	$x_{12} = 30$	$x_{13} = 36$	$x_{14} = 23$	$x_{1.} = 100$
	Drug B	$x_{21} = 8$	$x_{22} = 31$	$x_{23} = 25$	$x_{24} = 36$	$x_{2.} = 100$
	Drug C	$x_{31} = 13$	$x_{32} = 28$	$x_{33} = 28$	$x_{34} = 31$	$x_{3.} = 100$
		$x_{.1} = 32$	$x_{.2} = 89$	$x_{.3} = 89$	$x_{.4} = 90$	$n = x_{..} = 300$

FIGURE 10.29

Overage weedkiller product data set

Size of Container

Age of Product		Small	Medium	Large	
	Underage	$x_{11} = 15595$	$x_{12} = 25869$	$x_{13} = 10722$	$x_{1.} = 52186$
	Overage	$x_{21} = 612$	$x_{22} = 856$	$x_{23} = 1311$	$x_{2.} = 2779$
		$x_{.1} = 16207$	$x_{.2} = 26725$	$x_{.3} = 12033$	$n = x_{..} = 54965$

FIGURE 10.30

Building tile cracks data set

		Location		
		Buildings A	**Buildings B**	
Tile Condition	**Undamaged**	$x_{11} = 5594$	$x_{12} = 1917$	$x_{1.} = 7511$
	Cracked	$x_{21} = 406$	$x_{22} = 83$	$x_{2.} = 489$
		$x_{.1} = 6000$	$x_{.2} = 2000$	$n = x_{..} = 8000$

presented as a 2×2 contingency table. Notice that the column marginal frequencies

$$x_{.1} = 6000 \quad \text{and} \quad x_{.2} = 2000$$

are fixed and represent the sample sizes chosen for the two sets of buildings.

10.4.2 Testing for Independence

One of the most common test procedures applied to a two-way contingency table is a test of **independence** between the two categorizations. The exact interpretation of what independence means depends upon the specific contingency table under consideration, but it essentially can be taken to mean that the two factors that produce the two categorizations are *not associated* with each other. More technically it means that, conditional on being in any level of one of the categorizations, the sets of probabilities of being in the various levels of the other categorization are all the same. For example, being in level 1 or level 2 of the first categorization does not alter the chances of being in level 1 or level 2 of the second categorization.

In certain two-way contingency tables with either the row or column marginal frequencies fixed, the test for independence may more appropriately be considered to be a test for **homogeneity** between the probability distributions at each of the row or column levels. The null hypothesis in this case states that these probability distributions are all equal.

Tests of independence operate by taking independence to be the null hypothesis. A p-value is calculated using a chi-square test.

Example 29

Drug Allergies

In this example, the null hypothesis of independence between the type of drug and the type of reaction can be interpreted as meaning that the chances of the various kinds of allergic reaction are the same for each of the three drugs. This means that there is a set of probability values p_1, p_2, p_3, and p_4 that represent the probabilities of getting the four types of reaction *regardless* of which drug is used, as illustrated in Figure 10.31. Thus, the three types of drug can be considered to be equivalent in terms of the reactions that they cause.

If the null hypothesis of independence is rejected, then this implies that there is evidence to conclude that the three drugs have different sets of probability values for the four types of reaction. This means that the three types of drug *cannot* be considered to be equivalent in terms of the reactions that they cause. (Actually, two of the three drugs could still be equivalent, but all three cannot be the same.)

Example 58

Overage Weedkiller Product

Independence in this example can be interpreted as meaning that the overage proportions are identical for the three different sizes of container. Therefore, if p_s is the probability that a small container is overage, p_m is the probability that a medium container is overage, and p_l is the probability that a large container is overage, then the null hypothesis of independence

FIGURE 10.31

Independence for drug allergies example

No Independence

	Probability of hyperallergic		Probability of allergic		Probability of mildly allergic		Probability of no allergy		
Drug A	p_1^A	$+$	p_2^A	$+$	p_3^A	$+$	p_4^A	$=$	1
Drug B	p_1^B	$+$	p_2^B	$+$	p_3^B	$+$	p_4^B	$=$	1
Drug C	p_1^C	$+$	p_2^C	$+$	p_3^C	$+$	p_4^C	$=$	1

Independence

	Probability of hyperallergic		Probability of allergic		Probability of mildly allergic		Probability of no allergy		
All drugs	p_1	$+$	p_2	$+$	p_3	$+$	p_4	$=$	1

FIGURE 10.31

Independence for drug allergies example

FIGURE 10.32

Independence for overage weedkiller product example

No Independence

	Container Size		
	Small	**Medium**	**Large**
Probability of underage	$1 - p_s$	$1 - p_m$	$1 - p_l$
Probability of overage	p_s	p_m	p_l

Independence $H_0 : p_s = p_m = p_l$

	All containers
Probability of underage	$1 - p$
Probability of overage	p

can be written

$$H_0 : p_s = p_m = p_l$$

as illustrated in Figure 10.32. Thus, in this case, the test of independence can be thought of as a test of the equivalence of a set of binomial parameters. If the null hypothesis of independence is rejected, then there is enough evidence to conclude that the three overage proportions are not all equal.

Example 57
Building Tile Cracks

In this example, independence means that the probability that a tile on buildings A becomes cracked, p_A, is *equal* to the probability that a tile on buildings B becomes cracked, p_B. Consequently, testing for independence in this example can be viewed as being identical to the problem of testing the equivalence of two binomial parameters, as discussed in Section 10.2.2.

The null hypothesis of independence can be tested using either a Pearson or a likelihood ratio chi-square goodness of fit statistic, as described in the accompanying box. The summations in these statistics are taken to be over each of the $r \times c$ cells, and the expected cell frequencies are calculated as

$$e_{ij} = \frac{x_{i.}x_{.j}}{n}$$

Thus the expected cell frequency in the ij-cell is the product of the ith row marginal frequency and the jth column marginal frequency, divided by the total sample size n.

Testing for Independence in a Two-Way Contingency Table

The null hypothesis of independence between two categorizations based upon an $r \times c$ contingency table with cell frequencies x_{ij} can be performed using either the Pearson chi-square statistic

$$X^2 = \sum_{i=1}^{r}\sum_{j=1}^{c} \frac{(x_{ij} - e_{ij})^2}{e_{ij}}$$

or the likelihood ratio chi-square statistic

$$G^2 = 2\sum_{i=1}^{r}\sum_{j=1}^{c} x_{ij} \ln\left(\frac{x_{ij}}{e_{ij}}\right)$$

The expected cell frequencies are

$$e_{ij} = \frac{x_{i.}x_{.j}}{n}$$

where $x_{i.}$ is the ith row marginal frequency, $x_{.j}$ is the jth column marginal frequency, and n is the total sample size. A p-value can be calculated as either

$$p\text{-value} = P\left(\chi_{\nu}^2 \geq X^2\right) \qquad \text{or} \qquad p\text{-value} = P\left(\chi_{\nu}^2 \geq G^2\right)$$

where the degrees of freedom are

$$\nu = (r-1) \times (c-1)$$

A size α hypothesis test *accepts* the null hypothesis of independence if the chi-square statistic is *less* than the critical point $\chi_{\alpha,\nu}^2$, and *rejects* the null hypothesis of independence if the chi-square statistic is *greater* than $\chi_{\alpha,\nu}^2$.

One other point to notice is that the degrees of freedom of the chi-square distribution used to calculate the p-value is $(r-1) \times (c-1)$. Also, it is desirable to have the expected cell frequencies all larger than 5, although if one or two of them are less than 5 it should not be a big problem. If several are less than 5, then it is best to group some category levels to avoid this problem.

Example 29
Drug Allergies

Figure 10.33 shows the expected cell frequencies for this example and the calculation of the Pearson chi-square statistic $X^2 = 6.391$ (the likelihood ratio chi-square statistic is $G^2 = 6.450$). The appropriate degrees of freedom for calculating a p-value are $(r-1) \times (c-1) = 2 \times 3 = 6$, so that

$$p\text{-value} = P\left(\chi_6^2 \geq 6.391\right) \simeq 0.38$$

	Hyperallergic	Allergic	Mildly allergic	No allergy	
Drug A	$x_{11} = 11$ $e_{11} = \frac{100 \times 32}{300} = 10.67$	$x_{12} = 30$ $e_{12} = \frac{100 \times 89}{300} = 29.67$	$x_{13} = 36$ $e_{13} = \frac{100 \times 89}{300} = 29.67$	$x_{14} = 23$ $e_{14} = \frac{100 \times 90}{300} = 30.00$	$x_1. = 100$
Drug B	$x_{21} = 8$ $e_{21} = \frac{100 \times 32}{300} = 10.67$	$x_{22} = 31$ $e_{22} = \frac{100 \times 89}{300} = 29.67$	$x_{23} = 25$ $e_{23} = \frac{100 \times 89}{300} = 29.67$	$x_{24} = 36$ $e_{24} = \frac{100 \times 90}{300} = 30.00$	$x_2. = 100$
Drug C	$x_{31} = 13$ $e_{31} = \frac{100 \times 32}{300} = 10.67$	$x_{32} = 28$ $e_{32} = \frac{100 \times 89}{300} = 29.67$	$x_{33} = 28$ $e_{33} = \frac{100 \times 89}{300} = 29.67$	$x_{34} = 31$ $e_{34} = \frac{100 \times 90}{300} = 30.00$	$x_3. = 100$
	$x_{.1} = 32$	$x_{.2} = 89$	$x_{.3} = 89$	$x_{.4} = 90$	$n = x_{..} = 300$

Pearson chi-square statistic:
$$X^2 = \frac{(11.00 - 10.67)^2}{10.67} + \frac{(30.00 - 29.67)^2}{29.67} + \frac{(36.00 - 29.67)^2}{29.67} + \frac{(23.00 - 30.00)^2}{30.00}$$
$$+ \frac{(8.00 - 10.67)^2}{10.67} + \frac{(31.00 - 29.67)^2}{29.67} + \frac{(25.00 - 29.67)^2}{29.67} + \frac{(36.00 - 30.00)^2}{30.00}$$
$$+ \frac{(13.00 - 10.67)^2}{10.67} + \frac{(28.00 - 29.67)^2}{29.67} + \frac{(28.00 - 29.67)^2}{29.67} + \frac{(31.00 - 30.00)^2}{30.00}$$
$$= 6.391$$

FIGURE 10.33

Analysis of drug allergies data set

This large p-value implies that the null hypothesis of independence is plausible, and so there is no evidence to conclude that the three drugs are any different with respect to the types of allergic reaction that they cause.

Example 58

Overage Weedkiller Product

The estimates of the proportions of small, medium, and large containers that are overage are

$$\hat{p}_s = \frac{612}{16,207} = 0.038$$

$$\hat{p}_m = \frac{856}{26,725} = 0.032$$

$$\hat{p}_l = \frac{1311}{12,033} = 0.109$$

Is the difference between these estimates statistically significant?

Figure 10.34 shows the expected cell frequencies for this example and the calculation of the likelihood ratio chi-square statistic $G^2 = 930.8$. With $(2 - 1) \times (3 - 1) = 2$ degrees of freedom the p-value is

$$p\text{-value} = P\left(\chi_2^2 \geq 930.8\right) \simeq 0$$

and the null hypothesis of independence is clearly not plausible. In fact, with such a large chi-square statistic the survey provides extremely strong evidence that the overage proportions p_s, p_m, and p_l are not all equal. Moreover, the data suggest that the large containers have a considerably greater overage rate than the other types of container.

Suppose that the chemical company checks last year's records and finds that 26.3% of the weedkiller containers sold were the small size, 52.5% were the medium size, and 21.2%

	Small	**Medium**	**Large**	
Underage	$x_{11} = 15595$ $e_{11} = \frac{52186 \times 16207}{54965} = 15387.58$	$x_{12} = 25869$ $e_{12} = \frac{52186 \times 26725}{54965} = 25373.80$	$x_{13} = 10722$ $e_{13} = \frac{52186 \times 12033}{54965} = 11424.62$	$x_{1.} = 52186$
Overage	$x_{21} = 612$ $e_{21} = \frac{2779 \times 16207}{54965} = 819.42$	$x_{22} = 856$ $e_{22} = \frac{2779 \times 26725}{54965} = 1351.20$	$x_{23} = 1311$ $e_{23} = \frac{2779 \times 12033}{54965} = 608.38$	$x_{2.} = 2779$
	$x_{.1} = 16207$	$x_{.2} = 26725$	$x_{.3} = 12033$	$n = x_{..} = 54965$

Likelihood ratio chi-square statistic:
$$G^2 = 2 \times \left(15595 \ln\left(\frac{15595}{15387.58}\right) + 25869 \ln\left(\frac{25869}{25373.80}\right) + 10722 \ln\left(\frac{10722}{11424.62}\right) \right.$$
$$\left. + 612 \ln\left(\frac{612}{819.42}\right) + 856 \ln\left(\frac{856}{1351.20}\right) + 1311 \ln\left(\frac{1311}{608.38}\right) \right)$$
$$= 930.8$$

FIGURE 10.34

Analysis of overage weedkiller product data set

were the large size. This suggests that the overall proportion of sales that involve an overage product should be about equal to

$$\bar{p} = 0.263 p_s + 0.525 p_m + 0.212 p_l$$

This value can be estimated as

$$\hat{\bar{p}} = 0.263 \hat{p}_s + 0.525 \hat{p}_m + 0.212 \hat{p}_l$$
$$= (0.263 \times 0.038) + (0.525 \times 0.032) + (0.212 \times 0.109) = 0.050$$

with a variance

$$\text{Var}(\hat{\bar{p}}) = (0.263^2 \times \text{Var}(\hat{p}_s)) + (0.525^2 \times \text{Var}(\hat{p}_m)) + (0.212^2 \times \text{Var}(\hat{p}_l))$$

If the total number of products examined for each of the three sizes are considered to be fixed so that $\hat{p}_s$, $\hat{p}_m$, and $\hat{p}_l$ are considered to be estimates of three binomial parameters, then $\text{Var}(\hat{p}_s)$ can be estimated as

$$\frac{\hat{p}_s(1 - \hat{p}_s)}{x_{.1}} = \frac{x_{11} x_{21}}{x_{.1}^3} = \frac{15{,}595 \times 612}{16{,}207^3}$$

and similarly, $\text{Var}(\hat{p}_m)$ can be estimated as

$$\frac{25{,}869 \times 856}{26{,}725^3}$$

and $\text{Var}(\hat{p}_l)$ can be estimated as

$$\frac{10{,}722 \times 1311}{12{,}033^3}$$

FIGURE 10.35

Analysis of building tile cracks
data set

	Buildings A	Buildings B	
Undamaged	$x_{11} = 5594$ $e_{11} = \frac{7511 \times 6000}{8000} = 5633.25$	$x_{12} = 1917$ $e_{12} = \frac{7511 \times 2000}{8000} = 1877.75$	$x_{1.} = 7511$
Cracked	$x_{21} = 406$ $e_{21} = \frac{489 \times 6000}{8000} = 366.75$	$x_{22} = 83$ $e_{22} = \frac{489 \times 2000}{8000} = 122.25$	$x_{2.} = 489$
	$x_{.1} = 6000$	$x_{.2} = 2000$	$n = x_{..} = 8000$

Putting these all together gives

$$s.e.(\hat{\bar{p}}) = \sqrt{\text{Var}(\hat{\bar{p}})} = 0.0009$$

With $z_{0.005} = 2.576$, a two-sided 99% confidence interval for $\bar{p}$ is therefore

$$\bar{p} \in (0.050 - (2.576 \times 0.0009), 0.050 + (2.576 \times 0.0009)) = (0.048, 0.052)$$

In conclusion, this analysis predicts that somewhere between about 4.8% and 5.2% of the overall sales involve an overage product.

Example 57
Building Tile Cracks

Figure 10.35 shows the calculation of the expected cell frequencies for this example. The Pearson chi-square statistic is

$$X^2 = \frac{(5594.00 - 5633.25)^2}{5633.25} + \frac{(1917.00 - 1877.75)^2}{1877.75}$$
$$+ \frac{(406.00 - 366.75)^2}{366.75} + \frac{(83.00 - 122.25)^2}{122.25}$$
$$= 17.896$$

Compared with a chi-square distribution with $(2 - 1) \times (2 - 1) = 1$ degree of freedom, this result gives a p-value of

$$p\text{-value} = P\left(\chi_1^2 \geq 17.896\right) \simeq 0$$

Consequently, it is not plausible that p_A and p_B, the probabilities of tiles cracking on the two sets of buildings, are equal. This analysis is consistent with the confidence interval

$$p_A - p_B \in (0.0120, 0.0404)$$

calculated previously, since the confidence interval does not contain 0.

Finally, it is useful to know that for a 2×2 contingency table of this kind, a shortcut formula for the Pearson chi-square statistic is

$$X^2 = \frac{n(x_{11}x_{22} - x_{12}x_{21})^2}{x_{1.}x_{.1}x_{2.}x_{.2}}$$

FIGURE 10.36

Illustration of Simpson's paradox

		Internet sales	Telephone sales
Product A Sales	New customers	199 (11.10%)	63 (6.71%)
	Repeat customers	1594 (88.90%) $\overline{1793}$	876 (93.29%) $\overline{939}$
Product B Sales	New customers	243 (11.10%)	138 (9.98%)
	Repeat customers	1946 (88.90%) $\overline{2189}$	1245 (90.02%) $\overline{1383}$
Product C Sales	New customers	864 (16.15%)	1107 (15.90%)
	Repeat customers	4486 (83.85%) $\overline{5350}$	5855 (84.10%) $\overline{6962}$
Product D Sales	New customers	128 (38.32%)	180 (36.59%)
	Repeat customers	206 (61.68%) $\overline{334}$	312 (63.41%) $\overline{492}$
Total Sales	New customers	1434 (14.84%)	1488 (15.22%)
	Repeat customers	8232 (85.16%) $\overline{9666}$	8288 (84.78%) $\overline{9776}$

For this example, it can be checked that this value gives

$$X^2 = \frac{8000 \times ((5594 \times 83) - (1917 \times 406))^2}{7511 \times 6000 \times 489 \times 2000} = 17.896$$

as before.

Simpson's Paradox   When one analyzes categorical data in the form of contingency tables, it is important to consider the full extent of categorization that is possible. If the data set is collapsed over one or more categories, then the resulting contingency table may give misleading indications. This issue is exemplified in an unusual phenomenon known as Simpson's paradox.

Example 41
Internet Commerce

As an illustration of Simpson's paradox, consider the data set shown in Figure 10.36. A company sells four products either over the Internet or by telephone, and a manager investigates the company's sales to see whether they are from first-time customers or from repeat customers. It can be seen that for each of the four products, the proportion of the Internet sales that are from first-time customers is larger than the proportion of telephone sales that are from first-time customers. For example, there were 1793 sales of product A over the Internet, of which 199 or 11.10% were from first-time customers. However, out of 939 sales of product A by telephone, only 63 or 6.71% were from first-time costomers. Similarly, for product B the rate is 11.10% over the Internet but only 9.98% by telephone, for product C the rate is 16.15% over the Internet but only 15.90% by telephone, and for product D the rate is 38.32% over the Internet but only 36.59% by telephone. The slightly higher proportions of first-time customers from the Internet sales may be useful information for the manager.

However, if the manager had looked only at total sales, then it would have been seen that out of 9666 sales over the Internet, 1434 or 14.84% were from first-time customers, while out of 9776 sales by telephone, 1488 or 15.22% were from first-time customers. Rather surprisingly this provides the incorrect indication that the proportion of first-time customers is *lower* from the Internet than from telephone sales. This strange phenomenon has occurred as a result of looking at total sales instead of the sales broken down over each of the four products, and it is known as Simpson's paradox.

■ 10.4.3 Problems

10.4.1 Circuit Board Quality

A computer manufacturer has four suppliers of electrical circuit boards. Random samples of 200 circuit boards are taken from each of the suppliers and they are classified as being either acceptable or defective, as given in DS 10.4.1. Consider the problem of testing whether the defective rates are identical for all four suppliers.

(a) Calculate the expected cell frequencies.
(b) Calculate the Pearson chi-square statistic X^2.
(c) Calculate the likelihood ratio chi-square statistic G^2.
(d) What p-values do these chi-square statistics give?
(e) Is the null hypothesis that the defective rates are identical rejected at size $\alpha = 0.05$?
(f) Calculate a 95% two-sided confidence interval for the defective rate of supplier A.
(g) Calculate a 95% two-sided confidence interval for the difference between the defective rates of suppliers B and C.

10.4.2 Fertilizer Comparisons

Seedlings are grown without fertilizer or with one of two kinds of fertilizer. After a certain period of time a seedling's growth is classified into one of four categories, as given in DS 10.4.2. Test whether the seedlings' growth can be taken to be the same for all three sets of growing conditions.

10.4.3 Taste Tests for Soft Drink Formulations

DS 10.4.3 gives the results of a taste test in which a sample of 200 people in each of three age groups are asked which of three formulations of a soft drink they prefer. Test whether the preferences for the different formulations change with age.

10.4.4 Electric Motor Quality Tests

A factory has five production lines that assemble electric motors. A random sample of 180 motors is taken from each production line and is given a quality examination. The results are given in DS 10.4.4.

(a) Is there any evidence that the pass rates are any different for the five production lines?
(b) Construct a 95% two-sided confidence interval for the difference between the pass rates of production lines 1 and 2.

10.4.5 Customer Satisfaction Surveys

An air-conditioner supplier employs four technicians who visit customers to install and provide maintenance for their air-conditioner units. In an after-visit questionnaire the customers are asked to rate their satisfaction with the technician. DS 10.4.5 gives the ratings obtained by the four technicians over a period of time. Is there any evidence that some technicians are better than others in satisfying their customers?

10.4.6

Show that for a 2×2 contingency table the Pearson chi-square statistic can be written

$$X^2 = \frac{n(x_{11}x_{22} - x_{12}x_{21})^2}{x_{1\cdot}x_{\cdot 1}x_{2\cdot}x_{\cdot 2}}$$

10.4.7 Clinical Trial

DS 10.4.6 presents a 2×2 contingency table that compares two drugs with respect to the speed with which a patient recovers from an ailment. Let p_s be the probability that a patient recovers in less than one week if given the standard drug, and let p_n be the probability that a patient recovers in less than one week if given the new drug.

(a) Use the Pearson chi-square statistic to test whether there is any evidence that $p_s \neq p_n$.
(b) Construct a 99% two-sided confidence interval for $p_s - p_n$.

10.4.8 Reactive Ion Etching in Semiconductor Manufacturing

In the manufacture of semiconductors, reactive ion etching is a technique whereby the surface of the semiconductor is bombarded with ions to remove unwanted material and to leave the desired structure. In an experiment a set of semiconductors was produced by this technique, and each one was examined to see whether the desired structure was complete or incomplete and also whether the etch depth was satisfactory or unsatisfactory. It was found that in 1078 cases the structure was complete and the etch depth was satisfactory, in 544 cases the structure was complete and the etch depth was unsatisfactory, in 253 cases the structure was incomplete and the etch depth was satisfactory, and in 111 cases the structure was incomplete and the etch depth was unsatisfactory. Use the Pearson chi-square statistic to test whether the completeness of the structure is related to the etch depth, or whether these two factors can be considered to be independent of each other.

10.4.9 Consumer Warranty Purchases

A company's sales records show that an extended warranty was purchased on 38 out of 89 sales of a product of type A, an extended warranty was purchased on 62 out of 150 sales of a product of type B, and an extended warranty was purchased on 37 out of 111 sales of a product of type C. Do these data provide sufficient evidence to indicate that the probability of a customer purchasing the extended warranty is different for the three product types?

10.4.10 Asphalt Load Testing

An experiment was conducted to compare three types of asphalt. Samples of each type of asphalt were subjected to repeated loads at high temperatures, and the resulting cracking was analyzed. For type A, 57 samples were tested, of which 9 had severe cracking, 17 had medium cracking, and 31 had minor cracking. For type B, 49 samples were tested, of which 4 had severe cracking, 9 had medium cracking, and 36 had minor cracking. For type C, 90 samples were tested, of which 15 had severe cracking, 19 had medium cracking, and 56 had minor cracking. Does this experiment provide any evidence that the three types of asphalt are different with respect to cracking?

10.5 Case Study: Microelectronic Solder Joints

The data set in Figure 6.41 reveals that 451 out of 512 solder joints on an assembly were barrel shaped. With $z_{0.005} = 2.576$ and $\hat{p}_b = 451/512 = 0.881$, a 99% confidence level confidence interval for p_b, the probability that a solder joint will be barrel shaped for this production method can be calculated as

$$p_b \in \hat{p}_b \pm z_{0.005}\sqrt{\frac{\hat{p}_b(1 - \hat{p}_b)}{n}} = 0.881 \pm 2.576\sqrt{\frac{0.881(1 - 0.881)}{512}} = (0.844, 0.918)$$

A goodness of fit test can be performed to assess whether the data in Figure 6.41 are consistent with the supposition that using this production method there is a probability of 0.85 that a solder joint has a barrel shape, there is a probability of 0.03 that a solder joint has a cylinder shape, and there is a probability of 0.12 that a solder joint has an hourglass shape. The null hypothesis is therefore

$$H_0 : \; p_b = 0.85, \; p_c = 0.03, \; p_h = 0.12$$

and the expected values are

$$e_b = 512 \times 0.85 = 435.20, \; e_c = 512 \times 0.03 = 15.36, \; e_h = 512 \times 0.12 = 61.44$$

The Pearson chi-square statistic is

$$X^2 = \frac{(451 - 435.20)^2}{435.20} + \frac{(8 - 15.36)^2}{15.36} + \frac{(53 - 61.44)^2}{61.44} = 0.57 + 3.53 + 1.16 = 5.26$$

so that the p-value is $P(\chi_2^2 \geq 5.26) = 0.072$. Since the p-value falls in the range 1% to 10%, the researcher can conclude that there is some evidence that the shape probabilities are not as stated, but that the evidence is not overwhelming. In particular, the small number of hourglass-shaped solder joints observed in the data set is somewhat suspicious.

In an experiment to compare two different epoxy formulations for use in the underfill, the researcher prepares 40 assemblies using epoxy of type I and 40 assemblies using epoxy of type II. Each assembly is then subjected to 2000 temperature cycles before being tested to see whether it still functions properly. It was found that only 5 of the assemblies with type I epoxy failed, whereas 15 of the assemblies with type II epoxy failed.

The probability p_I that an assembly produced with underfill of type I epoxy fails within 2000 temperature cycles can therefore be estimated as

$$\hat{p}_I = \frac{5}{40} = 0.125$$

whereas the corresponding probability for assemblies produced with underfill of type II epoxy is

$$\hat{p}_{II} = \frac{15}{40} = 0.375$$

The hypotheses

$$H_0 : p_I = p_{II} \qquad \text{versus} \qquad H_A : p_I \neq p_{II}$$

can be tested with the test statistic

$$z = \frac{\hat{p}_I - \hat{p}_{II}}{\sqrt{\hat{p}(1 - \hat{p})\left(\frac{1}{n} + \frac{1}{m}\right)}} = \frac{0.125 - 0.375}{\sqrt{\frac{20}{80}\left(1 - \frac{20}{80}\right)\left(\frac{1}{40} + \frac{1}{40}\right)}} = 2.582$$

The p-value is $2 \times \Phi(-2.582) = 0.0098$ which is just less than 1%. Therefore, the researcher can conclude that this data set provides sufficient evidence to establish that the failure rates at 2000 temperature cycles are different for the two epoxy formulations, and clearly it is best to use epoxy type I for the underfill. Furthermore, it is interesting to note that if this data set is analyzed as a 2×2 contingency table, then the Pearson chi-square statistic is

$$X^2 = \frac{n(x_{11}x_{22} - x_{12}x_{21})^2}{x_{1.}x_{.1}x_{2.}x_{.2}} = \frac{80(5 \times 25 - 15 \times 35)^2}{20 \times 40 \times 60 \times 40} = 6.667$$

and the p-value $P(\chi_1^2 \geq 6.667)$ is again just less than 1%.

10.6   Supplementary Problems

10.6.1 Crystallization is an important step in the recovery and purification of biological molecules such as enzymes and proteins. The determination of conditions that produce suitable crystals is of great interest to molecular biologists. In one experiment, crystals had appeared within 24 hours in 27 out of 60 trials of a particular solution. Calculate a 95% two-sided confidence interval for the probability of crystallization within 24 hours using this method.

10.6.2 A consumer watchdog organization takes a random sample of 500 bags of flour made by a company, weighs them, and discovers that 18 of them are underweight. Suppose that legal action can be taken if it can be demonstrated that the proportion of bags sold by the company that are underweight is more than 1 in 40. Would you advise the watchdog organization that there are grounds for legal action?

10.6.3 A bank releases a new credit card that is to be targeted at a population group of about 1,000,000 customers. In a trial run the bank mails credit card applications to a random sample of 5000 customers within this group, and 384 of them request the credit card. If the bank goes ahead and mails application forms to all 1,000,000 customers in the target group, construct two-sided 99% confidence bounds on the total number of these customers who will request a card.

10.6.4 Upon checking hospital records, a hospital administrator notices that over a certain period of time, 443 out of 564 surgical operations performed in the *morning* turned out to be a "total success," whereas only 388 out of 545 surgical operations performed in the *afternoon* turned out to be a "total success." Does this substantiate the hypothesis that surgeons are less effective in the afternoon because they are more tired? How strong is the evidence that this data set provides to support this hypothesis? What other information would you like to know before making a judgment?

10.6.5 Householders are polled on whether they support a tax increase to build more schools. The householders are also asked whether their annual household income is above or below \$60,000. The results of the poll found 106 householders with an annual income above \$60,000 of whom 32 support the tax increase and 221 householders with an annual income below \$60,000 of whom 106 support the tax increase. Provide a two-sided analysis to investigate the evidence that the support for the tax increase depends upon the householder's income.

10.6.6 Archery Contest Scores

DS 10.6.1 contains a data set collected during an amateur archery contest. Calculate a X^2 statistic to assess whether it is appropriate to model the probabilities of a bull's-eye and a missed target as both being 10%.

10.6.7 Rush-Hour Car Accidents

DS 10.6.2 gives a data set of the number of car accidents occurring during evening rush-hour traffic in a certain city. Does it look like it's reasonable to model these with a Poisson distribution?

10.6.8 Random-Number Generation

A random-number generator is supposed to provide numbers that are uniformly distributed between 0 and 1. A total of $n = 10,000$ simulations are obtained, and they are classified as falling into one of ten intervals of length 0.1, as given in DS 10.6.3. Is there any evidence that the random-number generator is not operating correctly?

10.6.9 Metal Plate Thickness Variations

DS 10.6.4 gives a data set on the thicknesses of a set of $n = 200$ metal plates measured to an accuracy of 0.01 mm. Does the data set suggest that it is reasonable to model these thicknesses with a normal distribution? (The average plate thickness is $\bar{x} = 4.30$ mm and the sample standard deviation is $s = 0.12$ mm.)

10.6.10 Tire Sales

A garage sells tires of types A, B, and C, and this year's and last year's sales of the three types are given in DS 10.6.5. Is there evidence of a change in the preferences for the three types of tire between the two years?

10.6.11 Clinical Trial

DS 10.6.6 gives the results of a set of clinical trials involving three different medications. Test whether the three medications are equally effective at treating the infection.

10.6.12 Student Opinion Polls

DS 10.6.7 presents the results of an opinion poll conducted on students in the College of Engineering and students in the College of Arts and Sciences. Is there any evidence that opinions differ between students in these two colleges?

10.6.13 The dimensions of 3877 manufactured parts were examined and 445 were found to have a length outside a specified tolerance range.

   (a) Conduct a hypothesis test to investigate whether there is sufficient evidence to conclude that the probability of a part having a length outside the tolerance range is larger than 10%.

   (b) Construct a one-sided 99% confidence interval that provides a lower bound on the probability of a part having a length outside the tolerance range.

It was also found that out of the 445 parts that had a length outside the specified tolerance range, 25 also had a width outside a specified tolerance range. Furthermore, out of the 3432 parts that had a length inside the specified tolerance range, it was found that 161 had a width outside the specified tolerance range.

   (c) Use the Pearson chi-square statistic to test whether the acceptability of the length and the acceptability of the width of the parts are related to each other, or whether these two factors can be considered to be independent of each other.

10.6.14 Composite Material Properties

In a research report on the effect of moisture on a certain kind of composite material, it is reported that in 80% of cases the effect is minimal, in 15% of cases the effect is strong, and in the remaining 5% of cases the effect is severe. An experimenter tested these claims by subjecting 800 samples of the composite material to moisture, and the results are given in DS 10.6.8.

   (a) Perform a chi-square goodness of fit test to examine whether these experimental results are consistent with the claims made by the research report.

   (b) Use the experimental results to construct a 99% one-sided confidence interval that provides an upper bound on the probability of a severe moisture effect.

10.6.15 Chemical Preparation Methods

An experiment is performed to investigate the best way to produce a chemical solution. Three different preparation methods are considered, and various trials are conducted with each method. The resulting solutions are classified as being either too weak, satisfactory, or too strong, as given in DS 10.6.9. Perform a chi-square goodness of fit test to examine whether there is any evidence of a difference between the three preparation methods in terms of the quality of chemical solutions that they produce.

10.6.16 Metal Alloy Comparisons

Three types of a metal alloy were investigated to see how much damage they suffered when subjected to a high temperature. A total of 220 samples were obtained for each alloy, and after being subjected to the high temperature, the damage was classified as "none," "slight," "medium," or "severe." Out of the 220 samples of type I, 98 had no damage, 32 had slight damage, 48 had medium damage, and 42 had severe damage. Out of the 220 samples of type II, 52 had no damage, 27 had slight damage, 67 had medium damage, and 74 had severe damage. Out of the 220 samples of type III, 112 had no damage, 35 had slight damage, 41 had medium damage, and 32 had severe damage.

(a) Perform a hypothesis test to assess whether there is sufficient evidence to conclude that the three alloys are not all equivalent in terms of the damage that they suffer.

(b) Perform a hypothesis test to assess whether there is sufficient evidence to conclude that the probability of suffering severe damage is different for alloy type I and alloy type III.

(c) Construct a 99% confidence level two-sided confidence interval for the probability that a sample of alloy type II will not suffer any damage.

10.6.17 Company Sales Data

A company's orders are classified as coming from geographical area A, B, C, or D. Over a certain period, there were 119 orders from area A, 54 orders from area B, 367 orders from area C, and 115 orders from area D.

(a) It is claimed that the probability that an order is from area A is 0.25, the probability that an order is from area B is 0.10, the probability that an order is from area C is 0.40, and the probability that an order is from area D is 0.25. Are the data consistent with this claim?

(b) Construct a two-sided 99% confidence interval for the probability that an order is received from area C.

10.6.18 Concrete Breaking Loads

An experimenter obtained 84 samples of concrete. When the samples were each subjected to a load of size 115, a total of 17 of the samples broke while the other samples were unharmed. The remaining 67 samples were then subjected to a load of size 120, and 32 of them broke. Finally, the remaining 35 samples were subjected to a load of size 125, and 21 of them broke. This left 14 samples that survived the highest load. It is claimed that the breaking load of samples of this type of concrete is

normally distributed with a mean 120 and a standard deviation 4. Are the results of this experiment consistent with that claim?

10.6.19 Student Opinion Poll

When a random sample of 64 male students was asked their opinion on a proposal, 28 of them expressed support. Also, when a random sample of 85 female students was asked their opinion on the proposal, 31 of them expressed support.

(a) Use an appropriate hypothesis test to assess whether there is sufficient evidence to conclude that the support for the proposal is different for men and women.

(b) Construct a 99% two-sided confidence interval that illustrates the difference in support between men and women.

10.6.20 Clinical Trial

Patients were diagnosed as being either condition A or condition B before undergoing a treatment. The treatment was successful for 56 out of 94 patients classified as condition A, and the treatment was successful for 64 out of 153 patients classified as condition B.

(a) Perform a hypothesis test to assess whether there is sufficient evidence to conclude that the chance of success for patients with condition A is better than 50%.

(b) Construct a two-sided 99% confidence interval for the difference between the success probabilities for patients with condition A and with condition B.

(c) Perform a chi-square goodness of fit test to investigate whether there is sufficient evidence to conclude that the success probabilities are different for patients with condition A and with condition B. What is your conclusion?

10.6.21 Are the following statements true or false?

(a) Contingency tables can be used to summarize count frequencies for discrete data.

(b) The degrees of freedom used in a chi-square goodness of fit test are related to the number of cells being examined.

(c) Independence in a two-way contingency table implies that for each factor the different levels have equal probabilities.

(d) In a two-way contingency table analysis, the null hypothesis of independence is a more complicated model for the data than the alternative hypothesis.

(e) In a goodness of fit test an extremely high p-value suggests the possibility that the experimenter cheated and made up the data.

(f) For comparing two probabilities, either the methods of Section 10.2 or the methods of Section 10.4 can be used.

(g) Discrete data analysis can be referred to as categorical data analysis.

(h) A one-sided confidence interval that provides a lower bound on p can be used to obtain a one-sided confidence interval that provides an upper bound on $1 - p$.

(i) The margin of error in a political poll has a confidence level tacitly associated with it.

(j) The likelihood ratio chi-square statistic always provides values larger than the Pearson chi-square statistic, although the values are generally very close together.

10.6.22 Customer Satisfaction Surveys

In a customer satisfaction survey a random sample of 635 customers were asked their opinion on the service they received. A total of 485 of these customers replied that they were very satisfied.

(a) Construct a two-sided 95% confidence interval for the proportion of customers that are very satisfied.

(b) Is it safe to conclude that overall at least 75% of the customers are very satisfied? Use an appropriate hypothesis test.

10.6.23 Hospital Admission Rates

The records at the emergency rooms of five hospitals over a certain period of time were examined. Each patient was classified as either being "admitted to the hospital" or as "returned home," as given in DS 10.6.10.

(a) Is there any evidence to support the claim that the hospital admission rates differ between the five hospitals?

(b) Consider hospitals 3 and 4. Calculate a 95% two-sided confidence interval for the difference between the admission rates of these two hospitals.

(c) Consider hospital 1. Is there sufficient evidence to conclude that the admission rate for this hospital is larger than 10%?

10.6.24 Scouring Around Bridge Piers

After large floods and the consequent large river flows, there can be a reduction in the level of the riverbed around the piers of bridges that is known as scouring. This can be a serious problem if the foundations of the piers become exposed. The amount of scouring can depend on the shape of the pier and the consequent flows and vortexes that are generated. After a large flood, the data set in DS 10.6.11 was collected concerning the severity of the scouring around piers of different designs.

(a) Use a goodness of fit test to examine whether there is sufficient evidence to conclude that the pier design has any effect on the amount of scouring.

(b) Consider pier design 1. Are the data consistent with the contention that for this design the three levels of scouring are all equally likely?

(c) Show how to perform a two-sided hypothesis test of whether for pier design 3 the probability of a minimal scour depth is 25%.

(d) Let p_{1s} be the probability of severe scouring when pier design 1 is used, and let p_{2s} be the probability of severe scouring when pier design 2 is used. Construct a 99% two-sided confidence interval for $p_{1s} - p_{2s}$.

CHAPTER ELEVEN

The Analysis of Variance

Extensions to the methods presented in Chapter 9 for comparing two population means are made in this chapter, where the problem of comparing a set of three or more population means is considered. The basic ideas behind the statistical analysis are the same. The objective is to ascertain whether there is any evidence that the population means are unequal, and if there is evidence, to then ascertain which population means can be shown to be different and by how much.

In Chapter 9 a distinction was made between paired samples and independent samples. A similar distinction is appropriate in this chapter. A set of independent samples from a set of several populations is known as a **completely randomized design** and is analyzed with the statistical methodology known as the **analysis of variance,** or ANOVA for short. This topic is discussed in the first part of this chapter.

With three or more populations under consideration, the concept of pairing observations is known as **blocking**, which is a very important procedure for improving experimental designs so that they allow more sensitive statistical analyses. Experimental designs for comparing a set of several population means that incorporate blocking are known as **randomized block designs** and are discussed in the second part of this chapter.

11.1 One-Factor Analysis of Variance

11.1.1 One-Factor Layouts

Suppose that an experimenter is interested in k populations with unknown population means

$$\mu_1, \mu_2, \ldots, \mu_k$$

If only one population is of interest, $k = 1$, then the one sample inferences presented in Chapter 8 are applicable, and if $k = 2$, then the two sample comparisons discussed in Chapter 9 are appropriate. The one-factor analysis of variance methodology that is discussed in this section is appropriate for comparing three or more populations, that is, $k \geq 3$.

The experimenter's objective is to make inferences on the k unknown population means μ_i based upon a data set consisting of samples from each of the k populations, as illustrated in Figure 11.1. In this data set the observation

$$x_{ij}$$

represents the jth observation from the ith population. The sample from population i therefore consists of the n_i observations

$$x_{i1}, \ldots, x_{in_i}$$

If the sample sizes $n_1, \ldots, n_k$ are all equal, then the data set is referred to as being **balanced**, and if the sample sizes are unequal, then the data set is referred to as being **unbalanced**. The total sample size of the data set is

$$n_T = n_1 + \cdots + n_k$$

FIGURE 11.1

One factor layout

Sample from population 1 (factor level 1) $\cdots$ Sample from population i (factor level i) $\cdots$ Sample from population k (factor level k)

x_{11}	x_{i1}	x_{k1}
x_{12}	x_{i2}	x_{k2}
$\vdots$	$\vdots$	$\vdots$
x_{1n_1}	x_{in_i}	x_{kn_k}

Sample size n_1 Sample size n_i Sample size n_k

A data set of this kind is called a one-way or *one-factor* layout. The single factor is said to have k *levels* corresponding to the k populations under consideration. As in all experimental designs, care should be taken to ensure the integrity of the data set, so that, for example, there are no unseen biases between the k samples that may compromise the comparisons of the population means. Usually bias can be avoided by appropriate random sampling. If the experiment is performed by allocating a total of n_T "units" among the k populations, then it is appropriate to make this allocation in a random manner. With this in mind, one-factor layouts such as this are often referred to as *completely randomized designs*.

The analysis of variance is based upon the modeling assumption

$$x_{ij} = \mu_i + \epsilon_{ij}$$

where the **error terms** ϵ_{ij} are independently distributed as

$$\epsilon_{ij} \sim N(0, \sigma^2)$$

$$x_{1j} \sim N(\mu_1, \sigma^2) \qquad 1 \leq j \leq n_1$$

$$\hat{\mu}_1 = \bar{x}_{1.} = \frac{x_{11} + \cdots + x_{1n_1}}{n_1}$$

$$\vdots$$

$$x_{ij} \sim N(\mu_i, \sigma^2) \qquad 1 \leq j \leq n_i$$

$$\hat{\mu}_i = \bar{x}_{i.} = \frac{x_{i1} + \cdots + x_{in_i}}{n_i}$$

$$\vdots$$

$$x_{kj} \sim N(\mu_k, \sigma^2) \qquad 1 \leq j \leq n_k$$

$$\hat{\mu}_k = \bar{x}_{k.} = \frac{x_{k1} + \cdots + x_{kn_k}}{n_k}$$

FIGURE 11.2

Estimating the population (factor level) means

Thus, the observation x_{ij} consists of the fixed unknown population mean μ_i together with a random error term ϵ_{ij}, which is normally distributed with a mean of 0 and a variance of σ^2. Equivalently, x_{ij} can be thought of as just an observation from a

$$N(\mu_i, \sigma^2)$$

distribution. Notice that the unknown error variance σ^2 is taken to be the same in each of the k populations. The analysis of variance is therefore analogous to the *pooled variance* procedure for comparing two populations discussed in Section 9.3.2. A discussion of the importance and possible relaxation of these modeling assumptions is provided in Section 11.1.6.

Point estimates of the unknown population means μ_i are obtained in the obvious manner as the k sample averages, so that as Figure 11.2 shows,

$$\hat{\mu}_i = \bar{x}_{i.} = \frac{x_{i1} + \cdots + x_{in_i}}{n_i} \qquad 1 \leq i \leq k$$

An assessment of the evidence that there is a difference between some of the population means can be made by testing the null hypothesis

$$H_0 : \mu_1 = \cdots = \mu_k$$

which states that the population means are all equal. The alternative hypothesis states that at least two of the population means are not equal, and may be written

$$H_A : \mu_i \neq \mu_j \qquad \text{for some } i \text{ and } j$$

Acceptance of the null hypothesis indicates that there is no evidence that any of the population means are unequal. Rejection of the null hypothesis implies that there is evidence that at least some of the population means are unequal, and therefore that it is not plausible to assume that the population means are all equal.

Example 62

Collapse of Blocked Arteries

Atherosclerosis, or "hardening of the arteries," is a major contributor to ill health. It progresses by the accumulation of plaque on the interior wall of arteries that obstructs the blood flow through the arteries. This obstruction causes pressure differences within the blood flow, and eventually the artery collapses with possibly fatal consequences.

A mechanical engineer performs an experiment to investigate the collapsability of arteries due to the presence of stenosis (blocking). A silicone model of an artery is built and part of it is constricted as shown in Figure 11.3. The amount of stenosis is measured as

$$\text{stenosis} = 1 - \frac{d}{d_0}$$

and the engineer considers three levels

level 1: stenosis = 0.78

level 2: stenosis = 0.71

level 3: stenosis = 0.65

At each stenosis level, the engineer pumps a bloodlike liquid through the model artery and measures the flowrate at which the tube collapses. Various repetitions of the experiment are performed, and the resulting data set is shown in Figure 11.4.

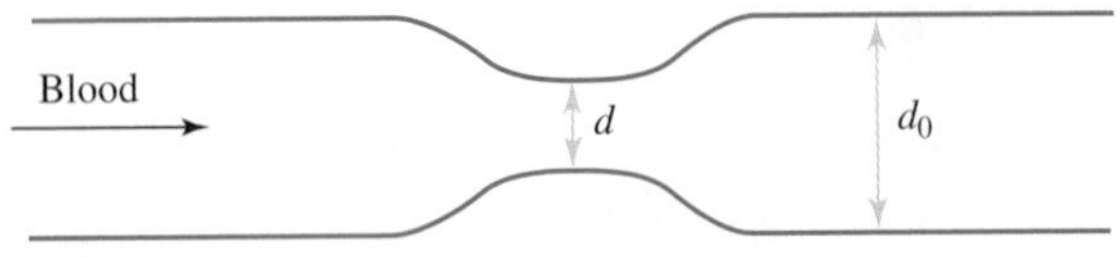

	Amount of Stenosis		
	Level 1	**Level 2**	**Level 3**
Flowrates in ml/s when artery model collapses	10.6	11.7	19.6
	9.7	12.7	15.1
	10.2	12.6	15.6
	14.3	16.6	19.2
	10.6	16.3	18.7
	11.0	14.7	13.0
	13.1	19.5	18.9
	10.8	16.8	18.0
	14.3	15.1	18.6
	10.4	15.2	16.6
	8.3	13.2	
		14.8	
		14.4	
		17.6	

Blood → d d_0

$$\text{Stenosis} = 1 - \frac{d}{d_0}$$

FIGURE 11.3

Stenosis in silicone model of an artery

$$\hat{\mu}_1 = \bar{x}_{1\cdot} = \frac{10.6 + \cdots + 8.3}{11} = 11.209$$

$$\hat{\mu}_2 = \bar{x}_{2\cdot} = \frac{11.7 + \cdots + 17.6}{14} = 15.086$$

$$\hat{\mu}_3 = \bar{x}_{3\cdot} = \frac{19.6 + \cdots + 16.6}{10} = 17.330$$

FIGURE 11.4

Data set for collapse of blocked arteries experiment

Notice that the factor under consideration is the amount of stenosis, which has $k = 3$ levels. Also, the sample sizes $n_1 = 11$, $n_2 = 14$, and $n_3 = 10$ are unequal, so that the data set is unbalanced. The unknown population mean μ_1 is the average flowrate at collapse for arteries with stenosis level 1, and similarly μ_2 and μ_3 are the average flowrates at collapse for arteries with stenosis levels 2 and 3.

As Figure 11.4 shows, these population means are estimated as

$$\hat{\mu}_1 = \bar{x}_{1.} = 11.209$$
$$\hat{\mu}_2 = \bar{x}_{2.} = 15.086$$
$$\hat{\mu}_3 = \bar{x}_{3.} = 17.330$$

Consequently, the data suggest that the average flowrate at collapse decreases as the amount of stenosis increases. Is this a statistically significant result? This question can be examined by testing the null hypothesis

$$H_0 : \mu_1 = \mu_2 = \mu_3$$

Rejection of this null hypothesis will allow the engineer to conclude that the average flowrate at collapse does depend upon the amount of stenosis.

Example 63

Roadway Base Aggregates

Immediately below the asphalt surface of a roadway is a layer of base material composed of a crushed stone or gravel aggregate. The resilient modulus of this aggregate is a measure of how the aggregate deforms when subjected to stress, and it is an important property affecting the manner in which the roadway responds to loads.

A construction engineer has four different suppliers of this aggregate material who obtain their raw materials from four different locations. The engineer would like to assess whether the aggregates from the four different locations have different values of resilient modulus. An experiment is performed by randomly selecting 10 samples of aggregate from each of the four locations and measuring their resilient modulus. The resulting data set is given in Figure 11.5.

FIGURE 11.5

Data set for roadway base aggregates experiment

Supplier

	Location 1	Location 2	Location 3	Location 4
Resilient modulus measurements	30,060	31,280	29,950	30,430
	28,740	30,380	29,190	28,120
	29,140	30,620	31,870	30,310
	29,090	29,650	30,010	31,650
	30,220	28,080	28,490	29,770
	31,120	27,920	31,600	33,100
	31,360	27,420	29,450	28,680
	28,300	28,860	32,890	31,730
	29,750	29,850	31,170	31,480
	33,350	30,550	29,470	32,960

$$\hat{\mu}_1 = \bar{x}_{1.} = \frac{30{,}060 + \cdots + 33{,}350}{10} = 30{,}113$$

$$\hat{\mu}_2 = \bar{x}_{2.} = \frac{31{,}280 + \cdots + 30{,}550}{10} = 29{,}461$$

$$\hat{\mu}_3 = \bar{x}_{3.} = \frac{29{,}950 + \cdots + 29{,}470}{10} = 30{,}409$$

$$\hat{\mu}_4 = \bar{x}_{4.} = \frac{30{,}430 + \cdots + 32{,}960}{10} = 30{,}823$$

The factor of interest here is the location from which the aggregate material is taken, and it has $k = 4$ levels. The data set is balanced with

$$n_1 = n_2 = n_3 = n_4 = 10$$

The unknown population means μ_1, μ_2, μ_3, and μ_4 represent the average resilient modulus values for aggregate taken from each of the four locations. The sample means are

$$\hat{\mu}_1 = \bar{x}_{1.} = 30,113$$
$$\hat{\mu}_2 = \bar{x}_{2.} = 29,461$$
$$\hat{\mu}_3 = \bar{x}_{3.} = 30,409$$
$$\hat{\mu}_4 = \bar{x}_{4.} = 30,823$$

The null hypothesis

$$H_0 : \mu_1 = \mu_2 = \mu_3 = \mu_4$$

states that the average resilient modulus values are the same for all four locations. If the null hypothesis is accepted, then the construction engineer has no reason to worry about which aggregate supplier is used since there is no evidence of a difference in their products with respect to average resilient modulus. However, if the null hypothesis is rejected, then the engineer may want to consider more carefully which supplier to use since the products differ from one supplier to another.

Example 64
Carpet Fiber Blends

A textile engineer performs an experiment to investigate which of $k = 6$ different carpet fiber blends produce the best quality carpet. The six different fiber blends considered are

type 1: nylon fiber type A

type 2: nylon fiber type B

type 3: nylon-acrylic fiber blend

type 4: polyester fiber blend type A

type 5: polyester fiber blend type B

type 6: polypropylene fiber

Various samples of carpets made with fibers of these six types are subjected to typical amounts of stress. After a suitable amount of time the worn carpets are taken to be examined by a set of experts who score the carpets on the amount of wear observed and on the general quality of the carpets.

The scores on each sample of carpet are averaged out to produce the data set shown in Figure 11.6. Higher scores indicate carpets that are in a better condition. The factor of interest here is the type of fiber employed in the carpet and it has $k = 6$ levels. The design is seen to be unbalanced, with the number of samples of carpet of each type ranging from 13 to 16.

The unknown population mean μ_i represents the average amount of wear for carpets made with fiber type i. These population means are estimated as

$$\hat{\mu}_1 = \bar{x}_{1.} = 5.26$$
$$\hat{\mu}_2 = \bar{x}_{2.} = 7.66$$
$$\hat{\mu}_3 = \bar{x}_{3.} = 5.10$$
$$\hat{\mu}_4 = \bar{x}_{4.} = 6.41$$
$$\hat{\mu}_5 = \bar{x}_{5.} = 4.80$$
$$\hat{\mu}_6 = \bar{x}_{6.} = 6.86$$

FIGURE 11.6

Data set for carpet fiber blends experiment

Carpet Fiber Blends

	Type 1	Type 2	Type 3	Type 4	Type 5	Type 6
Carpet scores	6.5	9.8	5.3	5.1	6.2	6.3
	5.4	7.6	4.6	6.7	4.8	7.5
	5.0	7.0	4.4	6.5	5.5	5.4
	5.4	8.9	5.9	6.5	5.1	7.2
	4.3	6.9	5.1	8.3	4.9	5.4
	4.6	7.7	5.8	4.9	4.8	7.9
	5.5	6.9	5.5	6.0	3.9	5.6
	4.0	8.3	5.2	5.5	4.7	6.5
	5.1	6.3	4.5	4.7	4.4	6.1
	7.0	9.7	6.5	6.7	5.2	9.2
	3.7	7.9	4.7	5.5	4.2	6.7
	5.9	8.0	4.9	7.5	4.5	7.8
	4.4	6.9	3.9	5.8	5.4	8.3
	5.0	5.5		7.6	3.6	6.8
	7.0	7.9		8.8		6.2
	5.3	7.3		6.5		

$$\hat{\mu}_1 = \bar{x}_{1.} = \frac{6.5 + \cdots + 5.3}{16} = 5.26$$

$$\hat{\mu}_2 = \bar{x}_{2.} = \frac{9.8 + \cdots + 7.3}{16} = 7.66$$

$$\hat{\mu}_3 = \bar{x}_{3.} = \frac{5.3 + \cdots + 3.9}{13} = 5.10$$

$$\hat{\mu}_4 = \bar{x}_{4.} = \frac{5.1 + \cdots + 6.5}{16} = 6.41$$

$$\hat{\mu}_5 = \bar{x}_{5.} = \frac{6.2 + \cdots + 3.6}{14} = 4.80$$

$$\hat{\mu}_6 = \bar{x}_{6.} = \frac{6.3 + \cdots + 6.2}{15} = 6.86$$

Is there any evidence that some fiber blends wear better than others? The null hypothesis

$$H_0 : \mu_1 = \mu_2 = \mu_3 = \mu_4 = \mu_5 = \mu_6$$

states that the average amount of wear is the same for all six fiber types. The rejection of this null hypothesis will allow the textile engineer to conclude that some fiber blends wear better than others.

11.1.2 Partitioning the Total Sum of Squares

Treatment Sum of Squares The assessment of the plausibility of the null hypothesis that the population means μ_i are all equal is based upon the variability among the population mean estimates $\hat{\mu}_i = \bar{x}_{i.}$. If the average of all of the data observations is denoted by

$$\bar{x}_{..} = \frac{n_1 \bar{x}_{1.} + \cdots + n_k \bar{x}_{k.}}{n_T} = \frac{x_{11} + \cdots + x_{kn_k}}{n_T}$$

then this variability can be measured by the statistic

$$\text{SSTr} = \sum_{i=1}^{k} n_i (\bar{x}_{i.} - \bar{x}_{..})^2$$

which is known as the **sum of squares for treatments**. The treatments are considered to be the different levels of the factor under consideration, so that this one factor layout with k levels can be thought of as a comparison of k treatments.

Notice that the sum of squares for treatments SSTr is formed as a weighted sum of the squared differences between the population mean estimates $\hat{\mu}_i = \bar{x}_{i.}$ and the overall mean $\bar{x}_{..}$. Generally speaking, as the variability among the population mean estimates $\hat{\mu}_i = \bar{x}_{i.}$ increases, the sum of squares for treatments SSTr also increases. Thus, the sum of squares for treatments SSTr should be thought of as a summary measure of the variability *between* the factor levels or treatments.

> *The sum of squares for treatments* SSTr *is a measure of the variability* between *the factor levels.*

It is useful to know that expanding the square in the expression for the sum of squares for treatments SSTr results in the alternative expression

$$\text{SSTr} = \sum_{i=1}^{k} n_i (\bar{x}_{i.} - \bar{x}_{..})^2 = \sum_{i=1}^{k} n_i \bar{x}_{i.}^2 - 2 \sum_{i=1}^{k} n_i \bar{x}_{i.} \bar{x}_{..} + \sum_{i=1}^{k} n_i \bar{x}_{..}^2$$

$$= \sum_{i=1}^{k} n_i \bar{x}_{i.}^2 - 2 n_T \bar{x}_{..}^2 + n_T \bar{x}_{..}^2 = \sum_{i=1}^{k} n_i \bar{x}_{i.}^2 - n_T \bar{x}_{..}^2$$

Error Sum of Squares   Another important consideration in the analysis of a one-factor layout is the amount of variability *within* the k factor levels. This variability can be attributed to the variance σ^2 of the error terms ϵ_{ij} and can be measured with the statistic

$$\text{SSE} = \sum_{i=1}^{k} \sum_{j=1}^{n_i} (x_{ij} - \bar{x}_{i.})^2$$

which is known as the **sum of squares for error**. Notice that within this expression the component

$$\sum_{j=1}^{n_i} (x_{ij} - \bar{x}_{i.})^2$$

measures the variability of the observations x_{ij} at the ith level of the factor about their own average $\bar{x}_{i.}$. The sum of squares for error SSE is then formed by adding up these components from each of the k levels of the factor.

It is important to notice that the sum of squares for error SSE measures only the variability *within* the k levels or treatments, and not between the k levels or treatments.

> *The sum of squares for error* SSE *is a measure of the variability* within *the factor levels.*

If the observations at the ith factor level

$$x_{i1}, \ldots, x_{in_i}$$

are replaced by the values

$$x_{i1} + c, \ldots, x_{in_i} + c$$

so that they are all shifted by an amount c, then the sum of squares for error SSE remains *unchanged* since the variability within the ith factor level is unchanged. However, the sample mean for the ith factor level changes so that the sum of squares for treatments SSTr will change.

Again, it is useful to know that when the square term is expanded, the sum of squares for error SSE can also be written as

$$
\text{SSE} = \sum_{i=1}^{k} \sum_{j=1}^{n_i} (x_{ij} - \bar{x}_{i\cdot})^2 = \sum_{i=1}^{k} \sum_{j=1}^{n_i} x_{ij}^2 - 2 \sum_{i=1}^{k} \sum_{j=1}^{n_i} x_{ij} \bar{x}_{i\cdot} + \sum_{i=1}^{k} \sum_{j=1}^{n_i} \bar{x}_{i\cdot}^2
$$

$$
= \sum_{i=1}^{k} \sum_{j=1}^{n_i} x_{ij}^2 - 2 \sum_{i=1}^{k} n_i \bar{x}_{i\cdot}^2 + \sum_{i=1}^{k} n_i \bar{x}_{i\cdot}^2 = \sum_{i=1}^{k} \sum_{j=1}^{n_i} x_{ij}^2 - \sum_{i=1}^{k} n_i \bar{x}_{i\cdot}^2
$$

Total Sum of Squares   If the k factor levels are ignored so that the data set is thought of as one big sample of size n_T, then the total variability within this sample can be measured by

$$
\text{SST} = \sum_{i=1}^{k} \sum_{j=1}^{n_i} (x_{ij} - \bar{x}_{..})^2
$$

which is known as the **total sum of squares**. SST is composed of the sum of the squared deviations of each observation x_{ij} about the overall sample average $\bar{x}_{..}$.

The total sum of squares SST *is a measure of the* total *variability in the data set.*

Notice that when the square term is expanded, the total sum of squares can be written

$$
\text{SST} = \sum_{i=1}^{k} \sum_{j=1}^{n_i} (x_{ij} - \bar{x}_{..})^2 = \sum_{i=1}^{k} \sum_{j=1}^{n_i} x_{ij}^2 - 2 \sum_{i=1}^{k} \sum_{j=1}^{n_i} x_{ij} \bar{x}_{..} + \sum_{i=1}^{k} \sum_{j=1}^{n_i} \bar{x}_{..}^2
$$

$$
= \sum_{i=1}^{k} \sum_{j=1}^{n_i} x_{ij}^2 - 2 n_T \bar{x}_{..}^2 + n_T \bar{x}_{..}^2 = \sum_{i=1}^{k} \sum_{j=1}^{n_i} x_{ij}^2 - n_T \bar{x}_{..}^2
$$

With the alternative expressions for SSTr, SSE, and SST, it can be easily seen that

$$
\text{SST} = \text{SSTr} + \text{SSE}
$$

In summary, it has been shown that the total variability in the data observations in a one-factor layout can be measured by the total sum of squares SST, which can be *partitioned* into two components, as illustrated in Figure 11.7. The first component, the sum of squares for treatments SSTr, measures the variability between the factor levels, while the second component, the sum of squares for error SSE, measures the variability within the factor levels.

FIGURE 11.7

Partition of total sum of squares for completely randomized one-factor layout

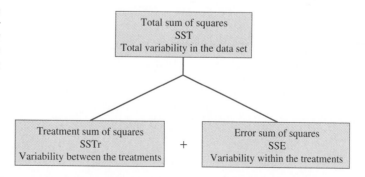

P-Value Considerations   The plausibility of the null hypothesis

$$H_0 : \mu_1 = \cdots = \mu_k$$

depends upon the relative sizes of the sum of squares for treatments SSTr and the sum of squares for error SSE.

> *The plausibility of the null hypothesis that the factor level means are all equal depends upon the* relative size *of the sum of squares for treatments* SSTr *to the sum of squares for error* SSE.

The mechanics of the calculation of a *p*-value for the null hypothesis are presented in an analysis of variance table, which is discussed in Section 11.1.3. Nevertheless, it is instructive to understand at an intuitive level how changes in the sum of squares for treatments SSTr and the sum of squares for error SSE affect the *p*-value.

A useful graphical representation of a one-factor layout is a series of boxplots, one for each factor level, drawn against the same axis. Figure 11.8 shows such a representation for a data set with $k = 3$ factor levels. The sample means $\bar{x}_{i\cdot}$ occur roughly at the center of each of the boxplots, and so location differences between the boxplots correspond to differences between the sample means $\bar{x}_{i\cdot}$. The sum of squares for treatments SSTr is therefore a measure of the variability in the locations of the different boxplots. On the other hand, the lengths of the boxplots provide an indication of the variability within the different factor levels, and so the sum of squares for error SSE depends upon the lengths of the boxplots, as illustrated in Figure 11.8.

Figure 11.9 shows how the *p*-value depends upon changes in the sum of squares for treatments SSTr and the sum of squares for error SSE. A base scenario is altered in one of four ways that correspond to increases or decreases in one of the sums of squares. In scenario I, the boxplots have the same lengths but they are farther apart. This corresponds to an increase

FIGURE 11.8

Interpretation of the sum of squares for treatments and the sum of squares for error

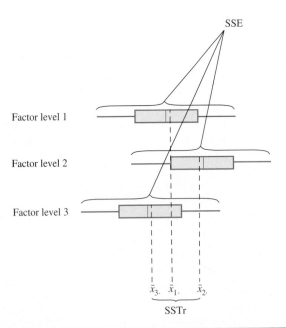

FIGURE 11.9

Dependence of p-value on the sum of squares for treatments and the sum of squares for error

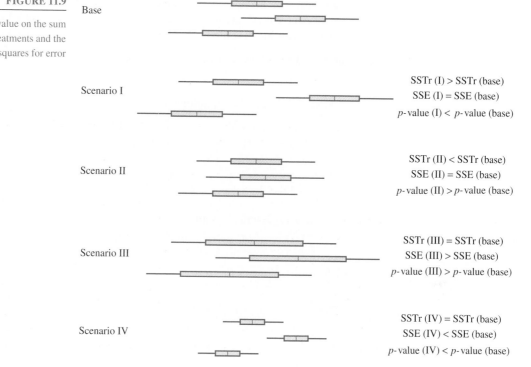

in the sum of squares for treatments SSTr, whereas the sum of squares for error SSE remains constant. The null hypothesis that the factor level means μ_i are all equal is less plausible in scenario I than in the base scenario, and so

$$p\text{-value (I)} < p\text{-value (base)}$$

In scenario II, the boxplots have the same lengths, but they are closer together than in the base scenario. This corresponds to a decrease in the sum of squares for treatments SSTr, whereas the sum of squares for error SSE remains constant. The null hypothesis that the factor level means μ_i are all equal is more plausible in scenario II than it is in the base scenario, and so

$$p\text{-value (II)} > p\text{-value (base)}$$

In scenario III, the boxplots have roughly the same central locations as in the base scenario, so that the sample means $\bar{x}_{i\cdot}$ are roughly the same, but the boxplots are longer than in the base scenario. This corresponds to an increase in the sum of squares for error SSE, while the sum of squares for treatments SSTr remains constant. This results in a decrease in the relative size of the sum of squares for treatments SSTr to the sum of squares for error SSE. Consequently, the null hypothesis that the factor level means μ_i are all equal is more plausible in scenario III than it is in the base scenario, so that

$$p\text{-value (III)} > p\text{-value (base)}$$

In scenario IV, the boxplots have roughly the same central locations as in the base scenario, but the boxplots are shorter than in the base scenario. This corresponds to a decrease in the sum of squares for error SSE, while the sum of squares for treatments SSTr remains constant.

There is an increase in the relative size of the sum of squares for treatments SSTr to the sum of squares for error SSE. Consequently, the null hypothesis that the factor level means μ_i are all equal is less plausible in scenario IV than it is in the base scenario, so that

$$p\text{-value (IV)} < p\text{-value (base)}$$

Sum of Squares Partition for One-Factor Layout

In a one factor layout the total variability in the data observations is measured by the total sum of squares SST, which is defined to be

$$\text{SST} = \sum_{i=1}^{k} \sum_{j=1}^{n_i} (x_{ij} - \bar{x}_{..})^2 = \sum_{i=1}^{k} \sum_{j=1}^{n_i} x_{ij}^2 - n_T \bar{x}_{..}^2$$

This value can be *partitioned* into two components

$$\text{SST} = \text{SSTr} + \text{SSE}$$

where the sum of squares for treatments

$$\text{SSTr} = \sum_{i=1}^{k} n_i (\bar{x}_{i.} - \bar{x}_{..})^2 = \sum_{i=1}^{k} n_i \bar{x}_{i.}^2 - n_T \bar{x}_{..}^2$$

measures the variability *between* the factor levels, and the sum of squares for error

$$\text{SSE} = \sum_{i=1}^{k} \sum_{j=1}^{n_i} (x_{ij} - \bar{x}_{i.})^2 = \sum_{i=1}^{k} \sum_{j=1}^{n_i} x_{ij}^2 - \sum_{i=1}^{k} n_i \bar{x}_{i.}^2$$

measures the variability *within* the factor levels. On an intuitive level, the plausibility of the null hypothesis that the factor level means μ_i are all equal depends upon the relative size of the sum of squares for treatments SSTr to the sum of squares for error SSE.

Example 62

Collapse of Blocked Arteries

Figure 11.10 shows boxplots of the flowrates at collapse for the three different stenosis levels. The boxplots provide a clear graphical presentation of the apparent decrease in the flowrates at collapse for the larger stenosis values.

FIGURE 11.10

Boxplots for collapse of blocked arteries data set

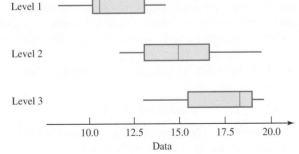

Hand calculations of the sums of squares are usually most easily performed using the alternative expressions. With $\bar{x}_{1.} = 11.209$, $\bar{x}_{2.} = 15.086$, and $\bar{x}_{3.} = 17.330$, and with

$$\bar{x}_{..} = \frac{10.6 + \cdots + 16.6}{35} = 14.509$$

and

$$\sum_{i=1}^{k} \sum_{j=1}^{n_i} x_{ij}^2 = 10.6^2 + \cdots + 16.6^2 = 7710.39$$

the total sum of squares is

$$\text{SST} = \sum_{i=1}^{k} \sum_{j=1}^{n_i} x_{ij}^2 - n_T \bar{x}_{..}^2 = 7710.39 - (35 \times 14.509^2) = 342.5$$

and the treatment sum of squares is

$$\begin{aligned}
\text{SSTr} &= \sum_{i=1}^{k} n_i \bar{x}_{i.}^2 - n_T \bar{x}_{..}^2 \\
&= (11 \times 11.209^2) + (14 \times 15.086^2) + (10 \times 17.330^2) - (35 \times 14.509^2) \\
&= 204.0
\end{aligned}$$

The sum of squares for error can then be calculated by subtraction to be

$$\text{SSE} = \text{SST} - \text{SSTr} = 342.5 - 204.0 = 138.5$$

In hand calculations of this kind it is important not to round numbers until the end of the calculation. This is because the sums of squares are generally obtained as the difference between two much larger numbers, and rounding in these numbers can lead to substantial inaccuracies in the resulting sums of squares.

Example 63	Figure 11.11 shows boxplots of the resilient modulus values of the aggregate samples from

Example 63
Roadway Base Aggregates

Figure 11.11 shows boxplots of the resilient modulus values of the aggregate samples from the four locations. A visual inspection of the boxplots reveals that there is not much difference between the four locations, although it looks as if location 2 may possibly have a smaller average resilient modulus value than the other three locations.

With $\bar{x}_{1.} = 30{,}113$, $\bar{x}_{2.} = 29{,}461$, $\bar{x}_{3.} = 30{,}409$, and $\bar{x}_{4.} = 30{,}823$, and with

$$\bar{x}_{..} = \frac{30{,}060 + \cdots + 32{,}960}{40} = 30{,}201.5$$

FIGURE 11.11

Boxplots for roadway base aggregates data set

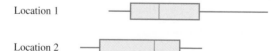
Location 1

Location 2

Location 3

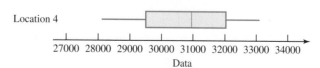
Location 4

27000 28000 29000 30000 31000 32000 33000 34000

Data

and

$$\sum_{i=1}^{k} \sum_{j=1}^{n_i} x_{ij}^2 = 30{,}060^2 + \cdots + 32{,}960^2 = 36{,}573{,}858{,}000$$

the total sum of squares is

$$\mathrm{SST} = \sum_{i=1}^{k} \sum_{j=1}^{n_i} x_{ij}^2 - n_T \bar{x}_{..}^2 = 36{,}573{,}858{,}000 - (40 \times 30{,}201.5^2)$$

$$= 88{,}633{,}910$$

and the treatment sum of squares is

$$\mathrm{SSTr} = \sum_{i=1}^{k} n_i \bar{x}_{i.}^2 - n_T \bar{x}_{..}^2$$

$$= (10 \times 30{,}113^2) + (10 \times 29{,}461^2) + (10 \times 30{,}409^2)$$
$$+ (10 \times 30{,}823^2) - (40 \times 30{,}201.5^2)$$

$$= 9{,}854{,}910$$

As before, the sum of squares for error can then be calculated by subtraction to be

$$\mathrm{SSE} = \mathrm{SST} - \mathrm{SSTr} = 88{,}633{,}910 - 9{,}854{,}910 = 78{,}779{,}000$$

Example 64
Carpet Fiber Blends

Figure 11.12 shows boxplots for the scores of the carpets made with the six types of fiber. There appear to be sizeable differences between the different fiber types, with types 2, 4, and 6 appearing to be better than types 1, 3, and 5.

The sums of squares can be calculated to be

$$\mathrm{SST} = 181.889$$
$$\mathrm{SSTr} = 97.219$$
$$\mathrm{SSE} = 84.670$$

FIGURE 11.12

Boxplots for carpet fiber blends data set

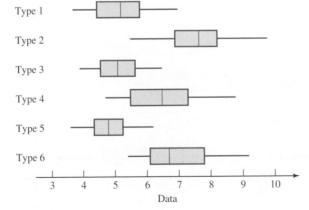

11.1.3 The Analysis of Variance Table

The sum of squares for treatments SSTr is said to have $k - 1$ degrees of freedom and the **mean squares for treatments** is defined to be

$$\text{MSTr} = \frac{\text{SSTr}}{\text{degrees of freedom}} = \frac{\text{SSTr}}{k - 1}$$

The sum of squares for error SSE has $n_T - k$ degrees of freedom (which is actually made up of $n_i - 1$ degrees of freedom from each of the k factor levels), and the **mean square error** is defined to be

$$\text{MSE} = \frac{\text{SSE}}{\text{degrees of freedom}} = \frac{\text{SSE}}{n_T - k}$$

The mean square error MSE is distributed as

$$\text{MSE} \sim \sigma^2 \frac{\chi^2_{n_T - k}}{n_T - k}$$

and since

$$E\left(\chi^2_{n_T - k}\right) = n_T - k$$

it follows that

$$E(\text{MSE}) = \sigma^2$$

so that the mean square error MSE is an unbiased point estimate of the error variance σ^2. Consequently, the mean square error MSE is sometimes written as $\hat{\sigma}^2$.

Notice that the estimate of the variance at the ith factor level is

$$s_i^2 = \frac{\sum_{j=1}^{n_i} (x_{ij} - \bar{x}_{i.})^2}{n_i - 1}$$

so that the sum of squares for error SSE is

$$\text{SSE} = \sum_{i=1}^{k} (n_i - 1)s_i^2$$

Consequently, the mean square error MSE is seen to be

$$\text{MSE} = \hat{\sigma}^2 = \frac{(n_1 - 1)s_1^2 + \cdots + (n_k - 1)s_k^2}{n_T - k}$$

so that it is a weighted average of the variance estimates from within each of the k factor levels.

It can be shown that

$$E(\text{MSTr}) = \sigma^2 + \frac{\sum_{i=1}^{k} n_i (\mu_i - \bar{\mu})^2}{k - 1}$$

where

$$\bar{\mu} = \frac{n_1 \mu_1 + \cdots + n_k \mu_k}{n_T}$$

Thus, the expected value of the mean square for treatments MSTr is equal to σ^2 plus a positive amount that depends upon the variability among the factor level means μ_i. Furthermore, if the factor level means μ_i are all equal (and consequently also all equal to $\bar{\mu}$), then

$$E(\text{MSTr}) = \sigma^2$$

In fact, if the factor level means μ_i are all equal, then it can be shown that the distribution of the mean square for treatments MSTr is

$$\text{MSTr} \sim \sigma^2 \frac{\chi_{k-1}^2}{k-1}$$

These results can be used to develop a method for calculating the p-value of the null hypothesis

$$H_0 : \mu_1 = \cdots = \mu_k$$

When this null hypothesis is true, then the **F-statistic**

$$F = \frac{\text{MSTr}}{\text{MSE}}$$

is distributed as a $\chi_{k-1}^2/(k-1)$ random variable divided by a $\chi_{n_T-k}^2/(n_T-k)$ random variable, which is an F_{k-1,n_T-k} random variable. In other words, when the null hypothesis is true, then

$$F = \frac{\text{MSTr}}{\text{MSE}} \sim F_{k-1,n_T-k}$$

The plausibility of the null hypothesis is therefore in doubt whenever the observed value of the F-statistic does not look like it is an observation from an F_{k-1,n_T-k} distribution.

Moreover, the mean square for treatments MSTr has a larger expected value when the factor level means μ_i are unequal than under the null hypothesis, and so in particular, it is large values of the F-statistic that cast doubt on the plausibility of the null hypothesis. The p-value is defined to be the probability of obtaining the observed data values or worse when the null hypothesis is true, and worse data sets are therefore data sets that give larger values of the F-statistic than the value observed. Consequently, the p-value is calculated as

$$p\text{-value} = P(X \geq F)$$

where the random variable X has an F_{k-1,n_T-k} distribution, as illustrated in Figure 11.13.

This one factor analysis of variance is presented in an **analysis of variance table** as shown in Figure 11.14. The table has three rows corresponding to the treatments (the factor levels), the errors, and the total of these two rows. The first column lists the degrees of freedom, the second column lists the sums of squares, and the third column lists the mean squares. Notice that the degrees of freedom in the "Total" row are $n_T - 1$, the sum of the degrees of freedom

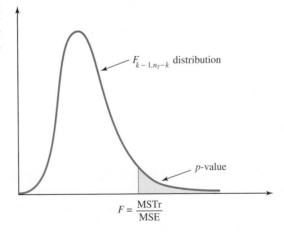

Source	Degrees of freedom	Sum of squares	Mean squares	F-statistic	p-value
Treatments	$k-1$	SSTr	$\text{MSTr} = \frac{\text{SSTr}}{k-1}$	$F = \frac{\text{MSTr}}{\text{MSE}}$	$P(F_{k-1,n_T-k} \geq F)$
Error	$n_T - k$	SSE	$\text{MSE} = \frac{\text{SSE}}{n_T-k}$		
Total	$n_T - 1$	SST			

FIGURE 11.14

Analysis of variance table for one-factor layout

for treatments and the degrees of freedom for error, but that a mean square is not given in the "Total" row. The penultimate column lists the F-statistic, calculated as the ratio of the two mean squares, and the final column contains the p-value.

F-Test for One-Factor Layout

In a one-factor layout with k levels and a total sample size n_T, the treatments have $k-1$ degrees of freedom and the error has $n_T - k$ degrees of freedom. Mean squares are obtained by dividing a sum of squares by its respective degrees of freedom so that

$$\text{MSTr} = \frac{\text{SSTr}}{k-1} \quad \text{and} \quad \text{MSE} = \frac{\text{SSE}}{n_T - k}$$

A p-value for the null hypothesis that the factor level means μ_i are all equal is calculated as

$$p\text{-value} = P(X \geq F)$$

where the F-statistic is

$$F = \frac{\text{MSTr}}{\text{MSE}}$$

and the random variable X has an F_{k-1,n_T-k} distribution.

Example 62
Collapse of Blocked Arteries

The degrees of freedom for treatments is $k - 1 = 3 - 1 = 2$ so that the mean square for treatments is

$$\text{MSTr} = \frac{\text{SSTr}}{2} = \frac{204.0}{2} = 102.0$$

With $n_T - k = 35 - 3 = 32$ degrees of freedom for error, the mean square error is

$$\text{MSE} = \frac{\text{SSE}}{32} = \frac{138.5}{32} = 4.33$$

This results in an F-statistic of

$$F = \frac{\text{MSTr}}{\text{MSE}} = \frac{102.0}{4.33} = 23.6$$

and a p-value of

$$p\text{-value} = P(X \geq 23.6) \simeq 0$$

Source	Degrees of freedom	Sum of squares	Mean squares	F-statistic	p-value
Stenosis level	2	204.02	102.01	23.57	0.000
Error	32	138.47	4.33		
Total	34	342.49			

Source	Degrees of freedom	Sum of squares	Mean squares	F-statistic	p-value
Locations	3	9854910	3284970	1.50	0.2307
Error	36	78779000	2188306		
Total	39	88633910			

Source	Degrees of freedom	Sum of squares	Mean squares	F-statistic	p-value
Carpet type	5	97.219	19.444	19.290	0.000
Error	84	84.670	1.008		
Total	89	181.889			

where the random variable X has an $F_{2,32}$ distribution. Consequently, the null hypothesis that the average flowrate at collapse is the same for all three amounts of stenosis is not plausible. This experiment has established that the flowrate at collapse does depend upon the amount of stenosis. Figure 11.15 shows the analysis of variance table for this data set.

Example 63

Roadway Base Aggregates

Figure 11.16 shows the analysis of variance table for this example. Notice that

$$\text{MSTr} = \frac{\text{SSTr}}{k-1} = \frac{9,854,910}{3} = 3,284,970$$

and

$$\text{MSE} = \frac{\text{SSE}}{n_T - k} = \frac{78,779,000}{36} = 2,188,306$$

The F-statistic is

$$F = \frac{\text{MSTr}}{\text{MSE}} = \frac{3,284,970}{2,188,306} = 1.50$$

so that the p-value is

$$p\text{-value} = P(X \geq 1.50) = 0.231$$

where the random variable X has an $F_{3,36}$ distribution. A p-value of this size indicates that the null hypothesis is plausible and so there is no reason for the engineer to conclude that the aggregate obtained from one location is different from the aggregate obtained from any other location.

Example 64

Carpet Fiber Blends

Figure 11.17 shows the analysis of variance table for the comparison of the different carpet fiber blends. For this data set

$$\text{MSTr} = \frac{\text{SSTr}}{k-1} = \frac{97.219}{5} = 19.444$$

and

$$MSE = \frac{SSE}{n_T - k} = \frac{84.670}{84} = 1.008$$

The F-statistic is therefore

$$F = \frac{MSTr}{MSE} = \frac{19.444}{1.008} = 19.290$$

so that the p-value is

$$p\text{-value} = P(X \geq 19.29) \simeq 0$$

where the random variable X has an $F_{5,84}$ distribution. Consequently, the textile engineer concludes that this experiment has established that some carpet fibers wear better than others. Which ones are better and by how much? This question is addressed in the next section.

11.1.4 Pairwise Comparisons of the Factor Level Means

When an analysis of variance is performed and the null hypothesis that the factor level means μ_i are all equal is rejected, the experimenter can follow up the analysis with pairwise comparisons of the factor level means to discover which ones have been shown to be different and by how much. These confidence intervals are often referred to as the Tukey multiple comparisons procedure. With k factor levels there are $k(k-1)/2$ pairwise differences

$$\mu_{i_1} - \mu_{i_2} \qquad 1 \leq i_1 < i_2 \leq k$$

Pairwise Comparisons of the Factor Level Means

When an analysis of variance is performed for a one-factor layout and the null hypothesis that the factor level means μ_i are all equal is rejected, it is useful to calculate the $1 - \alpha$ confidence level simultaneous confidence intervals

$$\mu_{i_1} - \mu_{i_2} \in \left(\bar{x}_{i_1 \cdot} - \bar{x}_{i_2 \cdot} - \frac{s\, q_{\alpha,k,\nu}}{\sqrt{2}} \sqrt{\frac{1}{n_{i_1}} + \frac{1}{n_{i_2}}}, \; \bar{x}_{i_1 \cdot} - \bar{x}_{i_2 \cdot} + \frac{s\, q_{\alpha,k,\nu}}{\sqrt{2}} \sqrt{\frac{1}{n_{i_1}} + \frac{1}{n_{i_2}}} \right)$$

for the $k(k-1)/2$ pairwise differences of the factor level means. The critical point $q_{\alpha,k,\nu}$ is the upper α point of the Studentized range distribution with parameters k and degrees of freedom $\nu = n_T - k$ (given in Table V), and $s = \hat{\sigma} = \sqrt{MSE}$. These simultaneous confidence intervals have an overall confidence level of $1 - \alpha$ and indicate which factor level means can be declared to be different, and by how much.

These confidence intervals are similar to the t-intervals based upon a pooled variance estimate for comparing two population means, which are discussed in Section 9.3.2. The difference is that the value $q_{\alpha,k,\nu}/\sqrt{2}$ is used in place of the critical point $t_{\alpha/2,\nu}$, and the reason for this is that the t-intervals have an *individual* confidence level of $1 - \alpha$ whereas this set of simultaneous confidence intervals have an *overall* confidence level of $1 - \alpha$. This overall confidence level has the interpretation that the experimenter has a confidence level of $1 - \alpha$ that *all* of the $k(k-1)/2$ confidence intervals contain their respective parameter value $\mu_{i_1} - \mu_{i_2}$, and this is achieved with a value of $q_{\alpha,k,\nu}/\sqrt{2}$ that is *larger* than the critical point $t_{\alpha/2,\nu}$.

If the confidence interval for the difference $\mu_{i_1} - \mu_{i_2}$ contains 0, then there is no evidence that the means at factor levels i_1 and i_2 are different. However, if the confidence interval for the difference $\mu_{i_1} - \mu_{i_2}$ does not contain 0, then there is evidence that the means at factor levels i_1 and i_2 are different. Furthermore, the confidence interval indicates by how much the factor level means have been shown to be different.

If the analysis of variance results in a rejection of the null hypothesis that the factor level means μ_i are all equal, then at least one pair of factor level means will have a confidence interval that does not contain 0. (Technically, this is generally the case as long as the critical point $q_{\alpha,k,\nu}$ is chosen with a value of α larger than the p-value calculated in the analysis of variance table.) If the analysis of variance results in an acceptance of the null hypothesis, with a p-value larger than 10%, then all of the confidence intervals will contain 0, since there is no evidence that any factor level means are different. In this situation there is usually no reason to construct the confidence intervals.

One property of these confidence intervals that the experimenter should be aware of is that as the number of factor levels k increases, the critical point $q_{\alpha,k,\nu}$ also increases so that the confidence intervals become longer and are consequently less sensitive.

Example 62 **Collapse of Blocked Arteries**	With $$s = \hat{\sigma} = \sqrt{\text{MSE}} = \sqrt{4.33} = 2.080$$ and the critical point $$q_{0.05,3,32} = 3.48$$

which can be estimated by interpolation in Table V, or obtained from a computer, the pairwise confidence intervals for the differences in the treatment means are

$$\mu_1 - \mu_2 \in \left(11.209 - 15.086 - \frac{2.080 \times 3.48}{\sqrt{2}} \sqrt{\frac{1}{11} + \frac{1}{14}}, \right.$$

$$\left. 11.209 - 15.086 + \frac{2.080 \times 3.48}{\sqrt{2}} \sqrt{\frac{1}{11} + \frac{1}{14}} \right)$$

$$= (-3.877 - 2.062, -3.877 + 2.062) = (-5.939, -1.814)$$

and

$$\mu_1 - \mu_3 \in \left(11.209 - 17.330 - \frac{2.080 \times 3.48}{\sqrt{2}} \sqrt{\frac{1}{11} + \frac{1}{10}}, \right.$$

$$\left. 11.209 - 17.330 + \frac{2.080 \times 3.48}{\sqrt{2}} \sqrt{\frac{1}{11} + \frac{1}{10}} \right)$$

$$= (-6.121 - 2.236, -6.121 + 2.236) = (-8.357, -3.884)$$

and

$$\mu_2 - \mu_3 \in \left(15.086 - 17.330 - \frac{2.080 \times 3.48}{\sqrt{2}} \sqrt{\frac{1}{14} + \frac{1}{10}}, \right.$$

$$\left. 15.086 - 17.330 + \frac{2.080 \times 3.48}{\sqrt{2}} \sqrt{\frac{1}{14} + \frac{1}{10}} \right)$$

$$= (-2.244 - 2.119, -2.244 + 2.119) = (-4.364, -0.125)$$

None of these three confidence intervals contains 0, and so the experiment has established that each of the three stenosis levels results in a different average flowrate at collapse.

Moreover, the average flowrate for a stenosis of 0.78 (level 1) is seen to be at least 1.8 ml/s lower than the average flowrate for a stenosis of 0.71 (level 2) (but no more than 5.9 ml/s lower) and at least 3.9 ml/s lower than the average flowrate for a stenosis of 0.65 (level 3) (but no more than 8.4 ml/s lower). Similarly, the average flowrate for a stenosis of 0.71 (level 2) is seen to be at least 0.1 ml/s lower than the average flowrate for a stenosis of 0.65 (level 3) (but no more than 4.4 ml/s lower).

Example 63 **Roadway Base** **Aggregates**	No evidence was found for any difference in the average resilient modulus values for the aggregates from the four locations, and so the experimenter knows that all of the confidence intervals for the pairwise differences $\mu_{i_1} - \mu_{i_2}$ will contain 0. Notice also that for balanced data sets such as this one with sample sizes all equal to n, the confidence intervals are

$$\mu_{i_1} - \mu_{i_2} \in \left(\bar{x}_{i_1.} - \bar{x}_{i_2.} - \frac{s\, q_{\alpha,k,\nu}}{\sqrt{n}}, \bar{x}_{i_1.} - \bar{x}_{i_2.} + \frac{s\, q_{\alpha,k,\nu}}{\sqrt{n}} \right)$$

which are all of the same length.

The largest sample mean is $\bar{x}_{4.} = 30{,}823$ and the smallest sample mean is $\bar{x}_{2.} = 29{,}461$. With

$$s = \hat{\sigma} = \sqrt{\text{MSE}} = \sqrt{2{,}188{,}305} = 1479$$

and a critical point

$$q_{0.05,4,36} = 3.81$$

the confidence interval for $\mu_4 - \mu_2$ is

$$\mu_4 - \mu_2 \in \left(30{,}823 - 29{,}461 - \frac{1479 \times 3.81}{\sqrt{10}}, 30{,}823 - 29{,}461 + \frac{1479 \times 3.81}{\sqrt{10}} \right)$$

$$= (1362 - 1782, 1362 + 1782) = (-420, 3144)$$

As expected, this confidence interval does contain 0 and so there is no evidence that μ_2 and μ_4 are unequal. However, notice that the difference in these means could be as large as 3144. Again, remember that acceptance of the null hypothesis does not prove that the average resilient modulus values are all equal. A more sensitive (powerful) test procedure and shorter confidence intervals are attained with larger sample sizes, as discussed in Section 11.1.5.

Example 64 **Carpet Fiber Blends**	Figure 11.18 shows the pairwise confidence intervals for this data set. Altogether there are $k(k-1)/2 = 6 \times 5/2 = 15$ confidence intervals, of which five contain 0 and ten do not contain 0.

It is useful to present these results schematically as in Figure 11.19. The sample means $\bar{x}_{i.}$ are written on a linear scale, and in this example they occur in the order 5, 3, 1, 4, 6, and then 2. Any two factor levels i_1 and i_2 for which the confidence interval for $\mu_{i_1} - \mu_{i_2}$ contains 0 are then joined by a line, and this is done in such a way that any two factor levels i_1 and i_2 for which the confidence interval for $\mu_{i_1} - \mu_{i_2}$ does not contain 0 are not joined by a line. In this case, the confidence intervals for $\mu_1 - \mu_3$, $\mu_1 - \mu_5$, and $\mu_3 - \mu_5$ all contain 0, and so one continuous line is drawn between $\bar{x}_{1.}$, $\bar{x}_{3.}$, and $\bar{x}_{5.}$. Also, the confidence intervals for $\mu_2 - \mu_6$ and $\mu_4 - \mu_6$ contain 0, and so lines are drawn between $\bar{x}_{2.}$ and $\bar{x}_{6.}$ and between $\bar{x}_{4.}$ and $\bar{x}_{6.}$. These lines are discontinuous at $\bar{x}_{6.}$ because the confidence interval for $\mu_2 - \mu_4$ does not contain 0 and so no line is required between $\bar{x}_{2.}$ and $\bar{x}_{4.}$.

FIGURE 11.18

Pairwise confidence intervals for the carpet fiber blends experiment

The critical point is $q_{0.05,6,84} = 4.12$ and $s = \hat{\sigma} = \sqrt{MSE} = 1.004$ so

$$\mu_i - \mu_j \in \bar{x}_{i.} - \bar{x}_{j.} \pm \frac{1.004 \times 4.12}{\sqrt{2}} \sqrt{\frac{1}{n_i} + \frac{1}{n_j}}$$

$\mu_1 - \mu_2$
$(-3.44, -1.37)$

$\mu_1 - \mu_3$ $\mu_2 - \mu_3$
$(-0.94, 1.25)$ $(1.47, 3.66)$
contains 0

$\mu_1 - \mu_4$ $\mu_2 - \mu_4$ $\mu_3 - \mu_4$
$(-2.19, -0.12)$ $(0.22, 2.28)$ $(-2.41, -0.22)$

$\mu_1 - \mu_5$ $\mu_2 - \mu_5$ $\mu_3 - \mu_5$ $\mu_4 - \mu_5$
$(-0.61, 1.53)$ $(1.79, 3.93)$ $(-0.83, 1.43)$ $(0.54, 2.68)$
contains 0   **contains 0**

$\mu_1 - \mu_6$ $\mu_2 - \mu_6$ $\mu_3 - \mu_6$ $\mu_4 - \mu_6$ $\mu_5 - \mu_6$
$(-2.66, -0.55)$ $(-0.25, 1.85)$ $(-2.87, -0.65)$ $(-1.50, 0.60)$ $(-3.15, -0.97)$
                   contains 0   **contains 0**

FIGURE 11.19

Schematic presentation of the results of the carpet fiber blends experiment

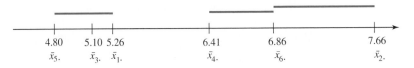

4.80   5.10 5.26   6.41   6.86   7.66
$\bar{x}_{5.}$ $\bar{x}_{3.}$ $\bar{x}_{1.}$ $\bar{x}_{4.}$ $\bar{x}_{6.}$ $\bar{x}_{2.}$

The diagram in Figure 11.19 is easily interpreted since factor levels joined by a line are not distinguishable (there is no evidence that their means are different), whereas factor levels that are not joined have means that have been shown to be different. One immediate observation is that as a group, fiber types 1, 3, and 5 are all "worse" than fiber types 2, 4, and 6. In other words, the first group has average scores that are known to be smaller than the average scores in the second group.

Also, fiber types 1, 3, and 5 are indistinguishable from one another. There is no evidence that their average scores are any different. In addition, fiber types 2 and 6 are indistinguishable, as are fiber types 4 and 6, but fiber type 2 is known to have an average score larger than that of fiber type 4. Is this last remark paradoxical? No, because a line joining two factor levels does not carry the interpretation that the two mean levels are equal. Rather it has the interpretation that there is no evidence of a difference between the two mean levels. Thus, there is evidence that $\mu_4 < \mu_2$, but the statements $\mu_2 = \mu_6$ and $\mu_4 = \mu_6$ are both plausible.

In summary, based solely on how well the carpets wear, the best fiber type is either type 2, the nylon fiber type B, or type 6, the polypropylene fiber. The textile engineer may want to investigate these two fiber types in greater detail to discover whether either one is much better than the other.

COMPUTER NOTE

Practice performing a one-factor analysis of variance on your computer package. Make sure that you can understand and interpret the analysis of variance table that the package presents you with. Find out how to obtain the follow-up pairwise confidence intervals. These are usually referred to as "Tukey intervals" or "pairwise intervals." Do not forget that you need to specify a confidence level for these.

11.1.5 Sample Size Determination

The sensitivity afforded by a one-factor analysis of variance depends upon the k sample sizes $n_1, \ldots, n_k$. Specifically, the power of the test of the null hypothesis that the factor level means are all equal increases as the sample sizes increase. This power is the probability that the null hypothesis is rejected when it is not true. In other words, for a given configuration of unequal factor level means μ_i, the probability of obtaining a small p-value in the analysis of variance table, so that the experimenter has evidence that the factor level means are different, increases as the sample sizes n_i increase.

An increase in the sample sizes also results in a decrease in the lengths of the pairwise confidence intervals. As before, the consideration of these confidence interval lengths is generally the most convenient manner in which to assess the sensitivity afforded by certain sample sizes. The pairwise confidence intervals have lengths

$$L = \sqrt{2}\, s q_{\alpha,k,\nu} \sqrt{\frac{1}{n_{i_1}} + \frac{1}{n_{i_2}}}$$

which are different if the sample sizes n_i are unequal. If the experimenter designs a balanced experiment with sample sizes all equal to n, then these lengths are all equal to

$$L = \frac{2 s q_{\alpha,k,\nu}}{\sqrt{n}}$$

As in other cases, notice that the confidence interval length L is inversely proportional to the square root of the sample size n

$$L \propto \frac{1}{\sqrt{n}}$$

so that in order to halve the confidence interval length, the sample size must be increased fourfold.

If, prior to experimentation, an experimenter decides that a confidence interval length no larger than L is required, then the necessary sample size is

$$n \simeq \frac{4 s^2 q_{\alpha,k,\nu}^2}{L^2}$$

The experimenter needs to estimate the value of $s = \hat{\sigma}$. Also, remember that the critical point $q_{\alpha,k,\nu}$ gets larger as the number of factor levels k increases, which results in a larger sample size required. This formula can also be used to determine the sample sizes that are required in a follow-up to an initial experiment in order to achieve increased sensitivity. In this case the values of s and $q_{\alpha,k,\nu}$ from the initial experiment can be used.

Example 63

Roadway Base Aggregates

For this balanced design, the lengths of the pairwise confidence intervals are all equal to

$$L = \frac{2 s q_{\alpha,k,\nu}}{\sqrt{n}} = \frac{2 \times 1479 \times 3.81}{\sqrt{10}} = 3564$$

Even though the analysis of variance indicates that there is no evidence that any of the aggregate types have different average values of resilient modulus, the pairwise confidence intervals indicate that the difference in average resilient modulus values between two locations could be as large as 3144. Therefore, the construction engineer decides to undertake additional sampling in order to increase the sensitivity of the comparisons between the different types of aggregate.

The engineer decides to aim for a confidence interval length of 2000. In this case, if all of the pairwise confidence intervals contain 0 so that there is no evidence of any difference between the aggregate types, the engineer will know that if there is a difference in the average values of resilient modulus, then it can be no larger than 2000. With values $s = 1479$ and $q_{0.05,4,36} = 3.81$ from the first experiment, the estimated sample size required is

$$n \simeq \frac{4s^2 q_{\alpha,k,\nu}^2}{L^2} = \frac{4 \times 1479^2 \times 3.81^2}{2000^2} = 31.8$$

Since the experimenter already has 10 observations of each aggregate type, this result implies that at least 22 more observations should be taken from each of the four aggregate types.

11.1.6 Model Assumptions

It is important to remember that the analysis of variance for a one-factor layout is based upon the modeling assumption that the observations are distributed independently with a normal distribution that has a common variance. The experimenter should be careful to notice whether any of these assumptions are violated in an obvious way.

The independence of the data observations can be judged from the manner in which a data set is collected. Where possible, the randomization of experimental units between the k factor levels helps to ensure independence. If the sample sizes are large enough, histograms of the data observations within each factor level may be constructed to indicate whether the distribution is obviously not normal. Often, prior experience with data observations of the kind under consideration tell the experimenter whether nonnormality is a problem. In any case, the analysis of variance is fairly robust to the distribution of the observations, so that it provides fairly accurate results as long as the distribution is not very different from a normal distribution. Nonparametric procedures for one-way layouts discussed in Section 15.3.1 provide alternative analysis techniques for data observations that clearly do not have a normal distribution.

The equality of the variances for each of the k factor levels can be judged from a comparison of the sample variances s_i^2 or from a comparison of the lengths of boxplots of the observations at each factor level. Again, the analysis of variance provides fairly accurate results unless the variances are very different. If the variances are different by more than a factor of five, say, then it is probably better for an experimenter to make pairwise comparisons between the k factor levels, one pair at a time, and use a small individual confidence level such as 90%. The pairwise comparisons can be made using the general procedure discussed in Section 9.3.1, which does not employ a pooled variance.

■ 11.1.7 Problems

11.1.1 Use Table IV to determine whether these probabilities are greater than 10%, between 5% and 10%, between 1% and 5%, or less than 1%. Then use a computer package to evaluate them exactly.

(a) $P(X \geq 4.2)$ where X has an $F_{4,15}$ distribution

(b) $P(X \geq 2.3)$ where X has an $F_{7,30}$ distribution

(c) $P(X \geq 31.7)$ where X has an $F_{3,10}$ distribution

(d) $P(X \geq 9.3)$ where X has an $F_{3,12}$ distribution

(e) $P(X \geq 0.9)$ where X has an $F_{6,60}$ distribution

11.1.2 Find the missing values in the analysis of variance table shown in Figure 11.20. What is the p-value?

11.1.3 Find the missing values in the analysis of variance table shown in Figure 11.21. What is the p-value?

11.1.4 A balanced experimental design has a sample size of $n = 12$ observations at each of $k = 7$ factor levels. The total sum of squares is SST $= 133.18$, and the sample averages are $\bar{x}_{1.} = 7.75$, $\bar{x}_{2.} = 8.41$, $\bar{x}_{3.} = 8.07$,

FIGURE 11.20

An analysis of variance table

Source	Degrees of freedom	Sum of squares	Mean squares	F-statistic	p-value
Treatments	?	?	111.4	?	?
Error	23	461.9	?		
Total	28	?			

FIGURE 11.21

An analysis of variance table

Source	Degrees of freedom	Sum of squares	Mean squares	F-statistic	p-value
Treatments	7	?	?	5.01	?
Error	?	?	3.62		
Total	29	?			

$\bar{x}_{4.} = 8.30$, $\bar{x}_{5.} = 8.17$, $\bar{x}_{6.} = 8.81$, and $\bar{x}_{7.} = 8.32$. Compute the analysis of variance table. (*Hint*: Calculate SSTr and then subtract it from SST to obtain SSE.) What is the p-value?

11.1.5 An experiment to compare $k = 4$ factor levels has $n_1 = 12$ and $\bar{x}_{1.} = 16.09$, $n_2 = 8$ and $\bar{x}_{2.} = 21.55$, $n_3 = 13$ and $\bar{x}_{3.} = 16.72$, and $n_4 = 11$ and $\bar{x}_{4.} = 17.57$. The total sum of squares is SST $= 485.53$. Compute the analysis of variance table. What is the p-value?

11.1.6 An experiment to compare $k = 3$ factor levels has $n_1 = 17$ and $\bar{x}_{1.} = 32.30$, $n_2 = 20$ and $\bar{x}_{2.} = 34.06$, and $n_3 = 18$ and $\bar{x}_{3.} = 32.02$. Also,

$$\sum_{i=1}^{3} \sum_{j=1}^{n_i} x_{ij}^2 = 59,843.21$$

Compute the analysis of variance table. What is the p-value?

11.1.7 A balanced experimental design has a sample size of $n = 14$ observations at each of $k = 4$ factor levels. The averages and variance estimates at each of the four factor levels are $\bar{x}_{1.} = 0.705$ and $s_1^2 = 0.766 \times 10^{-3}$, $\bar{x}_{2.} = 0.715$ and $s_2^2 = 1.276 \times 10^{-3}$, $\bar{x}_{3.} = 0.684$ and $s_3^2 = 2.608 \times 10^{-3}$, and $\bar{x}_{4.} = 0.692$ and $s_4^2 = 1.725 \times 10^{-3}$. Compute the analysis of variance table. What is the p-value?

11.1.8 A balanced experimental design has a sample size of $n = 11$ observations at each of $k = 3$ factor levels. The sample averages are $\bar{x}_{1.} = 48.05$, $\bar{x}_{2.} = 44.74$, and $\bar{x}_{3.} = 49.11$, and MSE $= 4.96$.
(a) Calculate pairwise confidence intervals for the factor level means with an overall confidence level of 95%.
(b) Make a diagram showing which factor level means are known to be different and which ones are indistinguishable.

(c) What additional sampling would you recommend to reduce the lengths of the pairwise confidence intervals to no more than 2.0 ?

11.1.9 A balanced experimental design has a sample size of $n = 6$ observations at each of $k = 6$ factor levels. The sample averages are $\bar{x}_{1.} = 136.3$, $\bar{x}_{2.} = 152.1$, $\bar{x}_{3.} = 125.7$, $\bar{x}_{4.} = 130.2$, $\bar{x}_{5.} = 142.3$, and $\bar{x}_{6.} = 128.0$, and MSE $= 15.95$.
(a) Calculate pairwise confidence intervals for the factor level means with an overall confidence level of 95%.
(b) Make a diagram showing which factor level means are known to be different and which ones are indistinguishable.
(c) What additional sampling would you recommend to reduce the lengths of the pairwise confidence intervals to no more than 10.0 ?

11.1.10 If each observation x_{ij} in a one-factor layout is replaced by the value $ax_{ij} + b$, for some constants a and b, what happens to the p-value in the analysis of variance table?

11.1.11 Consider the data set given in DS 11.1.1.
(a) Calculate the sample averages $\bar{x}_{i.}$
(b) Calculate the overall average $\bar{x}_{..}$
(c) Calculate the treatment sum of squares SSTr.
(d) Calculate $\sum_{i=1}^{k} \sum_{j=1}^{n_i} x_{ij}^2$
(e) Calculate the total sum of squares SST.
(f) Calculate the error sum of squares SSE.
(g) Complete the analysis of variance table. What is the p-value?
(h) Compute confidence intervals for the pairwise differences of the factor level means with an overall confidence level of 95%.

(i) Make a diagram showing which factor level means are known to be different and which ones are indistinguishable.

(j) If the experimenter wants confidence intervals for the pairwise differences of the factor level means with an overall confidence level of 95% that are no longer than 1.0, how much additional sampling would you recommend?

11.1.12 Complete parts (a)–(i) of Problem 11.1.11 for the data set in DS 11.1.2.

11.1.13 A factory has three production lines producing glass sheets that are all supposed to be of the same thickness. A quality inspector takes a random sample of $n = 30$ sheets from each production line and measures their thicknesses. The glass sheets from the first production line have a sample average of $\bar{x}_{1.} = 3.015$ mm with a sample standard deviation of $s_1 = 0.107$ mm. The glass sheets from the second production line have a sample average of $\bar{x}_{2.} = 3.018$ mm with a sample standard deviation of $s_2 = 0.155$ mm, while the glass sheets from the third production line have a sample average of $\bar{x}_{3.} = 2.996$ mm with a sample standard deviation of $s_3 = 0.132$ mm. What conclusions should the quality inspector draw?

11.1.14 A retail company is interested in examining whether the time taken for it to obtain approval for a credit card number by phone varies from one day of the week to another. The manager collects the following results. On Mondays, a sample of $n_1 = 20$ cases had a sample average of $\bar{x}_{1.} = 20.44$ seconds with a sample standard deviation of $s_1 = 1.338$ seconds. On Wednesdays, a sample of $n_2 = 15$ cases had a sample average of $\bar{x}_{2.} = 16.33$ seconds with a sample standard deviation of $s_2 = 1.267$ seconds, and on Fridays, a sample of $n_3 = 18$ cases had a sample average of $\bar{x}_{3.} = 15.34$ seconds with a sample standard deviation of $s_3 = 1.209$ seconds. What should the manager conclude?

11.1.15 Infrared Radiation Readings

The data set in DS 11.1.3 concerns the infrared radiation readings from an energy source measured by a particular meter with various levels of background radiation levels. The meter's manufacturers are interested in how the readings are affected by the background radiation levels. What would you advise them? Perform the calculations by hand.

11.1.16 Keyboard Layout Designs

DS 11.1.4 gives the times taken to perform a numerical task using three different kinds of keyboard layouts for the numerical keys. Perform hand calculations to investigate how the different layouts affect the time taken to perform the task.

11.1.17 Dispersion Polymerization

A chemical engineer is interested in how polymethyl methacrylate (PMMA) particles can be prepared by dispersion polymerization in the presence of a stabilizer polyvinyl pyrrolidone (PVP). DS 11.1.5 gives the average particle diameters achieved in various experiments with three different amounts of the stabilizer. How does the amount of stabilizer affect the average particle diameter? Use a computer package to obtain a graphical presentation of the data set, an analysis of variance table, and pairwise comparisons between the different stabilizer levels.

11.1.18 Computer Assembly Methods

A computer manufacturer has production lines that utilize three different assembly methods. These methods differ in the order in which tasks are performed and the consequent layout of the production lines. A quality inspector randomly observes 30 computers assembled under each of the three methods and records the assembly times. The resulting data set is given in DS 11.1.6. Is there any evidence that one assembly method is any quicker than the others? Use a computer package to obtain a graphical presentation of the data set, an analysis of variance table, and pairwise comparisons between the different assembly methods.

11.1.19 Three catalysts were compared to investigate how they affect the strength of a compound. When catalyst A was used, 8 samples of the compound were obtained and the strengths of these 8 samples had a sample average of 42.91 and a sample standard deviation of 5.33. When catalyst B was used, 11 samples of the compound were obtained and the strengths of these 11 samples had a sample average of 44.03 and a sample standard deviation of 4.01. When catalyst C was used, 10 samples of the compound were obtained and the strengths of these 10 samples had a sample average of 43.72 and a sample standard deviation of 5.10. Construct an analysis of variance table. Is there sufficient evidence to conclude that there is a difference between the three catalysts in terms of the strength of the compound?

11.1.20 The data set in DS 11.1.7 is a one-way layout to compare the quality scores of a product made from five different treatment methods.
  (a) Construct the ANOVA table.
  (b) Construct pairwise comparisons of the treatment effects and make a graphical presentation of what you find. Which treatment mean is largest? Which treatment mean is smallest?

11.1.21 In a one-way layout to compare five treatments an experiment gave $n_1 = 10$, $\bar{x}_{1.} = 46.09$, $n_2 = 9$, $\bar{x}_{2.} = 42.21$, $n_3 = 9$, $\bar{x}_{3.} = 55.32$, $n_4 = 7$, $\bar{x}_{4.} = 54.62$, $n_5 = 10$, $\bar{x}_{5.} = 38.79$, with $s = \hat{\sigma} = 4.33$. Construct pairwise comparisons of the treatment means μ_i, $1 \leq i \leq 5$, and draw a diagram to explain what you find. Which treatment mean μ_i is largest? Which treatment mean μ_i is smallest?

11.1.22 Metal Alloy Comparisons
An experiment was performed to compare four metal alloys. Various samples were made of each alloy, and their strengths were measured. The following data were obtained:
Type A : sample size 8, sample mean 10.50, sample standard deviation 1.02.
Type B : sample size 8, sample mean 9.22, sample standard deviation 0.86.
Type C : sample size 9, sample mean 6.32, sample standard deviation 1.13.
Type D : sample size 6, sample mean 11.39, sample standard deviation 0.98.
  (a) Construct the ANOVA table. What null hypothesis is tested by the ANOVA table? What bounds can you place on the p-value? What is your conclusion?
  (b) Make a graphical representation of the differences between the alloy types. Which alloy type is the strongest? Which alloy type is the weakest?

11.1.23 Durations of Investigatory Surgical Procedures
Thirty patients with a similar medical condition were randomly allocated among six physicians, so that each physician received exactly five patients. The times in minutes taken for the physicians to perform an investigatory surgical procedure on the patients were measured and the data set given in DS 11.1.8 was obtained. Construct an appropriate ANOVA table. What does the p-value in the ANOVA table tell you? Construct an appropriate graphical representation of the differences between the physicians. Which physician is the quickest? Which physician is the slowest?

11.1.24 E. Coli Colonies in Riverwater
Four samples of water were taken at five different locations along a river and measurements were obtained of the amount of E. Coli colonies within the water as given in DS 11.1.9. Investigate whether there is any evidence that the E. Coli pollution levels are different at the five locations. Make a graphical presentation that illustrates the differences between the E. Coli levels at the five locations. Which location has the highest E. Coli level? Which location has the smallest E. Coli level?

11.1.25 In a one-way layout, seven observations are taken from treatment 1 and their sum is 327.8, eight observations are taken from treatment 2 and their sum is 381.3, seven observations are taken from treatment 3 and their sum is 337.0, and six observations are taken from treatment 4 and their sum is 292.9. The ANOVA table gives a p-value of exactly 1%.
  (a) Write out the complete ANOVA table.
  (b) Calculate pairwise confidence intervals for the treatment means with an overall confidence level of 95%, and make a graphical representation of what you find.

11.2 Randomized Block Designs

11.2.1 One-Factor Layouts with Blocks

Recall that in the comparison of two population means discussed in Chapter 9, a distinction was made between independent samples and paired samples. In paired samples, comparisons are made within each pair of observations. A complete **randomized block design** is an extension of paired samples to accommodate the comparison of a set of k population means or factor levels.

A randomized block design consists of a set of b blocks, which each contain one observation from each of the k factor levels under consideration. The comparisons of the factor level

FIGURE 11.22

A randomized block design data set

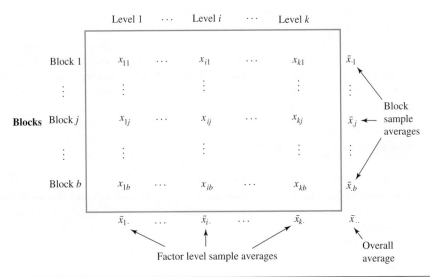

means are then based upon comparisons of the data observations *within* the blocks. In the same way that paired observations help eliminate the variation due to an extraneous source so that a more sensitive comparison can be made between two population means, a blocked design rather than an ordinary completely randomized design (discussed in Section 11.1) helps to eliminate variation due to the **blocking variable**, and thereby allows a more sensitive comparison of the set of k factor level means. The examples discussed in this section illustrate the use of a blocking variable. Typically a blocking variable may be the day on which an experiment is performed, the person or operator involved in an experiment, or the batch of raw material used in an experiment.

Figure 11.22 shows a data set obtained from a randomized block design to compare k factor levels. There are b blocks and the data set appears similar to a balanced one factor layout with b observations in each factor level. However, the difference with a randomized block design is that a data observation x_{ij} is known to be not from just the ith factor level, but also from the jth block.

The factor level sample averages are

$$\bar{x}_{i\cdot} = \frac{x_{i1} + \cdots + x_{ib}}{b} \qquad 1 \le i \le k$$

and the block sample averages are

$$\bar{x}_{\cdot j} = \frac{x_{1j} + \cdots + x_{kj}}{k} \qquad 1 \le j \le b$$

The total number of observations is $n_T = kb$, and the overall average is

$$\bar{x}_{\cdot\cdot} = \frac{\bar{x}_{1\cdot} + \cdots + \bar{x}_{k\cdot}}{k} = \frac{\bar{x}_{\cdot 1} + \cdots + \bar{x}_{\cdot b}}{b} = \frac{x_{11} + \cdots + x_{kb}}{kb}$$

It is usual to model the data observations as

$$x_{ij} = \mu_{ij} + \epsilon_{ij}$$

where μ_{ij} is the mean of the observation in the ij-cell, and the error terms ϵ_{ij} are independently distributed as

$$\epsilon_{ij} \sim N(0, \sigma^2)$$

Thus the error terms are taken to be independently normally distributed with zero means and a common variance σ^2. Equivalently, the data observation x_{ij} can be thought of as an observation from a

$$N(\mu_{ij}, \sigma^2)$$

distribution.

It is convenient to write the cell means as

$$\mu_{ij} = \mu + \alpha_i + \beta_j$$

In this expression, μ is referred to as the **overall average** of the cell means, the parameters

$$\alpha_1, \ldots, \alpha_k$$

are referred to as the **factor effects**, and the parameters

$$\beta_1, \ldots, \beta_b$$

are referred to as the **block effects**. The factor effects and the block effects are restricted to sum to 0, so that

$$\alpha_1 + \cdots + \alpha_k = 0 \qquad \text{and} \qquad \beta_1 + \cdots + \beta_b = 0$$

Notice that under this model and with the restrictions on the factor effects and block effects,

$$E(\bar{x}_{..}) = \frac{\sum_{i=1}^{k} \sum_{j=1}^{b} E(x_{ij})}{kb} = \frac{\sum_{i=1}^{k} \sum_{j=1}^{b} \mu_{ij}}{kb}$$

$$= \frac{\sum_{i=1}^{k} \sum_{j=1}^{b} (\mu + \alpha_i + \beta_j)}{kb} = \mu$$

Also,

$$E(\bar{x}_{i.}) = \frac{\sum_{j=1}^{b} E(x_{ij})}{b} = \frac{\sum_{j=1}^{b} \mu_{ij}}{b}$$

$$= \frac{\sum_{j=1}^{b} (\mu + \alpha_i + \beta_j)}{b} = \mu + \alpha_i$$

and

$$E(\bar{x}_{.j}) = \frac{\sum_{i=1}^{k} E(x_{ij})}{k} = \frac{\sum_{i=1}^{k} \mu_{ij}}{k}$$

$$= \frac{\sum_{i=1}^{k} (\mu + \alpha_i + \beta_j)}{k} = \mu + \beta_j$$

These results lead to the unbiased parameter estimates

$$\hat{\mu} = \bar{x}_{..}$$
$$\hat{\alpha}_i = \bar{x}_{i.} - \bar{x}_{..}$$
$$\hat{\beta}_j = \bar{x}_{.j} - \bar{x}_{..}$$

so that the cell mean estimates are

$$\hat{\mu}_{ij} = \hat{\mu} + \hat{\alpha}_i + \hat{\beta}_j = \bar{x}_{i.} + \bar{x}_{.j} - \bar{x}_{..}$$

FIGURE 11.23

Percentage reductions in heart rate
five minutes after drug is
administered

Drug Dosage Level

		Low	Medium	High	Averages
	1	12.2	16.5	25.4	18.033
	2	29.3	34.7	29.4	31.133
	3	14.1	26.1	34.0	24.733
Patient	4	18.5	33.3	21.7	24.500
	5	32.9	36.9	43.0	37.600
	6	15.3	32.6	32.1	26.667
	7	26.1	28.9	40.9	31.967
	8	22.8	28.6	31.1	27.500
Averages		21.400	29.700	32.200	27.767

For this model, the factor level means can be thought of as being

$$\mu + \alpha_1, \ldots, \mu + \alpha_k$$

so that the factor effect α_i is the difference between the ith factor level mean and the overall mean μ. The factor level means are all equal when the factor effects α_i are all equal to 0, and so the null hypothesis

$$H_0 : \alpha_1 = \cdots = \alpha_k = 0$$

implies that the k factor levels are indistinguishable. Rejection of this null hypothesis indicates that there is evidence that the k factor level means are not all equal. Finally, the block effect β_j measures the difference between the mean of the jth block level and the overall mean μ.

**Example 55
Heart Rate Reductions**

In a further evaluation of the effect of a new drug on inducing a temporary reduction in a patient's heart rate, three dosage levels of the new drug are compared. When the new drug was compared with the standard drug, a paired design was adopted in order to account for the variability in the drug efficacies from one patient to another. Similarly, a randomized block design is adopted to compare the three dosage levels of the new drug, with the blocks being the different patients.

The three dosage levels, low, medium, and high, are administered to each of the patients in a random order, with suitable precautions being made to ensure that there are no "carryover" effects from one dosage to the next. The data set consisting of the percentage reductions in the heart rate after 5 minutes is shown in Figure 11.23. There are $k = 3$ factor levels of interest (the dosage levels) and $b = 8$ blocks corresponding to eight patients who participated in the experiment.

Figure 11.24 provides a graphical presentation of the data set. Notice that the ordering of the x-axis is arbitrary because it depends upon the manner in which the eight patients are assigned the numbers 1 to 8. This graph, together with the factor level sample averages

$$\bar{x}_{1.} = 21.4, \qquad \bar{x}_{2.} = 29.7, \qquad \text{and} \qquad \bar{x}_{3.} = 32.2$$

suggest that the low dosage results in a smaller heart rate reduction than the medium and high dosages. Is this difference statistically significant? This question can be answered by testing the null hypothesis

$$H_0 : \alpha_1 = \alpha_2 = \alpha_3 = 0$$

FIGURE 11.24

Graphical presentation of the heart
rate reductions data set

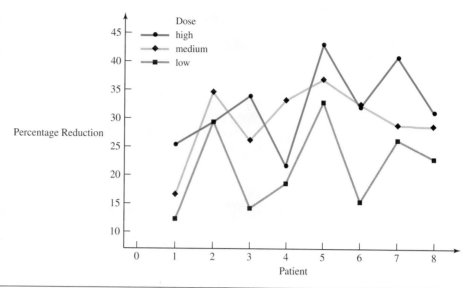

Acceptance of the null hypothesis implies that there is no evidence that the different dosage levels have any effect on the average heart rate reductions, whereas rejection of the null hypothesis indicates that there is evidence that the different dosage levels have different average effects on heart rate reduction.

The block sample averages $\bar{x}_{.j}$ are seen to be varied, and this indicates that the drug efficacy varies considerably from one patient to another. This finding underlines why a blocked design is appropriate. A similar (unblocked) completely randomized design would require a total of 24 patients split into three groups of eight patients, with each group receiving one (and only one) of the dosage levels. It is clear that the comparison of the dosage levels is then hindered by the variation between the groups due to the variabilities in the patients.

Example 65

Comparing Types of Wheat

Agricultural experimentation provides a classic example of the use of blocking. Consider an experiment to compare four types of wheat with five fields available for use. As Figure 11.25 shows, it is sensible to take the fields as blocks since growing conditions may vary from one field to another due to soil and drainage conditions. Each field is partitioned into four units, and the four types of wheat are randomly allocated to the four units within each field. If an experiment of this kind is run without blocking, so that different wheat types are allocated to different fields, then the variation between the different fields will hinder the comparisons between the wheat types.

Figure 11.26 shows a data set of crop yields from this blocked experiment, and Figure 11.27 provides a graphical presentation of the data set. Remember that in the graph the ordering of the x-axis (that is the numbering of the fields) is arbitrary. Wheat type 2 has the largest average yield, $\bar{x}_{2.} = 156.72$. Is the difference between wheat type 2 and the other wheat types statistically significant? This question can be addressed by testing the null hypothesis

$$H_0 : \alpha_1 = \alpha_2 = \alpha_3 = \alpha_4 = 0$$

which states that the four types of wheat produce equal yields on average.

Example 66

Infrared Radiation Meters

A research establishment has $k = 5$ infrared radiation meters and an experiment is run to investigate whether they are providing essentially equivalent readings. A total of $b = 15$ objects of different materials, temperatures, and sizes are used in the experiment, and radiation

FIGURE 11.25

Using fields as blocks in an experiment to compare wheat types

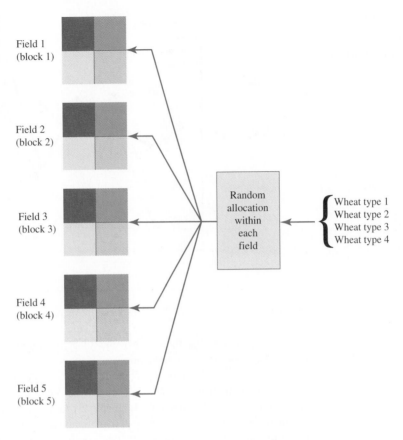

FIGURE 11.26

Crop yields for wheat experiment

Wheat Types

		Type 1	Type 2	Type 3	Type 4	Averages
	1	164.4	184.3	161.2	185.8	173.925
	2	145.0	142.1	110.8	135.4	133.325
Field	**3**	152.5	159.6	168.6	154.1	158.700
	4	138.5	137.2	134.9	123.2	133.450
	5	161.7	160.4	166.1	159.9	162.025
Averages		152.420	156.720	148.320	151.680	152.285

readings from each object are taken with each of the five meters. The resulting data set is given in Figure 11.28 and is presented graphically in Figure 11.29.

Notice that the different objects are taken to be the blocks, and the block sample averages $\bar{x}_{.j}$ indicate a wide variability in the radiation levels obtained from the 15 objects. As expected, this variability is much larger than the variability among the five meters, as evidenced by the meter sample averages $\bar{x}_{i.}$. Of course, it makes no sense to run this experiment without blocking on the objects, since using one meter on one set of objects and another meter on another set of objects does not allow a useful comparison of the two meters.

Acceptance of the null hypothesis

$$H_0 : \alpha_1 = \alpha_2 = \alpha_3 = \alpha_4 = \alpha_5 = 0$$

FIGURE 11.27

Graphical presentation of the crop yields for the wheat experiment

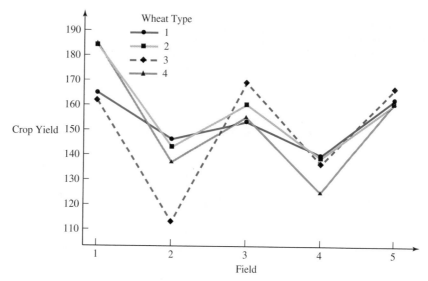

FIGURE 11.28

Infrared radiation readings

Infrared Radiation Meters

	Meter 1	Meter 2	Meter 3	Meter 4	Meter 5	Averages
1	2.00	2.01	1.89	2.09	2.04	2.006
2	2.27	2.24	2.44	2.40	2.38	2.346
3	1.25	1.03	1.20	1.19	1.33	1.200
4	2.97	2.91	2.99	3.05	3.07	2.998
5	3.97	3.72	3.75	3.90	3.81	3.830
6	1.60	1.67	1.79	1.77	1.83	1.732
7	2.27	2.06	2.16	2.28	2.37	2.228
Object 8	2.55	2.42	2.54	2.40	2.42	2.466
9	3.22	3.09	3.16	3.24	3.15	3.172
10	1.52	1.27	1.50	1.47	1.40	1.432
11	2.10	1.93	2.13	2.07	2.16	2.078
12	1.93	1.72	1.98	1.83	1.97	1.886
13	1.16	1.16	1.13	1.23	1.18	1.172
14	3.57	3.47	3.63	3.58	3.59	3.568
15	2.30	2.20	2.18	2.29	2.44	2.282
Averages	2.312	2.193	2.298	2.319	2.343	2.293

implies that there is no evidence of a difference between the average meter readings. However, rejection of this null hypothesis indicates that at least one of the meters needs recalibrating.

11.2.2 Partitioning the Total Sum of Squares

Total Sum of Squares As in a completely randomized one-factor layout, the total sum of squares

$$\text{SST} = \sum_{i=1}^{k} \sum_{j=1}^{b} (x_{ij} - \bar{x}_{..})^2 = \sum_{i=1}^{k} \sum_{j=1}^{b} x_{ij}^2 - kb\bar{x}_{..}^2$$

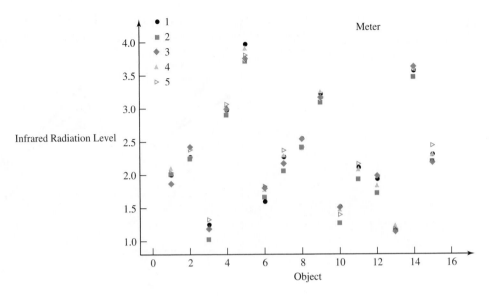

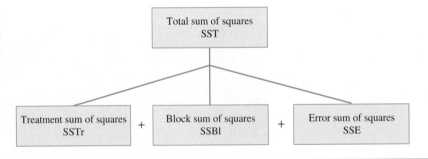

measures the total variability in the data observations. However, as Figure 11.30 shows, in randomized block designs SST is partitioned into three components, the treatment sum of squares SSTr, the block sum of squares SSBl, and the error sum of squares SSE, so that

$$SST = SSTr + SSBl + SSE$$

Treatment Sum of Squares   In a similar manner to the completely randomized one-factor layout, the treatment sum of squares

$$SSTr = \sum_{i=1}^{k} b(\bar{x}_{i\cdot} - \bar{x}_{\cdot\cdot})^2 = \sum_{i=1}^{k} b\bar{x}_{i\cdot}^2 - kb\bar{x}_{\cdot\cdot}^2$$

measures the variability between the k factor levels.

Block Sum of Squares   The block sum of squares SSBl measures the variability among the b blocks. It is based upon the comparison of the b block sample averages $\bar{x}_{\cdot j}$ and is defined to be

$$SSBl = \sum_{j=1}^{b} k(\bar{x}_{\cdot j} - \bar{x}_{\cdot\cdot})^2 = \sum_{j=1}^{b} k\bar{x}_{\cdot j}^2 - kb\bar{x}_{\cdot\cdot}^2$$

Notice that the form of the block sum of squares SSBl is similar to that of the treatment sum of squares SSTr except that it involves block sample averages $\bar{x}_{\cdot j}$ instead of factor level sample averages $\bar{x}_{i\cdot}$.

Error Sum of Squares   The error sum of squares SSE is defined to be the sum of the squared differences between the data observations x_{ij} and the estimated cell means $\hat{\mu}_{ij}$, so that

$$\text{SSE} = \sum_{i=1}^{k}\sum_{j=1}^{b}(x_{ij} - \hat{\mu}_{ij})^2 = \sum_{i=1}^{k}\sum_{j=1}^{b}(x_{ij} - \bar{x}_{i\cdot} - \bar{x}_{\cdot j} + \bar{x}_{\cdot\cdot})^2$$

When the square is expanded, or with subtraction from the total sum of squares as shown here, SST can be rewritten as

$$\text{SSE} = \text{SST} - \text{SSTr} - \text{SSBl}$$

$$= \left(\sum_{i=1}^{k}\sum_{j=1}^{b}x_{ij}^2 - kb\bar{x}_{\cdot\cdot}^2\right) - \left(\sum_{i=1}^{k}b\bar{x}_{i\cdot}^2 - kb\bar{x}_{\cdot\cdot}^2\right) - \left(\sum_{j=1}^{b}k\bar{x}_{\cdot j}^2 - kb\bar{x}_{\cdot\cdot}^2\right)$$

$$= \sum_{i=1}^{k}\sum_{j=1}^{b}x_{ij}^2 - \sum_{i=1}^{k}b\bar{x}_{i\cdot}^2 - \sum_{j=1}^{b}k\bar{x}_{\cdot j}^2 + kb\bar{x}_{\cdot\cdot}^2$$

Sum of Squares Partition for a Randomized Block Design

In a randomized block design with k factor levels and b blocks, the total sum of squares

$$\text{SST} = \sum_{i=1}^{k}\sum_{j=1}^{b}(x_{ij} - \bar{x}_{\cdot\cdot})^2 = \sum_{i=1}^{k}\sum_{j=1}^{b}x_{ij}^2 - kb\bar{x}_{\cdot\cdot}^2$$

measures the total variability in the data set. It can be partitioned into the treatment sum of squares

$$\text{SSTr} = \sum_{i=1}^{k}b(\bar{x}_{i\cdot} - \bar{x}_{\cdot\cdot})^2 = \sum_{i=1}^{k}b\bar{x}_{i\cdot}^2 - kb\bar{x}_{\cdot\cdot}^2$$

which measures the variability between the factor levels, the block sum of squares

$$\text{SSBl} = \sum_{j=1}^{b}k(\bar{x}_{\cdot j} - \bar{x}_{\cdot\cdot})^2 = \sum_{j=1}^{b}k\bar{x}_{\cdot j}^2 - kb\bar{x}_{\cdot\cdot}^2$$

which measures the variability between the blocks, and the error sum of squares

$$\text{SSE} = \sum_{i=1}^{k}\sum_{j=1}^{b}(x_{ij} - \bar{x}_{i\cdot} - \bar{x}_{\cdot j} + \bar{x}_{\cdot\cdot})^2$$

$$= \sum_{i=1}^{k}\sum_{j=1}^{b}x_{ij}^2 - \sum_{i=1}^{k}b\bar{x}_{i\cdot}^2 - \sum_{j=1}^{b}k\bar{x}_{\cdot j}^2 + kb\bar{x}_{\cdot\cdot}^2$$

Example 55

Heart Rate Reductions

With

$$\bar{x}_{..} = \frac{12.2 + \cdots + 31.1}{24} = 27.7667$$

and

$$\sum_{i=1}^{3} \sum_{j=1}^{8} x_{ij}^2 = 12.2^2 + \cdots + 31.1^2 = 20{,}065.76$$

the total sum of squares is

$$\text{SST} = 20{,}065.76 - (3 \times 8 \times 27.7667^2) = 1562.05$$

The treatment sum of squares is

$$\text{SSTr} = 8 \times (21.40^2 + 29.70^2 + 32.20^2) - (3 \times 8 \times 27.7667^2) = 511.41$$

and the block sum of squares is

$$\text{SSBl} = 3 \times (18.033^2 + \cdots + 27.500^2) - (3 \times 8 \times 27.7667^2) = 724.68$$

The error sum of squares can then be calculated by subtraction to be

$$\text{SSE} = \text{SST} - \text{SSTr} - \text{SSBl} = 1562.05 - 511.41 - 724.68 = 325.96$$

Example 65

Comparing Types of Wheat

For this example

$$\bar{x}_{..} = \frac{164.4 + \cdots + 159.9}{20} = 152.285$$

and

$$\sum_{i=1}^{4} \sum_{j=1}^{5} x_{ij}^2 = 164.4^2 + \cdots + 159.9^2 = 470{,}643.92$$

The total sum of squares is then

$$\text{SST} = 470{,}643.92 - (4 \times 5 \times 152.285^2) = 6829.51$$

The treatment sum of squares is

$$\text{SSTr} = 5 \times (152.42^2 + 156.72^2 + 148.32^2 + 151.68^2) - (4 \times 5 \times 152.285^2)$$
$$= 178.87$$

and the block sum of squares is

$$\text{SSBl} = 4 \times (173.925^2 + \cdots + 162.025^2) - (4 \times 5 \times 152.285^2) = 5274.10$$

As before, the error sum of squares can be calculated by subtraction to be

$$\text{SSE} = \text{SST} - \text{SSTr} - \text{SSBl} = 6829.5 - 178.9 - 5274.2 = 1376.44$$

Example 66

Infrared Radiation Meters

For this example the sums of squares can be shown to be

$$\text{SST} = 45.9536$$
$$\text{SSTr} = 0.2022$$
$$\text{SSBl} = 45.4816$$
$$\text{SSE} = 0.2698$$

11.2.3 The Analysis of Variance Table

As in completely randomized designs, mean squares for randomized block designs are obtained by dividing the sums of squares by the appropriate degrees of freedom. Treatments have $k - 1$ degrees of freedom, so that the mean squares for treatments is

$$\text{MSTr} = \frac{\text{SSTr}}{k - 1}$$

and blocks have $b - 1$ degrees of freedom so that the mean squares for blocks is

$$\text{MSBl} = \frac{\text{SSBl}}{b - 1}$$

For randomized block designs, the degrees of freedom for error are $(k - 1)(b - 1)$, so that the mean square error is

$$\text{MSE} = \frac{\text{SSE}}{(k - 1)(b - 1)}$$

Notice that the degrees of freedom for treatments, blocks, and error sum to $kb - 1$.

The mean square error MSE is distributed as

$$\text{MSE} \sim \sigma^2 \frac{\chi^2_{(k-1)(b-1)}}{(k - 1)(b - 1)}$$

so that, as before, the mean square error MSE is an unbiased point estimate of the error variance σ^2 and can be written as $\hat{\sigma}^2$. Also, it can be shown that

$$E(\text{MSTr}) = \sigma^2 + b \frac{\sum_{i=1}^{k} \alpha_i^2}{k - 1} \quad \text{and} \quad E(\text{MSBl}) = \sigma^2 + k \frac{\sum_{j=1}^{b} \beta_j^2}{b - 1}$$

Notice that under the null hypothesis

$$H_0 : \alpha_1 = \cdots = \alpha_k = 0$$

the expected mean square for treatments is

$$E(\text{MSTr}) = \sigma^2$$

and in fact when the null hypothesis is true, the F-statistic

$$F_T = \frac{\text{MSTr}}{\text{MSE}}$$

is an observation from an $F_{k-1,(k-1)(b-1)}$ distribution. Consequently the p-value for the null hypothesis is calculated as

$$p\text{-value} = P(X \geq F_T)$$

where the random variable X has a $F_{k-1,(k-1)(b-1)}$ distribution.

Also, it should be noted that in a similar fashion, with an F-statistic

$$F_B = \frac{\text{MSBl}}{\text{MSE}}$$

the p-value

$$p\text{-value} = P(X \geq F_B)$$

where the random variable X has a $F_{b-1,(k-1)(b-1)}$ distribution, measures the plausibility of the null hypothesis that the block effects β_j are all zero. In other words, this p-value measures the plausibility of the blocks being indistinguishable from each other.

Source	Degrees of freedom	Sum of squares	Mean squares	F-statistic	p-value
Treatments	$k - 1$	SSTr	$\text{MSTr} = \frac{\text{SSTr}}{k-1}$	$F_T = \frac{\text{MSTr}}{\text{MSE}}$	$P(F_{k-1,(k-1)(b-1)} \geq F_T)$
Blocks	$b - 1$	SSBl	$\text{MSBl} = \frac{\text{SSBl}}{b-1}$	$F_B = \frac{\text{MSBl}}{\text{MSE}}$	$P(F_{b-1,(k-1)(b-1)} \geq F_B)$
Error	$(k - 1)(b - 1)$	SSE	$\text{MSE} = \frac{\text{SSE}}{(k-1)(b-1)}$		
Total	$kb - 1$	SST			

FIGURE 11.31

The analysis of variance table for a randomized block design

The analysis of variance table for this randomized block design is shown in Figure 11.31. Notice that it consists of four rows corresponding to treatments, blocks, error, and their total. The F-statistic and p-value of primary interest to the experimenter are in the treatments row, and they indicate the evidence of a difference between the factor level means.

> **F-Tests for a Randomized Block Design**
>
> In a randomized block design with k factor levels and b blocks, the treatments have $k - 1$ degrees of freedom, the blocks have $b - 1$ degrees of freedom, and the error has $(k - 1)(b - 1)$ degrees of freedom. The mean squares are
>
> $$\text{MSTr} = \frac{\text{SSTr}}{k - 1}$$
>
> $$\text{MSBl} = \frac{\text{SSBl}}{b - 1}$$
>
> $$\text{MSE} = \frac{\text{SSE}}{(k - 1)(b - 1)}$$
>
> A p-value for the null hypothesis that the factor level means are all equal is calculated as
>
> $$p\text{-value} = P(X \geq F_T)$$
>
> where the F-statistic is
>
> $$F_T = \frac{\text{MSTr}}{\text{MSE}}$$
>
> and the random variable X has an $F_{k-1,(k-1)(b-1)}$ distribution. The p-value
>
> $$p\text{-value} = P(X \geq F_B)$$
>
> where the F-statistic is
>
> $$F_B = \frac{\text{MSBl}}{\text{MSE}}$$
>
> and the random variable X has an $F_{b-1,(k-1)(b-1)}$ distribution measures the plausibility of the blocks being indistinguishable from each other.

Source	Degrees of freedom	Sum of squares	Mean squares	F-statistic	p-value
Dose level	2	511.41	255.71	10.98	0.001
Person	7	724.68	103.53	4.45	0.009
Error	14	325.96	23.28		
Total	23	1562.05			

Example 55

Heart Rate Reductions

Figure 11.32 shows the analysis of variance table for this example. Notice that the degrees of freedom for treatments (dosage) is $k - 1 = 2$ and the mean square for treatments is

$$\text{MSTr} = \frac{\text{SSTr}}{2} = \frac{511.41}{2} = 255.71$$

Similarly, the degrees of freedom for blocks (person) is $b - 1 = 7$ and the mean square for blocks is

$$\text{MSBl} = \frac{\text{SSBl}}{7} = \frac{724.68}{7} = 103.53$$

Error has $(k - 1)(b - 1) = 14$ degrees of freedom and the mean square error is

$$\text{MSE} = \hat{\sigma}^2 = \frac{\text{SSE}}{14} = \frac{325.96}{14} = 23.28$$

The F-statistic for treatments is

$$F_T = \frac{\text{MSTr}}{\text{MSE}} = \frac{255.71}{23.28} = 10.98$$

which gives a p-value of

$$p\text{-value} = P(X \geq 10.98) \simeq 0.001$$

where the random variable X has an $F_{2,14}$ distribution. This low p-value indicates that the null hypothesis

$$H_0 : \alpha_1 = \alpha_2 = \alpha_3 = 0$$

is not plausible, and so the experiment provides evidence that the different dosage levels have different average effects on heart rate reduction. Are all three dosage levels significantly different? This question can be answered with the pairwise comparisons discussed in Section 11.2.4.

Finally, notice that the blocks (persons) have a p-value of 0.009. This value indicates that there is evidence of a block effect, which in this case is a difference in response from one patient to another. Thus, while it is clear that the dosage levels have different effects, the researchers should also notice that the heart rate reductions induced by the drugs vary considerably from one patient to another.

Example 65

Comparing Types of Wheat

Figure 11.33 shows the analysis of variance table for this example. Notice that

$$\text{MSTr} = \frac{\text{SSTr}}{k - 1} = \frac{178.87}{3} = 59.62$$

$$\text{MSBl} = \frac{\text{SSBl}}{b - 1} = \frac{5274.10}{4} = 1318.55$$

$$\text{MSE} = \hat{\sigma}^2 = \frac{\text{SSE}}{(k - 1)(b - 1)} = \frac{1376.44}{12} = 114.70$$

FIGURE 11.33

The analysis of variance table for the wheat comparisons experiment

Source	Degrees of freedom	Sum of squares	Mean squares	F-statistic	p-value
Wheat type	3	178.87	59.62	0.52	0.6766
Field	4	5274.10	1318.55	11.50	0.0005
Error	12	1376.44	114.70		
Total	19	6829.51			

FIGURE 11.34

The analysis of variance table for the infrared radiation meters experiment

Source	Degrees of freedom	Sum of squares	Mean squares	F-statistic	p-value
Meter	4	0.2022	0.0505	10.49	0.000
Object	14	45.4816	3.2487	674.29	0.000
Error	56	0.2698	0.0048		
Total	74	45.9536			

The F-statistic for treatments is

$$F_T = \frac{\text{MSTr}}{\text{MSE}} = \frac{59.62}{114.70} = 0.52$$

so that the p-value for the null hypothesis

$$H_0 : \alpha_1 = \alpha_2 = \alpha_3 = \alpha_4 = 0$$

is

$$p\text{-value} = P(X \geq 0.52) = 0.6766$$

where the random variable X has an $F_{3,12}$ distribution. This large p-value indicates that the experiment provides no evidence that the four types of wheat are in any way different with respect to their average yields. However, notice that the p-value of 0.0005 for blocks (fields) indicates that the fertility levels of the five fields are different. This finding confirms that it was sensible to run the experiment with fields as blocks.

Example 66
Infrared Radiation
Meters

Figure 11.34 shows the analysis of variance table for this experiment. Notice that

$$\text{MSTr} = \frac{\text{SSTr}}{k-1} = \frac{0.2022}{4} = 0.0505$$

$$\text{MSBl} = \frac{\text{SSBl}}{b-1} = \frac{45.4816}{14} = 3.2487$$

$$\text{MSE} = \hat{\sigma}^2 = \frac{\text{SSE}}{(k-1)(b-1)} = \frac{0.2698}{56} = 0.0048$$

The F-statistics are

$$F_T = \frac{\text{MSTr}}{\text{MSE}} = \frac{0.0505}{0.0048} = 10.49 \quad \text{and} \quad F_B = \frac{\text{MSBl}}{\text{MSE}} = \frac{3.2487}{0.0048} = 674.29$$

and both p-values are very small. Therefore, as well as the obvious difference in the radiation levels of the 15 objects, the experiment indicates a difference between the readings of the five radiation meters. This result immediately alerts the experimenters that at least one meter needs recalibrating, and the pairwise comparisons discussed in Section 11.2.4 show which meter this is.

COMPUTER NOTE

Find out how to obtain an analysis of variance table for a randomized block design with your computer software package. Computationally, it is equivalent to a two factor analysis of variance (without interactions), and so this is usually the way in which the computer handles randomized block designs. Two factor designs are discussed in Section 14.1.

11.2.4 Pairwise Comparisons of the Factor Level Means

For randomized block designs the interpretation and use of a set of simultaneous confidence intervals for the pairwise differences between the factor level means are the same as for a randomized one-way layout discussed in Section 11.1.4.

> **Pairwise Comparisons of Factor Level Means in a Randomized Block Design**
>
> When an analysis of variance is performed for a randomized block design and the null hypothesis that the factor level means are all equal is rejected, it is useful to calculate the $1 - \alpha$ confidence level simultaneous confidence intervals
>
> $$\mu_{i_1} - \mu_{i_2} \in \left(\bar{x}_{i_1.} - \bar{x}_{i_2.} - \frac{s\, q_{\alpha,k,\nu}}{\sqrt{b}}, \bar{x}_{i_1.} - \bar{x}_{i_2.} + \frac{s\, q_{\alpha,k,\nu}}{\sqrt{b}} \right)$$
>
> for the $k(k-1)/2$ pairwise differences of the factor level means. The critical point $q_{\alpha,k,\nu}$ is the upper α point of the Studentized range distribution with parameters k and degrees of freedom $\nu = (k-1)(b-1)$ (given in Table V), and $s = \hat{\sigma} = \sqrt{\mathrm{MSE}}$. These simultaneous confidence intervals have an overall confidence level of $1 - \alpha$ and they indicate which factor level means can be declared to be different, and by how much.

The pairwise confidence intervals are all of the same length

$$L = \frac{2s\, q_{\alpha,k,\nu}}{\sqrt{b}}$$

which is inversely proportional to the square root of the number of blocks b. Sample size determinations for randomized block designs simply involve the determination of the number of blocks that are to be used. The sensitivity of the comparison of the factor level means afforded by a given number of blocks b can be assessed by considering the resulting confidence interval length L.

Example 55

Heart Rate Reductions

To construct confidence intervals with an overall confidence level of 95%, the required critical point is

$$q_{0.05,3,14} = 3.70$$

so that

$$\frac{s\, q_{\alpha,k,\nu}}{\sqrt{b}} = \frac{\sqrt{23.28} \times 3.70}{\sqrt{8}} = 6.31$$

The confidence intervals are therefore

$$\mu_1 - \mu_2 \in (21.40 - 29.70 - 6.31, 21.40 - 29.70 + 6.31) = (-14.61, -1.99)$$
$$\mu_1 - \mu_3 \in (21.40 - 32.20 - 6.31, 21.40 - 32.20 + 6.31) = (-17.11, -4.49)$$
$$\mu_2 - \mu_3 \in (29.70 - 32.20 - 6.31, 29.70 - 32.20 + 6.31) = (-8.81, 3.81)$$

FIGURE 11.35

Schematic presentation of the results of the heart rate reductions example

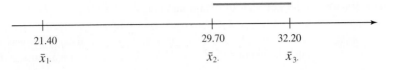

FIGURE 11.36

Schematic presentation of the results of the infrared radiation meters example

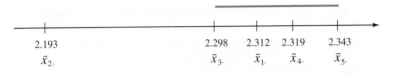

Thus, the medium dose induces an average heart rate reduction of somewhere between 2.0% and 14.6% more than the low dose, and the high dose induces an average heart rate reduction of somewhere between 4.5% and 17.1% more than the low dose. However, there is no evidence that the average heart rate reductions induced by the medium and high dosages are any different (see Figure 11.35). If they are different, then the difference is between -8.8% and 3.8%.

Example 66
Infrared Radiation Meters

With a confidence level of 95%, the critical point for this example is

$$q_{0.05,5,56} \simeq 4.0$$

so that

$$\frac{s\, q_{\alpha,k,\nu}}{\sqrt{b}} = \frac{\sqrt{0.0048} \times 4.0}{\sqrt{15}} = 0.072$$

The confidence intervals are therefore

$$\mu_{i_1} - \mu_{i_2} \in \left(\bar{x}_{i_1.} - \bar{x}_{i_2.} - 0.072, \bar{x}_{i_1.} - \bar{x}_{i_2.} + 0.072 \right)$$

where

$$\bar{x}_{1.} = 2.312$$
$$\bar{x}_{2.} = 2.193$$
$$\bar{x}_{3.} = 2.298$$
$$\bar{x}_{4.} = 2.319$$
$$\bar{x}_{5.} = 2.343$$

It can be seen that the confidence intervals for the comparisons of meters 1, 3, 4, and 5 all contain 0. However, meter 2 has an average reading significantly smaller than each of these four meters, since, for example, the confidence interval

$$\mu_1 - \mu_2 \in (2.312 - 2.193 - 0.072, 2.312 - 2.193 + 0.072) = (0.047, 0.191)$$

does not contain 0. This finding reveals that meter 2 is probably providing inaccurate radiation readings, and that it should be recalibrated in order to be consistent with the other four meters (see Figure 11.36).

11.2.5   Model Assumptions

It is important to remember that the analysis of variance of a randomized block design presented in this section is based upon the modeling assumption

$$x_{ij} = \mu + \alpha_i + \beta_j + \epsilon_{ij}$$

where the error terms ϵ_{ij} are independently distributed

$$\epsilon_{ij} \sim N(0, \sigma^2)$$

The assumption of independent error terms that are normally distributed with a common variance is similar to the assumption made in Section 11.1 for the analysis of a completely randomized design. An additional assumption of the analysis of a randomized block design is that there is no *interaction* between the factor levels and the blocks. In other words, it is assumed that the differences between the factor level effects are the same for each of the blocks.

The analysis of a randomized block design is similar to the analysis of a two-factor layout without replications, which is discussed in Section 14.1. Discussions of interactions and methods of checking the model assumptions through *residual analysis* are provided in Chapter 14. Also, it should be noted that a nonparametric method of analyzing a randomized block design, which does not require the assumption that the data observations are normally distributed, is discussed in Section 15.3.2.

■ 11.2.6   Problems

11.2.1 Find the missing values in the analysis of variance table shown in Figure 11.37. What are the p-values?

11.2.2 Find the missing values in the analysis of variance table shown in Figure 11.38. What are the p-values?

11.2.3 A randomized block design has $k = 4$ factor levels and $b = 10$ blocks. The total sum of squares is SST = 3736.64 and the block sum of squares is SSBl = 2839.97. The factor level sample averages are $\bar{x}_{1.} = 45.12$, $\bar{x}_{2.} = 42.58$, $\bar{x}_{3.} = 43.10$, and $\bar{x}_{4.} = 41.86$.

Compute the analysis of variance table. What are the p-values?

11.2.4 A randomized block design has $k = 5$ factor levels and $b = 15$ blocks. The total sum of squares is SST = 1947.89 and the block sum of squares is SSBl = 1527.12. The factor level sample averages are $\bar{x}_{1.} = 21.50$, $\bar{x}_{2.} = 22.51$, $\bar{x}_{3.} = 26.08$, $\bar{x}_{4.} = 21.03$, and $\bar{x}_{5.} = 23.43$. Compute the analysis of variance table. What are the p-values?

FIGURE 11.37

An analysis of variance table

Source	Degrees of freedom	Sum of squares	Mean squares	F-statistic	p-value
Treatments	?	?	?	3.02	?
Blocks	9	?	?	?	?
Error	?	?	1.12		
Total	39	64.92			

FIGURE 11.38

An analysis of variance table

Source	Degrees of freedom	Sum of squares	Mean squares	F-statistic	p-value
Treatments	?	?	3.77	3.56	?
Blocks	?	?	?	?	?
Error	?	51.92	?		
Total	63	122.47			

11.2.5 A randomized block design has $k = 3$ factor levels and $b = 8$ blocks. The block sum of squares is SSBl = 50.19, and the factor level sample averages are $\bar{x}_{1.} = 5.93$, $\bar{x}_{2.} = 4.62$, and $\bar{x}_{3.} = 4.78$. Also,

$$\sum_{i=1}^{3}\sum_{j=1}^{8} x_{ij}^2 = 691.44$$

(a) Compute the analysis of variance table. What is the p-value?

(b) Calculate pairwise confidence intervals for the factor level means with an overall confidence level of 95%.

11.2.6 Suppose that in a randomized block design the observations in the first block x_{i1} are replaced by the values $x_{i1} + c$ for some constant c. Which numbers change in the analysis of variance table?

11.2.7 Consider the data set given in DS 11.2.1.
(a) Calculate the factor level sample averages $\bar{x}_{i.}$
(b) Calculate the block sample averages $\bar{x}_{.j}$
(c) Calculate the overall average $\bar{x}_{..}$
(d) Calculate the treatment sum of squares SSTr.
(e) Calculate the block sum of squares SSBl.
(f) Calculate $\sum_{i=1}^{k}\sum_{j=1}^{b} x_{ij}^2$
(g) Calculate the total sum of squares SST.
(h) Calculate the error sum of squares SSE.
(i) Complete the analysis of variance table. What are the p-values?
(j) Compute confidence intervals for the pairwise differences of the factor level means with an overall confidence level of 95%.
(k) Make a diagram showing which factor level means are known to be different and which ones are indistinguishable.
(l) If the experimenter wants confidence intervals for the pairwise differences of the factor level means with an overall confidence level of 95% that are no longer than 2.0, how much additional sampling would you recommend?

11.2.8 Complete parts (a)–(k) of Problem 11.2.7 for the data set in DS 11.2.2. If the experimenter wants confidence intervals for the pairwise differences of the factor level means with an overall confidence level of 95% that are no longer than 4.0, how much additional sampling would you recommend?

11.2.9 Calciner Comparisons
DS 11.2.3 gives a data set of brightness measurements for $b = 7$ batches of kaolin processed through $k = 3$ calciners. The calciner operators are interested in whether their calciners are operating at different efficiencies. What would you advise them? Perform hand calculations.

11.2.10 Radar Detection of Airborne Objects
A randomized block design is used to compare $k = 3$ radar systems for detecting airborne objects. A total of $b = 8$ objects are flown toward the radar stations and the distances at detection by the radar systems are recorded in DS 11.2.4. Perform hand calculations to investigate whether there is evidence of any difference between the radar systems.

11.2.11 Golf Club Comparisons
A total of $b = 9$ golfers are each asked to hit a golf ball with each of $k = 4$ types of driver. DS 11.2.5 gives the straight line distances traveled by the balls. Are some drivers better than others? Perform hand calculations.

11.2.12 Production Line Assembly Methods
A manufacturing company wants to investigate which of three possible assembly methods for an electric motor is the most efficient. Since there is a fair amount of variability in the expertise of the workforce, it is decided to take workers as a blocking variable in the experiment. A random selection of $b = 10$ workers are trained in each of the $k = 3$ assembly methods, and then after a suitable amount of practice, they are timed under each method. The resulting data set is given in DS 11.2.6. Which assembly method is quickest? Use a computer package to obtain a graphical presentation of the data set and to help you with the analysis.

11.2.13 Realtor Commissions
A Realtor's office has five agents, and the manager wants to determine who is the best agent. A data set, given in DS 11.2.7, is collected that consists of the agents' commissions in each month of the previous year. Since house sales vary from one month to the next, the manager decides that it is sensible to take the months as blocking variables. Who is the best agent? Who is the worst agent? Use a computer package to obtain a graphical presentation of the data set and to help you with the analysis.

11.2.14 Cleanliness Scores for Detergent Comparisons
A chemical company runs a randomized block design to compare $k = 4$ different formulations of a detergent. A set of $b = 20$ pieces of cloth are dirtied in various ways and are then cut into four equal pieces and washed

in each of the four detergents. These different cloths are taken as the blocking variable. After washing, a grid is place over the cloth and each cell in the grid is evaluated on its cleanliness by a panel of judges. These evaluations are averaged to produce the scores given in DS 11.2.8. Higher scores represent cleaner cloths. Which detergent formulation is best? Use a computer package to obtain a graphical presentation of the data set and to help you with the analysis.

11.2.15 Consider the following ANOVA table for a randomized block design.

Source	df	SS	MS	F	p-value
Treatments	3	0.151	?	?	?
Blocks	?	?	0.054	?	?
Error	18	?	?		
Total	?	0.644			

(a) Find the missing values in the ANOVA table.

(b) Suppose that the treatment sample averages are $\bar{x}_{1.} = 0.810$, $\bar{x}_{2.} = 0.630$, $\bar{x}_{3.} = 0.797$, $\bar{x}_{4.} = 0.789$. At a 95% confidence level, is there sufficient evidence to conclude that treatment 2 has a smaller mean value than the other treatments?

11.2.16 A randomized block design has $k = 3$ factor levels and $b = 4$ blocks. The total sum of squares is SST $= 203.565$ and $s = 1.445$. The factor level sample

averages are $\bar{x}_{1.} = 107.68$ $\bar{x}_{2.} = 109.86$ and $\bar{x}_{3.} = 111.63$. Compute the analysis of variance table.

11.2.17 Consider a randomized block design with b blocks to compare k treatments. Suppose that each data value x_{ij} is replaced by $ax_{ij} + c$ for some constants $a > 0$ and c. For example, if temperature is the response variable, the units might be changed from Fahrenheit to Centigrade. Which values change in the ANOVA table, and how do they change?

11.2.18 A randomized block design has four factor levels with sample averages $\bar{x}_{1.} = 763.9$, $\bar{x}_{2.} = 843.9$, $\bar{x}_{3.} = 711.3$, and $\bar{x}_{4.} = 788.2$. There are 7 blocks and SSBl $= 13492.3$. Also, SSE $= 7052.8$. Compute the ANOVA table. What do the p-values tell you? Construct a graphical representation of the differences between the factor means.

11.2.19 Groundwater Pollution Levels
An experiment was conducted to measure the pollution level in the groundwater at four different locations. The pollution levels were measured at each of the locations at five different time points as given in DS 11.2.9. Is there any evidence to support the claim that the pollution levels are different at the four locations? Make a graphical presentation that illustrates the differences between the pollution levels at the four locations.

11.3 Case Study: Microelectronic Solder Joints

The researcher is interested in comparing the gold layer thicknesses on the bond pads of microelectronic assemblies manufactured by four different companies. A random selection of 10 bond pads is made from assemblies from each of the four companies, and the gold layer thicknesses are measured. The resulting data set is given in Figure 11.39, and it is represented by the boxplots in Figure 11.40.

In order to test whether there is a statistically significant difference between the four companies, the researcher conducts a one way analysis of variance which is shown in Figure 11.41. The p-value is equal to $P(F_{3,36} \geq 16.70)$ that is much less than 1%, and so the researcher can conclude that there is a difference between the companies in terms of the gold layer thicknesses that they have on their substrate bond pads.

The sample averages are $\bar{x}_A = 0.00714$, $\bar{x}_B = 0.00797$, $\bar{x}_C = 0.00878$, and $\bar{x}_D = 0.00795$, and $s = \sqrt{MSE} = 0.0005182$ so that

$$\frac{s \times q_{0.05,4,36}}{\sqrt{n}} = \frac{0.0005182 \times 3.81}{\sqrt{10}} = 0.000624$$

FIGURE 11.39

Data set of gold layer thicknesses
on substrate bond pads (microns)

Company A	Company B	Company C	Company D
0.0071	0.0084	0.0088	0.0084
0.0078	0.0088	0.0099	0.0089
0.0068	0.0077	0.0092	0.0081
0.0078	0.0077	0.0085	0.0083
0.0067	0.0082	0.0083	0.0070
0.0069	0.0079	0.0083	0.0086
0.0073	0.0083	0.0092	0.0075
0.0070	0.0079	0.0078	0.0078
0.0070	0.0075	0.0089	0.0076
0.0070	0.0073	0.0089	0.0073

FIGURE 11.40

Comparison of gold layer
thicknesses for each company

FIGURE 11.41

Analysis of variance table for
assessing the difference in gold
layer thicknesses between the four
companies

Source	Degrees of Freedom	Sum of Squares	Mean Squares	F-statistic	p-value
Companies	3	0.000013450	0.000004483	16.70	0.000
Error	36	0.000009666	0.000000268		
Total	39	0.000023116			

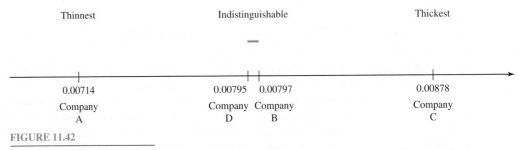

FIGURE 11.42

Schematic presentation of the results of the comparison of the gold layer thicknesses on the substrate bond pads for the four
companies

The differences between the sample averages are all larger than 0.000624 except for the
comparison of companies B and D, and consequently the researcher can conclude that while
there is no evidence of a difference between companies B and D, the data provides sufficient
evidence to conclude that company A has thinner gold layers on average than the other three
companies, while company C has thicker layers on average than the other three companies.
This result is illustrated in Figure 11.42.

11.4 Supplementary Problems

11.4.1 Biaxial Nanowire Tests

An experimenter measured the Young's modulus values of silicon carbide–silica biaxial nanowires. These synthesized nanowires are potentially useful for high-strength composites in which mechanical properties are critical. DS 11.4.1 gives a completely randomized design consisting of the Young's modulus measurements for samples of nanowires of four different types. Perform hand calculations to investigate whether there are any significant differences in the average Young's modulus of the four different types of nanowire.

11.4.2 Car Gas Efficiencies

A courier service buys four new cars of the same make and model, and they are driven under similar city and suburban conditions. Each time the cars are filled with gasoline, the car's mileage and amount of gasoline required to top off the tank are noted. At the end of a year, the company manager uses these data to calculate the gas mileages achieved by the cars between their fillups. This data set is given in DS 11.4.2. Are any of the cars getting a better gas mileage than the others? Use a computer package to obtain a graphical presentation of the data set, an analysis of variance table, and pairwise comparisons between the different cars.

11.4.3 Temperature Effect on Cement Curing

An experiment is conducted to assess the effect of the temperature during curing of a certain kind of cement. Six batches of the cement are prepared, and each batch is separated into five parts that are then cured at the five different temperature levels under consideration. The strengths of the cured cement samples are presented in DS 11.4.3. Perform hand calculations to investigate the effects of the different temperature levels.

11.4.4 Fertilizer Comparisons

A comparison of $k = 5$ fertilizers is made in a randomized block design that incorporates $b = 10$ fields of corn. Each field is partitioned into five equal areas, and the five fertilizers are randomly allocated to these areas. The resulting corn yields are given in DS 11.4.4. Which fertilizer is best? Use a computer package to obtain a graphical presentation of the data set and to help you with the analysis.

11.4.5 Red Blood Cell Adherence to Endothelial Cells

A hospital has $k = 4$ clinics where it can send blood samples for adhesion measurements of sickle red blood cells to endothelium layers. A hospital administrator decides to perform a randomized block experiment to investigate the consistency between the four clinics. A set of $b = 12$ blood samples is split into four parts, which are randomly sent to the four clinics. The numbers of adherent cells per square mm of endothelial monolayer reported by the clinics are given in DS 11.4.5. Are any of the clinics suspiciously different from the other clinics? Use a computer package to obtain a graphical presentation of the data set and to help you with the analysis.

11.4.6 Insertion Gains of Hearing Aids

The insertion gain of a hearing aid is defined to be the difference between the sound pressure level at the eardrum with and without the hearing aid in place. These sound pressure measurements can be made by a probe microphone placed in the ear canal of a subject. An experiment was conducted to investigate how the insertion gain of a particular hearing aid was affected by the elevation of the noise stimulus. Data were collected on the insertion gain for a constant noise stimulus placed at the horizontal level of the ear of a subject, placed above the horizontal level, and placed below the horizontal level. The data are listed in DS 11.4.6. What conclusions can you draw from this data set?

11.4.7 Air Resistance Drag for Road Vehicles

The reduction of drag from air resistance for large road vehicles is an important component of attempts to improve fuel efficiency. Wind tunnel tests were performed on models of four different vehicle designs, and the data set given in DS 11.4.7 was obtained. What conclusions can you draw from this data set about the drags of the four different designs?

11.4.8 Leather Shrinkage Measurements

When leather is used to produce furniture or clothes, its shrinkage is an important consideration. An experiment was conducted to discover the differences beween four different preparation methods in terms of the resulting shrinkages of the leather. Seven batches of leather were split into four parts each, which were then randomly assigned to each of the four preparation methods. The shrinkage measurements are given in DS 11.4.8. What conclusions can you draw from this data set about the differences between the four different preparation methods?

11.4.9   Are the following statements true or false?

(a) A one-way layout with only $k = 2$ treatments is equivalent to an independent samples design for comparing two means.

(b) In an ANOVA table the sum of squares for treatments can be negative if the only difference between the treatment levels can be explained by random variation.

(c) Large variability in the data within the treatments of a one-way layout produces a large sum of squares for error that makes it more difficult to identify a difference between the treatment means.

(d) In a randomized block design the degrees of freedom for error cannot be smaller than the degrees of freedom for blocks.

(e) An analysis of variance table shows how the total variability in a data set can be attributed to different sources.

(f) The distinction between randomized block designs and one-way layouts is analogous to the distinction between a paired two-sample data set and an unpaired two-sample data set.

(g) When a graphical representation is made of the differences between the treatments in a one-way layout, a line extending between two different treatments indicates that it has been proved that the two treatments are identical.

(h) A randomized block design has the advantage that it is not necessary to assume that the differences between the factor level effects are the same for each of the blocks.

11.4.10   Metal Alloy Hardness Tests

The data set in DS 11.4.9 is a one-way layout to compare the hardness measurements of three different alloys.

(a) Construct the ANOVA table. What conclusion can you draw from the ANOVA table about the differences between the alloys?

(b) Construct pairwise comparisons of the alloys and make a graphical representation of what you find. Which treatment mean is largest? Which treatment mean is smallest?

11.4.11   A randomized block design has four factor levels with sample averages $\bar{x}_{1.} = 11.43$, $\bar{x}_{2.} = 12.03$, $\bar{x}_{3.} = 14.88$, and $\bar{x}_{4.} = 11.76$. There are 9 blocks and SSBl $= 53.28$. Also, SSE $= 14.12$. Compute the ANOVA table. What do the p-values tell you? Construct a graphical representation of the differences between the factor means.

11.4.12   Aquatic Radon Levels

An investigation was conducted of the radon levels in five different rivers. Six samples of water were obtained from each river and the radon levels given in DS 11.4.10 were obtained. Construct an appropriate ANOVA table. What does the p-value in the ANOVA table tell you? Construct an appropriate graphical representation of the differences between the rivers. Which river has the highest radon level?

11.4.13   A completely randomized design to compare $k = 4$ factor levels has $n_1 = 11$ and $\bar{x}_{1.} = 213.7$, $n_2 = 8$ and $\bar{x}_{2.} = 206.3$, $n_3 = 13$ and $\bar{x}_{3.} = 205.7$, and $n_4 = 12$ and $\bar{x}_{4.} = 215.8$. The mean square error is MSE $= 11.23$. Calculate pairwise confidence intervals for the factor level means with an overall confidence level of 95%. Make a diagram showing which factor level means are known to be different and which ones are indistinguishable.

CHAPTER TWELVE

Simple Linear Regression and Correlation

Historically, the ideas of linear regression and correlation presented in this chapter have played a prominent part in statistical data analysis. When dealing with more than one variable, an experimenter is often interested in how a particular variable depends on one or more of the other variables. When the variables have a random component there will not be a deterministic relationship between them, but there may be an underlying structure that the experimenter can investigate. This **modeling** is often performed by finding a functional relationship between the *expected value* of a dependent variable and a set of explanatory or independent variables.

Linear regression is a modeling technique in which the expected value of a dependent variable is modeled as a **linear combination** of a set of explanatory variables. These linear models are easy to analyze and are applicable in many situations. **Simple linear regression**, discussed in this chapter, refers to a linear regression model with only one explanatory variable, while **multiple linear regression**, discussed in Chapter 13, refers to a linear regression model with two or more explanatory variables. A simple linear regression model is closely related to the calculation of a **correlation coefficient** to measure the degree of association between two variables, which is discussed at the end of this chapter.

12.1 The Simple Linear Regression Model

12.1.1 Model Definition and Assumptions

Consider a data set consisting of the paired observations

$$(x_1, y_1), \ldots, (x_n, y_n)$$

For example, x_i could represent the height and y_i could represent the weight of the ith person in a random sample of n adult males. Statistical modeling techniques can be used to investigate how the two variables, corresponding to the data values x_i and y_i, are related. In order to do this it is convenient to think of the data value y_i as being the observed value of a random variable Y_i, whose distribution depends upon a (fixed) value x_i.

In particular, with the *simple linear regression* model

$$y_i = \beta_0 + \beta_1 x_i + \epsilon_i$$

the observed value of the dependent variable y_i is composed of a linear function $\beta_0 + \beta_1 x_i$ of the explanatory variable x_i, together with an **error term** ϵ_i. The error terms $\epsilon_1, \ldots, \epsilon_n$ are generally taken to be independent observations from a $N(0, \sigma^2)$ distribution, for some **error variance** σ^2. This implies that the values $y_1, \ldots, y_n$ are observations from the independent random variables

$$Y_i \sim N(\beta_0 + \beta_1 x_i, \sigma^2)$$

as illustrated in Figure 12.1.

FIGURE 12.1

Simple linear regression model

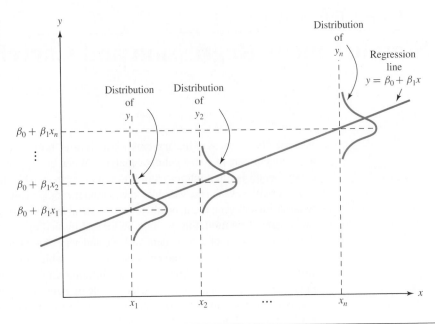

Notice that

$$E(Y_i) = \beta_0 + \beta_1 x_i$$

so that the expected value of the dependent variable is modeled as a linear function of the values of the explanatory variable. Thus, for the example of peoples' heights and weights, people with a height x are modeled as having weights with an expected value of $\beta_0 + \beta_1 x$.

In many cases the choice of which variable should be taken as the dependent variable y and which variable should be taken as the explanatory variable x is arbitrary. In an investigation of the relationship between peoples' heights and weights, either height or weight can sensibly be taken to be the dependent variable. However, if one variable can reasonably be thought of as depending upon and resulting from the specification of the other variable, then it is clearly sensible to model the former variable as the dependent variable y and the latter variable as the explanatory variable x.

The unknown parameters β_0 and β_1, which determine the relationship between the dependent variable and the explanatory variable, can be estimated from the data set. This procedure is referred to as "fitting a straight line to the data set." The parameter β_0 is known as the **intercept parameter**, and the parameter β_1 is known as the **slope parameter**. A third unknown parameter, the error variance σ^2, can also be estimated from the data set. As illustrated in Figure 12.2, the data values (x_i, y_i) lie closer to the line

$$y = \beta_0 + \beta_1 x$$

as the error variance σ^2 decreases.

The slope parameter β_1 is of particular interest since it indicates how the expected value of the dependent variable depends upon the explanatory variable x, as shown in Figure 12.3. If $\beta_1 = 0$ so that the line is flat, then changes in the explanatory variable x do not affect the distribution of the dependent variable. The two variables can then be interpreted as being unrelated. If $\beta_1 > 0$ so that the line slopes upward, then the dependent variable tends to

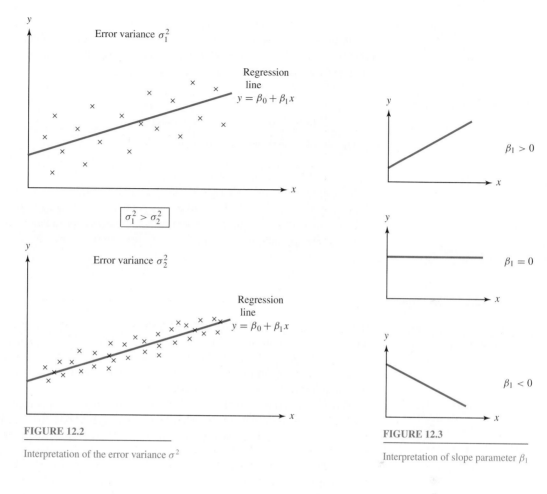

FIGURE 12.2

Interpretation of the error variance σ^2

FIGURE 12.3

Interpretation of slope parameter β_1

FIGURE 12.4

For this nonlinear relationship a simple linear regression model is not appropriate

increase in value as the explanatory variable x increases, and if $\beta_1 < 0$ so that the line slopes downward, then the dependent variable tends to decrease in value as the explanatory variable x increases. Clearly for the example of peoples' heights and weights, a positive value of β_1 would be expected.

It should always be remembered that it is ridiculous to fit a straight line to a data set that clearly does not exhibit a linear relationship. In other words, an experimenter should always look at a graph of a data set before fitting a line in order to assess whether this is a sensible thing to do. For example, the data set shown in Figure 12.4 clearly exhibits a quadratic (or at least nonlinear) relationship between the two variables, and it would make no sense to fit a straight line to the data set. Multiple linear regression methods for fitting quadratic and other more complicated models with more than one explanatory variable are discussed in Chapter 13, and variable transformations are discussed in Section 12.8.

Simple Linear Regression Model

The **simple linear regression** model

$$y_i = \beta_0 + \beta_1 x_i + \epsilon_i$$

fits a straight line through a set of paired data observations $(x_1, y_1), \ldots, (x_n, y_n)$. The error terms $\epsilon_1, \ldots, \epsilon_n$ are taken to be independent observations from a $N(0, \sigma^2)$ distribution. The three unknown parameters, the **intercept parameter** β_0, the **slope parameter** β_1, and the **error variance** σ^2, are estimated from the data set.

12.1.2 Examples

Example 67
Car Plant Electricity Usage

The manager of a car plant wishes to investigate how the plant's electricity usage depends upon the plant's production. The data set shown in Figure 12.5 is compiled and provides the plant's production and electrical usage for each month of the previous year. The electrical usage is in units of a million kilowatt-hours, and the production is measured as the value in million-dollar units of the cars produced in that month.

Figure 12.6 shows a graph of the data set, and the manager concludes that it is sensible to fit a straight line to the data points. The electricity usage is taken to be the dependent variable y, and the production is taken to be the explanatory variable x. The linear model

$$y = \beta_0 + \beta_1 x$$

	Production ($ million)	Electricity usage (million kWh)
January	4.51	2.48
February	3.58	2.26
March	4.31	2.47
April	5.06	2.77
May	5.64	2.99
June	4.99	3.05
July	5.29	3.18
August	5.83	3.46
September	4.70	3.03
October	5.61	3.26
November	4.90	2.67
December	4.20	2.53

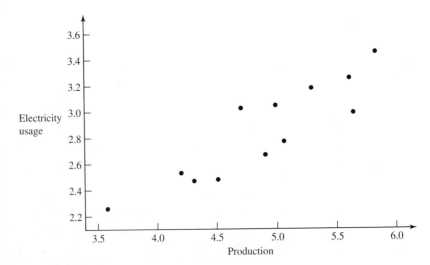

FIGURE 12.5

Car plant electricity usage data set

FIGURE 12.6

Scatter plot of the car plant electricity usage data set

will then allow a month's electrical usage to be estimated as a function of the month's production. The manager can estimate a month's electricity expenses once the month's production targets have been decided. Notice that for this example the slope parameter β_1 represents the expected increase in electricity usage resulting from an increase in production of $1 million.

Example 68
Nile River Flowrate

Figure 12.7 shows a map of the Nile River flowing from Sudan into Egypt and eventually into the Mediterranean Sea. The Aswan High Dam, situated at the northern end of Lake Nasser, controls the flow of water into the fertile regions bordering the Nile River in Egypt and to important metropolitan areas such as Cairo. The proper control of the water passing through the dam is critical to the well-being of agriculture, industry, and population centers.

The amount of water that the dam engineers can prudently allow to flow through the dam depends upon the amount of water reserves in Lake Nasser, which in turn depends upon the flow of water into the lake at Wadi Halfa. In this respect it is very useful to be able to predict the future Nile River flowrates at Wadi Halfa. Egyptian engineers have been measuring this inflow for over 100 years, and Figure 12.8 shows the inflows at Wadi Halfa for the months of January and February for the years 1874 to 1988. These inflows are defined to be the amount

FIGURE 12.7

The Nile River basin

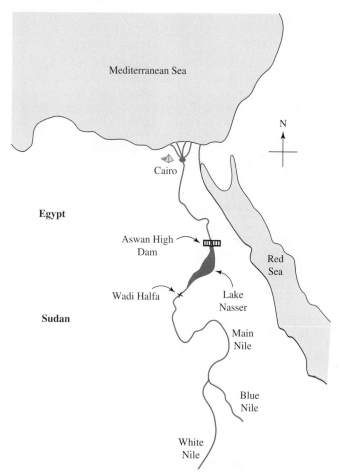

FIGURE 12.8

Nile River inflows in billions of cubic meters

Year	January	February	Year	January	February	Year	January	February
1874	6.41	4.48	1913	5.00	3.89	1951	3.71	2.77
1875	4.18	2.43	1914	4.34	3.20	1952	3.66	2.89
1876	5.43	4.18	1915	2.76	2.21	1953	3.94	3.00
1877	3.38	1.89	1916	2.89	1.91	1954	3.63	2.28
1878	5.47	3.81	1917	1.64	1.23	1955	3.23	1.91
1879	5.24	4.19	1918	3.98	2.74	1956	2.76	1.73
1880	4.57	2.90	1919	3.49	2.30	1957	3.12	1.75
1881	3.46	2.13	1920	4.74	3.79	1958	4.11	2.90
1882	7.07	5.36	1921	4.80	3.92	1959	3.59	2.89
1883	6.48	5.25	1922	3.28	2.42	1960	4.02	3.44
1884	4.45	3.01	1923	2.88	1.95	1961	2.63	1.80
1885	4.55	3.06	1924	3.28	2.16	1962	3.58	2.36
1886	4.66	3.46	1925	2.91	1.85	1963	3.68	2.64
1887	5.16	4.09	1926	3.14	2.01	1964	3.06	1.84
1888	4.81	3.33	1927	3.57	2.32	1965	4.34	3.37
1889	3.77	2.33	1928	3.25	2.14	1966	4.03	3.22
1890	4.64	3.04	1929	2.97	1.92	1967	5.13	3.96
1891	5.12	3.18	1930	3.60	2.50	1968	6.13	5.67
1892	2.86	1.70	1931	2.41	1.54	1969	4.58	3.61
1893	4.09	2.41	1932	3.28	2.18	1970	4.61	3.76
1894	5.24	3.19	1933	3.66	2.49	1971	4.94	4.28
1895	4.84	3.17	1934	2.46	1.67	1972	3.91	3.13
1896	6.14	4.80	1935	3.05	1.97	1973	4.03	3.19
1897	5.01	3.29	1936	3.84	3.14	1974	4.64	3.62
1898	6.25	4.67	1937	3.89	2.56	1975	4.45	3.73
1899	5.93	4.48	1938	3.77	2.62	1976	2.89	2.45
1900	6.44	4.54	1939	3.46	2.52	1977	3.57	2.57
1901	4.34	3.10	1940	3.11	2.15	1978	4.18	3.17
1902	5.58	4.31	1941	3.36	2.15	1979	4.71	3.93
1903	1.94	1.23	1942	4.02	3.28	1980	4.12	2.93
1904	3.84	2.39	1943	3.13	1.93	1981	5.76	3.53
1905	3.38	1.96	1944	2.24	1.53	1982	4.71	3.55
1906	3.62	2.24	1945	3.33	2.13	1983	3.63	2.82
1907	4.50	3.43	1946	3.15	1.89	1984	3.74	2.78
1908	3.85	2.50	1947	2.80	1.81	1985	3.94	2.85
1909	3.69	2.68	1948	2.65	1.68	1986	3.47	2.91
1910	4.29	2.89	1949	3.82	2.53	1987	3.88	4.01
1911	3.60	2.25	1950	4.33	3.67	1988	3.70	3.12
1912	4.53	3.43						

of water in billions of cubic meters that flows through a cross section of the Nile River at Wadi Halfa during the month under consideration.

Figure 12.9 shows summary statistics of the monthly inflows, and Figure 12.10 shows a scatter plot of this data set. The January inflows are seen to be generally slightly larger than the February inflows, and there does appear to be a fairly linear relationship between the inflows in the two months. It is sensible to fit the line

$$y = \beta_0 + \beta_1 x$$

where the explanatory variable x is the January inflow and the dependent variable y is the February inflow. This model is useful since at the end of January, when the January inflow

FIGURE 12.9

Summary statistics for the Nile River inflows

January inflow	February inflow
$n = 115$	$n = 115$
Sample mean = 4.0252	Sample mean = 2.8960
Sample standard deviation = 1.0228	Sample standard deviation = 0.9163
Maximum = 7.07	Maximum = 5.67
Upper quartile = 4.64	Upper quartile = 3.44
Median = 3.85	Median = 2.85
Lower quartile = 3.33	Lower quartile = 2.16
Minimum = 1.64	Minimum = 1.23

FIGURE 12.10

Scatter plot of the Nile River inflows data set

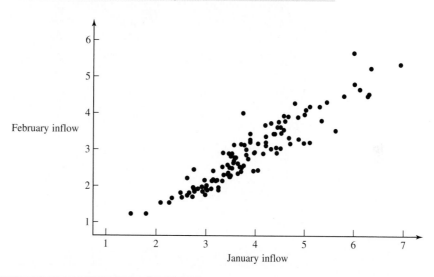

has been measured, the dam engineers can use it to estimate the inflow for the month ahead. The slope parameter β_1 indicates the expected increase in the February inflow for each unit increase in the January inflow.

Example 44
Army Physical Fitness Test

Recall that the Army Physical Fitness Test, discussed in Chapter 6, consists of 2 minutes of pushups followed by 2 minutes of situps and a 2-mile run. A data set of 84 run times for male Army officers was discussed in Chapter 6. Figure 12.11 shows this data set together with the number of pushups performed by the officers before their 2-mile run.

Figure 12.12 shows summary statistics of the number of pushups and the run times, and Figure 12.13 shows a scatter plot of the run times against the number of pushups. As might be expected, officers performing more pushups tend to have shorter run times, indicating a better all-around athletic ability. With the run time as the dependent variable y and the number of pushups as the explanatory variable x, the linear model

$$y = \beta_0 + \beta_1 x$$

allows run times to be predicted as a function of the number of pushups. This model is useful for medical researchers to help investigate how well people can perform different kinds of athletic tasks, and it can be used to identify officers whose performance on the test is in some way different from normal.

FIGURE 12.11

Army Physical Fitness Test data set

Number of pushups	Two-mile run time (seconds)	Number of pushups	Two-mile run time (seconds)	Number of pushups	Two-mile run time (seconds)
60	847	50	880	70	870
53	887	70	905	55	931
60	879	69	895	55	930
55	919	125	720	80	808
60	816	80	712	78	828
78	814	99	703	78	719
74	814	80	741	68	707
70	855	61	792	45	934
46	980	50	808	45	939
50	954	55	761	47	977
50	1078	60	785	88	896
59	1001	60	801	70	921
62	766	65	810	80	815
64	916	66	1013	79	838
51	798	58	882	71	854
66	782	62	861	50	1063
73	836	60	845	50	1024
78	837	90	865	73	780
80	791	78	883	78	813
70	838	86	881	78	850
70	853	86	921	63	902
79	840	69	816	60	906
93	740	74	837	78	865
80	763	40	1056	78	886
78	778	40	1034	78	881
70	855	78	774	50	825
60	875	70	821	61	821
78	868	78	850	62	832

FIGURE 12.12

Summary statistics for the Army
Physical Fitness Test

Pushups	Run time
$n = 84$	$n = 84$
Sample mean = 67.79	Sample mean = 857.7
Sample standard deviation = 14.31	Sample standard deviation = 81.98
Maximum = 125	Maximum = 1078
Upper quartile = 78	Upper quartile = 900.5
Median = 69.5	Median = 850
Lower quartile = 59.25	Lower quartile = 808.5
Minimum = 40	Minimum = 703

FIGURE 12.13

Scatter plot of the Army Physical
Fitness Test data set

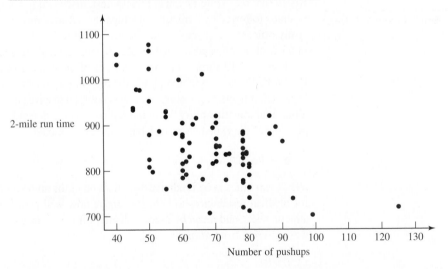

■ 12.1.3 Problems

12.1.1 Suppose that a dependent variable y is related to an explanatory variable x through a linear regression model with $\beta_0 = 4.2$, $\beta_1 = 1.7$, and with normally distributed error terms with a zero mean and a standard deviation of $\sigma = 3.2$.
  (a) What is the expected value of the dependent variable when $x = 10.0$?
  (b) How much does the expected value of the dependent variable change when the explanatory variable increases by 3?
  (c) What is the probability that the dependent variable is larger than 12.0 when $x = 5$?
  (d) What is the probability that the dependent variable is smaller than 17.0 when $x = 8$?
  (e) What is the probability that a dependent variable obtained at $x = 6$ is larger than a dependent variable obtained at $x = 7$?

12.1.2 Suppose that the purity of a chemical solution y is related to the amount of a catalyst x through a linear regression model with $\beta_0 = 123.0$, $\beta_1 = -2.16$, and with an error standard deviation $\sigma = 4.1$.
  (a) What is the expected value of the purity when the catalyst level is 20?
  (b) How much does the expected value of the purity change when the catalyst level increases by 10?
  (c) What is the probability that the purity is less than 60.0 when the catalyst level is 25?
  (d) What is the probability that the purity is between 30 and 40 when the catalyst level is 40?
  (e) What is the probability that the purity of a solution

with a catalyst level of 30 is smaller than the purity of a solution with a catalyst level of 27.5?

12.1.3 Consider the linear regression model $y = 5 + 0.9x$, with $\sigma = 1.4$, which relates the air content of a concrete sample x to the porosity of the sample y.
  (a) What is the expected value of the porosity for a concrete sample with an air content of 20?
  (b) How much does the expected value of the porosity decrease as the air content is reduced by 5?
  (c) What is the probability that a concrete sample with an air content of 25 has a porosity no larger than 30?
  (d) What is the probability that four independent samples of concrete each with an air content of 15 have an average porosity between 17 and 20?

12.1.4 Suppose that you wish to use simple linear regression to investigate the relationship between two variables. Why does the model you obtain depend on which variable is chosen as the y variable? How should you decide which of your two variables should be the y variable?

12.1.5 Consider the linear regression model

$$y = 675.30 - 5.87x$$

with $\sigma = 7.32$, which relates the viscosity y of a substance to the temperature x. If a sample of the substance is prepared at temperature 80, what is the probability that its viscosity is less than 220?

12.2 Fitting the Regression Line

12.2.1 Parameter Estimation

The regression line

$$y = \beta_0 + \beta_1 x$$

is fitted to the data points

$$(x_1, y_1), \ldots, (x_n, y_n)$$

by finding the line that is "closest" to the data points in some sense. There are many ways in which closeness can be defined, but the method most generally used is to consider the vertical deviations between the line and the data points

$$y_i - (\beta_0 + \beta_1 x_i) \qquad 1 \le i \le n$$

FIGURE 12.14

The least squares fit

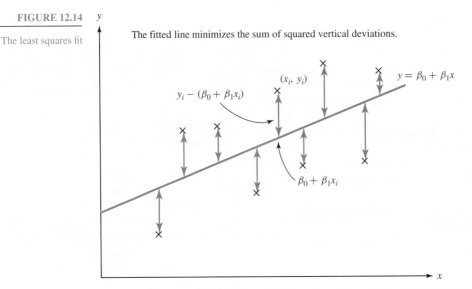

As Figure 12.14 illustrates, the fitted line is chosen to be the line that *minimizes* the sum of the squares of these vertical deviations

$$Q = \sum_{i=1}^{n}(y_i - (\beta_0 + \beta_1 x_i))^2$$

and this is referred to as the *least squares* fit.

Although this definition of closeness is in a sense arbitrary, with normally distributed error terms it results in parameter estimates $\hat{\beta}_0$ and $\hat{\beta}_1$ that are *maximum likelihood estimates*. This is because an error term ϵ_i has a probability density function

$$\frac{1}{\sqrt{2\pi}\sigma} e^{-\epsilon_i^2/2\sigma^2}$$

so that the joint density of the error terms $\epsilon_1, \ldots, \epsilon_n$ is

$$\left(\frac{1}{\sqrt{2\pi}\sigma}\right)^n e^{-\sum_{i=1}^{n} \epsilon_i^2/2\sigma^2}$$

This likelihood is maximized by minimizing

$$\sum_{i=1}^{n} \epsilon_i^2$$

which is equal to Q because

$$\epsilon_i = y_i - (\beta_0 + \beta_1 x_i)$$

The parameter estimates $\hat{\beta}_0$ and $\hat{\beta}_1$ are therefore the values that minimize the quantity Q. They are easily found by taking partial derivatives of Q with respect to β_0 and β_1 and setting the resulting expressions equal to 0. Since

$$\frac{\partial Q}{\partial \beta_0} = -\sum_{i=1}^{n} 2(y_i - (\beta_0 + \beta_1 x_i))$$

and

$$\frac{\partial Q}{\partial \beta_1} = -\sum_{i=1}^{n} 2x_i(y_i - (\beta_0 + \beta_1 x_i))$$

the parameter estimates $\hat{\beta}_0$ and $\hat{\beta}_1$ are thus the solutions to the equations

$$\sum_{i=1}^{n} y_i = n\beta_0 + \beta_1 \sum_{i=1}^{n} x_i$$

and

$$\sum_{i=1}^{n} x_i y_i = \beta_0 \sum_{i=1}^{n} x_i + \beta_1 \sum_{i=1}^{n} x_i^2$$

These equations are known as the **normal equations**.

The normal equations can be solved to give

$$\hat{\beta}_1 = \frac{n \sum_{i=1}^{n} x_i y_i - \left(\sum_{i=1}^{n} x_i\right)\left(\sum_{i=1}^{n} y_i\right)}{n \sum_{i=1}^{n} x_i^2 - \left(\sum_{i=1}^{n} x_i\right)^2}$$

and then $\hat{\beta}_0$ can be calculated as

$$\hat{\beta}_0 = \frac{\sum_{i=1}^{n} y_i}{n} - \hat{\beta}_1 \frac{\sum_{i=1}^{n} x_i}{n} = \bar{y} - \hat{\beta}_1 \bar{x}$$

Notice that with the notation

$$S_{XX} = \sum_{i=1}^{n} (x_i - \bar{x})^2 = \sum_{i=1}^{n} x_i^2 - n\bar{x}^2 = \sum_{i=1}^{n} x_i^2 - \frac{\left(\sum_{i=1}^{n} x_i\right)^2}{n}$$

and

$$S_{XY} = \sum_{i=1}^{n} (x_i - \bar{x})(y_i - \bar{y}) = \sum_{i=1}^{n} x_i y_i - n\bar{x}\bar{y} = \sum_{i=1}^{n} x_i y_i - \frac{\left(\sum_{i=1}^{n} x_i\right)\left(\sum_{i=1}^{n} y_i\right)}{n}$$

the parameter estimate $\hat{\beta}_1$ can be written as

$$\hat{\beta}_1 = \frac{S_{XY}}{S_{XX}}$$

The fitted regression line is

$$y = \hat{\beta}_0 + \hat{\beta}_1 x$$

For a specific value of the explanatory variable x^*, this equation provides a **fitted value**

$$\hat{y}|_{x^*} = \hat{\beta}_0 + \hat{\beta}_1 x^*$$

for the dependent variable y, as illustrated in Figure 12.15. For data points x_i, the fitted values are

$$\hat{y}_i = \hat{\beta}_0 + \hat{\beta}_1 x_i$$

The error variance σ^2 can be estimated by considering the deviations between the observed data values y_i and their fitted values $\hat{y}_i$. Specifically, the sum of squares for error SSE is defined

FIGURE 12.15

Fitted values from a regression line

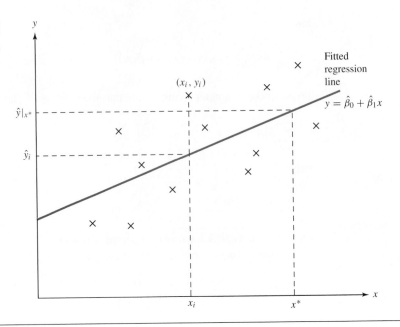

to be the sum of the squares of these deviations

$$\text{SSE} = \sum_{i=1}^{n}(y_i - \hat{y}_i)^2 = \sum_{i=1}^{n}(y_i - (\hat{\beta}_0 + \hat{\beta}_1 x_i))^2$$

and the estimate of the error variance is

$$\hat{\sigma}^2 = \frac{\text{SSE}}{n-2} = \frac{\sum_{i=1}^{n}(y_i - (\hat{\beta}_0 + \hat{\beta}_1 x_i))^2}{n-2}$$

Expanding the square in the expression for the sum of squares for error results in the expressions

$$\text{SSE} = \sum_{i=1}^{n} y_i^2 - \hat{\beta}_0 \sum_{i=1}^{n} y_i - \hat{\beta}_1 \sum_{i=1}^{n} x_i y_i$$

and consequently

$$\hat{\sigma}^2 = \frac{\sum_{i=1}^{n} y_i^2 - \hat{\beta}_0 \sum_{i=1}^{n} y_i - \hat{\beta}_1 \sum_{i=1}^{n} x_i y_i}{n-2}$$

which is a more convenient representation for computation purposes.

It is interesting to notice that a regression line can be fitted to the data points (x_i, y_i) and the error variance can be estimated using only the six quantities

$$n \qquad \sum_{i=1}^{n} x_i \qquad \sum_{i=1}^{n} y_i \qquad \sum_{i=1}^{n} x_i^2 \qquad \sum_{i=1}^{n} y_i^2 \qquad \sum_{i=1}^{n} x_i y_i$$

Parameter Estimates

The point estimates of the three unknown parameters in a simple linear regression model are

$$\hat{\beta}_1 = \frac{n \sum_{i=1}^{n} x_i y_i - \left(\sum_{i=1}^{n} x_i\right)\left(\sum_{i=1}^{n} y_i\right)}{n \sum_{i=1}^{n} x_i^2 - \left(\sum_{i=1}^{n} x_i\right)^2} = \frac{\sum_{i=1}^{n}(x_i - \bar{x})(y_i - \bar{y})}{\sum_{i=1}^{n}(x_i - \bar{x})^2} = \frac{S_{XY}}{S_{XX}}$$

$$\hat{\beta}_0 = \bar{y} - \hat{\beta}_1 \bar{x}$$

$$\hat{\sigma}^2 = \frac{\text{SSE}}{n-2} = \frac{\sum_{i=1}^{n}(y_i - (\hat{\beta}_0 + \hat{\beta}_1 x_i))^2}{n-2}$$

$$= \frac{\sum_{i=1}^{n} y_i^2 - \hat{\beta}_0 \sum_{i=1}^{n} y_i - \hat{\beta}_1 \sum_{i=1}^{n} x_i y_i}{n-2}$$

12.2.2 Examples

Example 67
Car Plant Electricity Usage

For this example $n = 12$ and

$$\sum_{i=1}^{12} x_i = 4.51 + \cdots + 4.20 = 58.62$$

$$\sum_{i=1}^{12} y_i = 2.48 + \cdots + 2.53 = 34.15$$

$$\sum_{i=1}^{12} x_i^2 = 4.51^2 + \cdots + 4.20^2 = 291.2310$$

$$\sum_{i=1}^{12} y_i^2 = 2.48^2 + \cdots + 2.53^2 = 98.6967$$

$$\sum_{i=1}^{12} x_i y_i = (4.51 \times 2.48) + \cdots + (4.20 \times 2.53) = 169.2532$$

The estimate of the slope parameter is therefore

$$\hat{\beta}_1 = \frac{n \sum_{i=1}^{n} x_i y_i - \left(\sum_{i=1}^{n} x_i\right)\left(\sum_{i=1}^{n} y_i\right)}{n \sum_{i=1}^{n} x_i^2 - \left(\sum_{i=1}^{n} x_i\right)^2}$$

$$= \frac{(12 \times 169.2532) - (58.62 \times 34.15)}{(12 \times 291.2310) - 58.62^2} = 0.49883$$

and the estimate of the intercept parameter is

$$\hat{\beta}_0 = \bar{y} - \hat{\beta}_1 \bar{x} = \frac{34.15}{12} - \left(0.49883 \times \frac{58.62}{12}\right) = 0.4090$$

The fitted regression line is thus

$$y = \hat{\beta}_0 + \hat{\beta}_1 x = 0.409 + 0.499x$$

which is shown together with the data points in Figure 12.16. If a production level of $5.5 million worth of cars is planned for next month, then the plant manager can predict that the

FIGURE 12.16

Fitted line plot for the car plant
electricity usage data set

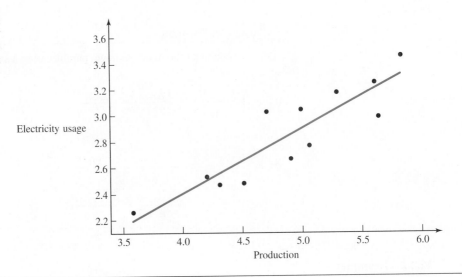

electricity usage will be

$$\hat{y}|_{5.5} = 0.409 + (0.499 \times 5.5) = 3.1535$$

With $\hat{\beta}_1 = 0.499$, this model predicts that the electricity usage increases by about half a million kilowatt-hours for every additional $1 million of production.

It must be remembered that it is best to use this regression model only for production values x that are close to the values x_i in the data set, that is, for production levels between about $3.5 million and $6 million. Using the model for production values x outside this range is known as **extrapolation** and may give inaccurate results. For example, suppose that production is going to increase to $8 million next month. Is the estimate of electrical usage

$$\hat{y}|_{8.0} = 0.409 + (0.499 \times 8.0) = 4.401$$

a good estimate? It may be inaccurate. Although it seems reasonable (based upon the data set available) to model electricity usage as a linear function of production for production values between $x = 3.5$ and $x = 6.0$, this linear relationship may not hold for production values greater than $x = 6.0$. Figure 12.17 illustrates the danger of extrapolation.

For this problem the error variance is estimated as

$$\hat{\sigma}^2 = \frac{\sum_{i=1}^{n} y_i^2 - \hat{\beta}_0 \sum_{i=1}^{n} y_i - \hat{\beta}_1 \sum_{i=1}^{n} x_i y_i}{n-2}$$

$$= \frac{98.6967 - (0.4090 \times 34.15) - (0.49883 \times 169.2532)}{10} = 0.0299$$

so that

$$\hat{\sigma} = \sqrt{0.0299} = 0.1729$$

Example 68

Nile River Flowrate

The statistics for this example are $n = 115$ and

$$\sum_{i=1}^{115} x_i = 6.41 + \cdots + 3.70 = 462.90$$

$$\sum_{i=1}^{115} y_i = 4.48 + \cdots + 3.12 = 333.04$$

FIGURE 12.17

Extrapolation dangers

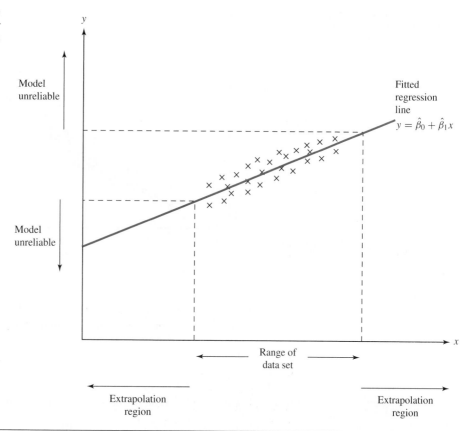

$$\sum_{i=1}^{115} x_i^2 = 6.41^2 + \cdots + 3.70^2 = 1982.5264$$

$$\sum_{i=1}^{115} y_i^2 = 4.48^2 + \cdots + 3.12^2 = 1060.2076$$

$$\sum_{i=1}^{115} x_i y_i = (6.41 \times 4.48) + \cdots + (3.70 \times 3.12) = 1440.2743$$

The estimate of the slope parameter is

$$\hat{\beta}_1 = \frac{(115 \times 1440.2743) - (462.90 \times 333.04)}{(115 \times 1982.5264) - 462.90^2} = 0.8362$$

and the estimate of the intercept parameter is

$$\hat{\beta}_0 = \frac{333.04}{115} - \left(0.8362 \times \frac{462.90}{115}\right) = -0.4698$$

The fitted regression line is therefore

$$y = \hat{\beta}_0 + \hat{\beta}_1 x = -0.470 + 0.836x$$

FIGURE 12.18

Fitted line plot for the Nile River inflows data set.

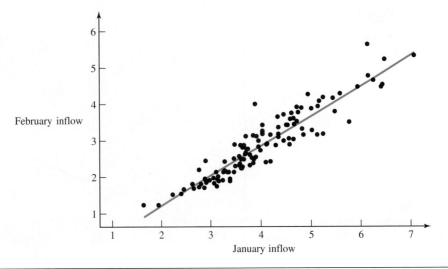

which is shown together with the data points in Figure 12.18. If the inflow in January is recorded as being 5 billion cubic meters, then the inflow for February can be estimated to be

$$\hat{y}|_5 = -0.470 + (0.836 \times 5) = 3.71$$

Each additional billion cubic meters of inflow in January results in an increase in the predicted February inflow of $\hat{\beta}_1 = 0.836$ billion cubic meters. Finally, the error variance is estimated as

$$\hat{\sigma}^2 = \frac{1060.2076 - (-0.4698 \times 333.04) - (0.8362 \times 1440.2743)}{113} = 0.1092$$

Example 44

Army Physical Fitness Test

In this case $n = 84$ and

$$\sum_{i=1}^{84} x_i = 60 + \cdots + 62 = 5694$$

$$\sum_{i=1}^{84} y_i = 847 + \cdots + 832 = 72{,}047$$

$$\sum_{i=1}^{84} x_i^2 = 60^2 + \cdots + 62^2 = 402{,}972$$

$$\sum_{i=1}^{84} y_i^2 = 847^2 + \cdots + 832^2 = 62{,}352{,}661$$

$$\sum_{i=1}^{84} x_i y_i = (60 \times 847) + \cdots + (62 \times 832) = 4{,}828{,}432$$

The estimate of the slope parameter is

$$\hat{\beta}_1 = \frac{(84 \times 4{,}828{,}432) - (5694 \times 72{,}047)}{(84 \times 402{,}972) - 5694^2} = -3.2544$$

FIGURE 12.19

Fitted line plot for the Army
Physical Fitness Test data set

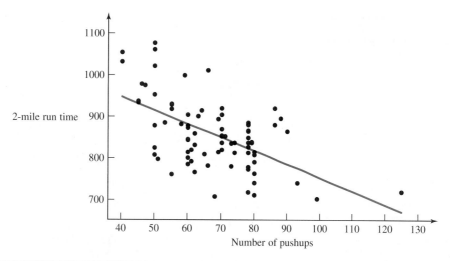

and the estimate of the intercept parameter is

$$\hat{\beta}_0 = \frac{72,047}{84} - \left(-3.2544 \times \frac{5694}{84}\right) = 1078.30$$

The fitted regression line is therefore

$$y = \hat{\beta}_0 + \hat{\beta}_1 x = 1078 - 3.254x$$

which is shown together with the data points in Figure 12.19. An officer who performs $x = 70$ pushups has an expected run time of

$$\hat{y}|_{70} = 1078 - (3.254 \times 70) = 850.22$$

seconds, which is 14 minutes and 10 seconds. The slope parameter $\hat{\beta}_1 = -3.254$ indicates that the expected run time *decreases* by 3.254 seconds for each additional pushup performed.

The error variance is estimated as

$$\hat{\sigma}^2 = \frac{62,352,661 - (1078.30 \times 72,047) - (-3.2544 \times 4,828,432)}{82} = 4606.42$$

and

$$\hat{\sigma} = \sqrt{4606.42} = 67.87$$

In summary, it can be seen that fitting straight lines to data sets is quite easy and has many uses. However, two prevalent abuses of this simple modeling technique should never be forgotten. These abuses are firstly, extrapolation, which was discussed in the example of electricity usage, and secondly, modeling nonlinear relationships with a straight line. With regard to the second abuse, although it is *always possible* to fit a straight line to a data set, this does not imply that it is *always appropriate* to fit a straight line to a data set. Figure 12.20 shows a straight line fitted to a data set exhibiting a nonlinear relationship, and the resulting linear model is clearly misleading and inaccurate.

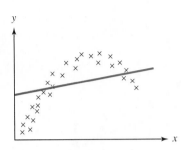

FIGURE 12.20

Inappropriate straight line fitted to a nonlinear relationship

---

■ 12.2.3 Problems

12.2.1 Show that the values

$$\hat{\beta}_1 = \frac{S_{XY}}{S_{XX}} \quad \text{and} \quad \hat{\beta}_0 = \bar{y} - \hat{\beta}_1 \bar{x}$$

satisfy the normal equations.

12.2.2 A data set has $n = 20$, $\sum_{i=1}^{20} x_i = 8.552$, $\sum_{i=1}^{20} y_i = 398.2$, $\sum_{i=1}^{20} x_i^2 = 5.196$, $\sum_{i=1}^{20} y_i^2 = 9356$, and $\sum_{i=1}^{20} x_i y_i = 216.6$. Calculate $\hat{\beta}_0$, $\hat{\beta}_1$, and $\hat{\sigma}^2$. What is the fitted value when $x = 0.5$?

12.2.3 A data set has $n = 30$, $\sum_{i=1}^{30} x_i = -67.11$, $\sum_{i=1}^{30} y_i = 1322.7$, $\sum_{i=1}^{30} x_i^2 = 582.0$, $\sum_{i=1}^{30} y_i^2 = 60,600$, and $\sum_{i=1}^{30} x_i y_i = -3840$. Calculate $\hat{\beta}_0$, $\hat{\beta}_1$, and $\hat{\sigma}^2$. What is the fitted value when $x = -2.0$?

12.2.4 Oil Well Drilling Costs
Estimating the costs of drilling oil wells is an important consideration for the oil industry. DS 12.2.1 contains the total costs and the depths of 16 offshore oil wells located in the Philippines (taken from "Identifying the major determinants of exploration drilling costs: A first approximation using the Philippine case" by Gary S. Makasiar, *Energy Exploration and Exploitation*, 1985).
(a) Fit a linear regression model with cost as the dependent variable and depth as the explanatory variable.
(b) What does your model predict as the cost increase for an additional depth of 1000 feet?
(c) What cost would you predict for an oil well of 10,000 feet depth?
(d) What is the estimate of the error variance?
(e) What could you say about the cost of an oil well of depth 20,000 feet?

(This problem is continued in Problems 12.3.3, 12.4.4, 12.5.3, 12.6.5, 12.7.1, and 12.9.3.)

12.2.5 Truck Unloading Times
A warehouse manager is interested in the possible improvements to labor efficiency if air-conditioning is installed in the warehouse. The data set given in DS 12.2.2 is collected which shows the times taken to unload a fully laden truck at various temperature levels.
(a) Fit a linear regression model with time as the dependent variable and temperature as the explanatory variable.
(b) What is the estimate of the error variance?
(c) Does the model suggest that the unloading time increases with temperature?
(d) What is the predicted unloading time when the temperature is 72° F?

(This problem is continued in Problems 12.3.4, 12.6.6, 12.7.2, and 12.9.4.)

12.2.6 VO2-max Aerobic Fitness Measurements
DS 12.2.3 is a data set concerning the aerobic fitness of a sample of 20 male subjects. An exercising individual breathes through an apparatus that measures the amount of oxygen in the inhaled air that is used by the individual. The maximum value per unit time of the utilized oxygen is then scaled by the person's body weight to come up with a variable VO2-max, which is a general indication of the aerobic fitness of the individual.
(a) Fit a linear regression model with VO2-max as the dependent variable and age as the explanatory variable.
(b) Does the sign of $\hat{\beta}_1$ surprise you? What does your model predict as the change in VO2-max for an additional five years of age?

(c) What VO2-max value would you predict for a 50-year-old man?

(d) What does the model tell you about the aerobic fitness of a 15-year-old boy?

(e) What is the estimate of the error variance?

(This problem is continued in Problems 12.3.5, 12.4.5, 12.5.4, 12.6.7, 12.7.3, and 12.9.5.)

12.2.7 Property Tax Appraisals

A Realtor collects the data set given in DS 12.2.4 concerning the sizes of a random selection of newly constructed houses in a certain area together with their appraised values for tax purposes.

(a) Fit a linear regression model with appraised value as the dependent variable and size as the explanatory variable.

(b) What is the predicted value of a house with 2600 square feet?

(c) How does the predicted value change as the size increases by 100 square feet?

(d) What is the estimate of the error variance?

(This problem is continued in Problems 12.3.6, 12.4.6, 12.5.5, 12.6.8, 12.7.4, and 12.9.6.)

12.2.8 Management of Computer Systems

During the installation of a large computer system, it is useful to know how long specific tasks will take, particularly programming changes. A great deal of effort is spent estimating the amount of time such tasks will take and learning how to effectively use such estimations. Having an accurate idea of the time required for these tasks is crucial for the effective planning and timely completion of the installation. The data set in DS 12.2.5 relates one expert's time estimates for programming changes to the actual time the tasks took.

(a) Fit a linear regression model with actual time as the dependent variable and estimated time as the explanatory variable.

(b) What is the predicted increase in the actual time when the estimated time increases by 1 hour? Is the expert underestimating or overestimating the times? If the estimated time is 7 hours, what is the predicted value of the actual time the task will take?

(c) What does your model tell you about the time it will take to perform a task that is estimated at 15 hours?

(d) What is the estimate of the error variance?

(This problem is continued in Problems 12.3.7, 12.4.7, 12.5.6, 12.6.9, 12.7.5, and 12.9.7.)

12.2.9 Vacuum Transducer Bobbin Resistances

The data set in DS 12.2.6 concerns the relationship between the temperature and resistance of vacuum transducer bobbins, which are used in the automobile industry.

(a) Fit a linear regression model with resistance as the dependent variable and temperature as the explanatory variable.

(b) What is the predicted resistance when the temperature is 69° F?

(c) How does the predicted resistance change as the temperature increases by 5 degrees?

(d) What is the estimate of the error variance?

(This problem is continued in Problems 12.3.8, 12.4.8, 12.5.7, 12.6.10, 12.7.6, and 12.9.8.)

12.3 Inferences on the Slope Parameter β_1

12.3.1 Inference Procedures

The slope parameter β_1 is of particular interest to an experimenter since it determines the nature of the relationship between the explanatory variable x and the dependent variable y. It is useful to be able to construct confidence intervals for the slope parameter and to test hypotheses that the slope parameter takes a certain value. Remember that β_1 is unknown and represents the slope of the true unknown regression line, whereas $\hat{\beta}_1$ is the estimate of the slope obtained by fitting a line to the data set.

The point estimate of the slope parameter is

$$\hat{\beta}_1 = \frac{S_{XY}}{S_{XX}}$$

Since

$$S_{XY} = \sum_{i=1}^{n}(x_i - \bar{x})(y_i - \bar{y}) = \sum_{i=1}^{n}(x_i - \bar{x})y_i - \bar{y}\sum_{i=1}^{n}(x_i - \bar{x})$$

$$= \sum_{i=1}^{n}(x_i - \bar{x})y_i - 0 = \sum_{i=1}^{n}(x_i - \bar{x})y_i$$

it follows that

$$\hat{\beta}_1 = \sum_{i=1}^{n} c_i y_i$$

where

$$c_i = \frac{x_i - \bar{x}}{S_{XX}}$$

Thus the point estimate $\hat{\beta}_1$ is the realization of the random variable

$$\sum_{i=1}^{n} c_i Y_i$$

where

$$Y_i \sim N(\beta_0 + \beta_1 x_i, \sigma^2) \qquad 1 \le i \le n$$

The notation $\hat{\beta}_1$ is generally used both for this random variable and for its realization, because this is unlikely to cause confusion. Notice then that

$$E(\hat{\beta}_1) = E\left(\sum_{i=1}^{n} c_i Y_i\right) = \sum_{i=1}^{n} c_i E(Y_i) = \sum_{i=1}^{n} c_i(\beta_0 + \beta_1 x_i)$$

$$= \beta_0 \frac{\sum_{i=1}^{n}(x_i - \bar{x})}{S_{XX}} + \beta_1 \frac{\sum_{i=1}^{n}(x_i - \bar{x})x_i}{S_{XX}}$$

$$= 0 + \beta_1 \frac{\sum_{i=1}^{n} x_i^2 - n\bar{x}^2}{S_{XX}} = \beta_1 \frac{S_{XX}}{S_{XX}} = \beta_1$$

so that $\hat{\beta}_1$ is an *unbiased* point estimate of the slope parameter β_1. Furthermore,

$$\text{Var}(\hat{\beta}_1) = \text{Var}\left(\sum_{i=1}^{n} c_i Y_i\right) = \sum_{i=1}^{n} c_i^2 \text{Var}(Y_i) = \sigma^2 \sum_{i=1}^{n} c_i^2$$

$$= \sigma^2 \sum_{i=1}^{n} \frac{(x_i - \bar{x})^2}{S_{XX}^2} = \frac{\sigma^2}{S_{XX}^2} \sum_{i=1}^{n}(x_i - \bar{x})^2 = \frac{\sigma^2}{S_{XX}^2} S_{XX} = \frac{\sigma^2}{S_{XX}}$$

so that the standard error of the point estimate $\hat{\beta}_1$ is

$$\text{s.e.}(\hat{\beta}_1) = \frac{\sigma}{\sqrt{S_{XX}}}$$

which can be estimated as

$$\frac{\hat{\sigma}}{\sqrt{S_{XX}}}$$

FIGURE 12.21

The slope parameter β_1 is estimated *more accurately* in scenario II than in scenario I because the data points are more spread out and S_{xx} is larger

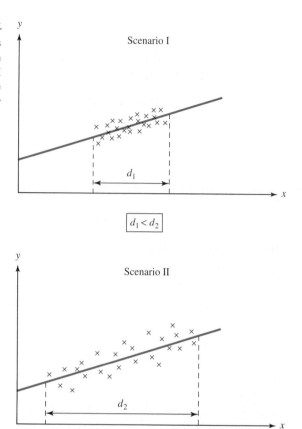

In fact, because $\hat{\beta}_1$ is a linear combination of normally distributed random variables Y_i, it also has a normal distribution, so that

$$\hat{\beta}_1 \sim N\left(\beta_1, \frac{\sigma^2}{S_{XX}}\right)$$

An interesting point to notice is that for a fixed value of the error variance σ^2, the variance of the slope parameter estimate decreases as the value of S_{XX} increases. This happens as the values of the explanatory variable x_i become more spread out, as illustrated in Figure 12.21. This result is intuitively reasonable because a greater spread in the values x_i provides a greater "leverage" for fitting the regression line, and therefore the slope parameter estimate $\hat{\beta}_1$ should be more accurate.

The normal distribution for the slope parameter estimate $\hat{\beta}_1$ can be standardized as

$$\frac{\hat{\beta}_1 - \beta_1}{\sqrt{\frac{\sigma^2}{S_{XX}}}} \sim N(0, 1)$$

Also, the estimate of the error variance is distributed

$$\hat{\sigma}^2 \sim \sigma^2 \frac{\chi^2_{n-2}}{n - 2}$$

Inferences on the Slope Parameter β_1

A two-sided confidence interval with a confidence level $1 - \alpha$ for the slope parameter in a simple linear regression model is

$$\beta_1 \in \left(\hat{\beta}_1 - t_{\alpha/2, n-2} \times \text{s.e.}(\hat{\beta}_1), \, \hat{\beta}_1 + t_{\alpha/2, n-2} \times \text{s.e.}(\hat{\beta}_1) \right)$$

which is

$$\beta_1 \in \left(\hat{\beta}_1 - \frac{\hat{\sigma} \, t_{\alpha/2, n-2}}{\sqrt{S_{XX}}}, \, \hat{\beta}_1 + \frac{\hat{\sigma} \, t_{\alpha/2, n-2}}{\sqrt{S_{XX}}} \right)$$

One-sided $1 - \alpha$ confidence level confidence intervals are

$$\beta_1 \in \left(-\infty, \, \hat{\beta}_1 + \frac{\hat{\sigma} \, t_{\alpha, n-2}}{\sqrt{S_{XX}}} \right) \qquad \text{and} \qquad \beta_1 \in \left(\hat{\beta}_1 - \frac{\hat{\sigma} \, t_{\alpha, n-2}}{\sqrt{S_{XX}}}, \, \infty \right)$$

The two-sided hypotheses

$$H_0 : \beta_1 = b_1 \quad \text{versus} \quad H_A : \beta_1 \neq b_1$$

for a fixed value b_1 of interest are tested with the t-statistic

$$t = \frac{\hat{\beta}_1 - b_1}{\sqrt{\frac{\hat{\sigma}^2}{S_{XX}}}}$$

The p-value is

$$p\text{-value} = 2 \times P(X > |t|)$$

where the random variable X has a t-distribution with $n - 2$ degrees of freedom. A size α test rejects the null hypothesis if $|t| > t_{\alpha/2, n-2}$.

The one-sided hypotheses

$$H_0 : \beta_1 \geq b_1 \quad \text{versus} \qquad H_A : \beta_1 < b_1$$

have a p-value

$$p\text{-value} = P(X < t)$$

and a size α test rejects the null hypothesis if $t < -t_{\alpha, n-2}$. The one-sided hypotheses

$$H_0 : \beta_1 \leq b_1 \quad \text{versus} \quad H_A : \beta_1 > b_1$$

have a p-value

$$p\text{-value} = P(X > t)$$

and a size α test rejects the null hypothesis if $t > t_{\alpha, n-2}$.

and if this estimate is used in place of the error variance σ^2, then the resulting expression has a t-distribution with $n - 2$ degrees of freedom

$$\frac{\hat{\beta}_1 - \beta_1}{\sqrt{\frac{\hat{\sigma}^2}{S_{XX}}}} \sim t_{n-2}$$

This distributional result implies that a two-sided confidence interval for the slope parameter β_1 can be constructed as

$$\beta_1 \in \left(\hat{\beta}_1 - \text{critical point} \times \text{s.e.}(\hat{\beta}_1), \ \hat{\beta}_1 + \text{critical point} \times \text{s.e.}(\hat{\beta}_1) \right)$$

where the critical point is taken from a t-distribution with $n-2$ degrees of freedom. Hypothesis tests concerning the value of the slope parameter β_1 are performed by comparing a t-statistic with a t-distribution with $n-2$ degrees of freedom. The confidence intervals and hypothesis testing procedures are outlined in the accompanying box.

Tests of the null hypothesis $H_0 : \beta_1 = 0$ are particularly interesting because if $\beta_1 = 0$, then the regression line is flat and the distribution of the dependent variable is not affected by the value of the explanatory variable. With $b_1 = 0$, the t-statistic in this case is simply

$$t = \frac{\hat{\beta}_1}{\text{s.e.}(\hat{\beta}_1)} = \frac{\hat{\beta}_1}{\sqrt{\frac{\hat{\sigma}^2}{S_{XX}}}} = \frac{\hat{\beta}_1 \sqrt{S_{XX}}}{\hat{\sigma}}$$

Computer output generally contains a two-sided p-value for this null hypothesis.

Computer output also usually contains a two-sided p-value for the null hypothesis $H_0 : \beta_0 = 0$ that the intercept parameter is equal to 0. Inferences on the intercept parameter are a special case of inferences on the fitted value of the regression line, which are discussed in Section 12.4. In most cases the null hypothesis $H_0 : \beta_0 = 0$ is not of any particular interest and its p-value given in the computer output is not important.

12.3.2   Examples

Example 67
Car Plant Electricity Usage

For this data set

$$S_{XX} = \sum_{i=1}^{12} x_i^2 - \frac{\left(\sum_{i=1}^{12} x_i \right)^2}{12} = 291.2310 - \frac{58.62^2}{12} = 4.8723$$

so that the standard error of the slope parameter estimate $\hat{\beta}_1$ is

$$\text{s.e.}(\hat{\beta}_1) = \frac{\hat{\sigma}}{\sqrt{S_{XX}}} = \frac{0.1729}{\sqrt{4.8723}} = 0.0783$$

The t-statistic for testing the null hypothesis $H_0 : \beta_1 = 0$ is

$$t = \frac{\hat{\beta}_1}{\text{s.e.}(\hat{\beta}_1)} = \frac{0.49883}{0.0783} = 6.37$$

The two-sided p-value is calculated as

$$p\text{-value} = 2 \times P(X > 6.37) \simeq 0$$

where the random variable X has a t-distribution with 10 degrees of freedom. This low p-value indicates that the null hypothesis is not plausible and so the slope parameter is known to be nonzero. In other words, it has been established that the distribution of electricity usage does depend on the level of production.

With $t_{0.005,10} = 3.169$, a 99% two-sided confidence interval for the slope parameter is

$$\beta_1 \in \left(\hat{\beta}_1 - \text{critical point} \times \text{s.e.}(\hat{\beta}_1), \ \hat{\beta}_1 + \text{critical point} \times \text{s.e.}(\hat{\beta}_1) \right)$$
$$= (0.49883 - 3.169 \times 0.0783, \ 0.49883 + 3.169 \times 0.0783) = (0.251, 0.747)$$

Thus the experimenter can be confident that within the range of the data set, the expected electricity usage increases by somewhere between a quarter of a million kilowatt-hours and three quarters of a million kilowatt-hours for every additional \$1 million of production.

If you analyze this data set with a statistical software package, you will find that a p-value of 0.314 is given for the intercept parameter β_0. In other words, the null hypothesis $H_0 : \beta_0 = 0$ is plausible. However, this is not important since β_0 is the expected electricity cost when there is no production ($x = 0$), which is extrapolating far below the data set. While it may be tempting to use a simplified regression model

$$y = \beta_1 x$$

with $\beta_0 = 0$, it is generally best to stay with the full fitted regression model

$$y = \hat{\beta}_0 + \hat{\beta}_1 x$$

Example 68

Nile River Flowrate

Here

$$S_{XX} = \sum_{i=1}^{115} x_i^2 - \frac{\left(\sum_{i=1}^{115} x_i\right)^2}{115} = 1982.5264 - \frac{462.9^2}{115} = 119.2533$$

and the standard error of the slope parameter estimate is

$$\text{s.e.}(\hat{\beta}_1) = \frac{\hat{\sigma}}{\sqrt{S_{XX}}} = \frac{\sqrt{0.1092}}{\sqrt{119.2533}} = 0.03026$$

The t-statistic for testing the null hypothesis $H_0 : \beta_1 = 0$ is

$$t = \frac{\hat{\beta}_1}{\text{s.e.}(\hat{\beta}_1)} = \frac{0.8362}{0.03026} = 27.63$$

and the two-sided p-value is

$$p\text{-value} = 2 \times P(X > 27.63) \simeq 0$$

where the random variable X has a t-distribution with 113 degrees of freedom. This low p-value indicates that the slope parameter is nonzero, so that the distribution of the February inflow has been shown to depend on the January inflow.

With $t_{0.005,113} = 2.620$, a 99% two-sided confidence interval for the slope parameter is

$$\beta_1 \in (0.8362 - 2.620 \times 0.03026, 0.8362 + 2.620 \times 0.03026) = (0.757, 0.915)$$

These values tell the dam engineers that an increase in the January inflow of 1 billion cubic meters results in an increase in the expected February inflow of between about 750 and 920 million cubic meters.

Example 44

Army Physical Fitness Test

With

$$S_{XX} = \sum_{i=1}^{84} x_i^2 - \frac{\left(\sum_{i=1}^{84} x_i\right)^2}{84} = 402,972 - \frac{5694^2}{84} = 17,000.14$$

the standard error of the slope parameter estimate is

$$\text{s.e.}(\hat{\beta}_1) = \frac{\hat{\sigma}}{\sqrt{S_{XX}}} = \frac{\sqrt{4606.42}}{\sqrt{17,000.14}} = 0.52054$$

The t-statistic for testing the null hypothesis $H_0 : \beta_1 = 0$ is

$$t = \frac{\hat{\beta}_1}{\text{s.e.}(\hat{\beta}_1)} = \frac{-3.2544}{0.52054} = -6.25$$

and the two-sided p-value is

$$p\text{-value} = 2 \times P(X > 6.25) \simeq 0$$

where the random variable X has a t-distribution with 82 degrees of freedom. Again, the low p-value indicates that the slope parameter is known to be nonzero, so that there is evidence that the distribution of the run times depends on the number of pushups performed.

The critical point $t_{0.005, 82} = 2.637$ gives a 99% two-sided confidence interval

$$\beta_1 \in (-3.254 - 2.637 \times 0.5205, -3.254 + 2.637 \times 0.5205) = (-4.63, -1.88)$$

Thus an additional pushup decreases the expected run time by somewhere between 1.88 and 4.63 seconds.

Example 69
Cranial
Circumferences

An anthropologist is interested in whether a fully grown human's cranial (skull) circumference is related to the length of the middle finger. Figure 12.22 shows a data set of these measurements obtained from $n = 20$ randomly selected people, together with a fitted regression line obtained with cranial circumference taken as the dependent variable and finger length taken as the explanatory variable.

Cranial circumference (cm)	Finger length (cm)
58.5	7.6
54.2	7.9
57.2	8.4
52.7	7.7
55.1	8.6
60.7	8.6
57.2	7.9
58.8	8.2
56.2	7.7
60.7	8.1
53.5	8.1
60.7	7.9
56.3	8.1
58.1	8.2
56.6	7.8
57.7	7.9
59.2	8.3
57.9	8.0
56.8	8.2
55.4	7.9

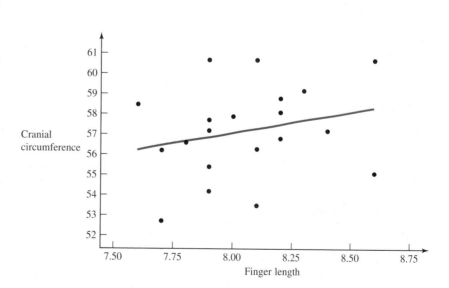

FIGURE 12.22

Fitted line plot and data set for the cranial circumferences example

When a line is fitted to the data, a slope parameter estimate $\hat{\beta}_1 = 2.086$ is obtained, which *suggests* that the expected cranial circumference increases by about 2 cm for each additional 1 cm of finger length. However, it can be shown that this point estimate has a standard error of

$$\text{s.e.}(\hat{\beta}_1) = 1.862$$

so that the t-statistic for testing the null hypothesis $H_0 : \beta_1 = 0$ is

$$t = \frac{\hat{\beta}_1}{\text{s.e.}(\hat{\beta}_1)} = \frac{2.086}{1.862} = 1.12$$

The two-sided p-value is therefore

$$p\text{-value} = 2 \times P(X > 1.12) = 0.277$$

where the random variable X has a t-distribution with 18 degrees of freedom.

This large p-value implies that the null hypothesis $H_0 : \beta_1 = 0$ is plausible and so there is not sufficient evidence to conclude that the distribution of the cranial circumference depends upon the finger length. In other words, the fitted regression line is not statistically significant. The anthropologist should conclude that no relationship has been established between the two variables. Of course, this does not prove that there is no relationship between the two variables. Perhaps a larger study with a larger sample size might find evidence of a relationship.

■ 12.3.3 Problems

12.3.1 In a simple linear regression analysis with $n = 18$ data points, an estimate $\hat{\beta}_1 = 0.522$ is obtained with s.e.$(\hat{\beta}_1) = 0.142$.
  (a) Calculate a two-sided 99% confidence interval for the slope parameter β_1.
  (b) Test the null hypothesis $H_0 : \beta_1 = 0$ against a two-sided alternative hypothesis.

12.3.2 In a simple linear regression analysis with $n = 22$ data points, an estimate $\hat{\beta}_1 = 56.33$ is obtained with s.e.$(\hat{\beta}_1) = 3.78$.
  (a) Calculate a two-sided 95% confidence interval for the slope parameter β_1.
  (b) Test the null hypothesis $H_0 : \beta_1 = 50.0$ against a two-sided alternative hypothesis.

12.3.3 Oil Well Drilling Costs
Consider the data set of oil well drilling costs given in DS 12.2.1.
  (a) What is the standard error of $\hat{\beta}_1$?
  (b) Construct a two-sided 95% confidence interval for the slope parameter β_1.
  (c) Test the null hypothesis $H_0 : \beta_1 = 0$. Interpret your answers.

12.3.4 Truck Unloading Times
Consider the data set of the times taken to unload a truck at a warehouse given in DS 12.2.2.
  (a) What is the standard error of $\hat{\beta}_1$?
  (b) Construct a two-sided 90% confidence interval for the slope parameter β_1.
  (c) Test the null hypothesis $H_0 : \beta_1 = 0$.
  (d) Does your analysis indicate that there is evidence that the trucks take longer to unload when the temperature is higher? Can a case be made that the installation of air-conditioning will improve worker efficiency?

12.3.5 VO2-max Aerobic Fitness Measurements
Consider the data set of aerobic fitness measurements given in DS 12.2.3.
  (a) What is the standard error of $\hat{\beta}_1$?
  (b) Construct and interpret a one-sided 95% confidence interval for the slope parameter β_1 that provides an upper bound.
  (c) Test the null hypothesis $H_0 : \beta_1 = 0$. Is it clear that on average aerobic fitness decreases with age?

12.3.6 Property Tax Appraisals
Consider the data set of appraised house values given in DS 12.2.4.
(a) What is the standard error of $\hat{\beta}_1$?
(b) Construct a two-sided 99% confidence interval for the slope parameter β_1.
(c) Test the null hypothesis $H_0 : \beta_1 = 0$. Interpret your answers.

12.3.7 Management of Computer Systems
Consider the data set of the times taken for programming changes given in DS 12.2.5.
(a) What is the standard error of $\hat{\beta}_1$?
(b) Construct a two-sided 95% confidence interval for the slope parameter β_1.
(c) Why is the null hypothesis $H_0 : \beta_1 = 1$ of particular interest? Test this null hypothesis. Interpret your answers.

12.3.8 Vacuum Transducer Bobbin Resistances
Consider the data set of vacuum transducer bobbin resistances given in DS 12.2.6.

(a) What is the standard error of $\hat{\beta}_1$?
(b) Construct a two-sided 99% confidence interval for the slope parameter β_1.
(c) Test the null hypothesis $H_0 : \beta_1 = 0$. Is it clear that resistance increases with temperature?

12.3.9 A simple linear regression is performed on 20 data pairs (x_i, y_i). It is found that $\hat{\beta}_1 = 54.87$ and s.e. $(\hat{\beta}_1) = 21.20$. Is the p-value for the two-sided hypothesis test of $H_0 : \beta_1 = 0$ (a) greater than 10%, (b) between 1% and 10%, or (c) less than 1%?

12.3.10 Ceramic Baking Procedures
Several samples of ceramic were made with different baking times. The densities of the samples were measured and the data set given in DS 12.3.1 was obtained. Use a simple linear regression model to investigate whether there is any evidence that the density of the ceramic is affected by the baking time.

12.4 Inferences on the Regression Line

12.4.1 Inference Procedures

For a particular value x^* of the explanatory variable, the true regression line

$$\beta_0 + \beta_1 x^*$$

specifies the *expected value* of the dependent variable or, in other words, the *expected response*. Thus, if the random variable $Y|_{x^*}$ represents the value of the dependent variable when the explanatory variable is equal to x^*, then

$$E(Y|_{x^*}) = \beta_0 + \beta_1 x^*$$

It is useful to be able to construct confidence intervals for this expected value.
The point estimate of the average response at x^* is

$$\hat{y}|_{x^*} = \hat{\beta}_0 + \hat{\beta}_1 x^* = (\bar{y} - \hat{\beta}_1 \bar{x}) + \hat{\beta}_1 x^* = \bar{y} + \hat{\beta}_1(x^* - \bar{x})$$

$$= \bar{y} + \sum_{i=1}^{n} \frac{(x_i - \bar{x})y_i}{S_{XX}}(x^* - \bar{x}) = \sum_{i=1}^{n} \left(\frac{1}{n} + \frac{(x_i - \bar{x})(x^* - \bar{x})}{S_{XX}} \right) y_i$$

This estimate is an observation from the random variable

$$\sum_{i=1}^{n} \left(\frac{1}{n} + \frac{(x_i - \bar{x})(x^* - \bar{x})}{S_{XX}} \right) Y_i$$

where

$$Y_i \sim N(\beta_0 + \beta_1 x_i, \sigma^2) \qquad 1 \le i \le n$$

Since it is a linear combination of normal random variables, this random variable is also normally distributed, and it can be shown that it has an expectation and variance of

$$\beta_0 + \beta_1 x^* \qquad \text{and} \qquad \sigma^2 \left(\frac{1}{n} + \frac{(x^* - \bar{x})^2}{S_{XX}} \right)$$

Thus, $\hat{\beta}_0 + \hat{\beta}_1 x^*$ is an *unbiased* estimator of the expected response $\beta_0 + \beta_1 x^*$ at x^*, and it has a standard error

$$\text{s.e.}(\hat{\beta}_0 + \hat{\beta}_1 x^*) = \sigma \sqrt{\frac{1}{n} + \frac{(x^* - \bar{x})^2}{S_{XX}}}$$

which can be estimated as

$$\text{s.e.}(\hat{\beta}_0 + \hat{\beta}_1 x^*) = \hat{\sigma} \sqrt{\frac{1}{n} + \frac{(x^* - \bar{x})^2}{S_{XX}}}$$

From this estimated standard error it follows that

$$\frac{(\hat{\beta}_0 + \hat{\beta}_1 x^*) - (\beta_0 + \beta_1 x^*)}{\text{s.e.}(\hat{\beta}_0 + \hat{\beta}_1 x^*)} \sim t_{n-2}$$

and this distributional result can be used to make inferences on the expected response as outlined in the box.

Inferences on the Expected Value of the Dependent Variable

A $1 - \alpha$ confidence level two-sided confidence interval for $\beta_0 + \beta_1 x^*$, the *expected value* of the dependent variable for a particular value x^* of the explanatory variable, is

$$\beta_0 + \beta_1 x^* \in (\hat{\beta}_0 + \hat{\beta}_1 x^* - t_{\alpha/2,n-2} \times \text{s.e.}(\hat{\beta}_0 + \hat{\beta}_1 x^*),$$
$$\hat{\beta}_0 + \hat{\beta}_1 x^* + t_{\alpha/2,n-2} \times \text{s.e.}(\hat{\beta}_0 + \hat{\beta}_1 x^*))$$

where

$$\text{s.e.}(\hat{\beta}_0 + \hat{\beta}_1 x^*) = \hat{\sigma} \sqrt{\frac{1}{n} + \frac{(x^* - \bar{x})^2}{S_{XX}}}$$

One-sided confidence intervals are

$$\beta_0 + \beta_1 x^* \in (-\infty, \hat{\beta}_0 + \hat{\beta}_1 x^* + t_{\alpha,n-2} \times \text{s.e.}(\hat{\beta}_0 + \hat{\beta}_1 x^*))$$

and

$$\beta_0 + \beta_1 x^* \in (\hat{\beta}_0 + \hat{\beta}_1 x^* - t_{\alpha,n-2} \times \text{s.e.}(\hat{\beta}_0 + \hat{\beta}_1 x^*), \infty)$$

Hypothesis tests on $\beta_0 + \beta_1 x^*$ can be performed by comparing the t-statistic

$$t = \frac{(\hat{\beta}_0 + \hat{\beta}_1 x^*) - (\beta_0 + \beta_1 x^*)}{\text{s.e.}(\hat{\beta}_0 + \hat{\beta}_1 x^*)}$$

with a t-distribution with $n - 2$ degrees of freedom.

Notice that the standard error of the point estimate $\hat{\beta}_0 + \hat{\beta}_1 x^*$ depends on the value of x^*. When $x^* = \bar{x}$ it takes its smallest value

$$\text{s.e.}(\hat{\beta}_0 + \hat{\beta}_1\bar{x}) = \frac{\hat{\sigma}}{\sqrt{n}}$$

and it increases in size as x^* moves farther away from $\bar{x}$. On an intuitive level this can be understood as indicating that the fitted regression line is the most accurate at the center of the data set and is less accurate toward the edges of the data set.

It is useful to draw **confidence bands** around the fitted regression line that are composed of individual confidence intervals for the expected response at different values of the explanatory variable. These bands are illustrated in the following examples. Since the confidence interval lengths are proportional to the standard error of $\hat{\beta}_0 + \hat{\beta}_1 x^*$, the bands are narrowest at $x^* = \bar{x}$ and become increasingly wide as x^* moves farther away from $\bar{x}$.

Notice that the standard error of $\hat{\beta}_0 + \hat{\beta}_1 x^*$ approaches 0 as the sample size n tends to infinity (this is the case since S_{XX} will also tend to infinity as n tends to infinity). This implies that the fitted regression line

$$y = \hat{\beta}_0 + \hat{\beta}_1 x$$

becomes an increasingly accurate estimate of the true regression line

$$y = \beta_0 + \beta_1 x$$

as more and more data observations (y_i, x_i) are available. Thus, the confidence bands shrink toward the fitted regression line as the sample size increases.

As a final point, notice that taking $x^* = 0$ provides inferences on the intercept parameter β_0. Hence, a $1 - \alpha$ confidence level two-sided confidence interval for β_0 is

$$\beta_0 \in \left(\hat{\beta}_0 - t_{\alpha/2,n-2}\hat{\sigma}\sqrt{\frac{1}{n} + \frac{\bar{x}^2}{S_{XX}}}, \; \hat{\beta}_0 + t_{\alpha/2,n-2}\hat{\sigma}\sqrt{\frac{1}{n} + \frac{\bar{x}^2}{S_{XX}}} \right)$$

and a two-sided p-value for the null hypothesis $H_0 : \beta_0 = 0$, contained in most computer output, is

$$p\text{-value} = 2 \times P(X > |t|)$$

where the random variable X has a t-distribution with $n - 2$ degrees of freedom, and the t-statistic is

$$t = \frac{\hat{\beta}_0}{\hat{\sigma}\sqrt{\frac{1}{n} + \frac{\bar{x}^2}{S_{XX}}}}$$

12.4.2 Examples

Example 67
Car Plant Electricity Usage

For this example

$$\text{s.e.}(\hat{\beta}_0 + \hat{\beta}_1 x^*) = \hat{\sigma}\sqrt{\frac{1}{n} + \frac{(x^* - \bar{x})^2}{S_{XX}}} = 0.1729 \times \sqrt{\frac{1}{12} + \frac{(x^* - 4.885)^2}{4.8723}}$$

FIGURE 12.23

Confidence bands for the car plant
electricity usage data set

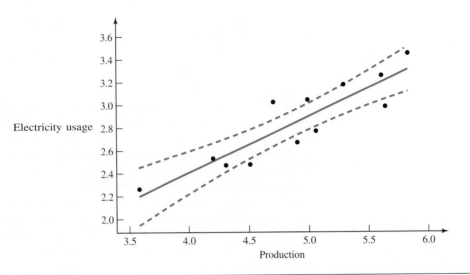

With $t_{0.025,10} = 2.228$, a 95% confidence interval for $\beta_0 + \beta_1 x^*$ is therefore

$$\beta_0 + \beta_1 x^* \in \left(0.409 + 0.499 x^* - 2.228 \times 0.1729 \times \sqrt{\frac{1}{12} + \frac{(x^* - 4.885)^2}{4.8723}}, \right.$$
$$\left. 0.409 + 0.499 x^* + 2.228 \times 0.1729 \times \sqrt{\frac{1}{12} + \frac{(x^* - 4.885)^2}{4.8723}} \right)$$

At $x^* = 5$ this is

$$\beta_0 + 5\beta_1 \in (0.409 + (0.499 \times 5) - 0.113, 0.409 + (0.499 \times 5) + 0.113)$$
$$= (2.79, 3.02)$$

This interval implies that with a monthly production of $5 million, the *expected* electricity usage is between about 2.8 and 3.0 kWh.

Figure 12.23 shows the confidence bands on the fitted regression line obtained from the individual 95% confidence intervals. As expected, they are narrowest at $\bar{x} = 4.885$ and they become wider toward the edges of the data set. It is convenient to think of the true regression line

$$y = \beta_0 + \beta_1 x$$

as lying somewhere *within* these confidence bands.

Example 68
Nile River Flowrate

The standard error of the fitted regression line is

$$\text{s.e.}(\hat{\beta}_0 + \hat{\beta}_1 x^*) = \sqrt{0.1092} \times \sqrt{\frac{1}{115} + \frac{(x^* - 4.0252)^2}{119.25}}$$

so that with $t_{0.025,113} = 1.9812$, a 95% confidence interval for the fitted regression line is

$$\beta_0 + \beta_1 x^* \in \left(-0.470 + 0.836 x^* - 1.9812 \times \sqrt{0.1092} \times \sqrt{\frac{1}{115} + \frac{(x^* - 4.0252)^2}{119.25}}, \right.$$
$$\left. -0.470 + 0.836 x^* + 1.9812 \times \sqrt{0.1092} \times \sqrt{\frac{1}{115} + \frac{(x^* - 4.0252)^2}{119.25}} \right)$$

FIGURE 12.24

Confidence bands for the Nile
River inflows data set

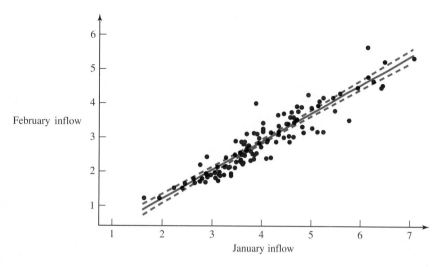

FIGURE 12.25

Confidence bands for the Army
Physical Fitness Test data set

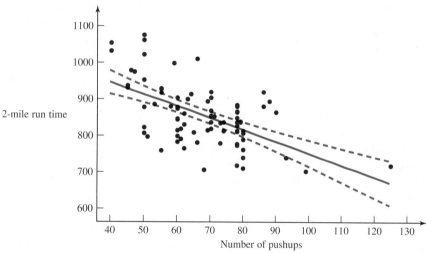

These confidence intervals produce the confidence bands shown in Figure 12.24. With $n = 115$ data observations, these bands appear fairly close together over the range of the data set so that the dam engineers can deduce that the regression line has been determined fairly accurately.

Example 44
Army Physical Fitness Test

For this example the standard error of the fitted regression line is

$$\text{s.e.}(\hat{\beta}_0 + \hat{\beta}_1 x^*) = 67.87 \times \sqrt{\frac{1}{84} + \frac{(x^* - 67.79)^2}{17{,}000}}$$

so that with $t_{0.025,82} = 1.9893$, the 95% confidence bands shown in Figure 12.25 are calculated as

$$\beta_0 + \beta_1 x^* \in \left(1078 - 3.254x^* - 1.9893 \times 67.87 \times \sqrt{\frac{1}{84} + \frac{(x^* - 67.79)^2}{17{,}000}}, \right.$$

$$\left. 1078 - 3.254x^* + 1.9893 \times 67.87 \times \sqrt{\frac{1}{84} + \frac{(x^* - 67.79)^2}{17{,}000}} \right)$$

For example, at $x^* = 70$ pushups the confidence interval is

$$\beta_0 + 70\beta_1 \in (1078 - (3.254 \times 70) - 14.91, 1078 - (3.254 \times 70) + 14.91)$$
$$= (835.3, 865.1)$$

This interval indicates that based upon the data set available, the *expected* 2-mile run time for an officer performing 70 pushups is known to lie somewhere between 13 minutes and 55 seconds and 14 minutes and 25 seconds.

■ 12.4.3 Problems

12.4.1 Show that the random variable

$$\sum_{i=1}^{n} \left(\frac{1}{n} + \frac{(x_i - \bar{x})(x^* - \bar{x})}{S_{XX}} \right) Y_i$$

where

$$Y_i \sim N(\beta_0 + \beta_1 x_i, \sigma^2), \quad 1 \leq i \leq n$$

has an expectation and variance

$$\beta_0 + \beta_1 x^* \quad \text{and} \quad \sigma^2 \left(\frac{1}{n} + \frac{(x^* - \bar{x})^2}{S_{XX}} \right)$$

12.4.2 In a simple linear regression analysis with $n = 17$ data points, the estimates $\hat{\beta}_0 = 12.08$, $\hat{\beta}_1 = 34.60$, and $\hat{\sigma}^2 = 17.65$ are obtained. If $S_{XX} = 1096$ and $\bar{x} = 53.2$, construct a two-sided 95% confidence interval for the expected response at $x^* = 40.0$. (This problem is continued in Problem 12.5.1.)

12.4.3 In a simple linear regression analysis with $n = 24$ data points, the estimates $\hat{\beta}_0 = -7.80$, $\hat{\beta}_1 = 2.23$, and $\hat{\sigma}^2 = 1.76$ are obtained. If $S_{XX} = 543.5$ and $\bar{x} = 10.8$, construct a two-sided 95% confidence interval for the expected response at $x^* = 13.6$. (This problem is continued in Problem 12.5.2.)

12.4.4 Oil Well Drilling Costs
Consider the data set of oil well costs given in DS 12.2.1. Construct a two-sided 95% confidence interval for the expected cost of oil wells with a depth of 9500 feet.

12.4.5 VO2-max Aerobic Fitness Measurements
Consider the data set of aerobic fitness measurements given in DS 12.2.3. Construct and interpret a two-sided 95% confidence interval for the expected VO2-max measurements of 50-year-old males.

12.4.6 Property Tax Appraisals
Consider the data set of appraised house values given in DS 12.2.4. Construct a two-sided 99% confidence interval

for the average appraised value of houses with a size of 3200 square feet.

12.4.7 Management of Computer Systems
Consider the data set of the times taken for programming changes given in DS 12.2.5. Construct a one-sided 95% confidence interval that provides an upper bound on the actual time on average for tasks that are estimated to take 7 hours.

12.4.8 Vacuum Transducer Bobbin Resistances
Consider the data set of vacuum transducer bobbin resistances given in DS 12.2.6. Construct a two-sided 99% confidence interval for the average resistance of a vacuum transducer bobbin at a temperature of 70° F.

12.4.9 A simple linear regression analysis is performed to determine how the strength of a substance depends on its density. Eight samples of the substance are prepared, with densities 11.2, 12.5, 13.4, 14.9, 16.0, 16.6. 17.9, and 20.1, and their strengths are found. The estimates $\hat{\beta}_0 = 75.32$, $\hat{\beta}_1 = 0.0674$, and $\hat{\sigma} = 0.0543$ are obtained. Calculate a 95% confidence interval for the average strength of substances with a density 15.

12.4.10 A linear regression model is fitted to the data

x	y
37.0	65.0
36.4	67.2
35.8	70.3
34.3	71.9
33.7	73.8
32.1	75.7
31.5	77.9

with x as the input variable and y as the output variable. Find $\hat{\beta}_0$, $\hat{\beta}_1$, and $\hat{\sigma}^2$. Construct a 99% confidence interval for the expected value of the output variable when the input variable is equal to 35.

12.5 Prediction Intervals for Future Response Values

12.5.1 Inference Procedures

One of the most important issues for an experimenter to investigate is a future value of the dependent variable y or, in other words, a *response* y obtained with a value x^* of the explanatory variable. It is useful to construct a **prediction interval** for this future response value $y|_{x^*}$, which like a confidence interval for an unknown parameter specifies a region for the future response value at a given confidence level.

Recall that under the modeling assumptions, a future response $y|_{x^*}$ is composed of a point on the regression line $\beta_0 + \beta_1 x^*$ together with an error term, which is taken to be an observation from a normal distribution with mean zero and variance σ^2. Consequently, when one makes inferences about a future response at x^*, there are two sources of uncertainty. The first is the uncertainty in the value of the regression line at x^* and the second is the variability in the error term. Thus, while the value on the fitted regression line

$$\hat{\beta}_0 + \hat{\beta}_1 x^*$$

serves as a point estimate of the expected response at x^*, the total variability is

$$\text{Var}(\hat{\beta}_0 + \hat{\beta}_1 x^*) + \sigma^2 = \sigma^2 \left(\frac{1}{n} + \frac{(x^* - \bar{x})^2}{S_{XX}} \right) + \sigma^2 = \sigma^2 \left(\frac{n+1}{n} + \frac{(x^* - \bar{x})^2}{S_{XX}} \right)$$

which is the variance of the fitted regression line together with the variance of the error term. This leads to the prediction intervals shown in the box.

Prediction Intervals for Future Response Values

A $1 - \alpha$ confidence level two-sided **prediction interval** for $y|_{x^*}$, a future value of the dependent variable for a particular value x^* of the explanatory variable, is

$$y|_{x^*} \in \left(\hat{\beta}_0 + \hat{\beta}_1 x^* - t_{\alpha/2, n-2} \hat{\sigma} \sqrt{\frac{n+1}{n} + \frac{(x^* - \bar{x})^2}{S_{XX}}}, \right.$$

$$\left. \hat{\beta}_0 + \hat{\beta}_1 x^* + t_{\alpha/2, n-2} \hat{\sigma} \sqrt{\frac{n+1}{n} + \frac{(x^* - \bar{x})^2}{S_{XX}}} \right)$$

One-sided prediction intervals are

$$y|_{x^*} \in \left(-\infty, \hat{\beta}_0 + \hat{\beta}_1 x^* + t_{\alpha, n-2} \hat{\sigma} \sqrt{\frac{n+1}{n} + \frac{(x^* - \bar{x})^2}{S_{XX}}} \right)$$

and

$$y|_{x^*} \in \left(\hat{\beta}_0 + \hat{\beta}_1 x^* - t_{\alpha, n-2} \hat{\sigma} \sqrt{\frac{n+1}{n} + \frac{(x^* - \bar{x})^2}{S_{XX}}}, \infty \right)$$

It is important to understand the difference between this prediction interval for a *future response* at x^* and the confidence interval on the *regression line* at x^* discussed in the previous section. Remember that the regression line at x^* indicates the expected or average response at x^*, and the length of a confidence interval for the regression line decreases to 0 as the sample size n becomes increasingly large.

The prediction interval for a future response discussed in this section is a prediction interval for one particular observation of the dependent variable obtained with the explanatory variable equal to x^*. It is necessarily *larger* than the confidence interval for the regression line. As the sample size n increases, the length of this prediction interval decreases, but not all the way to 0. The shortest prediction interval (in the limiting case when $n \to \infty$) is

$$y|_{x^*} \in \left(\hat{\beta}_0 + \hat{\beta}_1 x^* - t_{\alpha/2, n-2} \hat{\sigma}, \ \hat{\beta}_0 + \hat{\beta}_1 x^* + t_{\alpha/2, n-2} \hat{\sigma} \right)$$

since even though the regression line may be estimated with increasing precision, the error term component of $y|_{x^*}$ always has a fixed variance of σ^2.

Prediction bands for a future response can be drawn around a fitted regression line in a manner similar to the confidence bands for the regression line itself, as illustrated in the following examples. The prediction bands for a future response are always wider than the confidence bands for the regression line, and like the confidence bands for the regression line they are narrowest at $\bar{x}$ and become wider toward the edges of the data set.

12.5.2 Examples

Example 67
Car Plant Electricity Usage

With $t_{0.025, 10} = 2.228$, a 95% prediction interval for a future response $y|_{x^*}$ is

$$y|_{x^*} \in \left(0.409 + 0.499x^* - 2.228 \times 0.1729 \times \sqrt{\frac{13}{12} + \frac{(x^* - 4.885)^2}{4.8723}}, \right.$$

$$\left. 0.409 + 0.499x^* + 2.228 \times 0.1729 \times \sqrt{\frac{13}{12} + \frac{(x^* - 4.885)^2}{4.8723}} \right)$$

At $x^* = 5$ this interval is

$$y|_5 \in (0.409 + (0.499 \times 5) - 0.401, 0.409 + (0.499 \times 5) + 0.401)$$
$$= (2.50, 3.30)$$

This prediction interval indicates that if next month's production target is $5 million, then with 95% confidence next month's electricity usage will be somewhere between 2.5 and 3.3 kWh.

Recall that a 95% confidence interval for the regression line at $x^* = 5$ was calculated as $(2.8, 3.0)$. Thus, while the *expected* or *average* electricity usage in a month with $5 million of production is known to lie somewhere between 2.8 and 3.0 kWh, the electricity usage in a *particular* month with $5 million of production will lie somewhere between 2.5 and 3.3 kWh.

Figure 12.26 shows the prediction bands for a future response obtained from the individual 95% prediction intervals given above, together with the confidence bands on the regression line. As expected, both sets of bands are narrowest at $\bar{x} = 4.885$, and the prediction bands for a future response are wider than the confidence bands on the regression line.

FIGURE 12.26

Prediction bands and confidence
bands for the car plant electricity
usage data set

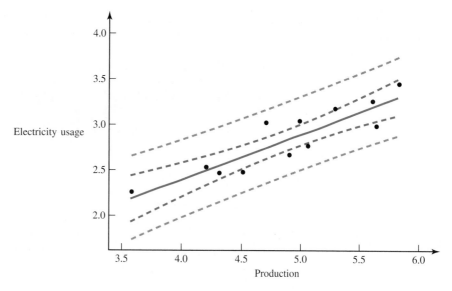

FIGURE 12.27

Prediction bands and confidence
bands for the Nile River inflows
data set

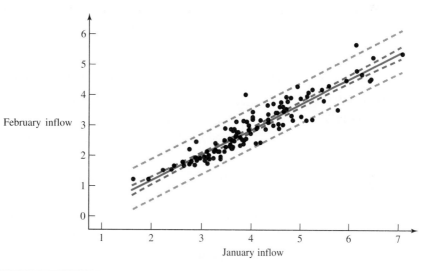

Example 68
Nile River Flowrate

With $t_{0.025,113} = 1.9812$, the 95% prediction bands for a future response, shown in Figure 12.27, are calculated as

$$y|_{x^*} \in \left(-0.470 + 0.836x^* - 1.9812 \times \sqrt{0.1092} \times \sqrt{\frac{116}{115} + \frac{(x^* - 4.0252)^2}{119.25}}, \right.$$

$$\left. -0.470 + 0.836x^* + 1.9812 \times \sqrt{0.1092} \times \sqrt{\frac{116}{115} + \frac{(x^* - 4.0252)^2}{119.25}} \right)$$

For example, the prediction interval at $x^* = 3$ is

$$y|_3 \in (-0.470 + (0.836 \times 3) - 0.660, -0.470 + (0.836 \times 3) + 0.660)$$
$$= (1.38, 2.70)$$

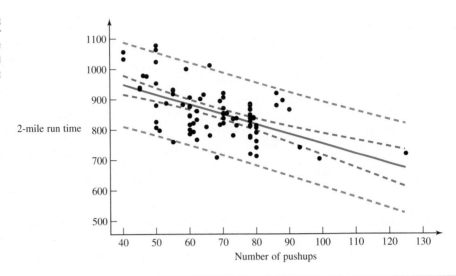

Thus, if in a particular year the January inflow is measured at 3 billion cubic meters, then February's inflow can be predicted as being somewhere between 1.38 and 2.70 cubic meters with 95% confidence.

Example 44
Army Physical Fitness Test

With $t_{0.025,82} = 1.9893$, the 95% prediction bands for a future response shown in Figure 12.28 are

$$y|_{x^*} \in \left(1078 - 3.254x^* - 1.9893 \times 67.87 \times \sqrt{\frac{85}{84} + \frac{(x^* - 67.79)^2}{17{,}000}}, \right.$$

$$\left. 1078 - 3.254x^* + 1.9893 \times 67.87 \times \sqrt{\frac{85}{84} + \frac{(x^* - 67.79)^2}{17{,}000}} \right)$$

For example, at $x^* = 70$ pushups the prediction interval for the 2-mile run time is

$$y|_{70} \in (1078 - (3.254 \times 70) - 135.83, 1078 - (3.254 \times 70) + 135.83)$$
$$= (714.4, 986.1)$$

Thus, although it was found in the last section that the *expected* 2-mile run time for an officer performing 70 pushups lies somewhere between 13 minutes and 55 seconds and 14 minutes and 25 seconds, this prediction interval indicates that a particular officer who just performed 70 pushups will have a run time between 11 minutes and 54 seconds and 16 minutes and 27 seconds, with 95% confidence.

■ 12.5.3 Problems

12.5.1 In a simple linear regression analysis with $n = 17$ data points, the estimates $\hat{\beta}_0 = 12.08$, $\hat{\beta}_1 = 34.60$, and $\hat{\sigma}^2 = 17.65$ are obtained. If $S_{XX} = 1096$ and $\bar{x} = 53.2$, construct a two-sided 95% prediction interval for a future response value at $x^* = 40.0$.

12.5.2 In a simple linear regression analysis with $n = 24$ data points, the estimates $\hat{\beta}_0 = -7.80$, $\hat{\beta}_1 = 2.23$, and $\hat{\sigma}^2 = 1.76$ are obtained. If $S_{XX} = 543.5$ and $\bar{x} = 10.8$, construct a two-sided 95% prediction interval for a future response value at $x^* = 13.6$.

12.5.3 Oil Well Drilling Costs

Consider the data set of oil well costs given in DS 12.2.1. Construct a two-sided 95% prediction interval for the cost of an oil well that is planned to have a depth of 9500 feet.

12.5.4 VO2-max Aerobic Fitness Measurements

Consider the data set of aerobic fitness measurements given in DS 12.2.3. Construct and interpret a two-sided 95% prediction interval for the VO2-max measurement of a new experimental subject who is a 50-year-old male.

12.5.5 Property Tax Appraisals

Consider the data set of appraised house values given in DS 12.2.4. Construct a two-sided 99% prediction interval for the appraised value of a newly constructed house with a size of 3200 square feet.

12.5.6 Management of Computer Systems

Consider the data set of the times taken for programming changes given in DS 12.2.5. A new programming change has been decided upon, and the expert estimates that it will take 7 hours to complete. Construct a one-sided 95% prediction interval that provides an upper bound on the actual time that the task will take.

12.5.7 Vacuum Transducer Bobbin Resistances

Consider the data set of vacuum transducer bobbin resistances given in DS 12.2.6. If a vacuum transducer bobbin is installed in a new car, construct a two-sided

99% prediction interval for the resistance of that vacuum transducer bobbin at a temperature of 70° F.

12.5.8 The amount of catalyst (x) and the yield (y) of a chemical experiment are analyzed using a simple linear regression model. There are 30 observations (x_i, y_i), and it is found that the fitted model is

$$y = 51.98 + 3.44x$$

Suppose that the sum of squared residuals is $\sum_{i=1}^{30} e_i^2 = 329.77$, and that $\sum_{i=1}^{30} x_i = 603.36$, $\sum_{i=1}^{30} x_i^2 = 12578.22$. If an additional experiment is planned with a catalyst level 22, give a range of values with 95% confidence for the yield that you think will be obtained for that experiment.

12.5.9 A linear regression model is fitted to the data

x	y
17.1	45.9
18.4	48.2
19.8	50.3
20.3	50.9
21.6	52.8
22.1	55.5
23.5	57.9

with x as the input variable and y as the output variable. Find $\hat{\beta}_0$, $\hat{\beta}_1$ and $\hat{\sigma}^2$. Construct a 99% prediction interval for an observation of the output variable when the input variable is equal to 20.

12.6 The Analysis of Variance Table

12.6.1 Sum of Squares Decomposition

As with a one-factor layout discussed in Chapter 11, an **analysis of variance** table can also be constructed for a simple linear regression analysis. This analysis of variance table is based upon the variability in the dependent variable y and provides a test of the null hypothesis

$$H_0 : \beta_1 = 0$$

which is equivalent to the similar test described in Section 12.3.

The total sum of squares

$$\text{SST} = \sum_{i=1}^{n}(y_i - \bar{y})^2$$

measures the total variability in the values of the dependent variable (this is also given the notation S_{YY}). It has $n - 1$ degrees of freedom. As Figure 12.29 shows, SST is partitioned into a **sum of squares for regression** SSR with 1 degree of freedom and a sum of squares for error SSE with $n - 2$ degrees of freedom. The sum of squares for regression

$$\text{SSR} = \sum_{i=1}^{n}(\hat{y}_i - \bar{y})^2$$

FIGURE 12.29

The sum of squares decomposition for a regression analysis

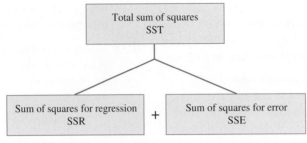

FIGURE 12.30

The sum of squares for a simple linear regression

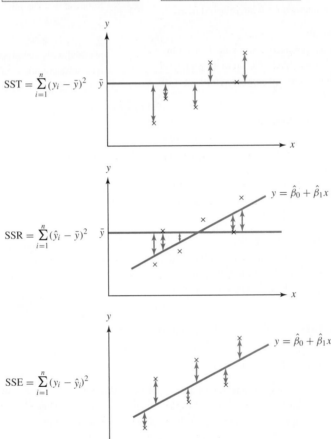

$$\text{SST} = \sum_{i=1}^{n} (y_i - \bar{y})^2 \quad \bar{y}$$

$$\text{SSR} = \sum_{i=1}^{n} (\hat{y}_i - \bar{y})^2 \quad \bar{y} \qquad y = \hat{\beta}_0 + \hat{\beta}_1 x$$

$$\text{SSE} = \sum_{i=1}^{n} (y_i - \hat{y}_i)^2 \qquad y = \hat{\beta}_0 + \hat{\beta}_1 x$$

measures the amount of variability in the dependent variable that is accounted for by the fitted regression line, and the sum of squares for error

$$\text{SSE} = \sum_{i=1}^{n} (y_i - \hat{y}_i)^2$$

measures the variability about the fitted regression line. These three sources of variability are illustrated in Figure 12.30. The mean squares are

$$\text{MSR} = \frac{\text{SSR}}{1} = \text{SSR} \qquad \text{and} \qquad \text{MSE} = \frac{\text{SSE}}{n-2} = \hat{\sigma}^2 \sim \sigma^2 \frac{\chi_{n-2}^2}{n-2}$$

Source	Degrees of freedom	Sum of squares	Mean squares	F-statistic	p-value
Regression	1	SSR	MSR = SSR	F = MSR/MSE	$P(F_{1,n-2} > F)$
Error	$n-2$	SSE	$\hat{\sigma}^2$ = MSE = SSE/$(n-2)$		
Total	$n-1$	SST			

FIGURE 12.31

The analysis of variance table for a simple linear regression analysis

An analysis of variance table can be constructed as shown in Figure 12.31. Under the null hypothesis

$$H_0 : \beta_1 = 0$$

the regression sum of squares is distributed

$$\text{SSR} \sim \chi_1^2$$

so that the F-statistic has an F-distribution

$$F = \frac{\text{MSR}}{\text{MSE}} \sim F_{1,n-2}$$

The (two-sided) p-value is then

$$p\text{-value} = P(X > F)$$

where the random variable X has an $F_{1,n-2}$ distribution.

This p-value is equal to the two-sided p-value obtained in Section 12.3 based upon the t-statistic

$$t = \frac{\hat{\beta}_1 \sqrt{S_{XX}}}{\hat{\sigma}}$$

This is because the F-statistic is the square of the t-statistic

$$t^2 = F$$

and the square of a random variable with a t-distribution with $n-2$ degrees of freedom has an $F_{1,n-2}$ distribution. Therefore, in computer output for simple linear regression problems the p-value given in the analysis of variance table and the p-value given for the slope parameter are always exactly the same.

The proportion of the total variability in the dependent variable y that is accounted for by the regression line is

$$R^2 = \frac{\text{SSR}}{\text{SST}} = 1 - \frac{\text{SSE}}{\text{SST}} = \frac{1}{1 + \frac{\text{SSE}}{\text{SSR}}}$$

which is known as the **coefficient of determination**. This coefficient takes a value between 0 and 1, and the closer it is to one the smaller is the sum of squares for error SSE in relation to the sum of squares for regression SSR. Thus, larger values of R^2 tend to indicate that the data points are closer to the fitted regression line, as shown in Figure 12.32. Nevertheless, a low value of R^2 should *not* necessarily be interpreted as implying that the fitted regression line is not appropriate or is not useful. A fitted regression line may be accurate and informative even though a small value of R^2 is obtained because of a large error variance σ^2.

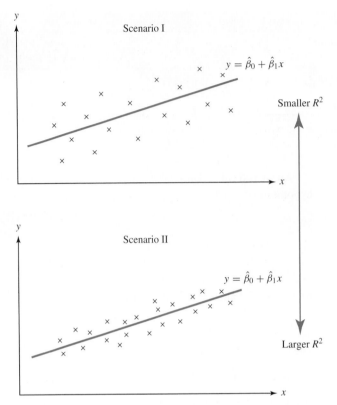

FIGURE 12.33

Analysis of variance table for the car plant electricity usage regression

Source	Degrees of freedom	Sum of squares	Mean squares	F-statistic	p-value
Regression	1	1.2124	1.2124	40.53	0.000
Error	10	0.2991	0.0299		
Total	11	1.5115			

12.6.2   Examples

Example 67

Car Plant Electricity Usage

The analysis of variance table is given in Figure 12.33. Notice that the F-statistic

$$F = \frac{\text{MSR}}{\text{MSE}} = \frac{1.2124}{0.0299} = 40.53$$

is the square of the t-statistic for the slope parameter $t = 6.37$. The coefficient of determination R^2 is reported as being 80.2%, which is

$$R^2 = \frac{\text{SSR}}{\text{SST}} = \frac{1.2124}{1.5115} = 0.802$$

Consequently, about 80% of the variability in the electricity usage can be accounted for by changes in the production levels.

Coefficient of Determination R^2

The total variability in the dependent variable, the total sum of squares

$$SST = \sum_{i=1}^{n}(y_i - \bar{y})^2$$

can be partitioned into the variability explained by the regression line, the regression sum of squares

$$SSR = \sum_{i=1}^{n}(\hat{y}_i - \bar{y})^2$$

and the variability about the regression line, the error sum of squares

$$SSE = \sum_{i=1}^{n}(y_i - \hat{y}_i)^2$$

The proportion of the total variability accounted for by the regression line is the **coefficient of determination**

$$R^2 = \frac{SSR}{SST} = 1 - \frac{SSE}{SST} = \frac{1}{1 + \frac{SSE}{SSR}}$$

which takes a value between 0 and 1.

Example 68
Nile River Flowrate

The analysis of variance table is given in Figure 12.34. Again, notice that the F-statistic

$$F = \frac{MSR}{MSE} = \frac{83.379}{0.109} = 763.250$$

is the square of the t-statistic for the slope parameter $t = 27.627$. The coefficient of determination R^2 is

$$R^2 = \frac{SSR}{SST} = \frac{83.379}{95.723} = 0.871$$

with $R = 0.933$. Thus, about 87% of the variability in the February inflows can be accounted for by changes in the January inflows.

Example 44
Army Physical Fitness Test

The analysis of variance table is given in Figure 12.35. The F-statistic is

$$F = \frac{MSR}{MSE} = \frac{180,051.10}{4606.42} = 39.09$$

which is the square of the t-statistic for the slope parameter $t = -6.25$. The coefficient of determination R^2 is reported as

$$R^2 = \frac{SSR}{SST} = \frac{180,051.10}{557,777.50} = 0.323$$

FIGURE 12.34

Analysis of variance table for the Nile River flowrate regression

Source	Degrees of freedom	Sum of squares	Mean squares	F-statistic	p-value
Regression	1	83.379	83.379	763.250	0.000
Error	113	12.344	0.109		
Total	114	95.723			

FIGURE 12.35

Analysis of variance table for the
Army Physical Fitness Test
regression

Source	Degrees of freedom	Sum of squares	Mean squares	F-statistic	p-value
Regression	1	180051.10	180051.10	39.09	0.000
Error	82	377726.40	4606.42		
Total	83	557777.50			

Therefore, about 32% of the variability in the 2-mile run time can be accounted for by the number of pushups performed.

■ 12.6.3 Problems

12.6.1 Calculate the missing values in the analysis of variance table for a simple linear regression analysis shown in Figure 12.36. What is the p-value? What is the coefficient of determination R^2?

12.6.2 Repeat Problem 12.6.1 for the analysis of variance table shown in Figure 12.37.

12.6.3 A data set has $n = 10$, $\sum_{i=1}^{10} y_i = 908.8$, $\sum_{i=1}^{10} y_i^2 = 83,470$, and $\hat{\sigma}^2 = 0.9781$. Compute the analysis of variance table and calculate the coefficient of determination R^2.

12.6.4 A data set has $n = 25$, $\sum_{i=1}^{25} x_i = 1356.25$, $\sum_{i=1}^{25} y_i = -6225$, $\sum_{i=1}^{25} x_i^2 = 97,025$, $\sum_{i=1}^{25} y_i^2 = 10,414,600$, and $\sum_{i=1}^{25} x_i y_i = -738,100$. Compute the analysis of variance table and calculate the coefficient of determination R^2.

12.6.5 Oil Well Drilling Costs
Consider the data set of oil well costs given in DS 12.2.1. Compute the analysis of variance table and calculate the coefficient of determination R^2. Check that the F-statistic

is the square of the t-statistic for testing $H_0 : \beta_1 = 0$, calculated earlier.

12.6.6 Truck Unloading Times
Consider the data set of the times taken to unload a truck at a warehouse given in DS 12.2.2. Compute the analysis of variance table and calculate the coefficient of determination R^2. Check that the F-statistic is the square of the t-statistic for testing $H_0 : \beta_1 = 0$, calculated earlier. What is the implication of the p-value in the analysis of variance table?

12.6.7 VO2-max Aerobic Fitness Measurements
Consider the data set of aerobic fitness measurements given in DS 12.2.3. Compute the analysis of variance table and calculate the coefficient of determination R^2. Explain the interpretation of the coefficient of determination.

12.6.8 Property Tax Appraisals
Consider the data set of appraised house values given in DS 12.2.4. Compute the analysis of variance table and calculate the coefficient of determination R^2. Check that

FIGURE 12.36

Analysis of variance table

Source	Degrees of freedom	Sum of squares	Mean squares	F-statistic	p-value
Regression	1	?	?	2.32	?
Error	?	576.51	?		
Total	34	?			

FIGURE 12.37

Analysis of variance table

Source	Degrees of freedom	Sum of squares	Mean squares	F-statistic	p-value
Regression	1	?	?	6.47	?
Error	19	?	?		
Total	?	474.80			

the F-statistic is the square of the t-statistic for testing $H_0 : \beta_1 = 0$, calculated earlier. What does the value of the coefficient of determination tell you about the way that houses are appraised?

12.6.9 Management of Computer Systems

Consider the data set of the times taken for programming changes given in DS 12.2.5. Compute the analysis of variance table and calculate the coefficient of determination R^2. Is the p-value in the analysis of variance table meaningful?

12.6.10 Vacuum Transducer Bobbin Resistances

Consider the data set of vacuum transducer bobbin resistances given in DS 12.2.6. Compute the analysis of variance table and calculate the coefficient of determination R^2. Check that the F-statistic is the square of the t-statistic for testing $H_0 : \beta_1 = 0$, calculated earlier.

12.7 Residual Analysis

12.7.1 Residual Analysis Methods

The **residuals** are defined to be

$$e_i = y_i - \hat{y}_i, \quad 1 \leq i \leq n$$

so that they are the differences between the observed values of the dependent variable y_i and the corresponding fitted values $\hat{y}_i$. A property of these residuals is that they sum to 0

$$\sum_{i=1}^{n} e_i = 0$$

(as long as the intercept parameter β_0 has not been removed from the model). The analysis of the residuals is an important tool for checking whether the fitted model is a good model and whether the modeling assumptions appear valid. This residual analysis can be used in conjunction with a visual examination of the data points and the fitted regression line for the simple linear regression problem discussed in this chapter. However, residual analysis becomes a much more important tool for the more complicated models and experimental designs discussed in Chapters 13 and 14.

Residual analysis can be used to

- identify data points that are outliers

- check whether the fitted model is appropriate

- check whether the error variance is constant

- check whether the error terms are normally distributed

A basic residual analysis technique for the simple linear regression problem is to plot the residuals e_i against the values of the explanatory variable x_i. Ideally, a nice random scatter plot such as the one in Figure 12.38 should be obtained, and if this is the case then there are no indications of any problems with the regression analysis. However, any patterns in the residual plot or any residuals with a large absolute value alert the experimenter to possible problems with the fitted regression model.

A data point (x_i, y_i) can be considered to be an outlier if it does not appear to be predicted well by the fitted model. Such data points lie far away from the fitted regression line relative to the other data points, and consequently their residuals have a *large absolute value*, as indicated in Figure 12.39. A convenient rule of thumb is to divide the residual by the estimated error standard deviation $\hat{\sigma}$, and to consider the data point to be a possible outlier

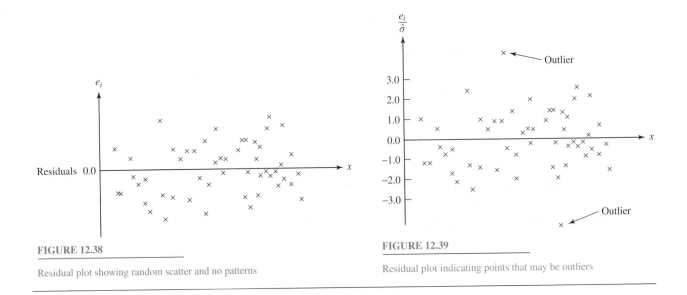

FIGURE 12.38

Residual plot showing random scatter and no patterns

FIGURE 12.39

Residual plot indicating points that may be outliers

if this "standardized residual" is larger in absolute value than 3. Computer packages also usually provide standardized residuals calculated from a more complicated but approximately equivalent formula, which is discussed in Section 13.3.1.

If a data point is identified as being a possible outlier, then it should be investigated further. An explanation may indicate that it should not be modeled together with the other data points and, if this is the case, it should then be removed from the data set and the regression line should be fitted again to the reduced data set.

If the residual plot shows positive and negative residuals grouped together as in Figure 12.40, then a linear model is not appropriate. As Figure 12.40 indicates, a nonlinear model is needed for such a data set. Also, if the residual plot shows a "funnel shape" as in Figure 12.41, so that the size of the residuals depends upon the value of the explanatory variable x, then the assumption of a constant error variance σ^2 is not valid.

The discovery of a nonconstant error variance may be interesting to the experimenter since it indicates that the variability in the distribution of the dependent variable y changes at different values of the explanatory variable x. This may be useful information about the variables under consideration. In practice, the fitted regression line and statistical inferences discussed in this chapter, which are based upon the assumption of a constant error variance, still provide fairly sensible results even if there is some variation in the error variance, although a more accurate analysis may be available using a *weighted* least squares approach.

Finally, a **normal probability plot** (also referred to as a **normal scores plot**) of the residuals can be used to check whether the error terms ϵ_i appear to be normally distributed. In such a plot, the residuals e_i are plotted against their "normal scores," which can be thought of as being their "expected values" if they have a normal distribution. Specifically, the normal score of the ith smallest residual is

$$\Phi^{-1}\left(\frac{i - \frac{3}{8}}{n + \frac{1}{4}}\right)$$

However, computer packages generally provide normal probability plots and it is not necessary to construct the normal scores oneself. If the main body of the points in a normal probability

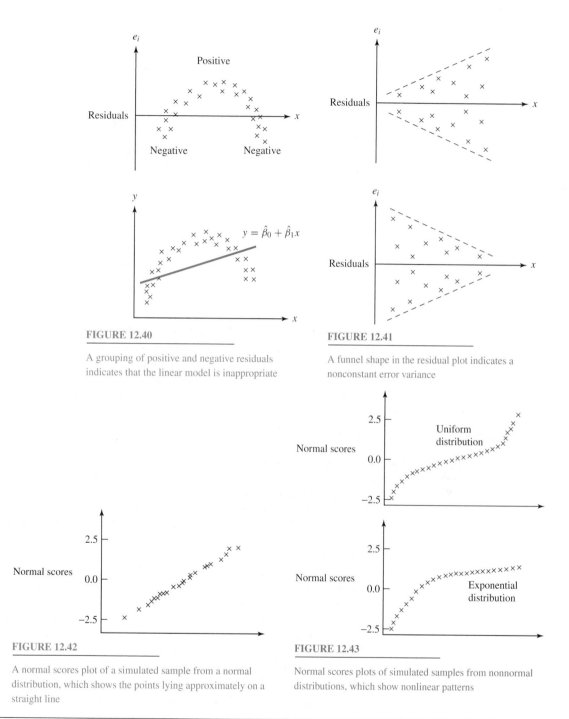

FIGURE 12.40

A grouping of positive and negative residuals indicates that the linear model is inappropriate

FIGURE 12.41

A funnel shape in the residual plot indicates a nonconstant error variance

FIGURE 12.42

A normal scores plot of a simulated sample from a normal distribution, which shows the points lying approximately on a straight line

FIGURE 12.43

Normal scores plots of simulated samples from nonnormal distributions, which show nonlinear patterns

plot lie approximately on a *straight line* as in Figure 12.42, then it is reasonable that the corresponding data points are observations from a normal distribution. Any obvious difference from a straight line in the normal probability plot, such as in Figure 12.43, indicates that the distribution is not normal.

FIGURE 12.44

Plot of the standardized residuals
for the Nile River inflows data set

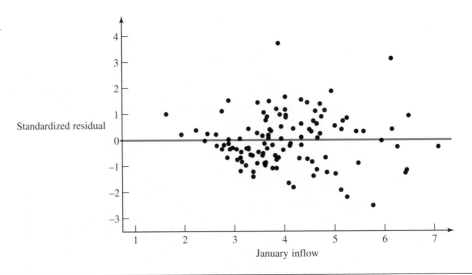

If the normal probability plot of the residuals exhibits a straight line, then the residuals can be taken to be observations from a normal distribution This in turn indicates that there is no reason to question the modeling assumption that the error terms ϵ_i are normally distributed. A lack of normality in the error terms suggests that the experimenter should be cautious about the fitted model and perhaps should adopt a different modeling approach. This may involve a variable transformation, some examples of which are discussed in Section 12.8.

12.7.2   Examples

Example 68
Nile River Flowrate

Figure 12.44 shows a plot of $e_i/\hat{\sigma}$ against the January inflow x_i. This residual plot raises a couple of points, but overall it indicates that the regression modeling appears to have been performed satisfactorily.

There appears to be at least one outlier. In 1987 the January inflow was $x = 3.88$ billion cubic meters while the February inflow was $y = 4.01$ billion cubic meters. For this January inflow the predicted February inflow is

$$\hat{y}|_5 = -0.470 + (0.836 \times 3.88) = 2.77$$

so that the residual is

$$e_i = y_i - \hat{y}_i = 4.01 - 2.77 = 1.24$$

This gives

$$\frac{e_i}{\hat{\sigma}} = \frac{1.24}{\sqrt{0.1092}} = 3.75$$

which is large enough to warrant investigation. Also, in 1968 the January inflow was $x = 6.13$ billion cubic meters while the February inflow was $y = 5.67$ billion cubic meters, which gives a residual of

$$e_i = y_i - \hat{y}_i = 5.67 - (-0.470 + (0.836 \times 6.13)) = 1.02$$

and

$$\frac{e_i}{\hat{\sigma}} = \frac{1.02}{\sqrt{0.1092}} = 3.07$$

FIGURE 12.45

Normal probability plot of the
standardized residuals for the Nile
River inflows data set

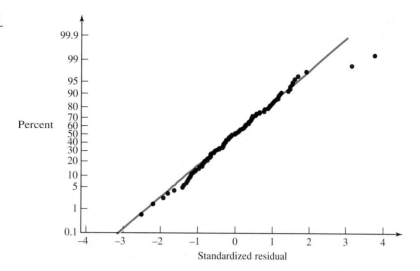

The dam engineers may wish to remove these two data points from the data set and to fit a regression line to the remaining 113 data points, although this should not affect the results very much.

The residual plot also suggests that there is possibly a very slight increase in the error variance as the value of the January inflow increases. This would imply that there is a slightly greater variability in the distribution of the February inflows when the expected February inflow is large, which seems quite a reasonable phenomenon.

Finally, the normal probability plot of the residuals in Figure 12.45 (in which the y-axis shows the cumulative distribution function of the normal scores) exhibits a fairly straight line and so there is no problem with the normality assumption for the error terms. Notice that the two top right points in the normal probability plot correspond to the two possible outliers identified above.

■ 12.7.3 Problems

12.7.1 Oil Well Drilling Costs
Consider the data set of oil well costs given in DS 12.2.1. Plot the residuals against the depth of the oil wells. Are there any points that are possible outliers? Does the residual plot have any patterns that suggest that the fitted regression model is not appropriate?

12.7.2 Truck Unloading Times
Consider the data set of the times taken to unload a truck at a warehouse given in DS 12.2.2. Plot the residuals against the temperature. Are there any points that might be considered to be outliers? Does the residual plot have any patterns that suggest that the fitted regression model is not appropriate? Construct and interpret a normal probability plot.

12.7.3 VO2-max Aerobic Fitness Measurements
Consider the data set of aerobic fitness measurements given in DS 12.2.3. Plot the residuals against the ages of the participants. Are there any obvious outliers? Does the residual plot cause you to have any misgivings about the regression analysis?

12.7.4 Property Tax Appraisals
Consider the data set of appraised house values given in DS 12.2.4. Plot the residuals against the house sizes. Are there any points that are obvious outliers? Construct and interpret a normal scores plot. What action do you think should be taken?

12.7.5 Management of Computer Systems
Consider the data set of the times taken for programming changes given in DS 12.2.5. Plot the residuals against the estimated times. What do you notice? Why do you think that this is?

12.7.6 Vacuum Transducer Bobbin Resistances

Consider the data set of vacuum transducer bobbin resistances given in DS 12.2.6. Plot the residuals against the temperature. Are there any points that are possible outliers? Does the residual plot have any patterns that suggest that the fitted regression model is not appropriate?

12.8 Variable Transformations

12.8.1 Intrinsically Linear Models

At first appearance it may seem restrictive to be able to fit only straight lines to data sets. However, many nonlinear relationships can be transformed into a linear relationship by transformations of one or both of the variables. The simple linear regression methods discussed in this chapter can then be used to fit a straight line to the transformed data values, and the fitted model can then be transformed back into the original variables.

Typically, transformations based on logarithms and reciprocals are the most useful, as illustrated in Figure 12.46. For example, the exponential model

$$y = \gamma_0 e^{\gamma_1 x}$$

for some parameter values γ_0 and γ_1 can be transformed to a linear format by taking logarithms of both sides

$$\ln(y) = \ln(\gamma_0) + \gamma_1 x$$

A linear regression model can then be fitted to the data points $(x_i, \ln(y_i))$. In the fitted linear model the slope parameter β_1 corresponds to γ_1, and the intercept parameter β_0 corresponds to $\ln(\gamma_0)$.

When models are transformed into a linear format, it should be remembered that the usual error assumptions are required for the linear relationship to which the linear regression model is fitted. Thus, for the exponential model example it is assumed that

$$\ln(y_i) = \ln(\gamma_0) + \gamma_1 x_i + \epsilon_i$$

where the error terms ϵ_i are independent observations from a $N(0, \sigma^2)$ distribution. This assumption can be checked in the usual manner through residual analysis. However, notice that in terms of the original exponential model it implies that

$$y = \gamma_0 e^{\gamma_1 x} \epsilon_i^*$$

so that there are *multiplicative* error terms $\epsilon_i^* = e^{\epsilon_i}$. Also, inferences on the original parameters γ_i must be made via corresponding inferences on the linear parameters β_i, as illustrated in the example in Section 12.8.2.

A model that can be transformed into a linear relationship is known as an **intrinsically linear** model. The extension of simple linear regression modeling techniques to intrinsically linear models makes it a general and useful modeling procedure. There are some models, however, that cannot be transformed into a linear format.

These models, such as

$$y = \gamma_0 + e^{\gamma_1 x}$$

for example, are not intrinsically linear and need to be handled with nonlinear regression techniques, which are discussed in Section 13.5.

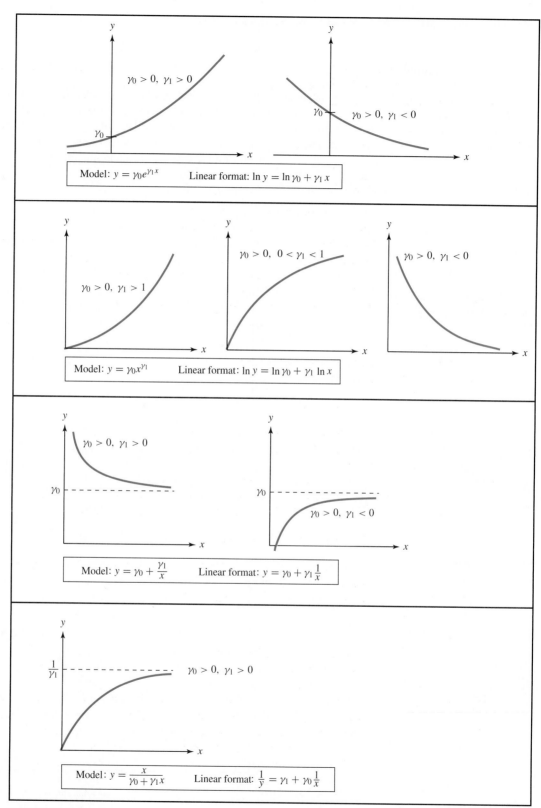

FIGURE 12.46

Some common intrinsically linear models

12.8.2   Example

**Example 63
Roadway Base
Aggregates**

When an aggregate material is subjected to different stress levels, the resilient modulus of the aggregate calculated at each stress level exhibits a nonlinear pattern of values. This can be seen from the plot in Figure 12.47 of $n = 16$ resilient modulus measurements M_i of an aggregate made at different values of the bulk stress variable θ_i. The data set indicates that there is a nonlinear relationship between the resilient modulus and the bulk stress, and it is clearly inappropriate to fit a straight line to the data set.

However, the intrinsically linear model

$$M = \gamma_0 \theta^{\gamma_1}$$

FIGURE 12.47

Nonlinear transformation for roadway base aggregates example

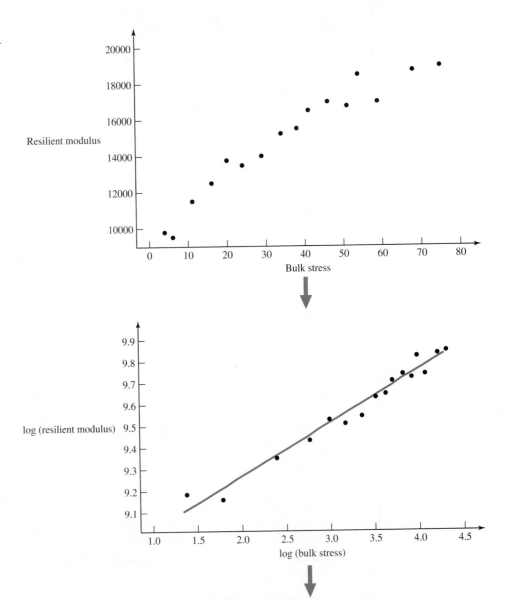

FIGURE 12.47 (*continued*)

Nonlinear transformation for roadway base aggregates example

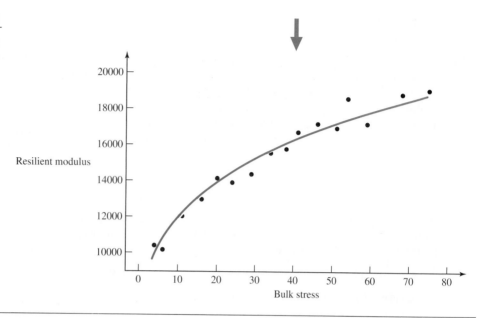

with two unknown parameters γ_0 and γ_1 is commonly used for this problem. With a logarithmic transformation, the linear relationship

$$y = \beta_0 + \beta_1 x$$

is obtained with $y = \ln(M)$, $x = \ln(\theta)$, $\beta_0 = \ln(\gamma_0)$, and $\beta_1 = \gamma_1$. The transformed variables $y_i = \ln(M_i)$ and $x_i = \ln(\theta_i)$ are plotted in Figure 12.47 and they exhibit a linear relationship, which confirms that this model is appropriate.

A straight line is fitted to the transformed data points (x_i, y_i) and the fitted regression line is

$$y = 8.761 + 0.2494x$$

This equation can be transformed back into a model in terms of the original variables

$$M = 6380 \, \theta^{0.2494}$$

which is shown together with the original data points in Figure 12.47.

The standard error of $\hat{\beta}_1$ is 0.0122, and so with $t_{0.025, 14} = 2.145$, a 95% confidence interval for $\beta_1 = \gamma_1$ is

$$\beta_1 = \gamma_1 \in (0.2494 - (2.145 \times 0.0122), 0.2494 + (2.145 \times 0.0122))$$
$$= (0.223, 0.276)$$

Also, the standard error of $\hat{\beta}_0$ is 0.04175, so that a 95% confidence interval for β_0 is

$$\beta_0 \in (8.761 - (2.145 \times 0.04175), 8.761 + (2.145 \times 0.04175))$$
$$= (8.67, 8.85)$$

Therefore with $\gamma_0 = e^{\beta_0}$, a 95% confidence interval for γ_0 is

$$\gamma_0 \in (e^{8.67}, e^{8.85}) = (5830, 6980)$$

■ 12.8.3   Problems

12.8.1 Make a plot of the data set given in DS 12.8.1. What intrinsically linear function should provide a good model for this data set? What transformation of the variables is needed? Fit a straight line to the transformed variables and write the fitted model back in terms of the original variables. What is the predicted value of the dependent variable y when $x = 2.0$?

12.8.2 Repeat Problem 12.8.1 for the data set given in DS 12.8.2.

12.8.3 Cell Growth
A bioengineer measures the growth rate of a substance by counting the number of cells N present at various times t as shown in DS 12.8.3. Fit the model

$$N = \gamma_0 e^{\gamma_1 t}$$

and calculate two-sided 95% confidence intervals for the unknown parameters γ_0 and γ_1.

12.8.4 Synthetic Human Arteries
In an experiment to investigate the suitability of using a silicone tube to model the behavior of a human artery, the data set in DS 12.8.4 is collected, which relates the pressure differential P across the walls of the tube to the cross-sectional area A of the tube.
(a) Show that the model

$$P = \gamma_0 A^{\gamma_1}$$

appears to provide a good fit to the data set.
(b) Make a suitable transformation of the variables and find point estimates for γ_0 and γ_1.
(c) Calculate two-sided 95% confidence intervals for γ_0 and γ_1.

12.8.5 An experimenter has data on the yield and the temperature of a chemical process and wishes to fit the model

$$\text{yield} = \gamma_0 e^{\gamma_1 \text{temperature}}$$

A linear regression model is fitted to the data $y = \ln(\text{yield})$ and $x = \text{temperature}$, with the results $n = 25$, $\hat{\beta}_0 = 2.628$, $\hat{\beta}_1 = 0.341$, and s.e. $(\hat{\beta}_1) = 0.025$. Find $\hat{\gamma}_0$, $\hat{\gamma}_1$, and calculate a 95% confidence interval for γ_1.

12.8.6 Explain how simple linear regression can be used to fit the model $e^{y/\gamma_0} = \gamma_1/x^2$. How would you find the parameter estimates $\hat{\gamma}_0$ and $\hat{\gamma}_1$?

12.8.7 Strengths of Cracked Plastic Compounds
A crack in a plastic compound affects the strength of the material. In order to investigate the relationship between the strength and the length of a crack, an experiment was conducted where cracked pieces of the plastic compound were subjected to increasing loads until the point at which they broke apart. DS 12.8.5 has the data set of the breakage loads and the lengths of the cracks. Show that a linear regression model with breakage load as the dependent variable and crack length as the explanatory variable is unsatisfactory because the residuals are first positive, then negative, and then positive again. Show that a better fit can be obtained with the model

$$\text{breakage load} = \gamma_0 e^{\gamma_1 \times (\text{crack length})}$$

Using this model, what is the expected breaking load when the crack length is 2.1?

12.9   Correlation Analysis

12.9.1   The Sample Correlation Coefficient

Recall that the correlation

$$\rho = \text{Corr}(X, Y) = \frac{\text{Cov}(X, Y)}{\sqrt{\text{Var}(X)\text{Var}(Y)}}$$

measures the strength of linear association between two jointly distributed random variables X and Y (see Section 2.5.4). Given a set of paired data observations

$$(x_i, y_i) \qquad 1 \le i \le n$$

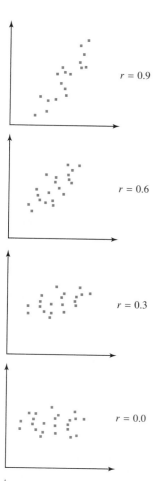

$r = 0.9$

$r = 0.6$

$r = 0.3$

$r = 0.0$

$r = -0.3$

$r = -0.6$

$r = -0.9$

FIGURE 12.48

Illustrations of the sample
correlation coefficient r

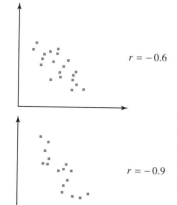

from this joint distribution, the correlation can be estimated by the **sample correlation coefficient** (also known as the Pearson product moment correlation coefficient)

$$r = \frac{S_{XY}}{\sqrt{S_{XX}}\sqrt{S_{YY}}} = \frac{\sum_{i=1}^{n}(x_i - \bar{x})(y_i - \bar{y})}{\sqrt{\sum_{i=1}^{n}(x_i - \bar{x})^2}\sqrt{\sum_{i=1}^{n}(y_i - \bar{y})^2}}$$

$$= \frac{\sum_{i=1}^{n} x_i y_i - n\bar{x}\bar{y}}{\sqrt{\sum_{i=1}^{n} x_i^2 - n\bar{x}^2}\sqrt{\sum_{i=1}^{n} y_i^2 - n\bar{y}^2}}$$

This sample correlation coefficient r takes a value between -1 and 1 and, as Figure 12.48 illustrates, it measures the strength of the *linear* association exhibited by the data points. A sample correlation coefficient $r = 0$ indicates that there is no linear association between the two variables, and their distributions can then be thought of as being independent of each other. A positive sample correlation coefficient $r > 0$ indicates that there is a positive association between the two variables, so that as one variable increases there is a tendency for the other variable to increase as well. On the other hand, a negative sample correlation coefficient $r < 0$ indicates that there is a negative association between the two variables, in which case as one variable increases there is a tendency for the other variable to decrease.

The closer the sample correlation coefficient is to either 1 or -1, the stronger is the linear association. This happens as the data points lie closer to a straight line. Technically, a sample correlation coefficient of $r = 1$ is obtained when the data points lie on a straight line with an upward slope, and a sample correlation coefficient of $r = -1$ is obtained when the data points lie on a straight line with a downward slope.

The sample correlation coefficient is clearly related to a regression line fitted to the data points. In fact, since

$$\hat{\beta}_1 = \frac{S_{XY}}{S_{XX}}$$

it follows that

$$r = \hat{\beta}_1 \sqrt{\frac{S_{XX}}{S_{YY}}}$$

so that the sample correlation coefficient is simply a scaled version of the estimated slope parameter $\hat{\beta}_1$. Moreover, it turns out that

$$r^2 = R^2$$

so that the square of the sample correlation coefficient is equal to the coefficient of determination of the regression model.

The sample correlation coefficient is a convenient way of summarizing the degree of association between two variables, whereas a more detailed regression analysis actually forms a linear model relating the two variables. Notice that the sample correlation coefficient is unchanged if the x and y variables are interchanged and are relabeled the y and x variables. This is in contrast to a regression analysis, which requires one of the variables to be designated the dependent variable and one the explanatory variable. In addition, the sample correlation coefficient is not affected (except for a possible sign change) by any linear transformations of the variables, so that if

$$x_i^* = ax_i + b \quad \text{and} \quad y_i^* = cy_i + d$$

for some constants a, b, c, and d, then the sample correlation coefficient of the data points (x_i, y_i) is equal to the sample correlation coefficient of the data points (x_i^*, y_i^*) if the constants

a and c are either both positive or both negative, and the two sample correlation coefficients are of the same magnitude but of different signs if the constants a and c have different signs.

A correlation $\rho = 0$ is particularly interesting since it implies that there is no linear association between the two variables, and consequently it is useful to be able to test the null hypothesis

$$H_0 : \rho = 0$$

against a two-sided alternative. Under the assumption that the X and Y random variables have a bivariate normal distribution, this hypothesis test can be performed by comparing the t-statistic

$$t = \frac{r\sqrt{n-2}}{\sqrt{1-r^2}}$$

with a t-distribution with $n - 2$ degrees of freedom. However, this is entirely equivalent to testing the null hypothesis

$$H_0 : \beta_1 = 0$$

with the t-statistic

$$t = \frac{\hat{\beta}_1}{\text{s.e.}(\hat{\beta}_1)} = \frac{\hat{\beta}_1\sqrt{S_{XX}}}{\hat{\sigma}}$$

since it can be shown that the two t-statistics are identical. Recall that this test is also equivalent to the test performed in the analysis of variance table for a regression problem, and so these are in fact three ways of performing the same test. However, the t-statistic written in terms of the sample correlation coefficient r is useful when only a correlation analysis is of interest rather than a full regression analysis.

It is important that an experimenter examines a scatter plot of the data points (x_i, y_i) before calculating a sample correlation coefficient since any nonlinear relationship between the two variables can render the sample correlation coefficient meaningless. This is because the sample correlation coefficient r is designed to measure the amount of *linear* association between two variables. Figure 12.49 shows how two variables with a nonlinear relationship might have a sample correlation coefficient close to 0, which is misleading in this case because there is a relationship between the two variables.

Association and Causality   A common pitfall in data interpretation is to mistake *association* for *causality*. If two variables are shown to be correlated, this does not in itself establish that there is a causal relationship between the two variables. It may, however, suggest that there

FIGURE 12.49

Misleading sample correlation coefficient for a nonlinear relationship

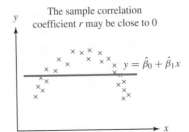

Sample Correlation Coefficient

The **sample correlation coefficient** r for a set of paired data observations (x_i, y_i) is

$$r = \frac{S_{XY}}{\sqrt{S_{XX}}\sqrt{S_{YY}}} = \frac{\sum_{i=1}^{n}(x_i - \bar{x})(y_i - \bar{y})}{\sqrt{\sum_{i=1}^{n}(x_i - \bar{x})^2}\sqrt{\sum_{i=1}^{n}(y_i - \bar{y})^2}}$$

$$= \frac{\sum_{i=1}^{n} x_i y_i - n\bar{x}\bar{y}}{\sqrt{\sum_{i=1}^{n} x_i^2 - n\bar{x}^2}\sqrt{\sum_{i=1}^{n} y_i^2 - n\bar{y}^2}}$$

It measures the strength of *linear* association between two variables and can be thought of as an estimate of the correlation ρ between the two associated random variables X and Y.

Under the assumption that the X and Y random variables have a bivariate normal distribution, a test of the null hypothesis

$$H_0 : \rho = 0$$

can be performed by comparing the t-statistic

$$t = \frac{r\sqrt{n - 2}}{\sqrt{1 - r^2}}$$

with a t-distribution with $n - 2$ degrees of freedom. In a regression framework, this test is equivalent to testing $H_0 : \beta_1 = 0$.

---

is a causal relationship between the two variables and motivate scientists to try to identify a mechanism for a causal relationship. Nevertheless, statistical data analysis in itself cannot establish causality.

For example, suppose that after a party some of the participants fall seriously ill. A doctor interviews all of the people who attended the party and finds out how ill they are and how much wine they consumed. Analysis of these data reveals a positive correlation between the level of illness and the amount of wine consumed. Does this prove that the wine was responsible for the illness?

The data analysis certainly suggests that the wine may be responsible for the illness, or at least may be a contributing factor. However, there are other possibilities. Suppose that the salted peanuts at the party are the real cause of the illness. Consequently, the more peanuts a person consumed, the more ill that person is likely to be, so that there is a causal relationship between peanuts and illness with a consequent positive association. Also, suppose that the more peanuts a person consumed, the thirstier the person is and so the greater is the person's wine consumption. Consequently, there is also a positive association between peanut consumption and wine consumption.

As Figure 12.50 illustrates, this scenario explains the positive association between wine consumption and illness, even though wine consumption in itself has nothing to do with the illness. In fact, *conditional* on the amount of peanuts consumed, wine consumption has nothing to do with illness. Therefore, even though there is a positive correlation between wine consumption and illness, it would be incorrect to use this result to infer that the wine consumption caused the illness. This example illustrates how the positive correlation exhibited between wine consumption and illness can be explained by the presence of a third variable, in this case peanuts, which is positively correlated with both wine consumption and illness.

FIGURE 12.50

A positive correlation between wine consumption and illness can be generated by their relationships with peanut consumption

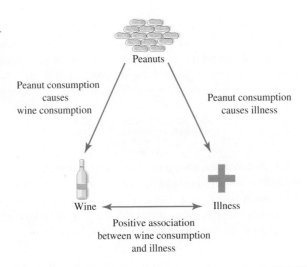

Peanuts

Peanut consumption causes wine consumption

Peanut consumption causes illness

Wine ← → Illness

Positive association between wine consumption and illness

12.9.2   Examples

Example 69
Cranial
Circumferences

The data set of finger lengths and cranial circumferences in Figure 12.22 gives

$$S_{YY} = 99.4574$$
$$S_{XY} = 3.10745$$
$$S_{XX} = 1.489505$$

The sample correlation coefficient is therefore

$$r = \frac{S_{XY}}{\sqrt{S_{XX}}\sqrt{S_{YY}}} = \frac{3.10745}{\sqrt{1.489505} \times \sqrt{99.4574}} = 0.255$$

This implies that the coefficient of determination for a regression analysis is $R^2 = r^2 = 0.065$. The t-statistic for testing the null hypothesis $H_0 : \rho = 0$ is

$$t = \frac{r\sqrt{n-2}}{\sqrt{1-r^2}} = \frac{0.255 \times \sqrt{18}}{\sqrt{1-0.255^2}} = 1.12$$

which, as expected, is equal to the t-statistic

$$t = \frac{\hat{\beta}_1}{\text{s.e.}(\hat{\beta}_1)}$$

obtained in Section 12.3, so that, as before, the p-value is 0.277. Consequently, it is plausible that the finger length and the cranial circumference are uncorrelated.

Example 68
Nile River Flowrate

For this data set the sample correlation coefficient is

$$r = \sqrt{R^2} = \sqrt{0.871} = 0.933$$

which confirms the strong positive linear association between the January and February inflows.

Example 44
Army Physical Fitness Test

For this data set the sample correlation coefficient is

$$r = -\sqrt{R^2} = -\sqrt{0.3228} = -0.568$$

with the number of pushups and the 2-mile run time being negatively correlated.

COMPUTER NOTE

Fortunately, the number crunching involved in a regression analysis can all be performed by a computer package and you need only enter the data into the package and specify the model that you want to fit. The computer will furnish you with the fitted regression line, p-values for the parameters, an analysis of variance table, coefficients of determination and correlation, and often much more additional information. Upon request, some packages will also compute confidence intervals for the regression line at specific values of the explanatory variable and prediction intervals for future response values. Some computer packages will draw confidence bands around the fitted regression line. Residuals and standardized residuals can also be obtained, and possible outliers are often brought to your attention.

The key to being a successful modeler is to make sure that you have a good understanding of what the computer output means. In addition, remember that one of the worst things that you can do is to haphazardly fit meaningless straight lines to data sets that exhibit a nonlinear relationship, so make good use of your package's graphics facilities to get a good feel for the data with which you are dealing.

■ 12.9.3 Problems

12.9.1 Show that if the data points lie on a straight line with

$$y_i = a + bx_i \qquad 1 \le i \le n$$

then the sample correlation coefficient r is equal to 1 if $b > 0$ and is equal to -1 if $b < 0$.

12.9.2 Show that if

$$x_i^* = ax_i + b \quad \text{and} \quad y_i^* = cy_i + d$$

for some constants a, b, c, and d, then the sample correlation coefficient of the data points (x_i, y_i) is equal to the sample correlation coefficient of the data points (x_i^*, y_i^*) except for a possible sign change.

For the data sets in Problems 12.9.3 through 12.9.9, what is the sample correlation coefficient r? Show that the t-statistic written in terms of the sample correlation coefficient $t = r\sqrt{n-2}/\sqrt{1-r^2}$ is equal to the t-statistic $t = \hat{\beta}_1/\text{s.e.}(\hat{\beta}_1)$ calculated earlier.

12.9.3 Oil Well Drilling Costs
The data set of oil well costs given in DS 12.2.1.

12.9.4 Truck Unloading Times
The data set of the times taken to unload a truck at a warehouse given in DS 12.2.2.

12.9.5 VO2-max Aerobic Fitness Measurements
The data set of aerobic fitness measurements given in DS 12.2.3.

12.9.6 Property Tax Appraisals
The data set of appraised house values given in DS 12.2.4.

12.9.7 Management of Computer Systems
The data set of the times taken for programming changes given in DS 12.2.5.

12.9.8 Vacuum Transducer Bobbin Resistances
The data set of vacuum transducer bobbin resistances given in DS 12.2.6.

12.9.9 Ceramic Baking Procedures
The data set of ceramic densities and baking times given in DS 12.3.1

12.9.10 The sample correlation coefficient between y and x is 0.3. If a simple linear regression is performed between y and x, what can you say about the sign (positive, 0, or negative) of the estimated slope parameter? What can you say about the p-value for testing whether the slope parameter is significant?

12.9.11 Suppose that variables A and B have a positive correlation. Provide a possible explanation for the claim that changes in B do not cause a change in A.

12.10 Case Study: Microelectronic Solder Joints

The researcher is interested in whether the heights of the solder joints have any influence on the reliability of the microelectronic assembly. Assemblies are constructed that are identical except for differences in the solder joint heights. All together, 3 assemblies are made with a height of 0.540 mm, 3 assemblies are made with a height of 0.555 mm, 3 assemblies are made with a height of 0.570 mm, and 3 assemblies are made with a height of 0.585 mm. These 12 assemblies are then subjected to temperature cycles, and after every 100 cycles they are tested to see whether they have failed. The resulting data set of the number of temperature cycles until failure is shown in Figure 12.51.

A graph of the data is shown in Figure 12.52 together with the fitted regression line that is

$$\text{cycles} = -983 + (9333 \times \text{height})$$

The positive slope of the fitted regression line seems to suggest that the reliability of the assemblies improves as the solder joint heights increase, but the researcher knows that the statistical significance of the regression line should be examined. The slope parameter has a standard error of 5532, so the t-statistic for testing whether the slope is 0 is

$$t = \frac{\hat{\beta}_1}{\text{s.e.}(\hat{\beta}_1)} = \frac{9333}{5532} = 1.69$$

This gives a p-value of $2 \times P(t_{10} \geq 1.69) = 0.122$, and since this is larger than 10%, the researcher concludes that the regression is not significant and that this data set does not provide sufficient evidence to establish that changing the solder joint heights in this range has any effect on the reliability of the assemblies. However, the researcher is aware that this analysis has not proved that there is no effect, and that collecting further data may provide evidence of a significant relationship between reliability and the solder joint heights.

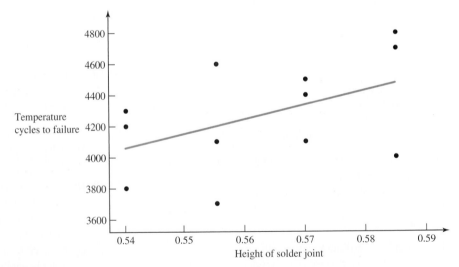

Solder joint height	Temperature cycles to failure
0.540	4300
0.540	3800
0.540	4200
0.555	4600
0.555	3700
0.555	4100
0.570	4400
0.570	4100
0.570	4500
0.585	4800
0.585	4000
0.585	4700

FIGURE 12.51

Data set of solder joint heights (mm) and temperature cycles to failure

FIGURE 12.52

Fitted regression line for temperature cycles to failure and solder joint height data set

12.11 Supplementary Problems

12.11.1 SAT Scores and Exam Completion Times

DS 12.11.1 gives the times taken by a class of $n = 14$ students to finish a math test together with the math SAT scores of the students. Perform the following linear regression analysis with time as the dependent variable and SAT as the explanatory variable.

(a) Calculate the point estimates $\hat{\beta}_0$, $\hat{\beta}_1$, and $\hat{\sigma}^2$.

(b) Calculate the sample correlation coefficient r.

(c) Calculate a two-sided 95% confidence interval for the slope parameter β_1.

(d) Is there any evidence that the time taken to finish the test depends upon the SAT score?

(e) How does the predicted time taken to finish the test change as the SAT score increases by 10 points?

(f) What is the predicted time for a student with an SAT score of 550? Construct a two-sided 95% confidence interval for the expected time for a student with an SAT score of 550. Construct a two-sided 95% prediction interval for the time taken by a student with an SAT score of 550.

(g) Plot the residuals against the SAT scores. Do there appear to be any problems with the regression analysis? Are there any points that are obvious outliers? Obtain a normal scores plot of the residuals.

12.11.2 Rolling Mill Scrap

The data used in Example 43 are taken from the data set given in DS 12.11.2, which relates the variable % scrap to the number of times that the metal plate is passed through the rollers. It is useful to be able to model the amount of discarded metal, % scrap, as a function of the number of passes through the rollers. Perform a linear regression analysis with % scrap as the dependent variable and the number of passes through the rollers as the explanatory variable.

(a) Fit the regression line and estimate the error variance.

(b) Calculate the sample correlation coefficient r.

(c) What is the evidence that the amount of scrap material depends upon the number of passes?

(d) Calculate a two-sided 95% confidence interval for the slope parameter β_1.

(e) How does the predicted amount of scrap change with each extra pass through the rollers?

(f) What is the predicted amount of scrap after seven passes through the rollers? Construct a two-sided

95% prediction interval for the amount of scrap after seven passes of a particular sheet through the rollers.

(g) Plot the residuals against the number of passes. Which points have values $e_i/\hat{\sigma}$ larger than 3 in absolute value? Do you think that a straight line regression model is appropriate? Obtain a normal scores plot of the residuals.

12.11.3 Friction Power Loss from Engine Bearings

DS 12.11.3 contains data from a set of automobile engines concerning the diameter of the bearing and the friction power loss that occurs in the bearing of the engine. Perform a linear regression analysis with power loss as the dependent variable and bearing diameter as the explanatory variable.

Can you conclude that there is a significant association between power loss and bearing diameter? What is the sample correlation coefficient? What can you predict for the power loss of a new engine with a bearing diameter of 25.0? Are there any data points with values $e_i/\hat{\sigma}$ larger than 3 in absolute value?

12.11.4 Double Pipe Heat Exchanger

A double pipe heat exchanger is a device that allows energy to be exchanged between a hot liquid and a cold liquid. It consists of two concentric pipes of different diameter. The hot liquid flows through the inner pipe, and the cold liquid flows in the gap between the two pipes, usually in the opposite direction to the hot liquid. DS 12.11.4 contains some data obtained from these heat exchangers, listing the energy lost by the hot liquid and the corresponding energy gained by the cold liquid. These values are not equal because of measurement errors, because not all of the energy lost by the hot liquid is aborbed by the cold liquid, and because the cold liquid can absorb energy from sources other than the hot liquid. Perform a linear regression analysis with energy gain as the dependent variable and energy loss as the explanatory variable. What is the fitted model and what is the sample correlation coefficient? What can you predict for the energy gained by the cold liquid when the energy lost by the hot liquid is 500?

12.11.5 Capacitance Discharge Pulse Times

Electrical capacitors can be used to deliver short high-voltage pulses. As the capacitor discharges, the time span of the resulting pulse is an important property. DS 12.11.5 contains measurements of the time spans of the discharges, in milliseconds, obtained for different capacitance values, in microfarads. Perform a linear regression analysis with pulse time as the dependent variable and capacitance as the explanatory variable. Is there a significant association between the pulse time and the capacitance value? What is the fitted model and what is the sample correlation coefficient? What can you predict for the pulse time of a discharge from a capacitance of 1700 microfarads? Which point has the largest residual in absolute value?

12.11.6 Deformability of Arteries with Atherosclerosis

DS 12.11.6 contains the circumferential strains S and lesion areas L of $n = 12$ specimens of human arteries with atherosclerosis. This disease is commonly referred to as "hardening of the arteries" due to the accumulation of plaque on the artery walls. The lesion area measures the amount of plaque that has accumulated on the artery walls and can be thought of as the disease severity. The circumferential strain is a measure of the deformability of the artery tissue. A model that relates the circumferential strain S to the lesion area L is helpful in assessing the potential impact of various levels of atherosclerosis.

(a) Show that the model

$$S = \gamma_0 + \frac{\gamma_1}{L}$$

appears to provide a good fit to the data set.

(b) After a suitable transformation of the variables, find point estimates for γ_0 and γ_1.

(c) Calculate two-sided 95% confidence intervals for γ_0 and γ_1.

(d) What is the predicted circumferential strain of an artery with a lesion area of 10.0 mm²?

12.11.7 Catalyst Effect on Chemical Process

DS 12.11.7 contains data on the measurements of the strength of a chemical solution for different amounts of a catalyst. It is decided to perform a linear regression analysis with strength as the dependent variable and amount of catalyst as the explanatory variable.

(a) Calculate the values of $\hat{\beta}_0$, $\hat{\beta}_1$, and $\hat{\sigma}$.

(b) Perform a two-sided hypothesis test of $H_0 : \beta_1 = 0$ and interpret the result.

(c) If a chemical solution is made with a catalyst amount of 21.0, obtain a range of values for the strength with 95% confidence.

(d) What is the residual for the point (28, 17)?

12.11.8

In a simple linear regression analysis with $n = 20$ data points, the estimates $\hat{\beta}_0 = 123.57$, $\hat{\beta}_1 = -3.90$, and $\hat{\sigma} = 11.52$ are obtained, with $\sum_{i=1}^{20} x_i = 856$, $\sum_{i=1}^{20} x_i^2 = 37636$, $\sum_{i=1}^{20} y_i = -869$, and $\sum_{i=1}^{20} y_i^2 = 55230$.

(a) Construct a two-sided 95% prediction interval for a future response value when the input value is 40.

(b) Compute the analysis of variance table and calculate the coefficient of determination R^2.

12.11.9 Bacteria Cultures

A simple linear regression model is to be fitted to the data set of temperature and yield values for bacteria cultivation in DS 12.11.8, with temperature as the explanatory variable and yield as the dependent variable.

(a) Calculate $\hat{\beta}_0$, $\hat{\beta}_1$, and $\hat{\sigma}^2$.

(b) Is there sufficient evidence to conclude that the yield depends upon the temperature?

(c) What is the residual for the first data point?

12.11.10

A simple linear regression analysis has $n = 20$ and $R^2 = 0.853$. What can you say about the p-value for the two-sided hypothesis test of $H_0 : \beta_1 = 0$?

12.11.11 Gold Extraction from Unprocessed Ore

Consider the data set in DS 12.11.9 from the mining industry that shows the amount of ore processed in various locations together with the resulting amount of gold that was obtained. A linear regression model is used to analyze the data with the amount of ore processed as the input variable and with the amount of gold obtained as the output variable.

(a) Calculate the fitted regression line.

(b) Perform a hypothesis test to assess whether it is a significant regression.

(c) What is the correlation between the amount of ore processed and the amount of gold obtained?

(d) What is the value of the coefficient of determination R^2?

(e) Calculate a 95% prediction interval for the amount of gold that will be obtained if 15.0 kilotons of ore are processed.

(f) Calculate the residual value for each data point.

12.11.12

Are the following statements true or false?

(a) In simple linear regression, the choice of which variable should be designated as the dependent variable and which variable should be designated as the explanatory variable is determined by considering the correlation between the variables.

(b) In regression modeling the dependent variable may be called the output variable.

(c) The standard fitted regression line is obtained from a least squares fit.

(d) In general, extrapolation should be avoided if possible, but it is not as risky as interpolation.

(e) Prediction bands are narrowest at the sample average of the input variable.

(f) A small value of the coefficient of determination indicates that the regression cannot be statistically significant.

(g) If two variables have a causal relationship, then they have an association.

(h) If two variables have an association, then they must have a causal relationship.

(i) A simple linear regression model has three unknown parameters.

(j) The solution of the "normal equations" provides the fitted intercept parameter and the fitted slope parameter.

(k) For very large sample sizes the confidence bands and the predictions bands will be virtually indistinguishable.

(l) Not every nonlinear model is intrinsically linear.

(m) The correlation between two variables can sometimes be reduced by changing the units that they are measured with.

12.11.13 Downloading Computer Files

The data set in DS 12.11.10 shows the size of some files and the times that they took to download from a website. A linear regression model is used to analyze the data with size as the input variable and time as the output variable.

(a) Find $\hat{\beta}_0$, $\hat{\beta}_1$, and $\hat{\sigma}^2$.

(b) Perform the hypothesis test to assess whether it is a significant regression.

(c) What is the expected download time for a file of size 6.00?

(d) What is the value of R^2?

(e) Calculate a 95% prediction interval for the downloading time of a file of size 6.00.

(f) What is the residual for the data point (4.56, 103)?

(g) What is the correlation between size and time?

(h) Can you predict the downloading time of a file of size 0.40?

12.11.14 Underwater Speed of Sound

Pressure changes in seawater affect the speed at which sound is transmitted, and the data set in DS 12.11.11 shows how the underwater speed of sound relates to the depth at which it is measured. Use a linear regression model to analyze the data with the depth as the input variable and with the speed of sound as the output variable.

(a) Show how to calculate the fitted regression line.

(b) Show how to calculate the estimate of the error variance.

(c) Perform a hypothesis test to assess whether it is a significant regression.

(d) Calculate a 95% confidence interval for the speed of sound at a depth of 4 meters.

(e) What is the value of the coefficient of determination R^2?

The following data sets can be used to practice the regression techniques discussed in this chapter.

12.11.15 Oxygen Purity in Chemical Distillation Processes

DS 12.11.12 contains the oxygen purities obtained from a chemical distillation process when the percentage of hydrocarbons that are present in the main condenser is varied.

12.11.16 Cotton Bale Productivities

DS 12.11.13 contains the cotton bales per acre harvested by a cotton farmer over a 30-year period. Year to year differences in the weather cause a substantial amount of variability in the productivity, but the farmer is interested in whether his attempts to improve soil fertility and irrigation methods have had a significant effect on his output.

12.11.17 Aerodynamics and Wing Design

DS 12.11.14 contains the wing lift generated in various trials of an airfoil design where the angle of inclination of the airfoil to the horizontal (the angle of attack) is varied over a small range. Can a linear model be effectively used to predict the wing lift generated by an angle of attack within this range?

12.11.18 Vapor Pressure Modeling for a Gas

DS 12.11.15 contains the vapor pressures of a gas at different temperatures (Kelvin). How do you feel about the linear regression model with temperature as the input variable and vapor pressure as the output variable? Theory suggests that within this temperature range the log of the vapor pressure can be modeled as a linear function of the reciprocal of the temperature. Do you prefer this model?

12.11.19 Static Breaking Strengths of Ropes

DS 12.11.16 contains the static breaking strengths of ropes of different diameters. This tensile strength is the maximum load that a rope can sustain before breaking. What model would you recommend for the relationship between breaking strength and rope diameter?

12.11.20 Protein Concentrations from Cell Lysing

In a biochemical experiment, various cultures of cells are grown and the number of cells are counted. Then the cells are lysed so that they release their contents and the resulting protein concentration is measured as given in DS 12.11.17.

12.11.21 Heart Tissue Elasticity

A sample of heart tissue was subjected to different stresses and the resulting strains are given in

DS 12.11.18. Compare the model

$$\text{strain} = \beta_0 + \beta_1 \text{ stress}$$

with the model

$$\text{strain} = \gamma_0 \text{ stress}^{\gamma_1}$$

12.11.22 Wind Speeds and Air Pressures of Hurricanes

DS 12.11.19 contains the maximum windspeeds and the minimum air pressures for the hurricanes and tropical storms occurring off the Atlantic seaboard of North America in 2004.

12.11.23 Thorax Width and Dry Mass of Wasps

DS 12.11.20 contains the thorax width and the dry mass of a random sample of male wasps.

CHAPTER THIRTEEN

Multiple Linear Regression and Nonlinear Regression

A **multiple linear regression** model is an extension of a simple linear regression model that allows the response variable y to be modeled as a linear function of *more than one* input variable x_i. The ideas and concepts behind multiple linear regression are similar to those discussed in Chapter 12 for simple linear regression, although the computations are considerably more complex. The best way to deal with multiple linear regression mathematically is with a matrix algebra approach, which is discussed in Section 13.3. Nevertheless, the discussion of multiple linear regression and the examples presented in Sections 13.1, 13.2, and 13.4 can be read without an understanding of Section 13.3.

Nonlinear regression models, discussed in Section 13.5, are models in which the expected value of the response variable y is not expressed as a linear combination of the parameters. Even though many of the ideas in nonlinear regression are similar to those in linear regression, the mathematical processes required to fit the models and to make statistical inferences are different.

13.1 Introduction to Multiple Linear Regression

13.1.1 The Multiple Linear Regression Model

Consider the problem of modeling a response variable y as a function of k input variables $x_1, \ldots, x_k$ based upon a data set consisting of the n sets of values

$$(y_1, x_{11}, x_{21}, \ldots, x_{k1})$$

$$\vdots$$

$$(y_n, x_{1n}, x_{2n}, \ldots, x_{kn})$$

Thus, y_i is the value taken by the response variable y for the ith observation, which is obtained with values $x_{1i}, x_{2i}, \ldots, x_{ki}$ of the k input variables $x_1, x_2, \ldots, x_k$.

In multiple linear regression, the value of the response variable y_i is modeled as

$$y_i = \beta_0 + \beta_1 x_{1i} + \cdots + \beta_k x_{ki} + \epsilon_i$$

which consists of a linear combination $\beta_0 + \beta_1 x_{1i} + \cdots + \beta_k x_{ki}$ of the corresponding values of the input variables together with an error term ϵ_i. The coefficients $\beta_0, \ldots, \beta_k$ are unknown parameters, and the error terms ϵ_i, $1 \leq i \leq n$, are taken to be independent observations from a $N(0, \sigma^2)$ distribution. Of course when $k = 1$ so that there is only one input variable, this model simplifies to the simple linear regression model discussed in Chapter 12.

The expected value of the response variable at $\mathbf{x} = (x_1, \ldots, x_k)$ is

$$E(Y|\mathbf{x}) = \beta_0 + \beta_1 x_1 + \cdots + \beta_k x_k$$

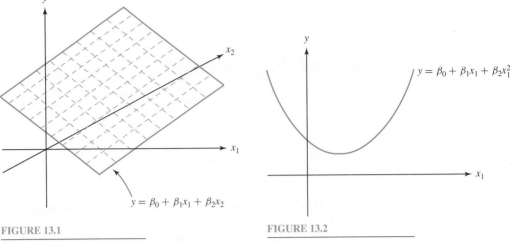

FIGURE 13.1

The expected value of the response variable for multiple linear regression model with two input variables is a plane

FIGURE 13.2

A quadratic regression model

For example, with $k = 2$ the expected values of the response variable lie on the plane

$$y = \beta_0 + \beta_1 x_1 + \beta_2 x_2$$

as illustrated in Figure 13.1. The parameter β_0 is referred to as the intercept parameter, and the parameter β_i determines how the ith input variable x_i influences the response variable when the other input variables are kept fixed. If $\beta_i > 0$, then the expected value of the response variable increases as the ith input variable increases (with the other input variables kept fixed), and if $\beta_i < 0$, then the expected value of the response variable decreases as the ith input variable increases (with the other input variables kept fixed). If $\beta_i = 0$, then the response variable is not influenced by changes in the ith input variable (with the other input variables kept fixed).

The multiple linear regression model can provide a rich variety of functional relationships by allowing some of the input variables x_i to be functions of other input variables. For example, with $k = 2$ and $x_2 = x_1^2$ the expected value of the response variable is modeled as

$$y = \beta_0 + \beta_1 x_1 + \beta_2 x_1^2$$

as illustrated in Figure 13.2. This equation is known as a **quadratic regression model** and allows the response variable to be related to a quadratic function of an input variable. More generally, a **polynomial regression model** of order k has $x_i = x_1^i$ so that the model is

$$y = \beta_0 + \beta_1 x_1 + \beta_2 x_1^2 + \cdots + \beta_k x_1^k$$

and the response variable is related to a polynomial function of the input variable x_1.

Another common model involving quadratic functions of two input variables x_1 and x_2 is

$$y = \beta_0 + \beta_1 x_1 + \beta_2 x_1^2 + \beta_3 x_2 + \beta_4 x_2^2 + \beta_5 x_1 x_2$$

The final term $x_1 x_2$ is referred to as the **cross product** or **interaction** term of the two input variables x_1 and x_2. This model is known as a **response surface model** and is illustrated in Figure 13.3.

The input variables x_i may be measurements of a continuous variable or a discrete variable, as illustrated in the examples in Section 13.2. In particular, a variable may be used to indicate

FIGURE 13.3

A typical response surface model

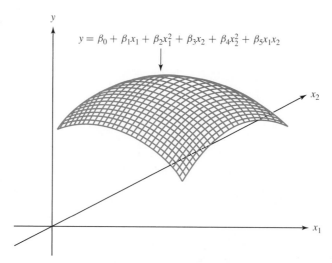

$$y = \beta_0 + \beta_1 x_1 + \beta_2 x_1^2 + \beta_3 x_2 + \beta_4 x_2^2 + \beta_5 x_1 x_2$$

the presence or absence of a certain effect or to designate at which of two levels a data observation is obtained. Such input variables are known as **indicator variables**. For example, if the response variable is a measurement obtained from a person, then the indicator variable x_i taking the values $x_i = 0$ if the person is male and $x_i = 1$ if the person is female can be used to investigate whether the response variable is related to the gender of the person. In other cases, indicator variables may be used to indicate whether an observation is taken before or after a change has occurred or to indicate from which of two machines an observation is taken, for example.

In summary, the key to deciding whether a particular model can be formulated as a multiple linear regression model is to determine whether the expected value of the response variable can be written as a linear combination of a set of (possibly related) input variables that can represent either continuous or discrete variables. As discussed in Chapter 12 with respect to the simple linear regression model, transformations of the response variable or the input variables can be employed to make the linear model appropriate. In practice, the multiple linear regression model can provide a general and useful way of exploring the relationship between a response variable and a set of input variables.

13.1.2 Fitting the Linear Regression Model

As in simple linear regression, the parameters $\beta_0, \ldots, \beta_k$ are estimated using the method of **least squares**. As Figure 13.4 illustrates for $k = 2$, this implies that the parameters are chosen so as to minimize the sum of squares of the (vertical) distances between the data observations y_i and their fitted values

$$\hat{\beta}_0 + \hat{\beta}_1 x_{1i} + \cdots + \hat{\beta}_k x_{ki}$$

With the normal distribution of the error terms ϵ_i the method of least squares produces the maximum likelihood estimates of the parameters.

The sum of squares is

$$Q = \sum_{i=1}^{n} (y_i - (\beta_0 + \beta_1 x_{1i} + \cdots + \beta_k x_{ki}))^2$$

FIGURE 13.4

Least squares fit for a multiple
linear regression model with two
input variables

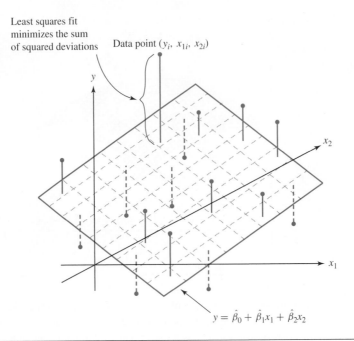

which can be minimized by taking derivatives with respect to each of the unknown parameters, setting the resulting expressions equal to 0, and then simultaneously solving the equations to obtain the parameter estimates $\hat{\beta}_i$. The derivative of Q with respect to β_0 is

$$\frac{\partial Q}{\partial \beta_0} = -2 \sum_{i=1}^{n} (y_i - (\beta_0 + \beta_1 x_{1i} + \cdots + \beta_k x_{ki}))$$

and the derivative of Q with respect to β_j, $1 \le j \le k$, is

$$\frac{\partial Q}{\partial \beta_j} = -2 \sum_{i=1}^{n} x_{ji} (y_i - (\beta_0 + \beta_1 x_{1i} + \cdots + \beta_k x_{ki}))$$

Setting these expressions equal to 0 results in the $k + 1$ equations

$$\sum_{i=1}^{n} y_i = n\beta_0 + \beta_1 \sum_{i=1}^{n} x_{1i} + \beta_2 \sum_{i=1}^{n} x_{2i} + \cdots + \beta_k \sum_{i=1}^{n} x_{ki}$$

$$\sum_{i=1}^{n} y_i x_{1i} = \beta_0 \sum_{i=1}^{n} x_{1i} + \beta_1 \sum_{i=1}^{n} x_{1i}^2 + \beta_2 \sum_{i=1}^{n} x_{2i} x_{1i} + \cdots + \beta_k \sum_{i=1}^{n} x_{ki} x_{1i}$$

$$\vdots$$

$$\sum_{i=1}^{n} y_i x_{ki} = \beta_0 \sum_{i=1}^{n} x_{ki} + \beta_1 \sum_{i=1}^{n} x_{1i} x_{ki} + \beta_2 \sum_{i=1}^{n} x_{2i} x_{ki} + \cdots + \beta_k \sum_{i=1}^{n} x_{ki}^2$$

The parameter estimates $\hat{\beta}_0, \hat{\beta}_1, \ldots, \hat{\beta}_k$ are the solutions to these equations, which are known as the **normal equations**.

It would be tedious to find the solutions to these equations by hand, yet it is a simple problem for a computer to handle. Mathematically, the best way to deal with this problem and any subsequent inferences concerning the regression model is to use matrix algebra.

This procedure is demonstrated in Section 13.3, although an understanding of the matrix formulation of the multiple linear regression analysis is not essential for the rest of this chapter.

13.1.3 Analysis of the Fitted Model

The fitted value of the response variable at the data point $(x_{1i}, \ldots, x_{ki})$ is

$$\hat{y}_i = \hat{\beta}_0 + \hat{\beta}_1 x_{1i} + \cdots + \hat{\beta}_k x_{ki}$$

and the ith residual is

$$e_i = y_i - \hat{y}_i$$

As in simple linear regression, the sum of squares for error is defined to be

$$\text{SSE} = \sum_{i=1}^{n} (y_i - \hat{y}_i)^2 = \sum_{i=1}^{n} e_i^2$$

The estimate of the error variance σ^2 is

$$\text{MSE} = \hat{\sigma}^2 = \frac{\text{SSE}}{n-k-1}$$

(where the degrees of freedom for error $n-k-1$ are equal to the sample size n minus the number of parameters estimated $k+1$), which is distributed as

$$\hat{\sigma}^2 \sim \sigma^2 \frac{\chi_{n-k-1}^2}{n-k-1}$$

Also, as with simple linear regression the total sum of squares

$$\text{SST} = \sum_{i=1}^{n} (y_i - \bar{y})^2$$

can be decomposed as

$$\text{SST} = \text{SSR} + \text{SSE}$$

where the regression sum of squares is

$$\text{SSR} = \sum_{i=1}^{n} (\hat{y}_i - \bar{y})^2$$

However, as the analysis of variance table in Figure 13.5 shows, there are now k degrees of freedom for regression and $n-k-1$ degrees of freedom for error. The p-value in the analysis

Source	Degrees of freedom	Sum of squares	Mean square	F-statistic	p-value
Regression	k	SSR	$\text{MSR} = \text{SSR}/k$	$F = \text{MSR}/\text{MSE}$	$P(F_{k,n-k-1} > F)$
Error	$n-k-1$	SSE	$\text{MSE} = \text{SSE}/(n-k-1)$		
Total	$n-1$	SST			

FIGURE 13.5

Analysis of variance table for a multiple linear regression analysis

of variance table is for the null hypothesis

$$H_0 : \beta_1 = \cdots = \beta_k = 0$$

(with the alternative hypothesis that at least one of these β_i is nonzero), which has the interpretation that the response variable is not related to any of the k input variables. A large p-value therefore indicates that there is no evidence that any of the input variables affects the distribution of the response variable y.

The coefficient of (multiple) determination

$$R^2 = \frac{\text{SSR}}{\text{SST}}$$

takes a value between 0 and 1 and indicates the amount of the total variability in the values of the response variable that is accounted for by the fitted regression model. The F-statistic in the analysis of variance table can be written as

$$F = \frac{(n - k - 1)R^2}{k(1 - R^2)}$$

Analysis of Variance Table for Multiple Linear Regression Problem

The analysis of variance table for a multiple linear regression problem provides a test of the null hypothesis

$$H_0 : \beta_1 = \cdots = \beta_k = 0$$

which implies that the response variable is not related to any of the k input variables. The p-value is

$$p\text{-value} = P(X > F)$$

where the random variable X has an F-distribution with degrees of freedom k and $n - k - 1$, and the F-statistic is

$$F = \frac{\text{SSR}}{k\hat{\sigma}^2} = \frac{(n - k - 1)R^2}{k(1 - R^2)}$$

A small p-value in the analysis of variance table indicates that the response variable is related to at least one of the input variables. Computer output provides the user with the parameter estimates

$$\hat{\beta}_0, \hat{\beta}_1, \ldots, \hat{\beta}_k$$

together with their standard errors

$$\text{s.e.}(\hat{\beta}_0), \text{s.e.}(\hat{\beta}_1), \ldots, \text{s.e.}(\hat{\beta}_k)$$

and these can be used to determine which input variables are needed in the regression model. If $\beta_i = 0$, then the distribution of the response variable does not directly depend on the input variable x_i, which can therefore be "dropped" from the model. Consequently, it is useful to test the hypotheses

$$H_0 : \beta_i = 0 \quad \text{versus} \quad H_A : \beta_i \neq 0$$

If the null hypothesis is accepted, then there is no evidence that the response variable is directly related to the input variable x_i, and it can therefore be dropped from the model. If the null hypothesis is rejected, then there is evidence that the response variable is related to the input variable, and it should therefore be kept in the model.

These hypotheses are tested with the t-statistics

$$t = \frac{\hat{\beta}_i}{\text{s.e.}(\hat{\beta}_i)}$$

which are compared to a t-distribution with $n - k - 1$ degrees of freedom. The (two-sided) p-value is therefore

$$p\text{-value} = 2 \times P(X > |t|)$$

where the random variable X has a t-distribution with $n - k - 1$ degrees of freedom.

Computer output for this analysis generally contains a list of p-values corresponding to each of the parameters $\beta_0, \beta_1, \ldots, \beta_k$. As with simple linear regression, the p-value for the intercept parameter β_0 is usually not important. For the other parameters, p-values larger than 10% indicate that the corresponding input variable can be dropped from the model, while p-values smaller than 1% indicate that the corresponding input variable should be kept in the model. However, p-values between 1% and 10% do not provide a clear indication, and how the corresponding input variables are dealt with can be left to the experimenter's judgment or particular requirements. It should also be noted that when one or more variables are removed from the model, the p-values of the remaining variables may change when the reduced model is fitted.

The more general hypotheses

$$H_0 : \beta_i = b_i \quad \text{versus} \quad H_A : \beta_i \neq b_i$$

for a fixed value of b_i of interest can be tested with the t-statistic

$$t = \frac{\hat{\beta}_i - b_i}{\text{s.e.}(\hat{\beta}_i)}$$

which is again compared to a t-distribution with $n - k - 1$ degrees of freedom. A confidence interval for β_i can be constructed as shown in the accompanying box.

Model fitting is performed by finding which subset of the k input variables is required to model the dependent variable y in the best and most succinct manner. The final model that the experimenter uses for inference problems should consist of input variables that each have p-values no larger than 10%, because otherwise some variables should be taken out of the model to simplify it. An important warning when fitting the model is that it is best to remove only one variable from the model at a time. This is because when one variable is removed from the model and the subsequent reduced model is fitted, the p-values of the remaining input variables in the model usually change.

Typically, model fitting may be performed in the following manner. First, an experimenter starts by fitting all k input variables. If all variables are needed in the model, then no reduction is necessary. If one or more input variables has a p-value larger than 10%, the variable with the largest p-value (smallest absolute value of the t-statistic) is removed. The reduced model with $k - 1$ input variables is then fitted, and the process is repeated. This is known as a **backwards elimination** modeling procedure and most computer packages will perform it automatically on request.

Alternatively, a **forward selection** modeling procedure can be employed whereby the model starts off without any variables, and input variables are then added one at a time until

---

Inferences on a Parameter β_i

Inferences on the parameter β_i in a multiple linear regression model, which indicates how the response variable is related to the ith input variable, are performed using $\hat{\beta}_i$ and s.e.$(\hat{\beta}_i)$. Specifically, the hypotheses

$$H_0 : \beta_i = b_i \quad \text{versus} \quad H_A : \beta_i \neq b_i$$

for a fixed value b_i of interest have a two-sided p-value

$$p\text{-value} = 2 \times P(X > |t|)$$

where the random variable X has a t-distribution with $n - k - 1$ degrees of freedom and the t-statistic is

$$t = \frac{\hat{\beta}_i - b_i}{\text{s.e.}(\hat{\beta}_i)}$$

A $1 - \alpha$ confidence level two-sided confidence interval for β_i can be constructed as

$$\beta_i \in (\hat{\beta}_i - t_{\alpha/2, n-k-1}\text{s.e.}(\hat{\beta}_i), \hat{\beta}_i + t_{\alpha/2, n-k-1}\text{s.e.}(\hat{\beta}_i))$$

---

no further additions are needed. Most computer packages can also implement a stepwise procedure that is a combination of the backward elimination and forward selection procedures, and that allows either the addition of new variables or the removal of variables already in the model.

Occasionally, an experimenter may wish to test the hypothesis that a certain subset of the parameters β_i are all equal to 0. Specifically, suppose that the null hypothesis

$$H_0 : \beta_{r+1} = \beta_{r+2} = \cdots = \beta_k = 0$$

is of interest. This hypothesis is tested by first fitting the *full* model with parameters $\beta_1, \ldots, \beta_k$ and finding the error sum of squares SSE_k, and then fitting the *reduced* model with parameters $\beta_1, \ldots, \beta_r$ and finding the new error sum of squares SSE_r. The F-statistic

$$F = \frac{(n - k - 1)(\text{SSE}_r - \text{SSE}_k)}{(k - r)\text{SSE}_k}$$

is calculated and the null hypothesis has a p-value

$$p\text{-value} = P(X > F)$$

where the random variable X has an F-distribution with parameters $k - r$ and $n - k - 1$.

13.1.4 Inferences on the Response Variable

Once a final model has been decided upon, it can be used to make inferences about the response variable at certain values of the input variables. Suppose that the final model includes the k input variables $x_1, \ldots, x_k$ which may be only a subset of the initial set of input variables that were investigated. The fitted value of the response variable at a specific set of values of the

FIGURE 13.6

Extrapolation of the fitted model
can occur even when both variables
x_1 and x_2 are within their ranges
from the data set

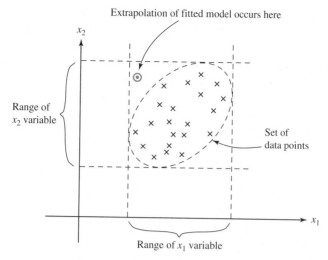

input variables x_i^* of interest is

$$\hat{y}|_{\mathbf{x}^*} = \hat{\beta}_0 + \hat{\beta}_1 x_1^* + \cdots + \hat{\beta}_k x_k^*$$

which is an estimate of the expected value of the response variable y at these values of the input variables.

A $1 - \alpha$ confidence level confidence interval for the expected value of the response variable at x_i^* is

$$\beta_0 + \beta_1 x_1^* + \cdots + \beta_k x_k^* \in (\hat{y}|_{\mathbf{x}^*} - t_{\alpha/2, n-k-1} \text{s.e.}(\hat{y}|_{\mathbf{x}^*}), \hat{y}|_{\mathbf{x}^*} + t_{\alpha/2, n-k-1} \text{s.e.}(\hat{y}|_{\mathbf{x}^*}))$$

which can be obtained from most computer packages. Prediction intervals for a future observation obtained at x_i^* can also be obtained.

When the fitted model is used, it is important to be aware of the danger of extrapolation that occurs when the model is used for values of the input variables *outside* the range of the data set. As Figure 13.6 shows, it is generally not sufficient to consider the ranges of each of the input variables x_i and to then consider the fitted model appropriate as long as the values of each of the input variables lie within its respective range. Extrapolation can occur even when each individual variable lies within its respective range of data values.

As a final point, it should always be remembered that as in simple linear regression, it is important to check whether the assumptions of the linear model appear to be appropriate. Residual plots and other techniques related to this issue are discussed in Section 13.4.

■ 13.1.5 Problems

13.1.1 The multiple linear regression model

$$y = \beta_0 + \beta_1 x_1 + \beta_2 x_2 + \beta_3 x_3$$

is fitted to a data set of $n = 30$ observations. The total sum of squares is SST = 108.9, and the error sum of squares is SSE = 12.4.

(a) What is the coefficient of determination R^2?

(b) Write out the analysis of variance table.

(c) What is the estimate of the error variance $\hat{\sigma}^2$?

(d) What is the p-value for the null hypothesis

$$H_0 : \beta_1 = \beta_2 = \beta_3 = 0$$

(e) If $\hat{\beta}_1 = 16.5$ and s.e.$(\hat{\beta}_1) = 2.6$, construct a two-sided 95% confidence interval for β_1.

13.1.2 The multiple linear regression model

$$y = \beta_0 + \beta_1 x_1 + \beta_2 x_2 + \beta_3 x_3 + \beta_4 x_4 + \beta_5 x_5 + \beta_6 x_6$$

is fitted to a data set of $n = 45$ observations. The total sum of squares is SST $= 11.62$, and the error sum of squares is SSE $= 8.95$.

(a) What is the coefficient of determination R^2?

(b) Write out the analysis of variance table.

(c) What is the estimate of the error variance $\hat{\sigma}^2$?

(d) What is the p-value for the null hypothesis

$$H_0 : \beta_1 = \beta_2 = \beta_3 = \beta_4 = \beta_5 = \beta_6 = 0$$

(e) If $\hat{\beta}_3 = 1.05$ and s.e.$(\hat{\beta}_3) = 0.91$, construct a two-sided 95% confidence interval for β_3.

13.1.3 The multiple linear regression model

$$y = \beta_0 + \beta_1 x_1 + \beta_2 x_2 + \beta_3 x_3 + \beta_4 x_4$$

is fitted to a data set of $n = 12$ observations.

(a) If $\hat{\beta}_2 = 132.4$ and s.e.$(\hat{\beta}_2) = 27.6$, construct a two-sided 95% confidence interval for β_2.

(b) What is the p-value for the null hypothesis $H_0 : \beta_2 = 0$?

13.1.4 The multiple linear regression model

$$y = \beta_0 + \beta_1 x_1 + \beta_2 x_2 + \beta_3 x_3$$

is fitted to a data set of $n = 15$ observations.

(a) If $\hat{\beta}_1 = 0.954$ and s.e.$(\hat{\beta}_1) = 0.616$, construct a two-sided 95% confidence interval for β_1.

(b) What is the p-value for the null hypothesis $H_0 : \beta_1 = 0$?

13.1.5 The multiple linear regression model

$$y = \beta_0 + \beta_1 x_1 + \beta_2 x_2 + \beta_3 x_3$$

is fitted to a data set of $n = 44$ observations. The parameter estimates $\hat{\beta}_1 = 11.64$, $\hat{\beta}_2 = 132.9$, and $\hat{\beta}_3 = 0.775$ are obtained with standard errors s.e.$(\hat{\beta}_1) = 1.03$, s.e.$(\hat{\beta}_2) = 22.8$, and s.e.$(\hat{\beta}_3) = 0.671$. Should any of the variables x_1, x_2, and x_3 be removed from the model?

13.1.6 The multiple linear regression model

$$y = \beta_0 + \beta_1 x_1 + \beta_2 x_2 + \beta_3 x_3 + \beta_4 x_4 + \beta_5 x_5 + \beta_6 x_6$$

is fitted to a data set of $n = 23$ observations and an error sum of squares of 1347.1 is obtained. The model

$$y = \beta_0 + \beta_1 x_1 + \beta_2 x_2$$

is then fitted to the data set and an error sum of squares of 1873.4 is obtained. Test the null hypothesis

$$H_0 : \beta_3 = \beta_4 = \beta_5 = \beta_6 = 0$$

13.1.7 The multiple linear regression model

$$y = \beta_0 + \beta_1 x_1 + \beta_2 x_2 + \beta_3 x_3 + \beta_4 x_4 + \beta_5 x_5$$

is fitted to a data set of $n = 19$ observations and an error sum of squares of 12.76 is obtained. The model

$$y = \beta_0 + \beta_4 x_4 + \beta_5 x_5$$

is then fitted to the data set and an error sum of squares of 28.33 is obtained. Test the null hypothesis

$$H_0 : \beta_1 = \beta_2 = \beta_3 = 0$$

13.1.8 The multiple linear regression model

$$y = \beta_0 + \beta_1 x_1 + \beta_2 x_2$$

is fitted to the data set given in DS 13.1.1.

(a) Write out the normal equations and show that they are satisfied by the parameter estimates $\hat{\beta}_0 = 7.280$, $\hat{\beta}_1 = -0.313$, and $\hat{\beta}_2 = -0.1861$.

(b) What is the fitted value of the expected value of the response variable when $x_1 = x_2 = 1$?

13.1.9 The multiple linear regression model

$$y = \beta_0 + \beta_1 x_1 + \beta_2 x_2$$

is fitted to a data set with $n = 20$ observations, and the parameter estimates $\hat{\beta}_0 = 104.9$, $\hat{\beta}_1 = 12.76$, and $\hat{\beta}_2 = 409.6$ are obtained.

(a) What is the fitted value of the expected value of the response variable when $x_1 = 10$ and $x_2 = 0.3$?

(b) If this fitted value has a standard error of 17.6, construct a two-sided 95% confidence interval for the expected value of the response variable at this point.

13.1.10 The multiple linear regression model

$$y = \beta_0 + \beta_1 x_1 + \beta_2 x_2 + \beta_3 x_3$$

is fitted to a data set with $n = 15$ observations, and the parameter estimates $\hat{\beta}_0 = 65.98$, $\hat{\beta}_1 = 23.65$, $\hat{\beta}_2 = 82.04$, and $\hat{\beta}_3 = 17.04$ are obtained.

(a) What is the fitted value of the expected value of the response variable when $x_1 = 1.5$, $x_2 = 1.5$, and $x_3 = 2.0$?

(b) If the fitted value from part (a) has a standard error of 2.55, construct a two-sided 95% confidence interval for the expected value of the response variable at this point.

13.1.11 The model

$$y = \beta_0 + \beta_1 x_1 + \beta_2 x_2 + \beta_3 x_3$$

is fitted to $n = 20$ data observations. Suppose that $\hat{\sigma} = 4.33$, $\sum_{i=1}^{20} y_i = -5.68$, and $\sum_{i=1}^{20} y_i^2 = 694.09$. Construct the ANOVA table and put bounds on the p-value. What hypothesis is being tested by the p-value? What proportion of the variability of the y variable is explained by the model?

13.1.12 The multiple linear regression model

$$y = \beta_0 + \beta_1 x_1 + \beta_2 x_2 + \beta_3 x_3 + \beta_4 x_4$$

is fitted to a data set of $n = 22$ observations. The total sum of squares is 45.76; the error sum of squares is 23.98.

(a) What is the coefficient of determination R^2?

(b) Write out the ANOVA table.

(c) What is the estimate of σ^2?

(d) What is the p-value for the null hypothesis

$$H_0 : \beta_1 = \beta_2 = \beta_3 = \beta_4 = 0$$

(e) If $\hat{\beta}_2 = 183.2$ and $s.e.(\hat{\beta}_2) = 154.3$, construct a two-sided 95% confidence interval for β_2.

13.2 Examples of Multiple Linear Regression

Skills at modeling with multiple linear regression models are developed primarily through experience. Every problem has its own unique characteristics due to the types of variables considered and the relationships between them. Rather than blindly throwing new variables into the model, a successful modeler always has an understanding and feeling for the variables under consideration.

13.2.1 Examples

Example 70
Chemical Yields

A chemical engineer measures the yield of a chemical obtained from a reaction performed at various temperatures, and the data set shown in Figure 13.7 is obtained. The data plot suggests that there is some curvature in the relationship between yield and temperature, and this is confirmed by the straight line fit (simple linear regression model with yield as the response variable and temperature as the input variable) shown in Figure 13.8, because the fitted line underestimates the yield at low and high temperatures and overestimates the yield at the middle temperatures.

The chemical engineer therefore decides to try the quadratic model

$$\text{yield} = \beta_0 + (\beta_1 \times \text{temperature}) + (\beta_2 \times \text{temperature}^2)$$

and a statistical software package is employed to give the fitted curve shown in Figure 13.9, which appears to provide a satisfactory fit. The equation of the fitted curve is

$$\text{yield} = 293 - (6.14 \times \text{temperature}) + (0.0411 \times \text{temperature}^2)$$

An F-statistic of 326.41 is given in the analysis of variance table in the engineer's computer output and the p-value for the null hypothesis

$$H_0 : \beta_1 = \beta_2 = 0$$

is very small so that the null hypothesis is not plausible. In other words, there is evidence of a relationship between the chemical yield and the temperature, which is obvious from the data plot.

Also, the computer output gives

$$\hat{\beta}_2 = 0.041086$$

Yield	Temperature °C
85	90
76	100
114	110
143	120
164	130
281	140
306	150
358	160
437	170
470	180
649	190
702	200

FIGURE 13.7

Chemical yields data set

FIGURE 13.8

Unsatisfactory linear model for the
chemical yields data set

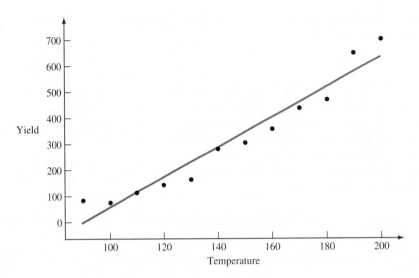

FIGURE 13.9

Quadratic model for the chemical
yields data set

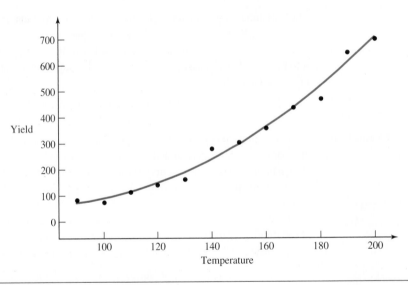

with

$$\text{s.e.}(\hat{\beta}_2) = 0.007570$$

so that the t-statistic for testing $H_0 : \beta_2 = 0$ is

$$t = \frac{\hat{\beta}_2}{\text{s.e.}(\hat{\beta}_2)} = \frac{0.041086}{0.007570} = 5.43$$

The p-value is therefore

$$p\text{-value} = 2 \times P(X > 5.43) \simeq 0$$

where the random variable X has a t-distribution with $n - k - 1 = 12 - 2 - 1 = 9$ degrees of
freedom. The small p-value indicates that the null hypothesis $H_0 : \beta_2 = 0$ is not plausible and
that the quadratic term in the model significantly improves the fit. In fact, with $t_{0.025,9} = 2.262$,

a two-sided 95% confidence interval for β_2 is

$$(0.041086 - (2.262 \times 0.007570), 0.041086 + (2.262 \times 0.007570)) = (0.024, 0.058)$$

The p-value for the intercept parameter β_0 is 0.089, but there is no reason to take it out of the model. In addition, the p-value for the linear parameter β_1 is 0.021. However, because yield has a significant quadratic dependence on temperature there is no reason to remove the linear term "$\hat{\beta}_1 \times$ temperature" from the model. Actually, because there is a quadratic term in the model, the presence or absence of the linear term and the intercept parameter depends on the scale used to measure temperature.

Finally, at a temperature of 155° the predicted value of the yield is

$$\text{yield} = 293 - (6.14 \times 155) + (0.0411 \times 155^2) = 328.3$$

This fitted value has a standard error of 11.72, and a 95% confidence interval for the expected value of the chemical yield obtained at this temperature is about $(301, 355)$. A 95% prediction interval for a future chemical yield obtained at a temperature of 155° is about $(260, 397)$. These can be obtained from a computer software package, using the analysis methods described in Section 13.3.

Example 71
Supermarket Deliveries

A supermarket manager decides to investigate how long it takes to unload deliveries from a truck. As illustrated in Figure 13.10, an initial consideration of the problem suggests that the unloading time could depend on both the volume of the delivery and the weight of the delivery. In addition, the supermarket has a day shift and a night shift, and the manager suspects that the unloading times may be different for the two shifts.

Ten deliveries are selected at random for both the day and night shifts, and the unloading times are measured together with the volumes and weights of the deliveries. The data set obtained is shown in Figure 13.11. Notice that the two shifts are designated by an indicator variable that takes the value 0 for the day shift and 1 for the night shift.

The multiple linear regression model employed is

$$y = \beta_0 + \beta_1 x_1 + \beta_2 x_2 + \beta_3 x_3$$

where y is the unloading time, x_1 is the volume of the delivery, x_2 is the weight of the delivery, and x_3 is the indicator variable designating the shift. When a computer is used to analyze this data set, an F-statistic of 60.99 is obtained in the analysis of variance table and the associated

FIGURE 13.10

Supermarket deliveries

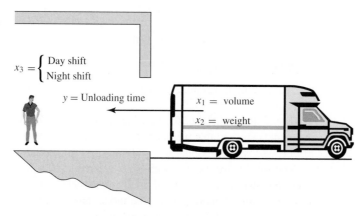

FIGURE 13.11

Supermarket deliveries data set

Volume of load (m³)	Weight of load (1000 kg)	Shift (0 day, 1 night)	Unloading time (minutes)
79	43	0	48
76	46	0	42
84	43	0	44
66	32	0	33
98	56	0	53
98	57	0	58
92	56	0	52
73	38	0	34
80	44	0	40
62	37	0	24
87	45	1	47
67	44	1	32
62	37	1	30
76	36	1	39
93	55	1	56
61	41	1	23
53	32	1	20
63	43	1	35
77	49	1	42
59	40	1	19

small p-value for the null hypothesis

$$H_0 : \beta_1 = \beta_2 = \beta_3 = 0$$

indicates that the unloading time is related to at least some of the input variables. However, when the coefficients β_i are considered individually, volume (β_1) has a small p-value, whereas weight (β_2) has a p-value of 0.949 and shift (β_3) has a p-value of 0.691. For example,

$$\hat{\beta}_2 = 0.0137$$

with

$$\text{s.e.}(\hat{\beta}_2) = 0.2109$$

which gives a t-statistic of

$$t = \frac{\hat{\beta}_2}{\text{s.e.}(\hat{\beta}_2)} = \frac{0.0137}{0.2109} = 0.07$$

The p-value of 0.949 for weight is thus obtained as

$$p\text{-value} = 2 \times P(X > 0.07)$$

where the random variable X has a t-distribution with $n - k - 1 = 20 - 3 - 1 = 16$ degrees of freedom.

The high p-values for the variables weight and shift suggest that they are not needed in the model. The variable weight has the largest p-value and so it is removed from the model first. The subsequent model with volume and shift should be fitted again.

The p-value for the shift variable now becomes 0.652 (notice that it has changed slightly after weight has been removed from the model), which again indicates that it should be removed from the model. Thus, the model has been reduced to a simple linear regression model involving just the variable volume.

FIGURE 13.12

Simple linear regression model for
time against volume for the
supermarket deliveries data set

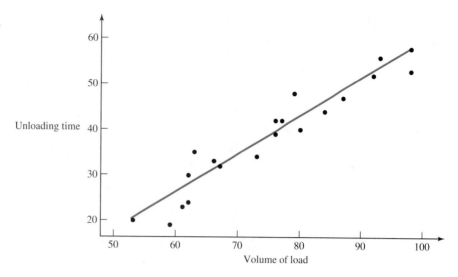

FIGURE 13.13

Simple linear regression model for
time against weight for the
supermarket deliveries data set

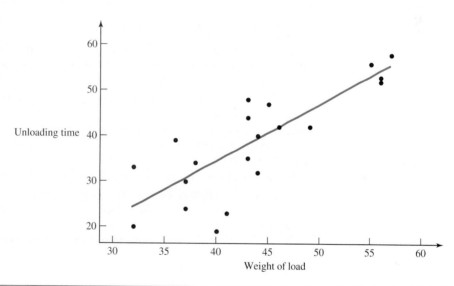

Figure 13.12 shows the final simple linear regression model together with a plot of the data. The manager has discovered that the best way to predict the unloading time of a delivery is with the model

$$\text{time} = -24.2 + (0.833 \times \text{volume})$$

and that neither the weight of the delivery nor the shift provides any additional useful information. In other words, the time it will take to unload a future load should be estimated from the volume of the load, and once the volume is known, the weight of the load and the shift are both irrelevant.

Does this mean that the unloading time and the weight of the delivery are uncorrelated? In fact it does not. Figure 13.13 shows a plot of the unloading times and the weights together with a simple linear regression model with unloading time as the response variable and weight

FIGURE 13.14

The input variables volume and
weight are positively correlated for
the supermarket deliveries data set

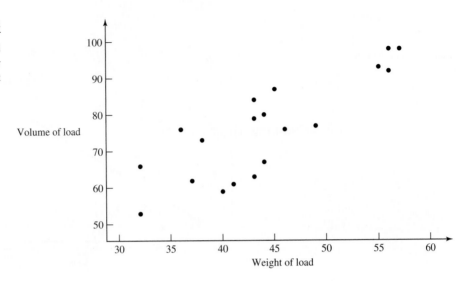

as the input variable. The slope parameter is significantly different from 0 (the t-statistic is
5.92) and the unloading time and weight are positively correlated.

Why then was weight dropped from the original model? Figure 13.14, which shows a
plot of the delivery volumes against their weights, provides an answer to this question. The
variables volume and weight are seen to be positively correlated, as would be expected.
Therefore, it turns out that once volume is in the model, weight serves no additional useful
purpose. In fact, the variable volume can be thought of as acting as a *surrogate* for the variable
weight.

When a model with both volume and weight is fitted, the mathematics of the fitting
procedure identify which of the two variables is needed to provide the best fit to the unloading
times. The correlation between the unloading times and the delivery volumes is $r = 0.96$,
whereas the correlation between the unloading times and the delivery weights is $r = 0.81$,
and the higher correlation for volume helps to explain why volume is preferred to weight in
the model.

Example 72
Turf Quality

A turf grower who provides high-quality turf for sporting purposes performs a small experiment to investigate how a particular variety of turf reacts to various amounts of water and
fertilizer. A total of $n = 14$ samples of turf are grown under different experimental conditions,
and at the end of the allotted time period they are each given a quality score based upon the
appearance of the turf, the health of the roots, the health of the grass blades, and the density of
the grass. The resulting data set is shown in Figure 13.15, where higher scores indicate better
quality grass.

The data set indicates that there is some curvature in the relationship between score and
the variables water and fertilizer, with too much water and fertilizer resulting in a drop in the
scores. Consequently, it should be useful to fit the response surface model

$$y = \beta_0 + \beta_1 x_1 + \beta_2 x_1^2 + \beta_3 x_2 + \beta_4 x_2^2 + \beta_5 x_1 x_2$$

where the response variable y is the score, x_1 is the water level, and x_2 is the fertilizer level.

Quality score	Water level	Fertilizer level
72	9	26
71	12	20
62	6	20
32	15	24
48	15	16
37	13	28
68	13	12
27	11	32
46	11	8
48	7	32
46	5	28
41	5	12
38	3	24
34	3	16

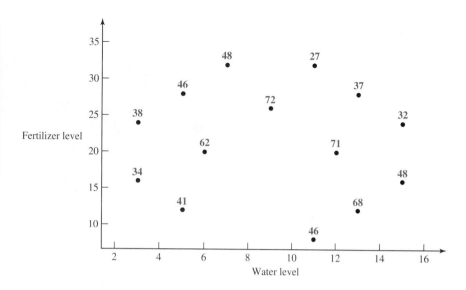

FIGURE 13.15

Turf quality data set and scatterplot

When this model is fitted, all of the parameters $\beta_1, \ldots, \beta_5$ have low p-values so that they are significantly different from 0, and therefore no terms need to be dropped from the model. The fitted model is

$$y = -144 + 24.2x_1 - 1.01x_1^2 + 11.5x_2 - 0.236x_2^2 - 0.270x_1x_2$$

which is shown in Figure 13.16. The fitted model is maximized at a water level of $x_1 = 9.4$ and a fertilizer level of $x_2 = 19.0$, which the turf grower can use as estimates of the optimum growing conditions.

There is a great deal of additional theory behind the design and analysis of experiments using response surfaces that the reader may wish to investigate. One question of importance is how should the *design points* (in the example, the values of the water and fertilizer levels used in the experiment) be chosen optimally? Furthermore, some experiments can be run in a *sequential* manner where the results of some experimental trials are used to indicate what additional experiments should be performed.

Example 44

Army Physical Fitness Test

Recall that the Army Physical Fitness Test consists of 2 minutes of pushups followed by 2 minutes of situps followed by a 2-mile run. Figure 13.17 shows the full data set including the number of situps performed, and Figure 13.18 gives summary statistics for the number of situps. How can a participant's run time be predicted from the number of pushups and the number of situps performed by the participant?

The multiple linear regression model

$$y = \beta_0 + \beta_1x_1 + \beta_2x_2$$

can be used where the response variable y is the run time, x_1 is the number of pushups, and x_2 is the number of situps. When this model is fitted, pushups are significant (p-value $= 0.0028$), but situps are not significant (p-value $= 0.3448$). This finding implies that situps

FIGURE 13.16

Fitted response surface for turf
quality example

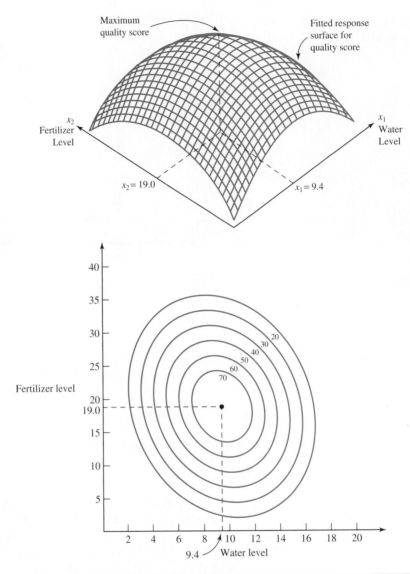

should be dropped from the model and that the run time should be predicted from the number of pushups using the simple linear regression model discussed in Chapter 12. Once the number of pushups performed by a participant is known, knowledge of the number of situps performed by the participant does not provide any further assistance in predicting the participant's run time.

However, as Figure 13.19 shows, by itself the variable situps is negatively correlated with the run time (in the simple linear regression of run time against situps, the t-statistic is -5.25 and the p-value is very small). Why then was situps dropped from the original model? This is because the variable situps is correlated with the variable pushups as shown in Figure 13.20. The correlation between run time and pushups is $r = -0.57$, whereas the correlation between run time and situps is $r = -0.50$, so that pushups is a marginally more effective predictor of run time than situps and acts as a *surrogate* for situps in the model.

2-mile run time (seconds)	Number of pushups	Number of situps
847	60	83
887	53	67
879	60	70
919	55	60
816	60	71
814	78	83
814	74	70
855	70	69
980	46	48
954	50	55
1078	50	48
1001	59	61
766	62	71
916	64	65
798	51	62
782	66	64
836	73	69
837	78	80
791	80	90
838	70	78
853	70	85
840	79	86
740	93	91
763	80	100
778	78	78
855	70	48
875	60	60
868	78	73
880	50	50
905	70	67
895	69	64
720	125	95
712	80	80
703	99	92
741	80	80
792	61	68
808	50	50
761	55	52
785	60	49
801	60	59
810	65	54
1013	66	82

2-mile run time (seconds)	Number of pushups	Number of situps
882	58	57
861	62	67
845	60	61
865	90	82
883	78	72
881	86	80
921	86	77
816	69	56
837	74	68
1056	40	45
1034	40	50
774	78	76
821	70	70
850	78	70
870	70	61
931	55	60
930	55	73
808	80	66
828	78	62
719	78	78
707	68	70
934	45	45
939	45	45
977	47	51
896	88	65
921	70	71
815	80	84
838	79	81
854	71	82
1063	50	55
1024	50	50
780	73	78
813	78	71
850	78	80
902	63	60
906	60	60
865	78	84
886	78	82
881	78	82
825	50	59
821	61	56
832	62	67

FIGURE 13.17

Data set for the Army Physical Fitness Test

Situps	
	$n = 84$
	Sample mean = 68.29
	Sample standard deviation = 13.05
	Maximum = 100
	Upper quartile = 80
	Median = 68.5
	Lower quartile = 59.25
	Minimum = 45

FIGURE 13.18

Summary statistics for the number of situps

An explanation from a physiological point of view may be that the ability to do situps is a rather specialized skill that depends primarily on the stomach muscles. The abilities to do pushups and to run fast are more likely to be reflections of general athletic ability or of how physically fit a person is. Consequently, for the purpose of modeling run times, pushups are the most important variable and situps do not provide any additional useful information.

FIGURE 13.19

A simple linear regression model of run time against situps for the Army Physical Fitness Test data set

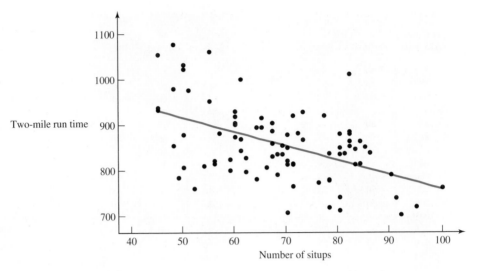

FIGURE 13.20

The input variables pushups and situps are positively correlated for the Army Physical Fitness Test data set

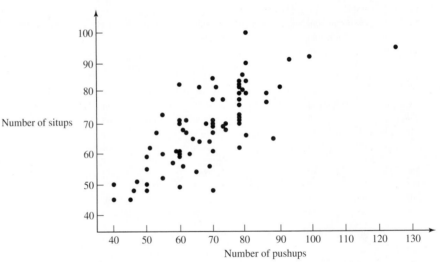

■   **13.2.2   Problems**

13.2.1   Competitive Pricing Policies

The data set given in DS 13.2.1 concerns the sales volume of a company, the price at which the company sells its product, and the price of a competing product for each of $n = 10$ quarter periods. Use this data set to fit the multiple linear regression model

$$y = \beta_0 + \beta_1 x_1 + \beta_2 x_2$$

with the response variable y as the sales volume and with x_1 as the product price and x_2 as the competitor's price.

(a) Plot the company's sales against its own price and against its competitor's price.

(b) Is the variable competitor's price needed in the regression model? What is the sample correlation coefficient between the competitor's price and sales? What is the sample correlation coefficient between the competitor's price and the company's price? Interpret your answers.

(c) What is the sample correlation coefficient between the company's price and its sales? What sales would you predict if the company priced its product at $10.0 next quarter?

13.2.2 Polymer Concentrations for Optimal Fiber Strengths

Two polymers are ingredients in the manufacture of a synthetic fiber. The data set given in DS 13.2.2 shows the results of an experiment conducted to measure the fiber strength resulting from different concentrations of the two polymers. The polymer concentrations examined represent their standard levels, coded as 0, together with one unit above and below the standard levels. Fit the response surface model

$$y = \beta_0 + \beta_1 x_1 + \beta_2 x_1^2 + \beta_3 x_2 + \beta_4 x_2^2 + \beta_5 x_1 x_2$$

where the response variable y is the fiber strength, x_1 is the concentration of polymer 1, and x_2 is the concentration of polymer 2. At what concentrations of the polymers would you estimate that the fiber strength is maximized?

13.2.3 Oil Well Drilling Costs

Consider again the problem of estimating the costs of drilling oil wells, which was originally discussed in Problem 12.2.4. The data set given in DS 13.2.3 contains the variables geology, downtime, and rig-index in addition to the variables depth and cost considered before. The variable geology is a score that measures the geological properties of the materials that have to be drilled through. Harder materials have larger scores, and so larger values of the geology variable indicate that harder materials have to be drilled through to complete the oil well. The variable downtime measures the number of hours that the drilling rig is idle due to factors such as inclement weather and interruptions for borehole and geological tests. The variable rig-index compares the daily rental costs of the drilling rig to the cost in 1980. Thus, an index of 1 implies that the rental costs are identical to those in 1980, and an index of 2 implies that the rental costs are twice what they were in 1980.

(a) Fit the multiple linear regression model

$$y = \beta_0 + \beta_1 x_1 + \beta_2 x_2 + \beta_3 x_3 + \beta_4 x_4$$

with the response variable y as the cost and with x_1 as the depth, x_2 as geology score, x_3 as the downtime, and x_4 as the rig-index, and make plots of cost against each of the four input variables.

(b) Explain why the variable geology should be removed from the model. Does this surprise you? What is the sample correlation coefficient between cost and geology? What is the sample correlation coefficient between depth and geology? Why do you think that geology is not needed in the model?

(c) Should any other variables be removed from the model? What is the final model that you would recommend for use? (This problem is continued in Problem 13.4.2.)

13.2.4 VO2-max Aerobic Fitness Measurements

The data set given in DS 13.2.4 extends the data set used in Problem 12.2.6 to include an individual's heart rate at rest, percentage body fat and weight, together with the variables age and VO2-max considered before. Fit a multiple linear regression model to assess whether the new input variables heart rate at rest, percentage body fat, and weight, together with age, can be used to provide an improved model for VO2-max. What model would you recommend? (This problem is continued in Problem 13.4.3.)

13.2.5 A categorical input variable has three levels. How can indicator variables be used to include it in a multiple linear regression model?

13.2.6 Consider a multiple linear regression of y on two input variables x_1 and x_2. The p-value for x_1 is less than 1%, and the p-value for x_2 is greater than 10%. Suppose that a simple linear regression is performed with y and one input variable x_2. What bounds can you put on the p-value for x_2 in the simple linear regression?

13.3 Matrix Algebra Formulation of Multiple Linear Regression

13.3.1 Matrix Representation

The linear model

$$y_i = \beta_0 + \beta_1 x_{1i} + \cdots + \beta_k x_{ki} + \epsilon_i$$

for $1 \leq i \leq n$ can be written in matrix form as

$$\mathbf{Y} = \mathbf{X}\boldsymbol{\beta} + \boldsymbol{\epsilon}$$

where $\mathbf{Y}$ is the $n \times 1$ vector of observed values of the response variable

$$\mathbf{Y} = \begin{pmatrix} y_1 \\ y_2 \\ \cdot \\ \cdot \\ \cdot \\ y_n \end{pmatrix}$$

the **design matrix** $\mathbf{X}$ is the $n \times (k+1)$ matrix containing the values of the input variables

$$\mathbf{X} = \begin{pmatrix} 1 & x_{11} & x_{21} & \cdots & x_{k1} \\ 1 & x_{12} & x_{22} & \cdots & x_{k2} \\ \cdot & \cdot & \cdot & \cdots & \cdot \\ \cdot & \cdot & \cdot & \cdots & \cdot \\ \cdot & \cdot & \cdot & \cdots & \cdot \\ 1 & x_{1n} & x_{2n} & \cdots & x_{kn} \end{pmatrix}$$

the parameter vector $\boldsymbol{\beta}$ is the $(k+1) \times 1$ vector

$$\boldsymbol{\beta} = \begin{pmatrix} \beta_0 \\ \beta_1 \\ \cdot \\ \cdot \\ \cdot \\ \beta_k \end{pmatrix}$$

and $\boldsymbol{\epsilon}$ is the $n \times 1$ vector containing the error terms

$$\boldsymbol{\epsilon} = \begin{pmatrix} \epsilon_1 \\ \epsilon_2 \\ \cdot \\ \cdot \\ \cdot \\ \epsilon_n \end{pmatrix}$$

Since the error terms ϵ_i have independent $N(0, \sigma^2)$ distributions, the vector of error terms $\boldsymbol{\epsilon}$ has a multivariate normal distribution with a zero mean vector and a covariance matrix $\sigma^2 \mathbf{I}_n$, where $\mathbf{I}_n$ is the $n \times n$ identity matrix

$$\boldsymbol{\epsilon} \sim N_n(\mathbf{0}, \sigma^2 \mathbf{I}_n)$$

Consequently, the response variable vector $\mathbf{Y}$ has the multivariate normal distribution

$$\mathbf{Y} \sim N_n(\mathbf{X}\boldsymbol{\beta}, \sigma^2 \mathbf{I}_n)$$

so that

$$E(\mathbf{Y}) = \mathbf{X}\boldsymbol{\beta}$$

Example 67
Car Plant Electricity
Usage

The car plant is in a southern location with a warm climate and for most of the year air-conditioning is required to cool the plant to a working temperature of 65° Fahrenheit. The manager therefore expects that in addition to the plant's production, the amount of air-conditioner use required should have a significant impact on the plant's electricity usage. In order to

FIGURE 13.21

Car plant electricity usage data set

	Electricity usage (million kWh)	Production ($ million)	Cooling degree days
January	2.48	4.51	0
February	2.26	3.58	0
March	2.47	4.31	13
April	2.77	5.06	56
May	2.99	5.64	117
June	3.05	4.99	306
July	3.18	5.29	358
August	3.46	5.83	330
September	3.03	4.70	187
October	3.26	5.61	94
November	2.67	4.90	23
December	2.53	4.20	0

investigate this the variable "cooling degrees days" (CDD) is calculated for each month as

$$CDD = \sum_{\text{days in month}} \max\{\text{average daily temperature} - 65, 0\}$$

This variable can be interpreted as being the degrees of cooling that are required during the month in order to reduce the temperature of the plant to $65°$ each day.

The full data set is shown in Figure 13.21 and the multiple linear regression model

$$y = \beta_0 + \beta_1 x_1 + \beta_2 x_2$$

is proposed, where the response variable y is the electricity usage, x_1 is the production level, and x_2 is CDD. Notice that $n = 12$ and $k = 2$. The response variable vector and the design matrix are

$$Y = \begin{pmatrix} 2.48 \\ 2.26 \\ 2.47 \\ 2.77 \\ 2.99 \\ 3.05 \\ 3.18 \\ 3.46 \\ 3.03 \\ 3.26 \\ 2.67 \\ 2.53 \end{pmatrix} \quad \text{and} \quad X = \begin{pmatrix} 1 & 4.51 & 0 \\ 1 & 3.58 & 0 \\ 1 & 4.31 & 13 \\ 1 & 5.06 & 56 \\ 1 & 5.64 & 117 \\ 1 & 4.99 & 306 \\ 1 & 5.29 & 358 \\ 1 & 5.83 & 330 \\ 1 & 4.70 & 187 \\ 1 & 5.61 & 94 \\ 1 & 4.90 & 23 \\ 1 & 4.20 & 0 \end{pmatrix}$$

and the parameter vector β is

$$\beta = \begin{pmatrix} \beta_0 \\ \beta_1 \\ \beta_2 \end{pmatrix}$$

The least squares estimates of the parameters β_i are found by minimizing the sum of squares Q, which can be written in matrix form as

$$Q = (Y - X\beta)'(Y - X\beta)$$

Notice that the $(k + 1) \times (k + 1)$ matrix $\mathbf{X'X}$ is

$$\mathbf{X'X} = \begin{pmatrix} n & \sum_{i=1}^{n} x_{1i} & \sum_{i=1}^{n} x_{2i} & \cdots & \sum_{i=1}^{n} x_{ki} \\ \sum_{i=1}^{n} x_{1i} & \sum_{i=1}^{n} x_{1i}^2 & \sum_{i=1}^{n} x_{1i} x_{2i} & \cdots & \sum_{i=1}^{n} x_{1i} x_{ki} \\ \sum_{i=1}^{n} x_{2i} & \sum_{i=1}^{n} x_{1i} x_{2i} & \sum_{i=1}^{n} x_{2i}^2 & \cdots & \sum_{i=1}^{n} x_{2i} x_{ki} \\ . & . & . & \cdots & . \\ . & . & . & \cdots & . \\ . & . & . & \cdots & . \\ \sum_{i=1}^{n} x_{ki} & \sum_{i=1}^{n} x_{1i} x_{ki} & \sum_{i=1}^{n} x_{2i} x_{ki} & \cdots & \sum_{i=1}^{n} x_{ki}^2 \end{pmatrix}$$

and the $(k + 1) \times 1$ vector $\mathbf{X'Y}$ is

$$\mathbf{X'Y} = \begin{pmatrix} \sum_{i=1}^{n} y_i \\ \sum_{i=1}^{n} y_i x_{1i} \\ . \\ . \\ . \\ \sum_{i=1}^{n} y_i x_{ki} \end{pmatrix}$$

so that the *normal equations* derived in Section 13.1.2 can be written as

$$\mathbf{X'X}\boldsymbol{\beta} = \mathbf{X'Y}$$

As long as the input variables are not linearly related, this matrix equation can be solved to give the vector of parameter estimates $\hat{\boldsymbol{\beta}}$ as

$$\hat{\boldsymbol{\beta}} = (\mathbf{X'X})^{-1}\mathbf{X'Y}$$

The $n \times 1$ vector of fitted values is

$$\hat{\mathbf{Y}} = \begin{pmatrix} \hat{y}_1 \\ \hat{y}_2 \\ . \\ . \\ . \\ \hat{y}_n \end{pmatrix} = \mathbf{X}\hat{\boldsymbol{\beta}}$$

and the $n \times 1$ vector of residuals is

$$\mathbf{e} = \begin{pmatrix} e_1 \\ e_2 \\ . \\ . \\ . \\ e_n \end{pmatrix} = \mathbf{Y} - \hat{\mathbf{Y}} = \begin{pmatrix} y_1 - \hat{y}_1 \\ y_2 - \hat{y}_2 \\ . \\ . \\ . \\ y_n - \hat{y}_n \end{pmatrix}$$

The sum of squares for error can then be calculated as

$$\text{SSE} = \mathbf{e'e} = \sum_{i=1}^{n} e_i^2 = \sum_{i=1}^{n} (y_i - \hat{y}_i)^2$$

and

$$\hat{\sigma}^2 = \text{MSE} = \frac{\text{SSE}}{n-k-1}$$

Example 67
Car Plant Electricity Usage

For this example

$$\sum_{i=1}^{12} x_{1i} = 4.51 + \cdots + 4.20 = 58.62$$

$$\sum_{i=1}^{12} x_{1i}^2 = 4.51^2 + \cdots + 4.20^2 = 291.231$$

$$\sum_{i=1}^{12} x_{2i} = 0 + 0 + 13 + \cdots + 23 + 0 = 1484$$

$$\sum_{i=1}^{12} x_{2i}^2 = 0^2 + 0^2 + 13^2 + \cdots + 23^2 + 0 = 392{,}028$$

$$\sum_{i=1}^{12} x_{1i}x_{2i} = (4.51 \times 0) + \cdots + (4.20 \times 0) = 7862.87$$

so that

$$\mathbf{X'X} = \begin{pmatrix} n & \sum_{i=1}^{n} x_{1i} & \sum_{i=1}^{n} x_{2i} \\ \sum_{i=1}^{n} x_{1i} & \sum_{i=1}^{n} x_{1i}^2 & \sum_{i=1}^{n} x_{1i}x_{2i} \\ \sum_{i=1}^{n} x_{2i} & \sum_{i=1}^{n} x_{1i}x_{2i} & \sum_{i=1}^{n} x_{2i}^2 \end{pmatrix} = \begin{pmatrix} 12.0 & 58.6 & 1484.0 \\ 58.6 & 291.2 & 7862.8 \\ 1484.0 & 7862.8 & 392{,}028.0 \end{pmatrix}$$

This result gives

$$(\mathbf{X'X})^{-1} = \begin{pmatrix} 6.82134 & -1.47412 & 3.74529 \times 10^{-3} \\ -1.47412 & 0.32605 & -9.5962 \times 10^{-4} \\ 3.74529 \times 10^{-3} & -9.5962 \times 10^{-4} & 7.6207 \times 10^{-6} \end{pmatrix}$$

Also,

$$\sum_{i=1}^{12} y_i = 2.48 + \cdots + 2.53 = 34.15$$

$$\sum_{i=1}^{12} y_i x_{1i} = (2.48 \times 4.51) + \cdots + (2.53 \times 4.20) = 169.2532$$

$$\sum_{i=1}^{12} y_i x_{2i} = (2.48 \times 0) + \cdots + (2.53 \times 0) = 4685.06$$

so that

$$\mathbf{X'Y} = \begin{pmatrix} \sum_{i=1}^{n} y_i \\ \sum_{i=1}^{n} y_i x_{1i} \\ \sum_{i=1}^{n} y_i x_{2i} \end{pmatrix} = \begin{pmatrix} 34.15 \\ 169.2532 \\ 4685.06 \end{pmatrix}$$

The parameter estimates are therefore

$$
\hat{\boldsymbol{\beta}} = \begin{pmatrix} \hat{\beta}_0 \\ \hat{\beta}_1 \\ \hat{\beta}_2 \end{pmatrix} = (\mathbf{X}'\mathbf{X})^{-1}\mathbf{X}'\mathbf{Y}
$$

$$
= \begin{pmatrix} 6.82134 & -1.47412 & 3.74529 \times 10^{-3} \\ -1.47412 & 0.32605 & -9.5962 \times 10^{-4} \\ 3.74529 \times 10^{-3} & -9.5962 \times 10^{-4} & 7.6207 \times 10^{-6} \end{pmatrix} \begin{pmatrix} 34.15 \\ 169.2532 \\ 4685.06 \end{pmatrix}
$$

$$
= \begin{pmatrix} 0.99 \\ 0.35 \\ 0.0012 \end{pmatrix}
$$

so that the fitted model is

$$
y = 0.99 + 0.35x_1 + 0.0012x_2
$$

The vector of fitted values is

$$
\hat{\mathbf{Y}} = \mathbf{X}\hat{\boldsymbol{\beta}} = \begin{pmatrix} 1 & 4.51 & 0 \\ 1 & 3.58 & 0 \\ 1 & 4.31 & 13 \\ 1 & 5.06 & 56 \\ 1 & 5.64 & 117 \\ 1 & 4.99 & 306 \\ 1 & 5.29 & 358 \\ 1 & 5.83 & 330 \\ 1 & 4.70 & 187 \\ 1 & 5.61 & 94 \\ 1 & 4.90 & 23 \\ 1 & 4.20 & 0 \end{pmatrix} \begin{pmatrix} 0.99 \\ 0.35 \\ 0.0012 \end{pmatrix} = \begin{pmatrix} 2.568 \\ 2.243 \\ 2.514 \\ 2.827 \\ 3.102 \\ 3.099 \\ 3.265 \\ 3.421 \\ 2.856 \\ 3.064 \\ 2.732 \\ 2.460 \end{pmatrix}
$$

The residuals are then

$$
\mathbf{e} = \mathbf{Y} - \hat{\mathbf{Y}} = \begin{pmatrix} 2.48 \\ 2.26 \\ 2.47 \\ 2.77 \\ 2.99 \\ 3.05 \\ 3.18 \\ 3.46 \\ 3.03 \\ 3.26 \\ 2.67 \\ 2.53 \end{pmatrix} - \begin{pmatrix} 2.568 \\ 2.243 \\ 2.514 \\ 2.827 \\ 3.102 \\ 3.099 \\ 3.265 \\ 3.421 \\ 2.856 \\ 3.064 \\ 2.732 \\ 2.460 \end{pmatrix} = \begin{pmatrix} -0.088 \\ 0.017 \\ -0.044 \\ -0.057 \\ -0.112 \\ -0.049 \\ -0.085 \\ 0.039 \\ 0.174 \\ 0.196 \\ -0.062 \\ 0.070 \end{pmatrix}
$$

and the sum of squares for error is

$$
\text{SSE} = \mathbf{e}'\mathbf{e} = (-0.088)^2 + \cdots + 0.070^2 = 0.1142
$$

The estimate of the error variance is therefore

$$
\hat{\sigma}^2 = \text{MSE} = \frac{\text{SSE}}{n - k - 1} = \frac{0.1142}{9} = 0.0127
$$

with $\hat{\sigma} = \sqrt{0.0127} = 0.113$.

The parameter estimates are unbiased

$$E(\hat{\boldsymbol{\beta}}) = \boldsymbol{\beta}$$

and they have a multivariate normal distribution with covariance matrix $\sigma^2(\mathbf{X'X})^{-1}$

$$\hat{\boldsymbol{\beta}} \sim N_{k+1}(\boldsymbol{\beta}, \sigma^2(\mathbf{X'X})^{-1})$$

The standard error of $\hat{\beta}_i$ is therefore σ multiplied by the square root of the ith diagonal element of the matrix $(\mathbf{X'X})^{-1}$.

At a set of values

$$\mathbf{x}^* = (1, x_1^*, x_2^*, \ldots, x_k^*)'$$

of the input variables, the fitted value of the response variable is

$$\hat{y}|_{\mathbf{x}^*} = \mathbf{x}^{*'}\hat{\boldsymbol{\beta}} = \hat{\beta}_0 + \hat{\beta}_1 x_1^* + \cdots + \hat{\beta}_k x_k^*$$

which is an estimate of

$$\mathbf{x}^{*'}\boldsymbol{\beta} = \beta_0 + \beta_1 x_1^* + \cdots + \beta_k x_k^*$$

A $1 - \alpha$ confidence level confidence interval for this expected value of the response variable at $\mathbf{x}^*$ is

$$\beta_0 + \beta_1 x_1^* + \cdots + \beta_k x_k^* \in (\hat{y}|_{\mathbf{x}^*} - t_{\alpha/2, n-k-1}\text{s.e.}(\hat{y}|_{\mathbf{x}^*}), \hat{y}|_{\mathbf{x}^*} + t_{\alpha/2, n-k-1}\text{s.e.}(\hat{y}|_{\mathbf{x}^*}))$$

where

$$\text{s.e.}(\hat{y}|_{\mathbf{x}^*}) = \hat{\sigma}\sqrt{\mathbf{x}^{*'}(\mathbf{X'X})^{-1}\mathbf{x}^*}$$

Also, a prediction interval for a future observation obtained at $\mathbf{x}^*$ takes into consideration the extra variability due to an error term ϵ and is given by

$$(\hat{y}|_{\mathbf{x}^*} - t_{\alpha/2, n-k-1}\text{s.e.}(\hat{y}|_{\mathbf{x}^*} + \epsilon), \hat{y}|_{\mathbf{x}^*} + t_{\alpha/2, n-k-1}\text{s.e.}(\hat{y}|_{\mathbf{x}^*} + \epsilon))$$

where

$$\text{s.e.}(\hat{y}|_{\mathbf{x}^*} + \epsilon) = \hat{\sigma}\sqrt{1 + \mathbf{x}^{*'}(\mathbf{X'X})^{-1}\mathbf{x}^*}$$

Finally, notice that

$$\hat{\mathbf{Y}} = \mathbf{X}\hat{\boldsymbol{\beta}} = \mathbf{X}(\mathbf{X'X})^{-1}\mathbf{X'Y} = \mathbf{HY}$$

so that the $n \times n$ "hat matrix"

$$\mathbf{H} = \mathbf{X}(\mathbf{X'X})^{-1}\mathbf{X'}$$

transforms the vector of response values $\mathbf{Y}$ to the vector of fitted values $\hat{\mathbf{Y}}$. The residual vector is therefore

$$\mathbf{e} = \mathbf{Y} - \hat{\mathbf{Y}} = \mathbf{Y} - \mathbf{HY} = (\mathbf{I}_n - \mathbf{H})\mathbf{Y}$$

and it has a covariance matrix $\sigma^2(\mathbf{I}_n - \mathbf{H})$. The variance of the ith residual e_i is thus

$$\text{Var}(e_i) = \sigma^2(1 - h_{ii})$$

where h_{ii} is the ith diagonal element of the matrix $\mathbf{H}$.

A **standardized residual** can therefore be calculated as

$$e_i^* = \frac{e_i}{\hat{\sigma}\sqrt{1 - h_{ii}}}$$

which is available from most computer packages. Data points that have a standardized residual with an absolute value larger than 3 do not fit the regression model very closely and the experimenter may want to treat them as being outliers. Notice that the values of h_{ii} are usually small, so that the standardized residuals can be approximated as

$$e_i^* \simeq \frac{e_i}{\hat{\sigma}}$$

as suggested in Chapter 12.

Example 67 **Car Plant Electricity** **Usage**	The diagonal elements of the matrix $(\mathbf{X}'\mathbf{X})^{-1}$ are used to calculate the standard errors of the parameter estimates as

$$\text{s.e.}(\hat{\beta}_0) = \hat{\sigma} \times \sqrt{6.82134} = \sqrt{0.0127} \times \sqrt{6.82134} = 0.294$$

$$\text{s.e.}(\hat{\beta}_1) = \hat{\sigma} \times \sqrt{0.32605} = \sqrt{0.0127} \times \sqrt{0.32605} = 0.0643$$

$$\text{s.e.}(\hat{\beta}_2) = \hat{\sigma} \times \sqrt{7.6207 \times 10^{-6}} = \sqrt{0.0127} \times \sqrt{7.6207 \times 10^{-6}} = 3.11 \times 10^{-4}$$

The t-statistic for testing the null hypothesis $H_0 : \beta_1 = 0$ is therefore

$$t = \frac{\hat{\beta}_1}{\text{s.e.}(\hat{\beta}_1)} = \frac{0.35}{0.0643} = 5.43$$

and the p-value is

$$p\text{-value} = 2 \times P(X > 5.43) \simeq 0.0004$$

where the random variable X has a t-distribution with $n - k - 1 = 9$ degrees of freedom. Similarly, the t-statistic for testing the null hypothesis $H_0 : \beta_2 = 0$ is

$$t = \frac{\hat{\beta}_2}{\text{s.e.}(\hat{\beta}_2)} = \frac{0.0012}{3.11 \times 10^{-4}} = 3.82$$

and the p-value is

$$p\text{-value} = 2 \times P(X > 3.82) \simeq 0.0042$$

These low p-values indicate that both production level and CDD should be kept in the model and that they are both useful in modeling the monthly electricity costs. The parameter estimate $\hat{\beta}_1 = 0.35$ indicates that when CDD remains constant, an extra 1 million in production is estimated to result in an additional 0.35 kWh of electricity usage. It is interesting to compare this with the coefficient $\hat{\beta}_1 = 0.499$ obtained in Chapter 12 without the variable CDD included in the model. The higher value obtained in Chapter 12 can be explained by the fact that this data set has a positive correlation between production levels and CDD. The parameter estimate $\hat{\beta}_2 = 0.0012$ indicates that for a fixed production level, an increase in CDD of $100°$ is estimated to result in an increase of 0.12 million kWh of electricity usage. Obviously, the model presented here with both production level and CDD as input variables is a better model than that calculated in Chapter 12 based only on the production level.

Suppose that next month a production level of \$5.5 million is predicted and that weather records suggest that there will be a CDD value of 150. How does the model allow the plant manager to forecast next month's electricity usage? At $\mathbf{x}^* = (1, 5.5, 150)$ the estimated electricity usage is

$$\hat{y}|_{\mathbf{x}^*} = \hat{\beta}_0 + 5.5\hat{\beta}_1 + 150\hat{\beta}_2 = 0.99 + (5.5 \times 0.35) + (150 \times 0.0012) = 3.09$$

Since

$$\mathbf{x}^{*'}(\mathbf{X}'\mathbf{X})^{-1}\mathbf{x}^*$$

$$= (1, 5.5, 150) \begin{pmatrix} 6.82134 & -1.47412 & 3.74529 \times 10^{-3} \\ -1.47412 & 0.32605 & -9.5962 \times 10^{-4} \\ 3.74529 \times 10^{-3} & -9.5962 \times 10^{-4} & 7.6207 \times 10^{-6} \end{pmatrix} \begin{pmatrix} 1 \\ 5.5 \\ 150 \end{pmatrix}$$

$$= 0.181$$

this fitted value has a standard error of

$$\text{s.e.}(\hat{y}|_{\mathbf{x}^*}) = \hat{\sigma}\sqrt{\mathbf{x}^{*'}(\mathbf{X}'\mathbf{X})^{-1}\mathbf{x}^*} = \sqrt{0.0127} \times \sqrt{0.181} = 0.0479$$

With a critical point $t_{0.025,9} = 2.262$, a 95% confidence interval for the average electricity usage in a month with $x_1 = 5.5$ and $x_2 = 150$ is

$$(3.09 - (2.262 \times 0.0479), 3.09 + (2.262 \times 0.0479)) = (2.98, 3.20)$$

Furthermore,

$$\hat{\sigma}\sqrt{1 + \mathbf{x}^{*'}(\mathbf{X}'\mathbf{X})^{-1}\mathbf{x}^*} = \sqrt{0.0127} \times \sqrt{1 + 0.181} = 0.122$$

so that a 95% prediction interval for the electricity usage for a future month with $x_1 = 5.5$ and $x_2 = 150$ is

$$(3.09 - (2.262 \times 0.122), 3.09 + (2.262 \times 0.122)) = (2.81, 3.37)$$

Finally, it can be shown that the first data observation has a residual

$$e_1 = -0.088029$$

and the first diagonal element of the hat matrix is

$$h_{11} = 0.156721$$

Consequently, the standardized residual for this point is

$$e_1^* = \frac{e_i}{\hat{\sigma}\sqrt{1 - h_{11}}} = \frac{-0.088029}{\sqrt{0.0127}\sqrt{1 - 0.156721}} = -0.851$$

■ 13.3.2 Problems

13.3.1 Consider fitting the multiple linear regression model

$$y = \beta_0 + \beta_1 x_1 + \beta_2 x_2$$

to the data set in DS 13.3.1.

(a) What is the 10×1 vector of observed values of the response variable $\mathbf{Y}$?

(b) What is the 10×3 design matrix $\mathbf{X}$?

(c) What is the 3×3 matrix $\mathbf{X}'\mathbf{X}$?

(d) What is the 3×3 matrix $(\mathbf{X}'\mathbf{X})^{-1}$?

(e) What is the 3×1 vector $\mathbf{X}'\mathbf{Y}$?

(f) Show that the parameter estimates are

$$\hat{\beta} = \begin{pmatrix} 0 \\ 1 \\ 29/30 \end{pmatrix}$$

(g) What is the 10×1 vector of predicted values of the response variable $\hat{\mathbf{Y}}$?

(h) What is the 10×1 vector of residuals $\mathbf{e}$?

(i) What is the sum of squares for error?

(j) Show that the estimate of the error variance is $\hat{\sigma}^2 = 17/15$.

(k) What is the standard error of $\hat{\beta}_1$? Of $\hat{\beta}_2$? Should either of the input variables be dropped from the model?

(l) What is the fitted value of the response variable when $x_1 = 1$ and $x_2 = 2$? What is the standard error of this fitted value? Construct a 95% confidence interval for the expected value of the response variable when $x_1 = 1$ and $x_2 = 2$.

(m) Construct a 95% prediction interval for a future value of the response variable obtained with $x_1 = 1$ and $x_2 = 2$.

13.3.2 Consider fitting the multiple linear regression model

$$y = \beta_0 + \beta_1 x_1 + \beta_2 x_2$$

to the data set in DS 13.3.2.

(a) What is the 8×1 vector of observed values of the response variable $\mathbf{Y}$?

(b) What is the 8×3 design matrix $\mathbf{X}$?

(c) What is the 3×3 matrix $\mathbf{X}'\mathbf{X}$?

(d) What is the 3×3 matrix $(\mathbf{X}'\mathbf{X})^{-1}$?

(e) What is the 3×1 vector $\mathbf{X}'\mathbf{Y}$?

(f) Calculate the parameter estimates $\hat{\boldsymbol{\beta}}$.

(g) What is the 8×1 vector of predicted values of the response variable $\hat{\mathbf{Y}}$?

(h) What is the 8×1 vector of residuals $\mathbf{e}$?

(i) Show that SSE = 20.0.

(j) What is the estimate of the error variance $\hat{\sigma}^2$?

(k) What is the standard error of $\hat{\beta}_1$? Of $\hat{\beta}_2$? Should either of the input variables be dropped from the model?

(l) What is the fitted value of the response variable when $x_1 = x_2 = 1$? What is the standard error of this fitted value? Construct a 95% confidence interval for the expected value of the response variable when $x_1 = x_2 = 1$.

(m) Construct a 95% prediction interval for a future value of the response variable when $x_1 = x_2 = 1$.

13.3.3 The multiple linear regression model

$$y = \beta_0 + \beta_1 x_1 + \beta_2 x_2 + \beta_3 x_3$$

is fitted to the data set in DS 13.3.3. Use matrix algebra to derive the parameter estimates $\hat{\beta}_0$, $\hat{\beta}_1$, $\hat{\beta}_2$, and $\hat{\beta}_3$.

13.4 Evaluating Model Adequacy

In this section various diagnostic tools are discussed that provide an experimenter with information about the adequacy of the regression model. These tools include residual plots to investigate whether the assumptions of the regression model appear to be met, and the investigation of individual data points that have large standardized residuals or that have an especially large influence on the fitted regression model. The tools also provide insight into how the input variables are related to one another and how they influence the model. A good modeler uses this kind of information to assess the strengths and limitations of a regression model.

13.4.1 Multicolinearity of the Input Variables

The correlation structure among a set of input variables has important consequences for the model fit and the interpretation of the model, and it is sensible for the modeler to calculate the sample correlations r between the input variables.

From a mathematical point of view, if two input variables are very highly correlated (that is, almost colinear), then the matrix $\mathbf{X}'\mathbf{X}$ is almost singular and there can be computational difficulties in obtaining the inverse matrix $\mathbf{X}'\mathbf{X}^{-1}$, which is used to calculate the parameter estimates $\hat{\beta}_i$. In general, computational difficulties and roundoff errors in fitting a model can be minimized by using standardized variables

$$\frac{x_i - \bar{x}_i}{s_i}$$

in place of x_i in the model, where $\bar{x}_i$ and s_i are the mean and the standard deviation of the values taken by the input variable x_i. However, a high correlation between two input variables can still present problems, and many computer packages will alert the user with a warning message and will automatically remove one of the variables from the model.

From a practical point of view, if two input variables x_i and x_j are highly correlated, then it suffices to have just one of them in the model. Little improvement in the model fit can be obtained by having both variables in the model. In this respect, it is important to remember that if a variable x_i is dropped from the model, then this does not necessarily imply that the variable x_i is not correlated with the response variable y. It may be that there is another variable x_j in the model that is correlated with the variable x_i and acts as a surrogate for it. This phenomenon was observed in Examples 71 and 44 discussed in Section 13.2.

The correlation structure among the input variables also affects the interpretation of the model parameters β_i. Remember that the parameter β_i represents the change in the expected value of the response variable y as the input variable x_i is increased by one unit and when *all the other input variables are kept fixed*. However, if the input variables are correlated, then a change in the variable x_i is in practice likely to be associated with changes in the other input variables so that the consequent change in the expected value of the response variable y may be quite different from β_i.

13.4.2 Residual Analysis

The concepts behind residual analysis for a multiple linear regression model are similar to those for a simple linear regression model discussed in Chapter 12. However, they are much more important for the multiple linear regression model because of the lack of good graphical representations of the data set and the fitted model. In simple linear regression a plot of the response variable against the input variable showing the data points and the fitted regression line provides a good graphical summary of the regression analysis. With the higher dimensions of a multiple linear regression model, similar plots cannot be obtained.

Plots of the residuals e_i against the fitted values $\hat{y}_i$ and against the input variables x_i, as suggested in Figure 13.22, may alert the experimenter to any problems with the regression model. As discussed in Chapter 12, if all is well with the regression model, then these residual plots should exhibit a *random scatter* of points. Any patterns in the residual plots should draw the experimenter's attention and should be investigated. For example, a funnel shape to a plot of the residuals e_i against the fitted values $\hat{y}_i$ indicates a lack of homogeneity of the error variance σ^2. A series of negative, positive, and then negative values in a plot of the residuals e_i against an input variable x_i suggests that the model can be improved with the addition of a quadratic term x_i^2.

In certain cases where the data observations are obtained sequentially over time, it is also prudent to plot the residuals against a time axis, which designates the order in which the observations are taken. A residual plot such as that shown in Figure 13.23 indicates that there is a lack of independence in the error terms, with adjacent observations being positively correlated. Normal probability plots can always be used to investigate whether there is any indication that the error terms are not normally distributed.

The **standardized residuals** can be used to identify individual data points that do not fit the model well. Typically, computer packages alert the experimenter to points that have a standardized residual with an absolute value larger than 2. However, remember that even in an ideal modeling situation, about 5% of the data points are expected to have a standardized residual with an absolute value larger than 2, and about 1% of the points are expected to have a standardized residual with an absolute value larger than 2.5. Points with large standardized residuals should be investigated by the experimenter to determine whether anything is strange about them. An experimenter may want to consider them as outliers, remove them from the data set, and refit the model to the remaining data set.

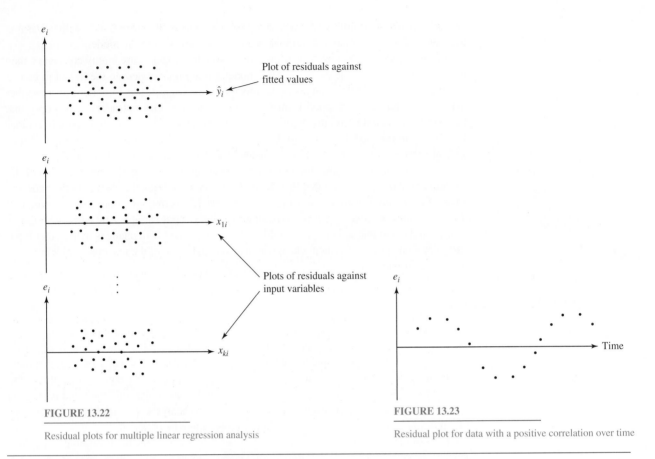

FIGURE 13.22

Residual plots for multiple linear regression analysis

FIGURE 13.23

Residual plot for data with a positive correlation over time

Example 71
Supermarket Deliveries

When a simple linear regression model is used to fit time against volume, a computer package may alert the user that the eighteenth observation has a standardized residual of $e_i^* = 2.03$. For this observation, an unloading time of 35 minutes was recorded for a load of 63 m^3, whereas the fitted time is

$$\hat{y}|_{63} = -24.16 + (0.8328 \times 63) = 28.306$$

This value gives a residual of

$$e_i = y_i - \hat{y}_i = 35.0 - 28.306 = 6.694$$

and with $h_{ii} = 0.093125$ the standardized residual is

$$e_i^* = \frac{e_i}{\hat{\sigma}\sqrt{1 - h_{ii}}} = \frac{6.694}{3.462\sqrt{1 - 0.093125}} = 2.03$$

However, this standardized residual is barely larger than 2, and even with an ideal model one out of $n = 20$ data points is expected to have a standardized residual larger than 2 anyhow, so the supermarket manager need not be overly concerned about this data point.

Similarly, when the unloading time is regressed on the weight of the load the twentieth observation is identified as having a standardized residual of $e_i^* = -2.18$. In this case, an unloading time of 19 minutes was recorded for a load of 40 tons, whereas the fitted time is

$$\hat{y}|_{40} = -16.08 + (1.250 \times 40) = 33.92$$

This value gives a residual of

$$e_i = y_i - \hat{y}_i = 19.0 - 33.92 = -14.92$$

and with $h_{ii} = 0.062221$ the standardized residual is

$$e_i^* = \frac{e_i}{\hat{\sigma}\sqrt{1 - h_{ii}}} = \frac{-14.92}{7.065\sqrt{1 - 0.062221}} = -2.18$$

However, as before, the experimenter does not need to be concerned about this data point.

Example 72
Turf Quality

Figure 13.24 shows plots of the residuals e_i against the fitted values $\hat{y}_i$, the water levels x_{1i}, and the fertilizer levels x_{2i}. Each residual plot exhibits a fairly random scatter of the points and they do not indicate that there are any problems with the regression model.

Finally, when an experimenter has a large enough data set, a good modeling procedure is to split the data set into two parts and to fit the model to just one of the two parts. The fitted model is then applied to the second part of the data set and residuals are calculated in the usual manner as the difference between the observed data points and the fitted values. In this case, however, standardized residuals are calculated as simply the residual e_i divided by the estimated standard error $\hat{\sigma}$. This modeling approach, known as **cross-validation**, is particularly reassuring when the model fitted from the first part of the data set provides a close fit to the data points in the second part of the data set.

13.4.3 Influential Points

Influential data points are points that have values of the input variables x_i that cause them to have an unusually large influence on the fitted model. It is prudent for the experimenter to be aware of these points and to be confident of their accuracy. In simple linear regression when there is only one input variable, the most influential points are those with the largest and smallest values of the input variable, as shown in Figure 13.25. More generally when there are two or more input variables, the most influential data points are those located on the edges of the collection of input variable values, as shown in Figure 13.26 for two input variables.

A good way to measure the influence of a data point is through the corresponding diagonal element of the hat matrix h_{ii}, which is often referred to as the **leverage value**. Larger leverage values h_{ii} indicate that the variable is more influential. The leverage values are nonnegative, and when there is an intercept and k input variables in the regression model, they sum to $k + 1$

$$\sum_{i=1}^{n} h_{ii} = k + 1$$

With n data points the average leverage value is therefore $(k + 1)/n$. A general rule is that a data point can be considered to be influential if $h_{ii} > 2(k + 1)/n$ and to be very influential if $h_{ii} > 3(k + 1)/n$.

FIGURE 13.24

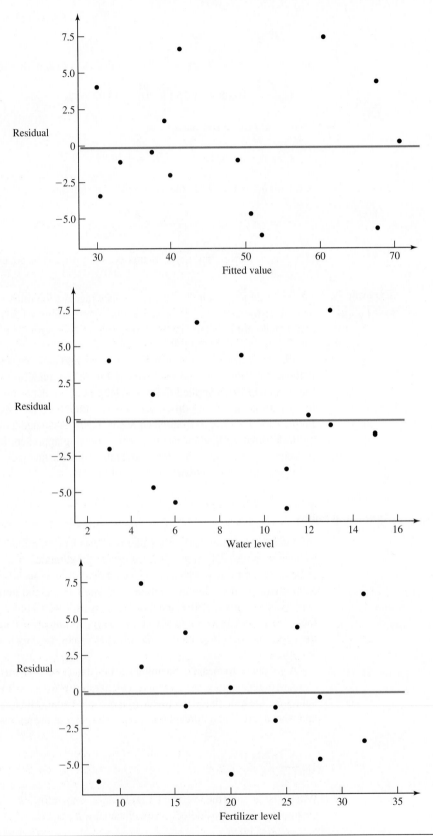

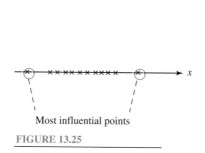

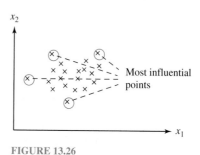

FIGURE 13.25

Influential data points for a simple linear regression analysis

FIGURE 13.26

Influential data points for a multiple linear regression analysis with two input variables

---

Example 44
Army Physical Fitness Test

Consider the linear regression model with run time as the dependent variable and with both pushups and situps as the input variables. Both the twenty-fourth and thirty-second observations may be identified as being points with a large influence on the fitted regression model. The twenty-fourth observation corresponds to an officer who performed 80 pushups and 100 situps, and the thirty-second observation corresponds to an officer (an age 39, 5'11", 149-lb major) who performed 125 pushups and 95 situps. These two points are easily identifiable in the plot of situps against pushups in Figure 13.20 as the topmost point and the rightmost point, respectively.

The twenty-fourth observation has a leverage $h_{24.24} = 0.119$ and the thirty-second observation has a leverage $h_{32.32} = 0.244$. With $k = 2$ input variables in the model

$$\frac{3(k+1)}{n} = \frac{9}{84} = 0.107$$

and so these two observations can be considered to be very influential. However, a check of the test records confirms the accuracy of these two data points and so no further action is required. Notice that in the simple linear regression model discussed in Chapter 12 with pushups as the only input variable, the thirty-second observation can be seen on the far right of Figure 12.19 and is clearly very influential in determining the slope of the fitted regression line.

COMPUTER NOTE

You will need to perform multiple linear regression analyses on a computer. However, this is easily done and only requires that you enter the data and specify the response variable, the input variables, and the linear model that you are interested in. As is always the case when using a statistical software package, the key is to have both a good understanding of what you are asking the computer to do and a good understanding of what the computer output is telling you.

Make good use of residual plots and other graphical devices where appropriate so that you have a good feeling for how well your model is doing. Make sure that you are aware of the correlations among your input variables and which data points have large standardized residuals or are very influential.

When you have a large number of input variables to investigate, you will want to start off using an automatic procedure to find good models for you. Stepwise procedures incorporating backwards elimination and forward selection procedures are useful. In addition, some software packages can implement a "best subsets" approach whereby regression models based upon all (or a specified number of) possible subsets of the collection of all of the input variables are analyzed. Some criterion such as the value of the coefficient of determination R^2 must be used to determine which models warrant further investigation. Often the statistic

$$C_r = \frac{\text{SSE}_r}{\hat{\sigma}^2} + 2(r+1) - n$$

is used as well, where $\hat{\sigma}^2$ is the estimate of the error variance from a full model with an intercept and k input variables, and SSE_r is the error sum of squares for the model under consideration with an intercept and r input variables. Values of C_r close to r indicate a reasonable model, while values of C_r much larger than r indicate a poor model.

■ **13.4.4   Problems**

13.4.1 Consider Example 70 and the data set in Figure 13.7.

(a) Make a plot of the residuals e_i against the fitted values $\hat{y}_i$. Make a plot of the residuals e_i against the temperature values x_i. Do either of these plots alert you to any problems with the regression analysis?

(b) Find the standardized residuals. Are any of them unusually large?

(c) Find the leverage values h_{ii}. Which points have the largest influence on the fitted regression line?

13.4.2   Oil Well Drilling Costs

Consider the modeling of oil well drilling costs described in Problem 13.2.3 and the data set in DS 13.2.3. Suppose that a model is used with cost as the dependent variable and with depth and downtime as input variables.

(a) Make plots of the residuals against the fitted values and against each of the two input variables. What do you find?

(b) Make a plot of the residuals against the variable geology. Why does this plot confirm that the variable geology is not needed in the regression model?

(c) Verify that there are no points with an unusually large influence on the regression model.

(d) Verify that there is one point with a standardized residual of 2.01.

13.4.3   VO2-max Aerobic Fitness Measurements

Consider the modeling of aerobic fitness described in Problem 13.2.4 and the data set in DS 13.2.4. Suppose that a model is used with VO2-max as the dependent variable and with heart rate at rest and percentage body fat as input variables.

(a) Make plots of the residuals against the fitted values and against each of the two input variables. Are you alerted to any problems with the regression model?

(b) Make a plot of the residuals against the variable weight. Why does this plot confirm that the variable weight is not needed in the regression model?

(c) Verify that there are no points with an unusually large influence on the regression model.

(d) Verify that there is one point with a standardized residual of -2.15.

13.4.4 In a multiple linear regression model, suppose that observation i has a positive residual. What can you say about how the leverage value h_{ii} of this observation will change if the response value y_i of the observation is increased?

13.5   Nonlinear Regression

13.5.1   Introduction

While linear regression modeling is very versatile and provides useful models for many important applications, it is sometimes necessary to fit a nonlinear model to a data set. Recall that an intrinsically linear model is one that can be transformed into a linear format, and it is generally best to perform such a transformation when possible. However, nonlinear regression techniques must be applied to models that cannot be transformed into a linear format.

In nonlinear regression a function

$$f(x_1, \ldots, x_k; \theta_1, \ldots, \theta_p)$$

is specified that relates the values of k input variables $x_1, \ldots, x_k$ to the expected value of a response variable y. This function depends upon a set of unknown parameters

$$\theta_1, \ldots, \theta_p$$

which are estimated by fitting the model to a data set. The function $f(\cdot)$ may be justified by a theoretical argument (as in the following example) or it may simply be chosen to provide a good fit to the available data.

As with linear regression, the criterion of least squares is generally employed to fit the model. With a data set

$$(y_1, x_{11}, x_{21}, \ldots, x_{k1})$$
$$\vdots$$
$$(y_n, x_{1n}, x_{2n}, \ldots, x_{kn})$$

consisting of n sets of values of the response variable and the k input variables, the parameter estimates

$$\hat{\theta}_1, \ldots, \hat{\theta}_p$$

are therefore chosen to minimize the sum of squares

$$\sum_{i=1}^{n} (y_i - f(x_{1i}, \ldots, x_{ki}; \theta_1, \ldots, \theta_p))^2$$

However, unlike linear regression, for a nonlinear regression problem there is in general no simple expression that can be used to calculate the parameter estimates, and in practice they need to be calculated by an iterative computer search procedure. These procedures generally require the user to specify "initial guesses" of the parameter values, which need to be suitably close to the actual values that minimize the sum of squares. The calculation of standard errors for the parameter estimates is also rather awkward for nonlinear regression problems and they are usually based on some general asymptotic arguments and should be treated as giving only a general idea of the sensitivity of the parameter estimates. Residual analysis can be used in a similar manner to linear regression problems to estimate the error variance and to assess whether the modeling assumptions appear to be appropriate.

13.5.2   Example

Example 73

Indoor Air Pollution Levels

This problem is discussed in the paper "Experimental Designs and Emission Rate Modeling for Chamber Experiments" by Anthony Hayter and Mary Dowling, *Atmospheric Environment* **27A**, 14, 2225–2234 (1993). In recent years heightened attention has been directed toward the problem of indoor pollution levels. In particular, indoor pollution levels may be the result of chemical emissions from objects such as carpets, paints, wallpaper, furniture, fabrics, and other general household or office appliances. An essential component of this investigation is the estimation of pollutant *emission rate profiles* for these polluting substances.

These emission rate profiles are estimated by collecting pollutant concentration level measurements from chamber studies as illustrated in Figure 13.27. A specimen of the sample under consideration is placed within a chamber, and a constant air flow is maintained through the chamber with polluted air being blown out in exchange for clean filtered air being drawn in.

FIGURE 13.27

Emission chamber experiment

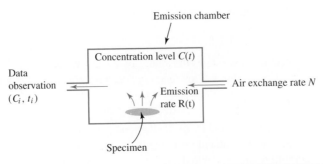

At various sampling times a portion of the exiting air is collected and the amount of pollutant is measured. This measurement presents an estimate of the pollutant concentration level within the chamber at that sampling time.

The data obtained from a chamber experiment consequently consist of paired observations (C_i, t_i) representing the chamber pollutant concentration levels C_i at times t_i. These values can be used to model the chamber pollutant concentration level time profile $C(t)$. This profile, in turn, can be used to estimate the sample emission rate time profile $R(t)$, which is the objective of the experiment.

The concentration level $C(t)$ and the emission rate $R(t)$ (measured per unit area of the emitting substance) are related by the following differential equation. If V represents the volume of the chamber, A represents the area of the emitting substance under investigation, and N represents the chamber air exchange rate (i.e., the proportion of the chamber's air that is replaced in unit time), then the conservation of pollutant mass results in the equation

$$VdC = ARdt - NVCdt$$

because during the small time interval dt, VdC is the change of pollutant mass within the chamber that is accounted for by the difference between the mass emitted from the specimen $ARdt$ and the mass leaving the chamber $NVCdt$. This lends to the differential equation

$$\frac{dC(t)}{dt} + NC(t) = aR(t)$$

where $a = A/V$.

Suppose that the emission rate $R(t)$ has an exponential decay so that it is given by

$$R(t; \theta_0, \theta) = \theta_0 e^{-\theta t}$$

for some unknown parameters θ_0 and θ. With the condition $C(0) = 0$, the differential equation above can then be solved to give

$$C(t; \theta_0, \theta) = \frac{a\theta_0 \left(e^{-\theta t} - e^{-Nt}\right)}{N - \theta}$$

For known values of the constants a and N, the unknown parameters θ_0 and θ are estimated by fitting this model to the data values (C_i, t_i) using nonlinear regression techniques. Notice that this model is not intrinsically linear because it cannot be transformed into a linear format.

Figure 13.28 shows an illustrative graph of this function $C(t; \theta_0, \theta)$. The concentration level of the pollutant in the chamber initially rises because the specimen emits pollutant at a rate greater than the rate at which it is blown out of the chamber. However, after a time

$$t_{\max} = \frac{\ln \theta - \ln N}{\theta - N}$$

FIGURE 13.28

Typical concentration level curve for emission chamber experiment

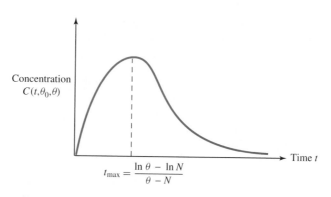

$$t_{max} = \frac{\ln \theta - \ln N}{\theta - N}$$

Time (hours)	Concentration (μg/m^3)
0.5	0.219
1.5	0.397
2.5	0.410
4.5	0.549
8.5	0.333
24.5	0.243
48.5	0.163
72.5	0.132
144.5	0.019
168.5	0.031
196.5	0.027
216.5	0.023
240.5	0.018

FIGURE 13.29

Data set of carpet formaldehyde emissions

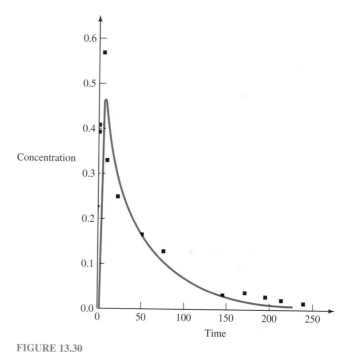

FIGURE 13.30

Fitted concentration level curve for carpet formaldehyde emissions

the concentration level in the chamber decreases because the rate of emission is smaller than the rate at which the pollutant is blown out of the chamber.

Figure 13.29 contains the data obtained from an experiment performed to measure formaldehyde emissions from a sample of carpet. A chamber of size $V = 0.053$ m^3 was used with a carpet sample of area $A = 0.0210$ m^2, so that $a = 0.40$ m^{-1}. A constant air exchange rate was maintained, which was measured to be $N = 1.01$ h^{-1}. The data set shows 13 concentration level observations that were taken within the first 250 hours.

A computer package can be used to fit the nonlinear model to the data set and to obtain the parameter estimates $\hat{\theta}_0 = 1.27$ and $\hat{\theta} = 0.024$. Figure 13.30 shows the fitted curve $C(t; 1.27, 0.024)$ superimposed on a plot of the data points. It can be seen that the model has done a fairly good job of fitting the data points, although it has underestimated the peak concentration level and has also consistently underestimated the concentration levels at times larger than 150 hours. Nevertheless, the fitted emission rate

$$R(t) = \hat{\theta}_0\, e^{-\hat{\theta}t} = 1.27\, e^{-0.024t}$$

should provide some useful information on the formaldehyde emissions from the carpet, at least within the first six days.

13.6   Supplementary Problems

13.6.1 Consider the data set in DS 13.6.1.

(a) Plot the response variable y against the input variable x and confirm that the quadratic model

$$y = \beta_0 + \beta_1 x + \beta_2 x^2$$

appears to be appropriate.

(b) Write out the analysis of variance table using the fact that SSE = 39.0.

(c) The parameter estimates are $\hat{\beta}_0 = 18.18$, $\hat{\beta}_1 = -44.90$, and $\hat{\beta}_2 = 44.08$. The standard error of $\hat{\beta}_2$ is s.e.$(\hat{\beta}_2) = 7.536$. Verify that the quadratic term is needed in the model.

(d) What is the fitted value of the response variable when $x = 1$? If this fitted value has a standard error of s.e.$(\hat{y}) = 1.005$, construct a 95% two-sided confidence interval for the expected value of the response variable when $x = 1$.

13.6.2 Use hand calculations to fit the multiple linear regression model

$$y = \beta_0 + \beta_1 x_1 + \beta_2 x_2$$

to the data set in DS 13.6.2.

(a) Write down the vector of observed values of the response variable $\mathbf{Y}$ and the design matrix $\mathbf{X}$.

(b) Calculate $\mathbf{X'X}$.

(c) Verify that

$$(\mathbf{X'X})^{-1} = \begin{pmatrix} 73/680 & -1/136 & 0 \\ -1/136 & 1/136 & 0 \\ 0 & 0 & 1/116 \end{pmatrix}$$

(d) Verify that $\hat{\beta}_0 = 4$, $\hat{\beta}_1 = -3$, and $\hat{\beta}_2 = 1$.

(e) Calculate the vector of predicted values of the response variable $\hat{\mathbf{Y}}$ and the vector of residuals $\mathbf{e}$.

(f) What is the sum of squares for error?

(g) Show that the estimate of the error variance is $\hat{\sigma}^2 = 54/7$.

(h) What is the standard error of $\hat{\beta}_1$? Of $\hat{\beta}_2$? Should either of the input variables be dropped from the model?

(i) What is the fitted value of the response variable when $x_1 = 2$ and $x_2 = -2$? What is the standard error of this fitted value?

(j) Construct a 95% prediction interval for a future value of the response variable obtained with $x_1 = 2$ and $x_2 = -2$.

13.6.3 Friction Power Loss from Engine Bearings

DS 13.6.3 contains an extension of the data set presented in DS 12.11.3 concerning the power loss that occurs in the bearing of an automobile engine. In addition to the diameter of the bearing, information is provided on the clearance and the length of the bearing.

(a) Perform a linear regression analysis with power loss as the dependent variable and bearing diameter, bearing clearance, and bearing length as the explanatory variables. Show that the bearing length has a p-value of 0.331 and can be removed from the model.

(b) What is the fitted model when the explanatory variables diameter and clearance are used? Calculate a 95% prediction interval for the power loss of an engine that has a bearing diameter of 25 and a bearing clearance of 0.07. Show that the largest standardized residual in absolute value is 2.46.

13.6.4 Bacteria Cultures

The data set in DS 13.6.4 shows the yields of a bacteria culture obtained for different amounts of an additive x_1 and for different growing temperatures x_2.

(a) Investigate the experimental design employed by looking at the values of the additive levels and the temperature levels used in the experiment.

(b) Fit the response surface model

$$y = \beta_0 + \beta_1 x_1 + \beta_2 x_1^2 + \beta_3 x_2 + \beta_4 x_2^2 + \beta_5 x_1 x_2$$

and show that one data point has a standardized residual of -3.01.

(c) Remove the data point with the large negative standardized residual and fit the response surface model to the remaining data points. Make a plot of the fitted model. What values of the additive and the temperature would you recommend to maximize the yield of the bacteria culture?

13.6.5 The regression model

$$y = -67.5 + 34.5x_1 - 0.44x_2 + 108.6x_3 + 55.8x_4$$

is obtained from $n = 44$ observations. The first observation is $y_1 = 288.9$, $x_1 = 12.3$, $x_2 = 143.4$, $x_3 = -7.2$, $x_4 = 14.4$ and it has a standardized residual -1.98 and a leverage value 0.0887. If $SST = 20554$, what is R^2?

13.6.6 The model $y = \beta_0 + \beta_1 x_1 + \beta_2 x_2 + \beta_3 x_3 + \beta_4 x_4 + \beta_5 x_5$ is fitted to a data set and $\hat{\beta}_3 = -5.602$ is obtained. What can you say about the sign (negative, zero, or positive) of the sample correlation coefficient between y and x_3?

13.6.7 Are the following statements true or false?

(a) It is necessary to perform diagnostic checks of the fit when the value of R^2 is large.

(b) Multiple linear regression is different from simple linear regression because more than one model is provided for the response variable.

13.6.8 The multiple linear regression model

$$y = \beta_0 + \beta_1 x_1 + \beta_2 x_2$$

is fitted to a data set of $n = 30$ observations.

(a) Suppose that $\hat{\beta}_1 = -45.2$ and $s.e.(\hat{\beta}_1) = 39.5$, and that $\hat{\beta}_2 = 3.55$ and $s.e.(\hat{\beta}_2) = 5.92$. Is it clear that both x_1 and x_2 should be removed from the model?

(b) Suppose that $\hat{\beta}_1 = -45.2$ and $s.e.(\hat{\beta}_1) = 8.6$, and that $\hat{\beta}_2 = 3.55$ and $s.e.(\hat{\beta}_2) = 0.63$. Is it clear that you would not want to remove either x_1 or x_2 from the model?

CHAPTER FOURTEEN

Multifactor Experimental Design and Analysis

Yields from a chemical process, for example, may depend upon a large number of different factors. The discussion in Chapter 11 on the analysis of variance concerns the relationship between a response variable of interest and various levels of a *single* factor of interest. However, in many situations like this chemical process, it is useful to simultaneously investigate the relationship between a response variable and *two or more* factors of interest. **Multifactor** experiments of this kind are considered in this chapter.

Experiments with two factors are considered in Section 14.1. The analysis of variance methodology is extended to analyze the structure of the relationship between the response variable and the two factors of interest. Of considerable importance is the assessment of any *interaction* between the two factors in the manner in which they influence the response variable. Extensions of the methodology to experiments with three or more factors are discussed in Section 14.2, together with the important research tool of **screening experiments**, which are used to determine which of a large number of factors have a significant influence on a response variable.

14.1 Experiments with Two Factors

The simplest multifactor experiment involves two factors. The analysis of a two factor experiment introduces the important concept of an interaction effect between the two factors, which is explained in Section 14.1.2.

14.1.1 Two-Factor Experimental Designs

A two-factor experiment can be used to investigate how a response variable of interest depends on two factors of interest. The following examples illustrate such experiments.

Example 72

Turf Quality

The manager of the turf growing company is interested in how the type of fertilizer and the temperature at which the grass is grown affect the quality of the turf. A two-factor experiment can be performed with *turf quality* as the response variable and with *fertilizer type* and *growing temperature* as the two factors.

Example 9

Car Body Assembly Line

In an automated car body assembly line, large multispot welding machines clamp two metal body parts together and weld them at various spots before passing them on to the next stage of construction. The strength and effectiveness of this welding procedure are clearly of considerable importance.

In one assembly line there are three supposedly similar machines in operation, and the manager wishes to investigate whether their performances are identical. In addition, the manager wishes to investigate various solder formulations that can be used by the machines. A two-factor experiment can be performed with the *strength of the weld* as the response variable and with *machines* and *solder type* as the two factors.

Example 74

Company Transportation Costs

A company has to transport some of its goods a considerable distance by road from the factory where they are produced to a port from where they are exported. The driving time required has important implications for the logistical planning of the transportation and for the company's transportation costs. A company manager realizes that the driving time may depend on the route taken by the driver and the time of day that the shipment leaves the factory. A two-factor experiment can be performed to investigate this dependency with *driving time* as the response variable and with *driving route* and *time of day* as the two factors.

Suppose that the two factors in a two-factor experiment are called factor A with a levels and factor B with b levels. As Figure 14.1 illustrates, there are therefore $a \times b$ different combinations of the two factors, which are known as **experimental configurations** or **cells**. In a complete balanced experimental design, the experimenter obtains n observations (measurements of the response variable) from each of the ab experimental configurations, so that a total of

$$n_T = abn$$

observations are obtained in the experiment. The n observations obtained at each experimental configuration are referred to as **replication measurements** obtained at the experimental configuration.

The data set obtained from the experiment therefore consists of the values

$$x_{ijk} \qquad 1 \leq i \leq a, 1 \leq j \leq b, 1 \leq k \leq n$$

as shown in Figure 14.2. Thus, x_{ijk} is the value of the kth measurement obtained at the ith level of factor A and the jth level of factor B. The cell averages, that is, the averages of the observations obtained at each of the ab experimental configurations, are denoted by

$$\bar{x}_{ij\cdot} = \frac{x_{ij1} + x_{ij2} + \cdots + x_{ijn}}{n} = \frac{1}{n} \sum_{k=1}^{n} x_{ijk}$$

It is also useful to consider the sample averages at each level of factor A. These are

$$\bar{x}_{1\cdot\cdot} \quad \bar{x}_{2\cdot\cdot} \quad \bar{x}_{3\cdot\cdot} \quad \cdots \quad \bar{x}_{a\cdot\cdot}$$

where

$$\bar{x}_{i\cdot\cdot} = \frac{1}{bn} \sum_{j=1}^{b} \sum_{k=1}^{n} x_{ijk}$$

FIGURE 14.2

Data observations and sample
averages for a two-factor
experiment

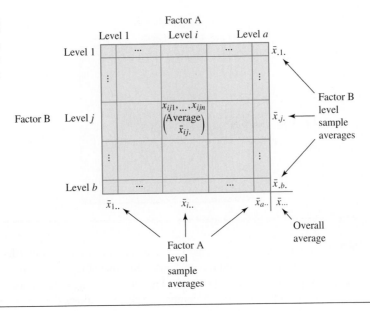

are the averages of all the bn data observations obtained at each of the levels of factor A. Similarly, the sample averages at each level of factor B are

$$\bar{x}_{.1.} \quad \bar{x}_{.2.} \quad \bar{x}_{.3.} \quad \ldots \quad \bar{x}_{.b.}$$

where

$$\bar{x}_{.j.} = \frac{1}{an} \sum_{i=1}^{a} \sum_{k=1}^{n} x_{ijk}$$

The average of all the abn data observations is

$$\bar{x}_{...} = \frac{1}{abn} \sum_{i=1}^{a} \sum_{j=1}^{b} \sum_{k=1}^{n} x_{ijk}$$

These sample averages are illustrated in Figure 14.2.

Two-Factor Experiments

Consider an experiment with two factors, A and B. If factor A has a levels and factor B has b levels, then there are ab experimental configurations or cells. In a complete balanced experiment n replicate measurements are taken at each experimental configuration, resulting in a total sample size of $n_T = abn$ data observations. The data observation x_{ijk} represents the value of the kth measurement obtained at the ith level of factor A and the jth level of factor B. The cell sample averages are denoted by $\bar{x}_{ij.}$, the sample averages at the levels of factor A are denoted by $\bar{x}_{i..}$, the sample averages at the levels of factor B are denoted by $\bar{x}_{.j.}$, and the overall sample average of all abn data observations is denoted by $\bar{x}_{...}$.

Example 72
Turf Quality

There are two fertilizers of interest, *fertilizer 1* and *fertilizer 2*, and two growing temperatures, *low* and *high*. Suppose that fertilizer type is considered to be factor A and growing temperature is considered to be factor B. Since each factor has two levels, $a = b = 2$ and there are $ab = 4$ experimental configurations.

FIGURE 14.3

Data set for the turf quality example

Temperature		Fertilizer 1	Fertilizer 2	
	Low	$x_{111} = 40$ $x_{112} = 24$ $x_{113} = 31$ $x_{114} = 21$ $(\bar{x}_{11\cdot} = 29.00)$	$x_{211} = 47$ $x_{212} = 37$ $x_{213} = 41$ $x_{214} = 41$ $(\bar{x}_{21\cdot} = 41.50)$	$\bar{x}_{\cdot 1\cdot} = 35.250$
	High	$x_{121} = 46$ $x_{122} = 53$ $x_{123} = 48$ $x_{124} = 46$ $(\bar{x}_{12\cdot} = 48.25)$	$x_{221} = 84$ $x_{222} = 76$ $x_{223} = 77$ $x_{224} = 79$ $(\bar{x}_{22\cdot} = 79.00)$	$\bar{x}_{\cdot 2\cdot} = 63.625$
		$\bar{x}_{1\cdot\cdot} = 38.625$	$\bar{x}_{2\cdot\cdot} = 60.250$	$\bar{x}_{\cdot\cdot\cdot} = 49.4375$

Sixteen turf samples are obtained, with four allocated to each experimental configuration in a random manner. This arrangement produces $n = 4$ replicate observations for each experimental configuration. The experiment is conducted by applying the appropriate fertilizer type to the turf samples and placing them in the appropriate growing temperatures. After a suitable amount of time, each turf sample is given a score relating to the quality of the grass. The resulting data set is shown in Figure 14.3. Recall that higher scores correspond to better quality grass.

Example 9
Car Body Assembly Line

Suppose that the different machines are taken to be factor A, which consequently has $a = 3$ levels. If four different solder formulations are to be investigated, then factor B has $b = 4$ levels. Consequently there are $ab = 12$ experimental configurations, and the experiment is run by conducting $n = 3$ separate weldings at each of these experimental configurations. The data set obtained from measuring the strengths of the weldings is shown in Figure 14.4.

Example 74
Company Transportation Costs

There are two different driving routes from the factory to the port, *route 1* and *route 2*, and the time of the day when the truck leaves the factory is classified as being either in the *morning*, the *afternoon*, or the *evening*. Driving route can be taken to be factor A with $a = 2$ levels, and period of day can be taken to be factor B with $b = 3$ levels.

Ten trucks leaving the factory in each of the three time periods are randomly chosen, and five of them are selected at random to take route 1 and the other five are selected to take route 2. The driving times of the trucks are recorded and are shown in Figure 14.5. Notice that there are $ab = 6$ experimental configurations with $n = 5$ replicate observations at each of the experimental configurations.

14.1.2 Models for Two-Factor Experiments

The unknown parameters of interest in a two-factor experiment are the **cell means**

$$\mu_{ij} \qquad 1 \leq i \leq a, 1 \leq j \leq b$$

FIGURE 14.4

Data set for the car body assembly line example

Machine

	Machine 1	Machine 2	Machine 3	
Solder 1	$x_{111} = 4.12$ $x_{112} = 4.21$ $x_{113} = 6.11$ $(\bar{x}_{11.} = 4.81)$	$x_{211} = 3.74$ $x_{212} = 6.40$ $x_{213} = 5.27$ $(\bar{x}_{21.} = 5.14)$	$x_{311} = 6.17$ $x_{312} = 3.79$ $x_{313} = 4.47$ $(\bar{x}_{31.} = 4.81)$	$\bar{x}_{.1.} = 4.92$
Solder 2	$x_{121} = 3.44$ $x_{122} = 3.32$ $x_{123} = 3.75$ $(\bar{x}_{12.} = 3.50)$	$x_{221} = 3.93$ $x_{222} = 4.41$ $x_{223} = 4.98$ $(\bar{x}_{22.} = 4.44)$	$x_{321} = 3.05$ $x_{322} = 4.03$ $x_{323} = 3.32$ $(\bar{x}_{32.} = 3.47)$	$\bar{x}_{.2.} = 3.80$
Solder 3	$x_{131} = 6.09$ $x_{132} = 4.46$ $x_{133} = 3.38$ $(\bar{x}_{13.} = 4.64)$	$x_{231} = 5.00$ $x_{232} = 5.15$ $x_{233} = 6.75$ $(\bar{x}_{23.} = 5.63)$	$x_{331} = 4.88$ $x_{332} = 4.49$ $x_{333} = 3.49$ $(\bar{x}_{33.} = 4.29)$	$\bar{x}_{.3.} = 4.85$
Solder 4	$x_{141} = 7.88$ $x_{142} = 6.75$ $x_{143} = 7.22$ $(\bar{x}_{14.} = 7.28)$	$x_{241} = 6.43$ $x_{242} = 5.87$ $x_{243} = 8.50$ $(\bar{x}_{24.} = 6.93)$	$x_{341} = 8.31$ $x_{342} = 7.10$ $x_{343} = 7.60$ $(\bar{x}_{34.} = 7.67)$	$\bar{x}_{.4.} = 7.30$
	$\bar{x}_{1..} = 5.06$	$\bar{x}_{2..} = 5.54$	$\bar{x}_{3..} = 5.06$	$\bar{x}_{...} = 5.22$

Solder (row label at left of the four Solder groups)

FIGURE 14.5

Driving times in minutes for the company transportation costs example

Route

Period of Day		Route 1	Route 2	
	Morning	$x_{111} = 490$ $x_{112} = 553$ $x_{113} = 489$ $x_{114} = 504$ $x_{115} = 519$ $(\bar{x}_{11.} = 511.0)$	$x_{211} = 485$ $x_{212} = 489$ $x_{213} = 475$ $x_{214} = 470$ $x_{215} = 459$ $(\bar{x}_{21.} = 475.6)$	$\bar{x}_{.1.} = 493.3$
	Afternoon	$x_{121} = 511$ $x_{122} = 490$ $x_{123} = 489$ $x_{124} = 492$ $x_{125} = 451$ $(\bar{x}_{12.} = 486.6)$	$x_{221} = 456$ $x_{222} = 460$ $x_{223} = 464$ $x_{224} = 485$ $x_{225} = 473$ $(\bar{x}_{22.} = 467.6)$	$\bar{x}_{.2.} = 477.1$
	Evening	$x_{131} = 435$ $x_{132} = 468$ $x_{133} = 463$ $x_{134} = 450$ $x_{135} = 444$ $(\bar{x}_{13.} = 452.0)$	$x_{231} = 406$ $x_{232} = 422$ $x_{233} = 459$ $x_{234} = 442$ $x_{235} = 464$ $(\bar{x}_{23.} = 438.6)$	$\bar{x}_{.3.} = 445.3$
		$\bar{x}_{1..} = 483.2$	$\bar{x}_{2..} = 460.6$	$\bar{x}_{...} = 471.9$

FIGURE 14.6

Cell means for a two-factor experiment

Factor A

		Level 1		Level i		Level a
	Level 1	μ_{11}	$\cdots$	μ_{i1}	$\cdots$	μ_{a1}
		$\vdots$				$\vdots$
Factor B	Level j	μ_{1j}		μ_{ij}		μ_{aj}
		$\vdots$				$\vdots$
	Level b	μ_{1b}	$\cdots$	μ_{ib}	$\cdots$	μ_{ab}

As illustrated in Figure 14.6, these are the unknown mean values at each of the ab experimental configurations, and they represent the expected values of the response variable at these experimental configurations.

A measurement obtained at the ith level of factor A and the jth level of factor B is taken to be an observation from a normally distributed random variable with a mean μ_{ij} and with some variance σ^2. Thus, an observation x_{ijk} is taken to be an observation from a

$$N(\mu_{ij}, \sigma^2)$$

distribution.

Equivalently, the observation can be written as

$$x_{ijk} = \mu_{ij} + \epsilon_{ijk}$$

where the **error term** ϵ_{ijk} is taken to be an observation from a

$$N(0, \sigma^2)$$

distribution. An observation taken at the ith level of factor A and the jth level of factor B can therefore be thought of as being composed of the unknown cell mean μ_{ij} together with an error term that is normally distributed with mean zero and some unknown variance σ^2. The error terms are assumed to be distributed independently of each other.

The cell means can be estimated by the sample averages within the cells, so that

$$\hat{\mu}_{ij} = \bar{x}_{ij\cdot}$$

However, the objective of the analysis of a two-factor experiment is not simply to estimate the ab cell means μ_{ij}, but rather to investigate the *structure* of these cell means. This structure provides an indication of how the expected value of the response variable depends on the two factors under consideration. Specifically, it is useful to consider three aspects of this relationship:

- the **interaction effect** between factors A and B
- the **main effect** of factor A
- the **main effect** of factor B

There is an *interaction* effect between factors A and B if the effect of one factor on the response variable is different at the various levels of the other factor. In other words, there is an interaction effect if the effects that different levels of factor A have on the response variable depend on which level of factor B is employed. Similarly, there is an interaction effect if the effects that different levels of factor B have on the response variable depend on which level of factor A is employed. At an intuitive level, it can be seen that the presence of an interaction between two factors implies that they *do not act independently* of each other on the response variable.

Suppose that there is no interaction effect between the two factors A and B. What then is the effect on the response variable of the different levels of factor A? This is the *main effect* of factor A. In other words, the main effect of factor A assesses how changing the levels of factor A influences the response variable. Notice that it makes sense to consider this main effect only when there is no interaction between the two factors, because if there is an interaction between the two factors, the effect that different levels of factor A have on the response variable depends on which level of factor B is employed. In a similar manner, when there is no interaction effect between the two factors A and B, the effect on the response variable of the different levels of factor B is the *main effect* of factor B.

A graphical illustration of the cell means μ_{ij} in the case where both factors have two levels ($a = b = 2$) helps to clarify the interpretation of the main effects and the interaction effect. In this case there are four experimental configurations and the four cell means μ_{11}, μ_{12}, μ_{21}, and μ_{22} can be plotted. Figure 14.7 shows a situation where there is no interaction effect and no main effects, so that the cell means are not affected by changes in the levels of the two factors. In other words, the expected value of the response variable is not influenced by either of the two factors considered.

Figure 14.8 shows a situation where there is no interaction effect and no main effect for factor B, but there is a main effect for factor A. In this case the values of the cell means depend upon which level of factor A is employed, but they are not affected by which level of factor B is employed. The interpretation of this model is that the response variable is influenced by factor A but not by factor B. Similarly, Figure 14.9 shows a situation where there is no interaction

FIGURE 14.7

Interpretation of models for two-factor experiments

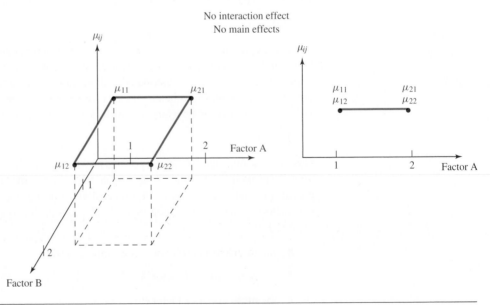

FIGURE 14.8

Interpretation of models for
two-factor experiments

No interaction effect
Factor A main effect
No factor B main effect

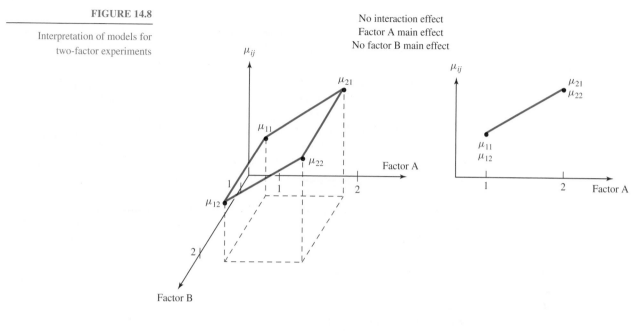

FIGURE 14.9

Interpretation of models for
two-factor experiments

No interaction effect
No factor A main effect
Factor B main effect

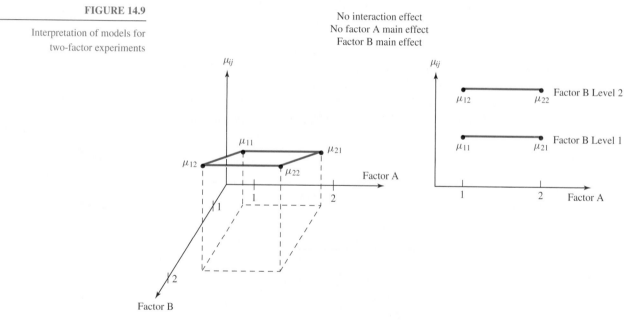

effect and no main effect for factor A, but there is a main effect for factor B, so that the response variable is influenced by factor B but not by factor A.

Figure 14.10 shows a situation where there is no interaction effect but there are main effects for factors A and B, and Figure 14.11 shows a situation where there is an interaction effect. The important distinction between these two cases is that when there is no interaction the plotted lines in the right-hand diagram are *parallel* to each other, but when there is an interaction the

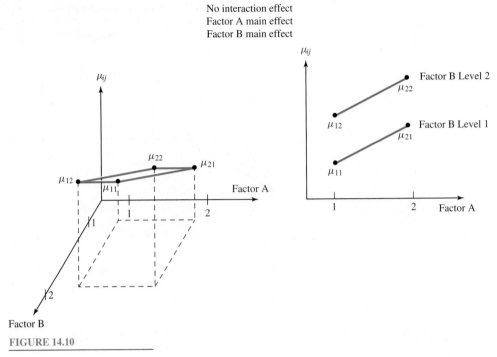

FIGURE 14.10

Interpretation of models for two-factor experiments

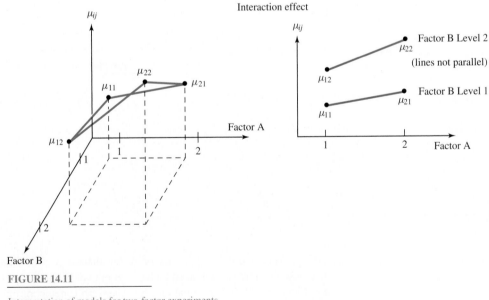

FIGURE 14.11

Interpretation of models for two-factor experiments

lines are not parallel. When the lines are parallel, the change in the cell means due to a change in the level of factor B is the same for both levels of factor A, because

$$\mu_{11} - \mu_{12} = \mu_{21} - \mu_{22}$$

Similarly,

$$\mu_{11} - \mu_{21} = \mu_{12} - \mu_{22}$$

so that the change in the cell means due to a change in the level of factor A is the same for both levels of factor B. This means that the changes in the expected value of the response variable resulting from changes in the levels of one factor are the same for all levels of the other factor. Hence the two factors influence the response variable in an independent manner. On the other hand, when there is an interaction effect and the lines are not parallel, it can be seen that

$$\mu_{11} - \mu_{12} \neq \mu_{21} - \mu_{22}$$

for example, so that the effect on the response variable due to a change in the level of factor B is different at the two levels of factor A.

Of course, in practice these plots of the cell means cannot be constructed because the cell means are unknown. Corresponding plots using the cell sample averages $\hat{\mu}_{ij} = \bar{x}_{ij\cdot}$ in place of the cell means can be made, but they are often difficult to interpret due to the error variability in the cell sample averages. Even if there is no interaction, the plots of the cell sample averages will generally have lines that are not parallel. However, hypothesis tests of the presence of interaction effects and factor main effects can be conducted, and they are discussed in Section 14.1.3.

The interpretation of interaction effects and factor main effects are illustrated in the following examples.

Example 72

Turf Quality

The interpretations of the various models including and excluding an interaction effect and factor main effects are shown in Figure 14.12. Suppose that fertilizer 1 produces on average higher quality turf than fertilizer 2 when a low growing temperature is used, but that the two fertilizers have identical effects when a high growing temperature is used. What model does this correspond to? Clearly this is an example of a model with an interaction effect, because

Fertilizer main effect	No	Yes	No	Yes	
Temperature main effect	No	No	Yes	Yes	
Interaction effect	No	No	No	No	Yes
	Quality of Turf				
Interpretation	Does not depend on either fertilizer or temperature.	Depends on fertilizer but not on temperature.	Depends on temperature but not on fertilizer.	Depends on both fertilizer and temperature. Independent influences.	Depends on both fertilizer and temperature. Influences not independent.

FIGURE 14.12

Model interpretation for the turf quality example

the effect of the level of factor A, fertilizer, depends on what level of factor B, growing temperature, is employed. If there is no interaction, then there is the same difference between the two fertilizers at each of the two growing temperatures.

Example 9
Car Body Assembly Line

In this example an interaction effect has the interpretation that the relative efficiencies of the four solder formulations are different for the three machines. This would appear to be an unlikely prospect because the three machines are supposed to be identical. If there is no interaction effect, then the presence of a factor A (machines) main effect indicates that the machines are actually operating differently, and the presence of a factor B (solder) main effect indicates that different solder formulations have different effects on the strength of the welding.

Example 74
Company Transportation Costs

Suppose that route 1 always takes on average 45 minutes longer to drive than route 2 regardless of the period of day when the truck leaves the factory. What model does this correspond to? Clearly there is no interaction effect, because the effect of factor A, the route taken, is the same for all levels of factor B, the period of day. There is also a factor A main effect because the average driving time is different for the two routes. Is there a factor B main effect? There may or may not be. If for each particular route driven the average driving times are the same for each period of day, then there is no factor B main effect. If these average driving times are different, then there is a factor B main effect.

The presence or absence of an interaction effect and main effects can be investigated mathematically by introducing parameters μ, α_i, β_j, and $(\alpha\beta)_{ij}$ so that the cell means μ_{ij} can be decomposed as

$$\mu_{ij} = \mu + \alpha_i + \beta_j + (\alpha\beta)_{ij}$$

From a technical point of view, in order for these parameters to be uniquely defined it is also necessary to impose the conditions

$$\sum_{i=1}^{a} \alpha_i = 0 \qquad \sum_{j=1}^{b} \beta_j = 0 \qquad \sum_{i=1}^{a} (\alpha\beta)_{ij} = 0 \qquad \sum_{j=1}^{b} (\alpha\beta)_{ij} = 0$$

The motivation behind this representation of the cell means is that it provides a clear indication of the presence or absence of the interaction and main effects. For example, if some of the parameters $(\alpha\beta)_{ij}$ are nonzero, then this indicates that there is an interaction effect between the two factors. On the other hand, there is no interaction effect between the two factors if all of the parameters $(\alpha\beta)_{ij}$ are equal to 0.

Consequently, the hypothesis testing problem

$$H_0 : (\alpha\beta)_{ij} = 0 \qquad 1 \le i \le a, 1 \le j \le b$$

versus

$$H_A : \text{some } (\alpha\beta)_{ij} \ne 0$$

can be used to assess the evidence that there is an interaction effect between the two factors. The null hypothesis corresponds to no interaction effect and the alternative hypothesis corresponds to the presence of an interaction effect. A large p-value leading to acceptance of the null hypothesis implies that it is plausible that there is no interaction effect, so that the two factors can be considered to act independently of each other. On the other hand, a small p-value leading to rejection of the null hypothesis establishes evidence of an interaction effect, so that it is known that the two factors do not influence the response variable in an independent manner.

If there is no interaction effect between the two factors, then the main effect of factor A can be investigated by considering the parameters α_i and the main effect of factor B can be investigated by considering the parameters β_j. If the parameters α_i are all equal to 0, then this implies that the expected value of the response variable does not depend upon which level of factor A is employed. In other words, the response variable does not depend on factor A. If some of the parameters α_i are nonzero, then the expected value of the response variable does depend upon which level of factor A is employed.

The hypothesis testing problem

$$H_0 : \alpha_i = 0 \quad 1 \leq i \leq a \qquad \text{versus} \qquad H_A : \text{some } \alpha_i \neq 0$$

can be used to assess the evidence that the response variable is influenced by factor A. The null hypothesis corresponds to the situation where the expected value of the response variable does not depend on the levels of factor A. Acceptance of the null hypothesis can therefore be interpreted as indicating that there is no evidence that the response variable depends on factor A. Rejection of the null hypothesis establishes evidence that the response variable does indeed depend on the levels of factor A.

Similarly, the hypothesis testing problem

$$H_0 : \beta_j = 0 \quad 1 \leq j \leq b \qquad \text{versus} \qquad H_A : \text{some } \beta_j \neq 0$$

can be used to assess the evidence that the response variable is influenced by factor B. Acceptance of the null hypothesis indicates that there is no evidence that the response variable depends on factor B, and rejection of the null hypothesis establishes evidence that the response variable does indeed depend on the levels of factor B.

In Section 14.1.3 it is shown how an analysis of variance approach can be used to test these hypotheses. First the null hypothesis of no interaction is tested, and if this hypothesis is accepted, then hypothesis tests of the factor A and factor B main effects can be performed. If the null hypothesis of no interaction is rejected, then the presence of an interaction effect has been established and the hypothesis tests of the factor A and factor B main effects are redundant and don't need to be considered.

Modeling Two-Factor Experiments

An observation x_{ijk} in a two-factor experiment is taken to consist of an unknown cell mean μ_{ij} together with an error term ϵ_{ijk} that is assumed to be an observation from a normal distribution with mean zero and unknown variance σ^2. The cell mean μ_{ij} represents the expected value of the response variable at that experimental configuration, and the structure of the cell means determines how the two factors affect the response variable. The presence of an interaction effect between the two factors indicates that the two factors do not operate independently of each other. If there is no interaction effect, then the presence or absence of main effects for each of the two factors indicates whether or not they influence the response variable.

14.1.3 Analysis of Variance Table

The analysis of variance table is based on a decomposition of the total sum of squares into sums of squares for the factor A and B main effects, the interaction effect, and an error sum of squares. Point estimates of the parameters μ, α_i, β_j, and $(\alpha\beta)_{ij}$ are first obtained in order to motivate the construction of these sums of squares.

Notice that

$$E(\bar{x}_{ij.}) = \frac{1}{n} \sum_{k=1}^{n} E(x_{ijk}) = \mu_{ij} = \mu + \alpha_i + \beta_j + (\alpha\beta)_{ij}$$

Also, the conditions on the summations of the parameters α_i, β_j, and $(\alpha\beta)_{ij}$ being equal to 0 implies that

$$E(\bar{x}_{i..}) = \frac{1}{bn} \sum_{j=1}^{b} \sum_{k=1}^{n} E(x_{ijk}) = \frac{1}{b} \sum_{j=1}^{b} \mu_{ij}$$

$$= \frac{1}{b} \sum_{j=1}^{b} (\mu + \alpha_i + \beta_j + (\alpha\beta)_{ij}) = \mu + \alpha_i$$

$$E(\bar{x}_{.j.}) = \frac{1}{an} \sum_{i=1}^{a} \sum_{k=1}^{n} E(x_{ijk}) = \frac{1}{a} \sum_{i=1}^{a} \mu_{ij}$$

$$= \frac{1}{a} \sum_{i=1}^{a} (\mu + \alpha_i + \beta_j + (\alpha\beta)_{ij}) = \mu + \beta_j$$

$$E(\bar{x}_{...}) = \frac{1}{abn} \sum_{i=1}^{a} \sum_{j=1}^{b} \sum_{k=1}^{n} E(x_{ijk}) = \frac{1}{ab} \sum_{i=1}^{a} \sum_{j=1}^{b} \mu_{ij}$$

$$= \frac{1}{ab} \sum_{i=1}^{a} \sum_{j=1}^{b} (\mu + \alpha_i + \beta_j + (\alpha\beta)_{ij}) = \mu$$

These expectations can be used to derive the following unbiased point estimates of the unknown parameters μ, α_i, β_j, and $(\alpha\beta)_{ij}$:

$$\hat{\mu} = \bar{x}_{...}$$

$$\hat{\alpha}_i = \bar{x}_{i..} - \hat{\mu} = \bar{x}_{i..} - \bar{x}_{...}$$

$$\hat{\beta}_j = \bar{x}_{.j.} - \hat{\mu} = \bar{x}_{.j.} - \bar{x}_{...}$$

$$(\hat{\alpha\beta})_{ij} = \bar{x}_{ij.} - \hat{\mu} - \hat{\alpha}_i - \hat{\beta}_j = \bar{x}_{ij.} - \bar{x}_{i..} - \bar{x}_{.j.} + \bar{x}_{...}$$

The sums of squares for factor main effects and for the interaction effect are based on the parameter estimates $\hat{\alpha}_i$, $\hat{\beta}_j$, and $(\hat{\alpha\beta})_{ij}$ as shown next.

Total Sum of Squares   The total sum of squares for a two-factor experiment is

$$\text{SST} = \sum_{i=1}^{a} \sum_{j=1}^{b} \sum_{k=1}^{n} (x_{ijk} - \bar{x}_{...})^2 = \sum_{i=1}^{a} \sum_{j=1}^{b} \sum_{k=1}^{n} x_{ijk}^2 - abn\bar{x}_{...}^2$$

which measures the total variability of all abn data observations about their overall mean value $\bar{x}_{...}$. As shown in Figure 14.13, this total sum of squares can be decomposed into a sum of squares for the factor A main effect SSA, a sum of squares for the factor B main effect SSB, a sum of squares for the interaction effect between factors A and B SSAB, and an error sum of squares SSE, so that

$$\text{SST} = \text{SSA} + \text{SSB} + \text{SSAB} + \text{SSE}$$

FIGURE 14.13

The sum of squares decomposition
for a two-factor experiment

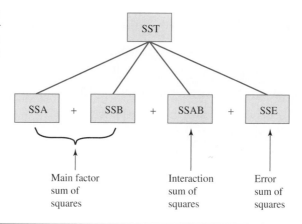

Sum of Squares for Factor A   The sum of squares for the factor A main effect SSA measures the presence or absence of a factor A main effect based on the estimated parameter values $\hat{\alpha}_i$. It is defined to be

$$\text{SSA} = bn \sum_{i=1}^{a} \hat{\alpha}_i^2 = bn \sum_{i=1}^{a} (\bar{x}_{i\cdot\cdot} - \bar{x}_{\cdots})^2 = bn \sum_{i=1}^{a} \bar{x}_{i\cdot\cdot}^2 - abn\bar{x}_{\cdots}^2$$

Sum of Squares for Factor B   The sum of squares for the factor B main effect SSB measures the presence or absence of a factor B main effect based on the estimated parameter values $\hat{\beta}_j$. It is defined to be

$$\text{SSB} = an \sum_{j=1}^{b} \hat{\beta}_j^2 = an \sum_{j=1}^{b} (\bar{x}_{\cdot j \cdot} - \bar{x}_{\cdots})^2 = an \sum_{j=1}^{b} \bar{x}_{\cdot j \cdot}^2 - abn\bar{x}_{\cdots}^2$$

Sum of Squares for Interaction   The sum of squares for the interaction effect between factors A and B SSAB measures the presence or absence of an interaction effect based on the estimated parameter values $(\hat{\alpha\beta})_{ij}$. It is defined to be

$$\text{SSAB} = n \sum_{i=1}^{a} \sum_{j=1}^{b} (\hat{\alpha\beta})_{ij}^2 = n \sum_{i=1}^{a} \sum_{j=1}^{b} (\bar{x}_{ij\cdot} - \bar{x}_{i\cdot\cdot} - \bar{x}_{\cdot j\cdot} + \bar{x}_{\cdots})^2$$

Error Sum of Squares   The sum of squares for error SSE measures the variability *within* each of the ab experimental configurations. It is defined to be

$$\text{SSE} = \sum_{i=1}^{a} \sum_{j=1}^{b} \sum_{k=1}^{n} (x_{ijk} - \bar{x}_{ij\cdot})^2 = \sum_{i=1}^{a} \sum_{j=1}^{b} \sum_{k=1}^{n} x_{ijk}^2 - n \sum_{i=1}^{a} \sum_{j=1}^{b} \bar{x}_{ij\cdot}^2$$

so that it is based on the variability of the data observations x_{ijk} about the cell sample averages $\bar{x}_{ij\cdot}$. It will be seen that the sum of squares for error is used to estimate the error variance σ^2.

These sum of squares are used to construct an analysis of variance table as shown in Figure 14.14. Notice that there are rows for the factor A main effect, the factor B main effect, the interaction effect (AB), and error, together with the total of these four components. Mean sums of squares are calculated from the sums of squares in the usual manner by dividing the

Source	Degrees of freedom	Sum of squares	Mean squares	F-statistics	p-value
Factor A	$a - 1$	SSA	$\text{MSA} = \frac{\text{SSA}}{a-1}$	$F_A = \frac{\text{MSA}}{\text{MSE}}$	$P(F_{a-1, ab(n-1)} > F_A)$
Factor B	$b - 1$	SSB	$\text{MSB} = \frac{\text{SSB}}{b-1}$	$F_B = \frac{\text{MSB}}{\text{MSE}}$	$P(F_{b-1, ab(n-1)} > F_B)$
AB interaction	$(a-1)(b-1)$	SSAB	$\text{MSAB} = \frac{\text{SSAB}}{(a-1)(b-1)}$	$F_{AB} = \frac{\text{MSAB}}{\text{MSE}}$	$P(F_{(a-1)(b-1), ab(n-1)} > F_{AB})$
Error	$ab(n-1)$	SSE	$\text{MSE} = \frac{\text{SSE}}{ab(n-1)}$		
Total	$abn - 1$	SST			

FIGURE 14.14

Analysis of variance table for a two-factor experiment

sums of squares by their respective degrees of freedom. For this two-factor experiment the degrees of freedom are

degrees of freedom for factor A main effect $= a - 1$

degrees of freedom for factor B main effect $= b - 1$

degrees of freedom for AB interaction effect $= (a - 1)(b - 1)$

degrees of freedom for error $= ab(n - 1)$

The total degrees of freedom are $abn - 1$, one less than the total number of data observations.

As in the analysis of variance tables discussed in previous chapters, the mean square error MSE is used as the estimate $\hat{\sigma}^2$ of the error variance. It has the scaled chi-square distribution

$$\hat{\sigma}^2 = \text{MSE} = \frac{\text{SSE}}{ab(n-1)} \sim \sigma^2 \frac{\chi^2_{ab(n-1)}}{ab(n-1)}$$

and is an unbiased estimate of the error variance

$$E(\text{MSE}) = \sigma^2$$

The other mean sums of squares have expectations

$$E(\text{MSA}) = \sigma^2 + \frac{bn}{a-1} \sum_{i=1}^{a} \alpha_i^2$$

$$E(\text{MSB}) = \sigma^2 + \frac{an}{b-1} \sum_{j=1}^{b} \beta_j^2$$

$$E(\text{MSAB}) = \sigma^2 + \frac{n}{(a-1)(b-1)} \sum_{i=1}^{a} \sum_{j=1}^{b} (\alpha\beta)_{ij}^2$$

so that their sizes relative to the mean square error MSE indicate whether or not the parameters α_i, β_j, and $(\alpha\beta)_{ij}$ can be taken to be 0. Three different F-statistics can be calculated, which can be used to test for the presence of an interaction effect and for the two-factor main effects.

Testing for Interaction Effects   The hypothesis testing problem

$$H_0 : (\alpha\beta)_{ij} = 0 \qquad 1 \leq i \leq a, 1 \leq j \leq b$$

versus

$$H_A : \text{some } (\alpha\beta)_{ij} \neq 0$$

is used to assess the evidence of an interaction effect and can be performed with the F-statistic

$$F_{AB} = \frac{\text{MSAB}}{\text{MSE}}$$

When the null hypothesis is true and there is no interaction effect, the statistic F_{AB} has an F-distribution with degrees of freedom $(a-1)(b-1)$ and $ab(n-1)$. Large values of the statistic F_{AB} suggest that the null hypothesis is not true, and a p-value can be calculated as

$$p\text{-value} = P(X > F_{AB})$$

where the random variable X has an F-distribution with degrees of freedom $(a-1)(b-1)$ and $ab(n-1)$.

If a small p-value is obtained in the test for an interaction effect, then the experimenter has established that the two factors do not operate independently of each other in the manner in which they influence the response variable. If a large p-value is obtained, however, so that there is no evidence of an interaction effect, then it is appropriate to perform hypothesis tests for the presence of main effects of the two factors.

Testing for Factor A Main Effects   The hypothesis testing problem

$$H_0 : \alpha_i = 0 \quad 1 \leq i \leq a \qquad \text{versus} \qquad H_A : \text{some } \alpha_i \neq 0$$

is used to assess the evidence of a factor A main effect and can be performed with the F-statistic

$$F_A = \frac{\text{MSA}}{\text{MSE}}$$

When the null hypothesis is true and there is no factor A main effect, the statistic F_A has an F-distribution with degrees of freedom $a-1$ and $ab(n-1)$. Large values of the statistic F_A suggest that the null hypothesis is not true, and a p-value can be calculated as

$$p\text{-value} = P(X > F_A)$$

where the random variable X has an F-distribution with degrees of freedom $a-1$ and $ab(n-1)$.

Testing for Factor B Main Effects   The hypothesis testing problem

$$H_0 : \beta_j = 0 \quad 1 \leq j \leq b \qquad \text{versus} \qquad H_A : \text{some } \beta_j \neq 0$$

is used to assess the evidence of a factor B main effect and can be performed with the F-statistic

$$F_B = \frac{\text{MSB}}{\text{MSE}}$$

When the null hypothesis is true and there is no factor B main effect, the statistic F_B has an F-distribution with degrees of freedom $b-1$ and $ab(n-1)$. Large values of the statistic F_B suggest that the null hypothesis is not true, and a p-value can be calculated as

$$p\text{-value} = P(X > F_B)$$

where the random variable X has an F-distribution with degrees of freedom $b-1$ and $ab(n-1)$.

COMPUTER NOTE

Computer packages can be used to construct the analysis of variance table for a two-factor experiment, and the p-values will be reported for each of the three hypothesis tests described above. However, the experimenter should remember to look first at the p-value for the hypothesis test concerning the interaction effect. If there is evidence of an interaction effect, then the p-values for the hypothesis tests of the factor A and factor B main effects are really not important.

FIGURE 14.15

The analysis of variance table for
the turf quality example

Source	Degrees of freedom	Sum of squares	Mean squares	F-statistic	p-value
Fertilizer	1	1870.6	1870.6	66.86	0.000
Temperature	1	3220.6	3220.6	115.11	0.000
Fertilizer*Temperature	1	333.1	333.1	11.90	0.005
Error	12	335.8	28.0		
Total	15	5759.9			

Example 72

Turf Quality

The analysis of variance table for this two-factor problem is shown in Figure 14.15. The overall average of the data observations is

$$\bar{x}_{...} = \frac{x_{111} + \cdots + x_{224}}{16} = \frac{40 + \cdots + 79}{16} = 49.4375$$

and

$$\sum_{i=1}^{2}\sum_{j=1}^{2}\sum_{k=1}^{4} x_{ijk}^2 = 40^2 + \cdots + 79^2 = 44,865$$

so that

$$SST = \sum_{i=1}^{2}\sum_{j=1}^{2}\sum_{k=1}^{4} x_{ijk}^2 - (2 \times 2 \times 4 \times \bar{x}_{...}^2)$$

$$= 44,865 - (16 \times 49.4375^2) = 5,759.9375$$

The cell sample averages are

$$\bar{x}_{11.} = 29.00, \quad \bar{x}_{12.} = 48.25, \quad \bar{x}_{21.} = 41.50, \quad \text{and} \quad \bar{x}_{22.} = 79.00$$

so that

$$\sum_{i=1}^{2}\sum_{j=1}^{2} \bar{x}_{ij.}^2 = 29.00^2 + 48.25^2 + 41.50^2 + 79.00^2 = 11,132.3125$$

Consequently, the sum of squares for error is

$$SSE = \sum_{i=1}^{2}\sum_{j=1}^{2}\sum_{k=1}^{4} x_{ijk}^2 - (4 \times \sum_{i=1}^{2}\sum_{j=1}^{2} \bar{x}_{ij.}^2)$$

$$= 44,865 - (4 \times 11,132.3125) = 335.75$$

With fertilizer level sample averages

$$\bar{x}_{1..} = 38.625 \quad \text{and} \quad \bar{x}_{2..} = 60.250$$

the sum of squares for the fertilizer main effect is

$$SSA = (2 \times 4 \times \sum_{i=1}^{2} \bar{x}_{i..}^2) - (2 \times 2 \times 4 \times \bar{x}_{...}^2)$$

$$= (2 \times 4 \times (38.625^2 + 60.250^2)) - (16 \times 49.4375^2) = 1870.5625$$

Similarly, the growing temperature level sample averages are

$$\bar{x}_{\cdot 1 \cdot} = 35.250 \quad \text{and} \quad \bar{x}_{\cdot 2 \cdot} = 63.625$$

so that the sum of squares for the growing temperature main effect is

$$SSB = (2 \times 4 \times \sum_{j=1}^{2} \bar{x}_{\cdot j \cdot}^2) - (2 \times 2 \times 4 \times \bar{x}_{\cdots}^2)$$

$$= (2 \times 4 \times (35.250^2 + 63.625^2)) - (16 \times 49.4375^2) = 3220.5625$$

The sum of squares for interaction SSAB can be calculated from its formula in terms of the sample averages, but it is now easiest to calculate it by subtracting the sums of squares already calculated from the total sum of squares, so that

$$SSAB = SST - SSA - SSB - SSE$$

$$= 5759.9375 - 1870.5625 - 3220.5625 - 335.75 = 333.0625$$

These sums of squares are seen to agree with the (rounded) values given in Figure 14.15, where the row in the analysis of variance table labeled "Fertizer*Temperature" corresponds to the interaction.

Notice that there are $ab(n - 1) = 2 \times 2 \times (4 - 1) = 12$ degrees of freedom for error so that the estimate of the error variance is

$$\hat{\sigma}^2 = MSE = \frac{SSE}{12} = \frac{335.75}{12} = 27.98$$

The estimate of the error standard deviation is therefore $\hat{\sigma} = \sqrt{27.98} = 5.29$.

Also, there is $(a - 1)(b - 1) = (2 - 1) \times (2 - 1) = 1$ degree of freedom for the AB interaction effect, so that

$$MSAB = \frac{SSAB}{1} = SSAB = 333.1$$

The F-statistic for testing the presence of an interaction effect is therefore

$$F_{AB} = \frac{MSAB}{MSE} = \frac{333.1}{27.98} = 11.90$$

and the p-value for the null hypothesis that there is no interaction effect is

$$p\text{-value} = P(X > 11.90) = 0.005$$

where the random variable X has an F-distribution with degrees of freedom 1 and 12. This small p-value indicates that there is evidence of an interaction effect between fertilizer and growing temperature. The presence of an interaction effect implies that it is not important to investigate the main effects of the fertilizer and the growing temperature.

The manager of the turf growing company has discovered from this experiment that the difference between the two fertilizers depends on which growing temperature is used. The plot of the cell sample averages shown in Figure 14.16 shows that fertilizer 2 always appears to produce higher quality turf than fertilizer 1 regardless of the growing temperature, but that while the increase in average quality is only about 10 units at the low growing temperature, the increase is about 30 units at the high growing temperature. This is the interaction effect that has been established. Notice also that of the four experimental configurations considered, the optimum configuration employs fertilizer 2 at the high growing temperature, where the average quality score of the turf is about 80.

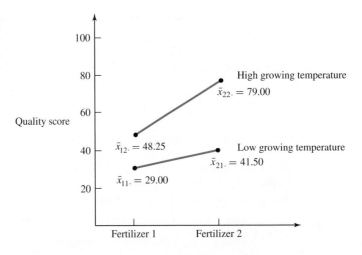

Source	Degrees of freedom	Sum of squares	Mean squares	F-statistic	p-value
Machine	2	1.81455	0.90728	0.98	0.3882
Solder	3	58.84646	19.61549	21.29	0.0001
Machine*Solder	6	3.95829	0.65972	0.72	0.6404
Error	24	22.11420	0.92142		
Total	35	86.73350			

Example 9

Car Body Assembly Line

Figure 14.17 contains the analysis of variance table for the two-factor experiment investigating welding strength. There are $ab(n - 1) = 3 \times 4 \times (3 - 1) = 24$ degrees of freedom for error and the mean square error is

$$\hat{\sigma}^2 = \text{MSE} = 0.92142$$

so that $\hat{\sigma} = \sqrt{0.92142} = 0.960$.

There are $(a - 1)(b - 1) = (3 - 1) \times (4 - 1) = 6$ degrees of freedom for interaction, and the mean sum of squares for interaction is

$$\text{MSAB} = 0.65972$$

The F-statistic for testing the presence of an interaction effect is therefore

$$F_{AB} = \frac{\text{MSAB}}{\text{MSE}} = \frac{0.65972}{0.92142} = 0.72$$

and the p-value is

$$p\text{-value} = P(X > 0.72) = 0.6404$$

where the random variable X has an F-distribution with degrees of freedom 6 and 24.

This large p-value indicates that there is no evidence of an interaction effect between the machines and the solder formulations. In other words, any difference between the solder formulations is the same for all three machines. Differences among the machines and among the solder formulations can now be investigated by considering the machine and solder main effects.

There are $a - 1 = 2$ degrees of freedom for the machine main effects, and the F-statistic for testing machine main effects is

$$F_A = \frac{\text{MSA}}{\text{MSE}} = \frac{0.90728}{0.92142} = 0.98$$

The corresponding p-value is

$$p\text{-value} = P(X > 0.98) = 0.3882$$

where the random variable X has an F-distribution with degrees of freedom 2 and 24. Again, this large p-value indicates that there is no evidence of a machine main effect, or in other words, there is no evidence that there is any difference between the three machines. This perhaps not unexpected result informs the experimenter that there is no evidence that the three machines are not operating in an identical fashion with respect to welding strength.

Finally, there are $b - 1 = 3$ degrees of freedom for the solder main effects, and the F-statistic for testing solder main effects is

$$F_B = \frac{\text{MSB}}{\text{MSE}} = \frac{19.61549}{0.92142} = 21.29$$

The corresponding p-value is

$$p\text{-value} = P(X > 21.29) = 0.0001$$

where the random variable X has an F-distribution with degrees of freedom 3 and 24. This very small p-value indicates that the null hypothesis that there are no solder main effects is not plausible, and so the experiment has provided evidence that the different solder formulations do indeed have different effects on the average welding strength.

In conclusion, this experiment has shown that while the three machines appear to be operating identically, the strength of the welds does depend upon which solder formulation is employed. An analysis of how the different solder formulations affect the welding strength should now be undertaken, and this is discussed in Section 14.1.4.

Example 74

Company Transportation Costs

The analysis of variance table for the two-factor experiment concerning driving times is shown in Figure 14.18. The p-value for the interaction effect is seen to be 0.430, which implies that there is no evidence of an interaction effect between the route taken and the period of day. However, the route main effect is seen to have a p-value of 0.004 and the period of day main effect is seen to have a p-value of 0.000, so that there is evidence that the driving time depends on both the route taken and the period of day. Nevertheless, the fact that there is no evidence of an interaction effect implies that these two factors influence driving time in an independent manner. For example, the difference between the average driving times for the two routes is the same regardless of the period of day.

The plot of the cell sample means in Figure 14.19 shows that route 1 tends to be longer than route 2 by about 20–25 minutes, and that the later in the day that the truck leaves the factory,

FIGURE 14.18

The analysis of variance table for the company transportation costs example

Source	Degrees of freedom	Sum of squares	Mean squares	F-statistic	p-value
Route	1	3830.7	3830.7	10.251	0.004
Period	2	11925.6	5962.8	15.956	0.000
Route*Period	2	653.6	326.8	0.874	0.430
Error	24	8968.8	373.7		
Total	29	25378.7			

FIGURE 14.19

A plot of the cell sample averages
for the company transportation
costs example

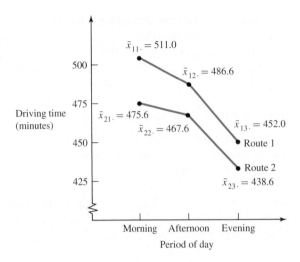

the shorter the driving time will be. (Notice that because there is no evidence of an interaction effect a plot of the *cell means* μ_{ij} could plausibly exhibit parallel lines, even though in the plot of the *cell sample averages* $\bar{x}_{ij.}$ in Figure 14.19 the lines are not quite parallel.) Analyses should be performed to compare the two different routes with each other and to compare the three different periods of day with each other, and these are discussed in Section 14.1.4.

14.1.4 Pairwise Comparisons of the Factor Level Means

If the analysis of a two-factor experiment reveals that there is no evidence of an interaction effect but that there is evidence of one or more factor main effects, then the experimenter can make pairwise comparisons of the factor level means to determine how the response variable depends upon the different levels of the factor or factors that have been found to be influential. Confidence intervals can be constructed for all of the pairwise comparisons of the factor level means in a fashion similar to that discussed in Section 11.1.4 for the comparisons of the treatment means in a one-factor experiment and in Section 11.2.4 for the comparisons of the treatment means in a randomized block experiment.

If there is evidence of a factor A main effect, then the effect of the a levels of factor A on the expected value of the response variable should be investigated. The expected value of the response variable at the a levels of factor A can be thought of as being represented by

$$\mu + \alpha_1, \ldots, \mu + \alpha_a$$

of which there are $a(a-1)/2$ pairwise comparisons

$$\alpha_{i_1} - \alpha_{i_2}$$

corresponding to the differences in the expectation of the response variable between levels i_1 and i_2 of factor A.

Using the critical point $q_{\alpha,a,\nu}$ given in Table V, a set of $1 - \alpha$ confidence level simultaneous confidence intervals for these pairwise differences are constructed as

$$\alpha_{i_1} - \alpha_{i_2} \in \left(\bar{x}_{i_1..} - \bar{x}_{i_2..} - \frac{s\, q_{\alpha,a,\nu}}{\sqrt{bn}}, \bar{x}_{i_1..} - \bar{x}_{i_2..} + \frac{s\, q_{\alpha,a,\nu}}{\sqrt{bn}} \right)$$

where $s = \hat{\sigma} = \sqrt{\text{MSE}}$ and the degrees of freedom for the critical point are the error degrees of freedom $\nu = ab(n-1)$.

If a confidence interval contains 0, then there is no evidence that the expected value of the response variable is different at the two corresponding levels of factor A. However, a confidence interval that does not contain 0 indicates how the expected value of the response variable differs between the two corresponding levels of factor A.

Notice that if these confidence intervals are constructed in a situation where there is no evidence of a factor A main effect, then each of the confidence intervals will contain 0 and there is no evidence that the expected value of the response variable is any different at any of the different levels of factor A. Also, if there is an interaction effect, then these confidence intervals are not interpretable since the different effects of the levels of factor A depend upon the level of factor B that is employed. Finally, it is worthwhile to point out that if factor A has only two levels so that $a = 2$, then the confidence interval constructed for the pairwise difference $\alpha_1 - \alpha_2$ is equivalent to the confidence interval constructed from a two-sample t-procedure with a pooled variance estimate discussed in Section 9.3.2. If there is evidence of a factor B main effect, then confidence intervals for the pairwise differences of the expected values of the response variable at the b different levels of factor B can be constructed in a similar manner as shown in the accompanying box.

The confidence intervals can be used to provide an indication of the sensitivity afforded by different sample sizes. Notice that the lengths of the confidence intervals are inversely proportional to the square root of the number of replications n for each experimental configuration. An experimenter may use the lengths of these confidence intervals to assess how large a sample size is required to achieve a certain amount of precision for comparing different factor levels.

Pairwise Comparisons of the Factor Level Means

If there is no evidence of an interaction effect in a two-factor experiment, then the confidence intervals

$$\alpha_{i_1} - \alpha_{i_2} \in \left(\bar{x}_{i_1\cdots} - \bar{x}_{i_2\cdots} - \frac{s\, q_{\alpha,a,\nu}}{\sqrt{bn}}, \bar{x}_{i_1\cdots} - \bar{x}_{i_2\cdots} + \frac{s\, q_{\alpha,a,\nu}}{\sqrt{bn}} \right)$$

can be used to investigate the interpretation of a factor A main effect, and the confidence intervals

$$\beta_{j_1} - \beta_{j_2} \in \left(\bar{x}_{\cdot j_1\cdot} - \bar{x}_{\cdot j_2\cdot} - \frac{s\, q_{\alpha,b,\nu}}{\sqrt{an}}, \bar{x}_{\cdot j_1\cdot} - \bar{x}_{\cdot bj_2\cdot} + \frac{s\, q_{\alpha,b,\nu}}{\sqrt{an}} \right)$$

can be used to investigate the interpretation of a factor B main effect. Each set of confidence intervals has an overall confidence level of $1 - \alpha$ and the critical point has $\nu = ab(n-1)$ degrees of freedom. Confidence intervals that contain 0 indicate that there is no evidence that the expected value of the response variable is different at the two corresponding levels of the factor, while confidence intervals that do not contain 0 indicate how the expected value of the response variable differs between the two corresponding levels of the factor.

Example 9
Car Body Assembly Line

In this experiment no evidence of an interaction effect between machines and solder formulations was found, but the presence of a solder effect was established. Pairwise comparisons of the $b = 4$ different solder formulations can be performed to identify how they affect the welding strength.

FIGURE 14.20

Comparisons of the solders for the
car body assembly line example

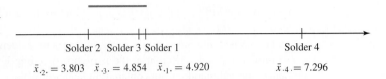

Solder 2   Solder 3 Solder 1   Solder 4

$\bar{x}_{.2.} = 3.803$ $\bar{x}_{.3.} = 4.854$ $\bar{x}_{.1.} = 4.920$ $\bar{x}_{.4.} = 7.296$

The sample averages for the four different solder formulations are

$$\bar{x}_{.1.} = 4.920, \quad \bar{x}_{.2.} = 3.803, \quad \bar{x}_{.3.} = 4.854, \quad \text{and} \quad \bar{x}_{.4.} = 7.296$$

Also, $q_{0.05,4,24} = 3.90$, so that

$$\frac{s\, q_{0.05,4,24}}{\sqrt{an}} = \frac{0.960 \times 3.90}{\sqrt{3 \times 3}} = 1.248$$

The confidence intervals for the pairwise differences of the solder effects are therefore

$$\beta_1 - \beta_2 \in (4.920 - 3.803 - 1.248, 4.920 - 3.803 + 1.248) = (-0.131, 2.365)$$
$$\beta_1 - \beta_3 \in (4.920 - 4.854 - 1.248, 4.920 - 4.854 + 1.248) = (-1.182, 1.314)$$
$$\beta_1 - \beta_4 \in (4.920 - 7.296 - 1.248, 4.920 - 7.296 + 1.248) = (-3.624, -1.128)$$
$$\beta_2 - \beta_3 \in (3.803 - 4.854 - 1.248, 3.803 - 4.854 + 1.248) = (-2.299, 0.197)$$
$$\beta_2 - \beta_4 \in (3.803 - 7.296 - 1.248, 3.803 - 7.296 + 1.248) = (-4.741, -2.245)$$
$$\beta_3 - \beta_4 \in (4.854 - 7.296 - 1.248, 4.854 - 7.296 + 1.248) = (-3.690, -1.194)$$

The three confidence intervals comparing solder formulations 1, 2, and 3 all contain 0 so there is no evidence of any difference among these three solders. However, none of the confidence intervals comparing solder formulation 4 with each of the other three solder formulations contains 0, so solder 4 has been established to have a different effect on welding strength than any of the other three solder formulations. This situation is shown graphically in Figure 14.20.

In summary, the experimenter can conclude that solder formulations 1, 2, and 3 appear to have the same effects on welding strength, but that solder formulation 4 has been shown to provide stronger welds than each of these other three solder formulations, with an increase in average strength of at least 1.1 units.

Example 74
Company
Transportation Costs

In this example there is no evidence of an interaction effect, but both route and period of day have been shown to influence the driving time, and so they should both be investigated further.

The sample averages for the two different routes are

$$\bar{x}_{1..} = 483.2 \quad \text{and} \quad \bar{x}_{2..} = 460.6$$

With $q_{0.05,2,24} = 2.92$, so that

$$\frac{s\, q_{0.05,2,24}}{\sqrt{bn}} = \frac{19.33 \times 2.92}{\sqrt{3 \times 5}} = 14.6$$

the confidence interval for the difference in the expected value of the driving time between the two routes is therefore

$$\alpha_1 - \alpha_2 \in (483.2 - 460.6 - 14.6, 483.2 - 460.6 + 14.6) = (8.0, 37.2)$$

Consequently, this simple experiment has demonstrated that route 1 takes on average at least 8 minutes longer to drive than route 2, and possibly as much as 37 minutes longer on average. Since there is no evidence of an interaction effect, remember that this result can be taken to hold regardless of the time of day that the truck leaves the factory.

If the morning is taken to be level 1, the afternoon to be level 2, and the evening to be level 3, the sample averages for the three different periods of day are

$$\bar{x}_{\cdot 1 \cdot} = 493.3, \quad \bar{x}_{\cdot 2 \cdot} = 477.1, \quad \text{and} \quad \bar{x}_{\cdot 3 \cdot} = 445.3$$

With $q_{0.05, 3, 24} = 3.53$, so that

$$\frac{s \, q_{0.05, 3, 24}}{\sqrt{an}} = \frac{19.33 \times 3.53}{\sqrt{2 \times 5}} = 21.6$$

the confidence intervals for the pairwise differences of the different periods of day are

$$\beta_1 - \beta_2 \in (493.3 - 477.1 - 21.6, 493.3 - 477.1 + 21.6) = (-5.4, 37.8)$$
$$\beta_1 - \beta_3 \in (493.3 - 445.3 - 21.6, 493.3 - 445.3 + 21.6) = (26.4, 69.6)$$
$$\beta_2 - \beta_3 \in (477.1 - 445.3 - 21.6, 477.1 - 445.3 + 21.6) = (10.2, 53.4)$$

This situation is summarized in Figure 14.21, and it is seen that while there is no evidence that the expected driving time is any different between the morning and afternoon periods, the experiment has established that trucks leaving in the evening have shorter expected driving times. The decrease in the expected driving time is seen to be at least 10 minutes between the evening and afternoon periods (and possibly as much as 53 minutes) and to be at least 26 minutes between the evening and morning periods (and possibly as much as 69 minutes).

14.1.5 Modeling Procedures and Residual Analysis

Residual analysis can be employed in the usual way to check on the model assumptions and specifically to investigate whether either of the factors influences the *variability* of the response variable. The construction of the residuals depends on the choice of which model is appropriate, and some general comments on the modeling procedure are given below.

Fitting Different Models Recall that for a two-factor experiment the model

$$\mu_{ij} = \mu + \alpha_i + \beta_j + (\alpha\beta)_{ij}$$

has been employed for the cell means μ_{ij}. This model allows for the two-factor main effects together with an interaction effect. If the experimenter finds evidence of an interaction effect, then this is the appropriate model for the data set under consideration.

If the experimenter finds no evidence of an interaction effect but finds evidence of both factor main effects, then this implies that the appropriate model for the data set under consideration is

$$\mu_{ij} = \mu + \alpha_i + \beta_j$$

Notice that this model does not include the interaction parameters $(\alpha\beta)_{ij}$.

Similarly, if the experimenter finds no evidence of an interaction effect or of a factor B main effect but finds evidence of a factor A main effect, then this implies that the appropriate model for the data set under consideration is

$$\mu_{ij} = \mu + \alpha_i$$

If there is a factor B main effect but no factor A main effect or interaction effect, then the appropriate model for the data set under consideration is

$$\mu_{ij} = \mu + \beta_j$$

If there is no evidence of an interaction effect or either of the two-factor main effects, then the appropriate model for the data set under consideration is

$$\mu_{ij} = \mu$$

which just indicates that the response variable has the same expected value in each of the experimental configurations.

COMPUTER NOTE   You will find that on your computer package you have the option to fit whichever model you wish to the data set. In fact it is important to make sure that you initially do fit a model with an interaction term (and not just with the two-factor main effects) so that you can investigate whether or not there is an interaction effect.

Experiments with One Observation Per Cell   Experiments with only one observation per cell are a special case of two-factor experiments and need to be handled a little differently. In this case $n = 1$ and there are no replicate observations at each experimental configuration. Notice that in this case the usual analysis would have $ab(n - 1) = 0$ degrees of freedom for error, and in fact the error sum of squares SSE would also be equal to 0.

With only one observation per cell it is not possible to investigate the presence or absence of an interaction effect, and an analysis of the factor main effects must be based on the *assumption* that there is no interaction effect. If this assumption is made, then the analysis of the factor main effects can proceed by taking the interaction sum of squares SSAB as the error sum of squares SSE with the corresponding degrees of freedom $(a - 1)(b - 1)$.

COMPUTER NOTE   You can analyze a two-factor experiment with only one observation per cell on your computer package by specifying a model with main effects but without an interaction effect, so that the model

$$\mu_{ij} = \mu + \alpha_i + \beta_j$$

is employed.

It should be clear that an experiment with only one observation per cell is a poorly designed one (although in some cases it may not be possible or practical to obtain replicates) and that replicate observations at each experimental configuration should be obtained whenever possible. The replications at the experimental configurations allow the investigation of the interaction effect between the two factors, which is usually of considerable interest.

Finally, notice that a two-factor experiment with only one observation per cell is similar to the randomized block experiments discussed in Section 11.2. In fact the data layout looks the same with the k factor levels in the randomized block experiment corresponding to the a levels of factor A in the two-factor experiment, and the b blocks in the randomized block experiment corresponding to the b levels of factor B in the two-factor experiment. Also, the analysis of variance tables are similar once the interaction sum of squares SSAB is used as the error sum of squares SSE in the two-factor experiment.

Nevertheless, the two experimental designs have different interpretations. In the randomized block experiment there is one factor of interest and a blocking variable is incorporated into the experiment to increase the precision with which this factor can be examined. In the two-factor experiment the objective is to investigate the influence that the two factors have on a response variable, although in this case there are unfortunately no replicate observations.

Unbalanced Experimental Designs   The analysis of a two-factor experiment described in this section considers the situation in which there are exactly n replicate observations for each of the ab experimental configurations. This case is known as a **balanced design**. In many experiments an unbalanced design occurs where there are not the same number of replications at each experimental configuration. This design may be intended by the experimenter or it may be as the result of some missing data observations in an experiment that was originally intended to be balanced.

The concepts behind the analysis of an unbalanced two-factor experiment are exactly the same as those for a balanced two-factor experiment. The interaction effect and the factor main effects are investigated in the same way with the analysis of variance table and their interpretations are the same.

COMPUTER NOTE

The mathematical analysis of an unbalanced experiment is more complicated than the analysis of a balanced experiment, and you may have to treat it differently on your computer package. For example, you may have to analyze it as a "linear model" rather than as a "two-factor analysis of variance."

Residual Analysis   The residuals for a particular model are calculated as

$$e_{ijk} = x_{ijk} - \hat{\mu}_{ij}$$

where the values $\hat{\mu}_{ij}$ are the fitted cell values for the model under consideration. Thus, the residuals are the differences between the data observations x_{ijk} and the fitted values under the model. You should find that the residuals can be calculated by your computer package upon request.

Residuals with a particularly large absolute value identify data observations that do not fit the model particularly well. The experimenter may want to investigate these outliers in more detail to see if there is an explanation for their behavior. The size of the residuals can be judged by dividing them by the estimated standard deviation $\hat{\sigma}$ to produce standardized residuals. The standardized residuals that are larger than 3 in absolute value are often considered to correspond to possible outliers.

A normal probability plot of the residuals (as discussed in Section 12.7.1) can be made to assess whether there is any indication that the residuals are not normally distributed. Evidence that the residuals are not normally distributed should cause the experimenter to be cautious about the analysis of the data set. In this case a transformation of the response variable may result in a model whose assumptions are better satisfied.

Finally, the residuals can be plotted against the factors in the experiment. These plots reveal whether there is any evidence that the *variability* in the response variable depends upon the different levels of the factor, as illustrated in Figure 14.22. This may be the case even though there is no interaction effect and there is no factor main effect so that the expected value of the response variable is unaffected by the different levels of the factor. The assessment of the effect of a factor on the variability of the response variable is often as important as the

FIGURE 14.22

A greater spread in the residual values at level 2 indicates that there is more variability in the response variable at level 2

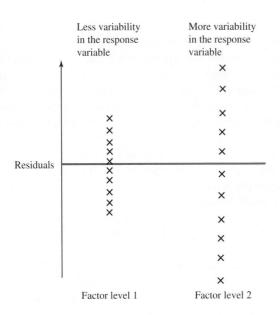

FIGURE 14.23

Residuals for the turf quality data set

Fertilizer

		Fertilizer 1	Fertilizer 2
Temperature	Low	$e_{111} = x_{111} - \bar{x}_{11.} = 11.00$ $e_{112} = x_{112} - \bar{x}_{11.} = -5.00$ $e_{113} = x_{113} - \bar{x}_{11.} = 2.00$ $e_{114} = x_{114} - \bar{x}_{11.} = -8.00$	$e_{211} = x_{211} - \bar{x}_{21.} = 5.50$ $e_{212} = x_{212} - \bar{x}_{21.} = -4.50$ $e_{213} = x_{213} - \bar{x}_{21.} = -0.50$ $e_{214} = x_{214} - \bar{x}_{21.} = -0.50$
	High	$e_{121} = x_{121} - \bar{x}_{12.} = -2.25$ $e_{122} = x_{122} - \bar{x}_{12.} = 4.75$ $e_{123} = x_{123} - \bar{x}_{12.} = -0.25$ $e_{124} = x_{124} - \bar{x}_{12.} = -2.25$	$e_{221} = x_{221} - \bar{x}_{22.} = 5.00$ $e_{222} = x_{222} - \bar{x}_{22.} = -3.00$ $e_{223} = x_{223} - \bar{x}_{22.} = -2.00$ $e_{224} = x_{224} - \bar{x}_{22.} = 0.00$

assessment of the effect of a factor on the expected value of the response variable, since it is generally desirable to employ levels of a factor that reduce the variability in the response variable.

Example 72
Turf Quality

The appropriate model for this data set has been found to contain an interaction effect and the fitted values are therefore the cell sample averages, so that

$$e_{ijk} = x_{ijk} - \hat{\mu}_{ij} = x_{ijk} - \bar{x}_{ij.}$$

These residuals are shown in Figure 14.23. Notice that the residual with the largest absolute value is

$$e_{111} = x_{111} - \bar{x}_{11.} = 40 - 29 = 11$$

but that with $\hat{\sigma} = 5.29$ this is not unusually large. The normal probability plot is shown in Figure 14.24, and the fairly linear relationship exhibited does not raise any alarms about the assumption of normality.

FIGURE 14.24

The normal probability plot of the residuals for the turf quality example

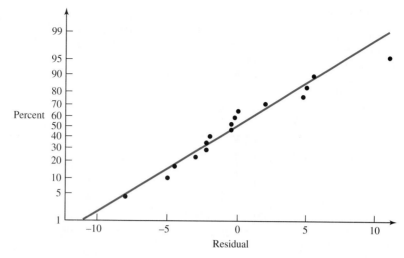

FIGURE 14.24

The normal probability plot of the residuals for the turf quality example

Example 9
Car Body Assembly Line

The analysis of this experiment did not reveal any evidence that the expected value of the welding strength is any different for the three machines on the factory floor. Is the *variability* of the welding strengths any different for the three machines? This is also an important question and can be addressed by making residual plots for each of the three machines.

The appropriate model for this data set has been found to be one that involves only main effects for the solder formulations, and under this model the fitted cell values are the solder level sample averages $\bar{x}_{.j.}$. The residuals are therefore calculated as

$$e_{ijk} = x_{ijk} - \hat{\mu}_{ij} = x_{ijk} - \bar{x}_{.j.}$$

For example, the residual for the first data observation is

$$e_{111} = x_{111} - \bar{x}_{.1.} = 4.120 - 4.920 = -0.800$$

The residual with the largest absolute value turns out to be

$$e_{233} = x_{233} - \bar{x}_{.3.} = 6.750 - 4.854 = 1.896$$

However, with $\hat{\sigma} = 0.960$ this is not unusually large.

Figure 14.25 shows a plot of the residuals for each of the three machines. There is no obvious difference in the amount of scatter of the three sets of residuals and so there is no evidence that the variability of the welding strengths is any different for the three machines. In addition, Figure 14.26 shows a plot of the residuals for the four different solder formulations. Solder formulation 4 (which provides the strongest welds) does not appear to produce a variability in weld strength that is markedly different from the other three solder formulations.

■ **14.1.6 Problems**

14.1.1 A two-factor experiment is conducted to compare different mixes of gasoline in different types of car. The response variable measures the driving characteristics of the car based on acceleration and other properties relating to the gasoline. Larger values of the response variable relate to improved driving characteristics. Factor A is the gasoline type that has $a = 2$ levels and factor B is the type of car that also has $b = 2$ levels. For each experimental configuration $n = 3$ replicate observations are taken. The resulting sums of squares are SSA = 96.33, SSB = 75.00, SSAB = 341.33, and SSE = 194.00. The cell sample averages are $\bar{x}_{11.} = 32.33$,

FIGURE 14.25

Plot of the residuals against the
machines for the car body
assembly line example

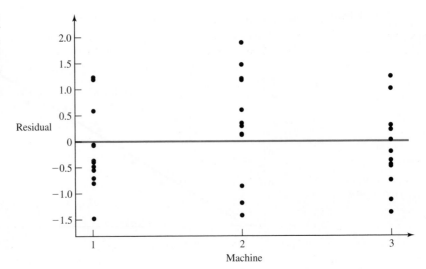

FIGURE 14.26

Plot of the residuals against the
solder types for the car body
assembly line example

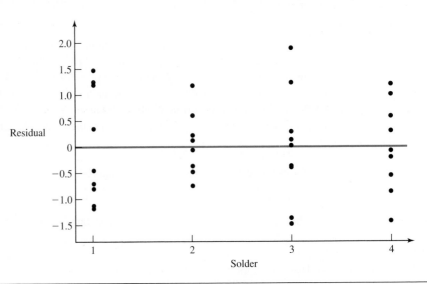

$\bar{x}_{12.} = 38.00$, $\bar{x}_{21.} = 37.33$, and $\bar{x}_{22.} = 21.67$. Construct the analysis of variance table. What are your conclusions from this experiment? Which type of gasoline is best?

14.1.2 A two-factor experiment is conducted to investigate how the tensile strength of a graphite-epoxy composite depends upon the formulation of the composite and the temperature of the composite. Four different composite formulations are considered in the experiment and they are taken to be factor A. Three different temperature levels are considered ($low = 1$, $medium = 2$, and $high = 3$) and they are taken to be factor B. The response variable is tensile strength and it is measured for two different samples at each experimental configuration. The sums of

squares SSA $= 160.61$, SSB $= 580.52$, SSAB $= 58.01$, and SSE $= 66.71$ are obtained. The cell sample averages are $\bar{x}_{11.} = 27.05$, $\bar{x}_{12.} = 18.80$, $\bar{x}_{13.} = 16.40$, $\bar{x}_{21.} = 25.50$, $\bar{x}_{22.} = 15.60$, $\bar{x}_{23.} = 8.25$, $\bar{x}_{31.} = 24.15$, $\bar{x}_{32.} = 17.90$, $\bar{x}_{33.} = 16.95$, $\bar{x}_{41.} = 30.00$, $\bar{x}_{42.} = 23.40$, and $\bar{x}_{43.} = 17.65$.

(a) Construct the analysis of variance table.

(b) Show that there is no evidence of an interaction effect but that there are main effects for both factors.

(c) Construct pairwise confidence intervals to compare the four different graphite-epoxy composite formulations. What conclusions can you draw?

(d) Construct pairwise confidence intervals to compare the three different temperature levels. What conclusions can you draw?

14.1.3 Indentation Measurements for Hardness Testing

The data set in DS 14.1.1 shows the results of an experiment performed to test the machinery used to measure the hardness of a material. The hardness of a material is measured by pushing a tip into the material with a specified force. The depth h of the resulting indentation is then measured. A laboratory purchases a machine that has three different tips that are supposed to provide identical results. The purpose of the experiment is to investigate whether the different tips can in fact be taken to produce identical results. Three types of material are used and four replicate measurements of indentation are obtained for each experimental configuration.

(a) Construct the analysis of variance table.

(b) Show that there is no evidence of an interaction effect and no evidence of a main effect for the tips, but that there is a main effect for material type. How do you interpret these results?

(c) Plot the residuals for each of the three tips. Why is it useful to do this? Does it tell you anything interesting?

14.1.4 Contact Lens Dimensions

The data set in DS 14.1.2 shows the results of an experiment performed to investigate how the deviation from specification of the width of a contact lens at its center point may depend on the type of material used to make the lens and the amount of magnification provided by the lens.

(a) Construct the analysis of variance table.

(b) Show that there is no evidence of an interaction effect and no evidence of a main effect for material type, but that there is a main effect for magnification level. How do you interpret these results?

(c) Plot the residuals for each of the four types of material. Which material would you recommend be employed to minimize the variability in the deviations from specification?

14.1.5 Silicon Mass Loss from Glass Leaching

When a specimen of glass is placed in a solution, leaching occurs whereby some of the constituents of the glass are absorbed into the solution. DS 14.1.3 contains the results of a two-factor experiment to investigate how the glass leaching depends on the glass composition and the acidity of the solution. The response variable measured is the mass loss of silicon. Analyze this data set.

14.1.6 Clinical Trial

The amount of improvement of a genetic disease is measured for 18 patients who are randomly assigned to the nine experimental configurations, with two patients to each configuration, corresponding to three dosage levels of each of two active ingredients of a drug. The experimental results are given in DS 14.1.4. Analyze this data set. What do you notice when you plot the residuals against the levels of ingredient B?

14.1.7 Optimal Mechanical Component Construction

A mechanical component can be made using three different designs and from two different material types. Six components are manufactured according to the six possible combinations of design and material, and DS 14.1.5 contains the lifetimes in hours of these components when they are employed. What does the data set tell you about the effectiveness of the different designs and materials?

14.2 Experiments with Three or More Factors

Experiments can be performed to investigate how a response variable depends on three or more factors of interest. The concepts behind the analysis of such an experiment are similar to those behind the analysis of a two-factor experiment, except that there are now a larger number of interaction terms that need to be considered. As with a two-factor experiment, the analysis is based upon the construction of an analysis of variance table and the consideration of the residuals.

If there are many factors that each have quite a few levels, then the number of experimental configurations can become extremely large so that a large amount of data collection is required to conduct the experiment. In this case, screening experiments are often performed whereby each factor is considered at only *two* levels. This provides a preliminary investigation into how the response variable depends on the factors of interest. Such an experiment with k factors is known as a 2^k design. In this section an example is given of a three-factor experiment followed by an example of a 2^4 experiment.

14.2.1 Three-Factor Experiments

Consider an experiment to investigate how a response variable depends on three factors of interest, A, B, and C, that have a levels, b levels, and c levels, respectively. There are therefore abc experimental configurations, and suppose that a complete balanced design is employed with n replicate observations obtained at each of the experimental configurations. The total number of data observations obtained is consequently

$$n_T = abcn$$

The data are of the form

$$x_{ijkl} \quad 1 \le i \le a, 1 \le j \le b, 1 \le k \le c, 1 \le l \le n$$

so that x_{ijkl} is the lth observation obtained at the ith level of factor A, the jth level of factor B, and the kth level of factor C. The sample averages are obtained in the obvious manner, so that the cell sample averages are

$$\bar{x}_{ijk\cdot} = \frac{1}{n} \sum_{l=1}^{n} x_{ijkl}$$

the overall average is

$$\bar{x}_{\cdots} = \frac{1}{abcn} \sum_{i=1}^{a} \sum_{j=1}^{b} \sum_{k=1}^{c} \sum_{l=1}^{n} x_{ijkl}$$

and, for instance, the averages at the a different levels of factor A are

$$\bar{x}_{i\cdots} = \frac{1}{bcn} \sum_{j=1}^{b} \sum_{k=1}^{c} \sum_{l=1}^{n} x_{ijkl}$$

while the averages at the ab different combinations of the levels of factors A and B are

$$\bar{x}_{ij\cdots} = \frac{1}{cn} \sum_{k=1}^{c} \sum_{l=1}^{n} x_{ijkl}$$

The unknown cell means can be denoted as

$$\mu_{ijk} \quad 1 \le i \le a, 1 \le j \le b, 1 \le k \le c$$

and they represent the expected value of the response variable at each of the abc experimental configurations. The data observations can then be modeled as

$$x_{ijkl} = \mu_{ijk} + \epsilon_{ijkl}$$

where the error terms ϵ_{ijkl} are taken to be independent observations, which are distributed

$$\epsilon_{ijkl} \sim N(0, \sigma^2)$$

for some unknown error variance σ^2.

As with two-factor experiments, the objective of the analysis of a three-factor experiment is to understand the *structure* of the cell means μ_{ijk}, so that the experimenter can interpret the manner in which the expected value of the response variable is influenced by the three factors under consideration. In this respect it is useful to decompose the cell means into a new set of parameters so that they are written

$$\mu_{ijk} = \mu + \alpha_i + \beta_j + \gamma_k + (\alpha\beta)_{ij} + (\alpha\gamma)_{ik} + (\beta\gamma)_{jk} + (\alpha\beta\gamma)_{ijk}$$

where these new parameters are conditioned to sum to 0 over each of their subscripts. Here the parameters $(\alpha\beta\gamma)_{ijk}$ represent a *three-way interaction* effect between the three factors. The presence of this three-way interaction effect indicates that there is a very complex relationship between the ways in which the three factors affect the expected value of the response variable. Specifically, it implies that the relationship between any two factors and the response variable depends upon the level of the additional factor.

Notice that there are now three different kinds of two-way interaction effects, which are the interaction effects discussed for two-factor experiments. The parameters $(\alpha\beta)_{ij}$ represent a two-way interaction effect between factors A and B, the parameters $(\alpha\gamma)_{ik}$ represent a two-way interaction effect between factors A and C, and the parameters $(\beta\gamma)_{jk}$ represent a two-way interaction effect between factors B and C. Finally, the parameters α_i, β_j, and γ_k represent the main effects for the factors A, B, and C, respectively.

With this representation of the cell means there are *seven* different hypotheses that can be tested corresponding to whether each of the seven different sets of parameters are equal to 0. The hypothesis that the parameters $(\alpha\beta\gamma)_{ijk}$ are all equal to 0 corresponds to there being no three-way interaction effect. Similarly, the three different hypotheses corresponding to whether the three different sets of parameters $(\alpha\beta)_{ij}$, $(\alpha\gamma)_{ik}$, and $(\beta\gamma)_{jk}$ are all equal to 0 test for the presence of two-way interaction effects. Also, the hypothesis that the parameters α_i are all equal to 0 tests for the presence of a factor A main effect, the hypothesis that the parameters β_j are all equal to 0 tests for the presence of a factor B main effect, and the hypothesis that the parameters γ_k are all equal to 0 tests for the presence of a factor C main effect.

An analysis of variance table is formed by decomposing the total sum of squares SST into sums of squares for the three factor main effects (SSA, SSB, SSC), the three two-way interaction effects (SSAB, SSAC, SSBC), the three-way interaction effect SSABC, and the sum of squares for error SSE, as shown in Figure 14.27, so that

$$SST = SSA + SSB + SSC + SSAB + SSAC + SSBC + SSABC + SSE$$

FIGURE 14.27

The sum of squares decomposition for a three-factor experiment

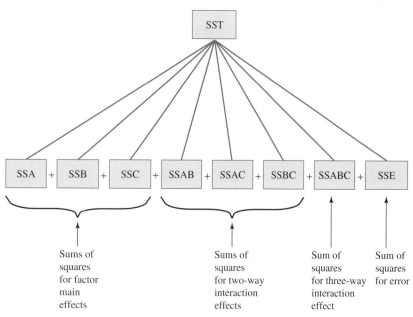

These sums of squares are based upon the sums of the squares of the corresponding parameters estimates and are given by

$$\text{SST} = \sum_{i=1}^{a}\sum_{j=1}^{b}\sum_{k=1}^{c}\sum_{l=1}^{n}(x_{ijkl} - \bar{x}_{....})^2$$

$$\text{SSE} = \sum_{i=1}^{a}\sum_{j=1}^{b}\sum_{k=1}^{c}\sum_{l=1}^{n}(x_{ijkl} - \bar{x}_{ijk.})^2$$

$$\text{SSA} = bcn\sum_{i=1}^{a}\hat{\alpha}_i^2 = bcn\sum_{i=1}^{a}(\bar{x}_{i...} - \bar{x}_{....})^2$$

$$\text{SSB} = acn\sum_{j=1}^{b}\hat{\beta}_j^2 = acn\sum_{j=1}^{b}(\bar{x}_{.j..} - \bar{x}_{....})^2$$

$$\text{SSC} = abn\sum_{k=1}^{c}\hat{\gamma}_k^2 = abn\sum_{k=1}^{c}(\bar{x}_{..k.} - \bar{x}_{....})^2$$

$$\text{SSAB} = cn\sum_{i=1}^{a}\sum_{j=1}^{b}(\hat{\alpha\beta})_{ij}^2 = cn\sum_{i=1}^{a}\sum_{j=1}^{b}(\bar{x}_{ij..} - \bar{x}_{i...} - \bar{x}_{.j..} + \bar{x}_{....})^2$$

$$\text{SSAC} = bn\sum_{i=1}^{a}\sum_{k=1}^{c}(\hat{\alpha\gamma})_{ik}^2 = bn\sum_{i=1}^{a}\sum_{k=1}^{c}(\bar{x}_{i.k.} - \bar{x}_{i...} - \bar{x}_{..k.} + \bar{x}_{....})^2$$

$$\text{SSBC} = an\sum_{j=1}^{b}\sum_{k=1}^{c}(\hat{\beta\gamma})_{jk}^2 = an\sum_{j=1}^{b}\sum_{k=1}^{c}(\bar{x}_{.jk.} - \bar{x}_{.j..} - \bar{x}_{..k.} + \bar{x}_{....})^2$$

$$\text{SSABC} = n\sum_{i=1}^{a}\sum_{j=1}^{b}\sum_{k=1}^{c}(\hat{\alpha\beta\gamma})_{ijk}^2$$

$$= n\sum_{i=1}^{a}\sum_{j=1}^{b}\sum_{k=1}^{c}(\bar{x}_{ijk.} - \bar{x}_{ij..} - \bar{x}_{i.k.} - \bar{x}_{.jk.} + \bar{x}_{i...} + \bar{x}_{.j..} + \bar{x}_{..k.} - \bar{x}_{....})^2$$

The degrees of freedom are $a-1$, $b-1$, and $c-1$ for the factor main effects, $(a-1)(b-1)$, $(a-1)(c-1)$, and $(b-1)(c-1)$ for the two-way interaction effects, and $(a-1)(b-1)(c-1)$ for the three-way interaction effect. There are $abc(n-1)$ degrees of freedom for error and as usual, the mean square error

$$\text{MSE} = \hat{\sigma}^2$$

is the estimate of the error variance. The complete analysis of variance table is shown in Figure 14.28, and notice that there are seven F-statistics that can be used to test the seven hypotheses of interest concerning the three-way interaction effect, the three two-way interaction effects, and the three main effects.

As with a two-factor experiment, interaction effects should be considered before main effects, and in this case the three-way interaction effect should be considered before the two-way interaction effects. If there is evidence of a three-way interaction effect, then there is a complex relationship between the three factors and the response variable, and the presence or absence of two-way interaction effects or main effects is redundant. If there is no evidence of a three-way interaction effect, then the three two-way interaction effects can be considered. Again, remember that the presence of a two-way interaction effect makes the tests of the main effects redundant for the two factors involved.

Source	Degrees of freedom	Sum of squares	Mean squares	F-statistics	p-value
Factor A	$a - 1$	SSA	$MSA = \frac{SSA}{a-1}$	$F_A = \frac{MSA}{MSE}$	$P(F_{a-1, abc(n-1)} > F_A)$
Factor B	$b - 1$	SSB	$MSB = \frac{SSB}{b-1}$	$F_B = \frac{MSB}{MSE}$	$P(F_{b-1, abc(n-1)} > F_B)$
Factor C	$c - 1$	SSC	$MSC = \frac{SSC}{c-1}$	$F_C = \frac{MSC}{MSE}$	$P(F_{c-1, abc(n-1)} > F_C)$
AB interaction	$(a - 1)(b - 1)$	SSAB	$MSAB = \frac{SSAB}{(a-1)(b-1)}$	$F_{AB} = \frac{MSAB}{MSE}$	$P(F_{(a-1)(b-1), abc(n-1)} > F_{AB})$
AC interaction	$(a - 1)(c - 1)$	SSAC	$MSAC = \frac{SSAC}{(a-1)(c-1)}$	$F_{AC} = \frac{MSAC}{MSE}$	$P(F_{(a-1)(c-1), abc(n-1)} > F_{AC})$
BC interaction	$(b - 1)(c - 1)$	SSBC	$MSBC = \frac{SSBC}{(b-1)(c-1)}$	$F_{BC} = \frac{MSBC}{MSE}$	$P(F_{(b-1)(c-1), abc(n-1)} > F_{BC})$
ABC interaction	$(a - 1)(b - 1)(c - 1)$	SSABC	$MSABC = \frac{SSABC}{(a-1)(b-1)(c-1)}$	$F_{ABC} = \frac{MSABC}{MSE}$	$P(F_{(a-1)(b-1)(c-1), abc(n-1)} > F_{ABC})$
Error	$abc(n - 1)$	SSE	$MSE = \frac{SSE}{abc(n-1)}$		
Total	$abcn - 1$	SST			

FIGURE 14.28

Analysis of variance table for a three-factor experiment

FIGURE 14.29

The automated application of grease to a shaft

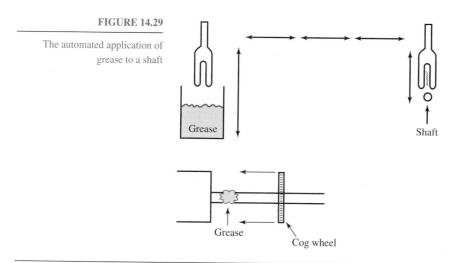

Finally, once a final model has been decided on, the residuals should be calculated and examined. Plots of the residuals against the factor levels, even for factors that are not included in the model, may indicate differences in the variability of the response variable at the different factor levels, which is important information for the experimenter to be aware of.

Example 75

Robot Assembly Methods

In an automated assembly line for the production of photocopiers, one of the many tasks is to apply grease to a metal shaft before a cog wheel is placed onto the shaft and pushed over the grease, as shown in Figure 14.29. The automatic application of the grease is achieved by a two pronged fork, which is dipped into a cup of grease and is then pushed over the shaft. It is important that this automated procedure apply neither too much grease nor too little grease to the shaft, and a three-factor experiment is conducted to investigate how the amount of grease deposited upon the shaft depends upon the type of fork used, the type of grease employed, and the temperature of the grease.

Suppose that two types of fork are used (one with longer and wider prongs than the other) and that this is taken to be factor A with $a = 2$ levels. Four different types of grease are considered, and this is taken to be factor B, which consequently has $b = 4$ levels. The temperature of the cup of grease can be controlled, and this is taken to be factor C with $c = 3$ levels (level 1 = low temperature, level 2 = medium temperature, level 3 = high temperature).

There are therefore $abc = 2 \times 4 \times 3 = 24$ experimental configurations, and $n = 5$ replicate observations are obtained at each of these configurations. In each case the weight of the grease applied to the shaft is carefully measured, and a total of

$$n_T = abcn = 2 \times 4 \times 3 \times 5 = 120$$

data observations are obtained which are shown in Figure 14.30.

FIGURE 14.30

Data set for the robot assembly methods example

Fork Type 1

		Grease type 1	Grease type 2	Grease type 3	Grease type 4
	Low	14.98	10.45	12.36	9.36
		11.67	9.54	15.47	8.89
		12.41	11.23	14.36	9.22
		13.39	10.68	16.35	11.29
		10.99	10.51	15.32	10.58
	Medium	11.17	12.51	13.55	10.20
		10.89	11.88	14.36	9.23
Temperature		12.48	11.95	13.14	9.02
		14.59	12.83	14.31	10.69
		11.43	12.82	14.38	9.30
	High	10.39	14.35	11.95	12.88
		11.67	13.88	11.43	10.98
		13.11	14.70	12.67	12.10
		13.20	12.18	10.68	11.00
		10.77	14.04	13.69	9.57

Fork Type 2

		Grease type 1	Grease type 2	Grease type 3	Grease type 4
	Low	7.55	10.59	15.47	14.41
		12.03	8.57	14.71	10.21
		14.03	10.42	12.89	9.94
		11.63	12.76	12.05	8.67
		11.87	13.10	15.19	11.64
	Medium	10.21	12.61	13.73	9.54
		9.06	9.57	19.36	10.85
Temperature		11.16	11.52	13.18	8.01
		14.06	10.91	13.26	12.59
		11.71	11.24	14.55	10.96
	High	8.47	11.00	13.98	11.38
		15.83	14.75	14.24	9.25
		13.60	15.29	13.34	9.75
		6.94	11.19	15.28	10.05
		10.18	12.39	10.39	13.24

FIGURE 14.31

The analysis of variance table for
the robot assembly methods
example

Source	Degrees of freedom	Sum of squares	Mean squares	F-statistic	p-value
Fork	1	0.626	0.626	0.22	0.642
Grease	3	173.714	57.905	20.15	0.000
Temperature	2	1.114	0.557	0.19	0.824
Fork*Grease	3	11.829	3.943	1.37	0.256
Fork*Temperature	2	0.145	0.072	0.03	0.975
Grease*Temperature	6	57.838	9.640	3.35	0.005
Fork*Grease*Temperature	6	14.635	2.439	0.85	0.536
Error	96	275.905	2.874		
Total	119	535.807			

The analysis of variance table for this experiment is shown in Figure 14.31. Notice first that the three-way interaction effect "Fork*Grease*Temperature" has an F-statistic

$$F_{ABC} = \frac{MSABC}{MSE} = \frac{2.439}{2.874} = 0.85$$

with a p-value of

$$p\text{-value} = P(X > 0.85) = 0.536$$

where the random variable X has an F-distribution with degrees of freedom 6 and 96. This large p-value indicates that there is no evidence of a three-way interaction effect.

Since there is no evidence of a three-way interaction effect, the p-values for the three two-way interaction effects can be considered. Notice that the two-way interaction effect "Grease*Temperature" has an F-statistic

$$F_{BC} = \frac{MSBC}{MSE} = \frac{9.640}{2.874} = 3.35$$

The corresponding p-value is

$$p\text{-value} = P(X > 3.35) = 0.005$$

where the random variable X has an F-distribution with degrees of freedom 6 and 96. This low p-value indicates that there is evidence of an interaction effect between the type of grease used and the temperature of the grease. In other words, there is evidence that the manner in which temperature changes affect the amount of grease deposited on the shaft depends upon the particular type of grease used.

The "Fork*Grease" two-way interaction has a p-value of 0.256 and the "Fork*Temperature" two-way interaction has a p-value of 0.975, and consequently there is no evidence of the presence of interaction effects between the type of fork and either of the two other factors.

Which factor main effects need to be considered? Since there is an interaction between the type of grease and the temperature, the tests of the main effects for these two factors are redundant. However, the main effect for the type of fork needs to be investigated. An F-statistic

$$F_A = \frac{MSA}{MSE} = \frac{0.626}{2.874} = 0.22$$

is obtained, and the p-value is

$$p\text{-value} = P(X > 0.22) = 0.642$$

where the random variable X has an F-distribution with degrees of freedom 1 and 96. There is thus no evidence of the presence for a main effect for the type of fork.

Grease

		Grease type 1	Grease type 2	Grease type 3	Grease type 4
Temperature	Low	$\bar{x}_{.11.} = 12.055$	$\bar{x}_{.21.} = 10.785$	$\bar{x}_{.31.} = 14.417$	$\bar{x}_{.41.} = 10.421$
	Medium	$\bar{x}_{.12.} = 11.676$	$\bar{x}_{.22.} = 11.784$	$\bar{x}_{.32.} = 14.382$	$\bar{x}_{.42.} = 10.039$
	High	$\bar{x}_{.13.} = 11.416$	$\bar{x}_{.23.} = 13.377$	$\bar{x}_{.33.} = 12.765$	$\bar{x}_{.43.} = 11.020$

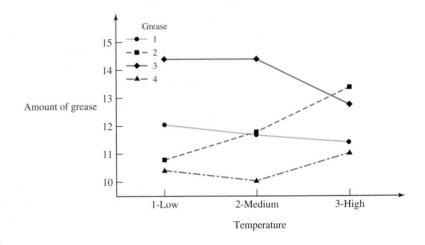

FIGURE 14.32

The sample averages for the grease-temperature combinations for the robot assembly methods example

In summary, the analysis conducted so far has shown the experimenter that the average amount of grease deposited on the shaft does not depend on which type of fork is used, but it does depend on the type of grease used and the temperature of the grease. Furthermore, the influences of the type of grease used and the temperature of the grease are not independent of each other. The effects of the different combinations of grease and temperature can be assessed from the sample averages $\bar{x}_{.jk.}$, which are shown in Figure 14.32.

Is it the case that it doesn't matter which type of fork is used? Let's look at a residual plot before jumping to this conclusion. The final model that has been decided on is

$$\mu_{ijk} = \mu + \beta_j + \gamma_k + (\beta\gamma)_{jk}$$

and this can be fitted to the data and residuals can be calculated in the usual manner. Figure 14.33 shows boxplots of the residuals for each of the two fork types, based on 60 residuals for each fork type. It is clear from these boxplots that the residuals tend to be closer to 0 for fork type 1 than for fork type 2, and this implies that the variability in the amount of grease deposited on the shaft is smaller for fork type 1 than for fork type 2. This finding indicates that it is preferable to use fork type 1 rather than fork type 2.

In conclusion, the objective of this experiment is to help design the best automated procedure to deposit the right amount of grease onto the shaft. It has been found that the grease type and the temperature can both be manipulated in order to make the expected amount of grease deposited equal to the required amount. Furthermore, it has been found that it is best to then

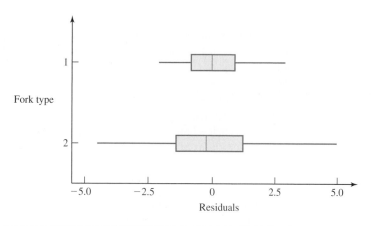

FIGURE 14.33

Boxplots of the residuals for the two fork types in the robot assembly methods example

use fork type 1 so that the variability of the amount of grease deposited about the required amount is as small as possible.

14.2.2 2^k Experiments

In an experiment to compare the effects of k factors on a response variable, where the levels of the factors are $a_1, \ldots, a_k$, the total number of different experimental configurations is

$$a_1 \times a_2 \times \cdots \times a_k$$

If n replicate observations are obtained at each experimental configuration, the total sample size required is

$$n_T = a_1 \times a_2 \times \cdots \times a_k \times n$$

This sample size becomes very large as the number of factors k increases, and in such circumstances it is therefore common to use screening experiments in which each factor is considered at only two levels so that the a_i are all equal to 2.

An experiment conducted with k factors each with two levels is known as a 2^k experimental design. There are 2^k experimental configurations and a total sample size of

$$n_T = 2^k \times n$$

is required if there are n replicate observations at each experimental configuration.

The usual type of model is employed in these experiments whereby the data observations are assumed to be composed of a cell mean together with an error term from a normal distribution. The cell means are then written as the sum of the k individual factor main effects together with all the possible two-way interaction effects plus higher order interaction effects. There are $k(k-1)/2$ two-way interaction effects, and in general there are

$$\binom{k}{r}$$

r way interaction effects, because this is the number of ways that a subset of r factors can be taken from the k factors.

For 2^k experiments the parameter estimates are easily obtainable as the differences between two sets of sample averages. The sums of squares for each set of parameters are obtained by taking the sums of the squares of the parameter estimates as was shown for three-factor experiments. An analysis of variance table is constructed to test for the presence of the various types of parameters in the usual manner, and all of the rows have one degree of freedom except for the error degrees of freedom, which are $2^k(n-1)$. The total degrees of freedom are $2^k n - 1$.

The analysis of variance table presents p-values for each of the hypotheses that can be tested in 2^k experiments. As usual, the higher order interactions should be analyzed first, and the presence of any interaction term makes redundant the analysis of any interactions or main effects formed from its component factors. In practice, there generally will not be any evidence of four-way interactions or higher order interactions, and three-way interaction effects may also be rare.

Example 70
Chemical Yields

An experiment is performed to investigate how the yield of a chemical reaction depends on four factors: temperature, pressure, amount of catalyst 1, and amount of catalyst 2. The four factors are each considered at two levels, a low level and a high level, which results in a 2^4 experimental design. There are 16 experimental configurations and $n = 2$ observations are taken at each of these experimental configurations so that the total sample size is $n_T = 32$. The data set obtained is shown in Figure 14.34.

The analysis of variance table is shown in Figure 14.35. Notice that there is a row for the four-way interaction effect

Temp*Pressure*Cat-1*Cat-2

FIGURE 14.34

Data set for the chemical yields example

Temperature	Pressure	Catalyst 1	Catalyst 2	Yield
Low	Low	Low	Low	548 521
Low	Low	Low	High	548 526
Low	Low	High	Low	580 568
Low	Low	High	High	562 526
Low	High	Low	Low	507 523
Low	High	Low	High	507 544
Low	High	High	Low	560 559
Low	High	High	High	538 533
High	Low	Low	Low	576 556
High	Low	Low	High	559 590
High	Low	High	Low	581 603
High	Low	High	High	608 586
High	High	Low	Low	542 547
High	High	Low	High	549 577
High	High	High	Low	583 588
High	High	High	High	550 535

Source	Degrees of freedom	Sum of squares	Mean squares	F-statistic	p-value
Temp	1	7200.0	7200.0	29.95	0.000
Pressure	1	2738.0	2738.0	11.39	0.004
Cat-1	1	3612.5	3612.5	15.03	0.001
Cat-2	1	338.0	338.0	1.41	0.253
Temp*Pressure	1	200.0	200.0	0.83	0.375
Temp*Cat-1	1	128.0	128.0	0.53	0.476
Temp*Cat-2	1	112.5	112.5	0.47	0.504
Pressure*Cat-1	1	50.0	50.0	0.21	0.654
Pressure*Cat-2	1	72.0	72.0	0.30	0.592
Cat-1*Cat-2	1	2178.0	2178.0	9.06	0.008
Temp*Pressure*Cat-1	1	162.0	162.0	0.67	0.424
Temp*Pressure*Cat-2	1	338.0	338.0	1.41	0.253
Temp*Cat-1*Cat-2	1	0.5	0.5	0.00	0.964
Pressure*Cat-1*Cat-2	1	450.0	450.0	1.87	0.190
Temp*Pressure*Cat-1*Cat-2	1	392.0	392.0	1.63	0.220
Error	16	3846.0	240.4		
Total	31	21817.5			

and that there are four rows for the three-way interaction effects

Temp*Pressure*Cat-1

Temp*Pressure*Cat-2

Temp*Cat-1*Cat-2

Pressure*Cat-1*Cat-2

There are also rows for the six two-way interaction effects and for the four individual factor main effects. Each of the interaction effects and main effects has one degree of freedom, while the degrees of freedom for error are $2^k(n-1) = 2^4 \times (2-1) = 16$.

The four-way interaction effect and each of the three-way interaction effects have large p-values, so there is no evidence of their presence. Notice that one of the six two-way interaction effects has a small p-value, and this is the interaction between the two catalysts (p-value = 0.008). This indicates that there is evidence of an interaction effect between the two catalysts, but that there is no evidence of any other interaction effects.

The tests of the main effects of catalyst 1 and catalyst 2 are now redundant because of the presence of their interaction effect, but the main effects for temperature and pressure should be examined. The temperature main effect has a very small p-value and the pressure main effect has a p-value of 0.004, and so there is evidence of the presence of main effects for both of these factors.

In summary, this experiment has revealed that the expected value of the chemical yield depends on all four of the factors. In addition, the effects of the changes in temperature and pressure are independent of each other and of the amounts of the two catalysts (since no interaction effects have been found involving temperature and pressure with each other or with the two catalysts). The effects of changes in the two catalysts are not independent of each other.

More specific information on these effects can be found by examining the appropriate sample averages. Figure 14.36 shows the sample averages for the two temperature levels, the two pressure levels, and the four combinations of the two catalysts. It can be seen that increasing the temperature increases the yield but that increasing the pressure decreases the

FIGURE 14.36

The sample averages for the
chemical yields example

Temperature

Low	High
$\bar{x}_{1\dots} = 540.62$	$\bar{x}_{2\dots} = 570.62$

Pressure

Low	High
$\bar{x}_{.1\dots} = 564.88$	$\bar{x}_{.2\dots} = 546.37$

Catalyst 1

Catalyst 2		Low	High
	Low	$\bar{x}_{..11.} = 540.00$	$\bar{x}_{..21.} = 577.75$
	High	$\bar{x}_{..12.} = 550.00$	$\bar{x}_{..22.} = 554.75$

yield. The interaction effect between the catalysts is evident since increasing the amount of catalyst 2 increases the yield at the low level of catalyst 1 but decreases the yield at the high level of catalyst 2.

This section provides a general introduction to 2^k experiments, which are very important tools for engineering research. A more detailed and in-depth description of the design and analysis of 2^k experiments can be found in more advanced books on the design of experiments. For example, some 2^k experiments are run without replicate observations so that $n = 1$. In this case the analysis proceeds on the assumption that some higher order interaction terms are 0 so that their sums of squares can be used as the error sum of squares (as in a two-factor experiment without replications). Alternatively, an analysis method can be based upon the construction of a normal probability plot of the parameters estimates.

Finally, 2^k experiments can be performed where data observations are obtained in only a subset of the 2^k experimental configurations. This serves to reduce the total sample size and is particularly useful when the number of factors k is large. Such an experimental design is known as a **fractional factorial design**. In these cases the parameters are said to be *confounded* with each other since they cannot be estimated independently of each other, and the analysis proceeds on the assumption that some of the higher order interaction effects can be taken to be 0.

■ 14.2.3 Problems

14.2.1 A three-factor experiment is performed to taste test different blends of a fruit juice mixture. The response variable scores how the participant liked the fruit juice based on the answers to a questionnaire. Factor A is the blend of fruit juice with $a = 3$ levels. Factor B is the gender of the participant with $b = 2$ levels, male and female. Factor C is the age of the participant with $c = 3$ age categories. In each experimental configuration $n = 5$ replicate observations are obtained.

   (a) Suppose that an interaction effect is found between factors A and B. How would you interpret this?

   (b) Suppose that an interaction effect is found between factors A and C. How would you interpret this?

   (c) Suppose that there are no interaction effects involving factor A, and that there is no factor A main effect. What does this tell you about the different blends of fruit juice?

   (d) The sums of squares are SSA = 90.65, SSB = 6.45, SSC = 23.44, SSAB = 17.82, SSAC = 24.09, SSBC = 24.64, SSABC = 27.87, and SSE = 605.40. Construct the analysis of variance table. What final model is appropriate to model the fruit juice scores? What is the interpretation of this model?

14.2.2 Optimal Rice Growing Conditions

A three-factor experiment is conducted to investigate the yields of different brands of rice grown in controlled greenhouse conditions. The response variable is the rice yield, and factor A is rice variety with $a = 3$ levels. Two levels of fertilizer are considered (factor B) together with two amounts of sunshine (factor C). There are $n = 4$ replicate observations per cell and the data set is given in DS 14.2.1.

(a) Construct the analysis of variance table and summarize what you learn from it.

(b) Is there any evidence that increasing the amount of fertilizer increases the expected yield?

(c) Is there any evidence that the effect of increasing the sunlight depends upon the amount of fertilizer employed?

(d) Is there any evidence that the effect of increasing the sunlight depends upon which variety of rice is used?

14.2.3 Effect of Gasoline Additives on Mileage

DS 14.2.2 contains the data set collected from a three-factor experiment to investigate how gas mileage depends on the amounts of two types of additive in the gasoline and the driving conditions.

(a) Construct the analysis of variance table. How do the three factors influence the gas mileage?

(b) Does the amount of additive B have an effect on the expected value of the gas mileage? Plot the residuals for each level of factor B. What do you learn?

14.2.4 Detection of Airborne Objects

DS 14.2.3 contains the data from a three-factor experiment to investigate the distance at detection for four different radar systems of two different aircraft flying at day and at night. Construct the analysis of variance table and summarize what you learn from it.

14.2.5 Digital-Weighing Machine Calibration

DS 14.2.4 contains the results of a 2^4 experiment conducted to investigate the accuracies of two digital weighing machines. The response variable is the deviation of the weight reading from the true weight of a particular object that is used throughout the experiment. The four factors are the machines, the temperature, the position of the weight on the surface of the weighing machine, and the angle of the weighing machine to the horizontal.

(a) What models imply that there is no difference between the expected values of the readings provided by the two machines?

(b) How would you interpret an interaction effect between temperature and machines?

(c) How would you interpret an interaction effect between position and angle?

(d) Construct the analysis of variance table and summarize what you learn from it.

(e) Is it fair to say that temperature has no influence on the readings provided by the machines? Make a plot of the residuals at the two temperature levels.

14.2.6 Golf Club Comparisons

DS 14.2.5 contains the results of a 2^4 experiment conducted to investigate the driving distances of two golfers using different clubs and balls and under different weather conditions. What do you learn from this experiment? Is there any evidence that the weather conditions have any effect on the driving distance?

14.2.7 An ANOVA table for an experiment with four factors A, B, C and D gave the p-values shown in Figure 14.37. Classify each term as being "significant," "not-significant," or "redundant."

Term	p-value
A	0.009
B	0.185
C	0.211
D	0.008
A*B	0.226
A*C	0.001
A*D	0.751
B*C	0.003
B*D	0.001
C*D	0.005
A*B*C	0.620
A*B*D	0.674
A*C*D	0.007
B*C*D	0.111
A*B*C*D	0.411

FIGURE 14.37

p-values obtained from an ANOVA analysis of a four factor experiment

14.2.8 A three-factor experiment ($3 \times 2 \times 2$) gave the following data:

When factor C is low

	A low	A middle	A high
B low	50	47	55
B high	40	38	47

When factor C is high

	A low	A middle	A high
B low	48	51	72
B high	42	40	64

Suppose that there is no three-way interaction effect, but that there is an A*C two-way interaction. Construct an interaction plot to demonstrate the A*C two-way interaction.

14.3   Supplementary Problems

14.3.1   Injection Molding Production

In an injection molding procedure plastic parts are produced by injecting the plastic material into a mold. Some variability is experienced in the resulting weight (or equivalently density) of the part. DS 14.3.1 contains the results of a two-factor experiment to investigate how the weight of the parts depends on the type of injection material and the pressure of injection.

(a) Construct the analysis of variance table.

(b) Show that there is no evidence of an interaction effect but that there are main effects for both factors.

(c) Construct pairwise confidence intervals to compare the three different types of injection material. What conclusions can you draw?

(d) Construct pairwise confidence intervals to compare the three different injection pressures. What conclusions can you draw?

14.3.2   Semiconductor Wafer Yields

Semiconductors (chips) are produced on wafers that contain 100 chips. The wafer yield is defined to be the proportion of these chips that are acceptable for use. One company has two different factory locations for producing chips and can use one of three different coatings for the wafers. The data set in DS 14.3.2 is the result of an experiment run to investigate how wafer yield depends on these two factors. What conclusions can you draw from this experiment?

14.3.3   Clinical Trial

DS 14.3.3 contains the recovery times in days for 16 patients allocated at random to a two-factor experiment to compare four drugs and two levels of severity of the illness. Analyze the data set and summarize your conclusions.

14.3.4   Effect of Furnace Operation on Metal Hardness

A three-factor experiment is performed to investigate how the hardness of a metal bar depends on which furnace is used to manufacture the bar and the location of the bar in the furnace. The response variable is the hardness of the metal, and one of the factors designates which of two furnaces is used for that metal bar. The other factors are the *layer* in the furnace (bottom, middle, or top) and the *position* in the furnace (front or back). The data set obtained is given in DS 14.3.4.

(a) What is the interpretation of a two-way interaction effect between layer and position?

(b) What is the interpretation of a two-way interaction effect between furnace and position?

(c) Construct the analysis of variance table and summarize what you learn from it.

14.3.5   Dispersion Polymerization

DS 14.3.5 contains the results of a 2^4 experiment that investigates the preparation of polymers by dispersion polymerization in an organic medium. The response variable is the mean diameter of the polymers in microns (m^{-6}). The four factors are monomer concentration, stabilizer concentration, catalyst concentration, and water concentration. Analyze the data set. What do you learn from the experiment?

14.3.6   Lathe Operation for Gas Cylinder Construction

In the production of low-pressure gas cylinders a lathe operator has to take the bottom and top parts of a cylinder, fit them with a footring and a collar, and then weld the halves together. An experiment was conducted to investigate the time taken to complete this process. Two different lathe designs were considered, and three skilled operators were used in the experiment. Six completion times were measured for each operator using both of the lathe designs. The times in seconds are given in DS 14.3.6. Is there any evidence of a difference in the efficiencies of the two lathe designs? Is the difference in the efficiencies of the two lathe designs the same for each of the operators?

14.3.7 Electric Motor Noise Levels

A company that manufactures large electric motors has to pay attention to the noise levels produced by the motors. Measurements of noise levels were obtained for three different motor speeds. These measurements were made at two positions, in front of the machine and at the side of the machine, and also for four different types of cooling, which involve different internal fans and ventilation covers. The data obtained are given in DS 14.3.7. Show that there is a three-way interaction effect between the factors speed, cooler, and position. Considering just the data obtained at position 1, make an interaction plot of the effects of speed and cooler at that position. Make a similar interaction plot for the data obtained at position 2. Compare the two interaction plots to investigate the three-way interaction.

CHAPTER FIFTEEN

Nonparametric Statistical Analysis

This chapter provides a discussion and illustration of the implementation of **nonparametric** or **distribution-free** statistical inference methodologies. The general distinction between these types of inference methods and the alternative *parametric* inference methods discussed in previous chapters is that a parametric inference approach is based on a **distributional assumption** for data observations, whereas a nonparametric inference approach provides answers that are not based on any distributional assumptions.

Most of the statistical methodologies discussed in the other chapters of this book would be considered parametric inference methods. For instance, the analysis of Example 14 concerning metal cylinder diameters in Chapters 7 and 8 is based upon the distributional assumption that the diameters are normally distributed with some unknown mean parameter μ and some unknown variance parameter σ^2. The parametric analysis is then based on obtaining estimates $\hat{\mu}$ and $\hat{\sigma}^2$ for these two unknown parameters. For other data sets other distributional assumptions may be appropriate. For example, measurements of the failure times of certain machine parts may be modeled with the Weibull or the gamma distribution. However, in all cases the parametric inference approach operates within the framework of an assumed distributional model, and it is based on the estimation of the unknown parameters of that model.

Obviously, the results of a parametric analysis are only as good as the validity of the assumptions on which they are based. If the assumption of a normal distribution is made, whereas in fact the unknown true distribution is skewed and asymmetric, then the corresponding inference results may in turn be misleading. Nonparametric inference methods have been developed with the objective of providing statistical inference methods that are free from any distributional assumptions. Thus the term "distribution-free method" may be considered to be synonymous with the term "nonparametric method" and may in fact be a more helpful term.

While nonparametric methods are thus valid under weaker assumptions than parametric methods, it should be realized that if the distributional assumption on which a parametric analysis is based is valid, then it will allow a more precise or more powerful analysis than the corresponding nonparametric method. Consequently, the comparison between nonparametric and parametric approaches to a problem can usefully be seen as a compromise between assumptions and precision. However, the more general validity of the nonparametric approaches provides the motivation for their widespread adoption in engineering and other sciences.

In this chapter the analysis of a single population is considered first, followed by the comparison of two populations and then the comparisons of three or more populations (one-way layouts). These types of problems have been considered in previous chapters where parametric methodologies such as t-tests, two-sample t-tests, and analysis of variance techniques have been considered. In this chapter alternative nonparametric inference methods are proposed and are compared with the parametric methodologies.

15.1 The Analysis of a Single Population

Chapter 8 (using the results presented in Chapters 6 and 7) contains a discussion of the analysis of a sample of data observations from a single population. Much of this analysis concerns the important case where the data observations can be taken to be normally distributed. In this section alternative analysis techniques that are not based on any distributional assumptions are discussed.

15.1.1 The Distribution Function

Consider a data set that consists of observations from some common unknown distribution function $F(x)$. A sensible starting point in the analysis of the data set is to obtain some information about the form of the distribution function $F(x)$. In this section the construction of a nonparametric estimate $\hat{F}(x)$ for the unknown distribution function is considered, which is based only on the assumption that the data represent independent, identically distributed observations from this common unknown distribution function.

For any value x_0 of interest, the estimate $\hat{F}(x)$ provides a point estimate of $F(x_0)$, the probability that an observation is no larger than a specific value x_0. $\hat{F}(x)$ can also be used to estimate specific quantiles $F^{-1}(p)$ of the distribution. The construction of confidence bands for the distribution function $F(x)$ provides an assessment of the accuracy of the nonparametric estimate $\hat{F}(x)$. In addition, these confidence bands can be used to test the plausibility that the true distribution function $F(x)$ is of a specific form $F_0(x)$.

The Empirical Cumulative Distribution Function

Given a data set $x_1, \ldots, x_n$ of independent observations from an unknown distribution with a cumulative distribution function $F(x)$, the empirical cumulative distribution function

$$\hat{F}(x) = \frac{\#x_i \leq x}{n}$$

provides an estimate of $F(x)$. The empirical cumulative distribution function is an increasing step function between the values of 0 and 1. A single data observation x_i causes the empirical distribution function to increase by an amount $1/n$ at x_i. If there are m data observations each taking the value x_i, then the function will increase by an amount m/n at x_i.

The Empirical Cumulative Distribution Function The empirical (or sample) cumulative distribution function provides a simple estimate of an unknown probability distribution. It can be considered to be the "best guess" of the true form of the underlying unknown distribution function $F(x)$. However, it should be pointed out that it is not necessarily expected that $F(x)$ looks exactly like the estimate $\hat{F}(x)$. For example, the true distribution function $F(x)$ may be considered to be smooth (corresponding to a continuous probability density function) rather than a step function, which corresponds to a discrete probability mass function. Nevertheless, the empirical cumulative distribution function $\hat{F}(x)$ should provide a close fit to the true distribution function $F(x)$ for reasonable sample sizes n.

Notice that if

$$S(x) = \#x_i \leq x$$

then

$$S(x) \sim B(n, F(x))$$

This binomial distribution for $S(x)$ implies that

$$E(\hat{F}(x)) = \frac{E(S(x))}{n} = \frac{n\,F(x)}{n} = F(x)$$

and that

$$\mathrm{Var}(\hat{F}(x)) = \frac{\mathrm{Var}(S(x))}{n^2} = \frac{n\,F(x)(1 - F(x))}{n^2} = \frac{F(x)(1 - F(x))}{n}$$

Consequently, for a given value x, the empirical cumulative distribution function $\hat{F}(x)$ is seen to be an unbiased estimator of $F(x)$, whose variance is a decreasing function of n.

The empirical cumulative distribution function $\hat{F}(x)$ becomes more accurate when it is based upon larger sample sizes. One reason for this is that for an underlying continuous distribution function $F(x)$, the estimate $\hat{F}(x)$ becomes smoother at larger sample sizes since the step sizes $1/n$ become smaller. In addition, since the variance of the empirical cumulative distribution function $\hat{F}(x)$ is a decreasing function of the sample size n, it becomes a more and more precise estimator of $F(x)$ for larger sample sizes.

Notice that as well as providing estimates of the cumulative distribution function $F(x)$ at various values of x, the empirical cumulative distribution function $\hat{F}(x)$ provides estimates of the **quantiles** $F^{-1}(p)$ of the distribution for various values of $p \in (0, 1)$. This is done by finding the x value (or values) for which $\hat{F}(x) = p$ or, equivalently, finding the values of x for which a proportion p of the data set is less than x.

It is interesting to compare this nonparametric approach to estimating the distribution function with a parametric approach. For a given distributional assumption, the parametric estimate will be the assumed distribution function evaluated at the estimated parameter values. For example, if the data set $x_1, \ldots, x_n$ is assumed to consist of independent, identically distributed observations from a normal distribution, and if $\Phi(x; \mu, \sigma)$ represents the cumulative distribution function of a normal distribution with mean μ and variance σ^2, then the distribution function may be estimated by $\hat{F}(x) = \Phi(x; \hat{\mu}, \hat{\sigma})$ for parameter estimates

$$\hat{\mu} = \bar{x} = \frac{\sum_{i=1}^{n} x_i}{n} \qquad \text{and} \qquad \hat{\sigma}^2 = s^2 = \frac{\sum_{i=1}^{n}(x_i - \bar{x})^2}{n - 1}$$

Finally, notice that attention is directed toward estimating the cumulative distribution function $F(x)$ of an unknown distribution rather than toward estimating the probability density function $f(x)$, although there is a one-to-one correspondence between these two. The reason for this is that the density function that corresponds to the empirical cumulative distribution function $\hat{F}(x)$ consists of point masses of amount $1/n$ at each of the data points. If the density (mass) function $f(x)$ is actually discrete, then for a large enough sample size (with consequently many repeat observations) this may be a good estimate of the probability mass function. However, for a continuous density function $f(x)$, this is not a useful representation. In this case an idea of the form of the density function $f(x)$ can best be obtained by grouping the data and drawing a histogram.

FIGURE 15.1	5.9	6.6	7.5	8.2	10.6	15.0	23.1	45.0
Execution times in seconds for the computer system performance example	6.0	6.7	7.5	8.9	11.8	16.2	33.0	45.8
	6.4	6.9	7.8	9.3	11.8	16.3	40.0	
	6.4	7.0	7.9	9.3	12.6	17.0	42.8	
	6.5	7.1	8.1	9.6	12.9	17.2	43.0	
	6.5	7.2	8.1	10.4	14.3	22.8	44.8	

FIGURE 15.2

Histogram of execution times for the computer system performance example

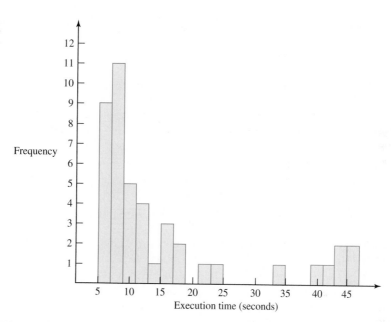

Example 76
Computer System
Performance

A computer system for recording, updating, and delivering maintenance requests for a large organization is being evaluated. The evaluation consists of recording the execution times for typical tasks that the system has to handle. At $n = 44$ randomly selected times a series of tasks is submitted to the system, and the average execution time for the tasks is calculated. The data set shown in Figure 15.1 is obtained, where the data values have been placed in increasing order. Since the series of tasks submitted to the computer is the same for each of the $n = 44$ data observations, the data set provides information on the variations in the execution times due to changes in the load on the computer system.

What can we learn from this data set about the distribution of the execution times? Figure 15.2 shows a histogram of the data set. The empirical cumulative distribution function is calculated in Figure 15.3 and it is plotted in Figure 15.4. The sample median execution time (see Section 6.3.2) can be taken to be the average of the twenty-second and twenty-third smallest execution times, which is

$$\frac{9.3 + 9.6}{2} = 9.45$$

seconds. The upper and lower sample quartiles (see Section 6.3.6) are 16.825 and 7.125 seconds, respectively. Since 35 out of the 44 observations are smaller than 20 seconds, the

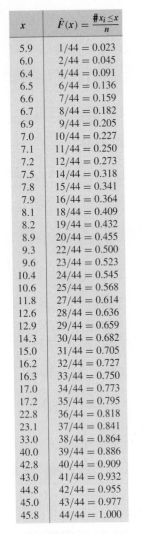

x	$\hat{F}(x) = \dfrac{\#x_i \leq x}{n}$
5.9	1/44 = 0.023
6.0	2/44 = 0.045
6.4	4/44 = 0.091
6.5	6/44 = 0.136
6.6	7/44 = 0.159
6.7	8/44 = 0.182
6.9	9/44 = 0.205
7.0	10/44 = 0.227
7.1	11/44 = 0.250
7.2	12/44 = 0.273
7.5	14/44 = 0.318
7.8	15/44 = 0.341
7.9	16/44 = 0.364
8.1	18/44 = 0.409
8.2	19/44 = 0.432
8.9	20/44 = 0.455
9.3	22/44 = 0.500
9.6	23/44 = 0.523
10.4	24/44 = 0.545
10.6	25/44 = 0.568
11.8	27/44 = 0.614
12.6	28/44 = 0.636
12.9	29/44 = 0.659
14.3	30/44 = 0.682
15.0	31/44 = 0.705
16.2	32/44 = 0.727
16.3	33/44 = 0.750
17.0	34/44 = 0.773
17.2	35/44 = 0.795
22.8	36/44 = 0.818
23.1	37/44 = 0.841
33.0	38/44 = 0.864
40.0	39/44 = 0.886
42.8	40/44 = 0.909
43.0	41/44 = 0.932
44.8	42/44 = 0.955
45.0	43/44 = 0.977
45.8	44/44 = 1.000

FIGURE 15.3

Calculation of the empirical
cumulative distribution function for the
computer system performance example

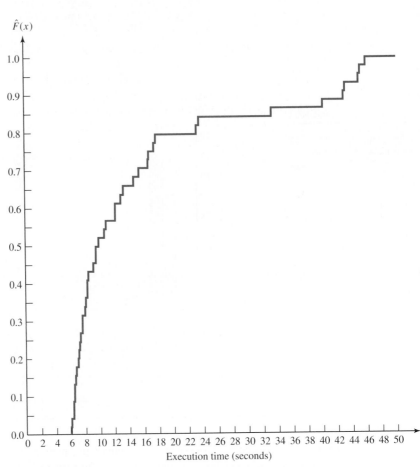

FIGURE 15.4

The empirical cumulative distribution function for the computer system performance example

probability of an execution time quicker than 20 seconds can be estimated as

$$\hat{F}(20) = \frac{\#x_i \leq 20}{44} = \frac{35}{44} = 0.795$$

Confidence Bands for the Distribution Function   Although the empirical cumulative distribution function $\hat{F}(x)$ may be considered to be the "best guess" of the true unknown distribution function $F(x)$, it is also useful to calculate some **confidence bands** for the distribution function $F(x)$. Confidence bands with a confidence level of $1 - \alpha$ have the property that there is is a confidence level of $1 - \alpha$ that the true distribution function $F(x)$ lies *completely* within the bands for all x values.

The confidence bands are constructed by taking the empirical cumulative distribution function $\hat{F}(x)$ and by shifting it up and down by a certain amount $d_{\alpha,n}$. Bands that exceed 1 are then truncated at 1, and bands below 0 are truncated at 0. For x values at which no truncation of the bands occurs, the point estimate $\hat{F}(x)$ is at the center of the confidence bands.

The shift amount $d_{\alpha,n}$ depends on both the confidence level $1 - \alpha$ and the precision of the empirical cumulative distribution function $\hat{F}(x)$ through the sample size n. Intuitively, it is clear that $d_{\alpha,n}$ should be a decreasing function of α since the width of the confidence bands should increase as the confidence level increases, and it is also clear that $d_{\alpha,n}$ should be a decreasing function of n because the width of the confidence bands should decrease as the precision of the empirical cumulative distribution function $\hat{F}(x)$ increases due to increases in the sample size n.

For sample sizes n larger than 40, the appropriate shift amount $d_{\alpha,n}$ is given by

$$\frac{d_\alpha}{\sqrt{n}}$$

where the values of d_α are

α	0.20	0.10	0.05	0.02	0.01
d_α	1.07	1.22	1.36	1.52	1.63

For smaller sample sizes the values of $d_{\alpha,n}$ are given in Table VI. They were originally calculated by the Russian probabilist Andrei Nikolaevich Kolmogorov in 1933, and the confidence bands are usually referred to as the Kolmogorov confidence bands.

Kolmogorov Confidence Bands

Confidence bands around an empirical cumulative distribution function $\hat{F}(x)$ with confidence level $1 - \alpha$ are constructed by adding and subtracting an amount $d_{\alpha,n}$ (see Table VI) to the empirical cumulative distribution function. The bands are truncated at 0 and 1.

Example 76
Computer System
Performance

Confidence bands with a confidence level of 95% are calculated with

$$d_{0.05,44} = \frac{d_{0.05}}{\sqrt{44}} = \frac{1.36}{\sqrt{44}} = 0.205$$

and are shown in Figure 15.5. The true unknown cumulative distribution function $F(x)$ lies completely within these confidence bands with a confidence level of 95%.

Testing Whether $F(x) = F_0(x)$ Consider the problem of assessing the evidence that the unknown cumulative distribution function $F(x)$ is equal to a specified distribution function $F_0(x)$. This problem may be formulated as a hypothesis testing problem with a null hypothesis

$$H_0 : F(x) = F_0(x)$$

FIGURE 15.5

Confidence bands for the computer system performance example

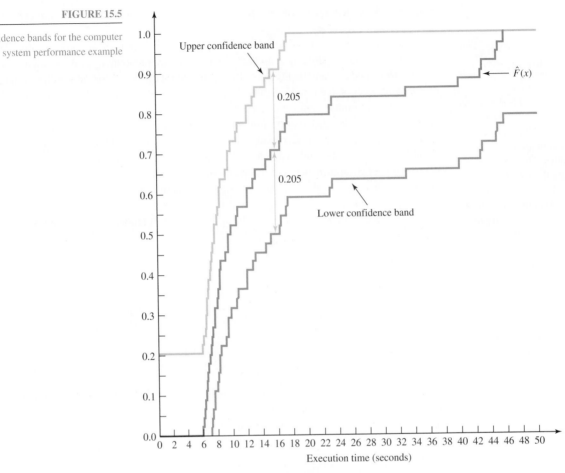

for all values of x, and with a general alternative hypothesis that $F(x) \neq F_0(x)$ for some value of x. Notice that this null hypothesis is subtly different from the usual kinds of hypotheses, which concern the values taken by certain parameters, such as $H_0 : \mu = \mu_0$. This null hypothesis requires that two functions be identical for all x values.

A size α hypothesis test can be performed by checking whether the specified distribution function $F_0(x)$ lies completely within the $1 - \alpha$ confidence bands for $F(x)$. The null hypothesis is accepted if $F_0(x)$ lies completely within the confidence bands and is rejected if $F_0(x)$ lies outside the confidence bands for some x values. The test can be done either graphically or by checking whether the value of the statistic

$$\max_x |\hat{F}(x) - F_0(x)|$$

is less than $d_{\alpha, n}$ in which case the null hypothesis is accepted, or is greater than $d_{\alpha, n}$ in which case the null hypothesis is rejected, as shown in Figure 15.6. If this latter approach is taken, then it is useful to notice that the maximum value of $|\hat{F}(x) - F_0(x)|$ must occur when x is equal to one of the data points x_i, at which point the empirical cumulative distribution function $\hat{F}(x)$ has a step. However, it is important to check the value of $|\hat{F}(x) - F_0(x)|$ at both the top and the bottom of the steps.

FIGURE 15.6

Using the confidence bands on the empirical cumulative distribution function to perform hypothesis tests regarding the distribution function

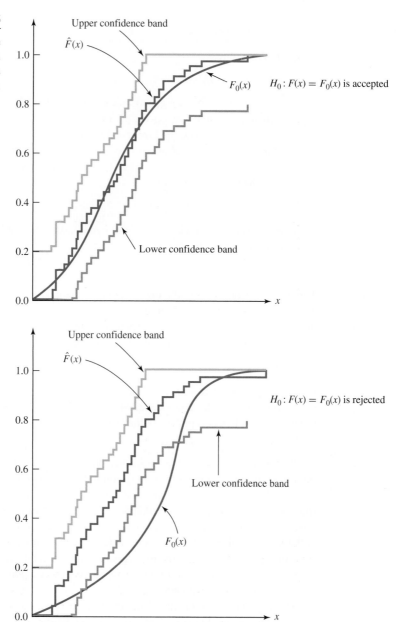

If the null hypothesis is accepted, then it is important not to fall into the trap of concluding that the hypothesis test has "proved" that the distribution function is $F_0(x)$. All that has happened is that $F_0(x)$ has been shown to be one of the many plausible distribution functions that lie within the confidence bands. Finally, recall that the chi-square goodness of fit tests described in Section 10.3.2 provide alternative methods of testing the distribution from which a sample is drawn, and probability plots such as a normal scores plot discussed in Section 12.7.1 can also be used in certain circumstances.

Testing Whether $F(x) = F_0(x)$

The null hypothesis

$$H_0 : F(x) = F_0(x)$$

that an unknown cumulative distribution function $F(x)$ is equal to a specified cumulative distribution function $F_0(x)$ is accepted at size α if $F_0(x)$ lies *completely* within the $1 - \alpha$ confidence level confidence bands for the empirical cumulative distribution function $\hat{F}(x)$ and is rejected if $F_0(x)$ lies outside the confidence bands for some x values. The hypothesis test can be done either graphically or by checking whether the value of the statistic

$$\max_x |\hat{F}(x) - F_0(x)|$$

is less than $d_{\alpha,n}$ in which case the null hypothesis is accepted, or is greater than $d_{\alpha,n}$ in which case the null hypothesis is rejected.

Example 76

Computer System Performance

The sample average execution time is about 15 seconds, and no observations are observed that are quicker than 5 seconds. Is it reasonable to model the execution times as being 5 seconds plus a random variable that has an exponential distribution with mean 10 seconds? This model corresponds to a cumulative distribution function

$$F_0(x) = 1 - e^{-0.1(x-5)}$$

for $x \geq 5$ (see Section 4.2), which is plotted together with the empirical cumulative distribution function and the confidence bands in Figure 15.7. Since the cumulative distribution function $F_0(x)$ lies completely within the confidence bands, it is a plausible model for the execution times.

It should be noted that the confidence bands constructed in this manner can be wide and imprecise. In the example concerning computer system execution times with a sample size of $n = 44$, the total length of the confidence band at a particular x value (where there is no truncation) is $2 \times d_{0.05,44} = 2 \times 0.205 = 0.410$. If a sample of size $n = 100$ is available, the confidence band length is still $2 \times d_{0.05,100} = 2 \times 1.36/\sqrt{100} = 0.272$. Confidence bands at the 0.99 confidence level are even wider. However, one useful point to notice concerning experimental design is that the experimenter knows in advance the precision afforded by a certain sample size. This information may be a useful component in the determination of how large a sample size n to obtain.

The apparent imprecision of the confidence bands is not due to any failings in their construction. In fact, for a given confidence level they are the best confidence bands available, and they efficiently summarize the information attainable from the given sample size and data set. In addition, it must be understood that they are **simultaneous** confidence bands. In other words, at the given confidence level they provide confidence bounds on $F(x)$ simultaneously for all values of x.

15.1.2   The Sign Test

The **sign test** is a basic nonparametric testing procedure that can be used to assess the plausibility of the null hypothesis

$$H_0 : F(x_0) = p_0$$

FIGURE 15.7

A plausible model for the execution
times in the computer system
performance example

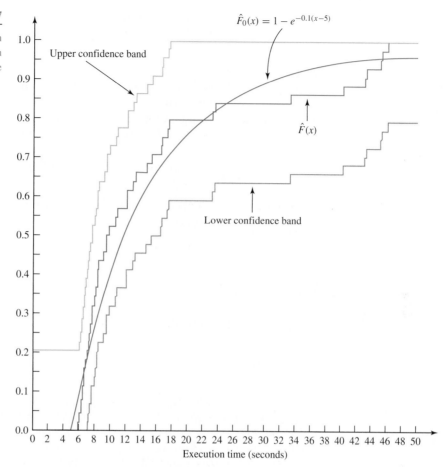

based upon a data set $x_1, \ldots, x_n$ of values that are assumed only to be independent and identically distributed observations from the unknown distribution $F(x)$. For certain fixed values of x_0 and p_0, this null hypothesis implies that the unknown underlying distribution function has the property that an observation no larger than x_0 occurs with a probability of p_0.

Consideration of the statement $F(x_0) = p_0$ is an example of how the experimenter can focus on a specific property of the unknown distribution function $F(x)$ and obtain an analysis that is more efficient than the more general overall estimation of the distribution function $F(x)$ for all values of x discussed in Section 15.1.1. A common example of this approach is to take $p_0 = 0.5$, in which case $F(x_0) = 0.5$ states that the value x_0 is the **median** of the distribution. The median, which will also be the **mean** of the distribution if it is a **symmetric** distribution, is often a useful summary statistic of the distribution with a meaningful interpretation to the experimenter.

As well as the median, inferences on the upper quartile or the lower quartile of the distribution may be of interest and are obtained by taking $p_0 = 0.75$ and $p_0 = 0.25$, respectively. In other cases the tails of the distribution may be of interest.

The first step in the implementation of the sign test is to *discard all data observations that are equal to x_0*. The sample size n is also reduced accordingly. The key aspect of the sign test is then that the remaining data observations are reduced to the single statistic

$$S(x_0) = \#x_i < x_0$$

which counts the number of the data observations that are no larger than x_0. This means that a particular data observation x_i influences the statistic $S(x_0)$ (and consequently influences the sign test) only through the consideration of whether it is *less* than x_0 or *greater* than x_0. Notice that the statistic $S(x_0)$ must take an integer value between 0 and n.

Consideration of the statistic $S(x_0)$ reveals that it has a binomial distribution with parameters n and $F(x_0)$

$$S(x_0) \sim B(n, F(x_0))$$

For reasonably large values of the sample size n it is also useful to consider the following approximate normal distribution for $S(x_0)/n$

$$S(x_0)/n \sim N\left(F(x_0), \frac{F(x_0)(1 - F(x_0))}{n}\right)$$

The assessment of the plausibility of the hypothesis $F(x_0) = p_0$ is based on whether the statistic $S(x_0)$ looks as if it could have come from a $B(n, p_0)$ distribution or, if the normal approximation is being employed, whether the statistic $S(x_0)/n$ looks as if it comes from a

$$N\left(p_0, \frac{p_0(1 - p_0)}{n}\right)$$

distribution.

A p-value can be obtained by comparing $S(x_0)$ with the *tails* of the $B(n, p_0)$ distribution or by comparing $S(x_0)/n$ with the *tails* of the $N(p_0, p_0(1 - p_0)/n)$ distribution. Either one-sided or two-sided tests can be performed depending on whether the experimenter is interested in alternative hypotheses of the form $F(x_0) > p_0$ or $F(x_0) < p_0$ in which case one-sided tests are appropriate, or of the form $F(x_0) \neq p_0$ in which case a two-sided test is appropriate. The details of the hypothesis tests are similar to the tests on a population proportion discussed in Section 10.1.2.

The Sign Test

The sign test for the null hypothesis

$$H_0 : F(x_0) = p_0$$

is based on the statistic

$$S(x_0) = \#x_i < x_0$$

which counts the number of the data observations that are no larger than x_0 (once any data observations equal to x_0 have been discarded). For the two-sided alternative hypothesis

$$H_A : F(x_0) \neq p_0$$

the exact p-value is

$$p\text{-value} = 2 \times P(X \geq S(x_0))$$

The Sign Test, continued

if $S(x_0) > np_0$, and

$$p\text{-value} = 2 \times P(X \le S(x_0))$$

if $S(x_0) < np_0$, where the random variable X has a $B(n, p_0)$ distribution. When np_0 and $n(1 - p_0)$ are both larger than 5, a normal approximation can be used to give a p-value of

$$p\text{-value} = 2 \times \Phi(-|z|)$$

where $\Phi(\cdot)$ is the standard normal cumulative distribution function and

$$z = \frac{S(x_0) - np_0}{\sqrt{np_0(1 - p_0)}}$$

In order to improve the normal approximation the value $S(x_0) - np_0 - 0.5$ may be used in the numerator of the z-statistic when $S(x_0) - np_0 > 0.5$, and the value $S(x_0) - np_0 + 0.5$ may be used in the numerator of the z-statistic when $S(x_0) - np_0 < 0.5$.

For the one-sided alternative hypothesis

$$H_A : F(x_0) < p_0$$

the exact p-value is

$$p\text{-value} = P(X \le S(x_0))$$

which can be approximated by

$$p\text{-value} = \Phi(z)$$

For the one-sided alternative hypothesis

$$H_A : F(x_0) > p_0$$

the exact p-value is

$$p\text{-value} = P(X \ge S(x_0))$$

which can be approximated by

$$p\text{-value} = 1 - \Phi(z)$$

The discussion above has dealt with the assessment of the plausibility of the statement $F(x_0) = p_0$ for fixed values of x_0 and p_0. An extension of this analysis is to consider only p_0 fixed, and then to consider the statement $F(x_0) = p_0$ for all values of x_0 and to "collect" those values of x_0 for which the statement is "acceptable" at a given error rate. Formally, if a collection is made of the x_0 values for which a size α hypothesis test of the statement $F(x_0) = p_0$ accepts, then the collection forms a confidence interval for the quantile $F^{-1}(p_0)$ with a confidence level equal to $1 - \alpha$.

An important special case concerns the construction of a confidence interval for the median of the distribution $F^{-1}(0.5)$, which can be obtained by taking $p_0 = 0.5$. Such a confidence interval is generally of more use to the experimenter than the calculation of the p-values for individual statements that the median takes a specific value. The confidence interval can be thought of as summarizing the plausible values for the median.

If for a data set $x_1, \ldots, x_n$ of size n the ordered values of the data observations are denoted by $x_{(1)} \leq \cdots \leq x_{(n)}$, then the confidence interval for the median generated by the sign test is of the form

$$F^{-1}(0.5) \in (x_{(m)}, x_{(n+1-m)})$$

for some integer value of m no larger than $(n + 1)/2$. The particular value of m depends on the confidence level required. Larger confidence levels result in smaller values of m, since for instance, the confidence interval $(X_{(10)}, X_{(n-9)})$ is shorter and has a smaller confidence level than the confidence interval $(X_{(9)}, X_{(n-8)})$. However, repeat observations (so that for instance $X_{(9)} = X_{(10)}$) may cause some of the confidence interval endpoints to remain unchanged when the confidence level changes. Strictly speaking, only confidence intervals with certain discrete confidence levels are available. For two-sided confidence intervals these confidence levels are $1 - 2p$, where p is a tail probability of the $B(n, 0.5)$ distribution.

Statistical software packages can be used to perform the sign test and to obtain confidence intervals for the median of a distribution, as illustrated in the following examples.

Example 45	Figure 15.8 shows the ordered values of the $n = 15$ data observations of % water pickup ob-

Example 45

Fabric Water Absorption Properties

Figure 15.8 shows the ordered values of the $n = 15$ data observations of % water pickup obtained with a roller pressure of 10 pounds per square inch. Could the median water pickup be 65%? Consider the hypotheses

$$H_0 : F(65) = 0.5 \quad \text{versus} \quad H_A : F(65) \neq 0.5$$

There are no data observations equal to 65, 2 data observations are larger than 65, and 13 data observations are smaller than 65, so that

$$S(65) = 13$$

The exact p-value is therefore

$$p\text{-value} = 2 \times P(X \geq 13) = 0.0074$$

where the random variable X has a $B(15, 0.5)$ distribution. This small p-value indicates that 65% is not a plausible value for the median value of the % water pickup.

If a confidence interval for the median % water pickup is of interest with a $1 - \alpha = 0.95$ confidence level, then a statistical software package can provide the confidence intervals with confidence levels either side of 95%. These intervals are

$$(56.70, 61.80)$$

with a confidence level of $1 - \alpha = 0.8815$, and

$$(55.80, 64.00)$$

with a confidence level of $1 - \alpha = 0.9648$. Notice that if $x_{(1)} \leq \cdots \leq x_{(15)}$ are the ordered values of the data observations, then these confidence intervals are $(x_{(5)}, x_{(11)})$ and $(x_{(4)}, x_{(12)})$.

FIGURE 15.8

Ordered % water pickup values for the fabric water absorption properties example

51.8	54.5	54.5	55.8	56.7	57.3	59.1	59.5
60.4	61.2	61.8	64.0	64.9	65.4	70.2	

<table>
<tr><td align="right">Example 76
Computer System
Performance</td><td>Consider a test that the median execution time is equal to 10 seconds. With the two-sided hypotheses.</td></tr>
</table>

$$H_0 : F(10) = 0.5 \quad \text{versus} \quad H_A : F(10) \neq 0.5$$

there are no data observations equal to 10, 21 data observations larger than 65, and 23 data observations smaller than 65, so that

$$S(10) = 23$$

and the exact p-value is

$$p\text{-value} = 2 \times P(X \geq 23) = 0.8804$$

where the random variable X has a $B(44, 0.5)$ distribution. Consequently, 10 seconds is clearly a plausible value for the median execution time. This is not surprising because 10 seconds is contained within the confidence intervals for the median execution time, which can be found to be

$$(x_{(17)}, x_{(28)}) = (8.10, 12.60)$$

with a confidence level of 0.9039 and

$$(x_{(16)}, x_{(29)}) = (7.90, 12.90)$$

with a confidence level of 0.9512.

Finally, consider the one-sided hypothesis testing problem

$$H_0 : F(7.5) \geq 0.5 \quad \text{versus} \quad H_A : F(7.5) < 0.5$$

The null hypothesis states that the probability of an execution time being shorter than 7.5 seconds is larger than 0.5, which in other words implies that the median execution time is no longer than 7.5 seconds. The alternative hypothesis states that the median execution time is longer than 7.5 seconds.

The first step in this analysis is to discard the two data observations equal to $x_0 = 7.5$. Of the remaining 42 data observations, 12 are smaller than 7.5 seconds and 30 are larger than 7.5 seconds, and so with

$$S(7.5) = 12$$

the p-value is

$$p\text{-value} = P(X \leq 12) = 0.004$$

where the random variable X has a $B(42, 0.5)$ distribution. This analysis therefore establishes that the median execution time is longer than 7.5 seconds.

The sign test is a particularly appropriate methodology for providing a simple basic analysis of *paired* data sets (see Section 9.2). Remember that the analysis of paired data sets $x_1, \ldots, x_n$ and $y_1, \ldots, y_n$ reduces to a one-sample problem through the consideration of the differences

$$z_1 = x_1 - y_1, \ldots, z_n = x_n - y_n$$

If $F(x)$ is defined to be the distribution function of these differences, then $z_1, \ldots, z_n$ can be considered to be independent, identically distributed observations from this distribution, and the sign test can be used to make inferences on the distribution function $F(x)$.

Specifically, the experimenter is usually interested in estimating the median of the distribution $F(x)$ and assessing whether or not it is 0. If the median is 0, then this implies that an observation x_i is equally likely to be less than or greater than the corresponding observation y_i, and this implies a certain equivalence between the process that generated the observations x_i and the process that generated the observations y_i. If the median is greater than 0, then this implies that the observations from the first population tend to be larger than the observations from the second population, and vice versa if the median is less than 0.

If the sign test is used to investigate whether the median of the distribution function $F(x)$ is 0, then the analysis depends on how many of the differences z_i are positive and how many are negative. In other words, the sign test is simply based on how many pairs there are with $x_i > y_i$ and how many there are with $x_i < y_i$.

Example 55 **Heart Rate Reductions**	Figure 9.13 contains the percentage reductions in heart rate for the standard drug x_i and the new drug y_i, together with the differences z_i for the $n = 40$ experimental subjects. The assessment using the sign test of whether the median reductions are the same for both drugs is based on the fact that 30 of the differences z_i are negative while only 10 are positive. In other words, 30 of the patients exhibited a larger reduction in heart rate using the new drug rather than with the standard drug, while only 10 patients exhibited a larger reduction in heart rate using the standard drug rather than with the new drug.

Formally, the two-sided hypothesis testing problem of interest is

$$H_0 : F(0) = 0.5 \quad \text{versus} \quad H_A : F(0) \neq 0.5$$

where $F(x)$ is the distribution function of the differences z_i. With

$$S(0) = 30$$

the exact p-value is

$$p\text{-value} = 2 \times P(X \geq 30) = 0.0022$$

where the random variable X has a $B(40, 0.5)$ distribution. This low p-value confirms that it is not plausible that the two drugs have equivalent effects.

Finally, it is interesting to note that the sign test is one of the oldest and most fundamental testing procedures. In fact, the sign test can be traced back to 1710 when John Arbuthnot, physician to Queen Anne, used data on infants in the city of London to investigate whether male births were more frequent than female births.

In addition, remember that gambling has provided much of the impetus behind the development of probability theory and statistical analysis, and in this context consider the following two questions:

■ If a coin is tossed 50 times and 30 heads and 20 tails are observed, what is the plausibility of the coin being fair?

■ If a six-sided die is rolled 100 times and a 6 is scored on 10 of the rolls, what is the plausibility of the die being fair?

These questions can both be easily answered with the use of the sign test. A common feature of the two questions is that the data are of a binary form taking one of two possible outcomes. In the first question an observation is recorded as "a head or a tail," and in the second question an observation is recorded as being " a 6 or not a 6." With this in mind, the key to an understanding of the application of the sign test toward the consideration of statements of the

FIGURE 15.9

Some symmetric probability
density functions

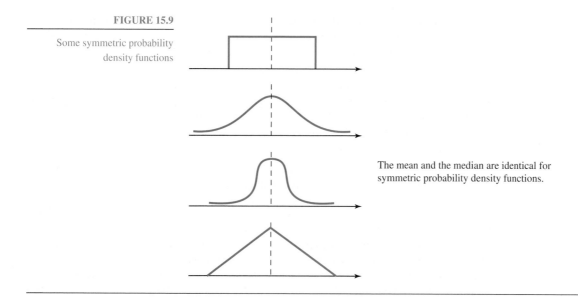

The mean and the median are identical for
symmetric probability density functions.

kind $F(x_0) = p_0$ is that the data observations x_i are being reduced to a binary form through
the determination of whether $x_i > x_0$ or whether $x_i < x_0$.

15.1.3 The Signed Rank Test

The **signed rank test** (often referred to as the Wilcoxon one-sample test procedure) can be
used to make inferences about the *median* of an unknown distribution function $F(x)$ under the
assumption that the probability density (mass) function is *symmetric*, as shown in Figure 15.9.
If the distribution is symmetric, then the median of the distribution is equal to the *mean* of
the distribution, and they can both be denoted by $\mu = F^{-1}(0.5)$. Recall that the sign test
discussed in Section 15.1.2 can also be used to provide inferences about the median without
the necessity of making any assumptions about the distribution function $F(x)$. However, the
motivation behind the use of the signed rank test is that it provides a more precise analysis in
situations where the distribution function can reasonably be taken to be symmetric.

The assumption of symmetry may be justified by an experimenter through observation or
through an understanding of the nature of the data. For example, a simple histogram of the data
may be sufficient to indicate that the assumption of symmetry is not inappropriate. In other
cases, the experimenter may expect that the randomness in the data is composed primarily of
a measurement error that can reasonably be expected to be distributed symmetrically.

The additional precision of the signed rank test over the sign test is achieved through
the consideration of the magnitudes of the deviations of the observed data values x_i from the
hypothesized median value μ_0. The sign test is performed by looking at whether an observation
x_i is less than μ_0 or is greater than μ_0. The signed rank test not only takes into account this
information but also considers

$$|x_i - \mu_0|$$

the magnitude of the distance between x_i and μ_0. The magnitudes provide additional information about the location of the median when the distribution is assumed to be symmetric.

Specifically, the signed rank test procedure for testing the plausibility of the statement
$\mu = \mu_0$ concerning the mean of a symmetric distribution, based on a data sample $x_1, \ldots, x_n$

of independent observations from this distribution, operates by first calculating the deviations of the data observations x_i from the hypothesized mean value μ_0

$$d_i(\mu_0) = x_i - \mu_0$$

As long as the data are not all to one side of μ_0, some of these deviations will be positive and some will be negative. The **ranks** $r_i(\mu_0)$ are then calculated by ranking the **absolute values** of the deviations $|\, d_i(\mu_0)\,|$. For example, if $n = 5$ and

$$|d_3(\mu_0)| < |d_2(\mu_0)| < |d_5(\mu_0)| < |d_1(\mu_0)| < |d_4(\mu_0)|$$

then $r_3(\mu_0) = 1$, $r_2(\mu_0) = 2$, $r_5(\mu_0) = 3$, $r_1(\mu_0) = 4$, and $r_4(\mu_0) = 5$. The signed rank test is based on the consideration of the statistic $S_+(\mu_0)$, which is the sum of the ranks of the observations x_i with positive deviations, $d_i(\mu_0) > 0$.

The sum of all of the ranks is

$$1 + 2 + \cdots + n = \frac{n(n+1)}{2}$$

and if μ_0 really is the mean of the distribution, then $S_+(\mu_0)$ can be expected to be about half of this value, that is about $n(n+1)/4$. However, if $S_+(\mu_0)$ is much larger than $n(n+1)/4$, then there is the suggestion that the mean μ is really larger than μ_0, as indicated in Figure 15.10. On the other hand, if $S_+(\mu_0)$ is much smaller than $n(n+1)/4$, then there is the suggestion that the mean μ is really smaller than μ_0.

Two practical points concerning the calculation of the statistic $S_+(\mu_0)$ involve data observations x_i that are equal to μ_0 and tied values of the magnitudes of the deviations $|d_i(\mu_0)|$. If an observation x_i is equal to μ_0, then it should be deleted from the data set and the sample

FIGURE 15.10

Interpretation of the signed rank test statistic $S_+(\mu_0)$

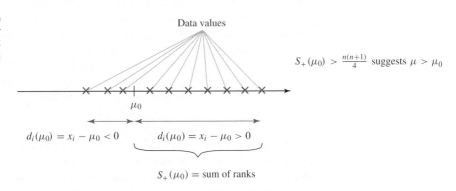

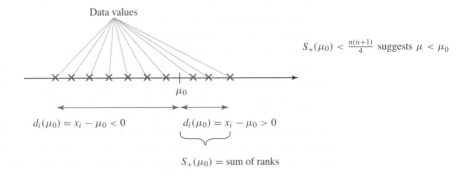

size n should be reduced by 1. If some of the absolute values of the deviations $|d_i(\mu_0)|$ are tied, which happens if there are repeat data observations x_i, or if some data values are equal amounts above and below μ_0, then it is appropriate to allocate each deviation the average of the corresponding ranks. For instance, if the fifth and sixth smallest deviation magnitudes $|d_i(\mu_0)|$ are equal, then they should both be ranked 5.5. If the fifth, sixth, and seventh smallest deviation magnitudes $|d_i(\mu_0)|$ are all equal, then they should all be ranked 6. This procedure ensures that the sum of the ranks always equals $n(n+1)/2$.

Either one-sided or two-sided p-values can be constructed for the plausibility of the statement $\mu = \mu_0$ by comparing the observed value of the statistic $S_+(\mu_0)$ with its distribution when $\mu = \mu_0$. This distribution can be calculated exactly for small sample sizes n, and a normal approximation to the distribution suffices for larger sample sizes.

If there are no ties in the absolute values of the deviations $|d_i(\mu_0)|$, and if $\mu = \mu_0$, then the statistic $S_+(\mu_0)$ can be written as

$$S_+(\mu_0) = (1 \times W_1) + (2 \times W_2) + \cdots + (n \times W_n)$$

where $W_1, \ldots, W_n$ are independent, identically distributed random variables that are equally likely to take the values 0 and 1. This representation of $S_+(\mu_0)$ follows from the fact that when $\mu = \mu_0$, so that the distribution is symmetric about μ_0, the deviation magnitude $|d_i(\mu_0)|$ corresponding to a particular rank j, say, is equally likely to be based upon an observation x_i that is greater than μ_0 or that is less than μ_0. Consequently, each of the ranks $1, \ldots, n$ is equally likely to contribute to $S_+(\mu_0)$ or not to contribute.

The exact distribution of the statistic $S_+(\mu_0)$ when $\mu = \mu_0$ is straightforward to calculate using this representation, although the calculations quickly become computationally intensive as the sample size n increases. Essentially, there are 2^n possible sets of values of the random variables $W_1, \ldots, W_n$ that are each equally likely with probability $1/2^n$. The exact discrete distribution of $S_+(\mu_0)$ is calculated by noticing that there is a probability of $m/2^n$ that $S_+(\mu_0) = x$, where m is the number of possible sets of values of $W_1, \ldots, W_n$ for which

$$(1 \times W_1) + (2 \times W_2) + \cdots + (n \times W_n) = x$$

Usually a computer package will calculate an exact p-value in this manner, although an approximate normal distribution for the statistic $S_+(\mu_0)$ may be used if the sample size is large. This normal approximation is determined by the mean and the variance of the statistic. When $\mu = \mu_0$, the mean value of $S_+(\mu_0)$ is $n(n+1)/4$, and since $\text{Var}(W_i) = 1/4$ the variance of $S_+(\mu_0)$ is

$$\text{Var}(S_+(\mu_0)) = (1^2 \times \text{Var}(W_1)) + (2^2 \times \text{Var}(W_2)) + \cdots + (n^2 \times \text{Var}(W_n))$$
$$= \frac{1^2 + 2^2 + \cdots + n^2}{4} = \frac{n(n+1)(2n+1)}{24}$$

Consequently, when $\mu = \mu_0$ the statistic

$$Z_+(\mu_0) = \frac{S_+(\mu_0) - n(n+1)/4}{\sqrt{n(n+1)(2n+1)/24}}$$

can be considered to have approximately a standard normal distribution.

In practice, a continuity correction of 0.5 helps increase the accuracy of the p-value calculations made by comparing the observed value of $Z_+(\mu_0)$ with the tail probabilities of the standard normal distribution. The amount 0.5 should be added to the numerator of $Z_+(\mu_0)$ if it is negative and subtracted from the numerator of $Z_+(\mu_0)$ if it is positive.

A positive value of $Z_+(\mu_0)$ suggests that the mean μ is larger than μ_0, and a negative value of $Z_+(\mu_0)$ suggests that the mean μ is smaller than μ_0. One-sided or two-sided

p-values can be obtained by comparing the observed value of $Z_+(\mu_0)$ with the tail probabilities of the standard normal distribution. For example, if $Z_+(\mu_0)$ is observed to be 2.00, then a one-sided p-value (for the alternative that the mean value is greater than μ_0) is calculated as $1 - \Phi(2.00) = 0.0228$, and a two-sided p-value is $2 \times 0.0228 = 0.0456$.

The Signed Rank Test

The **signed rank test** procedure for testing that the mean of a *symmetric* distribution is equal to a hypothesized value μ_0, based upon a data set $x_1, \ldots, x_n$, is performed by ranking the *absolute values* of the deviations $d_i(\mu_0) = x_i - \mu_0$. The test is based upon the consideration of the statistic $S_+(\mu_0)$, which is the sum of the ranks of the observations x_i with positive deviations $d_i(\mu_0) > 0$. One-sided or two-sided p-values can be calculated by comparing this statistic with its distribution when the mean is equal to μ_0.

The signed rank test procedure can be used to construct confidence intervals for the mean of a distribution when it is assumed to be symmetric. A $1 - \alpha$ confidence level confidence interval consists of the values μ_0 for which the null hypothesis $H_0 : \mu = \mu_0$ is accepted at size α. As with the sign test, only certain discrete values of the confidence level are attainable although they can often be found very close to typical values such as 95% and 99%. The use of the signed rank test is illustrated in the following examples.

Example 45

Fabric Water Absorption Properties

Consider the two-sided hypothesis testing problem that the median water pickup is 65%

$$H_0 : F(65) = 0.5 \quad \text{versus} \quad H_A : F(65) \neq 0.5$$

based on the assumption that the distribution of the % water pickup is symmetric. Figure 15.11 shows the deviations of the data values from $\mu_0 = 65$ together with the ranks of the absolute values. The two observations with positive deviations, $x = 65.4$ and $x = 70.2$, have ranks 2

FIGURE 15.11

Calculation of the signed rank test statistic for the fabric water absorption properties example

Data values x_i	Deviations $d_i(65) = x_i - 65$	Absolute deviations $\|d_i(65)\|$	Ranks $r_i(65)$
51.8	−13.2	13.2	15
54.5	−10.5	10.5 ⎫	13.5
54.5	−10.5	10.5 ⎭	13.5
55.8	−9.2	9.2	12
56.7	−8.3	8.3	11
57.3	−7.7	7.7	10
59.1	−5.9	5.9	9
59.5	−5.5	5.5	8
60.4	−4.6	4.6	6
61.2	−3.8	3.8	5
61.8	−3.2	3.2	4
64.0	−1.0	1.0	3
64.9	−0.1	0.1	1
65.4	0.4	0.4	2 ⎫ $S_+(65) = 2 + 7 = 9$
70.2	5.2	5.2	7 ⎭

and 7 respectively, so that the statistic employed by the signed rank test is

$$S_+(65) = 2 + 7 = 9$$

A statistical software package can be used to show that this statistic gives a p-value of 0.004, so that as with the analysis provided by the sign test, this finding establishes that 65% is not a plausible value for the median water pickup. Also, a 95% confidence interval for the median % water pickup constructed from the signed rank test is (56.80, 62.60), which is similar to the confidence interval obtained using the t-test in Section 8.1.3.

The signed rank test is particularly appropriate for analyzing paired data sets because the assumption of distributional symmetry is particularly appropriate in this case. Recall from Section 9.2.1 that if observations from one population are modeled as

$$x_i = \mu_A + \gamma_i + \epsilon_i^A$$

and if observations from the other population are modeled as

$$y_i = \mu_B + \gamma_i + \epsilon_i^B$$

then the differences z_i are

$$z_i = \mu_A - \mu_B + \epsilon_i^{AB}$$

where the error term is

$$\epsilon_i^{AB} = \epsilon_i^A - \epsilon_i^B$$

If the error terms ϵ_i^A and ϵ_i^B are identically distributed, then the error term ϵ_i^{AB} necessarily has a distribution that is symmetric about 0. Consequently, the differences z_i are symmetrically distributed about a mean value of $\mu_A - \mu_B$.

In other words, if the variations in the two populations under consideration can be assumed to be identical (although not necessarily symmetric), so that the distribution functions $F_A(x)$ and $F_B(x)$ are identical except for a location shift (that is, $F_A(x) = F_B(x + \delta)$ for some $\delta = \mu_B - \mu_A$), then the differences z_i necessarily have a symmetric distribution and it is appropriate to employ the signed rank test. This assumption that the population distributions are identical except for a location shift is often reasonable and therefore it is often appropriate to use the signed rank test to analyze paired data sets.

Example 55 **Heart Rate Reductions**	Figure 15.12 shows the differences in the heart rate reductions z_i together with the ranks of their absolute values. Notice that patient 20 has the smallest absolute value difference $|z_{20}| = 0.1$ and that this is given a rank of 1. Patient 13 has a difference $|z_{13}| = 0.3$, which is given a rank of 2. Patients 7, 8, and 15 then all have differences of either 0.5 or -0.5 and so they are each given the average of the ranks 3, 4, and 5, which is a ranking of 4 each. Patient 40 has a rank of 6, and patients 6 and 11 are then tied with $|z_6| = |z_{11}| = 0.8$, and they are each given a rank of 7.5.

The signed rank test that the differences z_i have a mean $\mu_0 = 0$ is based on the sum of the ranks of the positive differences z_i, which is

$$S_+(0) = 127.5$$

A computer package can be used to show that this statistic gives a very small p-value, and that a 95% confidence interval for the median of the distribution of the differences z_i based on the signed rank test is $(-3.85, -1.30)$. This interval implies that the new drug provides a reduction in a patient's heart rate of somewhere between 1% and 4% more on average than the

FIGURE 15.12

Calculation of the signed rank test
statistic for the heart rate
reductions example

Patient	Standard drug x_i	New drug y_i	Differences z_i	Absolute differences $|z_i|$	Rank $r_i(0)$
1	28.5	34.8	−6.3	6.3	30.5
2	26.6	37.3	−10.7	10.7	40.0
3	28.6	31.3	−2.7	2.7	21.0
4	22.1	24.4	−2.3	2.3	18.5
5	32.4	39.5	−7.1	7.1	35.0
6	33.2	34.0	−0.8	0.8	7.5
7	32.9	33.4	−0.5	0.5	4.0
8	27.9	27.4	0.5	0.5	4.0
9	26.8	35.4	−8.6	8.6	38.0
10	30.7	35.7	−5.0	5.0	28.0
11	39.6	40.4	−0.8	0.8	7.5
12	34.9	41.6	−6.7	6.7	33.0
13	31.1	30.8	0.3	0.3	2.0
14	21.6	30.5	−8.9	8.9	39.0
15	40.2	40.7	−0.5	0.5	4.0
16	38.9	39.9	−1.0	1.0	9.0
17	31.6	30.2	1.4	1.4	12.5
18	36.0	34.5	1.5	1.5	14.0
19	25.4	31.2	−5.8	5.8	29.0
20	35.6	35.5	0.1	0.1	1.0
21	27.0	25.3	1.7	1.7	15.0
22	33.1	34.5	−1.4	1.4	12.5
23	28.7	30.9	−2.2	2.2	17.0
24	33.7	31.9	1.8	1.8	16.0
25	33.7	36.9	−3.2	3.2	25.0
26	34.3	27.8	6.5	6.5	32.0
27	32.6	35.7	−3.1	3.1	24.0
28	34.5	38.4	−3.9	3.9	27.0
29	32.9	36.7	−3.8	3.8	26.0
30	29.3	36.3	−7.0	7.0	34.0
31	35.2	38.1	−2.9	2.9	22.0
32	29.8	32.1	−2.3	2.3	18.5
33	26.1	29.1	−3.0	3.0	23.0
34	25.6	33.5	−7.9	7.9	37.0
35	27.6	28.7	−1.1	1.1	10.0
36	25.1	31.4	−6.3	6.3	30.5
37	23.7	22.4	1.3	1.3	11.0
38	36.3	43.7	−7.4	7.4	36.0
39	33.4	30.8	2.6	2.6	20.0
40	40.1	40.8	−0.7	0.7	6.0

$$S_+(0) = 4 + 2 + 12.5 + 14 + 1 + 15 + 16 + 32 + 11 + 20 = 127.5$$

standard drug, and the results of this analysis are seen to be similar to the analysis provided by the sign test in Section 15.1.2 and the t-test in Section 9.2.2.

The signed rank test provides a useful middle ground between the sign test and a fully parametric test such as the t-test under a normal distributional assumption. The sign test is a very general procedure that requires no distributional assumptions. The signed rank test utilizes the additional information provided by an assumption that the distribution is symmetric and allows more precise inferences to be made about the mean of the distribution in this case. This assumption of symmetry can often be justified by an experimenter and provides a middle ground that is less restrictive than the exact specification of a symmetric distribution such as the normal distribution.

The choice of which inference method to adopt should be made on the basis of which set of assumptions are reasonable, and histograms or other data plots may be useful to investigate the form of the distribution. The experimenter should make use of any reasonable assumptions, so that for example, the signed rank test should be used in preference to the sign test when the assumption of symmetry can be justified. Remember that the central limit theorem implies that the t-test (or equivalently the z-test) can be used for large enough sample sizes regardless of the actual distribution of the data observations, although the sign test and the signed rank test in practice often provide as precise an inference as the t-test and may still be preferred.

Finally, it is worthwhile pointing out that an additional advantage of nonparametric test procedures, besides the fact that they require minimal distributional assumptions, is that they are generally less sensitive than parametric test procedures to "bad" data points. Such data points may be outliers, in the sense that they are really from a different distribution than the rest of the data set (due perhaps to some inadvertent change in the experimental conditions), or they may simply be miscoded or mistyped values.

The sign test is particularly robust to these kinds of errors, because if an observation of 4.00 is incorrectly recorded as 6.00 say, then any inferences about $F(x_0)$ for $x_0 < 4.00$ or $x_0 > 6.00$ are not affected because the statistic of interest, the number of data observations x_i on either side of x_0, is unchanged. At worst, a bad data point causes this statistic to be incorrect by an amount one.

The signed rank test is also quite robust to these kinds of errors. Parametric test procedures are generally less robust because an incorrect data point usually affects the estimates of each of the parameters. If a t-test is employed, for example, then an incorrect data point directly influences both the estimated mean and the estimated variance.

■ 15.1.4 Problems

15.1.1 Restaurant Service Times

(a) Construct the empirical cumulative distribution function for the data set of restaurant service times given in DS 6.1.4.

(b) Draw 95% confidence bands around the empirical cumulative distribution function.

(c) Is it plausible that the service times are normally distributed with a mean of 70 seconds and a standard deviation of 20 seconds?

(d) Is it plausible that the service times are exponentially distributed with a mean of 70 seconds?

(e) Consider the null hypothesis that the median service time is no longer than 65 seconds. What statistic is used by the sign test procedure to test this null hypothesis? What is the p-value?

(f) Test the null hypothesis in part (e) using the signed rank test.

(g) Use the sign test and the signed rank test to obtain 95% confidence intervals on the median service time.

15.1.2 Paving Slab Weights

(a) Construct the empirical cumulative distribution function for the data set of paving slab weights given in DS 6.1.7.

(b) Draw 95% confidence bands around the empirical cumulative distribution function.

(c) Is it plausible that the paving slab weights are normally distributed with a mean of 1.1 kg and a standard deviation of 0.05 kg? How about with a mean of 1.0 kg and a standard deviation of 0.05 kg?

(d) Consider the null hypothesis that the median paving slab weight is 1.1 kg. What statistic is used by the sign test procedure to test this null hypothesis? What is the p-value?

(e) Test the null hypothesis in part (d) using the signed rank test and the t-test. Compare your answers.

(f) Use the sign test, the signed rank test, and the t-test to obtain 95% confidence intervals for the median (or mean) paving slab weight. What assumptions are required by these three test procedures? Do the assumptions seem appropriate? How would you summarize your results?

15.1.3 Spray Painting Procedure

Construct the empirical cumulative distribution function for the data set of paint thicknesses given in DS 6.1.8. Draw 95% confidence bands around the empirical cumulative distribution function. Analyze the median (or mean) paint thickness using the sign test, the signed rank test, and the t-test. Pay particular attention to whether the median paint thickness is 0.2 mm. Which analysis method do you prefer? Summarize your conclusions.

15.1.4 Plastic Panel Bending Capabilities

Construct the empirical cumulative distribution function for the data set of deformity angles given in DS 6.1.9. Draw 95% confidence bands around the empirical cumulative distribution function. Analyze the median (or mean) deformity angle using the sign test, the signed rank test, and the t-test. Do you think that the assumptions behind these test procedures are valid? What evidence is there that the plastic can bend less than $9.5°$ on average before deforming?

15.1.5 Suppose that the data set in DS 15.1.1 consists of values that can be taken to be independent observations from a particular distribution. Consider testing whether the median of the distribution is equal to 18.0 against a two-sided alternative.

(a) What is the value of the test statistic used by the sign test?

(b) Write down an expression for the exact p-value using the sign test.

(c) Use the normal approximation to calculate the p-value using the sign test.

(d) What is the value of the test statistic used by the signed rank test?

(e) Use the normal approximation to calculate the p-value using the signed rank test.

15.1.6 Repeat Problem 15.1.5 using the data set in DS 15.1.2 and for the null hypothesis that the median of the distribution is equal to 40 against a two-sided alternative hypothesis.

15.1.7 Production Line Assembly Methods

Use the sign test and the signed rank test to analyze the paired data set of assembly times given in DS 9.2.1. Why might it be expected that the signed rank test is a better test procedure than the sign test for this paired data set? Do you find any evidence of a difference between the two assembly methods?

15.1.8 Red Blood Cell Adherence to Endothelial Cells

Use the sign test and the signed rank test to analyze the paired data set of adherent red blood cells given in DS 9.2.2. Do you find any evidence of a difference between the two stimulation conditions?

15.1.9 Calculus Teaching Methods

Use the sign test and the signed rank test to analyze the paired data set of calculus scores given in DS 9.2.4. Do you find any evidence of a difference between the two teaching methods? How much better is the new teaching method?

15.1.10 Radioactive Carbon Dating

Use the sign test and the signed rank test to analyze the paired data set given in DS 9.2.5 concerning the radioactive carbon dating methods. Do you find any evidence of a difference between the two dating methods?

15.1.11 Golf Ball Design

Use the sign test and the signed rank test to analyze the paired data set of golf shots given in DS 9.2.6. Do you find any evidence of a difference between the two ball types?

15.2 Comparing Two Populations

In this section the problem of comparing two distribution functions $F_A(x)$ and $F_B(x)$ based upon the analysis of two unpaired independent samples is discussed. Specifically, let $x_1, \ldots, x_n$ be independent, identically distributed observations from the distribution function $F_A(x)$, and let $y_1, \ldots, y_m$ be independent, identically distributed observations from the distribution function $F_B(x)$. In addition, suppose that the two samples are obtained independently of each other, as is the case with unpaired samples. Notice that the two sample sizes n and m need not be equal, although many experiments are designed to have equal sample sizes.

Two test procedures are considered in this section, the **Kolmogorov-Smirnov test** procedure and the **rank sum test** procedure. The Kolmogorov-Smirnov test addresses the general question of whether the two distribution functions $F_A(x)$ and $F_B(x)$ can be considered to be identical. The rank sum test procedure is based upon the assumption that the two distribution functions are identical except for a location shift, so that $F_A(x) = F_B(x + \delta)$ for some value δ, and it can be used to construct a confidence interval for and perform hypothesis tests of the location difference δ.

15.2.1 The Kolmogorov-Smirnov Test

The Kolmogorov-Smirnov test is a general test of the equivalence of two distribution functions $F_A(x)$ and $F_B(x)$, and it is based upon the comparison of the two empirical cumulative distribution functions $\hat{F}_A(x)$ and $\hat{F}_B(x)$. Recall from Section 15.1.1 that these are given by

$$\hat{F}_A(x) = \frac{\#x_i \leq x}{n} \qquad \text{and} \qquad \hat{F}_B(x) = \frac{\#y_i \leq x}{m}$$

As discussed in Section 15.1.1, $\hat{F}_A(x)$ and $\hat{F}_B(x)$ provide the best estimates of the true unknown distribution functions $F_A(x)$ and $F_B(x)$, respectively. The assessment of the plausibility that the two distribution functions $F_A(x)$ and $F_B(x)$ are identical is based on the consideration of how close together the two empirical cumulative distribution functions $\hat{F}_A(x)$ and $\hat{F}_B(x)$ appear to be.

The Kolmogorov-Smirnov Test

If $x_1, \ldots, x_n$ are independent, identically distributed observations from a distribution function $F_A(x)$ and $y_1, \ldots, y_m$ are independent, identically distributed observations from a distribution function $F_B(x)$, then the Kolmogorov-Smirnov test of the equivalence of two distribution functions $F_A(x)$ and $F_B(x)$ is based on the comparison of the two empirical cumulative distribution functions $\hat{F}_A(x)$ and $\hat{F}_B(x)$. The test statistic

$$M = \max_x |\hat{F}_A(x) - \hat{F}_B(x)|$$

which is the maximum vertical distance between the two empirical cumulative distribution functions, is used as a measurement of how close together the two empirical cumulative distribution functions are. For sample sizes n and m larger than 20, a size α hypothesis test of

$$H_0 : F_A(x) = F_B(x)$$

for all values of x is performed by comparing the value of the statistic M with the critical point

$$d_\alpha \sqrt{\frac{1}{n} + \frac{1}{m}}$$

where the values of d_α are given in Section 15.1.1. Values of the statistic M greater than the critical point cause the null hypothesis H_0 to be rejected at the size α confidence level, and values of M smaller than the critical point result in the null hypothesis being accepted.

Generally speaking, when using the Kolmogorov-Smirnov test, it is advisable to have sample sizes n and m both larger than 20. Critical points for use with sample sizes smaller than 20 can be found, although it must be remembered that with such small sample sizes the test procedure has very little power. In other words, with sample sizes less than 20 the null hypothesis $H_0 : F_A(x) = F_B(x)$ is unlikely to be rejected unless there is a substantial difference between the two distributions functions $F_A(x)$ and $F_B(x)$. Also, as a practical matter the maximum distance between $\hat{F}_A(x)$ and $\hat{F}_B(x)$ must occur at one of the data values x_i or y_i.

It is interesting and instructive to understand the relationship between the problem of comparing two empirical cumulative distribution functions, as discussed here, and the problem of comparing one empirical cumulative distribution function with a known distribution function $F_0(x)$, as discussed in Section 15.1.1. The connection between the two problems is that they become identical as one of the two sample sizes n or m in the two-sample problem increases to infinity.

In such a case, having an infinite number of observations from a distribution is conceptually equivalent to knowing the distribution exactly, so that the limiting value of the empirical cumulative distribution function can essentially be taken to be a known distribution function $F_0(x)$. Notice that the two-sample critical point $d_\alpha \sqrt{1/n + 1/m}$ reduces to $d_\alpha / \sqrt{n}$ as the sample size m tends to infinity, which is the critical point used in Section 15.1.1 for hypothesis testing and confidence band construction.

Consequently, it can be seen that the discussion in Section 15.1.1 is simply a special case of the two-sample problem discussed here, which is attained by letting one of the sample sizes grow to infinity. The two-sample test was first discussed by Smirnov in 1939, although it is usually referred to as the Kolmogorov-Smirnov test because of its similarity with the one-sample problem discussed by Kolmogorov in 1933.

Example 77
Petroleum Processing

In petroleum processing changes in the conditions of crude oil and natural gas may cause dissolved waxes to crystallize and to therefore affect the operation of processing equipment. If the waxes can be made to crystallize in a structure that is less likely to form a deposit, or if the waxes can be prevented from crystallizing altogether, then a problem that may be of serious concern to the petroleum industry can be mitigated. It is therefore important to study the crystallization properties of the dissolved waxes.

An experiment is performed whereby a solution of waxes is expanded rapidly through a nozzle, as shown in Figure 15.13, so that the wax particles can be collected on a plate. Scanning electron microscopy is then used to photograph the samples of wax particles so that their diameters can be measured. A question of importance is how the pre-expansion temperature of the solution affects the particle sizes.

FIGURE 15.13

The experimental apparatus for the petroleum processing example

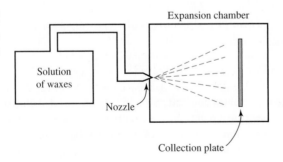

FIGURE 15.14

Data set of particle diameters for the petroleum processing example

Pre-expansion temperature of 80°C (temperature A)

9.00	9.50	9.75	10.50	10.50	11.50	12.00	12.25	12.75
13.25	14.25	14.75	15.25	17.00	18.25	18.25	18.50	18.75
19.25	19.75	20.25	20.75	22.00	22.75	22.75	23.00	23.00
23.25	26.00	26.00	29.00	31.75	32.50	33.00	36.50	38.00

Pre-expansion temperature of 170°C (temperature B)

4.00	4.17	4.75	4.83	5.17	5.33	5.50	5.67	5.67
5.67	6.17	6.17	6.67	6.67	7.33	7.60	7.67	7.67
7.75	8.17	8.50	8.50	8.67	8.67	8.67	9.17	9.17
9.33	9.44	9.67	10.00	10.00	10.25	10.67	10.83	12.16

FIGURE 15.15

Descriptive statistics and boxplots for the petroleum processing example

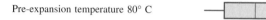

Pre-expansion temperature 80° C

Pre-expansion temperature 170° C

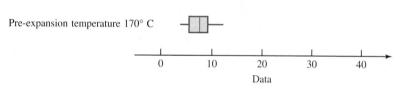

0   10   20   30   40

Data

Pre-expansion temperature 80° C
Sample size = 36
Sample mean = 19.88
Sample standard deviation = 7.88
Sample maximum = 38.00
Sample upper quartile = 23.19
Sample median = 19.00
Sample lower quartile = 12.88
Sample minimum = 9.00

Pre-expansion temperature 170° C
Sample size = 36
Sample mean = 7.68
Sample standard deviation = 2.09
Sample maximum = 12.16
Sample upper quartile = 9.29
Sample median = 7.71
Sample lower quartile = 5.67
Sample minimum = 4.00

In the first experiment a pre-expansion temperature of 80°C is used, and in a second experiment a pre-expansion temperature of 170°C is used. In either case 36 samples of particles are obtained, and the means of the particle diameters within each sample are calculated. The resulting data set is shown in Figure 15.14.

How does the pre-expansion temperature affect the particle sizes? This question can be answered by comparing the distribution function of particle sizes for a pre-expansion temperature of 80°C, $F_A(x)$, with the distribution function of particle sizes for a pre-expansion temperature of 170°C, $F_B(x)$. Figure 15.15 shows boxplots and sample statistics for the two data sets that indicate that the sample mean and the sample median of the particle sizes are much smaller at the higher pre-expansion temperature. In addition, notice that the sample standard deviation is also much smaller at the higher pre-expansion temperature.

These observations are confirmed by the plot of the two empirical cumulative distribution functions $\hat{F}_A(x)$ and $\hat{F}_B(x)$ in Figure 15.16. The empirical cumulative distribution function $\hat{F}_A(x)$ lies to the right of the empirical cumulative distribution function $\hat{F}_B(x)$, which indicates that the particle diameters tend to be larger at the lower pre-expansion temperature. In addition, the empirical cumulative distribution function $\hat{F}_B(x)$ increases quite sharply in comparison with the empirical cumulative distribution function $\hat{F}_A(x)$, which has a more gradual slope, and this confirms that there is more variability in the particle sizes at the lower pre-expansion temperature than at the higher pre-expansion temperature.

FIGURE 15.16

The empirical cumulative
distribution functions for the
petroleum processing example

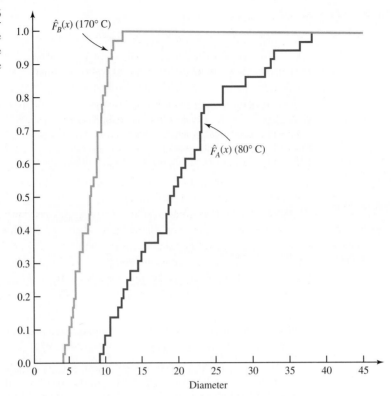

FIGURE 15.16

The empirical cumulative distribution functions for the petroleum processing example

Formally, the value of the Kolmogorov-Smirnov statistic is

$$M = \frac{30}{36} = 0.83$$

This value is the maximum vertical distance between the two empirical cumulative distribution functions $\hat{F}_A(x)$ and $\hat{F}_B(x)$, which occurs when $10.25 < x < 10.50$, in which case

$$\hat{F}_A(x) = \frac{3}{36} \quad \text{and} \quad \hat{F}_B(x) = \frac{33}{36}$$

and also when $10.83 < x < 11.50$, in which case

$$\hat{F}_A(x) = \frac{5}{36} \quad \text{and} \quad \hat{F}_B(x) = \frac{35}{36}$$

A size $\alpha = 0.01$ critical point is

$$d_{0.01} \sqrt{\frac{1}{n} + \frac{1}{m}} = 1.63 \times \sqrt{\frac{1}{36} + \frac{1}{36}} = 0.38$$

The value $M = 0.83$ is much larger than this critical point, which indicates that there is very strong evidence that the distributions of the particle sizes are different at the two different pre-expansion temperatures.

15.2.2   The Rank Sum Test

The rank sum test procedure is another test procedure for comparing two distribution functions $F_A(x)$ and $F_B(x)$ based on the analysis of two unpaired samples. Again let $x_1, \ldots, x_n$ be independent, identically distributed observations from the distribution function $F_A(x)$, and let $y_1, \ldots, y_m$ be independent, identically distributed observations from the distribution function $F_B(x)$. In addition, suppose that the two samples are obtained independently of each other, as is the case with unpaired samples. As before, the two sample sizes m and n need not be equal, although many experiments are designed to have equal sample sizes.

The rank sum test procedure is attributed to works of Wilcoxon in 1945 and Mann and Whitney in 1947. It is referred to as either the Wilcoxon rank sum test (to distinguish it from the Wilcoxon signed rank test) or the Mann-Whitney test. It is based upon the assumption that the two distribution functions are identical except for a difference in location, so that $F_A(x) = F_B(x + \delta)$ for some location difference δ, as illustrated in Figure 15.17. This is often a reasonable assumption to make, and it is valid if the observations from the two populations can be modeled as fixed population effects plus identically distributed error terms. The location difference δ can be interpreted as the difference between the means of the two populations $\mu_B - \mu_A$, or alternatively as the difference between the two population medians.

Under the assumption that $F_A(x) = F_B(x + \delta)$ the null hypothesis

$$H_0 : F_A(x) = F_B(x)$$

for all values of x is equivalent to testing whether the location difference δ is 0. The rank sum test procedure is typically applied to test this null hypothesis that the two distribution functions are identical. More generally, the rank sum test procedure can also be used to construct a confidence interval for the location difference δ.

The nonparametric nature of the analysis is evident from the fact that the actual shape of the distribution functions does not need to be specified. Under the assumption that the data observations are normally distributed, this problem can be analyzed using the two-sample t-test discussed in Section 9.3. For large enough sample sizes n and m, the central limit theorem also validates the test procedures discussed in Section 9.3. Nevertheless, in practice

FIGURE 15.17

The assumption of location shift for the rank sum test procedure

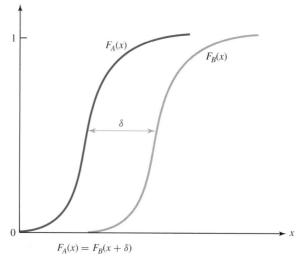

the rank sum test procedure often provides as precise an inference as the two-sample t-test without the distributional assumption.

The first step in the implementation of the rank sum test is to combine the two data samples $x_1, \ldots, x_n$ and $y_1, \ldots, y_m$ into one large sample and to rank the elements of this pooled sample from 1 to $m + n$. Tied observations are treated in the usual manner by assigning the averages of the corresponding rank values. The rank sum test is then based on the consideration of the statistic S_A, which is defined to be the sum of the ranks within the combined sample of the observations from population A. In other words, the statistic S_A is calculated as the sum of the ranks within the combined sample of the n observations $x_1, \ldots, x_n$. The results obtained from the rank sum test do not depend on which of the two populations is labeled population A.

In practice, instead of dealing directly with the statistic S_A it is more convenient to consider the statistic U_A defined by

$$U_A = S_A - \frac{n(n+1)}{2}$$

Notice that the statistic U_A can take any integer value between 0 and mn. For example, the smallest value of U_A occurs when each observation x_i is less than each observation y_i, so that the ranks of the observations x_i are $1, 2, \ldots, n$ in which case $S_A = n(n+1)/2$. On the other hand, the largest value of U_A occurs when each observation x_i is greater than each observation y_i, so that the ranks of the observations x_i are $m + 1, m + 2, \ldots, m + n$ in which case $S_A = mn + n(n+1)/2$.

If the two distribution functions $F_A(x)$ and $F_B(x)$ are identical, then the statistic U_A should be roughly equal to $mn/2$. However, if the statistic U_A is much larger than $mn/2$, this suggests that observations from population A tend to be larger than observations from population B. Conversely, if the statistic U_A is much smaller than $mn/2$, then this suggests that observations from population A tend to be smaller than observations from population B.

Computer packages will calculate an exact two-sided p-value for the plausibility of the null hypothesis $H_0 : F_A(x) = F_B(x)$ by comparing the observed value of the statistic U_A with its distribution under this null hypothesis. This discrete distribution can be found by considering the value of the statistic U_A for each of the different orderings of the n observations from population A and the m observations from population B. There are

$$\binom{n+m}{m}$$

of these orderings, and under the null hypothesis each has an equal probability of

$$\frac{1}{\binom{n+m}{m}}$$

Some minor modifications to the p-value calculation may be made if there are ties in the data observations.

However, for reasonably large sample sizes it is sufficient to calculate an approximate p-value by comparing the statistic

$$z = \frac{U_A - \frac{mn}{2}}{\sqrt{\frac{mn(m+n+1)}{12}}}$$

with the standard normal distribution. If $z > 0$ a two-sided p-value is $2 \times (1 - \Phi(z))$, and if $z < 0$ a two-sided p-value is $2 \times \Phi(z)$. As in other cases, a continuity correction of 0.5 can help increase the accuracy of these p-value calculations. The amount 0.5 should be added to the numerator of z if it is negative and subtracted from the numerator of z if it is positive.

In certain situations it may make sense to calculate a one-sided p-value. Remember that $\delta = \mu_B - \mu_A$, and so if, for example, the experimenter is interested only in detecting whether or not observations from population B tend to be *smaller* than observations from population A, then a null hypothesis of $H_0 : \delta \geq 0$ is appropriate with an alternative hypothesis of $H_A : \delta < 0$. Positive values of the statistic z suggest that $\delta < 0$, and a one-sided p-value can be calculated as $(1 - \Phi(z))$.

The Rank Sum Test

If $x_1, \ldots, x_n$ are independent, identically distributed observations from a distribution function $F_A(x)$ and $y_1, \ldots, y_m$ are independent, identically distributed observations from a distribution function $F_B(x)$, then under the assumption $F_A(x) = F_B(x + \delta)$ for some location difference δ, the **rank sum** test procedure can be used to test the null hypothesis

$$H_0 : F_A(x) = F_B(x)$$

for all values of x, which is equivalent to testing whether the location difference δ is 0. More generally, the rank sum test procedure can be used to construct a confidence interval for the location difference δ, which can be interpreted as the difference between the means of the two populations $\mu_B - \mu_A$, or alternatively as the difference between the two population medians.

The rank sum test is performed by combining the two data samples $x_1, \ldots, x_n$ and $y_1, \ldots, y_m$ into one large sample and ranking the elements of the pooled sample from 1 to $m + n$. The rank sum test is then based upon the consideration of the sum of the ranks within the combined sample of the observations from one of the populations.

If the null hypothesis

$$H_0 : F_A(x) = F_B(x + \delta_0)$$

is of interest for some fixed value of δ_0, then it suffices to apply the rank sum test to the modified data sets

$$x_1, \ldots, x_n \qquad \text{and} \qquad y_1 - \delta_0, \ldots, y_m - \delta_0$$

This is clear because if $F_A(x) = F_B(x + \delta_0)$, then the two distribution functions are identical except for a location difference of δ_0. Therefore, if one of the data sets is shifted by δ_0 in an appropriate manner, the assessment of the plausibility of the null hypothesis

$$H_0 : F_A(x) = F_B(x + \delta_0)$$

simplifies to the assessment of the plausibility that the two resulting samples are generated by identical distribution functions, which can be performed with the rank sum test.

Finally, under the assumption that $F_A(x) = F_B(x + \delta)$ for some location difference δ, the rank sum test procedure can be used to construct a confidence interval for δ. Specifically, a confidence interval for the location difference δ with a confidence level of $1 - \alpha$ consists of all the values δ_0 for which a two-sided size α test of the null hypothesis

$$H_0 : F_A(x) = F_B(x + \delta_0)$$

is accepted by the rank sum test. In other words, the confidence interval consists of all the values δ_0 for which the two-sided p-value of this null hypothesis is greater than α. Computer packages can be used to construct this confidence interval, and as with the sign test and the signed rank test, only certain discrete confidence levels are available.

Example 52

Kaolin Processing

Figure 15.18 shows how the $n = 12$ brightness measurements from calciner A and the $m = 12$ brightness measurements from calciner B are combined into one sample of 24 observations and ranked from 1 to 24, with tied values being assigned the average of the corresponding ranks. The sum of the ranks of the observations from calciner A is

$$S_A = 132$$

so that

$$U_A = S_A - \frac{n(n+1)}{2} = 132 - \frac{12 \times 13}{2} = 54$$

This result is smaller than

$$\frac{mn}{2} = \frac{12 \times 12}{2} = 72$$

so the data suggest that the brightness measurements from calciner A are smaller than the brightness measurements from calciner B. Notice that this finding is consistent with the

---

FIGURE 15.18

Calculation of the rank sum test statistic for the kaolin processing example

Calciner	Brightness measurement	Rank
A	87.4	1
A	88.4	2
A	89.0	3
B	89.2	4
A	90.5	5
B	90.7	6
A	90.8	7
B	91.5	8
B	91.7	9
A	91.9	10
B	92.0	11
B	92.6	12
A	92.8	13
A	93.0	14
A	93.1	15
B	93.2 ⎫	17
B	93.2 ⎬	17
A	93.2 ⎭	17
B	93.3	19
B	93.8	20
B	94.0	21
A	94.3 ⎫	22.5
A	94.3 ⎭	22.5
B	94.8	24

$$S_A = 1 + 2 + 3 + 5 + 7 + 10 + 13 + 14$$
$$+ 15 + 17 + 22.5 + 22.5 = 132$$

discussion in Section 9.3.4 where it is noted that the sample average 91.558 for calciner A is smaller than the sample average 92.500 for calciner B.

A statistical software package can be used to show that $U_A = 54$ gives a p-value of 0.3123 (or 0.3118 when a modification is made because of the ties in the data observations), and a confidence interval for δ with a confidence level of 95.4% is $(-2.7, 1.1)$ which, as expected, contains 0. Clearly, this analysis with the rank sum test procedure indicates that there is no evidence of a difference between the two calciners, which is consistent with the analysis performed in Section 9.3.4 using the two-sample t-test.

■ 15.2.3 Problems

15.2.1 Restaurant Service Times

Recall that DS 6.1.4 shows the service times of customers at a fast-food restaurant who were served between 2:00 and 3:00 on a Saturday afternoon, and that DS 9.3.5 shows the service times of customers at the fast-food restaurant who were served between 9:00 and 10:00 in the morning on the same day.

(a) Make a plot of the empirical cumulative distribution functions of these two data sets.

(b) What does a visual comparison of the two empirical cumulative distribution functions suggest about the difference between the two service time distributions?

(c) Use the Kolmogorov-Smirnov test to assess the evidence that the two distribution functions are different.

15.2.2 Paving Slab Weights

Recall that DS 6.1.7 shows the weights of a sample of paving slabs from manufacturer A and that DS 9.3.1 shows the weights of a sample of paving slabs from manufacturer B.

(a) Make a plot of the empirical cumulative distribution functions of these two data sets.

(b) What does a visual comparison of the two empirical cumulative distribution functions suggest about the difference between the distributions of the paving slabs weights for the two manufacturers?

(c) Use the Kolmogorov-Smirnov test to assess the evidence that the two distribution functions are different.

15.2.3 Heel-Strike Force on a Treadmill

DS 9.3.3 contains observations of heel-strike force for a runner on a treadmill with and without a damped feature activated. Use plots of the empirical cumulative distribution functions and the Kolmogorov-Smirnov test to investigate whether the damped feature is effective in reducing the heel-strike force.

15.2.4 Use the rank sum test procedure to analyze the two samples in DS 15.2.1.

(a) Combine the two samples and rank the observations.

(b) What is S_A?

(c) What is U_A?

(d) Is the value of U_A consistent with the observations from population A being larger or smaller than the observations from population B?

(e) Use a computer package to find a two-sided p-value for the null hypothesis that the two distribution functions are identical. Is the difference between the two distribution functions suggested in part (d) statistically significant?

15.2.5 Repeat Problem 15.2.4 using the data set in DS 15.2.2.

15.2.6 Use the rank sum test to analyze the data set in Figure 9.20 concerning Example 51 on acrophobia treatments. Let population A be with the standard treatment and population B be with the new treatment.

(a) Combine the two samples and rank the observations.

(b) What is S_A?

(c) What is U_A?

(d) Is the value of U_A consistent with the new treatment being better than the standard treatment?

(e) Use a computer package to find a one-sided p-value for the null hypothesis $H_0 : \mu_A \geq \mu_B$ versus the alternative hypothesis $H_A : \mu_A < \mu_B$.

(f) Are your conclusions from this analysis consistent with the analysis presented in Section 9.3.4 using the two-sample t-test?

15.2.7 Spray Painting Procedure

Recall that DS 6.1.8 contains a sample of paint thicknesses from production line A and that DS 9.3.2 contains a sample of paint thicknesses from production line B.

(a) Make a plot of the empirical cumulative distribution functions of these two data sets.

(b) What does a visual comparison of the two empirical cumulative distribution functions suggest about the difference between the distributions of the paint thicknesses from the two production lines?

(c) Use the Kolmogorov-Smirnov test to assess the evidence that the two distribution functions are different.

(d) Use the rank sum test to assess the evidence that the two distribution functions are different. Obtain a 95% confidence interval for the difference between the median paint thicknesses for the two production lines. On what assumption is the rank sum test based?

Do you think that it is a reasonable assumption in this case?

15.2.8 Bleaching Agents
Recall that DS 9.3.4 contains the results of an experiment to compare the bleaching effectiveness of two levels of hydrogen peroxide, a low level and a high level. Use the rank sum test to assess whether there is evidence of a difference between the low and high levels of hydrogen peroxide.

15.2.9 Clinical Trial
Use the rank sum test to analyze the clinical trial data in DS 9.3.6.

15.3 Comparing Three or More Populations

In Chapter 11 an analysis of variance approach is discussed for comparing three or more populations which is based upon the assumption that the error terms can be taken to be independent observations from a normal distribution. In this section nonparametric inference procedures for comparing three or more populations are discussed which are based only on the assumption that the error terms are independent observations from some common distribution. In other words, the assumption of normality is not required. The nonparametric approach can provide a useful inference, although it is not as good as the analysis of variance approach when the assumption of normality is reasonable.

In this section both one-way layouts and randomized block designs are considered. The nonparametric test procedure for a one-way layout is the **Kruskal-Wallis test** procedure, and the nonparametric test procedure for a randomized block design is the **Friedman test** procedure. Both of these nonparametric test procedures are based on an analysis of appropriate *ranks* of the data observations.

15.3.1 One-Way Layouts

Recall the one-way layout (one factor analysis of variance) discussed in Section 11.1 with data observations

$$x_{ij} = \mu_i + \epsilon_{ij} \qquad 1 \le i \le k, 1 \le j \le n_i$$

where the total sample size is

$$n_T = n_1 + \cdots + n_k$$

The F-test obtained in the analysis of variance table for the null hypothesis

$$H_0 : \mu_1 = \cdots = \mu_k$$

that the k factor level means are all equal is based upon the assumption that the error terms ϵ_{ij} are independent observations from a $N(0, \sigma^2)$ distribution. The Kruskal-Wallis test procedure is a nonparametric procedure for testing this same null hypothesis, which only requires that the error terms are identically distributed. Therefore, the test is appropriate as long as the error terms are independent observations from some common distribution, which need not be a normal distribution.

The first step in applying the nonparametric test procedure is to combine the k samples into one big sample and to rank the observations from 1 to n_T. Tied observations are handled in the usual manner by assigning the average of the corresponding ranks. Notice that the sum of the ranks is

$$1 + \cdots + n_T = \frac{n_T(n_T + 1)}{2}$$

and that the average of all the ranks is

$$\frac{n_T + 1}{2}$$

Let the rank of the observation x_{ij} be denoted by r_{ij} so that the averages of the ranks within each of the k populations are

$$\bar{r}_{1\cdot}, \ldots, \bar{r}_{k\cdot}$$

where

$$\bar{r}_{i\cdot} = \frac{r_{i1} + \cdots + r_{in_i}}{n_i}$$

The assessment of the plausibility of the null hypothesis that the k factor level means are all equal is based upon the consideration of how close the rank averages $\bar{r}_{i\cdot}$ are to the average rank value $(n_T + 1)/2$. The more widely spread they are around this average value, the less plausible is the null hypothesis.

The Kruskal-Wallis Test

The **Kruskal-Wallis test** procedure is a nonparametric test procedure for assessing the plausibility that the k factor level means in a one-way layout are all equal. Ranks r_{ij} of the data observations x_{ij} are obtained by ranking the combined sample of n_T data observations. The average ranks within the k populations are calculated as

$$\bar{r}_{i\cdot} = \frac{r_{i1} + \cdots + r_{in_i}}{n_i} \qquad 1 \le i \le k$$

and the test procedure is based upon the statistic

$$H = \frac{12}{n_T(n_T + 1)} \sum_{i=1}^{k} n_i \left(\bar{r}_{i\cdot} - \frac{n_T + 1}{2} \right)^2 = \frac{12}{n_T(n_T + 1)} \sum_{i=1}^{k} n_i \, \bar{r}_{i\cdot}^2 - 3(n_T + 1)$$

A p-value is calculated as

$$p\text{-value} = P(X > H)$$

where the random variable X has a chi-square distribution with $k - 1$ degrees of freedom.

Computer packages can be used to perform this nonparametric test procedure, and the test statistic H may be modified slightly if there are ties in the data observations x_{ij}.

FIGURE 15.19

Calculation of the ranks for the
blocked arteries example

Data values x_{ij}	Stenosis level	Rank r_{ij}
8.3	1	1.0
9.7	1	2.0
10.2	1	3.0
10.4	1	4.0
10.6 ⎱	1	5.5
10.6 ⎰	1	5.5
10.8	1	7.0
11.0	1	8.0
11.7	2	9.0
12.6	2	10.0
12.7	2	11.0
13.0	3	12.0
13.1	1	13.0
13.2	2	14.0
14.3 ⎱	1	15.5
14.3 ⎰	1	15.5
14.4	2	17.0
14.7	2	18.0
14.8	2	19.0
15.1 ⎱	2	20.5
15.1 ⎰	3	20.5
15.2	2	22.0
15.6	3	23.0
16.3	2	24.0
16.6 ⎱	2	25.5
16.6 ⎰	3	25.5
16.8	2	27.0
17.6	2	28.0
18.0	3	29.0
18.6	3	30.0
18.7	3	31.0
18.9	3	32.0
19.2	3	33.0
19.5	2	34.0
19.6	3	35.0

$$\bar{r}_{1.} = \frac{1 + 2 + 3 + 4 + 5.5 + 5.5 + 7 + 8 + 13 + 15.5 + 15.5}{11} = \frac{80}{11} = 7.273$$

$$\bar{r}_{2.} = \frac{9 + 10 + 11 + 14 + 17 + 18 + 19 + 20.5 + 22 + 24 + 25.5 + 27 + 28 + 34}{14} = \frac{279}{14} = 19.929$$

$$\bar{r}_{3.} = \frac{12 + 20.5 + 23 + 25.5 + 29 + 30 + 31 + 32 + 33 + 35}{10} = \frac{271}{10} = 27.100$$

Example 62

Collapse of Blocked Arteries

Recall that Figure 11.4 contains the values of the flowrate at collapse for tubes with three amounts of stenosis. Figure 15.19 shows how the ranks r_{ij} are calculated for the combined sample of $n_T = 35$ observations. The average ranks for the three stenosis levels are

$$\bar{r}_{1.} = \frac{80}{11} = 7.273$$

$$\bar{r}_{2.} = \frac{279}{14} = 19.929$$

$$\bar{r}_{3.} = \frac{271}{10} = 27.100$$

Notice that

$$\bar{r}_{1.} < \bar{r}_{2.} < \bar{r}_{3.}$$

so the data suggest that the flowrate at collapse increases from stenosis level 1 to level 2 and from stenosis level 2 to level 3, which is consistent with the comparisons of the sample averages $\bar{x}_{1.} = 11.209$, $\bar{x}_{2.} = 15.086$, and $\bar{x}_{3.} = 17.330$, with

$$\bar{x}_{1.} < \bar{x}_{2.} < \bar{x}_{3.}$$

The test statistic for the Kruskal-Wallis test procedure is

$$H = \frac{12}{n_T(n_T+1)} \sum_{i=1}^{k} n_i \, \bar{r}_{i.}^2 - 3(n_T+1)$$

$$= \frac{12}{35 \times 36} (11 \times 7.273^2 + 14 \times 19.929^2 + 10 \times 27.100^2) - 3 \times 36 = 20.44$$

and the p-value is

$$p\text{-value} = P(X > 20.44) \simeq 0$$

where the random variable X has a chi-square distribution with $k - 1 = 3 - 1 = 2$ degrees of freedom. Clearly there is substantial evidence that the amount of stenosis does affect the flowrate at collapse, which is the same conclusion as that reached in the analysis of this example in Section 11.1.

15.3.2 Randomized Block Designs

Recall the randomized block design discussed in Section 11.2 with data observations

$$x_{ij} = \mu_{ij} + \epsilon_{ij} \qquad 1 \le i \le k, 1 \le j \le b$$

where the cell means μ_{ij} are written as

$$\mu_{ij} = \mu + \alpha_i + \beta_j$$

The F-test obtained in the analysis of variance table for the null hypothesis

$$H_0 : \alpha_1 = \cdots = \alpha_k = 0$$

that the k factor levels are indistinguishable is based on the assumption that the error terms ϵ_{ij} are independent observations from a $N(0, \sigma^2)$ distribution. The Friedman test procedure is a nonparametric procedure for testing this same null hypothesis which requires only that the error terms are identically distributed. In other words, it is appropriate as long as the error terms are independent observations from some common distribution, which need not be a normal distribution.

The Friedman test operates by ranking the data observations *within each block*. Thus, in the first block the data observations

$$x_{11}, \ldots, x_{k1}$$

are ranked from 1 to k with ties being handled in the usual manner. In the second block the data observations

$$x_{12}, \ldots, x_{k2}$$

are also ranked from 1 to k, and similarly for each of the b blocks. The ranks r_{ij} consequently take values between 1 and k with an average value of

$$\frac{k+1}{2}$$

The averages of the ranks within each of the k populations are

$$\bar{r}_{1.}, \ldots, \bar{r}_{k.}$$

where

$$\bar{r}_{i.} = \frac{r_{i1} + \cdots + r_{ib}}{b}$$

and the assessment of the plausibility of the null hypothesis that the k factor level means are indistinguishable is based upon the consideration of how close the rank averages $\bar{r}_{i.}$ are to the average rank value $(k+1)/2$. The more variability that they have about this average value, the less plausible is the null hypothesis.

The Friedman Test

The **Friedman test** procedure is a nonparametric test procedure for assessing the plausibility that the k factor levels are indistinguishable in a randomized block design. Ranks r_{ij} of the data observations x_{ij} are obtained by ranking the data observations from one to k *within each of the b blocks*. The average ranks for the k factor levels are

$$\bar{r}_{i.} = \frac{r_{i1} + \cdots + r_{ib}}{b} \qquad 1 \le i \le k$$

and the test procedure is based upon the statistic

$$S = \frac{12\,b}{k(k+1)} \sum_{i=1}^{k} \left(\bar{r}_{i.} - \frac{k+1}{2} \right)^2 = \frac{12\,b}{k(k+1)} \sum_{i=1}^{k} \bar{r}_{i.}^2 - 3b(k+1)$$

A p-value is calculated as

$$p\text{-value} = P(X > S)$$

where the random variable X has a chi-square distribution with $k - 1$ degrees of freedom.

As with the Kruskal-Wallis test procedure, computer packages can be used to perform this nonparametric test procedure, and the test statistic S may be modified slightly if there are ties in the data observations x_{ij}.

Example 65
Comparing Types of Wheat

Recall that Figure 11.26 shows the crop yields from the experiment to compare $k = 4$ wheat types with $b = 5$ fields as blocks. Figure 15.20 shows how the ranks are calculated for the Friedman test. For example, in the first field

$$x_{31} < x_{11} < x_{21} < x_{41}$$

FIGURE 15.20

Calculation of the Friedman test ranks for the wheat comparisons data set

	Wheat Types			
	Type 1	Type 2	Type 3	Type 4
Field 1	$x_{11} = 164.4$ $r_{11} = 2$	$x_{21} = 184.3$ $r_{21} = 3$	$x_{31} = 161.2$ $r_{31} = 1$	$x_{41} = 185.8$ $r_{41} = 4$
Field 2	$x_{12} = 145.0$ $r_{12} = 4$	$x_{22} = 142.1$ $r_{22} = 3$	$x_{32} = 110.8$ $r_{32} = 1$	$x_{42} = 135.4$ $r_{42} = 2$
Field 3	$x_{13} = 152.5$ $r_{13} = 1$	$x_{23} = 159.6$ $r_{23} = 3$	$x_{33} = 168.6$ $r_{33} = 4$	$x_{43} = 154.1$ $r_{43} = 2$
Field 4	$x_{14} = 138.5$ $r_{14} = 4$	$x_{24} = 137.2$ $r_{24} = 3$	$x_{34} = 134.9$ $r_{34} = 2$	$x_{44} = 123.2$ $r_{44} = 1$
Field 5	$x_{15} = 161.7$ $r_{15} = 3$	$x_{25} = 160.4$ $r_{25} = 2$	$x_{35} = 166.1$ $r_{35} = 4$	$x_{45} = 159.9$ $r_{45} = 1$

so that $r_{11} = 2$, $r_{21} = 3$, $r_{31} = 1$, and $r_{41} = 4$. The average ranks for the four wheat types are

$$\bar{r}_{1.} = \frac{14}{5} = 2.8$$

$$\bar{r}_{2.} = \frac{14}{5} = 2.8$$

$$\bar{r}_{3.} = \frac{12}{5} = 2.4$$

$$\bar{r}_{4.} = \frac{10}{5} = 2.0$$

so that the statistic for the Friedman test procedure is

$$S = \frac{12\,b}{k(k+1)} \sum_{i=1}^{k} \bar{r}_{i.}^2 - 3b(k+1)$$

$$= \frac{12 \times 5}{4 \times 5} (2.8^2 + 2.8^2 + 2.4^2 + 2.0^2) - 3 \times 5 \times 5 = 1.32$$

and the p-value is

$$p\text{-value} = P(X > 1.32) = 0.724$$

where the random variable X has a chi-square distribution with $k - 1 = 4 - 1 = 3$ degrees of freedom. Clearly there is not sufficient evidence to conclude that there is any difference between the four wheat types with respect to their yields, which is the same result obtained in the analysis of this example in Section 11.2 using an analysis of variance approach.

■ **15.3.3 Problems**

15.3.1 Use the Kruskal-Wallis test procedure to analyze the data in DS 11.1.1.

(a) Find the ranks r_{ij}.

(b) What are the average ranks $\bar{r}_{1.}$, $\bar{r}_{2.}$, and $\bar{r}_{3.}$?

(c) What is the value of the test statistic H?

(d) Write down an expression for the p-value and use a computer package to evaluate it.

15.3.2 Use the Kruskal-Wallis test procedure to analyze the data in DS 11.1.2.

(a) Find the ranks r_{ij} and the average ranks $\bar{r}_{1\cdot}, \bar{r}_{2\cdot}, \bar{r}_{3\cdot}$, and $\bar{r}_{4\cdot}$.

(b) What is the value of the test statistic H?

(c) Write down an expression for the p-value and use a computer package to evaluate it.

15.3.3 Infrared Radiation Readings

The data set in DS 11.1.3 concerns the infrared radiation readings from an energy source measured by a particular meter with three different background radiation levels.

(a) Use the Kruskal-Wallis test procedure to investigate whether the radiation readings are affected by the background radiation level.

(b) Repeat the analysis using an analysis of variance table.

(c) What assumptions are required for the Kruskal-Wallis test procedure? What assumptions are required for the F-test in the analysis of variance table? Which analysis method do you prefer?

15.3.4 Keyboard Layout Designs

DS 11.1.4 contains the times taken to perform a task using three different keyboard layouts for the numerical keys. Use the Kruskal-Wallis test procedure to investigate whether the different layouts affect the time taken to perform a task.

15.3.5 Computer Assembly Methods

DS 11.1.6 contains the assembly times of computers for three different assembly methods. Use the Kruskal-Wallis test procedure to investigate whether there is any evidence that one assembly method is any quicker than the other methods.

15.3.6 Use the Friedman test procedure to analyze the data in DS 11.2.1.

(a) Find the ranks r_{ij}.

(b) What are the average ranks $\bar{r}_{1\cdot}, \bar{r}_{2\cdot}$, and $\bar{r}_{3\cdot}$?

(c) What is the value of the test statistic S?

(d) Write down an expression for the p-value and use a computer package to evaluate it.

15.3.7 Use the Friedman test procedure to analyze the data in DS 11.2.2.

(a) Find the ranks r_{ij} and the average ranks $\bar{r}_{1\cdot}, \bar{r}_{2\cdot}, \bar{r}_{3\cdot}$, and $\bar{r}_{4\cdot}$.

(b) What is the value of the test statistic S?

(c) Write down an expression for the p-value and use a computer package to evaluate it.

15.3.8 Calciner Comparisons

The data set in DS 11.2.3 concerns the brightness measurements for $b = 7$ batches of kaolin processed through $k = 3$ calciners.

(a) Use the Friedman test procedure to investigate whether the calciners are operating with different efficiencies.

(b) Repeat the analysis using an analysis of variance table.

(c) What assumptions are required for the Friedman test procedure? What assumptions are required for the F-test in the analysis of variance table? Which analysis method do you prefer?

15.3.9 Radar Detection of Airborne Objects

DS 11.2.4 contains distances at detection for three radar systems. Use the Friedman test procedure to investigate whether there is evidence of any difference between the radar systems.

15.3.10 Production Line Assembly Methods

The data set in DS 11.2.6 concerns an experiment to compare three different assembly methods for an electric motor. Use the Friedman test procedure to investigate whether there is evidence of any difference between the three assembly methods.

15.3.11 Realtor Commissions

DS 11.2.7 contains the commissions obtained by five agents in a realtor's office. Use the Friedman test procedure to investigate whether there is evidence of any real difference in the performances of the agents.

15.3.12 Cleanliness Scores for Detergent Comparisons

The data set in DS 11.2.8 concerns an experiment to compare four different formulations of a detergent. Use the Friedman test procedure to investigate whether there is evidence of any difference between the detergent formulations.

15.3.13 Durations of Investigatory Surgical Procedures

Use the appropriate nonparametric methodology from this section to analyze the data in DS 11.1.8.

15.3.14 E. Coli Colonies in Riverwater

Use the appropriate nonparametric methodology from this section to analyze the data in DS 11.1.9.

15.3.15 Groundwater Pollution Levels

Use the appropriate nonparametric methodology from this section to analyze the data in DS 11.2.9.

15.4   Supplementary Problems

15.4.1 Osteoporosis Patient Heights

(a) Construct the empirical cumulative distribution function for the data set of adult male heights given in DS 6.6.4.

(b) Draw 95% confidence bands around the empirical cumulative distribution function.

(c) Is it plausible that the heights are normally distributed with a mean of 70 inches and a standard deviation of 2 inches?

(d) Is it plausible that the heights are normally distributed with a mean of 71 inches and a standard deviation of 1 inch?

(e) Consider the null hypothesis that the median height is 70 inches. What statistic is used by the sign test procedure to test this null hypothesis? What is the p-value?

(f) Test the null hypothesis in part (e) using the signed rank test.

(g) Use the sign test, the signed rank test, and the t-test to obtain 95% confidence intervals on the median (or mean) service time. What assumptions are required by these three test procedures? Do the assumptions seem appropriate? How would you summarize your results?

15.4.2 Bamboo Cultivation

Construct the empirical cumulative distribution function for the data set of bamboo shoot heights given in DS 6.6.5. Draw 95% confidence bands around the empirical cumulative distribution function. Analyze the median (or mean) shoot height using the sign test, the signed rank test, and the t-test. Is there any evidence that the median (or mean) shoot height is less than 35 cm? Which analysis method do you prefer? Summarize your conclusions.

15.4.3 Tire Tread Wear

Use the sign test and the signed rank test to analyze the paired data set of tire wear given in DS 9.2.3. Do you find any evidence of a difference between the two types of tires? Do you think that the analysis with the signed rank test is better than the analysis with the sign test? Compare your results with an analysis using the t-test. What can you say about the difference in average wear for the two types of tires?

15.4.4 Video Display Designs

Use the sign test and the signed rank test to analyze the paired data set given in DS 9.6.1 concerning the assimilation of information from video monitors. Do you find any evidence of a difference between the two color types? Compare your results with an analysis based upon the t-test. Summarize your conclusions.

15.4.5 Consumer Complaints Division Reorganization

Recall that DS 9.6.4 contains data observations of waiting times for a consumer to speak to a company representative on a telephone complaints line both before and after a reorganization.

(a) Make a plot of the empirical cumulative distribution functions of these two data sets.

(b) What does a visual comparison of the two empirical cumulative distribution functions suggest about the difference between the distributions of the waiting times?

(c) Use the Kolmogorov-Smirnov test to assess the evidence that the two distribution functions are different. Does the reorganization appear to have been successful in affecting the times taken to answer calls?

15.4.6 Bamboo Cultivation

A researcher compares the bamboo shoot heights in DS 6.6.5 obtained under growing conditions A with the bamboo shoot heights in DS 9.6.3 obtained under growing conditions B. Recall that the growing conditions B allowed 10% more sunlight than growing conditions A.

(a) Make a plot of the empirical cumulative distribution functions of these two data sets.

(b) What does a visual comparison of the two empirical cumulative distribution functions suggest about the difference between the distributions of the bamboo shoot heights under the two growing conditions?

(c) Use the Kolmogorov-Smirnov test to assess the evidence that the two distribution functions are different.

(d) Use the rank sum test to assess the evidence that the two distribution functions are different. Obtain a 95% confidence interval for the difference between the median bamboo shoot heights for the two growing conditions. On what assumption is the rank sum test based? Do you think that it is a reasonable assumption in this case?

(e) Compare the results of the rank sum test with an analysis using a two-sample t-test. Which test procedure do you prefer in this case?

15.4.7 Use the rank sum test to analyze the data set in Figure 9.24 concerning Example 53 on kudzu pulping. Let population A be without the addition of anthraquinone and population B be with the addition of anthraquinone.

(a) Combine the two samples and rank the observations.
(b) What is S_A?
(c) What is U_A?
(d) Is the value of U_A consistant with the addition of anthraquinone increasing or decreasing the yield?
(e) Use a computer package to find a one-sided p-value for the null hypothesis $H_0 : \mu_A \geq \mu_B$ versus the alternative hypothesis $H_A : \mu_A < \mu_B$.
(f) Are your conclusions from this analysis consistent with the analysis presented in Section 9.3.4 using the two-sample t-test?

15.4.8 Biaxial Nanowire Tests
DS 11.4.1 contains Young's modulus measurements for four different types of nanowires. Use the Kruskal-Wallis test procedure to investigate whether there is any evidence of a difference in the types of nanowires.

(a) Find the ranks r_{ij}.
(b) What are the average ranks $\bar{r}_1.$, $\bar{r}_2.$, $\bar{r}_3.$, and $\bar{r}_4.$?
(c) What is the value of the test statistic H?
(d) Write down an expression for the p-value and use a computer package to evaluate it.

15.4.9 Car Gas Efficiencies
The data set in DS 11.4.2 concerns the gas mileages of four cars.

(a) Use the Kruskal-Wallis test procedure to investigate whether any of the cars are getting better gas mileages than the other cars.
(b) Repeat this analysis using an analysis of variance table.
(c) What assumptions are required for the Kruskal-Wallis test procedure? What assumptions are required for the F-test in the analysis of variance table? Which analysis method do you prefer?

15.4.10 Temperature Effect on Cement Curing
Use the Friedman test procedure to analyze the data set in DS 11.4.3 concerning cement strengths.

(a) Find the ranks r_{ij}.
(b) What are the average ranks $\bar{r}_1.$, $\bar{r}_2.$, $\bar{r}_3.$, $\bar{r}_4.$, and $\bar{r}_5.$?
(c) What is the value of the test statistic S?
(d) Write down an expression for the p-value and use a computer package to evaluate it.

15.4.11 Fertilizer Comparisons
DS 11.4.4 contains the results of an experiment to compare five fertilizers. Use the Friedman test procedure to investigate whether there is evidence of any difference between the fertilizers.

15.4.12 Red Blood Cell Adhesion to Endothelial Cells
The data set in DS 11.4.5 concerns the reports of $k = 4$ clinics for $b = 12$ samples of blood.

(a) Use the Friedman test procedure to investigate whether the clinics appear to be reporting similar results.
(b) Repeat the analysis using an analysis of variance table.
(c) What assumptions are required for the Friedman test procedure? What assumptions are required for the F-test in the analysis of variance table? Which analysis method do you prefer?

15.4.13 Soil Compressibility Tests
Recall the data set of soil compressibility measurements given in DS 6.6.6. Construct the empirical cumulative distribution function for this data set. Use the sign test and the signed rank test to investigate whether the engineers can conclude that the average soil compressibility is no larger than 25.5.

15.4.14 Ocular Motor Measurements
DS 9.6.5 contains the data from an experiment in which a group of 10 subjects had their ocular motor measurements recorded after they had been reading a book for an hour and also after they had been reading a computer screen for an hour. Use the sign test and the signed rank test to analyze the data set.

15.4.15 Engine Oil Viscosity
Oil viscosity values obtained from two engines are given in DS 9.6.6. Use the rank sum test to assess whether there is any evidence that the engines have different effects on the oil viscosity.

15.4.16 Insertion Gains of Hearing Aids
Data collected on the insertion gain of a hearing aid for a constant noise stimulus placed at the horizontal level of the ear of a subject, placed above the horizontal level,

and placed below the horizontal level are shown in DS 11.4.6. Use the Kruskal-Wallis test to analyze this data set.

15.4.17 Air Resistance Drag for Road Vehicles
Data from wind tunnel tests performed on models of four different vehicle designs are shown in DS 11.4.7. What conclusions can you draw from this data set about the drags of the four different designs using the Kruskal-Wallis test?

15.4.18 Leather Shrinkage Measurements
The shrinkage measurements of leather for four different preparation methods are given in DS 11.4.8. What conclusions can you draw from this data set about the differences between the four different preparation methods using the Friedman test?

For problems 15.4.19–15.4.37 use the appropriate nonparametric methodologies from this chapter to analyze the data sets.

15.4.19 Glass Fiber Reinforced Polymer Tensile Strengths
Data set in DS 6.6.7.

15.4.20 Infant Blood Levels of Hydrogen Peroxide
Data set in DS 6.6.8.

15.4.21 Paper Mill Operation of a Lime Kiln
Data set in DS 6.6.9.

15.4.22 River Salinity Levels
Data set in DS 6.6.10.

15.4.23 Dew Point Readings from Coastal Buoys
Data set in DS 6.6.11.

15.4.24 Brain pH Levels
Data set in DS 6.6.12.

15.4.25 Silicon Dioxide Percentages in Ocean Floor Volcanic Glass
Data set in DS 6.6.13.

15.4.26 Network Server Response Times
Data set in DS 6.6.14.

15.4.27 Flowrates in Urban Sewer Systems
Data set in DS 8.5.1.

15.4.28 Polymer Compound Densities
Data set in DS 8.5.2.

15.4.29 Reinforced Cement Strengths
Data set in DS 9.6.7.

15.4.30 Comparisons of Experimental Drug Therapies
Data set in DS 9.6.8.

15.4.31 Rubber Seal Curing Methods
Data set in DS 9.6.9.

15.4.32 Light and Dark Regimens for Plant Growth
Data set in DS 9.6.10.

15.4.33 Joystick Design for Spinal Cord Injury Patients
Data set in DS 9.6.11.

15.4.34 Ambient Air Carbon Monoxide Pollution Levels
Data set in DS 9.6.12.

15.4.35 Sphygmomanometer and Finger Monitor Systolic Blood Pressure Measurements
Data set in DS 9.6.13.

15.4.36 Metal Alloy Hardness Tests
Data set in DS 11.4.9.

15.4.37 Aquatic Radon Levels
Data set in DS 11.4.10.

CHAPTER SIXTEEN
Quality Control Methods

16.1 Introduction

Effective quality control methods have become increasingly important as industries strive to design and produce more reliable products more efficiently. Attention has been focused on the "quality" of a product or service, which is considered to be a general term denoting how well it meets the particular demands imposed upon it. Many quality control methods incorporate techniques involving probability and statistics, and they are discussed in this chapter.

The origins of quality control can be traced back to the implementation of control charts by Walter Shewhart in the 1920s. Interest in the area increased steadily from that beginning and received considerable attention in Japan due to the pioneering work of W. Edwards Deming. More recently the original quality control ideas together with related management principles and guidelines have formed a subject that is often referred to as **total quality management**.

Various aspects of the quality movement are related to probability and statistics. Foremost among these is the concentration on the use of proper experimental techniques, such as the statistical methodologies and experimental designs discussed in Chapters 8, 11, and 14. Another important quality control tool is **statistical process control** (SPC), which is discussed in Section 16.2. Statistical process control utilizes **control charts** to provide a continuous monitoring of a process. Control charts for the mean and variability of a variable of interest are considered in Section 16.3, and control charts for the number of defective items or the number of defects in an item are considered in Section 16.4. Finally, **acceptance sampling** procedures, which can be used to make decisions about the acceptability of batches of items, are discussed in Section 16.5.

16.2 Statistical Process Control

Consider a manufacturing organization that is involved in the production of a large number of a certain kind of product, such as a metal part, a computer chip, or a chemical solution. These products are manufactured using a *process* that typically involves the input of raw materials, a series of procedures, and possibly the involvement of one or more operators. Statistical process control concerns the continuous assessment of the various stages of such a process to ascertain the "quality" of the product as it passes through the process. A key component of this assessment is the use of control charts.

16.2.1 Control Charts

A control chart is a simple quality control tool in which certain measurements of products at a particular point in a manufacturing process are plotted against time. This simple graphical method allows a supervisor to detect when something unusual happens to the process.

The essential aspect of this tool is that measurements are taken successively over time, usually on a random sample of items, so that the supervisor has "real-time" information on the process. With this continuous assessment, any problems can be fixed as they occur. This procedure is in contrast to a less desirable scheme in which products are examined only at the end of the process, when it is too late to make any changes to the process, and there may be a large amount of wasted products or products of substandard quality.

Example 23
Piston Head
Construction

Consider the manufacture of a piston head that is designed to have a radius of 30.00 mm. A control chart can be used to monitor the actual values of the radius of the manufactured piston heads and to alert a supervisor if any changes in the process occur.

Suppose that every hour a random sample of n piston heads is taken and the radius of each one is measured. A simple control chart can be constructed by plotting the sample average of the radius measurements against time, as shown in Figure 16.1. In every manufacturing process there is some variability and consequently it is observed that there is some variation in the actual sizes of the manufactured pistons. Nevertheless, the control chart in Figure 16.1 exhibits a fairly random scatter of observations about the desired value of 30.00 mm, and so the process can be considered to be *in control*.

Generally speaking, a process can be considered to be in control if it does not exhibit any unusual changes. This does not mean that the process is perfect, or even good. In this example, a reduction in the variability of the radius measurements of the piston heads would be an improvement. However, if a process is observed to be in control, then this indicates that the process is performing fairly uniformly over time.

The control chart provides an easily interpretable graphical indication of when any changes in the process occur so that the process moves *out of control*. For example, the control chart in Figure 16.2 indicates that there has suddenly been an increase in the average radius of the piston heads. With the continuous monitoring provided by the control chart the supervisor can immediately investigate the reasons behind the radius increase and can take the appropriate corrective measures.

The control chart in Figure 16.3 also indicates that the process has moved out of control, in this case due to a sudden increase in the variability of the radius values. The control chart is obviously useful in detecting such a change so that corrective action can be taken before a large number of substandard piston heads have been produced.

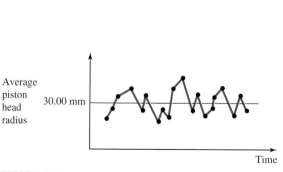

FIGURE 16.1

A control chart exhibiting random scatter

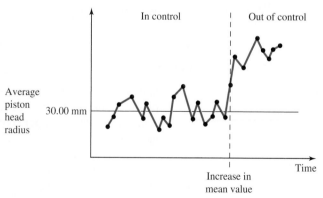

FIGURE 16.2

A control chart exhibiting a sudden increase in the mean value

FIGURE 16.3

FIGURE 16.3

A control chart exhibiting a sudden
increase in the variability

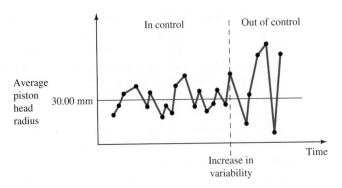

FIGURE 16.4

Control limits on a control chart

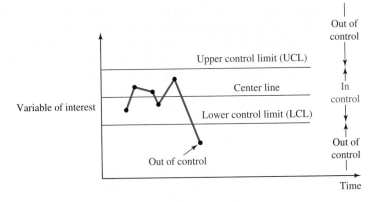

In summary, it is useful to interpret a control chart as providing an indication of whether the variability observed in the measurement of interest is consistent with the usual long-term variability in the process under consideration (process in control), or whether it is alerting the supervisor to some real change in the performance of the process (process out of control). Such a change may be due to such factors as a batch of raw material that is flawed in some manner, a temperature change at some part of the process, an instrument that needs some adjustment, or an operator error, and these factors can be corrected once they have been identified. In practice, experience with a process often implies that supervisors become very adept at identifying the causes behind various kinds of changes observed on the control charts.

16.2.2 Control Limits

In order to help judge whether a point on a control chart is indicative of the process having moved out of control, a control chart is drawn with a center line and two control limits. These control limits are the **upper control limit** (UCL) and the **lower control limit** (LCL) as shown in Figure 16.4.

The process is considered to be in control as long as all the points lie within the control limits. However, if a point lies outside the control limits, then this is taken to be evidence that the process has moved out of control.

It is useful to realize that this procedure is essentially performing a hypothesis test of whether the process is in control. It is useful to think of the null hypothesis as being

H_0 : process in control

with the alternative hypothesis

H_A : process out of control

When new observations on the process are taken, the null hypothesis is accepted as long as the point plotted on the control chart falls within the control limits. However, if the point lies outside the control limits, then the null hypothesis is rejected and there is evidence that the process is out of control.

Of course, it is clear that the control limits are consequently associated with a certain probability for a type I error, which corresponds to the size α of the hypothesis testing problem. This is the probability of observing a point outside the control limits when the process is actually still in control. The control limits are chosen so that this probability is very small.

Typically, "3-sigma" control limits are used, which are chosen to be three standard deviations σ above and below the center line, as illustrated in the following example. A random variable with a normal distribution has a probability of 0.9974 of taking a value within three standard deviations of its mean value, so that if the control chart measurements are normally distributed, 3-sigma control limits have a probability of a type I error of

$$\alpha = 1 - 0.9974 = 0.0026$$

Example 23
Piston Head
Construction

Suppose that experience with the process of manufacturing piston heads leads the supervisors to conclude that the in-control process produces piston heads with radius values that are normally distributed with a mean of $\mu_0 = 30.00$ mm and a standard deviation of $\sigma = 0.05$ mm. How should a control chart be constructed?

Suppose that the observations $x_1, \ldots, x_n$ represent the radius values of the random sample of n piston heads chosen at a particular time. The point plotted on the control chart is the observed value of the sample average

$$\bar{x} = \frac{x_1 + \cdots + x_n}{n}$$

When the process is in control this sample average is an observation from a distribution with a mean value of $\mu_0 = 30.00$ mm and a standard deviation of

$$\sigma_{\bar{X}} = \frac{\sigma}{\sqrt{n}} = \frac{0.05}{\sqrt{n}}$$

A 3-sigma control chart therefore has a center line at the "control value" $\mu_0 = 30.00$ mm together with control limits

$$\text{UCL} = \mu_0 + 3 \frac{\sigma}{\sqrt{n}} \qquad \text{and} \qquad \text{LCL} = \mu_0 - 3 \frac{\sigma}{\sqrt{n}}$$

If the sample size taken every hour is $n = 5$, then the control limits are

$$\text{UCL} = 30.00 + 3 \times \frac{0.05}{\sqrt{5}} = 30.067 \qquad \text{and} \qquad \text{LCL} = 30.00 - 3 \times \frac{0.05}{\sqrt{5}} = 29.933$$

as shown in Figure 16.5.

The probability of a type I error for this control chart, that is, the probability that a point on the control chart lies outside the control limits when the process is really still in control, is

$$1 - P(29.933 \leq \bar{X} \leq 30.067)$$

where the random variable $\bar{X}$ is normally distributed with a mean of 30.00 and a standard deviation of $0.05/\sqrt{5}$. This probability can be evaluated as

$$1 - P(-3 \leq Z \leq 3)$$

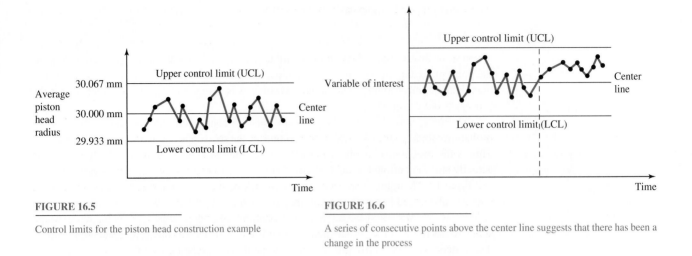

FIGURE 16.5

Control limits for the piston head construction example

FIGURE 16.6

A series of consecutive points above the center line suggests that there has been a change in the process

where the random variable Z has a standard normal distribution, which is

$$1 - 0.9974 = 0.0026$$

as mentioned previously.

It should be mentioned that even if a series of points all lie within the control limits there may still be reason to believe that the process has moved out of control. Remember that if the process is in control, then the points plotted on the control chart should exhibit random scatter about the center line. Any patterns observed in the control chart may be indications of an out-of-control process. For example, the last set of points on the control chart shown in Figure 16.6 all lie within the control limits but they are all above the center line. This suggests that the process may have moved out of control.

COMPUTER NOTE   A series of rules have been developed to help identify patterns in control charts that are symptomatic of an out-of-control process even though no individual point lies outside the control limits. These rules are often called the Western Electric rules, which is where they were first suggested. Most computer packages will implement these rules for you upon request.

16.2.3   Properties of Control Charts

In addition to the probability of a type I error α, it is also useful to consider the probability that the control chart indicates that the process is out of control when it really is out of control. Within the hypothesis testing framework this power is defined to be

$$\text{power} = P(\text{reject } H_0 \text{ when } H_0 \text{ is false}) = 1 - P(\text{Type II error}) = 1 - \beta$$

The actual value of the power depends upon how far the process has moved out of control.

Another point of interest relates to how long a control chart needs to be run before an out-of-control process is detected. The expected number of points that need to be plotted on a control chart before one of them lies outside the control limits and the process is determined to be out of control is known as the **average run length** (ARL).

If the process has moved out of control so that each point plotted has a probability of $1 - \beta$ of lying outside the control limits independent of the other points on the control chart, then the number of points that must be plotted before one of them lies outside the control limits has a *geometric distribution* with success probability $1 - \beta$ (see Section 3.2). The expected value of a random variable with a geometric distribution is the reciprocal of the success probability, so that in this case the average run length is

$$\text{ARL} = \frac{1}{1 - \beta}$$

Example 23
Piston Head
Construction

Suppose that the piston head manufacturing process has moved out of control due to a slight adjustment in some part of the machinery so that the piston heads have radius values that are now normally distributed with a mean value $\mu = 30.06$ mm, instead of the desired control value 30.00 mm, and with the same standard deviation $\sigma = 0.05$ mm as before. How good is the control chart at detecting this change?

The plotted points on the control chart are observations of the random variable $\bar{X}$, which is now normally distributed with a mean $\mu = 30.06$ mm and a standard deviation $0.05/\sqrt{5} = 0.0224$ mm. The probability that a point lies *within* the control limits is therefore

$$P(29.933 \leq \bar{X} \leq 30.067)$$

which can be written as

$$P\left(\frac{29.933 - 30.06}{0.0224} \leq Z \leq \frac{30.067 - 30.06}{0.0224}\right)$$

where the random variable Z has a standard normal distribution. This is

$$P(-5.680 \leq Z \leq 0.313) = \Phi(0.313) - \Phi(-5.680) = 0.622 - 0 = 0.622$$

The probability that a point lies outside the control limits is therefore

$$1 - \beta = 1 - 0.622 = 0.378$$

In other words, once the process has moved out of control in this manner, there is about a 40% chance that each point plotted on the control chart will alert the supervisor to the problem. The average run length in this case is

$$\text{ARL} = \frac{1}{1 - \beta} = \frac{1}{0.378} = 2.65$$

so the problem should be detected within 2 or 3 hours.

This section has provided a general introduction to the use of control charts and the motivation behind their use. Specific types of control charts for specific problems are considered in more detail in the subsequent sections. In addition to the consideration of the mean value of a measurement of interest, such as the mean piston head radius in the previous example, control charts can also be used to monitor changes in the variability of a measurement. In general, these control charts are referred to as $\bar{X}$-charts and R-charts respectively, and they are discussed in more detail in the next section. Control charts for the proportion of defective items in a batch, p-charts, and for the number of defects per item, c-charts, are discussed in Section 16.4.

■ 16.2.4 Problems

16.2.1 Suppose that when a process is in control a variable can be taken to be normally distributed with a mean of $\mu_0 = 10.0$ and a standard deviation of $\sigma = 0.2$, and that a control chart is to be implemented by plotting the average of $n = 4$ observations at each time point.

(a) What are the center line and control limits of a 3-sigma control chart that you would construct to monitor this process?

(b) If a sample average $\bar{x} = 9.5$ is obtained, should the process be declared to be out of control? What if $\bar{x} = 10.25$?

(c) If the process moves out of control so that the variable mean becomes $\mu = 10.15$ with $\sigma = 0.2$, what is the probability that an observation lies outside the control limits? What is the average run length for detecting this change?

16.2.2 A production process making chemical solutions is in control when the solution strengths have a mean of $\mu_0 = 0.650$ and a standard deviation of $\sigma = 0.015$. Suppose that it is a reasonable approximation to take the solution strengths as being normally distributed. At regular time intervals the solution strength is measured and is plotted on a control chart.

(a) What are the center line and control limits of a 3-sigma control chart that you would construct to monitor the strengths of the chemical solutions?

(b) If a randomly sampled solution had a strength of $x = 0.662$, would you take this as evidence that the production process had moved out of control? What if $x = 0.610$?

(c) If the production process moves out of control so that the chemical solution strengths have a mean $\mu = 0.630$ with $\sigma = 0.015$, what is the probability that a randomly sampled solution has a strength that lies outside the control limits? What is the average run length for detecting this change?

16.2.3 Suppose that a 2-sigma control chart is used to monitor a variable that can be taken to be normally distributed.

(a) What is the probability that an observation will lie outside the control limits when the process is still in control?

(b) If the mean of the variable increases to one standard deviation above the control value, what is the probability that an observation will lie outside the control limits? What is the average run length for detecting this change?

16.2.4 When a 3-sigma control chart is used, what is the average run length when the process is in control? In other words, if the process is in control, what is the expected number of points plotted on the control chart until one of them lies outside the control limits and provides the false indication that the process has moved out of control?

16.2.5 Suppose that a 3-sigma control chart is used. What is the probability that when the process is in control, the first eight points plotted lie on the same side of the center line but within the control limits? Do you think that this should be taken as evidence that the process has moved out of control?

16.3 Variable Control Charts

When the variable of interest is a continuous measurement, it is common practice to use two control charts in conjunction with each other. The first is an $\bar{X}$-chart, which is concerned with the *mean* value of the variable of interest, and the second is an R-chart, which is concerned with the *variation* among the measurements of the variable of interest. Put simply, the $\bar{X}$-chart looks for changes in the mean value μ and the R-chart looks for changes in the standard deviation σ of the variable measured.

These control charts can be constructed from a base set of data observations that are considered to be representative of the process when it is in control. This data set typically consists of a set of samples of size n taken at k different points in time, as shown in Figure 16.7. The sample size n is usually small, perhaps only 3, 4, or 5, but it may be as large as 20 in some cases. It should be a convenient number of observations that can be sampled at frequent

FIGURE 16.7

A data set for constructing $\bar{X}$-charts and R-charts

Sample size
n

$$\overbrace{}$$

Sample 1 $x_{11} \ldots x_{1n}$ Sample averages

Sample 2 $x_{21} \ldots x_{2n}$ $\bar{x}_i = \dfrac{x_{i1}+\cdots+x_{in}}{n}$

$\vdots$ Sample ranges

$$r_i = \max\{x_{i1}, \ldots, x_{in}\} - \min\{x_{i1}, \ldots, x_{in}\}$$

Sample k $x_{k1} \ldots x_{kn}$

---

intervals. Generally speaking, the control chart should be set up using data from at least $k = 20$ distinct points in time.

16.3.1 $\bar{X}$-Charts

The $\bar{X}$-chart consists of plots of the sample averages $\bar{x}_1, \ldots, \bar{x}_k$ against time. The sample averages

$$\bar{x}_i = \frac{x_{i1} + \cdots + x_{in}}{n}$$

are the average values of the n data values taken at each point in time.

The center line of the control chart is drawn at the overall average of the k sample averages

$$\bar{\bar{x}} = \frac{\bar{x}_1 + \cdots + \bar{x}_k}{k}$$

and the control limits are

$$\text{UCL} = \bar{\bar{x}} + 3\,\frac{\sigma}{\sqrt{n}} \quad \text{and} \quad \text{LCL} = \bar{\bar{x}} - 3\,\frac{\sigma}{\sqrt{n}}$$

where σ is the standard deviation of the individual data measurements.

There are various ways to estimate σ. Traditionally, a function of the *sample ranges* has been used with control charts since this was an easy calculation to perform before computers were widely employed. The range of the sample at the ith time point is

$$r_i = \max\{x_{i1}, \ldots, x_{in}\} - \min\{x_{i1}, \ldots, x_{in}\}$$

so that it is the difference between the largest data observation and the smallest data observation in the sample of size n.

The average sample range is then

$$\bar{r} = \frac{r_1 + \cdots + r_k}{k}$$

and the standard deviation is estimated as

$$\hat{\sigma} = \frac{\bar{r}}{d_2}$$

where the values of d_2, which depend only on the sample size n, are given in Figure 16.8. The control limits are therefore

$$\text{UCL} = \bar{\bar{x}} + A_2\bar{r} \quad \text{and} \quad \text{LCL} = \bar{\bar{x}} - A_2\bar{r}$$

	d_2	A_2
$n = 2$	1.128	1.880
$n = 3$	1.693	1.023
$n = 4$	2.059	0.729
$n = 5$	2.326	0.577
$n = 6$	2.534	0.483
$n = 7$	2.704	0.419
$n = 8$	2.847	0.373
$n = 9$	2.970	0.337
$n = 10$	3.078	0.308
$n = 11$	3.173	0.285
$n = 12$	3.258	0.266
$n = 13$	3.336	0.249
$n = 14$	3.407	0.235
$n = 15$	3.472	0.223
$n = 16$	3.532	0.212
$n = 17$	3.588	0.203
$n = 18$	3.640	0.194
$n = 19$	3.689	0.187
$n = 20$	3.735	0.180
$n = 21$	3.778	0.173
$n = 22$	3.819	0.167
$n = 23$	3.858	0.162
$n = 24$	3.895	0.157
$n = 25$	3.931	0.153

FIGURE 16.8

Values needed for the construction of $\bar{X}$-charts.

where

$$A_2 = \frac{3}{d_2\sqrt{n}}$$

which is also tabulated in Figure 16.8 for convenience.

	D_3	D_4
$n = 2$	0	3.267
$n = 3$	0	2.574
$n = 4$	0	2.282
$n = 5$	0	2.114
$n = 6$	0	2.004
$n = 7$	0.076	1.924
$n = 8$	0.136	1.864
$n = 9$	0.184	1.816
$n = 10$	0.223	1.777
$n = 11$	0.256	1.744
$n = 12$	0.283	1.717
$n = 13$	0.307	1.693
$n = 14$	0.328	1.672
$n = 15$	0.347	1.653
$n = 16$	0.363	1.637
$n = 17$	0.378	1.622
$n = 18$	0.391	1.608
$n = 19$	0.403	1.597
$n = 20$	0.415	1.585
$n = 21$	0.425	1.575
$n = 22$	0.434	1.566
$n = 23$	0.443	1.557
$n = 24$	0.451	1.548
$n = 25$	0.459	1.541

FIGURE 16.9

Values needed for the construction of R-charts.

The $\bar{X}$-Chart

An $\bar{X}$-chart consists of sample averages plotted against time and monitors changes in the mean value of a variable. The lines on the control chart can be determined from a set of k samples of size n, with the center line being taken as $\bar{\bar{x}}$, the overall average of the sample averages, and with control limits

$$\text{UCL} = \bar{\bar{x}} + A_2\bar{r} \quad \text{and} \quad \text{LCL} = \bar{\bar{x}} - A_2\bar{r}$$

where $\bar{r}$ is the average of the k sample ranges.

Once the lines on the control chart have been determined, the k sample averages $\bar{x}_i$ should be plotted. Since these data observations are taken to represent an in-control process, the points should all lie within the control limits. If some points lie outside the control limits, then this suggests that the process was not in control at those times. It is then generally best to remove these data points from the data set and to construct the control chart anew from the reduced data set.

The control chart can be used to monitor future observations. The sample average

$$\bar{x} = \frac{x_1 + \cdots + x_n}{n}$$

of a future sample of size n can be plotted on the control chart, and if it lies within the control limits, the process can be considered to be still in control. If it lies outside the control limits, then this can be taken as evidence that the mean value of the variable under consideration has moved from its control value. In practice, this $\bar{X}$-chart is used in conjunction with an R-chart discussed next, which monitors changes in the variability of the measurements.

16.3.2 R-Charts

The construction of an R-chart requires the values D_3 and D_4 that are given in Figure 16.9.

The R-Chart

An R-chart consists of sample ranges plotted against time and monitors changes in the variability of a measurement of interest. The lines on the control chart can be determined from a set of k samples of size n, with the center line being taken as $\bar{r}$, the average of the k sample ranges, and with control limits

$$\text{UCL} = D_4\bar{r} \quad \text{and} \quad \text{LCL} = D_3\bar{r}$$

Once the control chart lines have been determined, the sample ranges $r_1, \ldots, r_k$ can be plotted to check that they fall within the control limits. As with the $\bar{X}$-chart, if some points fall outside the control limits, they should be removed from the data set and the control chart should be reconstructed from the reduced data set. The ranges of future samples can then be plotted on the control chart to monitor for changes in the standard deviation of the variable of interest.

16.3.3 Modifications of $\bar{X}$-Charts and R-Charts

You will probably come across various modifications of the $\bar{X}$-chart and the R-chart that have the same motivation and interpretation but that are constructed differently. Recently the sample standard deviations are often used to estimate σ in place of the sample ranges r_i in the construction of the $\bar{X}$-chart. In addition, an "S-chart," which consists of sample standard deviations plotted against time, may be used in place of the R-chart for measuring process variability. Technically speaking, the sample standard deviations are a more efficient way of estimating σ than the sample ranges for most sample sizes. Most computer packages provide the option of using either approach, and it is unlikely that they will provide substantially different results.

In some circumstances only one observation is taken at each point in time so that $n = 1$. Then the process variability cannot be examined (although the time intervals could be grouped together to form larger time intervals where the sample sizes are larger than one). A control chart can still be constructed to monitor changes in the mean value of the process, although this will differ from the $\bar{X}$-chart due to the manner in which σ is estimated.

Finally, an alternative to the $\bar{X}$-chart is a CUSUM-chart where, for a control value μ_0 of the variable mean, the cumulative sums

$$\sum_{i=1}^{t} (\bar{x}_i - \mu_0)$$

are plotted against time. In other words, at time t the cumulative differences between the control value and the sample averages at time t and all previous times are plotted, rather than just the sample average at time t. While the interpretation of a CUSUM-chart is more complicated than the interpretation of the $\bar{X}$-chart, its use is motivated by the fact that it is more sensitive than the $\bar{X}$-chart to small changes in the variable mean from the control value, so that it can detect these changes with a smaller average run length.

16.3.4 Examples

Example 14
Metal Cylinder Production

Consider again the data set of 60 metal cylinder diameters given in Figure 6.5 and their summary statistics given in Figure 6.29. Suppose that these data values represent the measurements of a random sample of $n = 3$ cylinders taken every 30 minutes over a 10-hour run of a production process, as shown in Figure 16.10. How can this data set be used to construct a control chart for future runs of this production process?

As shown in Figure 16.10, the sample averages $\bar{x}_i$ and the sample ranges r_i are first constructed. For example, for the first sample the sample average is

$$\bar{x}_1 = \frac{x_{11} + x_{12} + x_{13}}{3} = \frac{50.08 + 49.78 + 50.02}{3} = 49.960$$

and the sample range is

$$r_1 = \max\{x_{11}, x_{12}, x_{13}\} - \min\{x_{11}, x_{12}, x_{13}\}$$
$$= \max\{50.08, 49.78, 50.02\} - \min\{50.08, 49.78, 50.02\}$$
$$= 50.08 - 49.78 = 0.30$$

FIGURE 16.10

Metal cylinder production data set

Sample number	Diameter measurements			Sample average $\bar{x}_i$	Sample range r_i
1	50.08	49.78	50.02	49.960	0.30
2	50.02	50.13	49.74	49.963	0.39
3	49.84	49.97	49.93	49.913	0.13
4	50.02	50.05	49.94	50.003	0.11
5	50.19	49.86	50.03	50.027	0.33
6	50.04	49.96	49.90	49.967	0.14
7	49.87	50.13	49.81	49.937	0.32
8	50.02	50.26	49.90	50.060	0.36
9	50.01	50.04	50.01	50.020	0.03
10	49.79	50.36	50.21	50.120	0.57
11	50.17	50.12	50.00	50.097	0.17
12	50.01	49.85	49.93	49.930	0.16
13	49.84	50.20	49.94	49.993	0.36
14	49.74	50.00	50.03	49.923	0.29
15	49.92	50.07	49.89	49.960	0.18
16	49.99	50.01	50.09	50.030	0.10
17	49.90	50.05	49.95	49.967	0.15
18	50.20	50.03	49.92	50.050	0.28
19	50.02	49.97	50.27	50.087	0.30
20	49.77	50.07	50.07	49.970	0.30

The center line on the $\bar{X}$-chart is drawn at

$$\bar{\bar{x}} = 50.00$$

which is the average of all 60 data observations (and similarly is the average of the 20 sample means), and the average sample range is

$$\bar{r} = \frac{r_1 + \cdots + r_{20}}{20} = \frac{0.30 + 0.39 + \cdots + 0.30}{20} = 0.2485$$

With $A_2 = 1.023$ for $n = 3$, the control limits for the $\bar{X}$-chart are

$$UCL = \bar{\bar{x}} + A_2\bar{r} = 50.00 + (1.023 \times 0.2485) = 50.25$$

and

$$LCL = \bar{\bar{x}} - A_2\bar{r} = 50.00 - (1.023 \times 0.2485) = 49.75$$

The resulting $\bar{X}$-chart with the 20 sample averages plotted is shown in Figure 16.11.

The R-chart has a center line at $\bar{r} = 0.2485$, and with $D_3 = 0$ and $D_4 = 2.574$ for $n = 3$, the control limits are

$$UCL = D_4\bar{r} = 2.574 \times 0.2485 = 0.6397$$

and

$$LCL = D_3\bar{r} = 0 \times 0.2485 = 0$$

Figure 16.12 shows the resulting R-chart with the 20 sample ranges plotted.

Notice that all of the points lie within the control limits for both the $\bar{X}$-chart and the R-chart, so that it is reasonable to consider the process to be in control while the data set was collected. However, as mentioned previously, this does not guarantee that the process

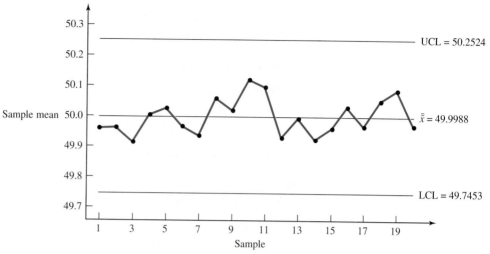

FIGURE 16.11

The $\bar{X}$-chart for the metal cylinder diameters example

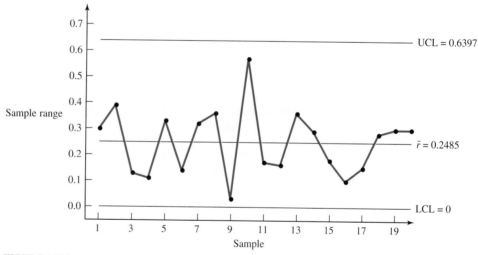

FIGURE 16.12

The R-chart for the metal cylinder diameters example

is working well. How well the process is working depends upon the actual specifications required for the metal cylinders. Other statistical techniques can be used to assess how well the production process is working, such as the analysis in Section 8.1.1, which revealed that a 95% two-sided confidence interval for the mean diameter of the cylinders is (49.964, 50.033).

These control charts can now be used to monitor future runs of the production line. Suppose that production has started again and after 30 minutes a random sample of $n = 3$ cylinders have diameters measured as

$$x_1 = 50.06, \ x_2 = 50.01, \text{ and } x_3 = 49.67$$

Is there any evidence that the process has moved out of control? The sample average is

$$\bar{x} = \frac{50.06 + 50.01 + 49.67}{3} = 49.913$$

and the sample range is

$$r = 50.06 - 49.67 = 0.39$$

and when plotted on their respective control charts, both of these points line within the control limits. Consequently, the process can be considered to be still in control.

What happens if the data values

$$x_1 = 50.34, \ x_2 = 50.48, \text{ and } x_3 = 50.16$$

are obtained? In this case the sample average is

$$\bar{x} = \frac{50.34 + 50.48 + 50.16}{3} = 50.327$$

and the sample range is

$$r = 50.48 - 50.16 = 0.32$$

While the sample range r lies within the control limits in the R-chart, the sample average lies above the upper control limit on the $\bar{X}$-chart. This finding alerts the supervisor that the process has moved out of control due to an increase in the mean value of the cylinder diameters. Suitable corrective action can be taken straightaway.

If the data values

$$x_1 = 49.66, \ x_2 = 50.47, \text{ and } x_3 = 50.03$$

are obtained, then the sample average is

$$\bar{x} = \frac{49.66 + 50.47 + 50.03}{3} = 50.053$$

and the sample range is

$$r = 50.47 - 49.66 = 0.81$$

In this case the sample average lies within the control limits on the $\bar{X}$-chart, but the sample range lies above the upper control limit on the R-chart. This finding indicates that the process has moved out of control due to an increase in the variability of the cylinder diameters.

Example 17

Milk Container Contents

The milk bottling company has decided to employ a control chart to provide continuous monitoring of the weight of its milk containers. During an 8-hour shift, a random sample of $n = 5$ milk containers is selected every 20 minutes and their weights are measured. The data set shown in Figure 16.13 is obtained.

FIGURE 16.13

Milk container contents data set

Sample number	Milk container weights					Sample average $\bar{x}_i$	Sample range r_i
1	2.106	1.965	2.081	2.079	2.129	2.0720	0.164
2	1.950	1.994	2.058	2.039	2.080	2.0242	0.130
3	2.095	2.043	2.158	2.005	2.085	2.0772	0.153
4	2.088	2.053	2.025	2.011	2.128	2.0610	0.117
5	1.980	2.080	2.189	2.024	2.072	2.0690	0.209
6	2.090	2.075	1.970	2.239	2.169	2.1086	0.269
7	2.147	2.124	2.069	2.077	1.999	2.0832	0.148
8	2.037	1.996	2.163	2.044	1.947	2.0374	0.216
9	2.072	2.083	2.033	2.103	2.018	2.0618	0.085
10	2.063	2.073	2.111	2.025	2.093	2.0730	0.086
11	2.045	2.162	2.084	2.032	2.079	2.0804	0.130
12	2.054	2.196	1.964	2.104	2.103	2.0842	0.232
13	1.883	1.833	2.288	2.119	1.957	2.0160	0.455
14	2.212	2.019	2.380	2.139	1.796	2.1092	0.584
15	2.131	1.986	1.785	2.233	2.058	2.0386	0.448
16	2.045	2.218	2.065	2.023	2.089	2.0880	0.195
17	2.048	2.054	2.005	1.969	1.995	2.0142	0.085
18	2.010	2.065	2.249	1.965	2.031	2.0640	0.284
19	2.022	1.996	1.970	1.946	2.146	2.0160	0.200
20	2.094	2.043	2.108	2.046	2.160	2.0902	0.117
21	2.086	2.015	2.097	2.121	2.008	2.0654	0.113
22	1.988	1.970	1.927	2.063	2.028	1.9952	0.136
23	2.031	2.016	2.088	2.096	2.121	2.0704	0.105
24	2.051	2.028	2.041	2.030	2.118	2.0536	0.090

The $\bar{X}$-chart and R-chart for this data set are shown in Figure 16.14. The $\bar{X}$-chart has a center line at

$$\bar{\bar{x}} = 2.0605$$

and the R-chart has a center line at

$$\bar{r} = 0.198$$

With $A_2 = 0.577$ for $n = 5$, the control limits for the $\bar{X}$-chart are

$$\text{UCL} = \bar{\bar{x}} + A_2\bar{r} = 2.0605 + (0.577 \times 0.198) = 2.175$$

and

$$\text{LCL} = \bar{\bar{x}} - A_2\bar{r} = 2.0605 - (0.577 \times 0.198) = 1.946$$

Also, with $D_3 = 0$ and $D_4 = 2.114$ for $n = 5$, the control limits for the R-chart are

$$\text{UCL} = D_4\bar{r} = 2.114 \times 0.198 = 0.419$$

and

$$\text{LCL} = D_3\bar{r} = 0 \times 0.198 = 0$$

Notice that while the points on the $\bar{X}$-chart all lie within the control limits, there are three consecutive points on the R-chart that lie above the upper control limit. This indicates that

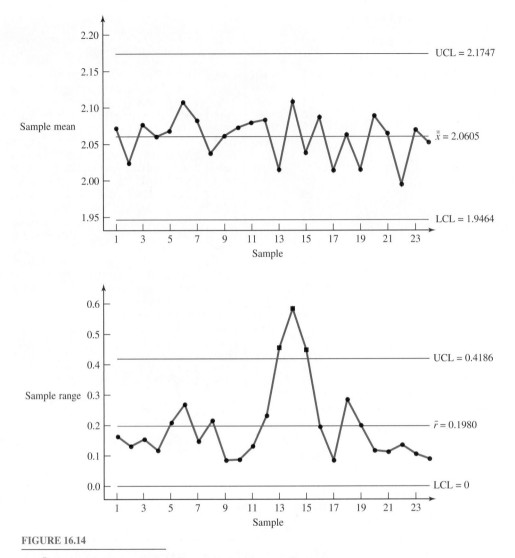

FIGURE 16.14

The $\bar{X}$-chart and the R-chart for the milk container weights example

this data set is not representative of an in-control process, and specifically it indicates that about 4 hours into the shift the process went out of control due to a sudden increase in the variability of the milk container weights. It would be instructive for the company to identify the reasons behind this (temporary) increase in variability so that the problem can be avoided in the future.

It is best to eliminate the three samples that have sample ranges above the control limit and to construct the control charts again. The control charts constructed from the reduced data set are shown in Figure 16.15, and it can be seen that all points now lie within the control limits. Notice that the control limits are also narrower than those obtained with the full data set. These control charts can now be used to monitor the weights of future samples of milk containers.

FIGURE 16.15

The $\bar{X}$-chart and the R-chart for the reduced data set of milk container weights

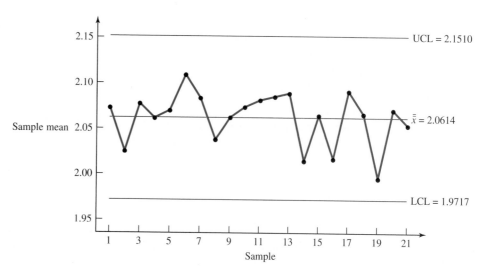

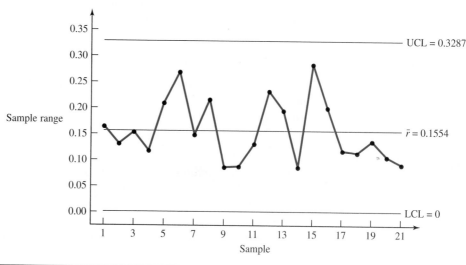

■ **16.3.5   Problems**

16.3.1 DS 16.3.1 contains the sample means $\bar{x}_i$ and the sample ranges r_i of samples of size $n = 4$ of a measurement of interest collected at $k = 20$ distinct time points from a production process.

(a) Construct an $\bar{X}$-chart and an R-chart.

(b) Is there any evidence that the process was out of control while the data values were collected? Should any of the samples be removed from the data set?

(c) If the control charts are used for monitoring future production runs, what conclusion would you draw if a sample $x_1 = 86.1$, $x_2 = 99.2$, $x_3 = 94.5$, and $x_4 = 90.7$ is obtained?

(d) What if $x_1 = 91.5$, $x_2 = 78.0$, $x_3 = 86.5$, and $x_4 = 82.4$?

(e) What if $x_1 = 95.3$, $x_2 = 92.1$, $x_3 = 89.6$, and $x_4 = 90.1$?

(f) What if $x_1 = 99.2$, $x_2 = 93.8$, $x_3 = 96.1$, and $x_4 = 94.2$?

16.3.2 DS 16.3.2 contains the sample means $\bar{x}_i$ and the sample ranges r_i of a variable measurement based upon samples of size $n = 5$, which are collected at $k = 25$ points in time.

(a) Use this data set to construct an $\bar{X}$-chart and an R-chart.

(b) Is there any evidence that the process was out of control while the data values were collected? Which samples should be removed from the data set?

(c) Construct control charts that can be employed to monitor future measurements.

16.3.3 Glass Sheet Thicknesses

Control charts are to be used to monitor the thicknesses of glass sheets. DS 16.3.3 contains the thicknesses in mm of random samples of $n = 4$ glass sheets collected at $k = 24$ different times.

(a) What $\bar{X}$-chart and R-chart would you recommend be used?

(b) What conclusion would you draw if in the future a sample $x_1 = 3.12$, $x_2 = 2.96$, $x_3 = 2.91$, and $x_4 = 2.88$ is obtained?

16.4   Attribute Control Charts

Whereas the $\bar{X}$-chart and the R-chart are appropriate for monitoring a continuous measurement of interest, attribute control charts can be used to monitor certain categorical measurements of the items of interest. Specifically a *p-chart* can be used to monitor the *proportion* of a sample that has a particular attribute. Typically, the proportion of defective items is of interest. A **c-chart** can be used to monitor the frequency of defects within an item.

16.4.1   *p*-Charts

A p-chart consists of a set of sample proportions plotted against time. If at a certain point in time a sample of size n items is obtained of which x possess a certain characteristic, the sample proportion

$$p = \frac{x}{n}$$

is plotted on the control chart. If this point lies within the control limits, then the process can be considered to be in control, and if it lies outside the control limits, then the process is known to be out of control due to a change in the proportion of items that possess the characteristic of interest.

A control chart can be constructed from a base set of data observations that are taken to be representative of the in-control process. Such a data set consists of a set of proportions $p_1, \ldots, p_k$ taken at k distinct points in time, where p_i is the proportion of the sample of size n taken at the ith time point that possess the characteristic of interest.

The center line of the control chart is the average proportion

$$\bar{p} = \frac{p_1 + \cdots + p_k}{k}$$

which is the proportion of all the items sampled that possess the characteristic. If $\bar{p}$ really is the probability that an item has the characteristic, then the number of such items in a future sample of size n has a binomial distribution with parameters n and $\bar{p}$. The standard error of the sample proportion with the characteristic is therefore

$$\sqrt{\frac{\bar{p}(1 - \bar{p})}{n}}$$

and consequently a 3-sigma control chart is constructed with control limits

$$\text{UCL} = \bar{p} + 3\sqrt{\frac{\bar{p}(1 - \bar{p})}{n}} \qquad \text{and} \qquad \text{LCL} = \bar{p} - 3\sqrt{\frac{\bar{p}(1 - \bar{p})}{n}}$$

Often the lower control limit turns out to be negative, in which case it should be taken to be 0. If any of the sample proportions $p_1, \ldots, p_k$ lie outside the control limits, then those samples should be removed from the data set and the control chart should be recalculated.

The p-Chart

A p-chart consists of sample proportions plotted against time, and it monitors changes in the proportion of items that possess a particular characteristic of interest. The lines on the control chart can be determined from a set of k sample proportions obtained from k samples of size n. The center line is taken to be $\bar{p}$, the average of the sample proportions, and the control limits are

$$\text{UCL} = \bar{p} + 3\sqrt{\frac{\bar{p}(1-\bar{p})}{n}} \quad \text{and} \quad \text{LCL} = \bar{p} - 3\sqrt{\frac{\bar{p}(1-\bar{p})}{n}}$$

Example 2
Defective Computer Chips

Consider again the data set in Figure 6.3, which contains the number of defective computer chips found in 80 boxes of chips, with each box containing a total of $n = 500$ chips. Suppose that these 80 boxes are obtained by taking every hundredth box off a production line. How can this data set be used to set up a control chart to monitor the number of defective chips in each box?

The sample proportions are

$$p_1 = \frac{1}{500} = 0.002, \; p_2 = \frac{3}{500} = 0.006, \ldots, p_{80} = \frac{1}{500} = 0.002.$$

These values have an average of

$$\bar{p} = 0.00615$$

which is the overall proportion of defective chips found. The control limits are consequently

$$\text{UCL} = \bar{p} + 3\sqrt{\frac{\bar{p}(1-\bar{p})}{n}}$$

$$= 0.00615 + \left(3 \times \sqrt{\frac{0.00615 \times (1 - 0.00615)}{500}} \right) = 0.0166$$

and

$$\text{LCL} = \bar{p} - 3\sqrt{\frac{\bar{p}(1-\bar{p})}{n}}$$

$$= 0.00615 - \left(3 \times \sqrt{\frac{0.00615 \times (1 - 0.00615)}{500}} \right) = -0.0043$$

Since a negative number is obtained, the lower control limit should be taken to be 0.

Figure 16.16 shows the control chart for this data set. All of the 80 sample proportions p_i lie within the control limits, and so the data set can be taken as being representative of an in-control process. The proportion of defective computer chips found in future samples from the production line can be plotted on the control chart to monitor for changes in the proportion of defective chips.

FIGURE 16.16

The p-chart for the defective
computer chips example

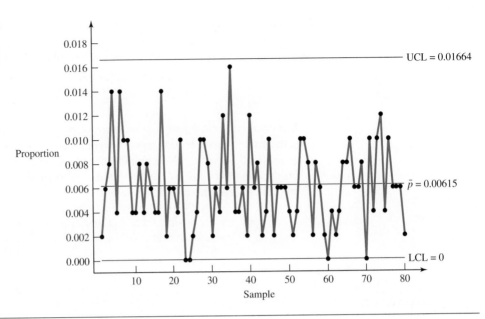

In fact, a point will lie outside the upper control limit if the proportion of defective chips is larger than 0.0166. This value corresponds to at least

$$0.0166 \times 500 = 8.3$$

defective chips in a box of $n = 500$ chips or, in other words, nine or more defective chips in a box. Thus, the control chart indicates that the process can be considered to have moved out of control if the random sampling (every hundredth box) produces a box containing nine or more defective chips.

If p is the true probability that a chip is defective, then each time a sample is taken this procedure is equivalent to testing the hypotheses

$$H_0 : p = \bar{p} \quad \text{versus} \quad H_A : p \neq \bar{p}$$

The number of defective chips X in a box of $n = 500$ chips is distributed

$$X \sim B(500, p)$$

and so this is a hypothesis testing problem concerning the success probability of a binomial random variable, which is discussed in Section 10.1.2. Notice that, in general, an observed proportion $\hat{p} = x/n$ falls *outside* the control limits of a p-chart constructed with $\bar{p} = p_0$ if the z-statistic

$$z = \frac{\hat{p} - p_0}{\sqrt{\frac{p_0(1 - p_0)}{n}}}$$

discussed in Section 10.1.2 has an absolute value larger than 3 (because the p-chart is constructed as a 3-sigma chart).

Sometimes an np-chart is used in place of a p-chart. An np-chart is similar to a p-chart except that the number of items possessing the characteristic of interest is plotted on the control chart instead of the proportion of items. The center line of an np-chart is $n\bar{p}$, and the control limits are

$$\text{UCL} = n\bar{p} + 3\sqrt{n\bar{p}(1-\bar{p})} \qquad \text{and} \qquad \text{LCL} = n\bar{p} - 3\sqrt{n\bar{p}(1-\bar{p})}$$

An np-chart and a p-chart look the same except that the scaling of the y-axis is different.

16.4.2   c-Charts

Sample number	Number of fractures
1	3
2	1
3	8
4	1
5	5
6	6
7	5
8	5
9	1
10	6
11	6
12	4
13	5
14	6
15	9
16	4
17	3
18	8
19	4
20	3
21	3
22	5
23	4
24	2
25	5
26	7
27	6
28	10
29	16
30	14
31	15
32	15

FIGURE 16.17

Steel girder fractures data set

Whereas a p-chart can be used when an item is classified as being either defective or satisfactory, a c-chart allows an item to be judged on the basis of a *count* of the number of flaws or defects that it contains. A c-chart is therefore a control chart where the number of defects found in a sample of items is plotted against time.

A base set of data observations corresponding to the number of defects observed in a set of k items sampled at k distinct times can be used to construct a c-chart. If these data values are $x_1, \ldots, x_k$, then the center line of the control chart is drawn at the sample average

$$\bar{x} = \frac{x_1 + \cdots + x_k}{k}$$

If the number of defects X in an item is modeled with a Poisson distribution with parameter λ, then the expected number of defects per item and the variance of the number of defects per item are both equal to λ (see Section 3.4). Consequently, λ can be estimated by $\hat{\lambda} = \bar{x}$, and the standard deviation of the number of defects X in an item is $\sqrt{\lambda}$, which can be estimated by $\sqrt{\bar{x}}$. Consequently, the 3-sigma control limits are chosen to be

$$\text{UCL} = \bar{x} + 3\sqrt{\bar{x}} \qquad \text{and} \qquad \text{LCL} = \bar{x} - 3\sqrt{\bar{x}}$$

As before, if the lower confidence bound turns out to be negative, it is taken to be 0.

The c-Chart

A c-chart consists of points corresponding to the number of defects found in sampled items plotted against time. The lines on the control chart can be determined from a set of data observations corresponding to the number of defects observed in a set of k sampled items. The center line is taken to be $\bar{x}$, the average of the data values, and the control limits are

$$\text{UCL} = \bar{x} + 3\sqrt{\bar{x}} \qquad \text{and} \qquad \text{LCL} = \bar{x} - 3\sqrt{\bar{x}}$$

In practice more than one item can be sampled at each time point, and in fact larger sample sizes produce a more sensitive control chart. However, the same number of items must be sampled at each time point with a c-chart. In this case the c-chart can be constructed using data values corresponding to the *total* number of defects found at each time point from the sampled items. A u-chart is a modification of a c-chart that allows for varying numbers of items to be sampled at different points in time.

Example 32

Steel Girder Fractures

A c-chart is to be constructed to monitor the number of hairline fractures occurring along the edges of steel girders. It is decided that a girder will be randomly selected every 15 minutes and that a 1-meter segment of the girder will then be randomly selected. The number of fractures within this segment will then be recorded. Figure 16.17 contains a data set consisting of 32 observations collected from an 8-hour shift which is to be used to construct the control chart.

FIGURE 16.18

The c-chart for the steel girder
fractures example

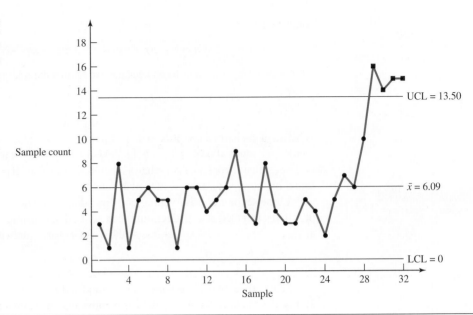

The average number of fractures found is

$$\bar{x} = \frac{3 + 1 + \cdots + 15}{32} = 6.094$$

which is the center line of the control chart. The control limits are

$$\text{UCL} = \bar{x} + 3\sqrt{\bar{x}} = 6.094 + (3 \times \sqrt{6.094}) = 13.50$$

and

$$\text{LCL} = \bar{x} - 3\sqrt{\bar{x}} = 6.094 - (3 \times \sqrt{6.094}) = -1.31$$

Since a negative value is obtained, the lower control limit can be taken to be 0.

Figure 16.18 shows the resulting control chart together with plots of the data values. It can be seen that the process appears to have moved out of control toward the end of the 8-hour shift. In fact the four data observations taken during the last hour are all above the upper control limit. This increase in the number of fractures should be investigated because it could provide some useful information on how to keep the number of fractures low in the future.

If the last four data points are removed from the data set, the sample average becomes

$$\bar{x} = \frac{3 + 1 + \cdots + 10}{28} = 4.821$$

and the control limits are

$$\text{UCL} = \bar{x} + 3\sqrt{\bar{x}} = 4.821 + (3 \times \sqrt{4.821}) = 11.41$$

and

$$\text{LCL} = \bar{x} - 3\sqrt{\bar{x}} = 4.821 - (3 \times \sqrt{4.821}) = -1.77$$

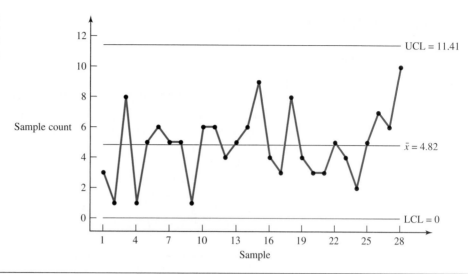

FIGURE 16.19

The c-chart for the reduced data set of steel girder fractures

which again can be taken to be 0. This control chart is shown in Figure 16.19 and all of the data points lie between the control limits. The chart can be used to monitor new girders, and the process will be determined to be out of control if a randomly sampled girder has 12 or more fractures within the randomly chosen 1-meter segment.

■ 16.4.3 Problems

16.4.1 Construct a p-chart from the data in DS 16.4.1, which are the number of defective items found in random samples of $n = 100$ items taken at $k = 30$ time points.
- **(a)** Is there any reason to believe that the process was out of control while this data set was collected?
- **(b)** How many defective items would need to be discovered in a future sample for the process to be considered to be out of control?

16.4.2 Spray Painting Procedure
Metal rods are spray painted by a machine and a p-chart is to be used to monitor the proportion of rods that are not painted correctly. These defective rods have either an incomplete coverage or a paint layer that is too thick in some places. The data set in DS 16.4.2 contains the number of defective rods found in random samples of $n = 400$ rods taken at $k = 25$ time points.
- **(a)** Why does it appear that the process was out of control at certain times while this data set was collected?
- **(b)** What p-chart would you recommend be used to monitor future samples of rods?
- **(c)** How many defective rods would need to be discovered in a future sample for the process to be considered to be out of control?

16.4.3 Construct a c-chart from the data in DS 16.4.3, which are the number of defects found in a randomly sampled product at $k = 24$ time points.
- **(a)** Should any observations be deleted from the data set?
- **(b)** How many defects would need to be discovered in a future randomly sampled product for the process to be considered to be out of control?

16.4.4 Paper Quality Assessment
A paper mill has decided to use a c-chart to monitor the number of imperfections in large paper sheets. The data set in DS 16.4.4 records the number of imperfections found in $k = 22$ sheets of paper sampled at different times.
- **(a)** Construct a c-chart. Why does it appear that the process was out of control at certain times while this data set was collected?
- **(b)** What c-chart would you recommend be used to monitor future samples of paper?
- **(c)** How many imperfections would need to be discovered in a future randomly sampled paper sheet for the process to be considered to be out of control?

16.5   Acceptance Sampling

16.5.1   Introduction

Acceptance sampling has traditionally been a partner of statistical process control and control charts in the area of statistical quality control. Products are shipped around in **batches** or **lots**, and the idea behind acceptance sampling is that a batch can be declared to be satisfactory or unsatisfactory on the basis of the number of defective items found within a random sample of items from the batch. Thus acceptance sampling provides a general check on the "quality" of the items within a batch.

Acceptance sampling can be performed by the **producer** of the products before they are shipped out. In addition, sampling may be performed by **customers** who receive the products in order to check that they are receiving high-quality materials. Many companies are accustomed to performing acceptance sampling on their incoming raw materials before taking delivery of them.

Notice that there is an important procedural distinction between statistical process control and acceptance sampling. The control charts used in statistical process control provide a real-time monitoring of a production process with the objective of avoiding the production of low-quality materials. Changes in the process can be identified almost immediately so that corrective actions can be taken. In contrast, acceptance sampling is performed after production has been completed. It does not allow any monitoring of the production process, and whole batches of products can be wasted if they are found to be unsatisfactory.

16.5.2   Acceptance Sampling Procedures

Consider a batch of N items, each one of which can be classified as being either satisfactory or defective. When N is very large, a 100% inspection scheme of the batch in which each item is examined is generally too expensive and time consuming. Therefore, a random sample of n of the items is chosen, and the number of defective items x in the sample is found.

The acceptance sampling procedure is based upon a rule whereby the batch is declared to be satisfactory (accept the batch) if the number of defective items x is no larger than a constant c, and the batch is declared to be unsatisfactory (reject the batch) if the number of defective items x is larger than c. Thus the batch is accepted if

$$x = 0, \ldots, c$$

and is rejected if

$$x = c + 1, \ldots, n$$

as shown in Figure 16.20.

The statistical properties of the acceptance sampling procedure can be determined by considering how the acceptance probability depends on the true proportion p of defective items in the batch. It is usual to define an **acceptable quality level** (AQL), p_0 say, so that a batch is considered to be acceptable as long as the actual proportion of defective items is no larger than the AQL, that is as long as

$$p \leq p_0$$

The batch is not considered to be acceptable if

$$p > p_0$$

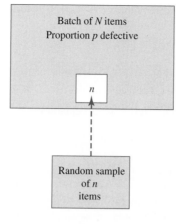

x = number of defective items in sample

$$x = 0, 1, 2, \ldots, c, c+1, \ldots, n$$

Accept batch   Reject batch

FIGURE 16.20

An acceptance sampling procedure

Acceptance sampling procedure

	Accept batch	Reject batch
H_0: proportion of defective items $p \leq p_0$	☺	Type I error ☹ Producer's risk
H_A: proportion of defective items $p > p_0$	Type II error ☹ Consumer's risk	☺

FIGURE 16.21

The risks associated with an acceptance sampling procedure

There are two kinds of risk associated with the acceptance sampling procedure, a **producer's risk** and a **consumer's risk**. The producer's risk is the probability that the batch is rejected when in fact the proportion of defective items p is smaller than the AQL p_0. This is the probability that the producer produces an acceptable batch of items, but that the batch is declared to be unsatisfactory by the acceptance sampling procedure. Obviously, it is desirable to have the producer's risk as small as possible.

In contrast, the consumer's risk is defined to be the probability that a batch is declared to be satisfactory by the acceptance sampling procedure when in fact the proportion of defective items p is larger than the AQL p_0. This is the probability that a consumer takes delivery of a batch that is really not acceptable. The amount of consumer's risk depends on the value of the actual proportion of defective items p, and again it is desirable to have it as small as possible.

The concepts of producer's risk and consumer's risk are analogous to the probabilities of a type I error and a type II error for a hypothesis testing problem, as shown in Figure 16.21. Consider the hypothesis testing problem

$$H_0 : p \leq p_0 \quad \text{versus} \quad H_A : p > p_0$$

The null hypothesis H_0 states that the batch is acceptable and the alternative hypothesis H_A states that the batch is not acceptable.

The size of the hypothesis test is

$\alpha = P(\text{Type I error})$

$\quad = P(\text{reject } H_0 \text{ when } H_0 \text{ is true})$

$\quad = P(\text{reject batch when batch is acceptable})$

$\quad = \text{producer's risk}$

Similarly, if the power of the hypothesis test is $1 - \beta$ for a proportion of defective items $p_1 > p_0$, then

$$\beta = 1 - \text{power}$$
$$= P(\text{Type II error})$$
$$= P(\text{accept } H_0 \text{ when } H_0 \text{ is false})$$
$$= P(\text{accept batch when batch is not acceptable})$$
$$= \text{consumer's risk}$$

Acceptance Sampling Procedures

An acceptance sampling procedure is based on a rule whereby a batch is declared to be satisfactory (the batch is accepted) if the number of defective items x found in a random sample from the batch is no larger than a constant c, and the batch is declared to be unsatisfactory (the batch is rejected) if the number of defective items x is larger than c.

The producer's risk is the probability that the batch is rejected when in fact the proportion of defective items is smaller than an acceptable quality level (AQL) p_0. The consumer's risk is the probability that a batch is declared to be satisfactory when in fact the proportion of defective items is larger than the acceptable quality level p_0.

The properties of an acceptance sampling procedure can be viewed graphically with an **operating characteristic curve**, which plots the probability of accepting a batch

$$P(\text{accept batch})$$

against the proportion of defective items p. As Figure 16.22 shows, both the producer's risk and the consumer's risk at a proportion p_1 can be found from the operating characteristic curve.

The operating characteristic curve is constructed from the distribution of the random variable X, which is the number of defective items found in the random sample of size n. Since sampling is performed without replacement, the random variable X has a hypergeometric distribution (see Section 3.3). If the proportion of defective items is p, so that the total number of defective items in the batch of size N is Np, then the probability mass function of the number of defective items in the sample is

$$P(X = x) = \frac{\binom{Np}{x} \times \binom{N(1-p)}{n-x}}{\binom{N}{n}}$$

The probability that the batch is accepted is

$$P(\text{accept batch}) = P(X = 0) + P(X = 1) + \cdots + P(X = c)$$

If the batch size N is much larger than the sample size n, as is usually the case, then recall that the hypergeometric distribution can be approximated by a binomial distribution, so that in this case the approximate distribution

$$X \sim B(n, p)$$

FIGURE 16.22

The operating characteristic curve
for an acceptance sampling plan

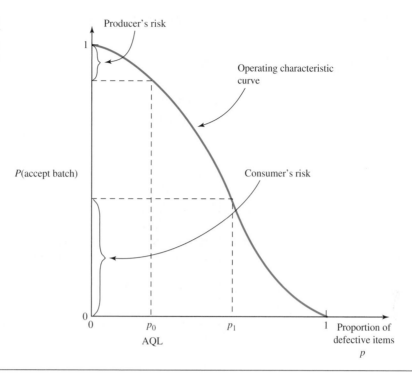

can be used. The acceptance probability can then be calculated as

$$P(\text{accept batch}) = \sum_{i=0}^{c} \binom{n}{i} p^i (1 - p)^{n-i}$$

Example 10
Fiber Coatings

Fibers are packaged in batches of $N = 500$ fibers and an acceptance sampling scheme is employed to check the acceptability of the coating on the fibers. An acceptable quality level of

$$\text{AQL} = p_0 = 0.05$$

is decided upon, and a random sample of $n = 20$ fibers is to be taken from a batch. A value of $c = 2$ is used so that the batch is considered to be acceptable if no more than two of the fibers in the sample are found to have an unsatisfactory coating. What are the operating characteristics of this acceptance sampling procedure?

With the binomial approximation, the probability of accepting a batch when the true proportion of defective fibers is p is

$$P(\text{accept batch}) = \sum_{i=0}^{2} \binom{20}{i} p^i (1 - p)^{20-i}$$
$$= (1 - p)^{20} + 20 \, p(1 - p)^{19} + 190 \, p^2 (1 - p)^{18}$$

The operating characteristic curve obtained from this formula is shown in Figure 16.23.

When $p = p_0 = 0.05$, the probability that the batch is accepted is

$$P(\text{accept batch}) = 0.95^{20} + (20 \times 0.05 \times 0.95^{19}) + (190 \times 0.05^2 \times 0.95^{18})$$
$$= 0.92$$

FIGURE 16.23

The operating characteristic curve
for the fiber coatings acceptance
sampling plan

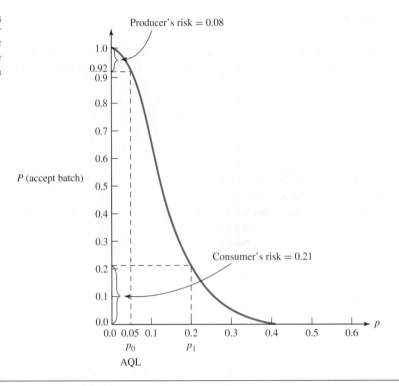

The producer's risk is the probability that the batch is rejected when $p = p_0 = 0.05$ and is therefore

producer's risk $= 1 - 0.92 = 0.08$

What is the consumer's risk when a proportion $p_1 = 0.2$ of the fibers have an unsatisfactory coating? This is the probability that the batch is accepted when $p = p_1 = 0.2$, which is

consumer's risk $= P(\text{accept batch})$
$$= 0.8^{20} + (20 \times 0.2 \times 0.8^{19}) + (190 \times 0.2^2 \times 0.8^{18}) = 0.21$$

as shown in Figure 16.23.

Many handbooks are available to assist in the implementation of acceptance sampling procedures. These handbooks indicate what sample sizes n should be taken and what values of c should be used in order to obtain a desired operating characteristic curve.

It is also worth pointing out that more sophisticated sampling schemes such as *double* or *sequential* sampling plans have been developed. In a double sampling scheme two constants $c_1 < c_2$ are employed. An initial random sample is obtained and the batch is accepted if the number of defective items is smaller than c_1 and the batch is rejected if the number of defective items is larger than c_2. However, if the number of defective items lies between c_1 and c_2, then a second random sample is taken. In this latter case, a final decision is then based upon the number of defective items found in the combination of the two samples. Sampling plans such as this have been developed with the objective of minimizing the average amount of sampling that needs to be performed while maintaining the desired properties of the operating characteristic curve.

■ 16.5.3 Problems

16.5.1 A random sample of size $n = 5$ is taken from a batch of size $N = 50$, and the batch is accepted if the number of defective items found is no larger than $c = 2$. Use the hypergeometric distribution to calculate each exact value.
  (a) The producer's risk for an acceptable quality level of $p_0 = 0.06$
  (b) The consumer's risk if the proportion of defective items is $p_1 = 0.20$
  Recompute these risks using the binomial approximation to the distribution of the number of defective items found in the sample.

16.5.2 Electrical fuses are sold in boxes of $N = 20$ fuses, and an acceptance sampling procedure is implemented with a random sample of size $n = 3$ and a value $c = 1$. Calculate and compare each value exactly using the hypergeometric distribution and approximately using the binomial distribution.
  (a) The producer's risk for an acceptable quality level of $p_0 = 0.10$
  (b) The consumer's risk if the proportion of defective items is $p_1 = 0.20$

16.5.3 An acceptance sampling procedure has $n = 40$ and $c = 8$ for a very large batch size N so that the binomial distribution can be used to calculate the probability of accepting the batch. Calculate the probability of accepting the batch for various values of the proportion of defective items p, and sketch the operating characteristic curve.
  (a) What is the producer's risk for an acceptable quality level of $p_0 = 0.01$?
  (b) What is the consumer's risk if the proportion of defective items is $p_1 = 0.25$?

16.5.4 Ceramic tiles are shipped in very large batches and an acceptance sampling procedure to monitor the number of cracked tiles in a batch has $n = 50$ and $c = 10$. Use the binomial approximation to calculate the probability of accepting the batch of tiles for various values of the proportion of cracked tiles p, and sketch the operating characteristic curve.
  (a) What is the producer's risk for an acceptable quality level of $p_0 = 0.01$?
  (b) What is the consumer's risk if the proportion of defective items is $p_1 = 0.10$?
  (c) How do the producer's and consumer's risks change if a value of $c = 9$ is used?

16.5.5 Suppose that an acceptance plan has $n = 30$ and that the binomial approximation can be used to calculate the probability of accepting a batch. What value of c should be employed so that the consumer's risk is minimized while the producer's risk is no larger than 5% for an acceptable quality limit of $p_0 = 0.10$?

16.6 Supplementary Problems

16.6.1 A company that manufactures soap bars finds that the bars tend to be underweight if too much air is blown into the soap solution so that its density is too low. It is decided to set up a control chart to monitor the density of the soap solution, and at regular time intervals the soap solution density is measured and plotted on a control chart. Experience indicates that the process is in control when the solution density has a mean value of $\mu_0 = 1250$ and a standard deviation of $\sigma = 12$. Suppose that it is a reasonable approximation to take the soap solution density as being normally distributed.
  (a) What are the center line and control limits of a 3-sigma control chart that you would construct to monitor the soap solution density?
  (b) If a randomly sampled solution had a density of 1300 would you take this as evidence that the production

process had moved out of control? What if the density is 1210?
  (c) If the soap production process moves out of control so that the soap solution density has a mean $\mu = 1240$ with $\sigma = 12$, what is the probability that a randomly sampled solution has a density that lies outside the control limits? What is the average run length for detecting this change?

16.6.2 Paper Quality Assessment
  A paper mill has decided to implement a control chart to monitor the weight of the paper that it is producing. DS 16.6.1 contains the weights in g/m^2 of $n = 3$ random paper samples collected at $k = 22$ different points in time.
  (a) Does it look as though the paper producing process was in control when this data set was collected?

(b) What control charts would you suggest are used to monitor future runs of paper production?

(c) What conclusion would you draw if in the future a sample $x_1 = 76.01$, $x_2 = 73.42$, and $x_3 = 72.61$ is obtained?

(d) What if $x_1 = 77.82$, $x_2 = 79.04$, and $x_3 = 75.83$?

16.6.3 Date Code Legibility

A factory that packages food products in metal cans has installed an ink jet to spray a date code onto the bottom of the cans. Sometimes the process does not work correctly and the date codes are either missing or illegible. A p-chart is to be constructed based on the data set in DS 16.6.2, which records the number of defective date codes found in random samples of $n = 250$ cans taken at $k = 25$ time points.

(a) Does it appear that the process was out of control at any time while this data set was collected?

(b) What p-chart would you recommend be used to monitor future samples of date codes?

(c) How many defective date codes would need to be discovered in a future sample for the process to be considered to be out of control?

16.6.4 Fabric Flaws

In the textile industry c-charts can be used to monitor the number of flaws occurring in segments of fabric. Construct a c-chart from the data set in DS 16.6.3, which records the number of flaws found in $k = 25$ fabric segments sampled at different times.

(a) Should you remove any points from the data set? What c-chart would you recommend be used to monitor future samples of fabric?

(b) How many flaws would need to be discovered in a future randomly selected fabric segment for the process to be considered to be out of control?

16.6.5 An acceptance sampling procedure is being developed to check whether batteries have the required voltage. The batteries are shipped in very large batches, and the voltages of a random sample of $n = 50$ batteries are to be measured. What value of c should be used if it is decided that the producer's risk should be no larger than 2.5% for an acceptable quality limit of $p_0 = 0.06$? In this case, what is the consumer's risk if the proportion of defective batteries is $p_1 = 0.30$?

CHAPTER SEVENTEEN

Reliability Analysis and Life Testing

Reliability analysis is an important component of engineering work that draws on the areas of probability and statistics. As the name suggests, reliability analysis is concerned with the investigation of the failure rates of components and systems, which are typically represented as probabilities. Life testing is a general term used to describe the experimentation and statistical analysis performed to investigate failure rates. Some of the topics in probability and statistics that have been discussed in this book and that relate directly to the areas of reliability and life testing are brought together in this chapter and are discussed more explicitly within this framework.

In Section 17.1 the basic rules for calculating the probabilities of combinations of independent events discussed in Chapter 1 are used to calculate the reliabilities of complex systems. In Section 17.2 the problem of modeling failure rates is considered. Probability distributions that are used to model failure times are typically the exponential distribution, the gamma distribution, the Weibull distribution, and the lognormal distribution. Some of the statistical analysis techniques used in life testing are discussed in Section 17.3. Statistical analysis can be employed to make inferences on the failure rates under these distributional assumptions or to make general inferences on the distribution of failure times without any modeling assumptions. Data sets of failure times often contain censored data observations where the exact failure time is known only to lie within a certain region and is not known exactly. The product limit estimator discussed in Section 17.3 allows the estimation of the failure time distribution when some of the data observations are censored in this way.

17.1 System Reliability

The reliability of a component, which can be denoted by r, can in general be thought of as the probability that it performs a certain task. The complement of this probability is therefore the probability that the component fails to perform the required task, or the probability that the component "fails." In Section 17.2 probability distributions are employed to model how these reliabilities vary with time. In this section the probability calculations that can be used to find the overall reliability of a system of components are discussed.

A complex system may consist of a large number of components each with its own reliability value. A system reliability diagram, such as that shown in Figure 17.1, can be used to show how the failure of the various components affects the overall status of the system. The overall system operates successfully only if it is possible to progress from one side of the diagram to the other side by passing only through components that have not failed. If this is not possible, then the system has failed. If the reliabilities of the individual components are known, then the question of interest is what is the overall reliability of the system or, in other words, what is the probability that the system operates successfully?

This question is typically answered by assuming that the failures of the components are independent of each other and by then applying the basic rules of probability theory. These

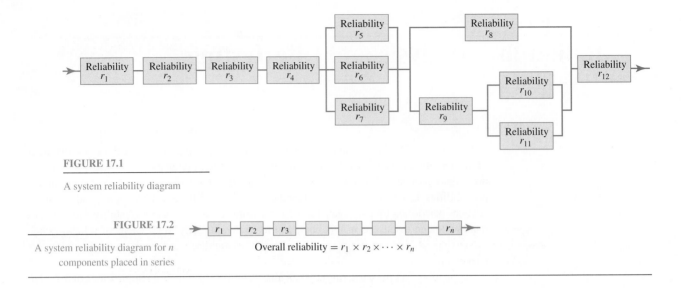

FIGURE 17.1

A system reliability diagram

FIGURE 17.2

A system reliability diagram for n components placed in series

Overall reliability $= r_1 \times r_2 \times \cdots \times r_n$

probability rules can be used to calculate the reliabilities of systems consisting of **components in series** and in **parallel**, and the reliabilities of more complex systems can be calculated by decomposing them into an appropriate collection of *modules*.

17.1.1 Components in Series

A set of components is considered to be in *series* if the system works only if each one of the components works. In other words, the system fails whenever one or more of the components has failed. The system reliability diagram for a system of n components in series is shown in Figure 17.2. If the reliabilities of the components are $r_1, \ldots, r_n$, then the multiplication law for independent events discussed in Section 1.5.2 implies that the overall system reliability is therefore

$$r = r_1 \times \cdots \times r_n$$

Components in Series

If a system consists of n components with independent reliabilities $r_1, \ldots, r_n$ placed in **series**, then the overall system reliability is

$$r = r_1 \times \cdots \times r_n$$

Example 6

Satellite Launching

Suppose that the satellite is launched on a rocket that has four thrusters, each of which must work in order for the launch to be successful. Suppose furthermore that each thruster is estimated to have a reliability of 0.995, which implies that they fail on average one time in 200 attempts. Since each thruster must work for the launch to be successful, the system reliability diagram has the four thrusters in series as shown in Figure 17.3. The overall system reliability is therefore

$$r = r_1 \times r_2 \times r_3 \times r_4 = 0.995^4 = 0.980150$$

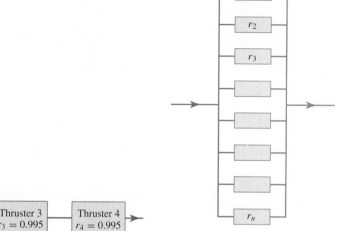

Overall reliability
$$= 1 - [(1 - r_1) \times \cdots \times (1 - r_n)]$$

FIGURE 17.4

A system reliability diagram for n components placed in parallel

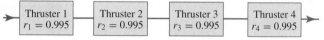

| Thruster 1 $r_1 = 0.995$ | Thruster 2 $r_2 = 0.995$ | Thruster 3 $r_3 = 0.995$ | Thruster 4 $r_4 = 0.995$ |

Overall reliability $= r_1 \times r_2 \times r_3 \times r_4 = 0.995^4 = 0.980$

FIGURE 17.3

The system reliability diagram for the thrusters in the satellite launching example

17.1.2 Components in Parallel

A set of components is considered to be in *parallel* if the system works whenever at least one of the components works. In other words, the system fails only when all of the components have failed. The system reliability diagram for a system of n components in parallel is shown in Figure 17.4. If the reliabilities of the components are $r_1, \ldots, r_n$ and these are independent, then the probability that all n components fail is

$$(1 - r_1) \times \cdots \times (1 - r_n)$$

This implies that the overall reliability of the system is

$$r = 1 - [(1 - r_1) \times \cdots \times (1 - r_n)]$$

Notice that adding additional components in parallel *increases* the reliability of the system. When components are in parallel, they can be considered as acting as "backups" for each other.

Components in Parallel

If a system consists of n components with independent reliabilities $r_1, \ldots, r_n$ placed in **parallel**, then the overall system reliability is

$$r = 1 - [(1 - r_1) \times \cdots \times (1 - r_n)]$$

Example 6
Satellite Launching

Recall that the satellite launch requires a computer and that three computers are available, with computers 2 and 3 acting as backups for computer 1. Consequently, the system reliability

FIGURE 17.5

The system reliability diagram for the computers in the satellite launching example

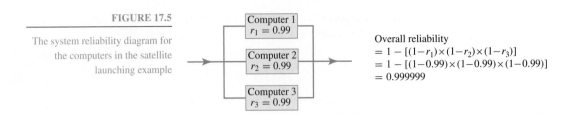

Overall reliability
$= 1 - [(1-r_1) \times (1-r_2) \times (1-r_3)]$
$= 1 - [(1-0.99) \times (1-0.99) \times (1-0.99)]$
$= 0.999999$

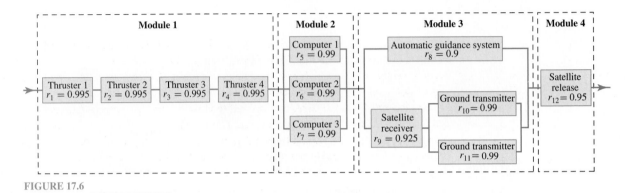

FIGURE 17.6

The system reliability diagram for the satellite launching example

diagram shown in Figure 17.5 has the three computers placed in parallel. Each computer is estimated to malfunction with a probability of 0.01, so that the reliabilities are

$$r_1 = r_2 = r_3 = 0.99$$

The overall system reliability is therefore

$$r = 1 - [(1 - r_1) \times (1 - r_2) \times (1 - r_3)] = 1 - 0.01^3 = 0.999999$$

and the probability of a system failure is

$$1 - r = 1 - 0.999999 = 10^{-6}$$

as calculated in Section 1.5.3.

17.1.3 Complex Systems

The overall reliability of a complex system of components can be calculated by decomposing the system into a series of modules. Each module should consist of a fairly simple system of components in series or in parallel. The reliabilities of each of the modules can then be calculated, and the overall system reliability is found by combining the module reliabilities in the appropriate manner. These ideas are illustrated in the following example.

Example 6
Satellite Launching

The system reliability diagram in Figure 17.1 actually corresponds to the system reliability diagram for a satellite launch shown in Figure 17.6. Notice that it can be decomposed into four modules placed in series.

The first module corresponds to the four thrusters operating successfully, and this has been shown to have an overall reliability of

$$r_{m1} = 0.980150$$

The second module corresponds to at least one of the three computers working, which has been shown to have an overall reliability of

$$r_{m2} = 0.999999$$

The third module is concerned with whether the rocket can be maneuvered into the correct orbit for releasing the satellite. This maneuvering can be performed by an automatic guidance system on the rocket, which has a reliability of $r_8 = 0.9$. If this automatic guidance system fails, then as a backup the rocket can be maneuvered from the ground. In order for this to work successfully the receiver on the satellite must function properly and at least one of two ground transmitters must be operational. The satellite receiver has an estimated reliability of $r_9 = 0.925$ and the ground transmitters have estimated reliabilities of $r_{10} = r_{11} = 0.99$. Notice how the third module is therefore composed of the automatic guidance system placed in *parallel* with the ground-based system of maneuvering, and that the ground-based system of maneuvering consists of the two ground transmitters placed in *parallel* with each other and in *series* with the satellite receiver.

The reliability of the third module can be calculated by decomposing it into smaller modules as shown in Figure 17.7. Module 5 consists of the two ground transmitters placed in parallel so that it has a reliability of

$$r_{m5} = 1 - [(1 - r_{10}) \times (1 - r_{11})] = 1 - 0.01^2 = 0.9999$$

Module 6 then consists of the satellite receiver and module 5 placed in series so that its reliability is

$$r_{m6} = r_9 \times r_{m5} = 0.925 \times 0.9999 = 0.924908$$

FIGURE 17.7

The decomposition of module 3 in the satellite launching example

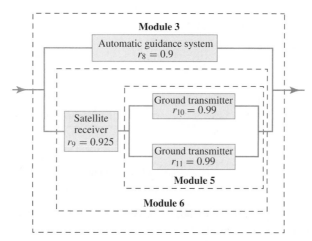

Finally, module 3 is composed of the automatic guidance system placed in parallel with module 6 so that its reliability is

$$r_{m3} = 1 - [(1 - r_8) \times (1 - r_{m6})] = 1 - [(1 - 0.9) \times (1 - 0.924908)] = 0.992491$$

Module 4 in Figure 17.6 corresponds to a successful release of the satellite from the rocket, which has an estimated reliability of $r_{12} = 0.95$. The complete system consists of modules 1–4 placed in series, so the overall reliability of the system is therefore

$$r = r_{m1} \times r_{m2} \times r_{m3} \times r_{m4}$$
$$= 0.980150 \times 0.999999 \times 0.992491 \times 0.95 = 0.924$$

Consequently, this reliability analysis estimates that the chance of a successful satellite deployment is 92.4%.

COMPUTER NOTE   As a final comment in this section, it is worthwhile to point out that many systems are considerably more complicated than the ones that we have considered, and engineers are consequently often forced to resort to computer *simulation* methods to evaluate the reliabilities of complex systems. In these simulations the reliabilities can be allowed to be time dependent and the components can be randomly assigned as being operational or as having failed. The system reliability can be estimated as the proportion of the simulations in which the overall system is operational.

■ 17.1.4   Problems

17.1.1 Calculate the reliability of the system diagram in Figure 17.8.

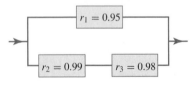

FIGURE 17.8

17.1.2 Calculate the reliability of the system diagram in Figure 17.9.

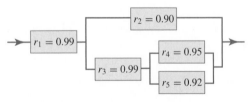

FIGURE 17.9

17.1.3 (a) If four identical components are placed in *series*, how large must their individual reliabilities be in order for the overall system reliability to be at least 0.95?

(b) If four identical components are placed in *parallel*, how large must their individual reliabilities be in order for the overall system reliability to be at least 0.95?

(c) Provide general answers for parts (a) and (b) assuming that n identical components are used and that an overall system reliability r is required.

17.1.4 (a) Three components with reliabilities $r_1 = 0.92$, $r_2 = 0.95$, and $r_3 = 0.975$ are placed in series. A fourth component with reliability $r_4 = 0.96$ can be placed in parallel with any one of these three components. How should the fourth component be placed in order to maximize the overall system reliability?

(b) Does your answer to part (a) change if the value of r_4 changes? In general, how does your answer depend upon the values of r_1, r_2, and r_3?

17.1.5 Calculate the reliability of the system diagram in Figure 17.10.

17.1.6 Calculate the reliability of the system diagram in Figure 17.11.

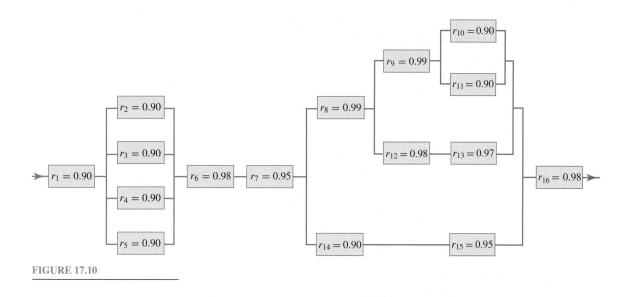

FIGURE 17.10

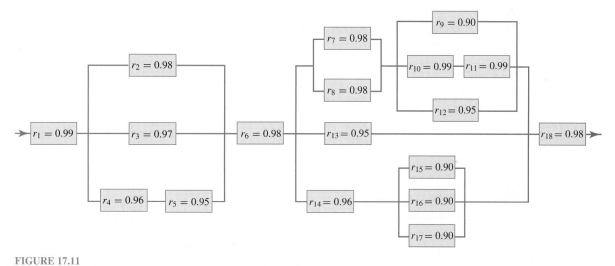

FIGURE 17.11

17.2   Modeling Failure Rates

It is often important to be able to model the reliability of a component as a function of time, $r(t)$. This can be accomplished by assigning a probability density function for the time to failure of the component. The **hazard rate** is then a particularly useful tool for assessing how the risk of failure varies with time.

17.2.1   Time to Failure

Suppose that the random variable T measures the time to failure of a component. A probability density function $f(t)$ for this random variable provides the probability distribution of the failure time, and the cumulative distribution function $F(t)$ can be interpreted as the probability

FIGURE 17.12

Components with exponential
failure times in series

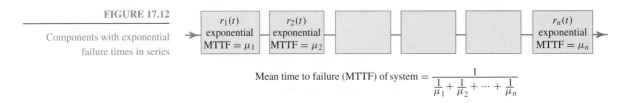

that the component has failed by time t. Conversely, the reliability function

$$r(t) = 1 - F(t)$$

is the probability that the component has not failed by time t. The reliability function is sometimes written as $\bar{F}(t)$ or may be referred to as the survival function $S(t)$. The **mean time to failure** (MTTF) is the expectation of the time to failure T

$$E(T) = \int_0^\infty t\, f(t)\, dt$$

Failure times are typically modeled with the exponential distribution, the gamma distribution, the Weibull distribution, and the lognormal distribution.

The Exponential Distribution   The exponential distribution with parameter λ is discussed in Section 4.2. The probability of a failure by time t is

$$F(t) = 1 - e^{-\lambda t}$$

so that the reliability function is

$$r(t) = e^{-\lambda t}$$

The mean time to failure is $1/\lambda$.

The exponential distribution is often used to model failure rates, and one main advantage of doing so is that it allows mathematically tractable solutions to many reliability problems. For example, suppose that a system consists of n components in series, and that the ith component has a failure time that can be modeled with an exponential distribution with a mean time to failure μ_i, as shown in Figure 17.12. What is the distribution of the time to failure of the system?

The ith component has a failure time that is exponentially distributed with parameter $\lambda_i = 1/\mu_i$, and so the probability that it is still working at time t is

$$r_i(t) = e^{-\lambda_i t}$$

The system is still working at time t only if all of the n components are still working at time t, and if the failure times of the components are independent, then this implies that the reliability function of the system is

$$r(t) = e^{-\lambda_1 t} \times \cdots \times e^{-\lambda_n t} = e^{-\lambda t}$$

where

$$\lambda = \lambda_1 + \cdots + \lambda_n$$

This implies that the failure time of the system is exponentially distributed with parameter λ, so that the mean time to failure of the system is

$$\frac{1}{\lambda} = \frac{1}{\lambda_1 + \cdots + \lambda_n} = \frac{1}{\frac{1}{\mu_1} + \cdots + \frac{1}{\mu_n}}$$

Notice that if the mean time to failure for each component is μ^*, then the mean time to failure of the system is

$$\frac{\mu^*}{n}$$

The Gamma Distribution   The gamma distribution with parameters k and λ is discussed in Section 4.3. The exponential distribution is obtained as a special case when $k = 1$. The mean time to failure is k/λ, but there is no simple expression for the reliability function $r(t)$.

The Weibull Distribution   The Weibull distribution with parameters a and λ is discussed in Section 4.4. It is perhaps the most widely used probability distribution for modeling failure times. The exponential distribution is obtained as a special case when $a = 1$. The probability of a failure by time t is

$$F(t) = 1 - e^{-(\lambda t)^a}$$

so that the reliability function is

$$r(t) = e^{-(\lambda t)^a}$$

The mean time to failure is

$$\frac{1}{\lambda} \Gamma\left(1 + \frac{1}{a}\right)$$

where $\Gamma(x)$ is the gamma function.

The Lognormal Distribution   The lognormal distribution with parameters μ and σ^2 is discussed in Section 5.4.1. Recall that if the random variable T has a lognormal distribution, then the transformed random variable $\ln(T)$ has a normal distribution with mean μ and variance σ^2. The probability of a failure by time t is

$$F(t) = \Phi\left(\frac{\ln(t) - \mu}{\sigma}\right)$$

so that the reliability function is

$$r(t) = 1 - \Phi\left(\frac{\ln(t) - \mu}{\sigma}\right)$$

The mean time to failure is

$$e^{\mu + \sigma^2/2}$$

17.2.2   The Hazard Rate

The hazard rate $h(t)$ of a failure time distribution $f(t)$ measures the instantaneous rate of failure at time t conditional on the component still being operational at time t. It is sometimes also referred to as the **failure rate** and intuitively it represents the chance of a component that has not failed by time t suddenly failing. Thus if $h(t_1)$ is larger than $h(t_2)$, then a component that is still operating at time t_1 is more likely to suddenly fail than a component that is still operating at time t_2.

The hazard rate is calculated as

$$h(t) = \frac{f(t)}{r(t)} = \frac{f(t)}{1 - F(t)}$$

Notice that the hazard rate is uniquely determined by the failure time probability density function $f(t)$. Similarly, the failure time probability density function $f(t)$ can be calculated from a hazard rate using the expression

$$f(t) = h(t) \, e^{-\int_0^t h(x)dx}$$

The Hazard Rate

The **hazard rate** $h(t)$ of a failure time distribution $f(t)$, which is also known as the **failure rate**, is calculated as

$$h(t) = \frac{f(t)}{r(t)} = \frac{f(t)}{1 - F(t)}$$

and can be interpreted as the chance that a component that has not failed by time t suddenly fails.

A common form of hazard rate for an engineering component is the "bathtub" shape shown in Figure 17.13. In this case the hazard rate initially decreases, remains fairly constant for a period of time, and then increases again. The hazard rate has high values initially because some components fail very early due to errors in their manufacture or construction. These failures are often referred to as failures in the "burn-in" phase of the components. If a component survives this initial time period, then it can be expected to last a reasonable period of time so that the hazard rate remains fairly low over the middle portion. Eventually the age of the components and their wear and tear lead to an increase in the hazard rate at later times. Interestingly, human mortality follows this "bathtub" pattern. Child mortality rates are quite

FIGURE 17.13

A typical bathtub shape for a hazard rate

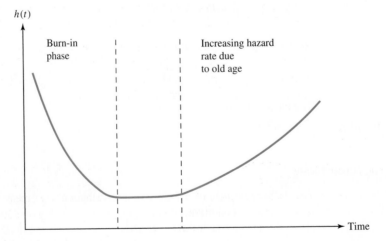

high, but having survived infancy a person enjoys a fairly low and constant mortality risk for a substantial period of time. However, with the onset of old age the mortality rate starts to increase again.

In practice it is difficult to find a probability model that has a hazard rate with the "bathtub" shape. Consequently, most probability models assume that the burn-in phase of the components has been passed so that it is reasonable to use a model with an increasing hazard rate. Such models are generally referred to as **increasing failure rate** (IFR) models and have the intuitively reasonable interpretation that the longer a component has been operating without failing, the larger is the chance of its sudden failure.

The Exponential Distribution   The hazard rate for an exponential distribution with parameter λ is

$$h(t) = \frac{f(t)}{r(t)} = \frac{\lambda \, e^{-\lambda t}}{e^{-\lambda t}} = \lambda$$

This constant hazard rate is related to the *memoryless* property of the exponential distribution discussed in Section 4.2.2, and it implies that any two components that are operating successfully are equally likely to suddenly fail, regardless of how much older one component is than the other. This is an unrealistic result for most situations, but nevertheless the simplicity and mathematical tractability of the exponential distribution cause it to be widely used in reliability analysis.

The Weibull Distribution   The hazard rate for a Weibull distribution with parameters a and λ is

$$h(t) = \frac{f(t)}{r(t)} = \frac{a \, \lambda^a \, t^{a-1} \, e^{-(\lambda t)^a}}{e^{-(\lambda t)^a}} = a \, \lambda^a \, t^{a-1}$$

This formula provides an increasing failure rate for values of the parameter a larger than one, and the larger the value of the parameter a, the steeper the increase in the hazard rate.

Example 33	Recall that the lifetime of a bacterium at a certain high temperature is modeled with a Weibull
Bacteria Lifetimes	distribution with parameters $a = 2$ and $\lambda = 0.1$. The hazard rate is therefore

$$h(t) = a \, \lambda^a \, t^{a-1} = 2 \times 0.1^2 \times t^{2-1} = 0.02 \, t$$

so that it increases linearly in time. Thus, for example, a bacterium still alive at a time $2t$ is twice as likely to suddenly die as a bacterium still alive at a time t.

No simple expressions can be obtained for the hazard rates of the gamma distribution and the lognormal distribution. The gamma distribution has an increasing failure rate when the parameter k is larger than 1, although the lognormal distribution typically has a hump-shaped hazard rate that first increases and then decreases.

■ **17.2.3 Problems**

17.2.1 A component has an exponential failure time distribution with a mean time to failure of 225 hours.
  (a) What is the probability that the component is still operating after 250 hours?
  (b) What is the probability that the component fails within 150 hours?

  (c) If three of these components are placed in series, what is the probability that the system is still operating after 100 hours?

17.2.2 A component has an exponential failure time distribution with a mean time to failure of 35 days.

(a) What is the probability that the component is still operating after 35 days?

(b) What is the probability that the component fails within 40 days?

(c) If six of these components are placed in series, what is the probability that the system is still operating after 5 days?

17.2.3 Four components with mean times to failure of 125 minutes, 60 minutes, 150 minutes, and 100 minutes are placed in series. If the failure times are exponentially distributed, what is the mean time to failure of the system?

17.2.4 A component has a constant hazard rate of 0.2.

(a) What is the probability that the component is still operating at time 4.0?

(b) What is the probability that the component fails before time 6.0?

17.2.5 A component has a lognormal failure time distribution with parameters $\mu = 2.5$ and $\sigma = 1.5$.

(a) What is the probability that the component is still operating at time 40?

(b) What is the probability that the component fails before time 10?

(c) What is the mean time to failure of the component?

(d) What is the median failure time?

17.2.6 A component has a lognormal failure time distribution with parameters $\mu = 3.0$ and $\sigma = 0.5$.

(a) What is the probability that the component is still operating at time 50?

(b) What is the probability that the component fails before time 40?

(c) What is the mean time to failure of the component?

(d) What is the median failure time?

17.2.7 A component has a Weibull failure time distribution with parameters $a = 3.0$ and $\lambda = 0.25$.

(a) What is the probability that the component is still operating at time 5?

(b) What is the probability that the component fails before time 3?

(c) What is the median failure time?

(d) What is the hazard rate?

(e) How much more likely to suddenly fail is a component operating at time 5 compared with a component operating at time 3?

17.2.8 A component has a Weibull failure time distribution with parameters $a = 4.5$ and $\lambda = 0.1$.

(a) What is the probability that the component is still operating at time 12?

(b) What is the probability that the component fails before time 8?

(c) What is the median failure time?

(d) What is the hazard rate?

(e) How much more likely to suddenly fail is a component operating at time 12 compared with a component operating at time 8?

17.3   Life Testing

The failure time distribution $f(t)$ can be investigated by measuring the failure times $t_1, \ldots, t_n$ of a set of n components. Inferences can be made on the failure time distribution $f(t)$ by considering these observations to be independent observations from the distribution. Some model fitting techniques are discussed in Section 17.3.1, and censored data observations, where the exact failure time is unknown, are discussed in Section 17.3.2.

17.3.1   Model Fitting

Some model fitting methodologies for reliability distributions are illustrated with the exponential distribution and the lognormal distribution.

Exponential Distribution   If the failure times $t_1, \ldots, t_n$ are considered to be observations from an exponential distribution, then the parameter λ can be estimated as

$$\hat{\lambda} = \frac{n}{t_1 + \cdots + t_n} = \frac{1}{\bar{t}}$$

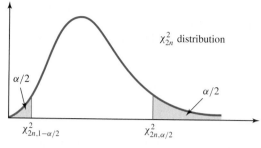

FIGURE 17.14

Critical points of a χ^2_{2n} distribution

| 0.6 | 8.4 | 12.8 | 4.3 | 1.3 | 8.3 | 18.1 | 20.6 | 21.0 | 17.8 | 3.3 |
| 6.2 | 9.7 | 35.2 | 13.9 | 2.6 | 5.2 | 5.6 | 13.7 | 3.8 |

FIGURE 17.15

Electrical discharge times in seconds

This is the maximum likelihood estimate of λ and the estimate obtained from the method of moments (see Section 7.4). The mean failure time $\mu = 1/\lambda$ is then estimated as

$$\hat{\mu} = \frac{1}{\hat{\lambda}} = \bar{t}$$

which is simply the sample average.

A $1 - \alpha$ confidence level confidence interval for the mean failure time can be constructed as

$$\mu \in \left(\frac{2n\bar{t}}{\chi^2_{2n,\alpha/2}}, \frac{2n\bar{t}}{\chi^2_{2n,1-\alpha/2}} \right)$$

where $\chi^2_{2n,\alpha/2}$ is the *upper* $\alpha/2$ critical point and $\chi^2_{2n,1-\alpha/2}$ is the *lower* $\alpha/2$ critical point of a chi-square distribution with $2n$ degrees of freedom, as shown in Figure 17.14 and tabulated in Table II. This confidence interval can be rewritten in terms of the parameter λ as

$$\lambda = \left(\frac{\chi^2_{2n,1-\alpha/2}}{2n\bar{t}}, \frac{\chi^2_{2n,\alpha/2}}{2n\bar{t}} \right)$$

Confidence Interval for the Mean Failure Time of an Exponential Distribution

If $\bar{t}$ is the sample average of a set of n data observations from an exponential distribution, then

$$\mu \in \left(\frac{2n\bar{t}}{\chi^2_{2n,\alpha/2}}, \frac{2n\bar{t}}{\chi^2_{2n,1-\alpha/2}} \right)$$

is a $1 - \alpha$ confidence level confidence interval for the mean failure time.

Example 78

Electrical Discharges

A metal plate is administered a certain electrical charge and the time taken for it to suddenly discharge is measured. The experiment is performed $n = 20$ times and the resulting data set is shown in Figure 17.15.

t_i	$\ln(t_i)$
0.610	−0.494
0.718	−0.331
0.877	−0.131
0.646	−0.437
1.012	0.012
0.586	−0.534
0.542	−0.612
0.596	−0.518
0.631	−0.460
0.888	−0.119
0.872	−0.137
0.667	−0.405
0.804	−0.218
0.735	−0.308
0.997	−0.003
0.535	−0.625
0.853	−0.159
0.641	−0.445
0.738	−0.304
0.838	−0.177
0.716	−0.334
0.671	−0.399
0.933	−0.069
0.436	−0.830
0.793	−0.232

FIGURE 17.16

Reaction times in seconds

The exponential distribution is often employed to model the time until an electrical discharge occurs since the constant hazard rate (memoryless property) is often appropriate. It is reasonable to consider the chance of a sudden discharge to be independent of the time that has elapsed since the electrical charge was administered.

The mean discharge time is estimated as

$$\hat{\mu} = \bar{t} = \frac{0.6 + \cdots + 3.8}{20} = 10.62$$

and with $\chi^2_{40,0.025} = 59.34$ and $\chi^2_{40,0.975} = 24.43$, a 95% confidence interval for the mean discharge time is

$$\mu \in \left(\frac{2 \times 20 \times 10.62}{59.34}, \frac{2 \times 20 \times 10.62}{24.43} \right)$$
$$= (7.16, 17.39)$$

Lognormal Distribution   A method of moments approach is typically used to fit a lognormal distribution to a data set $t_1, \ldots, t_n$ of failure times. This can be performed by taking the logarithms of the failure times

$$\ln(t_1), \ldots, \ln(t_n)$$

which can be considered to be observations from a normal distribution with mean μ and standard deviation σ. The sample average and sample standard deviation of the logarithms of the failure times can then be used as estimates of μ and σ.

Example 40

Testing Reaction Times

Figure 17.16 contains $n = 25$ reaction times t_i measured in an experiment. If a lognormal distribution is to be used to model the reaction time, what parameter values μ and σ are appropriate? The logarithms of the reaction times $\ln(t_i)$ are also shown in Figure 17.16 and these have a sample average

$$\overline{\ln(t)} = \frac{\ln(t_1) + \cdots + \ln(t_n)}{n} = -0.331$$

and a sample standard deviation

$$\sqrt{\frac{\sum_{i=1}^{n} (\ln(t_i) - \overline{\ln(t)})^2}{n-1}} = 0.210$$

so that $\hat{\mu} = -0.331$ and $\hat{\sigma} = 0.210$.

The $(1 - \alpha)$ quantile of a lognormal distribution is $e^{\mu + \sigma z_\alpha}$, so with $z_{0.10} = 1.282$ the 90th percentile of the distribution of reaction times can be estimated as

$$e^{\hat{\mu} + \hat{\sigma} 1.282} = e^{-0.331 + (0.210 \times 1.282)} = 0.940$$

Fitting a Weibull distribution or a gamma distribution to a data set is not so straightforward, and various methods have been proposed based on method of moments approaches or regression techniques. Details of these model fitting methods can be found in more detailed reliability textbooks.

17.3.2   Censored Data

Whereas the previous discussion in this chapter considered data sets comprised of n exact failure times $t_1, \ldots, t_n$, attention is now directed to circumstances where a data observation t_i is not known exactly but where a bound on its value is available. Such observations are known as **censored observations**.

Censored observations occur very commonly when measurements are taken of the failure times of a particular experimental item. Whereas an exact observation t_i records the exact failure time of an item, a censored observation arises when an item's failure time is not observed exactly. In particular, an observation is said to be *right-censored* if all that is known is that failure had not occurred by the given time, and an observation is said to be *left-censored* if all that is known is that failure occurred some time before the given time.

There are various "censoring mechanisms" that result in censored observations rather than exact observations. A common reason for an observation to be right-censored is that the experimenter loses track of the item before it fails or that the experiment is concluded before the particular item fails. Left-censored observations can arise if a particular item fails before it has begun to be monitored.

Example 1	
Machine Breakdowns	

15	11*
22*	20
39	17
31	17
12*	13*
8*	24
24*	23*
26	19*
21*	19
18	27
24	23
28	5*
14*	19
18*	25
17	9*
22*	16*
32	24
21*	15
17*	

FIGURE 17.17

Data set for machine breakdowns example

Example 1

Machine Breakdowns

Suppose that a manager is interested in how often a particular component of a machine needs to be replaced. This component may be something like a fan-belt, a cartridge, or a battery which needs to be changed with some regularity. Over a period of time, the manager monitors various machines and collects the data set of component failure times in days shown in Figure 17.17, where an asterisk represents a right-censored observation.

In certain cases the manager is able to observe how long a component lasts before it needs to be replaced, and this provides an exact observation of a failure time. Thus, for example, the first data observation 15 is for a particular component that is observed to last exactly 15 days before it needed to be replaced.

However, at regularly scheduled maintenance overhauls of the machines the components are replaced regardless of whether or not they have failed. This results in some right-censored observations. For example, the second data observation 22* is for a particular component that had been working successfully for 22 days before it was replaced during a maintenance overhaul. All that the manager knows about the failure time of this component is that it would have been more than 22 days.

The presence of censored observations necessarily complicates the analysis of a data set. However, since the censored observations provide information about the underlying probability distribution of interest, it is inefficient and misleading to reduce the data set to a collection of exact observations by ignoring the censored observations. Consequently, it is important to consider analysis techniques that can incorporate the information provided by the censored observations.

The following discussion considers the nonparametric analysis of data sets with right-censored observations. First, the estimation of an unknown reliability function $r(t)$ based upon a data set of exact and right-censored, independent, identically distributed observations is addressed using the **product limit estimator**. Finally, the construction of confidence intervals for the reliability function $r(t)$ is considered.

The Product Limit Estimator   In Section 15.1.1, the empirical distribution function

$$\hat{F}(t) = \frac{\#t_i \leq t}{n}$$

was proposed as a nonparametric estimator of an unknown distribution function $F(t)$ based upon a sample $t_1, \ldots, t_n$ of independent, identically distributed random variables from that distribution. If some of the data observations are censored, however, then clearly the empirical distribution function cannot be formed, since the quantity $\#t_i \leq t$ is in general not known exactly. For example, in the estimation of $F(10)$, it is not known whether a right-censored observation at time 9 corresponds to an exact observation less than 10 or greater than 10.

Consequently, a more sophisticated estimator is needed, and this is provided by the **product limit estimator**, which was first proposed by Edward Kaplan and Paul Meier in 1958. The product limit estimator is most easily discussed in terms of the reliability function

$$r(t) = 1 - F(t)$$

rather than the distribution function $F(t)$ itself.

The product limit estimator $\hat{r}(t)$ provides an estimate of the reliability function $r(t)$ when there are some right-censored data observations. Its construction is illustrated in the following example.

Example 1

Machine Breakdowns

5*	20
8*	21*
9*	21*
11*	22*
12*	22*
13*	23
14*	23*
15	24
15	24
16*	24
17	24*
17	25
17	26
17*	27
18	28
18*	31
19	32
19	39
19*	

FIGURE 17.18

The ordered data set for the machine breakdowns example

The first step in the construction of the product limit estimator $\hat{r}(t)$ is to order the data observations. The ordered machine component failure times are shown in Figure 17.18.

Next, a list is formed of the distinct times t_i at which exact failures are observed. For this data set these values are

$$t_1 = 15, t_2 = 17, t_3 = 18, t_4 = 19, t_5 = 20, t_6 = 23, t_7 = 24, t_8 = 25,$$
$$t_9 = 26, t_{10} = 27, t_{11} = 28, t_{12} = 31, t_{13} = 32, t_{14} = 39$$

It is also necessary to record the number of exact failures d_i observed at these times, which are

$$d_1 = 2, d_2 = 3, d_3 = 1, d_4 = 2, d_5 = 1, d_6 = 1, d_7 = 3, d_8 = 1, d_9 = 1,$$
$$d_{10} = 1, d_{11} = 1, d_{12} = 1, d_{13} = 1, d_{14} = 1$$

Finally, the number of observations n_i, censored or exact, that are recorded at or later than each of the times t_i should be calculated. For this example these are

$$n_1 = 30, n_2 = 27, n_3 = 23, n_4 = 21, n_5 = 18, n_6 = 13, n_7 = 11, n_8 = 7,$$
$$n_9 = 6, n_{10} = 5, n_{11} = 4, n_{12} = 3, n_{13} = 2, n_{14} = 1$$

Notice that the smallest seven observations are all censored, so that t_1, the first time at which an exact failure is observed, is equal to the eighth observation, which is 15. The next distinct time at which an exact failure is observed is then $t_2 = 17$. There are two items observed to fail at exactly $t_1 = 15$, and so $d_1 = 2$. Similarly, $d_2 = 3$ since three items are observed to fail at exactly $t_2 = 17$. All except the first seven censored observations have failure times (either exact or censored) that are equal to or are larger than $t_1 = 15$, and so $n_1 = 30$. Of these 30 observations, three are smaller than $t_2 = 17$, and so $n_2 = 30 - 3 = 27$.

The product limit estimator $\hat{r}(t)$ is a step function with steps at the times t_i, and with the step heights depending upon the values of the d_i and the n_i. It is given by the formula

$$\hat{r}(t) = \prod_{j | t_j < t} \left(\frac{n_j - d_j}{n_j} \right)$$

The product is over all the j values for which $t_j < t$, and if $t \leq t_1$, then $\hat{r}(t)$ is taken to be 1.

The product limit estimator $\hat{r}(t)$ is easily calculated in a sequential manner as shown.

$$0 < t \leq t_1 = 15 \Rightarrow \hat{r}(t) = 1$$
$$t_1 = 15 < t \leq t_2 = 17 \Rightarrow \hat{r}(t) = 1 \times (n_1 - d_1)/n_1 = 0.933$$
$$t_2 = 17 < t \leq t_3 = 18 \Rightarrow \hat{r}(t) = 0.933 \times (n_2 - d_2)/n_2 = 0.830$$
$$t_3 = 18 < t \leq t_4 = 19 \Rightarrow \hat{r}(t) = 0.830 \times (n_3 - d_3)/n_3 = 0.794$$
$$t_4 = 19 < t \leq t_5 = 20 \Rightarrow \hat{r}(t) = 0.794 \times (n_4 - d_4)/n_4 = 0.718$$
$$t_5 = 20 < t \leq t_6 = 23 \Rightarrow \hat{r}(t) = 0.718 \times (n_5 - d_5)/n_5 = 0.678$$
$$t_6 = 23 < t \leq t_7 = 24 \Rightarrow \hat{r}(t) = 0.678 \times (n_6 - d_6)/n_6 = 0.626$$
$$t_7 = 24 < t \leq t_8 = 25 \Rightarrow \hat{r}(t) = 0.626 \times (n_7 - d_7)/n_7 = 0.455$$
$$t_8 = 25 < t \leq t_9 = 26 \Rightarrow \hat{r}(t) = 0.455 \times (n_8 - d_8)/n_8 = 0.390$$
$$t_9 = 26 < t \leq t_{10} = 27 \Rightarrow \hat{r}(t) = 0.390 \times (n_9 - d_9)/n_9 = 0.325$$
$$t_{10} = 27 < t \leq t_{11} = 28 \Rightarrow \hat{r}(t) = 0.325 \times (n_{10} - d_{10})/n_{10} = 0.260$$
$$t_{11} = 28 < t \leq t_{12} = 31 \Rightarrow \hat{r}(t) = 0.260 \times (n_{11} - d_{11})/n_{11} = 0.195$$
$$t_{12} = 31 < t \leq t_{13} = 32 \Rightarrow \hat{r}(t) = 0.195 \times (n_{12} - d_{12})/n_{12} = 0.130$$
$$t_{13} = 32 < t \leq t_{14} = 39 \Rightarrow \hat{r}(t) = 0.130 \times (n_{13} - d_{13})/n_{13} = 0.065$$
$$t_{14} = 39 < t \Rightarrow \hat{r}(t) = 0.065 \times (n_{14} - d_{14})/n_{14} = 0.000$$

Figure 17.19 shows a plot of the product limit estimator.

The product limit estimator $\hat{r}(t)$ provides point estimates of the probability of surviving at least until a given time t. In this case $\hat{r}(20) = 0.718$ so that the manager would estimate that

FIGURE 17.19

The product limit estimator of the reliability function for the machine breakdowns example

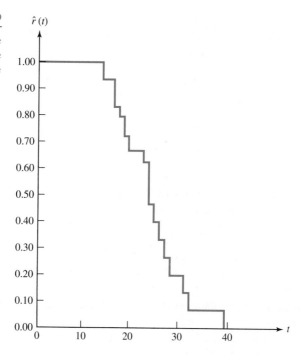

about 72% of components last for at least 20 days, and $\hat{r}(30) = 0.195$ so that the manager would estimate that about 20% of components last for at least 30 days. The median lifetime can be estimated as $\hat{r}^{-1}(0.5) = 24$, so that the manager expects about half of the components to last less than 24 days and about half to last more than 24 days. The construction of confidence intervals around these point estimates is discussed below.

One aspect of the product limit estimator is that if the largest observation is a censored one, then the estimator $\hat{r}(t)$ technically never reaches 0. Consequently, it is usual to define the product limit estimator to be 0 after the largest observation.

Notice that if the censored observations are discarded and ignored, then the empirical cumulative distribution function provides an estimate of the distribution function that is a step function with steps at the values t_i. The heights of the steps would be d_i/n, where n is the sample size. The product limit estimator is also a step function with steps at the values t_i. However, the step heights depend not only on the number of exact observations d_i, but also on the censored data observations through the quantities n_i.

An estimate of the survival function $r(t)$ can also be constructed when some of the data observations are left-censored, or when some observations are doubly censored, so that they are both right-censored and left-censored. Again, the survival function $r(t)$ is estimated by a step function with steps at the observed times of exact failures. However, in these cases there is not a simple form for the step heights and a computer program is needed to construct the estimator $\hat{r}(t)$.

Finally, it is interesting to consider how the presence of censored data affects a parametric approach to the estimation of the distribution function or reliability function. If a parametric density function $f(t; \boldsymbol{\theta})$ is assumed, with a corresponding distribution function $F(t; \boldsymbol{\theta})$, then the parameters $\boldsymbol{\theta}$ are generally estimated by the method of maximum likelihood.

For a given data set, the likelihood consists of the product of terms of the form $f(t_i; \boldsymbol{\theta})$ for an observation of an exact failure at time t_i, and of terms of the form $(1 - F(t_i; \boldsymbol{\theta}))$ for a right-censored observation at time t_i. The parameter estimates $\hat{\boldsymbol{\theta}}$ are the values that maximize the likelihood, although there is usually no simple closed form expressions for these estimates when there are censored data.

Confidence Intervals for $\hat{r}(t)$ For a data set with right-censored observations, a confidence interval for the survival function $r(t)$ at a particular value of t, based on the product limit estimator $\hat{r}(t)$ at t, may be calculated as

$$r(t) \in \left(\hat{r}(t) - z_{\alpha/2} \sqrt{\text{Var}(\hat{r}(t))}, \ \hat{r}(t) + z_{\alpha/2} \sqrt{\text{Var}(\hat{r}(t))} \right)$$

with an approximate confidence level of $1 - \alpha$. The critical point $z_{\alpha/2}$ is the upper $\alpha/2$ point of the standard normal distribution, and $\text{Var}(\hat{r}(t))$ is an estimate of the variance of $\hat{r}(t)$, which is given by

$$\text{Var}(\hat{r}(t)) = \hat{r}(t)^2 \times \sum_{j|t_j < t} \frac{d_j}{n_j(n_j - d_j)}$$

This expression for the variance is known as Greenwood's formula, and again the sum is over all the j values for which $t_j < t$.

Note that a confidence interval cannot be constructed for $r(t)$ if $t \leq t_1$, in which case $\hat{r}(t) = 1$, nor if t is greater than the largest data observation, in which case $\hat{r}(t) = 0$.

Example 1

Machine Breakdowns

For the data set of machine component failure times, the variances $\text{Var}(\hat{r}(t))$ can be calculated to be

$$
\begin{aligned}
t_1 &= 15 < t \le t_2 &= 17 \Rightarrow \text{Var}(\hat{r}(t)) = 0.00207 \\
t_2 &= 17 < t \le t_3 &= 18 \Rightarrow \text{Var}(\hat{r}(t)) = 0.00483 \\
t_3 &= 18 < t \le t_4 &= 19 \Rightarrow \text{Var}(\hat{r}(t)) = 0.00567 \\
t_4 &= 19 < t \le t_5 &= 20 \Rightarrow \text{Var}(\hat{r}(t)) = 0.00722 \\
t_5 &= 20 < t \le t_6 &= 23 \Rightarrow \text{Var}(\hat{r}(t)) = 0.00794 \\
t_6 &= 23 < t \le t_7 &= 24 \Rightarrow \text{Var}(\hat{r}(t)) = 0.00929 \\
t_7 &= 24 < t \le t_8 &= 25 \Rightarrow \text{Var}(\hat{r}(t)) = 0.01196 \\
t_8 &= 25 < t \le t_9 &= 26 \Rightarrow \text{Var}(\hat{r}(t)) = 0.01241 \\
t_9 &= 26 < t \le t_{10} &= 27 \Rightarrow \text{Var}(\hat{r}(t)) = 0.01214 \\
t_{10} &= 27 < t \le t_{11} &= 28 \Rightarrow \text{Var}(\hat{r}(t)) = 0.01115 \\
t_{11} &= 28 < t \le t_{12} &= 31 \Rightarrow \text{Var}(\hat{r}(t)) = 0.00944 \\
t_{12} &= 31 < t \le t_{13} &= 32 \Rightarrow \text{Var}(\hat{r}(t)) = 0.00701 \\
t_{13} &= 32 < t \le t_{14} &= 39 \Rightarrow \text{Var}(\hat{r}(t)) = 0.00387
\end{aligned}
$$

For example, if $t_5 = 20 < t \le t_6 = 23$, then

$$
\begin{aligned}
\text{Var}(\hat{r}(t)) &= \hat{r}(t)^2 \times \sum_{j=1}^{5} \frac{d_j}{n_j(n_j - d_j)} \\
&= 0.678^2 \times \left(\frac{2}{30 \times 28} + \frac{3}{27 \times 24} + \frac{1}{23 \times 22} + \frac{2}{21 \times 19} + \frac{1}{18 \times 17} \right) \\
&= 0.00794
\end{aligned}
$$

A confidence interval for $r(22)$ with a confidence level of 0.95, so that $z_{\alpha/2} = 1.96$, is thus calculated as

$$
r(22) \in \left(0.678 - 1.96\sqrt{0.00794}, \, 0.678 + 1.96\sqrt{0.00794} \right) = (0.503, 0.853)
$$

Consequently, the manager can be about 95% certain that between 50% and 85% of the machine components will last at least 22 days. Notice that this statement is based upon a sample of only 37 observations of which 17 are censored. Obviously, greater precision can be achieved with a larger sample size.

Individual confidence intervals with confidence levels of 0.95 can be calculated for other times as

$$
\begin{aligned}
t_1 &= 15 < t \le t_2 &= 17 \Rightarrow r(t) \in (0.844, 1.000) \\
t_2 &= 17 < t \le t_3 &= 18 \Rightarrow r(t) \in (0.694, 0.966) \\
t_3 &= 18 < t \le t_4 &= 19 \Rightarrow r(t) \in (0.646, 0.942) \\
t_4 &= 19 < t \le t_5 &= 20 \Rightarrow r(t) \in (0.551, 0.885) \\
t_5 &= 20 < t \le t_6 &= 23 \Rightarrow r(t) \in (0.503, 0.853) \\
t_6 &= 23 < t \le t_7 &= 24 \Rightarrow r(t) \in (0.437, 0.815) \\
t_7 &= 24 < t \le t_8 &= 25 \Rightarrow r(t) \in (0.241, 0.669) \\
t_8 &= 25 < t \le t_9 &= 26 \Rightarrow r(t) \in (0.172, 0.608) \\
t_9 &= 26 < t \le t_{10} &= 27 \Rightarrow r(t) \in (0.109, 0.541) \\
t_{10} &= 27 < t \le t_{11} &= 28 \Rightarrow r(t) \in (0.053, 0.467) \\
t_{11} &= 28 < t \le t_{12} &= 31 \Rightarrow r(t) \in (0.005, 0.385) \\
t_{12} &= 31 < t \le t_{13} &= 32 \Rightarrow r(t) \in (0.000, 0.294) \\
t_{13} &= 32 < t \le t_{14} &= 39 \Rightarrow r(t) \in (0.000, 0.187)
\end{aligned}
$$

Note that confidence intervals with an upper limit greater than one are truncated at 1, and confidence intervals with a lower limit less than 0 are truncated at 0.

■ 17.3.3 Problems

17.3.1 A set of $n = 30$ components are tested and their average lifetime is $\bar{t} = 132.4$ hours.
   (a) If the lifetimes are modeled with an exponential distribution, construct a 99% confidence interval for the mean time to failure.
   (b) The components are advertised as having "an average lifetime of at least 150 hours." Do you think that this is a plausible claim?

17.3.2 Vibration Robustness of Electrical Circuits
   The failure times in hours of $n = 20$ identical electrical circuits subjected to an intense vibration are given in DS 17.3.1.
   (a) If the failure times are modeled with an exponential distribution, construct a 95% confidence interval for the mean time to failure.
   (b) Do you think that it is plausible that the mean time to failure is 14 hours?

17.3.3 Thirty computer chips are tested in a sequential manner. A chip is placed in a circuit and when it fails it is immediately replaced by another chip. The final chip fails 176.5 hours after the experiment was started.

   (a) If the failure times of the computer chips are modeled with an exponential distribution, construct a 99% confidence interval for the mean time to failure.
   (b) Do you think that it is plausible that the mean time to failure is 10 hours?

17.3.4 Virus Survival Times
   The survival times in hours of a virus under certain conditions are given in DS 17.3.2.
   (a) If the survival times are modeled with a lognormal distribution, estimate the parameters μ and σ.
   (b) Use these point estimates to estimate the probability that a virus will survive for more than 10 hours under these conditions.

17.3.5 DS 17.3.3 contains a data set of failure times, where an asterisk represents a right-censored observation.
   (a) Construct and graph the product limit estimator of the reliability function.
   (b) Construct a 95% confidence interval for the probability that a component operates until at least time 100.

17.4 Supplementary Problems

17.4.1 (a) A set of n identical components with reliabilities 0.90 are placed in parallel. What value of n is needed to ensure that the overall system reliability is at least 0.995?
   (b) In general, how many identical components with reliabilities r_i need to be placed in parallel to ensure an overall system reliability of at least r?

17.4.2 Calculate the reliability of the system diagram in Figure 17.20.

FIGURE 17.20

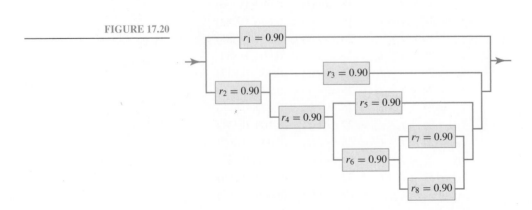

17.4.3 The germination time of a seed in days is modeled as having a constant hazard rate of 0.31.

    (a) What is the probability that a seed has not germinated after 6 days?

    (b) What is the probability that a seed germinates within 2 days?

17.4.4 The failure time of a light bulb in days is modeled as a Weibull distribution with parameters $a = 2.5$ and $\lambda = 0.01$.

    (a) What is the probability that a light bulb is still operating after 120 days?

    (b) What is the probability that a light bulb fails within 50 days?

    (c) What is the median failure time of a light bulb?

    (d) What is the hazard rate?

    (e) How much more likely to suddenly fail is a light bulb operating after 120 days compared with a light bulb operating after 100 days?

17.4.5 Conveyor Belt Malfunctions

The times in hours that a conveyor belt operates before a mechanical malfunction occurs are given in DS 17.4.1.

    (a) If the failure times are modeled with an exponential distribution, construct a 95% confidence interval for the mean time to failure.

    (b) Do you think that it is plausible that the mean time to failure is one week?

17.4.6 Concrete Stress Test

The times in minutes taken by $n = 25$ samples of concrete to fracture when subjected to a certain stress are given in DS 17.4.2.

    (a) If the failure times are modeled with a lognormal distribution, estimate the parameters μ and σ.

    (b) Use these point estimates to estimate the probability that a fracture will not occur within 15 minutes.

17.4.7 Electric Motor Reliabilities

DS 17.4.3 contains a data set of failure times in days of a certain type of electric motor. Exact failure times are observed when the motor fails to operate correctly and needs replacing. Right-censored observations, denoted with an asterisk, occur when a motor is replaced while it is still operating successfully.

    (a) Construct and graph the product limit estimator of the reliability function for the electric motors.

    (b) Construct a 95% confidence interval for the probability that an electric motor operates successfully for at least 200 days.

TABLES

Table I: Cumulative Distribution Function of the Standard Normal Distribution

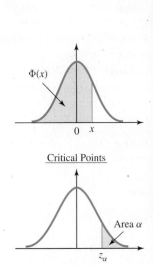

$\Phi(x)$

Critical Points

Area α

z_α

α	z_α
0.10	1.282
0.05	1.645
0.025	1.960
0.01	2.326
0.005	2.576

x	0.00	0.01	0.02	0.03	0.04	0.05	0.06	0.07	0.08	0.09
−3.4	0.0003	0.0003	0.0003	0.0003	0.0003	0.0003	0.0003	0.0003	0.0003	0.0002
−3.3	0.0005	0.0005	0.0005	0.0004	0.0004	0.0004	0.0004	0.0004	0.0004	0.0003
−3.2	0.0007	0.0007	0.0006	0.0006	0.0006	0.0006	0.0006	0.0005	0.0005	0.0005
−3.1	0.0010	0.0009	0.0009	0.0009	0.0008	0.0008	0.0008	0.0008	0.0007	0.0007
−3.0	0.0013	0.0013	0.0013	0.0012	0.0012	0.0011	0.0011	0.0011	0.0010	0.0010
−2.9	0.0019	0.0018	0.0017	0.0017	0.0016	0.0016	0.0015	0.0015	0.0014	0.0014
−2.8	0.0026	0.0025	0.0024	0.0023	0.0023	0.0022	0.0021	0.0021	0.0020	0.0019
−2.7	0.0035	0.0034	0.0033	0.0032	0.0031	0.0030	0.0029	0.0028	0.0027	0.0026
−2.6	0.0047	0.0045	0.0044	0.0043	0.0041	0.0040	0.0039	0.0038	0.0037	0.0036
−2.5	0.0062	0.0060	0.0059	0.0057	0.0055	0.0054	0.0052	0.0051	0.0049	0.0048
−2.4	0.0082	0.0080	0.0078	0.0075	0.0073	0.0071	0.0069	0.0068	0.0066	0.0061
−2.3	0.0107	0.0104	0.0102	0.0099	0.0096	0.0094	0.0091	0.0089	0.0087	0.0084
−2.2	0.0139	0.0136	0.0132	0.0129	0.0125	0.0122	0.0119	0.0116	0.0113	0.0110
−2.1	0.0179	0.0174	0.0170	0.0166	0.0162	0.0158	0.0154	0.0150	0.0146	0.0143
−2.0	0.0228	0.0222	0.0217	0.0212	0.0207	0.0202	0.0197	0.0192	0.0188	0.0183
−1.9	0.0287	0.0281	0.0274	0.0268	0.0262	0.0256	0.0250	0.0244	0.0239	0.0233
−1.8	0.0359	0.0352	0.0344	0.0336	0.0329	0.0322	0.0314	0.0307	0.0301	0.0294
−1.7	0.0446	0.0436	0.0427	0.0418	0.0409	0.0401	0.0392	0.0384	0.0375	0.0367
−1.6	0.0548	0.0537	0.0526	0.0516	0.0505	0.0495	0.0485	0.0475	0.0465	0.0455
−1.5	0.0668	0.0655	0.0643	0.0630	0.0618	0.0606	0.0594	0.0582	0.0571	0.0559
−1.4	0.0808	0.0793	0.0778	0.0764	0.0749	0.0735	0.0722	0.0708	0.0694	0.0681
−1.3	0.0968	0.0951	0.0934	0.0918	0.0901	0.0885	0.0869	0.0853	0.0838	0.0823
−1.2	0.1151	0.1131	0.1112	0.1093	0.1075	0.1056	0.1038	0.1020	0.1003	0.0985
−1.1	0.1357	0.1335	0.1314	0.1292	0.1271	0.1251	0.1230	0.1210	0.1190	0.1170
−1.0	0.1587	0.1562	0.1539	0.1515	0.1492	0.1469	0.1446	0.1423	0.1401	0.1379
−0.9	0.1841	0.1814	0.1788	0.1762	0.1736	0.1711	0.1685	0.1660	0.1635	0.1611
−0.8	0.2119	0.2090	0.2061	0.2033	0.2005	0.1977	0.1949	0.1922	0.1894	0.1867
−0.7	0.2420	0.2389	0.2358	0.2327	0.2296	0.2266	0.2236	0.2206	0.2177	0.2148
−0.6	0.2743	0.2709	0.2676	0.2643	0.2611	0.2578	0.2546	0.2514	0.2483	0.2451
−0.5	0.3085	0.3050	0.3015	0.2981	0.2946	0.2912	0.2877	0.2843	0.2810	0.2776
−0.4	0.3446	0.3409	0.3372	0.3336	0.3300	0.3264	0.3228	0.3192	0.3156	0.3121
−0.3	0.3821	0.3783	0.3745	0.3707	0.3669	0.3632	0.3594	0.3557	0.3520	0.3483
−0.2	0.4207	0.4168	0.4129	0.4090	0.4052	0.4013	0.3974	0.3936	0.3897	0.3859
−0.1	0.4602	0.4562	0.4522	0.4483	0.4443	0.4404	0.4364	0.4325	0.4286	0.4247
−0.0	0.5000	0.4960	0.4920	0.4880	0.4840	0.4801	0.4761	0.4721	0.4681	0.4641

(Continued on next page)

Table I: (*Continued*)

x	0.00	0.01	0.02	0.03	0.04	0.05	0.06	0.07	0.08	0.09
0.0	0.5000	0.5040	0.5080	0.5120	0.5160	0.5199	0.5239	0.5279	0.5319	0.5339
0.1	0.5398	0.5438	0.5478	0.5517	0.5557	0.5596	0.5636	0.5675	0.5714	0.5753
0.2	0.5793	0.5832	0.5871	0.5910	0.5948	0.5987	0.6026	0.6064	0.6103	0.6141
0.3	0.6179	0.6217	0.6255	0.6293	0.6331	0.6368	0.6406	0.6443	0.6480	0.6517
0.4	0.6554	0.6591	0.6628	0.6664	0.6700	0.6736	0.6772	0.6808	0.6844	0.6879
0.5	0.6915	0.6950	0.6985	0.7019	0.7054	0.7088	0.7123	0.7157	0.7190	0.7224
0.6	0.7257	0.7291	0.7324	0.7357	0.7389	0.7422	0.7454	0.7486	0.7517	0.7549
0.7	0.7580	0.7611	0.7642	0.7673	0.7704	0.7734	0.7764	0.7794	0.7823	0.7852
0.8	0.7881	0.7910	0.7939	0.7967	0.7995	0.8023	0.8051	0.8078	0.8106	0.8133
0.9	0.8159	0.8186	0.8212	0.8238	0.8264	0.8289	0.8315	0.8340	0.8365	0.8389
1.0	0.8413	0.8438	0.8461	0.8485	0.8508	0.8531	0.8554	0.8577	0.8599	0.8621
1.1	0.8643	0.8665	0.8686	0.8708	0.8729	0.8749	0.8770	0.8790	0.8810	0.8830
1.2	0.8849	0.8869	0.8888	0.8907	0.8925	0.8944	0.8962	0.8980	0.8997	0.9015
1.3	0.9032	0.9019	0.9066	0.9082	0.9099	0.9115	0.9131	0.9147	0.9162	0.9177
1.4	0.9192	0.9207	0.9222	0.9236	0.9251	0.9265	0.9278	0.9292	0.9306	0.9319
1.5	0.9332	0.9345	0.9357	0.9370	0.9382	0.9394	0.9406	0.9418	0.9429	0.9441
1.6	0.9452	0.9463	0.9474	0.9484	0.9495	0.9505	0.9515	0.9525	0.9535	0.9545
1.7	0.9554	0.9564	0.9573	0.9582	0.9591	0.9599	0.9608	0.9610	0.9625	0.9633
1.8	0.9641	0.9649	0.9656	0.9664	0.9671	0.9678	0.9686	0.9693	0.9699	0.9706
1.9	0.9713	0.9719	0.9726	0.9732	0.9738	0.9744	0.9750	0.9756	0.9761	0.9767
2.0	0.9772	0.9778	0.9783	0.9788	0.9793	0.9798	0.9803	0.9808	0.9812	0.9817
2.1	0.9821	0.9826	0.9830	0.9834	0.9838	0.9842	0.9846	0.9850	0.9854	0.9857
2.2	0.9861	0.9864	0.9868	0.9871	0.9875	0.9878	0.9881	0.9884	0.9887	0.9890
2.3	0.9893	0.9896	0.9898	0.9901	0.9904	0.9906	0.9909	0.9911	0.9913	0.9916
2.4	0.9918	0.9920	0.9922	0.9925	0.9927	0.9929	0.9931	0.9932	0.9934	0.9936
2.5	0.9938	0.9940	0.9941	0.9943	0.9945	0.9946	0.9948	0.9949	0.9951	0.9952
2.6	0.9953	0.9955	0.9956	0.9957	0.9959	0.9960	0.9961	0.9962	0.9963	0.9964
2.7	0.9965	0.9966	0.9967	0.9968	0.9969	0.9970	0.9971	0.9972	0.9973	0.9974
2.8	0.9974	0.9975	0.9976	0.9977	0.9977	0.9978	0.9979	0.9979	0.9980	0.9981
2.9	0.9981	0.9982	0.9982	0.9983	0.9984	0.9984	0.9985	0.9985	0.9986	0.9986
3.0	0.9987	0.9987	0.9987	0.9988	0.9988	0.9989	0.9989	0.9989	0.9990	0.9990
3.1	0.9990	0.9991	0.9991	0.9991	0.9992	0.9992	0.9992	0.9992	0.9993	0.9993
3.2	0.9993	0.9993	0.9994	0.9994	0.9994	0.9994	0.9994	0.9995	0.9995	0.9995
3.3	0.9995	0.9995	0.9995	0.9996	0.9996	0.9996	0.9996	0.9996	0.9996	0.9997
3.4	0.9997	0.9997	0.9997	0.9997	0.9997	0.9997	0.9997	0.9997	0.9997	0.9998

Table II: Critical Points of the Chi-Square Distribution

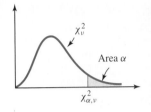

Degrees of freedom ν	α									
	0.995	0.99	0.975	0.95	0.90	0.10	0.05	0.025	0.01	0.005
1	0.000	0.000	0.001	0.004	0.016	2.706	3.841	5.024	6.635	7.879
2	0.010	0.020	0.051	0.103	0.211	4.605	5.991	7.378	9.210	10.597
3	0.072	0.115	0.216	0.352	0.584	6.251	7.815	9.348	11.345	12.838
4	0.207	0.297	0.484	0.711	1.064	7.779	9.488	11.143	13.277	14.860
5	0.412	0.554	0.831	1.145	1.610	9.236	11.071	12.833	15.086	16.750
6	0.676	0.872	1.237	1.635	2.204	10.645	12.592	14.449	16.812	18.548
7	0.989	1.239	1.690	2.167	2.833	12.017	14.067	16.013	18.475	20.278
8	1.344	1.646	2.180	2.733	3.490	13.362	15.507	17.535	20.090	21.955
9	1.735	2.088	2.700	3.325	4.168	14.684	16.919	19.023	21.666	23.589
10	2.156	2.558	3.247	3.940	4.865	15.987	18.307	20.483	23.209	25.188
11	2.603	3.053	3.816	4.575	5.578	17.275	19.675	21.920	24.725	26.757
12	3.074	3.571	4.404	5.226	6.304	18.549	21.026	23.337	26.217	28.299
13	3.565	4.107	5.009	5.892	7.042	19.812	22.362	24.736	27.688	29.819
14	4.075	4.660	5.629	6.571	7.790	21.064	23.685	26.119	29.141	31.319
15	4.601	5.229	6.262	7.261	8.547	22.307	24.996	27.488	30.578	32.801
16	5.142	5.812	6.908	7.962	9.312	23.542	26.296	28.845	32.000	34.267
17	5.697	6.408	7.564	8.672	10.085	24.769	27.587	30.191	33.409	35.718
18	6.265	7.015	8.231	9.390	10.865	25.989	28.869	31.526	34.805	37.156
19	6.844	7.633	8.907	10.117	11.651	27.204	30.144	32.852	36.191	38.582
20	7.434	8.260	9.591	10.851	12.443	28.412	31.410	34.170	37.566	39.997
21	8.034	8.897	10.283	11.591	13.240	29.615	32.671	35.479	38.932	41.401
22	8.643	9.542	10.982	12.338	14.042	30.813	33.924	36.781	40.289	42.796
23	9.260	10.196	11.689	13.091	14.848	32.007	35.172	38.076	41.638	44.181
24	9.886	10.856	12.401	13.848	15.659	33.196	36.415	39.364	42.980	45.559
25	10.520	11.524	13.120	14.611	16.473	34.382	37.652	40.646	44.314	46.928
26	11.160	12.198	13.844	15.379	17.292	35.563	38.885	41.923	45.642	48.290
27	11.808	12.879	14.573	16.151	18.114	36.741	40.113	43.194	46.963	49.645
28	12.461	13.565	15.308	16.928	18.939	37.916	41.337	44.461	48.278	50.993
29	13.121	14.257	16.017	17.708	19.768	39.087	42.557	45.722	49.588	52.336
30	13.787	14.954	16.791	18.493	20.599	40.256	43.773	46.979	50.892	53.672
40	20.707	22.164	24.433	26.509	29.051	51.805	55.758	59.342	63.691	66.766
50	27.991	29.707	32.357	34.764	37.689	63.167	67.505	71.420	76.154	79.490
60	35.534	37.485	40.482	43.188	46.459	74.397	79.082	83.298	88.379	91.952
70	43.275	45.442	48.758	51.739	55.329	85.527	90.531	95.023	100.425	104.215
80	51.172	53.540	57.153	60.391	64.278	96.578	101.879	106.629	112.329	116.321
90	59.196	61.754	65.647	69.126	73.291	107.565	113.145	118.136	124.116	128.299
100	67.328	70.065	74.222	77.929	82.358	118.498	124.342	129.561	135.807	140.169

Table III: Critical Points of the *t*-Distribution

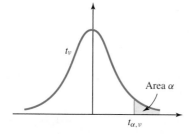

Degrees of freedom ν	α						
	0.10	0.05	0.025	0.01	0.005	0.001	0.0005
1	3.078	6.314	12.706	31.821	63.657	318.31	636.62
2	1.886	2.920	4.303	6.965	9.925	22.326	31.598
3	1.638	2.353	3.182	4.541	5.841	10.213	12.924
4	1.533	2.132	2.776	3.747	4.604	7.173	8.610
5	1.476	2.015	2.571	3.365	4.032	5.893	6.869
6	1.440	1.943	2.447	3.143	3.707	5.208	5.959
7	1.415	1.895	2.365	2.998	3.499	4.785	5.408
8	1.397	1.860	2.306	2.896	3.355	4.501	5.041
9	1.383	1.833	2.262	2.821	3.250	4.297	4.781
10	1.372	1.812	2.228	2.764	3.169	4.144	4.587
11	1.363	1.796	2.201	2.718	3.106	4.025	4.437
12	1.356	1.782	2.179	2.681	3.055	3.930	4.318
13	1.350	1.771	2.160	2.650	3.012	3.852	4.221
14	1.345	1.761	2.145	2.624	2.977	3.787	4.140
15	1.341	1.753	2.131	2.602	2.947	3.733	4.073
16	1.337	1.746	2.120	2.583	2.921	3.686	4.015
17	1.333	1.740	2.110	2.567	2.898	3.646	3.965
18	1.330	1.734	2.101	2.552	2.878	3.610	3.922
19	1.328	1.729	2.093	2.539	2.861	3.579	3.883
20	1.325	1.725	2.086	2.528	2.845	3.552	3.850
21	1.323	1.721	2.080	2.518	2.831	3.527	3.819
22	1.321	1.717	2.074	2.508	2.819	3.505	3.792
23	1.319	1.714	2.069	2.500	2.807	3.485	3.767
24	1.318	1.711	2.064	2.492	2.797	3.467	3.745
25	1.316	1.708	2.060	2.485	2.787	3.450	3.725
26	1.315	1.706	2.056	2.479	2.779	3.435	3.707
27	1.314	1.703	2.052	2.473	2.771	3.421	3.690
28	1.313	1.701	2.048	2.467	2.763	3.408	3.674
29	1.311	1.699	2.045	2.462	2.756	3.396	3.659
30	1.310	1.697	2.042	2.457	2.750	3.385	3.646
40	1.303	1.684	2.021	2.423	2.704	3.307	3.551
60	1.296	1.671	2.000	2.390	2.660	3.232	3.460
120	1.289	1.658	1.980	2.358	2.617	3.160	3.373
∞	1.282	1.645	1.960	2.326	2.576	3.090	3.291

Table IV: Critical Points of the F-Distribution

$$F_{\nu_1, \nu_2} \sim \frac{\chi^2_{\nu_1}/\nu_1}{\chi^2_{\nu_2}/\nu_2}$$

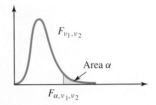

F_{ν_1, ν_2}

Area α

F_{α, ν_1, ν_2}

$\alpha = 0.10$

ν_2 \ ν_1	1	2	3	4	5	6	7	8	9	10	12	15	20	24	30	40	60	120	∞
1	39.86	49.50	53.59	55.84	57.24	58.20	58.90	59.44	59.85	60.20	60.70	61.22	61.74	62.00	62.27	62.53	62.79	63.05	63.33
2	3.53	9.00	9.16	9.24	9.29	9.33	9.35	9.37	9.38	9.39	9.41	9.42	9.44	9.45	9.46	9.47	9.47	9.48	9.49
3	5.54	5.46	5.39	5.34	5.31	5.28	5.27	5.25	5.24	5.23	5.22	5.20	5.18	5.18	5.17	5.16	5.15	5.14	5.13
4	4.54	4.32	4.19	4.11	4.05	4.01	3.98	3.95	3.94	3.92	3.90	3.87	3.84	3.83	3.82	3.80	3.79	3.78	3.76
5	4.06	3.78	3.62	3.52	3.45	3.40	3.37	3.34	3.32	3.30	3.27	3.24	3.21	3.19	3.17	3.16	3.14	3.12	3.10
6	3.78	3.46	3.29	3.18	3.11	3.05	3.01	2.98	2.96	2.94	2.90	2.87	2.84	2.82	2.80	2.78	2.76	2.74	2.72
7	3.59	3.26	3.07	2.96	2.88	2.83	2.78	2.75	2.72	2.70	2.67	2.63	2.59	2.58	2.56	2.54	2.51	2.49	2.47
8	3.46	3.11	2.92	2.81	2.73	2.67	2.62	2.59	2.56	2.54	2.50	2.46	2.42	2.40	2.38	2.36	2.34	2.32	2.29
9	3.36	3.01	2.81	2.69	2.61	2.55	2.51	2.47	2.44	2.42	2.38	2.34	2.30	2.28	2.25	2.23	2.21	2.18	2.16
10	3.28	2.92	2.73	2.61	2.52	2.46	2.41	2.38	2.35	2.32	2.28	2.24	2.20	2.18	2.16	2.13	2.11	2.08	2.06
11	3.23	2.86	2.66	2.54	2.45	2.39	2.34	2.30	2.27	2.25	2.21	2.17	2.12	2.10	2.08	2.05	2.03	2.00	1.97
12	3.18	2.81	2.61	2.48	2.39	2.33	2.28	2.24	2.21	2.19	2.15	2.10	2.06	2.04	2.01	1.99	1.96	1.93	1.90
13	3.14	2.76	2.56	2.43	2.35	2.28	2.23	2.20	2.16	2.14	2.10	2.05	2.01	1.98	1.96	1.93	1.90	1.88	1.85
14	3.10	2.73	2.52	2.39	2.31	2.24	2.19	2.15	2.12	2.10	2.05	2.01	1.96	1.94	1.91	1.89	1.86	1.83	1.80
15	3.07	2.70	2.49	2.36	2.27	2.21	2.16	2.12	2.09	2.06	2.02	1.97	1.92	1.90	1.87	1.85	1.82	1.79	1.76
16	3.05	2.67	2.46	2.33	2.24	2.18	2.13	2.09	2.06	2.03	1.99	1.94	1.89	1.87	1.84	1.81	1.78	1.75	1.72
17	3.03	2.64	2.44	2.31	2.22	2.15	2.10	2.06	2.03	2.00	1.96	1.91	1.86	1.84	1.81	1.78	1.75	1.72	1.69
18	3.01	2.62	2.42	2.29	2.20	2.13	2.08	2.04	2.00	1.98	1.93	1.89	1.84	1.81	1.78	1.75	1.72	1.69	1.66
19	2.99	2.61	2.40	2.27	2.18	2.11	2.06	2.02	1.98	1.96	1.91	1.86	1.81	1.79	1.76	1.73	1.70	1.67	1.63
20	2.97	2.59	2.38	2.25	2.16	2.09	2.04	2.00	1.96	1.94	1.89	1.84	1.79	1.77	1.74	1.71	1.68	1.64	1.61
21	2.96	2.57	2.36	2.23	2.14	2.08	2.02	1.98	1.95	1.92	1.87	1.83	1.78	1.75	1.72	1.69	1.66	1.62	1.59
22	2.95	2.56	2.35	2.22	2.13	2.06	2.01	1.97	1.93	1.90	1.86	1.81	1.76	1.73	1.70	1.67	1.64	1.60	1.57
23	2.94	2.55	2.34	2.21	2.11	2.05	1.99	1.95	1.92	1.89	1.84	1.80	1.74	1.72	1.69	1.66	1.62	1.59	1.55
24	2.93	2.54	2.33	2.19	2.10	2.04	1.98	1.94	1.91	1.88	1.83	1.78	1.73	1.70	1.67	1.64	1.61	1.57	1.51
25	2.92	2.53	2.32	2.18	2.09	2.02	1.97	1.93	1.89	1.87	1.82	1.77	1.72	1.69	1.66	1.63	1.59	1.56	1.52
26	2.91	2.52	2.31	2.17	2.08	2.01	1.96	1.92	1.88	1.86	1.81	1.76	1.71	1.68	1.65	1.61	1.58	1.54	1.50
27	2.90	2.51	2.30	2.17	2.07	2.00	1.95	1.91	1.87	1.85	1.80	1.75	1.70	1.67	1.64	1.60	1.57	1.53	1.49
28	2.89	2.50	2.29	2.16	2.06	2.00	1.94	1.90	1.87	1.84	1.79	1.74	1.69	1.66	1.63	1.59	1.56	1.52	1.48
29	2.89	2.50	2.28	2.15	2.06	1.99	1.93	1.89	1.86	1.83	1.78	1.73	1.68	1.65	1.62	1.58	1.55	1.51	1.47
30	2.88	2.49	2.28	2.14	2.05	1.98	1.93	1.88	1.85	1.82	1.77	1.72	1.67	1.64	1.61	1.57	1.54	1.50	1.46
40	2.84	2.44	2.23	2.09	2.00	1.93	1.87	1.83	1.79	1.76	1.71	1.66	1.61	1.57	1.54	1.51	1.47	1.42	1.35
60	2.79	2.39	2.18	2.04	1.95	1.87	1.82	1.77	1.74	1.71	1.66	1.60	1.54	1.51	1.48	1.44	1.40	1.35	1.29
120	2.75	2.35	2.13	1.99	1.90	1.82	1.77	1.72	1.68	1.65	1.60	1.55	1.48	1.45	1.41	1.37	1.32	1.26	1.18
∞	2.71	2.30	2.08	1.94	1.85	1.77	1.72	1.67	1.63	1.60	1.55	1.49	1.42	1.38	1.34	1.30	1.24	1.17	1.10

(Continued on next page)

Table IV: (*Continued*)

		$\alpha = 0.05$																	
								ν_1											
ν_2		**1**	**2**	**3**	**4**	**5**	**6**	**7**	**8**	**9**	**10**	**12**	**15**	**20**	**24**	**30**	**40**	**60**	**120**	**∞**
	1	161.44	199.50	215.69	224.57	230.16	233.98	236.78	238.89	240.55	241.89	243.91	245.97	248.02	249.04	250.07	251.13	252.18	253.27	254.31
	2	18.51	19.00	19.16	19.25	19.30	19.33	19.35	19.37	19.39	19.40	19.41	19.43	19.45	19.45	19.46	19.47	19.48	19.49	19.50
	3	10.13	9.55	9.28	9.12	9.01	8.94	8.89	8.85	8.81	8.79	8.74	8.70	8.66	8.64	8.62	8.59	8.57	8.55	8.53
	4	7.71	6.94	6.59	6.39	6.20	6.16	6.09	6.04	6.00	5.96	5.91	5.86	5.80	5.77	5.75	5.72	5.69	5.66	5.63
	5	6.61	5.79	5.41	5.19	5.05	4.95	4.88	4.82	4.77	4.74	4.68	4.62	4.56	4.53	4.50	4.46	4.43	4.40	4.36
	6	5.99	5.14	4.76	4.53	4.39	4.28	4.21	4.15	4.10	4.06	4.00	3.94	3.87	3.84	3.81	3.77	3.74	3.70	3.67
	7	5.59	4.74	4.35	4.12	3.97	3.87	3.79	3.73	3.68	3.64	3.57	3.51	3.44	3.41	3.38	3.34	3.30	3.27	3.23
	8	5.52	4.46	4.07	3.84	3.69	3.58	3.50	3.44	3.39	3.35	3.28	3.22	3.15	3.12	3.08	3.04	3.01	2.97	2.93
	9	5.12	4.26	3.86	3.63	3.48	3.37	3.29	3.23	3.18	3.14	3.07	3.01	2.94	2.90	2.86	2.83	2.79	2.75	2.71
	10	4.96	4.10	3.71	3.48	3.33	3.22	3.14	3.07	3.02	2.98	2.91	2.85	2.77	2.74	2.70	2.66	2.62	2.58	2.54
	11	4.84	3.98	3.59	3.36	3.20	3.09	3.01	2.95	2.90	2.85	2.79	2.72	2.65	2.61	2.57	2.53	2.49	2.45	2.40
	12	4.75	3.89	3.49	3.26	3.11	3.00	2.91	2.85	2.80	2.75	2.69	2.62	2.54	2.51	2.47	2.43	2.38	2.34	2.30
	13	4.67	3.81	3.41	3.18	3.03	2.92	2.83	2.77	2.71	2.67	2.60	2.53	2.46	2.42	2.38	2.34	2.30	2.25	2.21
	14	4.60	3.74	3.34	3.11	2.96	2.85	2.76	2.70	2.65	2.60	2.53	2.46	2.39	2.35	2.31	2.27	2.22	2.18	2.13
	15	4.54	3.68	3.29	3.06	2.90	2.79	2.71	2.64	2.59	2.54	2.48	2.40	2.33	2.29	2.25	2.20	2.16	2.11	2.07
	16	4.49	3.63	3.24	3.01	2.85	2.74	2.66	2.59	2.54	2.49	2.42	2.35	2.28	2.24	2.19	2.15	2.11	2.06	2.01
	17	4.45	3.59	3.20	2.96	2.81	2.70	2.61	2.55	2.49	2.45	2.38	2.31	2.23	2.19	2.15	2.10	2.06	2.01	1.96
	18	4.41	3.55	3.16	2.93	2.77	2.66	2.58	2.51	2.46	2.41	2.34	2.27	2.19	2.15	2.11	2.06	2.02	1.97	1.92
	19	4.38	3.52	3.13	2.90	3.74	2.63	2.54	2.48	2.42	2.38	2.31	2.23	2.16	2.11	2.07	2.03	1.98	1.93	1.88
	20	4.35	3.49	3.10	2.87	2.71	2.60	2.51	2.45	2.30	2.35	2.28	2.20	2.12	2.08	2.04	1.99	1.95	1.90	1.84
	21	4.32	3.47	3.07	2.84	2.68	2.57	2.49	2.42	2.37	2.32	2.25	2.18	2.10	2.05	2.01	1.96	1.92	1.87	1.81
	22	4.30	3.44	3.05	2.82	2.66	2.55	2.46	2.40	2.34	2.30	2.23	2.15	2.07	2.03	1.98	1.94	1.89	1.84	1.78
	23	4.28	3.42	3.03	2.80	2.64	2.53	2.44	2.37	2.32	2.27	2.20	2.13	2.05	2.01	1.96	1.91	1.86	1.81	1.76
	24	4.26	3.40	3.01	2.78	2.62	2.51	2.42	2.36	2.30	2.25	2.18	2.11	2.03	1.98	1.94	1.89	1.84	1.79	1.73
	25	4.24	3.39	2.99	2.76	2.60	2.49	2.40	2.34	2.28	2.24	2.16	2.09	2.01	1.96	1.92	1.87	1.82	1.77	1.71
	26	4.23	3.37	2.98	2.74	2.59	2.47	2.39	2.32	2.27	2.22	2.15	2.07	1.99	1.95	1.90	1.85	1.80	1.75	1.69
	27	4.21	3.35	2.96	2.73	2.57	2.46	2.37	2.31	2.25	2.20	2.13	2.06	1.97	1.93	1.88	1.84	1.79	1.73	1.67
	28	4.20	3.34	2.95	2.71	2.56	2.45	2.36	2.29	2.24	2.19	2.12	2.04	1.96	1.91	1.87	1.82	1.77	1.71	1.65
	29	4.18	3.33	2.93	3.70	2.55	2.43	2.35	2.28	2.22	2.18	2.10	2.03	1.94	1.90	1.85	1.81	1.75	1.70	1.64
	30	4.17	3.32	2.92	2.69	2.53	2.42	2.33	2.27	2.21	2.16	2.09	2.01	1.93	1.89	1.84	1.79	1.74	1.68	1.62
	40	4.08	3.23	2.84	2.61	2.45	2.34	2.25	2.18	2.12	2.08	2.00	1.92	1.84	1.79	1.74	1.69	1.64	1.58	1.51
	60	4.00	3.15	2.76	2.53	2.37	2.25	2.17	2.10	2.04	1.99	1.92	1.84	1.75	1.70	1.65	1.59	1.53	1.47	1.39
	120	3.92	3.09	2.68	2.45	2.29	2.18	2.09	2.02	1.96	1.91	1.83	1.75	1.66	1.61	1.55	1.50	1.43	1.35	1.25
	∞	3.84	3.00	2.60	2.37	2.21	2.10	2.01	1.94	1.88	1.83	1.75	1.67	1.57	1.52	1.46	1.39	1.32	1.22	1.00

(*Continued on next page*)

Table IV: (*Continued*)

$\alpha = 0.01$

ν_2 \ ν_1	1	2	3	4	5	6	7	8	9	10	12	15	20	24	30	40	60	120	∞
1	4052	4999	5403	5625	5764	5859	5929	5981	6023	6055	6107	6157	6209	6235	6260	6287	6312	6339	6366
2	98.51	99.00	99.17	99.25	99.30	99.33	99.35	99.38	99.39	99.40	99.41	99.43	99.44	99.45	99.47	99.47	99.48	99.49	99.50
3	34.12	30.82	29.46	28.71	28.24	27.91	27.67	27.49	27.35	27.23	27.05	26.87	26.69	26.60	26.51	26.41	26.32	26.22	26.13
4	21.20	18.00	16.69	15.98	15.52	15.21	14.98	14.80	14.66	14.55	14.37	14.20	14.02	13.93	13.84	13.75	13.65	13.56	13.46
5	16.26	13.27	12.06	11.39	10.97	10.67	10.46	10.29	10.16	10.05	9.89	9.72	9.55	9.47	9.38	9.29	9.20	9.11	9.02
6	13.75	10.92	9.78	9.15	8.75	8.47	8.26	8.10	7.98	7.87	7.72	7.56	7.40	7.31	7.23	7.14	7.06	6.97	6.88
7	12.25	9.55	8.45	7.85	7.46	7.19	6.99	6.84	6.72	6.62	6.47	6.31	6.16	6.07	5.99	5.91	5.82	5.74	5.65
8	11.26	8.65	7.59	7.01	6.63	6.37	6.18	6.03	5.91	5.81	5.67	5.52	5.36	5.28	5.20	5.12	5.03	4.95	4.86
9	10.56	8.02	6.99	6.42	6.06	5.80	5.61	5.47	5.35	5.26	5.11	4.96	4.81	4.73	4.65	4.57	4.48	4.40	4.31
10	10.04	7.56	6.55	5.99	5.64	5.39	5.20	5.06	4.94	4.85	4.71	4.56	4.41	4.33	4.25	4.17	4.08	4.00	3.91
11	9.65	7.21	6.22	5.67	5.32	5.07	4.89	4.74	4.63	4.54	4.40	4.25	4.10	4.02	3.94	3.86	3.78	3.69	3.60
12	9.33	6.93	5.95	5.41	5.06	4.82	4.64	4.50	4.39	4.30	4.16	4.01	3.86	3.78	3.70	3.62	3.54	3.45	3.36
13	9.07	6.70	5.74	5.21	4.86	4.62	4.44	4.30	4.19	4.10	3.96	3.82	3.66	3.59	3.51	3.43	3.34	3.25	3.17
14	8.86	6.51	5.56	5.04	4.69	4.46	4.28	4.14	4.03	3.94	3.80	3.66	3.51	3.43	3.35	3.27	3.18	3.09	3.00
15	8.68	6.36	5.42	4.89	4.56	4.32	4.14	4.00	3.89	3.80	3.67	3.52	3.37	3.29	3.21	3.13	3.05	2.96	2.87
16	8.53	6.23	5.29	4.77	4.44	4.20	4.03	3.89	3.78	3.69	3.55	3.41	3.26	3.18	3.10	3.02	2.93	2.84	2.75
17	8.40	6.11	5.18	4.67	4.34	4.10	3.93	3.79	3.68	3.59	3.46	3.31	3.16	3.08	3.00	2.92	2.83	2.75	2.65
18	8.29	6.01	5.09	4.58	4.25	4.01	3.84	3.71	3.60	3.51	3.37	3.23	3.08	3.00	2.92	2.84	2.75	2.66	2.57
19	8.19	5.93	5.01	4.50	4.17	3.94	3.77	3.63	3.52	3.43	3.30	3.15	3.00	2.92	2.84	2.76	2.67	2.58	2.49
20	8.10	5.85	4.94	4.43	4.10	3.87	3.70	3.56	3.46	3.37	3.23	3.09	2.94	2.86	2.78	2.69	2.61	2.52	2.42
21	8.02	5.78	4.87	4.37	4.04	3.81	3.64	3.51	3.40	3.31	3.17	3.03	2.88	2.80	2.72	2.64	2.55	2.46	2.36
22	7.95	5.72	4.82	4.31	3.99	3.76	3.59	3.45	3.35	3.26	3.12	2.98	2.83	2.75	2.67	2.58	2.50	2.40	2.31
23	7.88	5.66	4.76	4.26	3.94	3.71	3.54	3.41	3.30	3.21	3.07	2.93	2.78	2.70	2.62	2.54	2.45	2.35	2.26
24	7.82	5.61	4.72	4.22	3.90	3.67	3.50	3.36	3.26	3.17	3.03	2.89	2.74	2.66	2.58	2.49	2.40	2.31	2.21
25	7.77	5.57	4.68	4.18	3.85	3.63	3.46	3.32	3.22	3.13	2.99	2.85	2.70	2.62	2.54	2.45	2.36	2.27	2.17
26	7.72	5.53	4.64	4.14	3.82	3.59	3.42	3.29	3.18	3.09	2.96	2.81	2.66	2.58	2.50	2.42	2.33	2.23	2.13
27	7.68	5.49	4.60	4.11	3.78	3.56	3.39	3.26	3.15	3.06	2.93	2.78	2.63	2.55	2.47	2.38	2.29	2.20	2.10
28	7.64	5.45	4.57	4.07	3.75	3.53	3.36	3.23	3.12	3.03	2.90	2.75	2.60	2.52	2.44	2.35	2.26	2.17	2.06
29	7.60	5.42	4.54	4.04	3.73	3.50	3.33	3.20	3.09	3.00	2.87	2.73	2.57	2.49	2.41	2.33	2.23	2.14	2.03
30	7.56	5.39	4.51	4.02	3.70	3.47	3.30	3.17	3.07	2.98	2.84	2.70	2.55	2.47	2.39	3.30	2.31	2.11	2.01
40	7.31	5.18	4.31	3.83	3.51	3.29	3.12	2.99	2.89	2.80	2.66	2.52	2.37	2.29	2.20	2.11	2.02	1.92	1.80
60	7.08	4.98	4.13	3.65	3.34	3.12	2.95	2.82	2.72	2.63	2.50	2.35	2.20	2.12	2.03	1.94	1.84	1.73	1.60
120	6.85	4.79	3.95	3.48	3.17	2.96	2.79	2.66	2.56	2.47	2.34	2.19	2.03	1.95	1.86	1.76	1.66	1.53	1.38
∞	6.63	4.61	3.78	3.32	3.02	2.80	2.64	2.51	2.41	2.32	2.18	2.04	1.88	1.79	1.70	1.59	1.47	1.32	1.00

Table V: Critical Points $q_{\alpha,k,v}$ of the Studentized Range Distribution

Degrees of freedom v	$\alpha = 0.05$ k																		
	2	3	4	5	6	7	8	9	10	11	12	13	14	15	16	17	18	19	20
5	3.64	4.60	5.22	5.67	6.03	6.33	6.58	6.80	6.99	7.17	7.32	7.47	7.60	7.72	7.83	7.93	8.03	8.12	8.21
6	3.46	4.34	4.90	5.30	5.63	5.90	6.12	6.32	6.49	6.65	6.79	6.92	7.03	7.14	7.24	7.34	7.43	7.51	7.59
7	3.34	4.16	4.68	5.06	5.36	5.61	5.82	6.00	6.16	6.30	6.43	6.55	6.66	6.76	6.85	6.94	7.02	7.10	7.17
8	3.26	4.04	4.53	4.89	5.17	5.40	5.60	5.77	5.92	6.05	6.18	6.29	6.39	6.48	6.57	6.65	6.73	6.80	6.87
9	3.20	3.95	4.41	4.76	5.02	5.24	5.43	5.59	5.74	5.87	5.98	6.09	6.19	6.28	6.36	6.44	6.51	6.58	6.64
10	3.15	3.88	4.33	4.65	4.91	5.12	5.30	5.46	5.60	5.72	5.83	5.93	6.03	6.11	6.19	6.27	6.34	6.40	6.47
11	3.11	3.82	4.26	4.57	4.82	5.03	5.20	5.35	5.49	5.61	5.71	5.81	5.90	5.98	6.06	6.13	6.20	6.27	6.33
12	3.08	3.77	4.20	4.51	4.75	4.95	5.12	5.27	5.39	5.51	5.61	5.71	5.80	5.88	5.95	6.02	6.09	6.15	6.21
13	3.06	3.73	4.15	4.45	4.69	4.88	5.05	5.19	5.32	5.43	5.53	5.63	5.71	5.79	5.86	5.93	5.99	6.05	6.11
14	3.03	3.70	4.11	4.41	4.64	4.83	4.99	5.13	5.25	5.36	5.46	5.55	5.64	5.71	5.79	5.85	5.91	5.97	6.03
15	3.01	3.67	4.08	4.37	4.59	4.78	4.94	5.08	5.20	5.31	5.40	5.49	5.57	5.65	5.72	5.78	5.85	5.90	5.96
16	3.00	3.65	4.05	4.33	4.56	4.74	4.90	5.03	5.15	5.26	5.35	5.44	5.52	5.59	5.66	5.73	5.79	5.84	5.90
17	2.98	3.63	4.02	4.30	4.52	4.70	4.86	4.99	5.11	5.21	5.31	5.39	5.47	5.54	5.61	5.67	5.73	5.79	5.84
18	2.97	3.61	4.00	4.28	4.49	4.67	4.82	4.96	5.07	5.17	5.27	5.35	5.43	5.50	5.57	5.63	5.69	5.74	5.79
19	2.96	3.59	3.98	4.25	4.47	4.65	4.79	4.92	5.04	5.14	5.23	5.31	5.39	5.46	5.53	5.59	5.65	5.70	5.75
20	2.95	3.58	3.96	4.23	4.45	4.62	4.77	4.90	5.01	5.11	5.20	5.26	5.36	5.43	4.49	5.55	5.61	5.66	5.71
24	2.92	3.53	3.90	4.17	4.37	4.54	4.68	4.81	4.92	5.01	5.10	5.16	5.25	5.32	5.38	5.44	5.49	5.55	5.59
30	2.89	3.49	3.85	4.10	4.30	4.46	4.60	4.72	4.82	4.92	5.00	5.08	5.15	5.21	5.27	5.33	5.38	5.43	5.47
40	2.86	3.44	3.79	4.04	4.23	4.39	4.52	4.63	4.73	4.82	4.90	4.98	5.04	5.11	5.16	5.22	5.27	5.31	5.36
60	2.83	3.40	3.74	3.98	4.16	4.31	4.44	4.55	4.65	4.73	4.81	4.88	4.94	5.00	5.06	5.11	5.15	5.20	5.24
120	2.80	3.36	3.68	3.92	4.10	4.24	4.36	4.47	4.56	4.64	4.71	4.78	4.84	4.90	4.95	5.00	5.04	5.09	5.13
∞	2.77	3.31	3.63	3.86	4.03	4.17	4.29	4.39	4.47	4.55	4.62	4.68	4.74	4.80	4.85	4.89	4.93	4.97	5.01

Degrees of freedom v	$\alpha = 0.01$ k																		
	2	3	4	5	6	7	8	9	10	11	12	13	14	15	16	17	18	19	20
5	5.70	6.97	7.80	8.42	8.91	9.32	9.67	9.97	10.2	10.5	10.7	10.9	11.1	11.2	11.4	11.6	11.7	11.8	11.9
6	5.24	6.33	7.03	7.56	7.97	8.32	8.61	8.87	9.10	9.30	9.49	9.65	9.81	9.95	10.1	10.2	10.3	10.4	10.5
7	4.95	5.92	6.54	7.01	7.37	7.68	7.94	8.17	8.37	8.55	8.71	8.86	9.00	9.12	9.24	9.35	9.46	9.55	9.65
8	4.74	5.63	6.20	6.63	6.96	7.24	7.47	7.68	7.87	8.03	8.18	8.31	8.44	8.55	8.66	8.76	8.85	8.94	9.03
9	4.60	5.43	5.96	6.35	6.66	6.91	7.13	7.32	7.49	7.65	7.78	7.91	8.03	8.13	8.23	8.32	8.41	8.49	8.57
10	4.48	5.27	5.77	6.14	6.43	6.67	6.87	7.05	7.21	7.36	7.48	7.60	7.71	7.81	7.91	7.99	8.07	8.15	8.22
11	4.39	5.14	5.62	5.97	6.25	6.48	6.67	6.84	6.99	7.13	7.25	7.36	7.46	7.56	7.65	7.73	7.81	7.88	7.95
12	4.32	5.04	5.50	5.84	6.10	6.32	6.51	6.67	6.81	6.94	7.06	7.17	7.26	7.36	7.44	7.52	7.59	7.66	7.73
13	4.26	4.96	5.40	5.73	5.98	6.19	6.37	6.53	6.67	6.79	6.90	7.01	7.10	7.19	7.27	7.34	7.42	7.48	7.55
14	4.21	4.89	5.32	5.63	5.88	6.08	6.26	6.41	6.54	6.66	6.77	6.87	6.96	7.05	7.12	7.20	7.27	7.33	7.39
15	4.17	4.83	5.25	5.56	5.80	5.99	6.16	6.31	6.44	6.55	6.66	6.76	6.84	6.93	7.00	7.07	7.14	7.20	7.26
16	4.13	4.78	5.19	5.49	5.72	5.92	6.08	6.22	6.35	6.46	6.56	6.66	6.74	6.82	6.90	6.97	7.03	7.09	7.15
17	4.10	4.74	5.14	5.43	5.66	5.85	6.01	6.15	6.27	6.38	6.48	6.57	6.66	6.73	6.80	6.87	6.94	7.00	7.05
18	4.07	4.70	5.09	5.38	5.60	5.79	5.94	6.08	6.20	6.31	6.41	6.50	6.58	6.65	6.72	6.79	6.85	6.91	6.96
19	4.05	4.67	5.05	5.33	5.55	5.73	5.89	6.02	6.14	6.25	6.34	6.43	6.51	6.58	6.65	6.72	6.78	6.84	6.89
20	4.02	4.64	5.02	5.29	5.51	5.69	5.84	5.97	6.09	6.19	6.29	6.37	6.45	6.52	6.59	6.65	6.71	6.76	6.82
24	3.96	4.54	4.91	5.17	5.37	5.54	5.69	5.81	5.92	6.02	6.11	6.19	6.26	6.33	6.39	6.45	6.51	6.56	6.61
30	3.89	4.45	4.80	5.05	5.24	5.40	5.54	5.65	5.76	5.85	5.93	6.01	6.08	6.14	6.20	6.26	6.31	6.36	6.41
40	3.82	4.37	4.70	4.93	5.11	5.27	5.39	5.50	5.60	5.69	5.77	5.84	5.90	5.96	6.02	6.07	6.12	6.17	6.21
60	3.76	4.28	4.60	4.82	4.99	5.13	5.25	5.36	5.45	5.53	5.60	5.67	5.73	5.79	5.84	5.89	5.93	5.98	6.02
120	3.70	4.20	4.50	4.71	4.87	5.01	5.12	5.21	5.30	5.38	5.44	5.51	5.56	5.61	5.66	5.71	5.75	5.79	5.83
∞	3.64	4.12	4.40	4.60	4.76	4.88	4.99	5.08	5.16	5.23	5.29	5.35	5.40	5.45	5.49	5.54	5.57	5.61	5.65

Table VI: Critical Points $d_{\alpha,n}$ for the Kolmogorov and Kolmogorov-Smirnov Procedures

n	$\alpha = 0.20$	$\alpha = 0.10$	$\alpha = 0.05$	$\alpha = 0.02$	$\alpha = 0.01$
10	0.323	0.369	0.409	0.457	0.489
11	0.308	0.352	0.391	0.437	0.468
12	0.296	0.338	0.375	0.419	0.449
13	0.285	0.325	0.361	0.404	0.432
14	0.275	0.314	0.349	0.390	0.418
15	0.268	0.304	0.338	0.377	0.404
16	0.258	0.295	0.327	0.366	0.392
17	0.250	0.286	0.318	0.355	0.381
18	0.244	0.279	0.309	0.346	0.371
19	0.237	0.271	0.301	0.337	0.361
20	0.232	0.265	0.294	0.329	0.362
21	0.226	0.259	0.287	0.321	0.344
22	0.221	0.253	0.281	0.314	0.337
23	0.216	0.247	0.275	0.307	0.330
24	0.212	0.242	0.269	0.301	0.323
25	0.208	0.238	0.264	0.295	0.317
26	0.204	0.233	0.259	0.290	0.311
27	0.200	0.229	0.254	0.284	0.305
28	0.197	0.226	0.250	0.279	0.300
29	0.193	0.221	0.246	0.275	0.295
30	0.190	0.218	0.242	0.270	0.290
31	0.187	0.214	0.238	0.266	0.285
32	0.184	0.211	0.234	0.262	0.281
33	0.182	0.208	0.231	0.258	0.277
34	0.179	0.205	0.227	0.254	0.273
35	0.177	0.202	0.224	0.251	0.269
36	0.174	0.199	0.221	0.247	0.265
37	0.172	0.196	0.218	0.244	0.262
38	0.170	0.194	0.215	0.241	0.258
39	0.168	0.191	0.213	0.238	0.255
40	0.165	0.189	0.210	0.235	0.252
Approximation for $n > 40$:	$\dfrac{1.07}{\sqrt{n}}$	$\dfrac{1.22}{\sqrt{n}}$	$\dfrac{1.36}{\sqrt{n}}$	$\dfrac{1.52}{\sqrt{n}}$	$\dfrac{1.63}{\sqrt{n}}$

ANSWERS TO ODD-NUMBERED PROBLEMS

Chapter 1   Probability Theory

1.1   Probabilities

1.1.1 $S = \{(\text{head, head, head}), (\text{head, head, tail}), (\text{head, tail, head}),$ (head, tail, tail), (tail, head, head), (tail, head, tail), (tail, tail, head), (tail, tail, tail)$\}$

1.1.3 $S = \{0,1,2,3,4\}$

1.1.5 $S = \{(\text{on time, satisfactory}), (\text{on time, unsatisfactory}), (\text{late, satisfactory}), (\text{late, unsatisfactory})\}$

1.1.7 (a) $p = 0.5$ (b) $p = \frac{2}{3}$ (c) $\frac{1}{3}$

1.1.9 $0 \leq P(V) \leq 0.39$ $P(IV) = P(V) = 0.195$

1.2   Events

1.2.1 (a) $P(b) = 0.15$ (b) $P(A) = 0.50$ (c) $P(A') = 0.50$

1.2.3 $\frac{124}{1461}$ and $\frac{113}{1461}$

1.2.5 $P(\text{at least one score is a prime number}) = \frac{8}{9}$

$P(\text{neither score prime}) = \frac{1}{9}$

1.2.7 $P(\spadesuit \text{ or } \clubsuit) = \frac{1}{2}$

1.2.9 (a) $P(\text{Terica is winner}) = \frac{1}{4}$

(b) $P(\text{Terica is winner or runner up}) = \frac{1}{2}$

1.2.11 (a) $P(\text{both assembly lines are shut down}) = 0.02$
(b) $P(\text{neither assembly line is shut down}) = 0.74$
(c) $P(\text{at least one assembly line is at full capacity}) = 0.71$
(d) $P(\text{exactly one assembly line at full capacity}) = 0.52$
The complement of *"neither assembly line is shut down"* consists of the outcomes

$$\{(S, S), (S, P), (S, F), (P, S), (F, S)\}.$$

The complement of *"at least one assembly line is at full capacity"* consists of the outcomes

$$\{(S, S), (S, P), (P, S), (P, P)\}.$$

1.3   Combinations of Events

1.3.1 The event A contains the outcome 0 while the empty set does not contain any outcomes.

1.3.5 Yes   No

1.3.7 $P(B) = 0.4$

1.3.9 Yes
$P(A \cup B \cup C) = \frac{3}{4}$

1.3.11 $P(O' \cap S') = 0.11$

1.3.13 0.19

1.4   Conditional Probability

1.4.1 (a) $P(A \mid B) = 0.1739$ (b) $P(C \mid A) = 0.59375$
(c) $P(B \mid A \cap B) = 1$ (d) $P(B \mid A \cup B) = 0.657$
(e) $P(A \mid A \cup B \cup C) = 0.3636$
(f) $P(A \cap B \mid A \cup B) = 0.1143$

1.4.3 (a) $P(A\heartsuit \mid \text{red suit}) = \frac{1}{26}$ (b) $P(\text{heart} \mid \text{red suit}) = \frac{1}{2}$
(c) $P(\text{red suit} \mid \text{heart}) = 1$ (d) $P(\text{heart} \mid \text{black suit}) = 0$
(e) $P(\text{King} \mid \text{red suit}) = \frac{1}{13}$
(f) $P(\text{King} \mid \text{red picture card}) = \frac{1}{3}$

1.4.5 $P(\text{shiny} \mid \text{red}) = \frac{3}{8}$

$P(\text{dull} \mid \text{red}) = \frac{5}{8}$

1.4.9 (a) 0.512   (b) 0.619
(c) 0   (d) 0.081

1.4.11 (a) 0.9326   (b) 0.9756

1.4.13 (a) 0.02   (b) 0.775

1.4.15 $P(\text{delay}) = 0.3568$

1.5   Probabilities of Event Intersections

1.5.1   (a) P(both cards are picture cards) $= \frac{132}{2652}$

   (b) P(both cards are from red suits) $= \frac{650}{2652}$

   (c) P(one card is from a red suit and one is from black suit) $= \frac{26}{51}$

1.5.3   (a)   No, they are not independent.
   (b)   Yes, they are independent.
   (c)   No, they are not independent.
   (d)   Yes, they are independent.
   (e)   No, they are not independent.

1.5.5 P(all 4 cards are hearts) $= \frac{1}{256}$
   P(all 4 cards are from red suits) $= \frac{1}{16}$
   P(all 4 cards from different suits) $= \frac{3}{32}$

1.5.7 P(message gets through the network) $= 0.98096$

1.5.9 P(no broken bulbs) $= 0.5682$
   P(no more than one broken bulb in the sample) $= 0.9260$

1.5.11 P(drawing 2 green balls) $= 0.180$
   P(two balls different colors) $= 0.656$

1.5.13 $p = 0.5$ (a fair coin)

1.5.15   (a) $\frac{1}{32}$ (b) $\frac{5}{9}$
   (c) $\frac{3}{8}$ (d) $\frac{13}{34}$

1.6   Posterior Probabilities

1.6.1   (a) P(positive blood test) $= 0.0691$
   (b) P(disease | positive blood test) $= 0.1404$
   (c) P(no disease | negative blood test) $= 0.9997$

1.6.3   (a) P(Section I) $= \frac{55}{100}$ (b) P(grade is A) $= \frac{21}{100}$
   (c) P(A | Section I) $= \frac{10}{55}$ (d) P(Section I | A) $= \frac{10}{21}$

1.6.5   (a)   0.4579
   (b) P(A | did not fail) $= 0.7932$

1.6.7   (a) $P(H \mid L) == 0.224$
   (b) $P(M \mid L') = 0.551$

1.7   Counting Techniques

1.7.1   (a) $7! = 5040$
   (b) $8! = 40320$
   (c) $4! = 24$
   (d) $13! = 6,227,020,800$

1.7.3   (a) $C_2^6 == 15$ (b) $C_4^8 = 70$
   (c) $C_2^5 = 10$ (d) $C_6^{14} = 3003$

1.7.5   24

1.7.7   1,860,480
   15504

1.7.13   48
   72

1.7.15   (a)   27720

1.7.17   168,168,000

1.7.19   3,628,800
   252

1.7.21   (a)   0.082
   (b)   0.049

Chapter 2   Random Variables

2.1   Discrete Random Variables

2.1.1   (a) $P(X = 4) = 0.21$
   (c) $F(0) = 0.08$
      $F(1) = 0.19$
      $F(2) = 0.46$
      $F(3) = 0.79$
      $F(4) = 1.00$

2.1.3

x_i	1	2	3	4	5	6	8	9	10
p_i	$\frac{1}{36}$	$\frac{2}{36}$	$\frac{2}{36}$	$\frac{3}{36}$	$\frac{2}{36}$	$\frac{4}{36}$	$\frac{2}{36}$	$\frac{1}{36}$	$\frac{2}{36}$
$F(x_i)$	$\frac{1}{36}$	$\frac{3}{36}$	$\frac{5}{36}$	$\frac{8}{36}$	$\frac{10}{36}$	$\frac{14}{36}$	$\frac{16}{36}$	$\frac{17}{36}$	$\frac{19}{36}$

x_i	12	15	16	18	20	24	25	30	36
p_i	$\frac{4}{36}$	$\frac{2}{36}$	$\frac{1}{36}$	$\frac{2}{36}$	$\frac{2}{36}$	$\frac{2}{36}$	$\frac{1}{36}$	$\frac{2}{36}$	$\frac{1}{36}$
$F(x_i)$	$\frac{23}{36}$	$\frac{25}{36}$	$\frac{26}{36}$	$\frac{28}{36}$	$\frac{30}{36}$	$\frac{32}{36}$	$\frac{33}{36}$	$\frac{35}{36}$	1

2.1.5

x_i	-5	-4	-3	-2	-1	0	1	2	3	4	6	8	10	12
p_i	$\frac{1}{36}$	$\frac{1}{36}$	$\frac{2}{36}$	$\frac{2}{36}$	$\frac{3}{36}$	$\frac{3}{36}$	$\frac{2}{36}$	$\frac{5}{36}$	$\frac{1}{36}$	$\frac{4}{36}$	$\frac{3}{36}$	$\frac{3}{36}$	$\frac{3}{36}$	$\frac{3}{36}$
$F(x_i)$	$\frac{1}{36}$	$\frac{2}{36}$	$\frac{4}{36}$	$\frac{6}{36}$	$\frac{9}{36}$	$\frac{12}{36}$	$\frac{14}{36}$	$\frac{19}{36}$	$\frac{20}{36}$	$\frac{24}{36}$	$\frac{27}{36}$	$\frac{30}{36}$	$\frac{33}{36}$	1

2.1.7 (a)

x_i	0	1	2	3	4	6	8	12
p_i	0.061	0.013	0.195	0.067	0.298	0.124	0.102	0.140

(b)

x_i	0	1	2	3	4	6	8	12
$F(x_i)$	0.061	0.074	0.269	0.336	0.634	0.758	0.860	1.000

(c) The most likely value is 4.
$P(\text{not shipped}) = 0.074$

2.1.9

x_i	1	2	3	4
p_i	$\frac{2}{5}$	$\frac{3}{10}$	$\frac{1}{5}$	$\frac{1}{10}$
$F(x_i)$	$\frac{2}{5}$	$\frac{7}{10}$	$\frac{9}{10}$	1

2.1.11 (a) The state space is $\{3, 4, 5, 6\}$.
(b) $P(X = 3) = \frac{1}{20}$
$P(X = 4) = \frac{3}{20}$
$P(X = 5) = \frac{6}{20}$
$P(X = 6) = \frac{1}{2}$
$P(X \leq 3) = \frac{1}{20}$
$P(X \leq 4) = \frac{4}{20}$
$P(X \leq 5) = \frac{10}{20}$
$P(X \leq 6) = 1$

2.2 Continuous Random Variables

2.2.1 (a) Continuous   (b) Discrete
(c) Continuous   (d) Continuous
(e) Discrete
(f) This depends on what level of accuracy to which it is measured.
It could be considered to be either discrete or continuous.

2.2.3 (a) $c = -\frac{1}{8}$
(b) $P(-1 \leq X \leq 1) = \frac{69}{128}$
(c) $F(x) = \frac{x^2}{128} + \frac{15x}{64} + \frac{7}{16}$
for $-2 \leq x \leq 0$
$F(x) = -\frac{x^2}{16} + \frac{3x}{8} + \frac{7}{16}$
for $0 \leq x \leq 3$

2.2.5 (a) $A = 1$
$B = -1$
(b) $P(2 \leq X \leq 3) = 0.0855$
(c) $f(x) = e^{-x}$
for $x \geq 0$

2.2.7 (a) $A = -0.25$
$B = 0.361$
(b) $P(X > 2) = 0.5$
(c) $f(x) = \frac{1.08}{3x+2}$
for $0 \leq x \leq 10$

2.2.9 (a) $A = 1.0007$
$B = -125.09$
(b) $P(X \leq 10) = 0.964$
(c) $P(X \geq 30) = 0.002$
(d) $f(r) = \frac{375.3}{(r+5)^4}$
for $0 \leq r \leq 50$

2.2.11 (a) $A = \frac{4}{819}$
(b) $F(x) = \frac{4}{819}\left(65x^2 - \frac{x^4}{4} - 4000\right)$
for $10 \leq x \leq 11$
(c) 0.283

2.3 The Expectation of a Random Variable

2.3.1 $E(X) = 2.48$

2.3.3 With replacement:

$$E(X) = 0.5$$

Without replacement:

$$E(X) = 0.5$$

2.3.5 $E(X) = \$8.77$

2.3.7 $E(\text{net winnings}) = \$1.31$

2.3.9 $E(\text{winnings}) = \$0.09$

2.3.11 (a) $E(X) = 2.67$ (b) 2.83

2.3.13 $E(X) = 0.9977$
0.6927

2.3.17 (a) $E(X) = 10.418234$
(b) 10.385

2.4 The Variance of a Random Variable

2.4.1 (a) $E(X) = \frac{11}{6}$ (b) $\text{Var}(X) = \frac{341}{36}$

2.4.3 $\sigma = 1$

2.4.5 (a) $\text{Var}(X) = 0.25$ (b) $\sigma = 0.5$
(c) 4.43 (d) 0.99
5.42

2.4.7 (a) $\text{Var}(X) = 0.0115$ (b) $\sigma = 0.107$
(c) 0.217 (d) 0.184
$x = 0.401$

2.4.9 (a) $\text{Var}(X) = 18.80 - 2.44^2 = 12.8$
(b) $\sigma = 3.58$ (c) 0.50 (d) 2.43
2.93

2.4.11 $P(109.55 \le X \le 112.05) \ge 0.84$

2.4.13 (a) 0.275
(b) 10.69
10.07

2.4.15 $E(X) = \$1.55$
The standard deviation is $\$1.96$.

2.4.17 (a) $E(X) = 3.74$
(b) The standard deviation is 0.844.

2.5 Jointly Distributed Random Variables

2.5.3 (a) $A = -\frac{1}{125}$
(b) $P(0 \le X \le 1, 4 \le Y \le 5) = \frac{9}{100}$
(c) $f_X(x) = \frac{2(3-x)}{25}$
for $-2 \le x \le 3$
$f_Y(y) = \frac{y}{10}$
for $4 \le x \le 6$
(d) The random variables X and Y are independent.
(e) $f_{X|Y=5}(x)$ is equal to $f_X(x)$

2.5.5 (a) $A = 0.00896$
(b) $P(1.5 \le X \le 2, 1 \le Y \le 2) = 0.158$
(c) $f_X(x) = 0.00896 \, (e^{x+3} - e^{2x-3} - e^x + e^{2x})$
for $1 \le x \le 2$
$f_Y(y) = 0.00896 \, (e^{2+y} + 0.5e^{4-y} - e^{1+y} - 0.5e^{2-y})$
for $0 \le y \le 3$
(d) No, the random variables X and Y are not independent.
(e) $f_{X|Y=0}(x) = \frac{e^x + e^{2x}}{28.28}$

2.5.7 (a)

X\Y	0	1	2	p_{i+}
0	$\frac{4}{16}$	$\frac{4}{16}$	$\frac{1}{16}$	$\frac{9}{16}$
1	$\frac{4}{16}$	$\frac{2}{16}$	0	$\frac{6}{16}$
2	$\frac{1}{16}$	0	0	$\frac{1}{16}$
p_{+j}	$\frac{9}{16}$	$\frac{6}{16}$	$\frac{1}{16}$	1

(b) See the table above.
(c) No, the random variables X and Y are not independent.

(d) $E(X) = \frac{1}{2}$
$\text{Var}(X) = 0.3676$
The random variable Y has the same mean and variance as X.
(e) $\text{Cov}(X, Y) = -\frac{1}{8}$
(f) $\text{Corr}(X, Y) = -\frac{1}{3}$
(g) $P(Y = 0|X = 0) = \frac{4}{9}$
$P(Y = 1|X = 0) = \frac{4}{9}$
$P(Y = 2|X = 0) = \frac{1}{9}$
$P(Y = 0|X = 1) = \frac{2}{3}$
$P(Y = 1|X = 1) = \frac{1}{3}$
$P(Y = 2|X = 1) = 0$

2.5.9 (a) $P(\text{same score}) = 0.80$
(b) $P(X < Y) = 0.07$
(c)

x_i	1	2	3	4
p_{i+}	0.12	0.20	0.30	0.38

$E(X) = 2.94$
$\text{Var}(X) = 1.0564$

(d)

y_j	1	2	3	4
p_{+j}	0.14	0.21	0.30	0.35

$E(Y) = 2.86$
$\text{Var}(Y) = 1.1004$

(e) The scores are not independent.

(f) $P(Y = 1 | X = 3) = \frac{1}{30}$

$P(Y = 2 | X = 3) = \frac{3}{30}$

$P(Y = 3 | X = 3) = \frac{24}{30}$

$P(Y = 4 | X = 3) = \frac{2}{30}$

(g) $\text{Cov}(X, Y) = 0.8816$

(h) $\text{Corr}(X, Y) = 0.82$

2.6 Combinations and Functions of Random Variables

2.6.1 (a) $E(3X + 7) = 13$

$\text{Var}(3X + 7) = 36$

(b) $E(5X - 9) = 1$

$\text{Var}(5X - 9) = 100$

(c) $E(2X + 6Y) = -14$

$\text{Var}(2X + 6Y) = 88$

(d) $E(4X - 3Y) = 17$

$\text{Var}(4X - 3Y) = 82$

(e) $E(5X - 9Z + 8) = -54$

$\text{Var}(5X - 9Z + 8) = 667$

(f) $E(-3Y - Z - 5) = -4$

$\text{Var}(-3Y - Z - 5) = 25$

(g) $E(X + 2Y + 3Z) = 20$

$\text{Var}(X + 2Y + 3Z) = 75$

(h) $E(6X + 2Y - Z + 16) = 14$

$\text{Var}(6X + 2Y - Z + 16) = 159$

2.6.3 $E(Y) = 3\mu$

$\text{Var}(Y) = 9\sigma^2$

$E(Z) = 3\mu$

$\text{Var}(Z) = 3\sigma^2$

2.6.5 $18.75

$25.72

2.6.7 $\frac{10}{13}$

$\frac{120}{169}$

$\frac{10}{13}$

2.6.9 (a) $A = \frac{3}{4}$

(b) $f_V(v) = \frac{1}{2}(\frac{3}{4\pi})^{2/3} v^{-1/3} - \frac{3}{16\pi}$

for $0 \le v \le \frac{32\pi}{3}$.

(c) $E(V) = \frac{32\pi}{15}$

2.6.11 (a) The return has an expectation of $100, a standard deviation of $20, and a variance of 400.

(b) The return has an expectation of $100, a standard deviation of $30, and a variance of 900.

(c) The total return has an expectation of $100 and a variance of 325, so that the standard deviation is $18.03.

(d) $x = 692

2.6.13 (a) The mean is 66.

The standard deviation is 5.63.

(b) The mean is 64.8.

The standard deviation is 6.61.

2.6.15 (a) The mean is $\mu = 65.90$.

The standard deviation is 0.143.

(b) The mean is 527.2.

The standard deviation is 0.905.

2.6.17 The mean is 280.

The standard deviation is 15.3.

2.6.19 The mean is 43.33.

The standard deviation is 1.22.

2.6.21 $n \ge 32$

Chapter 3   Discrete Probability Distributions

3.1 The Binomial Distribution

3.1.1 (a) $P(X = 3) = 0.0847$

(b) $P(X = 6) = 0.0004$

(c) $P(X \le 2) = 0.8913$

(d) $P(X \ge 7) = 3.085 \times 10^{-5}$

(e) $E(X) = 1.2$

(f) $\text{Var}(X) = 1.056$

3.1.3 $X \sim B(6, 0.5)$

x_i	0	1	2	3	4	5	6
p_i	0.0156	0.0937	0.2344	0.3125	0.2344	0.0937	0.0156

$E(X) = 3$

$\sigma = 1.22$

$X \sim B(6, 0.7)$

x_i	0	1	2	3	4	5	6
p_i	0.0007	0.0102	0.0595	0.1852	0.3241	0.3025	0.1176

$E(X) = 4.2$
$\sigma = 1.12$

3.1.5 (a) $P\left(B\left(8, \frac{1}{2}\right) = 5\right) = 0.2187$
(b) 0.3721
(c) 0.2326
(d) 0.4682

3.1.7 0.35
0.0013
$p = 0.5$

3.1.9 $X \sim B(18, 0.6)$
(a) 0.3789   (b) 0.0013

3.2   The Geometric and Negative Binomial Distributions

3.2.1 (a) $P(X = 4) = 0.0189$ (b) $P(X = 1) = 0.7$
(c) $P(X \leq 5) = 0.9976$ (d) $P(X \geq 8) = 0.0002$

3.2.5 (a) 0.0678   (b) 0.0136   (c) 11.11   (d) 33.33

3.2.7 (a) 0.1406   (b) 0.0584
The expected number of cards drawn before the fourth heart is obtained is 16.

If the first two cards are spades, then the probability that the first heart card is obtained on the fifth drawing is the same as the probability in part (a).

3.2.9 (a) 0.01536   (b) 0.116

3.2.11 0.123

3.3   The Hypergeometric Distribution

3.3.1 (a) $P(X = 4) = \frac{5}{11}$
(b) $P(X = 5) = \frac{2}{11}$
(c) $P(X \leq 3) = \frac{23}{66}$

3.3.3 (a) $\frac{90}{221}$ (b) $\frac{25}{442}$ (c) $\frac{139}{442}$

3.3.5 $\frac{55}{833}$
The expected value is 3 and the variance is $\frac{30}{17}$.

3.3.7 (a) $\frac{7}{33}$ (b) $\frac{7}{165}$

3.3.9 (a) 0.431   (b) 0.375

3.4   The Poisson Distribution

3.4.1 (a) $P(X = 1) = 0.1304$
(b) $P(X \leq 3) = 0.6025$
(c) $P(X \geq 6) = 0.1054$
(d) $P(X = 0|X \leq 3) = 0.0677$

3.4.5 0.9735

3.4.7 0.7576

3.5   The Multinomial Distribution

3.5.1 (a) 0.0416   (b) 0.7967

3.5.3 (a) 0.0502   (b) 0.0670   (c) 0.1577
The expected number of misses is 0.96.

3.5.5 0.0212

Chapter 4   Continuous Probability Distributions

4.1   The Uniform Distribution

4.1.1 (a) $E(X) = 2.5$
(b) $\sigma = 3.175$
(c) The upper quartile is 5.25.
(d) $P(0 \leq X \leq 4) = \frac{4}{11}$

4.1.3 (a) The expectations are: The variances are:
                                6   4.2
                                4   3.2
                                5   3.75
                                5   3.75
            (b) 0.0087

4.1.5 (a) $\frac{2}{3}$

(b) $P(\text{difference} \leq 0.0005 \mid \text{fits in hole}) = \frac{1}{4}$

4.2 The Exponential Distribution

4.2.3 (a) $E(X) = 5$
(b) $\sigma = 5$
(c) The median is 3.47.
(d) $P(X \geq 7) = 0.2466$
(e) 0.6703

4.2.5 $F(x) = \frac{1}{2}e^{-\lambda(\theta-x)}$
for $-\infty \leq x \leq \theta$, and
$F(x) = 1 - \frac{1}{2}e^{-\lambda(x-\theta)}$
for $\theta \leq x \leq \infty$.
(a) $P(X \leq 0) = 0.0012$
(b) $P(X \geq 1) = 0.9751$

4.2.7 (a) $\lambda = 1.8$
(b) $E(X) = 0.5556$
(c) $P(X \geq 1) = 0.1653$
(d) A Poisson distribution with parameter 7.2.
(e) $P(X \geq 4) = 0.9281$

4.2.9 (a) $P(X \geq 1.5) = 0.301$
(b) 0.217

4.2.11 (a) $P(X \leq 6) = 0.699$
(b) 0.180

4.2.13 0.355

4.3 The Gamma Distribution

4.3.1 $\Gamma(5.5) = 52.34$

4.3.3 (a) $f(3) = 0.2055$ (b) $f(3) = 0.0227$
$F(3) = 0.3823$ $F(3) = 0.9931$
$F^{-1}(0.5) = 3.5919$ $F^{-1}(0.5) = 1.3527$
(c) $f(3) = 0.2592$
$F(3) = 0.6046$
$F^{-1}(0.5) = 2.6229$

4.3.5 (a) A gamma distribution with parameters $k = 4$ and $\lambda = 2$.
(b) $E(X) = 2$ (c) $\sigma = 1$
(d) The probability can be calculated as

$$P(X \geq 3) = 0.1512$$

where the random variable X has a gamma distribution with parameters $k = 4$ and $\lambda = 2$.

The probability can also be calculated as

$$P(Y \leq 3) = 0.1512$$

where the random variable Y has a Poisson distribution with parameter

$$2 \times 3 = 6$$

which counts the number of imperfections in a 3 meter length of fiber.

4.3.7 (a) The expectation is $E(X) = 62.86$
the variance is $\text{Var}(X) = 89.80$
and the standard deviation is 9.48.
(b) 0.3991

4.4 The Weibull Distribution

4.4.3 (a) 0.5016
(b) 0.6780
0.3422
(c) $P(0.5 \leq X \leq 1.5) = 0.5023$

4.4.5 (a) 0.8000   (b) 4.5255   (c) 31.066
0.0888   91.022
(d) $P(3 \leq X \leq 5) = 0.0722$

4.4.7 0.0656

4.5 The Beta Distribution

4.5.1 (a) $A = 60$
(b) $E(X) = \frac{4}{7}$
$\text{Var}(X) = \frac{3}{98}$
(c) This is a beta distribution with $a = 4$ and $b = 3$.

4.5.3 (a) $f(0.5) = 1.9418$ (b) $f(0.5) = 0.7398$
$F(0.5) = 0.6753$ $F(0.5) = 0.7823$
$F^{-1}(0.5) = 0.5406$ $F^{-1}(0.5) = 0.4579$

(c) $f(0.5) = 0.6563$
$F(0.5) = 0.9375$
$F^{-1}(0.5) = 0.3407$

4.5.5 (a) $E(X) = 0.7579$
$\text{Var}(X) = 0.0175$
(b) From the computer $P(X \geq 0.9) = 0.1368$.

Chapter 5   The Normal Distribution

5.1   Probability Calculations Using the Normal Distribution

5.1.1 (a) 0.9099 (b) 0.5871
  (c) 0.6521 (d) 0.4249
  (e) 0.2960 (f) $x = 0.1257$
  (g) $x = -0.5828$ (h) $x = 0.3989$

5.1.3 (a) $P(X \leq 10.34) = 0.5950$
  (b) $P(X \geq 11.98) = 0.0807$
  (c) $P(7.67 \leq X \leq 9.90) = 0.4221$
  (d) $P(10.88 \leq X \leq 13.22) = 0.2555$
  (e) $P(|X - 10| \leq 3) = 0.9662$
  (f) $x = 11.2415$
  (g) $x = 12.4758$
  (h) $x = 0.6812$

5.1.5 $\mu = 1.1569$
  $\sigma = 4.5663$

5.1.9 (a) 0.0478 (b) 0.0062 (c) $c = 0.3091$.

5.1.11 (a) 4.3809 (b) $c = 0.1538$
  4.2191

5.1.13 (a) 0.2398 (b) 0.4298
  (c) 0.0937 (d) 24.56
  (e) 25.66

5.1.15 (a) The probability of being outside the range is 0.0956.
  (b) $\sigma = 0.304$

5.1.17 $\mu = 93.09$

5.2   Linear Combinations of Normal Random Variables

5.2.1 (a) 0.6360 (b) 0.6767 (c) 0.4375
  (d) 0.0714 (e) 0.8689 (f) 0.1315

5.2.3 (a) 0.3830 (b) 0.8428
  (c) $n \geq 27$.

5.2.5 0.7777

5.2.7 (a) \$1050
  (b) $0.0002y^2 + 0.0003(1000 - y)^2$
  (c) The variance is minimized with $y = 600$.
  0.1807

5.2.9 (a) $N(22.66, 4.312 \times 10^{-3})$ (b) $x = 22.704$
  $x = 22.616$

5.2.11 (a) 0.843 (b) $c = 0.624$

5.2.13 0.645

5.2.15 $n \geq 22$

5.2.17 (a) $E(X) = 1268000$
  The standard deviation is 11,180.
  (b) $E(X) = \mu = 63400$
  The standard deviation is 456.4.

5.3   Approximating Distributions with the Normal Distribution

5.3.1 (a) The exact probability is 0.3823.
  The normal approximation is 0.3650.
  (b) The exact probability is 0.9147.
  The normal approximation is 0.9090.
  (c) The exact probability is 0.7334.
  The normal approximation is 0.7299.
  (d) The exact probability is 0.6527.
  The normal approximation is 0.6429.

5.3.3 The required probability is equal to
  0.2358 for $n = 100$
  0.2764 for $n = 200$
  0.3772 for $n = 500$
  0.4934 for $n = 1000$
  and 0.6408 for $n = 2000$.

5.3.5 (a) A normal distribution can be used with
$$\mu = 1200 \quad \text{and} \quad \sigma^2 = 1200$$
  (b) 0.0745

5.3.7 0.9731

5.3.9 0.7210

5.3.11 0.7824

5.3.13 0.92

5.3.15 (a) 0.138 (b) Very close to 1.

5.4   Distributions Related to the Normal Distribution

5.4.1 (a) $E(X) = 10.23$
  (b) $\text{Var}(X) = 887.69$
  (c) 9.13 (d) 1.21 (e) 7.92
  (f) $P(5 \leq X \leq 8) = 0.1136$

5.4.5 (a) $\chi^2_{0.10,9} = 14.68$
  (b) $\chi^2_{0.05,20} = 31.41$
  (c) $\chi^2_{0.01,26} = 45.64$
  (d) $\chi^2_{0.90,50} = 39.69$
  (e) $\chi^2_{0.95,6} = 1.635$

5.4.7 (a) $t_{0.10,7} = 1.415$ (b) $t_{0.05,19} = 1.729$
(c) $t_{0.01,12} = 2.681$ (d) $t_{0.025,30} = 2.042$
(e) $t_{0.005,4} = 4.604$

5.4.9 (a) $F_{0.10,9,10} = 2.347$ (b) $F_{0.05,6,20} = 2.599$
(c) $F_{0.01,15,30} = 2.700$ (d) $F_{0.05,4,8} = 3.838$
(e) $F_{0.01,20,13} = 3.665$

5.4.13 $P(F_{5,20} \geq 4.00) = 0.011$

5.4.15 (a) $P(F_{10,50} \geq 2.5) = 0.016$
(b) $P(\chi^2_{17} \leq 12) = 0.200$
(c) $P(t_{24} \geq 3) = 0.003$
(d) $P(t_{14} \geq -2) = 0.967$

5.4.17 (a) $P(t_{16} \leq 1.9) = 0.962$
(b) $P(\chi^2_{25} \geq 42.1) = 0.018$
(c) $P(F_{9,14} \leq 1.8) = 0.844$
(d) $P(-1.4 \leq t_{29} \leq 3.4) = 0.913$

Chapter 6   Descriptive Statistics

6.3   Sample Statistics

6.3.1 The sample mean is $\bar{x} = 155.95$.
The sample median is 159.
The sample trimmed mean is 156.50.
The sample standard deviation is $s = 18.43$.
The upper sample quartile is 169.5.
The lower sample quartile is 143.25.

6.3.3 The sample mean is $\bar{x} = 37.08$.
The sample median is 35.
The sample trimmed mean is 36.35.
The sample standard deviation is $s = 8.32$.
The upper sample quartile is 40.
The lower sample quartile is 33.5.

6.3.5 The sample mean is $\bar{x} = 69.35$.
The sample median is 66.
The sample trimmed mean is 67.88.
The sample standard deviation is $s = 17.59$.

The upper sample quartile is 76.
The lower sample quartile is 61.

6.3.7 The sample mean is $\bar{x} = 12.211$.
The sample median is 12.
The sample trimmed mean is 12.175.
The sample standard deviation is $s = 2.629$.
The upper sample quartile is 14.
The lower sample quartile is 10.

6.3.9 The sample mean is $\bar{x} = 0.23181$.
The sample median is 0.220.
The sample trimmed mean is 0.22875.
The sample standard deviation is $s = 0.07016$.
The upper sample quartile is 0.280.
The lower sample quartile is 0.185.

6.3.11 $x = 13$

Chapter 7   Statistical Estimation and Sampling Distributions

7.2   Properties of Point Estimates

7.2.1 (a) $\text{bias}(\hat{\mu}_1) = 0$
The point estimate $\hat{\mu}_1$ is unbiased.
$\text{bias}(\hat{\mu}_2) = 0$
The point estimate $\hat{\mu}_2$ is unbiased.
$\text{bias}(\hat{\mu}_3) = 9 - \frac{\mu}{2}$
(b) $\text{Var}(\hat{\mu}_1) = 6.2500$
$\text{Var}(\hat{\mu}_2) = 9.0625$
$\text{Var}(\hat{\mu}_3) = 1.9444$
The point estimate $\hat{\mu}_3$ has the smallest variance.
(c) $\text{MSE}(\hat{\mu}_1) = 6.2500$
$\text{MSE}(\hat{\mu}_2) = 9.0625$
$\text{MSE}(\hat{\mu}_3) = 1.9444 + (9 - \frac{\mu}{2})^2$
This is equal to 26.9444 when $\mu = 8$.

7.2.3 (a) $\text{Var}(\hat{\mu}_1) = 2.5$
(b) The value $p = 0.6$ produces the smallest variance which is
$\text{Var}(\hat{\mu}) = 2.4$.
(c) The relative efficiency is $\frac{2.4}{2.5} = 0.96$.

7.2.5 (a) $a_1 + \ldots + a_n = 1$
(b) $a_1 = \ldots = a_n = \frac{1}{n}$

7.2.7 $\text{MSE}(\hat{\mu}) = \frac{\sigma^2}{4} + \frac{(\mu_0 - \mu)^2}{4}$
$\text{MSE}(X) = \sigma^2$

7.2.9 The standard deviation is $\sqrt{29.49} = 5.43$.

7.3   Sampling Distributions

7.3.1 The relative efficiency is $\frac{n_1}{n_2}$.

7.3.3 (a) 0.4418 (b) 0.7150

7.3.5 (a) $c = 45.46$ (b) $c = 50.26$

7.3.7 (a) $c = 0.4552$ (b) $c = 0.6209$

7.3.9 $\hat{\mu} = 974.3$
$\text{s.e.}(\hat{\mu}) = 3.594$

7.3.11 $\hat{p} = 0.22$
s.e.$(\hat{p}) = 0.0338$

7.3.13 $\hat{\mu} = 69.35$
s.e.$(\hat{\mu}) = 1.244$

7.3.15 $\hat{\mu} = 12.211$
s.e.$(\hat{\mu}) = 0.277$

7.3.17 $\hat{\mu} = 0.23181$
s.e.$(\hat{\mu}) = 0.00810$

7.3.19 0.8022
0.9178

7.3.21 0.324

7.3.23 0.017

7.3.25 0.103

7.3.27 0.747

7.3.29 (a) 0.7626   (b) 0.5416

7.4   Constructing Parameter Estimates

7.4.1 $\hat{\lambda} = 5.63$
s.e.$(\hat{\lambda}) = 0.495$

Chapter 8   Inferences on a Population Mean

8.1   Confidence Intervals

8.1.1 (52.30, 54.54)

8.1.3 (431.9, 441.1)
(430.9, 442.1)
(428.9, 444.1)

8.1.5 (0.0272, 0.0384)

8.1.7 A sample size of about $n = 64$ should be sufficient.

8.1.9 An additional sample of at least 8 observations should be sufficient.

8.1.11 An additional sample of at least 27 observations should be sufficient.

8.1.13 $c = 0.761$

8.1.15 $c = 420.0$

8.1.17 (2.464, 3.040)

8.1.19 At 95% confidence the confidence interval is (11.66, 12.76).

8.1.21 At 95% confidence the confidence interval is (0.2157, 0.2480).

8.1.23 90%

8.1.25 (a) (5219.6, 5654.8)
(b) It can be estimated that an additional 16 chemical solutions would need to be measured.

8.2   Hypothesis Testing

8.2.1 (a) The p-value is $2 \times P(t_{17} \geq 1.04) = 0.313$.
(b) The p-value is $P(t_{17} \leq -2.75) = 0.0068$.

8.2.3 (a) The p-value is $2 \times \Phi(-1.34) = 0.180$.
(b) The p-value is $\Phi(-2.45) = 0.007$.

8.2.5 (a) The null hypothesis is accepted when $|t| \leq 1.684$.
(b) The null hypothesis is rejected when $|t| > 2.704$.
(c) The null hypothesis is rejected at size $\alpha = 0.10$ and accepted at size $\alpha = 0.01$.
(d) The p-value is $2 \times P(t_{40} \geq 2.066) = 0.045$.

8.2.7 (a) The null hypothesis is accepted when $|t| \leq 1.753$.
(b) The null hypothesis is rejected when $|t| > 2.947$.
(c) The null hypothesis is rejected at size $\alpha = 0.10$ and accepted at size $\alpha = 0.01$.
(d) The p-value is $2 \times P(t_{15} \geq 1.931) = 0.073$.

8.2.9 (a) The null hypothesis is accepted when $t \leq 1.296$.
(b) The null hypothesis is rejected when $t > 2.390$.
(c) The null hypothesis is rejected at size $\alpha = 0.01$ and consequently also at size $\alpha = 0.10$.
(d) The p-value is $P(t_{60} \geq 3.990) = 0.0001$.

8.2.11 There is sufficient evidence to conclude that the machine is miscalibrated.

8.2.13 There is sufficient evidence to conclude that the chemical plant is in violation of the working code.

8.2.15 There is sufficient evidence to conclude that the average corrosion rate of chilled cast iron of this type is larger than 2.5.

8.2.17 There is sufficient evidence to conclude that the average number of calls taken per minute is less than 13 so that the manager's claim is false.

8.2.19 There is not sufficient evidence to conclude that the spray painting machine is not performing properly.

8.2.21 There is not sufficient evidence to conclude that the average voltage of the batteries from the production line is at least 238.5.

8.2.23 There is sufficient evidence to conclude that the average length of the components is not 82.50.

8.2.25 There is sufficient evidence to conclude that the average breaking strength is not 7.000.

8.2.27 The hypotheses are $H_0 : \mu \geq 25$ versus $H_A : \mu < 25$.

8.2.29 (a) The experiment does not provide sufficient evidence to conclude that the average time to toxicity of salmon fillets under these storage conditions is more than 11 days.

(b) (9.40, 14.55)

Chapter 9   Comparing Two Population Means

9.2   Analysis of Paired Samples

9.2.1 There is *not* sufficient evidence to conclude that the new assembly method is any quicker on average than the standard assembly method.

9.2.3 There is sufficient evidence to conclude that the new tires are better than the standard tires in terms of the average reduction in tread depth.

9.2.5 There is *not* sufficient evidence to conclude that the two laboratories are any different in the datings that they provide.

9.2.7 There is not sufficient evidence to conclude that procedures A and B give different readings on average.
The reviewer's comments are plausible.

9.2.9 There is *not* sufficient evidence to conclude that the addition of the surfactant has an effect on the amount of uranium-oxide removed from the water.

9.3   Analysis of Independent Samples

9.3.3 (a) $(-76.4, 21.8)$ (b) $(-83.6, 29.0)$
(c) The null hypothesis is accepted at size $\alpha = 0.01$.
The p-value is $2 \times P(t_{23} \geq 1.56) = 0.132$.

9.3.5 (a) $(-\infty, -0.0014)$
(b) The null hypothesis is rejected at size $\alpha = 0.01$ and consequently is also rejected at size $\alpha = 0.05$.
The p-value is $P(t_{26} \leq -4.95) = 0.000$.

9.3.7 (a) The null hypothesis is rejected at size $\alpha = 0.01$.
(b) $(-\infty, -9.3)$
(c) There is sufficient evidence to conclude that the synthetic fiber bundles have an average breaking strength larger than the wool fiber bundles.

9.3.9 (a) The p-value is $2 \times \Phi(-1.92) = 0.055$.
(b) (4.22, 18.84)

9.3.11 (a) The p-value is $2 \times \Phi(-2.009) = 0.045$.
(b) A 90% two-sided confidence interval is (0.16, 1.56).
A 95% two-sided confidence interval is (0.02, 1.70).
A 99% two-sided confidence interval is $(-0.24, 1.96)$.

9.3.13 Equal sample sizes of at least 47 can be recommended.

9.3.15 Additional sample sizes of at least $96 - 41 = 55$ from each population can be recommended.

9.3.17 There is sufficient evidence to conclude that the paving slabs from company A weigh more on average than the paving slabs from company B.
There is also more variability in the weights of the paving slabs from company A.

9.3.19 There is sufficient evidence to conclude that the damped feature is effective in reducing the heel-strike force.

9.3.21 There is not sufficient evidence to conclude that the average service times are any different at these two times of day.

9.3.23 There is sufficient evidence to conclude that on average medicine A provides a higher response than medicine B.

9.3.25 $\bar{x}_B < 149.9$

Chapter 10   Discrete Data Analysis

10.1   Inferences on a Population Proportion

10.1.1 (a) (0.127, 0.560)
(b) (0.179, 0.508)
(c) (0, 0.539)
(d) The exact p-value is 0.110.
The approximate p-value is 0.077.

10.1.3 (a) Let p be the probability that a value produced by the random number generator is a zero, and consider the hypotheses

$$H_0 : p = 0.5 \text{ versus } H_A : p \neq 0.5$$

The p-value is 0.018.

(b) $(0.4995, 0.5110)$

(c)   An additional sample size of about 215,500 would be required.

10.1.5   There is more support for foul play from the second experiment than from the first.

10.1.7   There is sufficient evidence to conclude that the jurors are not being randomly selected.

10.1.9   The required sample size for the worst case scenario is $n \geq 9604$. If $p \geq 0.75$ the required sample size can be calculated as $n \geq 7203$.

10.1.11 $(0.494, 0.723)$
An additional sample size of 512 would be required.

10.1.13 $(0.385, 0.815)$

10.1.15   The required sample size for the worst case scenario is 385 householders.
If $p \leq 1/3$ then the required sample size can be calculated as 342 householders.

10.1.17   The standard confidence interval is $(0.161, 0.557)$.
The alternative confidence interval is $(0.195, 0.564)$.

10.1.19 $(0.085, 0.211)$

10.1.21   The margin of error was calculated with 95% confidence under the worst case scenario where the estimated probability could be close to 0.5.

10.2   Comparing Two Population Proportions

10.2.1   (a) $(-0.195, 0.413)$
(b) $(-0.122, 0.340)$
(c) $(-0.165, 1)$
(d)   The p-value is 0.365.

10.2.3   (a) $(-0.124, 0.331)$
(b)   The p-value is 0.251.
There is *not* sufficient evidence to conclude that one radar system is any better than the other radar system.

10.2.5 $(-1, -0.002)$
The p-value is 0.050.
There is some evidence that the presence of seed crystals increases the probability of crystallization within 24 hours but it is not overwhelming.

10.2.7 $(-0.017, 0.008)$
The p-value is 0.479.
There is *not* sufficient evidence to conclude that there is a difference in the operating standards of the two production lines.

10.2.9 $(-0.051, 0.027)$
The p-value is 0.549.
There is *not* sufficient evidence to conclude that there is a difference in the quality of the computer chips from the two suppliers.

10.2.11   There is sufficient evidence to conclude that machine A is better than machine B.

10.2.13   (a)   There is not sufficient evidence to conclude that the probability of an insulator of this type having a dielectric breakdown strength below the specified threshold level is larger at 250 degrees Centigrade than it is at 180 degrees Centigrade.
(b) $(-0.269, 0.117)$

10.2.15 $(-0.172, 0.036)$
The confidence interval contains zero so there is not sufficient evidence to conclude that the failure rates due to operator misuse are different for the two products.

10.3   Goodness of Fit Tests for One-way Contingency Tables

10.3.1   (a)   The expected cell frequencies are $e_i = 83.33$.
(b)   The Pearson chi-square statistic is 4.36.
(c)   The likelihood ratio chi-square statistic is 4.44.
(d)   The p-values are 0.499 and 0.488.
A size $\alpha = 0.01$ test of the null hypothesis that the die is fair is accepted.
(e) $(0.159, 0.217)$

10.3.3   (a)   There is a fairly strong suggestion that the supposition is not plausible although the evidence is not completely overwhelming.
(b) $(0.425, 0.598)$

10.3.5   (a)   These probability values are plausible.
(b) $(0.358, 0.531)$

10.3.7   It is not reasonable to model the number of arrivals with a Poisson distribution with mean $\lambda = 7$.

10.3.9   It is plausible that the pearl oyster diameters have a uniform distribution between 0 and 10 mm.

10.3.13   The data set is consistent with the proposed genetic theory.

10.3.15   (a)   There is not sufficient evidence to conclude that the probability that a solution has normal acidity is not 0.80.
(b)   The data is not consistent with the claimed probabilities.

10.3.17   It is plausible that the number of shark attacks per year follows a Poisson distribution with mean 2.5.

10.4   Testing for Independence in Two-way Contingency Tables

10.4.1 (a) The expected cell frequencies are

	Acceptable	Defective
Supplier A	186.25	13.75
Supplier B	186.25	13.75
Supplier C	186.25	13.75
Supplier D	186.25	13.75

(b) The Pearson chi-square statistic is $X^2 = 7.087$.
(c) The likelihood ratio chi-square statistic is $G^2 = 6.889$.
(d) The p-values are 0.069 and 0.076.

(e) The null hypothesis that the defective rates are identical for the four suppliers is accepted at size $\alpha = 0.05$.
(f) $(0.020, 0.080)$ (g) $(-0.086, 0.026)$

10.4.3 There is *not* sufficient evidence to conclude that the preferences for the different formulations change with age.

10.4.5 There is sufficient evidence to conclude that some technicians are better than others in satisfying their customers.

10.4.7 (a) There is sufficient evidence to conclude that $p_s \neq p_n$.
(b) $(-0.375, -0.085)$

10.4.9 There is not sufficient evidence to conclude that the probability of a customer purchasing the extended warranty is different for the three product types.

Chapter 11   The Analysis of Variance

11.1   One-Factor Analysis of Variance

11.1.1 (a) $P(X \geq 4.2) = 0.0177$
(b) $P(X \geq 2.3) = 0.0530$
(c) $P(X \geq 31.7) \leq 0.0001$
(d) $P(X \geq 9.3) = 0.0019$
(e) $P(X \geq 0.9) = 0.5010$

11.1.3

Source	df	SS	MS	F	p-value
Treatments	7	126.95	18.136	5.01	0.0016
Error	22	79.64	3.62		
Total	29	206.59			

11.1.5

Source	df	SS	MS	F	p-value
Treatments	3	162.19	54.06	6.69	0.001
Error	40	323.34	8.08		
Total	43	485.53			

11.1.7

Source	df	SS	MS	F	p-value
Treatments	3	0.0079	0.0026	1.65	0.189
Error	52	0.0829	0.0016		
Total	55	0.0908			

11.1.9 (a) $\mu_1 - \mu_2 \in (-22.8, -8.8)$
$\mu_1 - \mu_3 \in (3.6, 17.6)$
$\mu_1 - \mu_4 \in (-0.9, 13.1)$
$\mu_1 - \mu_5 \in (-13.0, 1.0)$
$\mu_1 - \mu_6 \in (1.3, 15.3)$
$\mu_2 - \mu_3 \in (19.4, 33.4)$
$\mu_2 - \mu_4 \in (14.9, 28.9)$
$\mu_2 - \mu_5 \in (2.8, 16.8)$
$\mu_2 - \mu_6 \in (17.1, 31.1)$
$\mu_3 - \mu_4 \in (-11.5, 2.5)$
$\mu_3 - \mu_5 \in (-23.6, -9.6)$
$\mu_3 - \mu_6 \in (-9.3, 4.7)$
$\mu_4 - \mu_5 \in (-19.1, -5.1)$

$\mu_4 - \mu_6 \in (-4.8, 9.2)$
$\mu_5 - \mu_6 \in (7.3, 21.3)$
(c) An additional sample size of 6 observations from each factor level can be recommended.

11.1.11 (a) $\bar{x}_{1.} = 5.633$
$\bar{x}_{2.} = 5.567$
$\bar{x}_{3.} = 4.778$
(b) $\bar{x}_{..} = 5.326$
(c) $SSTR = 4.076$
(d) $\sum_{i=1}^{k} \sum_{j=1}^{n_i} x_{ij}^2 = 791.30$
(e) $SST = 25.432$
(f) $SSE = 21.356$
(g)

Source	df	SS	MS	F	p-value
Treatments	2	4.076	2.038	2.29	0.123
Error	24	21.356	0.890		
Total	26	25.432			

(h) $\mu_1 - \mu_2 \in (-1.04, 1.18)$
$\mu_1 - \mu_3 \in (-0.25, 1.97)$
$\mu_2 - \mu_3 \in (-0.32, 1.90)$
(j) An additional sample size of 36 observations from each factor level can be recommended.

11.1.13

Source	df	SS	MS	F	p-value
Treatments	2	0.0085	0.0042	0.24	0.787
Error	87	1.5299	0.0176		
Total	89	1.5384			

$\mu_1 - \mu_2 \in (-0.08, 0.08)$
$\mu_1 - \mu_3 \in (-0.06, 0.10)$
$\mu_2 - \mu_3 \in (-0.06, 0.10)$

There is *not* sufficient evidence to conclude that there is a difference between the three production lines.

11.1.15

Source	df	SS	MS	F	p-value
Treatments	2	0.0278	0.0139	1.26	0.299
Error	30	0.3318	0.0111		
Total	32	0.3596			

$\mu_1 - \mu_2 \in (-0.15, 0.07)$
$\mu_1 - \mu_3 \in (-0.08, 0.14)$
$\mu_2 - \mu_3 \in (-0.04, 0.18)$

There is *not* sufficient evidence to conclude that the radiation readings are affected by the background radiation levels.

11.1.17

Source	df	SS	MS	F	p-value
Treatments	2	0.4836	0.2418	7.13	0.001
Error	93	3.1536	0.0339		
Total	95	3.6372			

$\mu_1 - \mu_2 \in (-0.01, 0.22)$
$\mu_1 - \mu_3 \in (0.07, 0.29)$
$\mu_2 - \mu_3 \in (-0.03, 0.18)$

There is sufficient evidence to conclude that the average particle diameter is larger at the low amount of stabilizer than at the high amount of stabilizer.

11.1.19

Source	df	SS	MS	F	p-value
Treatments	2	5.981	2.990	0.131	0.878
Error	26	593.753	22.837		
Total	28	599.734			

There is not sufficient evidence to conclude that there is a difference between the catalysts in terms of the strength of the compound.

11.1.21 With a 95% confidence level the pairwise confidence intervals that contain 0 are:
$\mu_1 - \mu_2$
$\mu_2 - \mu_5$
$\mu_3 - \mu_4$

It can be inferred that the largest mean is either μ_3 or μ_4 and that the smallest mean is either μ_2 or μ_5.

11.1.23

Source	df	SS	MS	F	p-value
Physician	5	1983.8	396.8	15.32	0.000
Error	24	621.6	25.9		
Total	29	2605.4			

The p-value of 0.000 implies that there is sufficient evidence to conclude that the times taken by the physicians for the investigatory surgical procedures are different.
The slowest physician is either physician 1, physician 4, or physician 5.
The quickest physician is either physician 2, physician 3, or physician 6.

11.1.25 (a)

Source	df	SS	MS	F	p-value
Treatments	3	13.77	4.59	4.72	0.01
Error	24	23.34	0.97		
Total	27	37.11			

(b) $\mu_1 - \mu_2 \in (-2.11, 0.44)$
$\mu_1 - \mu_3 \in (-2.63, 0.01)$
$\mu_1 - \mu_4 \in (-3.36, -0.62)$
$\mu_2 - \mu_3 \in (-1.75, 0.79)$
$\mu_2 - \mu_4 \in (-2.48, 0.18)$
$\mu_3 - \mu_4 \in (-2.04, 0.70)$

There is sufficient evidence to establish that μ_4 is larger than μ_1.

11.2 Randomized Block Designs

11.2.1

Source	df	SS	MS	F	p-value
Treatments	3	10.15	3.38	3.02	0.047
Blocks	9	24.53	2.73	2.43	0.036
Error	27	30.24	1.12		
Total	39	64.92			

11.2.3

Source	df	SS	MS	F	p-value
Treatments	3	58.72	19.57	0.63	0.602
Blocks	9	2839.97	315.55	10.17	0.0000
Error	27	837.96	31.04		
Total	39	3736.64			

11.2.5 (a)

Source	df	SS	MS	F	p-value
Treatments	2	8.17	4.085	8.96	0.0031
Blocks	7	50.19	7.17	15.72	0.0000
Error	14	6.39	0.456		
Total	23	64.75			

(b) $\mu_1 - \mu_2 \in (0.43, 2.19)$
$\mu_1 - \mu_3 \in (0.27, 2.03)$
$\mu_2 - \mu_3 \in (-1.04, 0.72)$

11.2.7 (a) $\bar{x}_{1.} = 6.0617$
$\bar{x}_{2.} = 7.1967$
$\bar{x}_{3.} = 5.7767$
(b) $\bar{x}_{.1} = 7.4667$
$\bar{x}_{.2} = 5.2667$
$\bar{x}_{.3} = 5.1133$
$\bar{x}_{.4} = 7.3300$
$\bar{x}_{.5} = 6.2267$
$\bar{x}_{.6} = 6.6667$
(c) $\bar{x}_{..} = 6.345$
(d) $SSTr = 6.7717$
(e) $SSBl = 15.0769$
(f) $\sum_{i=1}^{3} \sum_{j=1}^{6} x_{ij}^2 = 752.1929$
(g) $SST = 27.5304$
(h) $SSE = 5.6818$

(i)

Source	df	SS	MS	F	p-value
Treatments	2	6.7717	3.3859	5.96	0.020
Blocks	5	15.0769	3.0154	5.31	0.012
Error	10	5.6818	0.5682		
Total	17	27.5304			

(j) $\mu_1 - \mu_2 \in (-2.33, 0.05)$
$\mu_1 - \mu_3 \in (-0.91, 1.47)$
$\mu_2 - \mu_3 \in (0.22, 2.61)$

(l) An additional 3 blocks can be recommended.

11.2.9

Source	df	SS	MS	F	p-value
Treatments	2	17.607	8.803	2.56	0.119
Blocks	6	96.598	16.100	4.68	0.011
Error	12	41.273	3.439		
Total	20	155.478			

$\mu_1 - \mu_2 \in (-1.11, 4.17)$
$\mu_1 - \mu_3 \in (-0.46, 4.83)$
$\mu_2 - \mu_3 \in (-1.99, 3.30)$

There is *not* sufficient evidence to conclude that the calciners are operating at different efficiencies.

11.2.11

Source	df	SS	MS	F	p-value
Treatments	3	3231.2	1,077.1	4.66	0.011
Blocks	8	29256.1	3,657.0	15.83	0.000
Error	24	5545.1	231.0		
Total	35	38032.3			

$\mu_1 - \mu_2 \in (-8.20, 31.32)$
$\mu_1 - \mu_3 \in (-16.53, 22.99)$
$\mu_1 - \mu_4 \in (-34.42, 5.10)$
$\mu_2 - \mu_3 \in (-28.09, 11.43)$
$\mu_2 - \mu_4 \in (-45.98, -6.46)$
$\mu_3 - \mu_4 \in (-37.65, 1.87)$

There is sufficient evidence to conclude that driver 4 is better than driver 2.

11.2.13

Source	df	SS	MS	F	p-value
Treatments	4	8.462×10^8	2.116×10^8	66.55	0.000
Blocks	11	19.889×10^8	1.808×10^8	56.88	0.000
Error	44	1.399×10^8	3.179×10^6		
Total	59	29.750×10^8			

$\mu_1 - \mu_2 \in (4372, 8510)$
$\mu_1 - \mu_3 \in (4781, 8919)$
$\mu_1 - \mu_4 \in (5438, 9577)$
$\mu_1 - \mu_5 \in (-3378, 760)$
$\mu_2 - \mu_3 \in (-1660, 2478)$
$\mu_2 - \mu_4 \in (-1002, 3136)$
$\mu_2 - \mu_5 \in (-9819, -5681)$
$\mu_3 - \mu_4 \in (-1411, 2727)$
$\mu_3 - \mu_5 \in (-10228, -6090)$
$\mu_4 - \mu_5 \in (-10886, -6748)$

There is sufficient evidence to conclude that either agent 1 or agent 5 is the best agent.
The worst agent is either agent 2, 3 or 4.

11.2.15 (a)

Source	df	SS	MS	F	p-value
Treatments	3	0.151	0.0503	5.36	0.008
Blocks	6	0.324	0.054	5.75	0.002
Error	18	0.169	0.00939		
Total	27	0.644			

(b) $\mu_2 - \mu_1 \in (-0.326, -0.034)$
$\mu_2 - \mu_3 \in (-0.313, -0.021)$
$\mu_2 - \mu_4 \in (-0.305, -0.013)$

None of these confidence intervals contains 0 so there is sufficient evidence to conclude that treatment 2 has a smaller mean value than each of the other treatments.

11.2.17 The new analysis of variance table is

Source	df	SS	MS	F	p-value
Treatments	same	a^2 SSTr	a^2 MSTr	same	same
Blocks	same	a^2 SSBl	a^2 MSBl	same	same
Error	same	a^2 SSE	a^2 MSE		
Total	same	a^2 SST			

11.2.19

Source	df	SS	MS	F	p-value
Locations	3	3.893	1.298	0.49	0.695
Time	4	472.647	118.162	44.69	0.000
Error	12	31.729	2.644		
Total	19	508.270			

The p-value of 0.695 implies that there is not sufficient evidence to conclude that the pollution levels are different at the four locations.
The confidence intervals for all of the pairwise comparisons contain 0, so the graphical representation has one line joining all four sample means.

Chapter 12   Simple Linear Regression and Correlation

12.1   The Simple Linear Regression Model

12.1.1 (a) 21.2   (b) 5.1   (c) 0.587
(d) 0.401   (e) 0.354

12.1.3 (a) 23.0
(b) The expected value of the porosity decreases by 4.5.

(c) 0.963
(d) 0.968

12.1.5 0.975

12.2 Fitting the Regression Line

12.2.3 $\hat{\beta}_0 = 39.5$
$\hat{\beta}_1 = -2.04$
$\hat{\sigma}^2 = 17.3$
43.6

12.2.5 (a) $\hat{\beta}_0 = 36.19$ (b) $\hat{\sigma}^2 = 70.33$
$\hat{\beta}_1 = 0.2659$
(c) Yes.   (d) 55.33

12.2.7 (a) $\hat{\beta}_0 = -29.59$ (b) 173.1
$\hat{\beta}_1 = 0.07794$
(c) 7.794   (d) $\hat{\sigma}^2 = 286$

12.2.9 (a) $\hat{\beta}_0 = 12.864$ (b) 68.42
$\hat{\beta}_1 = 0.8051$
(c) 4.03   (d) $\hat{\sigma}^2 = 3.98$

12.3 Inferences on the Slope Parameter $\hat{\beta}_1$

12.3.1 (a) $(0.107, 0.937)$ (b) The p-value is 0.002.

12.3.3 (a) $s.e.(\hat{\beta}_1) = 0.08532$ (b) $(0.820, 1.186)$
(c) The p-value is 0.000.

12.3.5 (a) $s.e.(\hat{\beta}_1) = 0.1282$ (b) $(-\infty, -0.115)$
(c) The (two-sided) p-value is 0.017.

12.3.7 (a) $s.e.(\hat{\beta}_1) = 0.2829$ (b) $(1.041, 2.197)$
(c) If $\beta_1 = 1$ then the actual times are equal to the estimated times except for a constant difference of β_0.
The p-value is 0.036.

12.3.9 The p-value is 0.019.

12.4 Inferences on the Regression Line

12.4.3 $(21.9, 23.2)$

12.4.5 $(33.65, 41.02)$

12.4.7 $(-\infty, 10.63)$

12.4.9 $(76.284, 76.378)$

12.5 Prediction Intervals for Future Response Values

12.5.1 $(1386, 1406)$

12.5.3 $(5302, 9207)$

12.5.5 $(165.7, 274.0)$

12.5.7 $(63.48, 74.96)$

12.5.9 $\hat{\beta}_1 = 1.851$
$\hat{\beta}_0 = 13.875$
$\hat{\sigma}^2 = 0.494$
$(47.864, 53.941)$

12.6 The Analysis of Variance Table

12.6.1

Source	df	SS	MS	F	p-value
Regression	1	40.53	40.53	2.32	0.137
Error	33	576.51	17.47		
Total	34	617.04			

$R^2 = 0.066$

12.6.3

Source	df	SS	MS	F	p-value
Regression	1	870.43	870.43	889.92	0.000
Error	8	7.82	0.9781		
Total	9	878.26			

$R^2 = 0.991$

12.6.5

Source	df	SS	MS	F	p-value
Regression	1	10.71×10^7	10.71×10^7	138.29	0.000
Error	14	1.08×10^7	774,211		
Total	15	11.79×10^7			

$R^2 = 0.908$

12.6.7

Source	df	SS	MS	F	p-value
Regression	1	397.58	397.58	6.94	0.017
Error	18	1031.37	57.30		
Total	19	1428.95			

$R^2 = 0.278$

12.6.9

Source	df	SS	MS	F	p-value
Regression	1	411.26	411.26	32.75	0.000
Error	30	376.74	12.56		
Total	31	788.00			

$R^2 = 0.522$

The p-value is not very meaningful because it tests the null hypothesis that the actual times are unrelated to the estimated times.

12.7   Residual Analysis

12.7.1 There is no suggestion that the fitted regression model is not appropriate.

12.7.3 There is a possible suggestion of a slight reduction in the variability of the VO2-max values as age increases.

12.7.5 The variability of the actual times increases as the estimated time increases.

12.8   Variable Transformations

12.8.1 The model

$$y = \gamma_0\, e^{\gamma_1 x}$$

is appropriate.
A linear regression can be performed with $\ln(y)$ as the dependent variable and with x as the input variable.

$\hat{\gamma}_0 = 9.12$
$\hat{\gamma}_1 = 0.28$
$\hat{\gamma}_0\, e^{\hat{\gamma}_1 \times 2.0} = 16.0$

12.8.3 $\hat{\gamma}_0 = 8.81$
$\hat{\gamma}_1 = 0.523$
$\gamma_0 \in (6.84, 11.35)$
$\gamma_1 \in (0.473, 0.573)$

12.8.5 $\hat{\gamma}_0 = 13.85$
$\hat{\gamma}_1 = 0.341$
$(0.289, 0.393)$

12.8.7 $\hat{\gamma}_0 = 12.775$
$\hat{\gamma}_1 = -0.5279$

When the crack length is 2.1 the expected breaking load is 4.22.

12.9   Correlation Analysis

12.9.3 The sample correlation coefficient is $r = 0.95$.

12.9.5 The sample correlation coefficient is $r = -0.53$.

12.9.7 The sample correlation coefficient is $r = 0.72$.

12.9.9 The sample correlation coefficient is $r = 0.431$.

12.9.11 The variables A and B may both be related to a third surrogate variable C.

Chapter 13   Multiple Linear Regression and Nonlinear Regression

13.1   Introduction to Multiple Linear Regression

13.1.1 (a) $R^2 = 0.89$

(b)

Source	df	SS	MS	F	p-value
Regression	3	96.5	32.17	67.4	0.000
Error	26	12.4	0.477		
Total	29	108.9			

(c) $\hat{\sigma}^2 = 0.477$
(d) The p-value is 0.000.
(e) $(11.2, 21.8)$

13.1.3 (a) $(67.1, 197.7)$
(b) The p-value is 0.002.

13.1.5 Variable x_3 should be removed from the model.

13.1.7 The p-value is 0.013.

13.1.9 (a) $\hat{y} = 355.38$ (b) $(318.24, 392.52)$

13.1.11

Source	df	SS	MS	F	p-value
Regression	3	392.495	130.832	6.978	0.003
Error	16	299.982	18.749		
Total	19	692.477			

The p-value in the analysis of variance table is for the null hypothesis

$$H_0 : \beta_1 = \beta_2 = \beta_3 = 0.$$

The proportion of the variability of the y variable that is explained by the model is

$$R^2 = 56.7\%.$$

13.2   Examples of Multiple Linear Regression

13.2.1 (b) The variable competitor's price has a p-value of 0.216 and is not needed in the model.
The sample correlation coefficient between the competitor's price and the sales is $r = -0.91$.
The sample correlation coefficient between the competitor's price and the company's price is $r = 0.88$.
(c) The sample correlation coefficient between the company's price and the sales is $r = -0.96$.

Using the model

$$\text{sales} = 107.4 - (3.67 \times \text{company's price})$$

the predicted sales are $107.4 - (3.67 \times 10.0) = 70.7$.

13.2.3 (a) $\hat{\beta}_0 = -3,238.6$
$\hat{\beta}_1 = 0.9615$
$\hat{\beta}_2 = 0.732$
$\hat{\beta}_3 = 2.889$
$\hat{\beta}_4 = 389.9$

(b) The variable geology has a p-value of 0.737 and is not needed in the model.
The sample correlation coefficient between the cost and geology is $r = 0.89$.

The sample correlation coefficient between the depth and geology is $r = 0.92$.
The variable geology is not needed in the model because it is highly correlated with the variable depth which is in the model.

(c) The variable rig-index can also be removed from the model.
A final model

$$\text{cost} = -3011 + (1.04 \times \text{depth}) + (2.67 \times \text{downtime})$$

can be recommended.

13.2.5   Two indicator variables x_1 and x_2 are needed.

13.3   Matrix Algebra Formulation of Multiple Linear Regression

13.3.1 (a)

$$Y = \begin{pmatrix} 2 \\ -2 \\ 4 \\ -2 \\ 2 \\ -4 \\ 1 \\ 3 \\ 1 \\ -5 \end{pmatrix}$$

(b)

$$X = \begin{pmatrix} 1 & 0 & 1 \\ 1 & 0 & -1 \\ 1 & 1 & 4 \\ 1 & 1 & -4 \\ 1 & -1 & 2 \\ 1 & -1 & -2 \\ 1 & 2 & 0 \\ 1 & 2 & 0 \\ 1 & -2 & 3 \\ 1 & -2 & -3 \end{pmatrix}$$

(c)

$$X'X = \begin{pmatrix} 10 & 0 & 0 \\ 0 & 20 & 0 \\ 0 & 0 & 60 \end{pmatrix}$$

(d)

$$(X'X)^{-1} = \begin{pmatrix} 0.1000 & 0 & 0 \\ 0 & 0.0500 & 0 \\ 0 & 0 & 0.0167 \end{pmatrix}$$

(e)

$$X'Y = \begin{pmatrix} 0 \\ 20 \\ 58 \end{pmatrix}$$

(g)

$$\hat{Y} = \begin{pmatrix} 0.967 \\ -0.967 \\ 4.867 \\ -2.867 \\ 0.933 \\ -2.933 \\ 2.000 \\ 2.000 \\ 0.900 \\ -4.900 \end{pmatrix}$$

(h)

$$e = \begin{pmatrix} 1.033 \\ -1.033 \\ -0.867 \\ 0.867 \\ 1.067 \\ -1.067 \\ -1.000 \\ 1.000 \\ 0.100 \\ -0.100 \end{pmatrix}$$

(i) $SSE = 7.933$

(k) $s.e.(\hat{\beta}_1) = 0.238$
$s.e.(\hat{\beta}_2) = 0.137$
Both input variables should be kept in the model.

(l) The fitted value is 2.933.
The standard error is 0.496.
The confidence interval is (1.76, 4.11).

(m) The prediction interval is (0.16, 5.71).

13.3.3

$$Y = \begin{pmatrix} 10 \\ 0 \\ -5 \\ 2 \\ 3 \\ -6 \end{pmatrix}$$

$$X = \begin{pmatrix} 1 & -3 & 1 & 3 \\ 1 & -2 & 1 & 0 \\ 1 & -1 & 1 & -5 \\ 1 & 1 & -6 & 1 \\ 1 & 2 & -3 & 0 \\ 1 & 3 & 6 & 1 \end{pmatrix}$$

$$X'X = \begin{pmatrix} 6 & 0 & 0 & 0 \\ 0 & 28 & 0 & 0 \\ 0 & 0 & 84 & -2 \\ 0 & 0 & -2 & 36 \end{pmatrix}$$

$$(X'X)^{-1} = \begin{pmatrix} 0.16667 & 0 & 0 & 0 \\ 0 & 0.03571 & 0 & 0 \\ 0 & 0 & 0.01192 & 0.00066 \\ 0 & 0 & 0.00066 & 0.02781 \end{pmatrix}$$

$$X'Y = \begin{pmatrix} 4 \\ -35 \\ -52 \\ 51 \end{pmatrix}$$

$$\hat{\beta} = \begin{pmatrix} 0.6667 \\ -1.2500 \\ -0.5861 \\ 1.3841 \end{pmatrix}$$

13.4   Evaluating Model Accuracy

13.4.1 (a) There is a slight suggestion of a greater variability in the yields at higher temperatures.
(b) There are no unusually large standardized residuals.
(c) The points $(90, 85)$ and $(200, 702)$ have leverage values $h_{ii} = 0.547$.

13.4.3 (a) The residual plots do not indicate any problems.
(b) If it were beneficial to add the variable weight to the model, then there would be some pattern in this residual plot.
(c) The observation with VO2-max = 23 has a standardized residual of -2.15.

Chapter 14   Multifactor Experimental Design and Analysis

14.1   Experiments with Two Factors

14.1.1

Source	df	SS	MS	F	p-value
Fuel	1	96.33	96.33	3.97	0.081
Car	1	75.00	75.00	3.09	0.117
Fuel*Car	1	341.33	341.33	14.08	0.006
Error	8	194.00	24.25		
Total	11	706.66			

14.1.3

Source	df	SS	MS	F	p-value
Tip	2	0.1242	0.0621	1.86	0.175
Material	2	14.1975	7.0988	212.31	0.000
Tip*Material	4	0.0478	0.0120	0.36	0.837
Error	27	0.9028	0.0334		
Total	35	15.2723			

14.1.5

Source	df	SS	MS	F	p-value
Glass	2	3.134	1.567	0.32	0.732
Acidity	1	18.201	18.201	3.72	0.078
Glass*Acidity	2	83.421	41.711	8.52	0.005
Error	12	58.740	4.895		
Total	17	163.496			

14.1.7

Source	df	SS	MS	F	p-value
Design	2	3.896×10^3	1.948×10^3	0.46	0.685
Material	1	0.120×10^3	0.120×10^3	0.03	0.882
Error	2	8.470×10^3	4.235×10^3		
Total	5	12.487×10^3			

14.2   Experiments with Three or More Factors

14.2.1

Source	df	SS	MS	F	p-value
Drink	2	90.65	45.32	5.39	0.007
Gender	1	6.45	6.45	0.77	0.384
Age	2	23.44	11.72	1.39	0.255
Drink*Gender	2	17.82	8.91	1.06	0.352
Drink*Age	4	24.09	6.02	0.72	0.583
Gender*Age	2	24.64	12.32	1.47	0.238
Drink*Gender*Age	4	27.87	6.97	0.83	0.511
Error	72	605.40	8.41		
Total	89	820.36			

14.2.3

Source	df	SS	MS	F	p-value
Add-A	2	324.11	162.06	8.29	0.003
Add-B	2	5.18	2.59	0.13	0.877
Conditions	1	199.28	199.28	10.19	0.005
Add-A*Add-B	4	87.36	21.84	1.12	0.379
Add-A*Conditions	2	31.33	15.67	0.80	0.464
Add-B*Conditions	2	2.87	1.44	0.07	0.930
Add-A*Add-B*Conditions	4	21.03	5.26	0.27	0.894
Error	18	352.05	19.56		
Total	35	1023.21			

14.2.5

Source	df	SS	MS	F	p-value
Machine	1	387.1	387.1	3.15	0.095
Temp	1	29.5	29.5	0.24	0.631
Position	1	1271.3	1271.3	10.35	0.005
Angle	1	6865.0	6685.0	55.91	0.000
Machine*Temp	1	43.0	43.0	0.35	0.562
Machine*Position	1	54.9	54.9	0.45	0.513
Machine*Angle	1	1013.6	1013.6	8.25	0.011
Temp*Position	1	67.6	67.6	0.55	0.469
Temp*Angle	1	8.3	8.3	0.07	0.798
Position*Angle	1	61.3	61.3	0.50	0.490
Machine*Temp*Position	1	21.0	21.0	0.17	0.685
Machine*Temp*Angle	1	31.4	31.4	0.26	0.620
Machine*Position*Angle	1	13.7	13.7	0.11	0.743
Temp*Position*Angle	1	17.6	17.6	0.14	0.710
Machine*Temp*Position*Angle	1	87.5	87.5	0.71	0.411
Error	16	1964.7	122.8		
Total	31	11937.3			

14.2.7 A redundant
B redundant
C redundant
D redundant
A*B not significant
A*C redundant
A*D redundant
B*C significant
B*D significant
C*D redundant
A*B*C not significant
A*B*D not significant
A*C*D significant
B*C*D not significant
A*B*C*D not significant

Chapter 15   Nonparametric Statistical Analysis

15.1   The Analysis of a Single Population

15.1.1 (c) It is not plausible. (d) It is not plausible.
(e) The p-value is 0.064.
(f) The p-value is 0.001.
(g) The confidence interval from the sign test is (65.0, 69.0).
The confidence interval from the signed rank test is (66.0, 69.5).

15.1.3 The p-values for the hypotheses

$$H_0 : \mu = 0.2 \quad \text{versus} \quad H_A : \mu \neq 0.2$$

are 0.004 for the sign test,
0.000 for the signed rank test,
and 0.000 for the t-test.
Confidence intervals for μ with a confidence level of at least 95% are
(0.207, 0.244) for the sign test,
(0.214, 0.244) for the signed rank test,
and (0.216, 0.248) for the t-test.

There is sufficient evidence to conclude that the median paint thickness is larger than 0.2 mm.

15.1.5 (a) $S(18.0) = 14$
(b) The exact p-value is 0.115.
(c) 0.116 (d) $S_+(18.0) = 37$ (e) 0.012

15.1.7 It is reasonable to assume that the differences of the data have a symmetric distribution in which case the signed rank test can be used.
The p-values for the hypotheses

$$H_0 : \mu_A - \mu_B = 0 \quad \text{versus} \quad H_A : \mu_A - \mu_B \neq 0$$

are 0.296 for the sign test and 0.300 for the signed rank test.
Confidence intervals for $\mu_A - \mu_B$ with a confidence level of at least 95% are
(−1.0, 16.0) for the sign test and
(−6.0, 17.0) for the signed rank test.

There is not sufficient evidence to conclude that there is a difference between the two assembly methods.

15.1.9 The p-values for the hypotheses

$$H_0 : \mu_A - \mu_B = 0 \quad \text{versus} \quad H_A : \mu_A - \mu_B \neq 0$$

are 0.003 for the sign test and 0.002 for the signed rank test. Confidence intervals for $\mu_A - \mu_B$ with a confidence level of at least 95% are
$(-13.0, -1.0)$ for the sign test and
$(-12.0, -3.5)$ for the signed rank test.

15.2 Comparing Two Populations

15.2.1 (c) The Kolmogorov-Smirnov statistic is $M = 0.2006$.
There is sufficient evidence to conclude that the two distribution functions are different.

15.2.3 The Kolmogorov-Smirnov statistic is $M = 0.40$.
There is sufficient evidence to conclude that the two distribution functions are different.

15.2.5 (b) $S_A = 245$ (c) $U_A = 140$
(d) The value of U_A is consistent with the observations from population A being larger than the observations from population B.
(e) The p-value is 0.004.
There is sufficient evidence to conclude that the observations

15.1.11 The p-values for the hypotheses

$$H_0 : \mu_A - \mu_B = 0 \quad \text{versus} \quad H_A : \mu_A - \mu_B \neq 0$$

are 0.541 for the sign test and 0.721 for the signed rank test. Confidence intervals for $\mu_A - \mu_B$ with a confidence level of at least 95% are
$(-13.6, 7.3)$ for the sign test and
$(-6.6, 6.3)$ for the signed rank test.
There is not sufficient evidence to conclude that there is a difference between the two ball types.

from population A tend to be larger than the observations from population B.

15.2.7 (c) The Kolmogorov-Smirnov statistic is $M = 0.218$.
There is some evidence that the two distribution functions are different, although the evidence is not overwhelming.
(d) $S_A = 6555.5$
$U_A = 3705.5$
The value of U_A is consistent with the observations from production line A being larger than the observations from production line B.
The two-sided p-value is 0.027.
A 95% confidence interval for the difference in the population medians is $(0.003, 0.052)$.

15.3 Comparing Three or More Populations

15.3.1 (b) $\bar{r}_{1.} = 16.6$
$\bar{r}_{2.} = 15.5$
$\bar{r}_{3.} = 9.9$
(c) $H = 3.60$
(d) The p-value is $P(\chi_2^2 > 3.60) = 0.165$.

15.3.3 (a) $H = 1.84$
The p-value is 0.399.
There is not sufficient evidence to conclude that the radiation readings are affected by the background radiation level.

15.3.5 $H = 25.86$
The p-value is about 0.000.
There is sufficient evidence to conclude that the computer assembly times are affected by the different assembly methods.

15.3.7 (a) $\bar{r}_{1.} = 2.250$
$\bar{r}_{2.} = 1.625$
$\bar{r}_{3.} = 3.500$
$\bar{r}_{4.} = 2.625$
(b) $S = 8.85$ (c) The p-value is 0.032.

15.3.9 $S = 12.25$
The p-value is 0.002.
There is sufficient evidence to conclude that there is a difference between the radar systems.

15.3.11 $S = 37.88$
The p-value is about 0.000.
There is sufficient evidence to conclude that there is a difference in the performances of the agents.

Chapter 16   Quality Control Methods

16.2 Statistical Process Control

16.2.1 (a) The center line is 10.0 and the control limits are 9.7 and 10.3.
(b) The process is declared to be out of control at $\bar{x} = 9.5$ but not at $\bar{x} = 10.25$.
(c) The probability that an observation lies outside the control limits is 0.0668.
The average run length for detecting the change is 5.0.

16.2.3 (a) The probability that an observation lies outside the control limits is 0.0456.

(b) The probability that an observation lies outside the control limits is 0.1600.
The average run length for detecting the change is 6.25.

16.2.5 The probability that all eight points lie on the same side of the center line and within the control limits is 0.0076.

16.3 Variable Control Charts

16.3.1 (a) The $\bar{X}$-chart has a center line at 91.33 and control limits at 87.42 and 95.24.
The R-chart has a center line at 5.365 and control limits at 0 and 12.24.
(b) No
(c) The process can be declared to be out of control due to an increase in the variability.
(d) The process can be declared to be out of control due to an increase in the variability and a decrease in the mean value.

(e) There is no evidence that the process is out of control.
(f) The process can be declared to be out of control due to an increase in the mean value.

16.3.3 (a) The $\bar{X}$-chart has a center line at 2.993 and control limits at 2.801 and 3.186.
The R-chart has a center line at 0.2642 and control limits at 0 and 0.6029.
(b) There is no evidence that the process is out of control.

16.4 Attribute Control Charts

16.4.1 The p-chart has a center line at 0.0500 and control limits at 0.0000 and 0.1154.
(a) No
(b) It is necessary that $x \geq 12$.

16.4.3 The c-chart has a center line at 12.42 and control limits at 1.85 and 22.99.
(a) No
(b) At least 23.

16.5 Acceptance Sampling

16.5.1 (a) With $p_0 = 0.06$ there would be 3 defective items in the batch of $N = 50$ items.
The producer's risk is 0.0005.
(b) With $p_1 = 0.20$ there would be 10 defective items in the batch of $N = 50$ items.
The consumer's risk is 0.952.

Using a binomial approximation these probabilities are estimated to be 0.002 and 0.942.

16.5.3 (a) The producer's risk is 0.000.
(b) The consumer's risk is 0.300.

16.5.5 $c = 6$

Chapter 17   Reliability Analysis and Life Testing

17.1 System Reliability

17.1.1 $r = 0.9985$

17.1.3 (a) If r_1 is the individual reliability then it is necessary that $r_1 \geq 0.9873$.

(b) If r_1 is the individual reliability then it is necessary that $r_1 \geq 0.5271$.
(c) $r_1 \geq r^{1/n}$ $r_1 \geq 1 - (1 - r)^{1/n}$

17.1.5 $r = 0.9017$

17.2 Modeling Failure Rates

17.2.1 (a) 0.329   (b) 0.487   (c) 0.264

17.2.3 24.2 minutes

17.2.5 (a) 0.214   (b) 0.448
(c) 37.5   (d) 12.2.

17.2.7 (a) 0.142   (b) 0.344   (c) 3.54
(d) $h(t) = 0.0469 \times t^2$ (e) 2.78

17.3 Life Testing

17.3.1 (a) (86.4, 223.6)
(b) The value 150 is within the confidence interval, so the claim is plausible.

17.3.3 (a) (3.84, 9.93)

(b) The value 10 is not included within the confidence interval, and so it is not plausible that the mean time to failure is 10 hours.

17.3.5 (b) (0.457, 0.833)

INDEX

Boldface indicates references to tables.